Amazing and Aesthetic Aspects of Analysis

$$\frac{\pi^2}{6} = \frac{1}{1^2} + \frac{1}{2^2} + \frac{1}{3^2} + \frac{1}{4^2} + \cdots$$

$$= \frac{2^2}{2^2 - 1} \cdot \frac{3^2}{3^2 - 1} \cdot \frac{5^2}{5^2 - 1} \cdot \frac{7^2}{7^2 - 1} \cdot \frac{11^2}{11^2 - 1} \cdots$$

$$= \cfrac{1}{0^2 + 1^2 - \cfrac{1^4}{1^2 + 2^2 - \cfrac{2^4}{2^2 + 3^2 - \cfrac{3^4}{3^2 + 4^2 - \cfrac{4^4}{4^2 + 5^2 - \cdots}}}}}$$

$$= \frac{1}{\text{Probability a natural number is square free}}$$

$$= \frac{1}{\text{Probability two natural numbers are coprime}}$$

Paul Loya

Amazing and Aesthetic
Aspects of Analysis

 Springer

Paul Loya
Department of Mathematics
Binghamton University
Binghamton, NY
USA

ISBN 978-1-4939-9245-4   ISBN 978-1-4939-6795-7   (eBook)
https://doi.org/10.1007/978-1-4939-6795-7

Printed on acid-free paper

This Springer imprint is published by the registered company Springer Science+Business Media, LLC part of Springer Nature
The registered company address is: 233 Spring Street, New York, NY 10013, U.S.A.

Preface

I have truly enjoyed writing this book on some amazing and aesthetic aspects of analysis. Admittedly, some of the writing is too overdone (e.g., overuse of colloquial language and abundant alliteration at times). But what can I say? I was having fun. The sections of the book with an ★ are meant to be optional or just for fun and don't interfere with other sections, besides perhaps other starred sections. Most of the quotations that you'll find in these pages are taken from the website http://www-gap.dcs.st-and.ac.uk/history/Quotations/.

The contents of this book are based on lectures I have given to Binghamton University students taking our fall semester undergraduate real analysis course from 2003 to about 2006. The audience consisted of math majors, including actuarial students, as well as students from the fields of chemistry, computer science, economics, and physics, among others. In order to interest such a diverse body of students, I wanted to write a book that not only teaches the fundamentals of analysis, but also shows its usefulness, beauty, and excitement. I also wanted a book that is personal, in which the students come with me on a journey through some amazing and aesthetic aspects of analysis. There are no derivatives or integrals in this book. This is on purpose, because I wanted to focus on the "elementary" limiting processes only, those directly involving sequences and continuity, without the "higher" technology of calculus. The student completing Chapters 1 through 4 of this book will have mastered the fundamental arts of analysis and can move on to the "higher" arts, like the Lebesgue theory of integration, which I usually teach in our spring semester real analysis course.

Besides giving an appreciation of the amazing and aesthetic aspects of analysis (of course!), the overarching goals of this textbook are similar to those of any advanced math textbook, regardless of the subject:

GOALS OF THIS TEXTBOOK. THE STUDENT WILL BE ABLE TO...

- Comprehend and write mathematical reasonings and proofs.
- Wield the language of mathematics in a precise and effective manner.
- State the fundamental ideas, axioms, definitions, and theorems on which real analysis is built and flourishes.

- Articulate the need for abstraction and the development of mathematical tools and techniques in a general setting.

The objectives of this book make up the framework of how these goals will be accomplished, and more or less follow the chapter headings:

OBJECTIVES OF THIS TEXTBOOK. THE STUDENT WILL BE ABLE TO...

- Identify the interconnections between set theory and mathematical statements and proofs (Chapter 1).
- State the fundamental axioms of the natural, integer, and real number systems and how the completeness axiom of the real number system distinguishes that system from the rational system in a powerful way (Chapter 2).
- Apply the rigorous ε-N definition of convergence for sequences and series and recognize monotone and Cauchy sequences (Chapter 3).
- Apply the rigorous ε-δ definition of limits and continuity for functions and apply the fundamental theorems of continuous functions (Chapter 4).
- Analyze the convergence properties of an infinite series, product, or continued fraction (mainly Chapters 5–8).
- Identify series, product, and continued fraction formulas for the various elementary functions and constants (Throughout!).

In one semester, I usually review parts of Chapters 1 and 2, then cover most of Chapters 3 and 4, and end with some applications from Chapters 5–8.

Although not a history book (though I do give tiny history bites throughout the book) nor a "little" book like Herbert Westren Turnbull's book *The Great Mathematicians*, in the words of Turnbull, I do hope ...

> *If this little book perhaps may bring to some, whose acquaintance with mathematics is full of toil and drudgery, a knowledge of those great spirits who have found in it an inspiration and delight, the story has not been told in vain. There is a largeness about mathematics that transcends race and time: mathematics may humbly help in the market-place, but it also reaches to the stars. To one, mathematics is a game (but what a game!) and to another it is the handmaiden of theology. The greatest mathematics has the simplicity and inevitableness of supreme poetry and music, standing on the borderland of all that is wonderful in Science, and all that is beautiful in Art. Mathematics transfigures the fortuitous concourse of atoms into the tracery of the finger of God.*

Herbert Westren Turnbull (1885–1961). Quoted from (243, p. 141)

I'd like to thank Brett Bernstein and Ye Li for looking over the notes and giving many valuable suggestions, with special thanks to Ye Li for writing up solutions to many problems (which will eventually be available as a student and instructor's guide) and for pushing me to finally get this book into print. Thanks also to Dikran Karagueuzian and Dennis Pixton for using the book when they taught real analysis. The editors at Springer have been wonderful to work with; my thanks to them all. Also, thanks to the many people throughout the world who have emailed me about the book with encouragement and comments, including Jeremiah Goertz, Scott Lindstrom, Zbigniew Szewczak and Fabio Ricci. There are many others to whom I owe thanks, but due to situations beyond my control, I've either lost their emails or was not able to reply. *Please accept my sincerest gratitude to you all.* I thank my wife, Deborah, as well as my children, Melodie, Blaise, Theo, and Harmonie, for their continued support. Amid the difficulties, I thank and dedicate this book to Jesus, my Lord, Savior, and friend, for allowing me to complete this work.

Soli Deo Gloria

Binghamton, NY, USA   Paul Loya

A Word to the Student

One can imagine mathematics as a movie with exciting scenes, action, plots, etc. There are a couple things you can do. First, you can simply sit back and watch the movie playing out. Second, you can take an active role in shaping the movie. A mathematician does both at times, but is more an actor than an observer. I recommend that you be an actor in the great mathematics movie. To do so, I recommend that you read this book with a pencil and paper at hand, writing down definitions, working through examples, filling in any missing details, and of course doing exercises (even the ones that are not assigned).[1] Of course, please feel free to mark up the book as much as you wish with remarks and highlighting and even corrections if you find a typo or error. (Just let me know if you find one!) The sections with an ★ are optional and are meant to showcase some of the most breathtaking scenes in this *Amazing and Aesthetic Aspects of Analysis.*

---

[1]There are many footnotes in this book. Most are quotations from famous mathematicians and others are remarks that I might make to you if I were reading the book with you. All footnotes may be ignored if you wish!

Some of the Most Beautiful Formulas in the World

In addition to the formulas involving $\pi^2/6$ on the front page of this book, below are more of the main characters we'll meet on our journey, Φ (the golden ratio), log 2, π, γ (the Euler–Mascheroni constant), $\zeta(z)$ (the zeta function), $\sqrt{2}$, and e. Indicated are a section and page number where we prove the formula, most of which are proved in different ways on other pages.

$$\Phi = \frac{1+\sqrt{5}}{2} = \sqrt{1+\sqrt{1+\sqrt{1+\sqrt{1+\sqrt{1+\sqrt{1+\cdots}}}}}} \quad \text{(Section 3.3, p. 177)}$$

$$\Phi = 1 + \cfrac{1}{1+\cfrac{1}{1+\cfrac{1}{1+\ddots}}} \quad \text{(Section 3.4, p. 193)}$$

$$\sqrt{2} = 1 + \cfrac{1}{2+\cfrac{1}{2+\cfrac{1}{2+\ddots}}} \quad \text{(Section 3.4, p. 192)}$$

$$\log 2 = 1 - \frac{1}{2} + \frac{1}{3} - \frac{1}{4} + \frac{1}{5} - \frac{1}{6} + \cdots \quad \text{(Section 4.7, p. 311)}$$

$$\frac{2}{\pi} = \sqrt{\frac{1}{2}} \cdot \sqrt{\frac{1}{2}+\frac{1}{2}\sqrt{\frac{1}{2}}} \cdot \sqrt{\frac{1}{2}+\frac{1}{2}\sqrt{\frac{1}{2}+\frac{1}{2}\sqrt{\frac{1}{2}}}} \cdots \quad \text{(Section 5.1, p. 382)}$$

$$\frac{\pi}{2} = \frac{1}{1} \cdot \frac{2}{1} \cdot \frac{2}{3} \cdot \frac{4}{3} \cdot \frac{4}{5} \cdot \frac{6}{5} \cdot \frac{6}{7} \cdot \frac{8}{7} \cdots \quad \text{(Section 5.1, p. 389)}$$

$$\pi = \frac{\left(1+\dfrac{1}{1\cdot 3}\right)\left(1+\dfrac{1}{3\cdot 5}\right)\left(1+\dfrac{1}{5\cdot 7}\right)\left(1+\dfrac{1}{7\cdot 9}\right)\cdots}{\dfrac{1}{1\cdot 3}+\dfrac{1}{3\cdot 5}+\dfrac{1}{5\cdot 7}+\dfrac{1}{7\cdot 9}+\cdots}\quad\text{(Section 5.1, p. 390)}$$

$$\frac{\pi}{4} = \frac{1}{1} - \frac{1}{3} + \frac{1}{5} - \frac{1}{7} + \cdots \quad \text{(Section 5.2, p. 400)}$$

$$\gamma = \frac{\zeta(2)}{2} - \frac{\zeta(3)}{3} + \frac{\zeta(4)}{4} - \frac{\zeta(5)}{5} + \cdots \quad \text{(Section 6.8, p. 517)}$$

$$\log 2 = \frac{2}{1+\sqrt{2}} \cdot \frac{2}{1+\sqrt{\sqrt{2}}} \cdot \frac{2}{1+\sqrt{\sqrt{\sqrt{2}}}} \cdot \frac{2}{1+\sqrt{\sqrt{\sqrt{\sqrt{2}}}}} \cdots \quad \text{(Section 7.1, p. 538)}$$

$$\sin \pi z = \pi z \left(1-\frac{z^2}{1^2}\right)\left(1-\frac{z^2}{2^2}\right)\left(1-\frac{z^2}{3^2}\right)\left(1-\frac{z^2}{4^2}\right)\left(1-\frac{z^2}{5^2}\right)\cdots \quad \text{(Section 7.3, p. 547)}$$

$$\sqrt{2} = \frac{2}{1}\cdot\frac{2}{3}\cdot\frac{6}{5}\cdot\frac{6}{7}\cdot\frac{10}{9}\cdot\frac{10}{11}\cdots \quad \text{(Section 7.3, p. 554)}$$

$$\frac{\pi}{\sin \pi z} = \frac{1}{z} - \frac{2z}{z^2-1^2} + \frac{2z}{z^2-2^2} - \frac{2z}{z^2-3^2} + \frac{2z}{z^2-4^2} - \cdots \quad \text{(Section 7.4, p. 560)}$$

$$\zeta(z) = \frac{2^z}{2^z-1}\cdot\frac{3^z}{3^z-1}\cdot\frac{5^z}{5^z-1}\cdot\frac{7^z}{7^z-1}\cdot\frac{11^z}{11^z-1}\cdot\frac{13^z}{13^z-1}\cdot\frac{17^z}{17^z-1}\cdots \quad \text{(Section 7.6, p. 566)}$$

$$\log 2 = \cfrac{1}{1+\cfrac{1^2}{1+\cfrac{2^2}{1+\cfrac{3^2}{1+\cfrac{4^2}{1+\ddots}}}}}\quad\text{(Section 8.2, p. 600)}$$

$$\frac{4}{\pi} = 1+\cfrac{1^2}{2+\cfrac{3^2}{2+\cfrac{5^2}{2+\cfrac{7^2}{2+\ddots}}}}\quad\text{(Section 8.2, p. 602)}$$

$$e = 1+1+\cfrac{2}{2+\cfrac{3}{3+\cfrac{4}{4+\cfrac{5}{5+\ddots}}}}\quad\text{(Section 8.2, p. 607)}$$

Contents

Part I
Some Standard Curriculum

Chapter 1
Very Naive Set Theory, Functions, and Proofs

Mathematics is not a deductive science—that's a cliché. When you try to prove a theorem, you don't just list the hypotheses, and then start to reason. What you do is trial and error, experimentation, guesswork.
Paul R. Halmos (1916–2006), I Want to Be a Mathematician [99].

One of the goals of this text is to get you proving mathematical statements in real analysis. Set theory provides a safe environment in which to learn about math statements, "if … then," "if and only if," etc., and to learn the logic behind proofs. Since this is an introductory book on analysis, our treatment of sets is "*very* naive," in the sense that we actually don't define sets rigorously, only informally; we are mostly interested in how "they work," not really what they are.

The students at a university, the people in your family, your pets, the food in your refrigerator, are all examples of sets of objects. Mathematically, a set is defined by some property or attribute that an object must have or must not have; if an object has the property, then it's in the set. For example, the collection of all registered students at a university who are signed up for real analysis forms a set. (A student is either signed up for real analysis or is not.) For an example of a property that cannot be used to define a set, try to answer the following question proposed by Bertrand Russell (1872–1970) in 1918 [206, p. 101]:

> **A puzzle for the student:** *A barber in a local town puts up a sign saying that he shaves only those people who do not shave themselves.* "*Who, then, shaves the barber?*"

Try to answer the question. (Does the barber shave himself or does someone else shave him?) In any case, the idea of a set is perhaps the most fundamental idea in all of mathematics. Sets can be combined to form other sets, and the study of such operators is called the *algebra of sets*, which we cover in Section 1.1. In Section 1.2, we look at the relationship between set theory and the language of mathematics.

© Paul Loya 2017
P. Loya, *Amazing and Aesthetic Aspects of Analysis*,
https://doi.org/10.1007/978-1-4939-6795-7_1

Second to sets in fundamental importance is the idea of a function, which we cover in Section 1.3. In order to illustrate relevant examples of sets, we shall presume elementary knowledge of the real numbers. A thorough discussion of real numbers is left for the next chapter.

This chapter is short, since we do not want to spend too much time on set theory so as to start real analysis ASAP. In the words of Paul Halmos [98, p. vi], "general set theory is pretty trivial stuff really, but, if you want to be a mathematician, you need some, and here it is; read it, absorb it, and forget it."

CHAPTER 1 OBJECTIVES: THE STUDENT WILL BE ABLE TO ...

- Manipulate and create new sets from old ones using the algebra of sets.
- Identify the interconnections between set theory and math statements/proofs.
- Define functions and the operations of functions on sets.

1.1   The Algebra of Sets and the Language of Mathematics

In this section, we study sets and various operations on sets, referred to as the *algebra of sets*, from the "*very* naive" (informal, intuitive) viewpoint. We shall see that the algebra of sets is indispensable in many branches of mathematics such as the study of topology in later chapters. Set theory also provides a lot of language by which mathematics and logic are built.

1.1.1   Sets and Intervals

A **set** is a collection of definite, well-distinguished objects, also called **elements** or **members,** which are usually defined by a conditional statement or simply by listing the set's elements. Intuitively, a set can be thought of as a polyethylene bag containing various objects. All sets and objects that we deal with have the property that there is a definite "yes" or "no" answer to whether an object is in a set. If there is no definite answer, paradoxes can arise, as seen in the barber paradox; see also Problem 3 for another puzzle.

Example 1.1 Consider the set of letters in the word *analysis,*

$$A = \{a, n, a, l, y, s, i, s\} = \{a, n, l, y, s, i\} = \{i, s, y, l, n, a\}.$$

Here, the order of the elements in a set and presence of duplicates are immaterial. Sets for which we can list some of the elements include

$$\mathbb{N} = \{1, 2, 3, 4, 5, 6, 7, \dots\} \quad \text{and} \quad \mathbb{Z} = \{\dots, -2, -1, 0, 1, 2, \dots\},$$

the **natural numbers** and **integers,** respectively.

Example 1.2 The **rational numbers** can be written in terms of a conditional statement:

$$\mathbb{Q} := \left\{ x \in \mathbb{R} \, ; \, x = \frac{a}{b}, \text{ where } a, b \in \mathbb{Z} \text{ and } b \neq 0 \right\}.$$

Here, the symbol ":=" means that the symbol on the left is by definition equal to the expression on the right, and we usually read ":=" as "equals by definition".[1] Thus,

---

:= means "is by definition equal to."

The symbol $\in$ means "belongs to" or "is a member of." The symbol $\mathbb{R}$ denotes the set of real numbers, and the semicolon should be read "such that." So, $\mathbb{Q}$ is the set of all real numbers x such that x can be written as a ratio $x = a/b$, where a and b are integers with b not zero.

Example 1.3 The **empty set** is a set with no elements—think of an empty clear plastic bag. We denote this empty set by $\varnothing$. (In the next subsection, we prove that there is only one empty set.)

Example 1.4 Intervals provide many examples of sets defined by conditional statements. Let a and b be real numbers with $a \leq b$. Then the set

$$\{ x \in \mathbb{R} \, ; \, a < x < b \}$$

is called an **open interval** and is often denoted by (a, b). If $a = b$, then there are no real numbers between a and b, so $(a, a) = \varnothing$. The set

$$\{ x \in \mathbb{R} \, ; \, a \leq x \leq b \}$$

is called a **closed interval** and is denoted by $[a, b]$. There are also half-open and half-closed intervals,

$$\{ x \in \mathbb{R} \, ; \, a < x \leq b \}, \qquad \{ x \in \mathbb{R} \, ; \, a \leq x, < b \},$$

called **left-half-open** and **right-half-open** intervals and denoted by $(a, b]$ and $[a, b)$, respectively. The points a and b are called the **endpoints** of the intervals. There are also infinite intervals. The sets

$$\{ x \in \mathbb{R} \, ; \, x < a \}, \qquad \{ x \in \mathbb{R} \, ; \, a < x \}$$

are open intervals, denoted by $(-\infty, a)$ and (a, ∞), respectively, and

$$\{ x \in \mathbb{R} \, ; \, x \leq a \}, \qquad \{ x \in \mathbb{R} \, ; \, a \leq x \}$$

---

[1] "The errors of definitions multiply themselves according as the reckoning proceeds; and lead men into absurdities, which at last they see but cannot avoid, without reckoning anew from the beginning." Thomas Hobbes (1588–1679) [172].

are closed intervals, denoted by $(-\infty, a]$ and $[a, \infty)$, respectively. Note that the sideways eight symbol ∞ for **infinity**, introduced in 1655 by John Wallis (1616–1703) [45, p. 44], is just that, a symbol, and is not to be taken to be a real number. The real line is itself an interval, namely $\mathbb{R} = (-\infty, \infty)$.

1.1.2 Subsets and "If ... Then" Statements

If a belongs to a set A, then we usually say that a is in A, and we write $a \in A$, and if a does not belong to A, then we write $a \notin A$. If each element of a set A is also an element of a set B, we write $A \subseteq B$ and say that A is a **subset** of, or contained in, B. If A is not a subset of B, we write $A \nsubseteq B$. To say that two sets A and B are the same just means that they contain exactly the same elements; in other words, every element in A is also in B (that is, $A \subseteq B$) and also every element in B is also in A (that is, $B \subseteq A$). Thus, we define

$$A = B \quad \text{means that } A \subseteq B \text{ and } B \subseteq A.$$

Here's a picture to consider:

Fig. 1.1 The *left-hand* side displays a subset. The *right-hand* side deals with complements, which we'll look at in Section 1.1.3

Example 1.5 $\mathbb{N} \subseteq \mathbb{Z}$, since every natural number is also an integer, and $\mathbb{Z} \subseteq \mathbb{R}$, since every integer is also a real number, but $\mathbb{R} \nsubseteq \mathbb{Z}$, because not every real number is an integer.

Stated another way, $A \subseteq B$ means that if we take an element of A, that element must also belong to B, that is, the following is a true statement:

$$\text{If } x \in A, \text{ then } x \in B.$$

This statement is an example of a **conditional statement**, or **implication**, and in general is any statement that looks like

$$\text{If } P, \text{ then } Q,$$

where P and Q are statements. Here, P is called the **hypothesis**, and Q is called the **conclusion**, and the implication can be written out more fully as

"If P is a true statement, then Q is also a true statement."

In the subset example above, P is "$x \in A$" and Q is "$x \in B$." We will discuss conditional statements quite a bit in Section 1.2. An important point concerning conditional statements is made in the following example.

Example 1.6 Consider the following statement made by a student Joe:

If my professor cancels class on Friday, then on Friday I'm going fishing.

Obviously, Joe told the truth if indeed the class was canceled and he went fishing, and he lied if the class was canceled, yet Joe stayed in his dorm all Friday. Now what if class was not canceled but Joe still went fishing? Did Joe make a true or false statement? Everyone would certainly agree that Joe's statement is not false, regardless of whether he went fishing. Indeed, Joe said that *if* the professor canceled class, *then* he would go fishing; Joe said nothing about what he would do if the professor did not cancel class. In mathematics, statements are either true or false,[2] so, since Joe's statement is not false, it must therefore be true! This is the standing *convention* mathematicians take for any "if ... then" statement. Thus, given statements P and Q, we consider the statement "If P, then Q" to be true if the statement P is true and the statement Q is also true, and we also regard "If P, then Q" as true if the statement P is false, whether or not the statement Q is true or false. We consider "If P, then Q" to be false if and only if P is true and Q is false. Finally, we consider "If P, then Q" to be false only in the case that P is true and Q is false.

Using the logic in this example, we claim that $\varnothing$, a set with nothing in it, is a subset of every set. To see this, let A be a set. We must verify that the following statement is true:

$$\text{If } x \in \varnothing, \text{ then } x \in A.$$

However, the hypothesis "$x \in \varnothing$" is always false, since $\varnothing$ has nothing in it! Therefore, by our *convention*, the statement "if $x \in \varnothing$, then $x \in A$" is true! Thus, $\varnothing \subseteq A$. We can also see that there is only one empty set, for suppose that $\varnothing'$ is another empty set. Then the same argument that we just made for $\varnothing$ shows that $\varnothing'$ is also a subset of every set. Now to say that $\varnothing = \varnothing'$, we must show that $\varnothing \subseteq \varnothing'$ and $\varnothing' \subseteq \varnothing$. But $\varnothing \subseteq \varnothing'$ holds because $\varnothing$ is a subset of every set, and $\varnothing' \subseteq \varnothing$ holds because $\varnothing'$ is a subset of every set. Therefore, $\varnothing = \varnothing'$.

The fact that we regard "If P, then Q" to be true when P is false has implications when we do proofs:

Remark: Since for a false statement P, we always consider a statement "If P, then Q" to be true, regardless of the validity of the statement Q, if we want to prove that a statement "If P, then Q" is true, we usually start the proof by assuming P is true, then under that assumption, deducing that Q is also true.

---

[2]Later in your math career you will find some "neither true nor false" statements (perhaps a better wording is "neither provable nor refutable") such as, e.g., the continuum hypothesis ... but that is another story! There is no such thing as a "neither statement" in this book.

The following theorem states an important law of sets.

Transitive law

Theorem 1.1 *If $A \subseteq B$ and $B \subseteq C$, then $A \subseteq C$.*

Proof Suppose that $A \subseteq B$ and $B \subseteq C$ (as detailed in our remark, we are assuming "P," that $A \subseteq B$ and $B \subseteq C$, is true). We need to prove that $A \subseteq C$, which by definition means that if $x \in A$, then $x \in C$ is a true statement. So, let x be in A (we are again following our remark, with now "P" being that $x \in A$). We will show that x is also in C. Since x is in A and $A \subseteq B$, we know that x is also in B. Now $B \subseteq C$, and therefore x is also in C. In conclusion, we have proved that if $x \in A$, then $x \in C$, which is exactly what we wanted to prove.   ∎

The **power set** of a given set A is the collection consisting of all subsets of A, which we usually denote by $\mathscr{P}(A)$.

Example 1.7 For the set $\{e, \pi\}$, we have

$$\mathscr{P}(\{e, \pi\}) = \big\{\varnothing, \{e\}, \{\pi\}, \{e, \pi\}\big\}.$$

Before moving on to set operations, we remark that we can see that $\varnothing$ is a subset of every set by invoking the *contrapositive*. Consider again the statement that $A \subseteq B$, meaning that

(1) If $x \in A$, then $x \in B$,

must be a true statement. This statement is equivalent (that is, either both statements are true or both are false) to the **contrapositive** statement

(2) If $x \notin B$, then $x \notin A$.

Indeed, suppose that statement (1) holds, that is, $A \subseteq B$. We shall prove that statement (2) holds. So, let us assume that some x has the property that $x \notin B$ is true; is it true that $x \notin A$? Well, the object x is either in A or it's not. If $x \in A$, then since $A \subseteq B$, we must have $x \in B$. However, we know that $x \notin B$, and so $x \in A$ is not a valid option, and therefore $x \notin A$. Assume now that statement (2) holds; we shall prove that statement (1) holds. So, let $x \in A$. We must prove that $x \in B$. Well, either $x \in B$ or it's not. If $x \notin B$, then we know that $x \notin A$. However, we are given that $x \in A$, so $x \notin B$ is not a valid option. Therefore, the other option $x \in B$ must be true. Therefore, (1) and (2) really say the same thing. We now prove that $\varnothing \subseteq A$ for every set A using the contrapositive. Assume that $x \notin A$; we must prove that $x \notin \varnothing$. But this last statement is true because $\varnothing$ does not contain anything, so $x \notin \varnothing$ is certainly true. Thus, $\varnothing \subseteq A$.

1.1.3 Unions, "or" Statements, Intersections, and Set Differences

Given two sets A and B, their **union**, denoted by $A \cup B$, is the set of elements that are in A or B:

$$A \cup B = \{x \, ; \, x \in A \text{ or } x \in B\}.$$

By "or" we always allow the option that x could be in *both* A and B; in logic, this is called the *inclusive or*. In standard English, "or" can refer to the inclusive or, but it often means "one or the other, but not both," called the *exclusive or*. Only by context or social standards do we know which "or" is meant.

Example 1.8 Your parents offer to buy you a new laptop *or* a new smartphone as a reward for getting an A in calculus *or* making the dean's list. For most people, the first "or" was an exclusive or (since laptops and smartphones are rather expensive), while the second "or" was definitely an inclusive or.

In math, please remember that "or" by *convention* refers to the inclusive or. Thus, $A \cup B$ is the set of elements that are in A or B or in both A and B.

Example 1.9
$$\{0, 1, e, i\} \cup \{e, i, \pi, \sqrt{2}\} = \{0, 1, e, i, \pi, \sqrt{2}\}.$$

The **intersection** of two sets A and B, denoted by $A \cap B$, is the set of elements that are in both A and B:

$$A \cap B = \{x \, ; \, x \in A \text{ and } x \in B\}.$$

(Here, "and" means just what you think it means.)

Example 1.10
$$\{0, 1, e, i\} \cap \{e, i, \pi, \sqrt{2}\} = \{e, i\}.$$

If the sets A and B have no elements in common, then $A \cap B = \varnothing$, and the sets are said to be **disjoint**. Here are some properties of unions and intersections, the proofs of which we leave mostly to the reader.

> **Theorem 1.2** *Unions and intersections are commutative and associative in the sense that if A, B, and C are sets, then*
>
> *(1) $A \cup B = B \cup A$ and $A \cap B = B \cap A$.*
> *(2) $(A \cup B) \cup C = A \cup (B \cup C)$ and $(A \cap B) \cap C = A \cap (B \cap C)$.*

Proof Consider the proof that $A \cup B = B \cup A$. By definition of equality of sets, we must show that $A \cup B \subseteq B \cup A$ and $B \cup A \subseteq A \cup B$. To prove that $A \cup B \subseteq$

$B \cup A$, let x be in $A \cup B$. Then by definition of union, $x \in A$ or $x \in B$. This of course is the same thing as $x \in B$ or $x \in A$. Therefore, x is in $B \cup A$. The proof that $B \cup A \subseteq A \cup B$ is similar. Therefore, $A \cup B = B \cup A$. We leave the proof that $A \cap B = B \cap A$ to the reader. We also leave the proof of *(2)* to the reader. ∎

Our last operation on sets is the **set difference** $A \setminus B$ (read "A take away B" or the "**complement** of B in A"), which is the set of elements of A that do not belong to B. Thus,

$$A \setminus B = \{x \,;\, x \in A \text{ and } x \notin B\}.$$

Example 1.11

$$\{0, 1, e, i\} \setminus \{e, i, \pi, \sqrt{2}\} = \{0, 1\}.$$

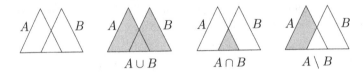

$A \cup B \qquad\qquad A \cap B \qquad\qquad A \setminus B$

Fig. 1.2 Visualization of the various set operations. Here, A and B are overlapping *triangles*

Figure 1.2 shows some pictorial representations of union, intersection, and set difference. Such pictures are called **Venn diagrams** after John Venn (1834–1923), who introduced them.

In any given situation, we are usually working with subsets of some underlying set X (our "universe"). Given any subset A of X, we denote $X \setminus A$, the set of elements in X that are outside of A, by A^c, called the **complement** of A; see Fig. 1.1 on p. 6. Therefore,

$$A^c = X \setminus A = \{x \in X \,;\, x \notin A\}.$$

Example 1.12 Let us take our "universe" to be $\mathbb{R}$. Then,

$$(-\infty, 1]^c = \{x \in \mathbb{R} \,;\, x \notin (-\infty, 1]\} = (1, \infty),$$

and

$$[0, 1]^c = \{x \in \mathbb{R} \,;\, x \notin [0, 1]\} = (-\infty, 0) \cup (1, \infty).$$

In any given situation, the set X will always be clear from context, either because it is stated what X is, or because we are working in, say a section dealing with only real numbers, so $\mathbb{R}$ is by default the universal set. Otherwise, we assume that X is "there" but simply not stated.

1.1.4 Arbitrary Unions and Intersections

We can also consider arbitrary (finite or infinite) unions and intersections. Let I be a nonempty set and assume that for each $\alpha \in I$, there corresponds a set A_α. The sets A_α where $\alpha \in I$ are said to be a **family** of sets **indexed** by I, which we often denote by $\{A_\alpha \,;\, \alpha \in I\}$. An index set that shows up quite often is $I = \mathbb{N}$; in this case, we usually call $\{A_n \,;\, n \in \mathbb{N}\}$ a **sequence** of sets.

Example 1.13 The sets $A_1 = [0, 1]$, $A_2 = [0, 1/2]$, $A_3 = [0, 1/3]$, and in general,

$$A_n = \left[0, \frac{1}{n}\right] = \left\{x \in \mathbb{R}\,;\, 0 \leq x \leq \frac{1}{n}\right\},$$

form a family of sets indexed by $\mathbb{N}$ (or a sequence of sets). Here's a picture of these sets:

Fig. 1.3 The sequence of sets $A_n = [0, 1/n]$ for $n \in \mathbb{N}$

How do we define the union of all the sets A_α in a family $\{A_\alpha \,;\, \alpha \in I\}$? Consider the case of two sets A and B. We can write

$$A \cup B = \{x \,;\, x \in A \text{ or } x \in B\}$$
$$= \{x \,;\, x \text{ is in at least one of the sets on the left-hand side}\}.$$

With this as motivation, we define the union of all the sets A_α as

$$\boxed{\bigcup_{\alpha \in I} A_\alpha = \{x \,;\, x \in A_\alpha \text{ for at least one } \alpha \in I\}.}$$

To simplify notation, we sometimes just write $\bigcup A_\alpha$ or $\bigcup_\alpha A_\alpha$ for the left-hand side.

Example 1.14 For the sequence $\{A_n \,;\, n \in \mathbb{N}\}$, where $A_n = [0, 1/n]$, by staring at Fig. 1.3, we see that

$$\bigcup_{n \in \mathbb{N}} A_n = \{x \,;\, x \in [0, 1/n] \text{ for at least one } n \in \mathbb{N}\} = [0, 1].$$

We how do we define the intersection of all the sets A_α in a family $\{A_\alpha \,;\, \alpha \in I\}$? Consider the case of two sets A and B. We can write

$$A \cap B = \{x \; ; \; x \in A \text{ and } x \in B\}$$
$$= \{x \; ; \; x \text{ is in every set on the left-hand side}\}.$$

With this as motivation, we define the intersection of all the sets A_α as

$$\boxed{\bigcap_{\alpha \in I} A_\alpha = \{x \; ; \; x \in A_\alpha \text{ for every } \alpha \in I\}.}$$

To simplify notation, we sometimes just write $\bigcap A_\alpha$ or $\bigcap_\alpha A_\alpha$ for the left-hand side.

Example 1.15 For the sequence $A_n = [0, 1/n]$ in Fig. 1.3, we have

$$\bigcap_{n \in \mathbb{N}} A_n = \{x \; ; \; x \in [0, 1/n] \text{ for every } n \in \mathbb{N}\} = \{0\}.$$

If $I = \{1, 2, \ldots, N\}$ is a finite set of natural numbers, or if $I = \mathbb{N}$, then it's common to use the notation

$$\boxed{\bigcup_{n=1}^{N} A_n \text{ instead of } \bigcup_{\alpha \in \{1,2,\ldots,N\}} A_\alpha \quad \text{and} \quad \bigcup_{n=1}^{\infty} A_n \text{ instead of } \bigcup_{\alpha \in \mathbb{N}} A_\alpha,}$$

with similar notation for intersections.

> **Theorem 1.3** *Let A be a set and $\{A_\alpha\}$ a family of sets. Then unions and intersections distribute in the sense that*
>
> $$A \cap \bigcup_\alpha A_\alpha = \bigcup_\alpha (A \cap A_\alpha), \qquad A \cup \bigcap_\alpha A_\alpha = \bigcap_\alpha (A \cup A_\alpha)$$
>
> *and satisfy the Augustus De Morgan (1806–1871) laws:*
>
> $$A \setminus \bigcup_\alpha A_\alpha = \bigcap_\alpha (A \setminus A_\alpha), \qquad A \setminus \bigcap_\alpha A_\alpha = \bigcup_\alpha (A \setminus A_\alpha).$$

Proof We shall prove the first distributive law and leave the second one to the reader. We need to show that $A \cap \bigcup_\alpha A_\alpha = \bigcup_\alpha (A \cap A_\alpha)$, which means that

$$A \cap \bigcup_\alpha A_\alpha \subseteq \bigcup_\alpha (A \cap A_\alpha) \quad \text{and} \quad \bigcup_\alpha (A \cap A_\alpha) \subseteq A \cap \bigcup_\alpha A_\alpha. \qquad (1.1)$$

To prove the first inclusion, let $x \in A \cap \bigcup_\alpha A_\alpha$; we must show that $x \in \bigcup_\alpha (A \cap A_\alpha)$. The statement $x \in A \cap \bigcup_\alpha A_\alpha$ means that $x \in A$ and $x \in \bigcup_\alpha A_\alpha$, which means, by the definition of union, that $x \in A$ and $x \in A_\alpha$ for some α. Hence, $x \in A \cap A_\alpha$ for some α, which is to say that $x \in \bigcup_\alpha (A \cap A_\alpha)$. Consider now the second inclusion in (1.1). To prove this, let $x \in \bigcup_\alpha (A \cap A_\alpha)$. This means that $x \in A \cap A_\alpha$ for some α. Therefore, by definition of intersection, $x \in A$ and $x \in A_\alpha$ for some α. This means that $x \in A$ and $x \in \bigcup_\alpha A_\alpha$, which is to say that $x \in A \cap \bigcup_\alpha A_\alpha$. Having established both inclusions in (1.1), we've proved the equality of the sets.

We shall prove the first De Morgan law and leave the second to the reader. We need to show that $A \setminus \bigcup_\alpha A_\alpha = \bigcap_\alpha (A \setminus A_\alpha)$, which means that

$$A \setminus \bigcup_\alpha A_\alpha \subseteq \bigcap_\alpha (A \setminus A_\alpha) \quad \text{and} \quad \bigcap_\alpha (A \setminus A_\alpha) \subseteq A \setminus \bigcup_\alpha A_\alpha. \tag{1.2}$$

To prove the first inclusion, let $x \in A \setminus \bigcup_\alpha A_\alpha$. This means that $x \in A$ and $x \notin \bigcup_\alpha A_\alpha$. For x not to be in the union, it must be that $x \notin A_\alpha$ for any α whatsoever (because if x happened to be in some A_α, then x would be in the union $\bigcup_\alpha A_\alpha$, which we know x is not). Hence, $x \in A$ and $x \notin A_\alpha$ for all α, or in other words, $x \in A \setminus A_\alpha$ for all α, which means that $x \in \bigcap_\alpha (A \setminus A_\alpha)$. We now prove the second inclusion in (1.2). So, let $x \in \bigcap_\alpha (A \setminus A_\alpha)$. This means that $x \in A \setminus A_\alpha$ for all α. Therefore, $x \in A$ and $x \notin A_\alpha$ for all α. Since x is not in any A_α, it follows that $x \notin \bigcup_\alpha A_\alpha$. Therefore, $x \in A$ and $x \notin \bigcup_\alpha A_\alpha$, and hence $x \in A \setminus \bigcup_\alpha A_\alpha$. In summary, we have established both inclusions in (1.2), which proves the equality of the sets. ∎

If $A = X$, the universe in which we are working, then the De Morgan laws are

$$\left(\bigcup_\alpha A_\alpha \right)^c = \bigcap_\alpha A_\alpha, \quad \left(\bigcap_\alpha A_\alpha \right)^c = \bigcup_\alpha A_\alpha^c.$$

The best way to remember De Morgan's laws is the versions expressed in plain English:

> *The complement of a union is the intersection of the complements and the complement of an intersection is the union of the complements.*

For a family $\{A_\alpha\}$ consisting of just two sets B and C, the distributive and De Morgan laws are

$$A \cap (B \cup C) = (A \cap B) \cup (A \cap C), \quad A \cup (B \cap C) = (A \cup B) \cap (A \cup C)$$

and

$$A \setminus (B \cup C) = (A \setminus B) \cap (A \setminus C), \quad A \setminus (B \cap C) = (A \setminus C) \cup (A \setminus C).$$

Here are some exercises in which we ask you to prove statements concerning sets. In Problem 2, it is very helpful to draw Venn diagrams to see why the statements are true. Here is some useful advice as you do problems: if you can't see how to prove something after some effort, take a break and come back to the problem later.[3]

▶ **Exercises 1.1**

1. Prove that $\varnothing = \{x \,;\, x \neq x\}$. Is the following statement true, false, or neither: if $x \in \varnothing$, then real analysis is everyone's favorite class.

2. For sets A, B, C, prove some of the following statements:

   (a) $A \cup \varnothing = A$ and $A \cap \varnothing = \varnothing$.
   (b) $A \setminus B = A \cap B^c$.
   (c) $(A^c)^c = A$.
   (d) $A \cap B = A \setminus (A \setminus B)$.
   (e) $B \cap (A \setminus B) = \varnothing$.
   (f) If $A \subseteq B$, then $B = A \cup (B \setminus A)$.
   (g) $A \cup B = A \cup (B \setminus A)$.
   (h) $A \subseteq A \cup B$ and $A \cap B \subseteq A$.
   (i) If $A \cap B = A \cap C$ and $A \cup C = A \cup B$, then $B = C$.
   (j) $(A \setminus B) \setminus C = (A \setminus C) \setminus (B \setminus C)$
   (k) $(A \cap B) \cup C = A \cap (B \cup C)$ if and only if $C \subseteq A$.

3. **(Russell's paradox)**[4] Define a "thing" to be any collection of items. The reason that we use the word "thing" is that these things are not sets. Let

$$\mathscr{B} = \{\text{"things" } A \,;\, A \notin A\},$$

   that is, $\mathscr{B}$ is the collection of all "things" that do not contain themselves. Questions: Is $\mathscr{B}$ a "thing"? Is $\mathscr{B} \in \mathscr{B}$ or is $\mathscr{B} \notin \mathscr{B}$? Is $\mathscr{B}$ a set?

4. Find

$$(a)\ \bigcup_{n=1}^{\infty} \left(0, \frac{1}{n}\right), \quad (b)\ \bigcap_{n=1}^{\infty} \left(0, \frac{1}{n}\right), \quad (c)\ \bigcup_{n=1}^{\infty} \left[\frac{1}{2^n}, \frac{1}{2^{n-1}}\right),$$

$$(d)\ \bigcap_{n=1}^{\infty} \left[\frac{1}{2^n}, \frac{1}{2^{n-1}}\right), \quad (e)\ \bigcap_{\alpha \in \mathbb{R}} (\alpha, \infty), \quad (f)\ \bigcap_{\alpha \in (0,\infty)} \left[1, 1 + \frac{1}{\alpha}\right].$$

---

[3]"Finally, two days ago, I succeeded—not on account of my hard efforts, but by the grace of the Lord. Like a sudden flash of lightning, the riddle was solved. I am unable to say what was the conducting thread that connected what I previously knew with what made my success possible." Carl Friedrich Gauss (1777–1855) [72].

[4]"The point of philosophy is to start with something so simple as not to seem worth stating, and to end with something so paradoxical that no one will believe it." Bertrand Russell (1872–1970).

1.2 Set Theory and Mathematical Statements

As already mentioned, set theory provides a comfortable environment in which to do proofs and to learn the ins and outs of mathematical statements.[5] In this section we give a brief account of the various ways mathematical statements can be worded using the background of set theory.

1.2.1 More on "if ... Then" Statements

We begin by exploring different ways of saying "if ... then." Consider again the statement that $A \subseteq B$:

$$\text{If } x \in A, \text{ then } x \in B.$$

Other common ways to write this are

$$x \in A \text{ implies } x \in B \quad \text{ and } \quad x \in A \Longrightarrow x \in B;$$

that is, x belongs to A implies that x also belongs to B. Here,

> $\Longrightarrow$ is the common symbol for "implies."

Another way to say this is

$$x \in A \text{ only if } x \in B;$$

that is, the object x belongs to A only if it follows that x also belongs to B. Here is yet one more way to write the statement:

$$x \in B \text{ if } x \in A;$$

that is, the object x belongs to B if, or given that, the object x belongs to A. Finally, we also know that the **contrapositive** statement says the same thing:

$$\text{If } x \notin B, \text{ then } x \notin A.$$

Thus, "If $x \in A$, then $x \in B$" can also be written as

---

[5]"Another advantage of a mathematical statement is that it is so definite that it might be definitely wrong; and if it is found to be wrong, there is a plenteous choice of amendments ready in the mathematicians' stock of formulae. Some verbal statements have not this merit; they are so vague that they could hardly be wrong, and are correspondingly useless." Lewis Richardson (1881–1953). Mathematics of War and Foreign Politics.

$$x \in A \implies x \in B; \quad x \in A \text{ implies } x \in B; \quad \text{Given } x \in A, x \in B;$$
$$x \in A \text{ only if } x \in B; \quad x \in B \text{ if } x \in A; \quad \text{If } x \notin B, \text{ then } x \notin A. \tag{1.3}$$

We now consider each of these set statements in more generality. First of all, a *statement* in the mathematical sense is a statement that is either true or false, but never both, much in the same way that we work with only sets and objects such that any given object is either in or not in a given set, but never both. A mathematical statement always has **hypotheses** or **assumptions**, and a **conclusion**. Almost always there are **hidden assumptions**, that is, assumptions that are not stated, but taken for granted, because the context makes it clear what these assumptions are. Whenever you read a mathematical statement, make sure that you fully understand the hypotheses or assumptions (including hidden ones) and the conclusion. For the statement "If $x \in A$, then $x \in B$," the assumption is $x \in A$ and the conclusion is $x \in B$. The "if–then" wording means: if the assumptions ($x \in A$) are true, then the conclusion ($x \in B$) is also true, or stated another way, given that the assumptions are true, the conclusion follows. Let P denote the statement that $x \in A$, and Q the statement that $x \in B$. Then rewriting the statements (1.3) in terms of P's and Q's, we see that the following statements are equivalent[6]:

$$\boxed{\begin{array}{ll} \text{If } P, \text{ then } Q; \quad P \implies Q; \quad P \text{ implies } Q; \quad \text{given } P, Q \text{ holds}; \\ \quad P \text{ only if } Q; \quad Q \text{ if } P; \quad \text{if not } Q, \text{ then not } P. \end{array}} \tag{1.4}$$

Each of these statements is for P being $x \in A$ and Q being $x \in B$, but as you can probably guess, they work for any mathematical statements P and Q. Let us consider statements concerning real numbers.

Example 1.16 Each of the following statements is equivalent to "$x > 5 \implies x^2 > 100$":

$$\text{If } x > 5, \text{ then } x^2 > 100; \quad x > 5 \text{ implies } x^2 > 100; \quad \text{Given } x > 5, x^2 > 100;$$
$$x > 5 \text{ only if } x^2 > 100; \quad x^2 > 100 \text{ if } x > 5; \quad \text{If } x^2 \leq 100, \text{ then } x \leq 5.$$

The hidden assumptions are that x represents a real number and that the real numbers satisfy all the axioms you think they do. For the last statement, we used that "not $x^2 > 100$" (that is, it's *not* true that $x^2 > 100$" is the same as "$x^2 \leq 100$" and that "not $x > 5$" is the same as $x \leq 5$." Of course, any one (and hence every one) of the six statements is false. For instance, $x = 6 > 5$ is true, but $x^2 = 36$, which is not greater than 100.

Example 1.17 Each of the following statements is equivalent to "$x^2 = 2 \implies x$ is irrational":

---

[6] P implies Q is sometimes translated as "P is sufficient for Q" in the sense that the truth of P is sufficient or enough or ample to imply that Q is also true. Moreover, P implies Q is also translated "Q is necessary for P" because Q is necessarily true given that P is true. However, we shall not use this language in this book.

If $x^2 = 2$, then x is irrational; $x^2 = 2$ implies x is irrational;

Given $x^2 = 2$, x is irrational ; $x^2 = 2$ only if x is irrational ;

x is irrational if $x^2 = 2$;   If x is rational, then $x^2 \neq 2$.

Again, the hidden assumptions are that x represents a real number and that the real numbers satisfy all their usual properties. Any one (and hence every one) of these six statements is of course true (since we have been told since high school that $\pm\sqrt{2}$ are irrational; we shall prove this fact in Section 2.6 on p. 81).

As these two examples show, it is very important to remember that none of the statements in (1.4) assert that P or Q is true; they simply state that *if* P is true, *then* Q is also true.

1.2.2 Converse Statements and "if and only if" Statements

Given a statement P implies Q, the reverse statement Q implies P is called the **converse** statement. For example, back to set theory, the converse of the statement

$$\text{If } x \in A, \text{ then } x \in B; \text{ that is, } A \subseteq B,$$

is just the statement that

$$\text{If } x \in B, \text{ then } x \in A; \text{ that is, } B \subseteq A.$$

These set theory statements make it clear that the converse of a true statement may not be true, for $\{e, \pi\} \subseteq \{e, \pi, i\}$, but $\{e, \pi, i\} \nsubseteq \{e, \pi\}$. Let us consider examples with real numbers.

Example 1.18 The statement "If $x^2 = 2$, then x is irrational" is true, but its converse statement, "If x is irrational, then $x^2 = 2$," is false.

Statements for which the converse is equivalent to the original statement are called "if and only if" statements.

Example 1.19 Consider the statement "If $x = -5$, then $2x + 10 = 0$." This statement is true. Its converse statement is "If $2x + 10 = 0$, then $x = -5$." By solving the equation $2x + 10 = 0$, we see that the converse statement is also true.

The implication $x = -5 \implies 2x + 10 = 0$ can be written

$$2x + 10 = 0 \text{ if } x = -5, \tag{1.5}$$

while the implication $2x + 10 = 0 \implies x = -5$ can be written

$$2x + 10 = 0 \text{ only if } x = -5. \tag{1.6}$$

Combining the two statements (1.5) and (1.6) into one statement, we get

$$2x + 10 = 0 \text{ } if \text{ and } only \text{ } if \text{ } x = -5.$$

This statement is often written using a double arrow,

$$2x + 10 = 0 \Longleftrightarrow x = -5,$$

and is often stated as $2x + 10 = 0$ *is equivalent to* $x = -5$. We regard the statements $2x + 10 = 0$ and $x = -5$ as equivalent because if one statement is true, then so is the other one, whence the wording "is equivalent to." In summary, if the statements

$$Q \text{ if } P \text{ (that is, } P \Longrightarrow Q) \quad \text{and} \quad Q \text{ only if } P \text{ (that is, } Q \Longrightarrow P)$$

are either both true or both false, then we write

$$\boxed{Q \text{ if and only if } P \quad \text{or} \quad Q \Longleftrightarrow P.}$$

Also, if you are asked to prove a statement "Q if and only if P," then you have to prove both the "if" statement "Q if P" (that is, $P \Longrightarrow Q$) and the "only if" statement "Q only if P" (that is, $Q \Longrightarrow P$).

The if and only if notation $\Longleftrightarrow$ comes in quite handy in proofs whenever we want to move from one statement to an equivalent one.

Example 1.20 In the proof of Theorem 1.3 on p. 12, we wanted to show that $A \cap \bigcup_\alpha A_\alpha = \bigcup_\alpha (A \cap A_\alpha)$, which means that $A \cap \bigcup_\alpha A_\alpha \subseteq \bigcup_\alpha (A \cap A_\alpha)$ and $\bigcup_\alpha (A \cap A_\alpha) \subseteq A \cap \bigcup_\alpha A_\alpha$, or

$$x \in A \cap \bigcup_\alpha A_\alpha \Longrightarrow x \in \bigcup_\alpha (A \cap A_\alpha) \quad \text{and} \quad x \in \bigcup_\alpha (A \cap A_\alpha) \Longrightarrow x \in A \cap \bigcup_\alpha A_\alpha.$$

That is, we wanted to prove that

$$x \in A \cap \bigcup_\alpha A_\alpha \Longleftrightarrow x \in \bigcup_\alpha (A \cap A_\alpha).$$

We can prove this quickly and easily using $\Longleftrightarrow$:

$$x \in A \cap \bigcup_\alpha A_\alpha \Longleftrightarrow x \in A \text{ and } x \in \bigcup_\alpha A_\alpha \Longleftrightarrow x \in A \text{ and } x \in A_\alpha \text{ for some } \alpha$$

$$\Longleftrightarrow x \in A \cap A_\alpha \text{ for some } \alpha$$

$$\Longleftrightarrow x \in \bigcup_\alpha (A \cap A_\alpha).$$

Just make sure that if you use $\Longleftrightarrow$, *the expressions to the immediate left and right of* $\Longleftrightarrow$ *are indeed equivalent.*

1.2.3 Negations and Logical Quantifiers

We already know that a statement and its contrapositive are always equivalent: "if P, then Q" is equivalent to "if not Q, then not P." Therefore, it is important to know how to "not" something, that is, find the **negation** of a statement. Here, for any statement S, "not S" is the statement that S is false. A common notation for "not S" is $\neg S$, but we'll use "not" instead of $\neg$. Sometimes the negation is obvious.

Example 1.21 Under the assumption that x represents a real number, the negation of the statement that $x > 5$ is $x \leq 5$, and the negation of the statement that x is irrational is that x is rational.

Some statements are not so easy to negate, especially when there are **logical quantifiers**, such as the **universal quantifier**

$$\text{"for every"} = \text{"for each"} = \text{"for all" (denoted by } \forall),$$

and the **existential quantifier**

$$\text{"for some"} = \text{"there exists"} = \text{"there is"} = \text{"for at least one" (denoted by } \exists).$$

The equal signs represent the fact that we mathematicians consider "for every" as another way of saying "for all," "for some" as another way of saying "there exists," and so forth. Working under the assumption that the numbers under consideration are real, consider the statement

$$\text{For every } x, x^2 \geq 0. \tag{1.7}$$

What is the negation of this statement? One way to find out is to think of this in terms of set theory. Let $A = \{x \in \mathbb{R} \, ; \, x^2 \geq 0\}$. Then the statement (1.7) is just that $A = \mathbb{R}$. It is obvious that the negation of the statement $A = \mathbb{R}$ is just $A \neq \mathbb{R}$. Now this means that there must exist some real number x such that $x \notin A$. In order for x to not be in A, it must be that $x^2 < 0$. Therefore, $A \neq \mathbb{R}$ just means there is a real number x such that $x^2 < 0$. Hence, the negation of (1.7) is just

$$\text{For at least one } x, x^2 < 0.$$

Thus, the "for every" statement (1.7) becomes a "there is" statement. In general, the negation of a statement of the form

"For every x, P" is the statement "For at least one x, not P."

Similarly, the negation of a "there is" statement becomes a "for every" statement. Explicitly, the negation of

"For at least one x, Q" is the statement "For every x, not Q."

Example 1.22 With the understanding that x represents a real number, the negation of "There is an x such that $x^2 = 2$" is "For every x, $x^2 \neq 2$".

Using sets, one can also find the negation of other statements. For instance, De Morgan's laws for complements of unions and intersections can be used to see that

the negation of "P or Q" is the statement "not P and not Q, "

and the negation of "P and Q" is "not P or not Q."

▶ **Exercises 1.2**

1. In this problem, all numbers are understood to be real. Write down the contrapositive and converse of the following statement:

$$\text{If } x^2 - 2x + 10 = 25, \text{ then } x = 5,$$

and determine which (if any) of the three statements are true.
2. Here are some more set theory proofs to brush up on.

   (a) Prove that $A = A \cup B$ if and only if $B \subseteq A$.
   (b) Prove that $A = A \cap B$ if and only if $A \subseteq B$.

3. Write the negation of the following statements, where x represents an integer.

   (a) For every x, $2x + 1$ is odd.
   (b) There is an x such that $2^x + 1$ is prime.[7]
   (c) If $x^2 - 5x + 6 = 0$, then $x = 2$ or $x = 3$.

1.3   What Are Functions?

In high school, we learned that a function is a "rule that assigns to each input exactly one output," like a machine:

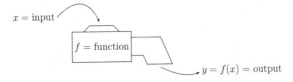

$x = $ input

$f = $ function

$y = f(x) = $ output

---

[7] An integer exceeding 1 that is not prime.

In practice, what usually comes to mind is a formula, such as

$$p(x) = x^2 - 3x + 10.$$

In fact, to Gottfried Leibniz (1646–1716), who in 1692 (or as early as 1673) introduced the word "function" [236, p. 272], and to all the mathematicians of the seventeenth century, a function was always associated with some type of analytic expression, "a formula." However, because of necessity related to problems in mathematical physics, the notion of function was generalized throughout the years, and in this section we present the modern view of what a function is; see [127] or [150, 151] for some history.

1.3.1 (Cartesian) Product

If A and B are sets, their **(Cartesian) product**, denoted by $A \times B$, is the set of all 2-tuples (or ordered pairs) whose first element is in A and second element is in B (see Problem 6). Explicitly,

$$A \times B = \{(a, b) \, ; \, a \in A, \, b \in B\}.$$

We use the adjective "ordered" because we distinguish between ordered pairs, e.g., $(e, \pi) \neq (\pi, e)$, but as sets we regard them as equal, $\{e, \pi\} = \{\pi, e\}$. Of course, one can also define the product of any finite number of sets

$$A_1 \times A_2 \times \cdots \times A_m$$

as the set of all m-tuples $(a_1, \ldots, a_m)$, where $a_k \in A_k$ for each $k = 1, \ldots, m$.

Example 1.23 Of particular interest is m-dimensional Euclidean space (see Section 2.9 starting on p. 118):

$$\mathbb{R}^m := \underbrace{\mathbb{R} \times \cdots \times \mathbb{R}}_{(m \text{ times})}.$$

1.3.2 Functions

Let X and Y be sets. Informally, we say that a function f from X to Y, denoted by $f : X \longrightarrow Y$, is a rule that associates to each element $x \in X$, a single element $y \in Y$. We usually think of a function as a "machine," or if we think of X and Y abstractly as "blobs" of points, then we usually draw an abstract function as associating points in X to points in Y as follows:

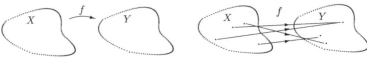

where on the right we draw an arrow from a point $x \in X$ to the associated point $y \in Y$. Mathematically, we define a **function** from X to Y as a subset f of the product $X \times Y$ such that each element $x \in X$ appears exactly once as the first entry of an ordered pair in the subset f. Explicitly, for each $x \in X$ there is a unique $y \in Y$ such that $(x, y) \in f$.

Example 1.24 For instance,

$$p = \{(x, y) ; \; x \in [0, 1] \text{ and } y = x^2 - 3x + 10\} \subseteq [0, 1] \times \mathbb{R} \qquad (1.8)$$

defines a function $p : [0, 1] \longrightarrow \mathbb{R}$, since p is an example of a subset of $[0, 1] \times \mathbb{R}$ such that each real number $x \in [0, 1]$ appears exactly once as the first entry of an ordered pair in p; e.g., the real number 1 appears as the first entry of $(1, 1^2 - 3 \cdot 1 + 10) = (1, 8)$, and there is no other ordered pair in the set (1.8) with 1 as the first entry. Thus, p satisfies the mathematical definition of a function, as you thought it should!

If $f : X \longrightarrow Y$ is a function, then we say that f **maps X to Y**, and f **is a map from X to Y**. If $Y = X$, so that $f : X \longrightarrow X$, we say that f is a **function on X**. For a function $f : X \longrightarrow Y$, the **domain** of f is X, the **codomain** or **target** of f is Y, and the **range** of f, sometimes denoted by $R(f)$, is the set of all elements in Y that occur as the second entry of an ordered pair in f. If $(x, y) \in f$ (recall that f is a set of ordered pairs), then we call the second entry y the **value** or **image** of the function at x and we write $y = f(x)$, and sometimes we write

$$x \mapsto y = f(x).$$

Using this $f(x)$ notation, which by the way was introduced in 1734 by Leonhard Euler (1707–1783) [183], [35, p. 443], we have

$$f = \{(x, y) \in X \times Y ; \; y = f(x)\} = \{(x, f(x)) ; \; x \in X\} \subseteq X \times Y, \qquad (1.9)$$

and

$$R(f) = \{y \in Y ; \; y = f(x) \text{ for some } x \in X\} = \{f(x) ; \; x \in X\}.$$

Figure 1.4 shows the familiar **graph** illustration of domain, codomain, and range, for the trig function $\tau : \mathbb{R} \longrightarrow \mathbb{R}$ given by $\tau(x) = \sin x$.

Also using this $f(x)$ notation, we can return to our previous ways of thinking of functions. For instance, we can say "let $p : [0, 1] \longrightarrow \mathbb{R}$ be the function $p(x) = x^2 - 3x + 10$" or "let $p : [0, 1] \longrightarrow \mathbb{R}$ be the function $x \mapsto x^2 - 3x + 10$," by which we mean of course the set (1.8). In many situations in this book, we are dealing with a fixed codomain such as functions whose codomain is $\mathbb{R}$, which we call **real-valued functions**. Then we can omit the codomain and simply say, "let p be the function

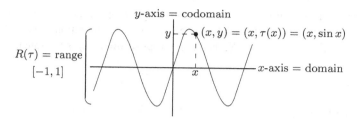

Fig. 1.4 The function $\tau : \mathbb{R} \longrightarrow \mathbb{R}$ defined by $\tau(x) = \sin x$

$x \mapsto x^2 - 3x + 10$ for $x \in [0, 1]$." In this case, we again mean the set (1.8). We shall also deal quite a bit with complex-valued functions, that is, functions whose codomain is $\mathbb{C}$. Then if we say, "let f be a complex-valued function on $[0, 1]$," we mean that $f : [0, 1] \longrightarrow \mathbb{C}$ is a function. Here are some more examples.

Example 1.25 Consider the function $s : \mathbb{N} \longrightarrow \mathbb{R}$ defined by

$$s = \left\{ \left(n, \frac{(-1)^n}{n} \right) ; n \in \mathbb{N} \right\} \subseteq \mathbb{N} \times \mathbb{R}.$$

We usually define $s(n) = \frac{(-1)^n}{n}$ by s_n and write $\{s_n\}$ for the function s, and we call $\{s_n\}$ a **sequence**. We shall study sequences in great depth in Chapter 3.

Example 1.26 Here is a function $a : \mathbb{R} \longrightarrow \mathbb{R}$ defined "piecewise":

$$a(x) = \begin{cases} x & \text{if } x \geq 0; \\ -x & \text{if } x < 0. \end{cases}$$

Of course, $a(x)$ is usually denoted by $|x|$ and is called the **absolute value function**.

Example 1.27 Here's an example of a "pathological function," the **Dirichlet function**, named after Lejeune Dirichlet (1805–1859), which is the function $\mathcal{D} : \mathbb{R} \longrightarrow \mathbb{R}$ defined by

$$\mathcal{D}(x) = \begin{cases} 1 & \text{if } x \text{ is rational;} \\ 0 & \text{if } x \text{ is irrational.} \end{cases}$$

This function was introduced in 1829 in Dirichlet's study of Fourier series and was the first function (1) not given by an analytic expression and (2) not continuous anywhere [127, p. 292]. See Section 4.3, p. 261, for more on continuous functions. Here's an attempted graph of Dirichlet's function:

In elementary calculus, you often encountered composition of functions when learning, for instance, the chain rule. Here is the precise definition of composition. If $f : X \longrightarrow Y$ and $g : Z \longrightarrow X$, then the **composition** $f \circ g$ is the function

$$f \circ g : Z \longrightarrow Y$$

defined by $(f \circ g)(z) := f(g(z))$ for all $z \in Z$. As a set of ordered pairs, $f \circ g$ is given by (do you see why?)

$$f \circ g = \{(z, y) \in Z \times Y ; \text{ for some } x \in X, (z, x) \in g \text{ and } (x, y) \in f\} \subseteq Z \times Y.$$

Also, when learning about the exponential or logarithmic functions, you probably encountered inverse functions. Here are some definitions related to this area. A function $f : X \longrightarrow Y$ is called **one-to-one** or **injective** if for each $y \in R(f)$, there is exactly one $x \in X$ with $y = f(x)$. Another way to state this is

$$\boxed{f \text{ is one-to-one means: If } f(x_1) = f(x_2), \text{ then } x_1 = x_2.} \qquad (1.10)$$

In terms of the contrapositive, we have

$$\boxed{f \text{ is one-to-one means: If } x_1 \neq x_2, \text{ then } f(x_1) \neq f(x_2).} \qquad (1.11)$$

In case $f : X \longrightarrow Y$ is injective, the **inverse** map f^{-1} is the map with domain $R(f)$ and codomain X:

$$f^{-1} : R(f) \longrightarrow X$$

defined by $f^{-1}(y) := x$, where $y = f(x)$. The function f is called **onto** or **surjective** if $R(f) = Y$; that is,

$$\boxed{f \text{ is onto means: For every } y \in Y \text{ there is an } x \in X \text{ such that } y = f(x).} \qquad (1.12)$$

A one-to-one and onto map is called a **bijection**. Here is a picture to consider:

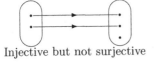

Injective but not surjective

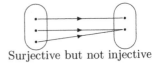

Surjective but not injective

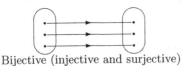

Bijective (injective and surjective)

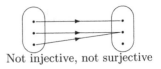

Not injective, not surjective

Here are some examples.

Example 1.28 Let $f : \mathbb{R} \longrightarrow \mathbb{R}$ be defined by $f(x) = x^2$. Then f is not one-to-one, because, e.g., (see the condition (1.11)) $2 \neq -2$ yet $f(2) = f(-2)$. This function is also not onto, because it fails (1.12): e.g., for $y = -1 \in \mathbb{R}$, there is no $x \in \mathbb{R}$ such that $-1 = f(x)$. (However, note that if we change the codomain to $[0, \infty)$, then f is onto.)

Example 1.29 In elementary calculus, we learn that the exponential function

$$\exp : \mathbb{R} \longrightarrow (0, \infty), \qquad f(x) = e^x,$$

is both one-to-one and onto, that is, a bijection, with inverse

$$\log : (0, \infty) \longrightarrow \mathbb{R}, \qquad f^{-1}(x) = \log x.$$

Here, $\log x$ denotes the "natural logarithm," which in many calculus courses is denoted by $\ln x$, with $\log x$ denoting the base 10 logarithm; however, in this book and in most advanced math texts, $\log x$ denotes the natural logarithm. In Section 3.7, starting on p. 216, and in Section 4.7, starting on p. 300, we shall define and study the exponential and logarithmic functions.

1.3.3 Images and Inverse Images

Functions act on sets as follows. Given a function $f : X \longrightarrow Y$ and a set $A \subseteq X$, we define

$$\boxed{f(A) = \{f(x)\,;\ x \in A\} = \{y \in Y\,;\ y = f(x) \text{ for some } x \in A\},}$$

and call this set the **image** of A under f. Thus,

$$y \in f(A) \iff y = f(x) \text{ for some } x \in A.$$

Given a set $B \subseteq Y$, we define

$$\boxed{f^{-1}(B) = \{x \in X\,;\ f(x) \in B\},}$$

and call this set the **inverse image** or **preimage** of B under f. Thus,

$$x \in f^{-1}(B) \iff f(x) \in B.$$

Warning: The notation f^{-1} in the preimage $f^{-1}(B)$ is merely notation and *does not* represent the inverse function of f. (Indeed, the function may not have an inverse, so the inverse function may not even be defined.)

Example 1.30 Let $f(x) = x^2$ with domain $\mathbb{R}$ as in Example 1.28. The following picture (Fig. 1.5) gives examples of images and inverse images:

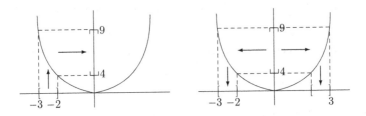

Fig. 1.5 *Left* the function $f(x) = x^2$ takes all the points in $[-3, -2]$ to the set $[4, 9]$, so $f([-3, -2]) = [4, 9]$. *Right* $f^{-1}([4, 9])$ consists of every point in $\mathbb{R}$ that f brings inside of $[4, 9]$, so $f^{-1}([4, 9]) = [-3, -2] \cup [2, 3]$

Here are more examples: You are invited to check that

$$f((1, 2]) = (1, 4], \quad f^{-1}([-4, -1)) = \varnothing, \quad f^{-1}((1, 4]) = [-2, -1) \cup (1, 2].$$

The following theorem gives the main properties of images and inverse images.

Theorem 1.4 *Let $f : X \longrightarrow Y$, let $B, C \subseteq Y$, and let $\{B_\alpha\}$ be a family of subsets of Y, and let $\{A_\alpha\}$ a family of subsets of X. Then*

$$f^{-1}(C \setminus B) = f^{-1}(C) \setminus f^{-1}(B), \quad f^{-1}\left(\bigcup_\alpha B_\alpha\right) = \bigcup_\alpha f^{-1}(B_\alpha),$$

$$f^{-1}\left(\bigcap_\alpha B_\alpha\right) = \bigcap_\alpha f^{-1}(B_\alpha), \quad f\left(\bigcup_\alpha A_\alpha\right) = \bigcup_\alpha f(A_\alpha).$$

Proof Using the definition of inverse image and set difference, we have

$$\begin{aligned}
x \in f^{-1}(C \setminus B) &\iff f(x) \in C \setminus B \iff f(x) \in C \text{ and } f(x) \notin B \\
&\iff x \in f^{-1}(C) \text{ and } x \notin f^{-1}(B) \\
&\iff x \in f^{-1}(C) \setminus f^{-1}(B).
\end{aligned}$$

Thus, $f^{-1}(C \setminus B) = f^{-1}(C) \setminus f^{-1}(B)$.

Using the definition of inverse image and union, we have

$$x \in f^{-1}\left(\bigcup_\alpha B_\alpha\right) \iff f(x) \in \bigcup_\alpha B_\alpha \iff f(x) \in B_\alpha \text{ for some } \alpha$$

$$\iff x \in f^{-1}(B_\alpha) \text{ for some } \alpha$$

$$\iff x \in \bigcup_\alpha f^{-1}(B_\alpha).$$

Thus, $f^{-1}\left(\bigcup_\alpha B_\alpha\right) = \bigcup_\alpha f^{-1}(B_\alpha)$. The proof of the last two properties in this theorem are similar enough to the proof just presented that we leave their verification to the reader. ∎

We end this section with some definitions needed for the exercises. Let X be a set and let A be any subset of X. The **characteristic function** of A is the function $\chi_A : X \longrightarrow \mathbb{R}$ defined by

$$\chi_A(x) = \begin{cases} 1 & \text{if } x \in A; \\ 0 & \text{if } x \notin A. \end{cases}$$

The sum and product of two characteristic functions χ_A and χ_B are the functions $\chi_A + \chi_B : X \longrightarrow \mathbb{R}$ and $\chi_A \cdot \chi_B : X \longrightarrow \mathbb{R}$ defined by

$$(\chi_A + \chi_B)(x) = \chi_A(x) + \chi_B(x) \quad \text{and} \quad (\chi_A \cdot \chi_B)(x) = \chi_A(x) \cdot \chi_B(x), \quad \text{for all } x \in X.$$

Of course, the sum and product of *any* functions $f : X \longrightarrow \mathbb{R}$ and $g : X \longrightarrow \mathbb{R}$ are defined in the same way. We can also replace $\mathbb{R}$ by, say, $\mathbb{C}$, or by any set Y as long as "+" and "·" are defined on Y. Given any constant $c \in \mathbb{R}$, we denote by the same letter the function $c : X \longrightarrow \mathbb{R}$ defined by $c(x) = c$ for all $x \in X$. This is the **constant function** c. For instance, 0 is the function defined by $0(x) = 0$ for all $x \in X$. The **identity map** on X is the map defined by $I(x) = x$ for all $x \in X$. Finally, we say that two functions $f : X \longrightarrow Y$ and $g : X \longrightarrow Y$ are **equal** if $f = g$ as subsets of $X \times Y$, which holds if and only if $f(x) = g(x)$ for all $x \in X$.

▶ **Exercises 1.3**

1. Which of the following subsets of $\mathbb{R}^2$ define functions from $\mathbb{R}$ to $\mathbb{R}$?

   (a) $A_1 = \{(x, y) \in \mathbb{R} \times \mathbb{R} ; x^2 = y\}$,   (b) $A_2 = \{(x, y) \in \mathbb{R} \times \mathbb{R} ; x = \sin y\}$,

   (c) $A_3 = \{(x, y) \in \mathbb{R} \times \mathbb{R} ; y = \sin x\}$,   (d) $A_4 = \{(x, y) \in \mathbb{R} \times \mathbb{R} ; x = 4y - 1\}$.

   (Assume well-known properties of trig functions.) Of those sets that do define functions, find the range of the function. Is the function one-to-one; is it onto?

2. Let $f(x) = 1 - x^2$. Find

   $$f([1, 4]), \quad f([-1, 0] \cup (2, 10)), \quad f^{-1}([-1, 1]), \quad f^{-1}([5, 10]), \quad f(\mathbb{R}), \quad f^{-1}(\mathbb{R}).$$

3. If $f : X \longrightarrow Y$ and $g : Z \longrightarrow X$ are bijective, prove that $f \circ g$ is a bijection and $(f \circ g)^{-1} = g^{-1} \circ f^{-1}$.

4. Let $f : X \longrightarrow Y$ be a function.

   (a) Given any subset $B \subseteq Y$, prove that $f(f^{-1}(B)) \subseteq B$.
   (b) Prove that $f(f^{-1}(B)) = B$ for all subsets B of Y if and only if f is surjective.
   (c) Given any subset $A \subseteq X$, prove that $A \subseteq f^{-1}(f(A))$.
   (d) Prove that $A = f^{-1}(f(A))$ for all subsets A of X if and only if f is injective.

5. Let $f : X \longrightarrow Y$ be a function. Show that f is one-to-one if and only if there is a function $g : Y \longrightarrow X$ such that $g \circ f$ is the identity map on X. Show that f is onto if and only if there is a function $h : Y \longrightarrow X$ such that $f \circ h$ is the identity map on Y.

6. (**Definition of ordered pair**) Given objects x, y, we define the **ordered pair** (x, y) by
$$(x, y) := \{\{x\}, \{x, y\}\}.$$

   The main property of ordered pair is the following result: We have $(a, b) = (x, y)$ if and only if $a = x$ and $b = y$. Prove this.

7. (Cf. [164]) In this problem we give various applications of characteristic functions to prove statements about sets. First, prove at least two the following assertions. (a) $\chi_X = 1, \chi_\varnothing = 0$; (b) $\chi_A \cdot \chi_B = \chi_B \cdot \chi_A = \chi_{A \cap B}$ and $\chi_A \cdot \chi_A = \chi_A$; (c) $\chi_{A \cup B} = \chi_A + \chi_B - \chi_A \cdot \chi_B$; (d) $\chi_{A^c} = 1 - \chi_A$; (e) $\chi_A = \chi_B$ if and only if $A = B$. Here are some applications of these properties. Prove the distributive law

$$A \cup (B \cap C) = (A \cup B) \cap (A \cup C)$$

by showing that the characteristic functions of each side are equal as functions. Then invoke (e) to demonstrate equality of sets. Prove the nonobvious equality

$$(A \cap B^c) \cap (C^c \cap A) = A \cap (B \cup C)^c.$$

Here's a harder question: Consider the sets $(A \cup B) \cap C$ and $A \cup (B \cap C)$. When, if ever, are they equal? When is one set a subset of the other?

Chapter 2
Numbers, Numbers, and More Numbers

I believe there are 15, 747, 724, 136, 275, 002, 577, 605, 653, 961, 181, 555, 468, 044, 717, 914, 527, 116, 709, 366, 231, 425, 076, 185, 631, 031, 296 protons in the universe and the same number of electrons.
Arthur Eddington (1882–1944), "The Philosophy of Physical Science." Cambridge, 1939.

This chapter is on the study of numbers. Of course, we all have a working understanding of the real numbers and we use many aspects of these numbers in everyday life: tallying up tuition and fees, figuring out how much we have left on our food cards, etc. We have accepted from our childhood all the properties of numbers that we use every day. In this chapter, we *prove* many of these properties.

In everyday life, what usually comes to mind when we think of *numbers* are the counting, or natural, numbers $1, 2, 3, 4, \ldots$ We shall study the natural numbers and their properties in Sections 2.1 and 2.2. These numbers have been used from the beginning. The Hindus became the first to systematically use "zero" and "negative" integers [34], [35, p. 220]; for example, Brahmagupta (598–670) gave arithmetic rules for multiplying and dividing with such numbers (although he mistakenly believed that $0/0 = 0$). We study the integers in Sections 2.3–2.5. Everyday life forces us to talk about fractions, for example, $2/3$ of a pizza "two pieces of a pizza that is divided into three equal parts." Such fractions (and their negatives and zero) make up the so-called rational numbers, which are called *rational* not because they are "sane" or "comprehensible," but simply because they are *ratios* of integers. It was a shock to the ancient Greeks, who discovered that the rational numbers are not enough to describe geometry. They noticed that according to the Pythagorean theorem, the length of the hypotenuse of a triangle with sides of length 1 is $\sqrt{1^2 + 1^2} = \sqrt{1 + 1} = \sqrt{2}$. We shall prove that $\sqrt{2}$ is "irrational," which simply means "not rational," that is, not a ratio of integers. In fact, we'll see that "most" numbers that you encountered in high school are irrational:

© Paul Loya 2017
P. Loya, *Amazing and Aesthetic Aspects of Analysis*,
https://doi.org/10.1007/978-1-4939-6795-7_2

> *Square roots, and more generally, nth roots, roots of polynomials, and values of trigonometric and logarithmic functions, are "mostly" irrational!*

You'll have to wait for this mouthwatering subject until Section 2.6! In Section 2.7, we study the all-important property of the real numbers called the completeness property, which in some sense says that real numbers can describe any length whatsoever. In the optional Section 2.8, we construct the real numbers via Dedekind cuts. In Sections 2.9 and 2.10, we leave the one-dimensional real line and discuss m-dimensional space and the complex numbers (which is really just two-dimensional space). Finally, in Section 2.11, we define "most" using cardinality and show that "most" real numbers are not only irrational, they are transcendental.

CHAPTER 2 OBJECTIVES: THE STUDENT WILL BE ABLE TO . . .

- State the fundamental axioms of the natural, integer, and real number systems.
- Explain how the completeness axiom of the real number system distinguishes this system from the rational number system in a powerful way.
- Prove statements about numbers from basic axioms including induction.
- Define $\mathbb{R}^m$ and $\mathbb{C}$ and the norms on these spaces.
- Explain cardinality and how "most" real numbers are irrational or even transcendental.

2.1 The Natural Numbers

The numbers we encounter most often in "everyday life" are the counting numbers, or the positive **whole numbers**

These numbers are the **natural numbers** and have been around since time immemorial. We shall study the essential properties of these fundamental numbers.

2.1.1 Axioms for the Natural Numbers

The *set*, or collection, of natural numbers is denoted by $\mathbb{N}$. We all know that if two natural numbers are added, we obtain a natural number; for example, $3 + 4 = 7$. Similarly, if two natural numbers are multiplied, we get a natural number. We say that the natural numbers are **closed** under addition and multiplication. Thus, if a, b are in $\mathbb{N}$, then using the familiar notation for addition and multiplication, $a + b$ and $a \cdot b$ are also in $\mathbb{N}$.[1] The following properties of $+$ and $\cdot$ are also familiar.

---

[1] By the way, $+$ and $\cdot$ are functions, as we studied on p. 22 in Section 1.3, from $\mathbb{N} \times \mathbb{N}$ to $\mathbb{N}$. These are not arbitrary functions but must satisfy the properties (**A**), (**M**), and (**D**) listed.

Addition satisfies

(A1) $a + b = b + a$ (commutative law);
(A2) $(a + b) + c = a + (b + c)$ (associative law).

By the associative law, we may "drop" parentheses in sums of more than two numbers:

$$a + b + c \text{ is unambiguously defined as } (a + b) + c = a + (b + c).$$

Multiplication satisfies

(M1) $a \cdot b = b \cdot a$ (commutative law);
(M2) $(a \cdot b) \cdot c = a \cdot (b \cdot c)$ (associative law);
(M3) there is a natural number, denoted by 1 "one," such that

$$1 \cdot a = a = a \cdot 1 \quad \text{(existence of multiplicative identity)}.$$

By the associative law for multiplication, we may "drop" parentheses:

$$a \cdot b \cdot c \text{ is unambiguously defined as } (a \cdot b) \cdot c = a \cdot (b \cdot c).$$

Addition and multiplication are related by

(D) $a \cdot (b + c) = (a \cdot b) + (a \cdot c)$ (distributive law).

We sometimes drop the dot $\cdot$ and just write ab for $a \cdot b$. The natural numbers are also ordered in the sense that you can compare the magnitude of any two of them; for example, $2 < 5$, because five is three greater than two, or two is three less than five. This inequality relationship satisfies the following law of **trichotomy**. Given any natural numbers a and b, *exactly one* of the following (in)equalities holds:

(O1) $a = b$;
(O2) $a < b$, which by definition means that $b = a + c$ for some natural number c;
(O3) $b < a$, which by definition means that $a = b + c$ for some natural number c.

Thus, $2 < 5$, because $5 = 2 + c$, where $c = 3$. Of course, we write $a \leq b$ if $a < b$ or $a = b$. There are similar meanings for the opposite inequalities ">" and "$\geq$." The inequality signs $<$ and $>$ are called **strict**. There is one more property of the natural numbers, called **induction**. Let M be a subset of $\mathbb{N}$.

(I) Suppose that M contains 1 and that M has the following property: If n belongs to M, then $n + 1$ also belongs to M. Then M contains all natural numbers.

The statement that $M = \mathbb{N}$ is "obvious" with a little thought. M contains 1. Because 1 belongs to M, by **(I)**, we know that $1 + 1 = 2$ also belongs to M. Because 2 belongs to M, by **(I)** we know that $2 + 1 = 3$ also belongs to M. Assuming that we can continue this process indefinitely makes it clear that we should have $M = \mathbb{N}$.

Everyday experience convinces us that the counting numbers satisfy properties **(A)**, **(M)**, **(D)**, **(O)**, and **(I)**. Consider commutativity of addition. In grade school, we learned that one way to understand $a + b$ is to "start at a and move b units"; then the commutative law for addition is "obvious" by taking an example such as

$2 + 3 =$ start at 2 and move 3 units $3 + 2 =$ start at 3 and move 2 units

Of course, an example doesn't prove that a law holds in general. Thus, mathematically, we will assume, or take on faith, the existence of a set $\mathbb{N}$ with operations $+$ and $\cdot$ that satisfy properties **(A)**, **(M)**, **(D)**, **(O)**, and **(I)**.[2] From these properties alone we shall *prove* many well-known properties of these numbers that we have accepted since grade school. It is quite satisfying to see that many of the well-known properties about numbers that are memorized (or even those that are not so well known) can in fact be proved from a basic set of axioms!

> **Rules of the game:** Henceforth, we are allowed to prove statements using only facts that we know are true because those facts are **(1)** given to us in a set of axioms, or **(2)** proved by us previously in this book, by your teacher in class, or by you in an exercise, or **(3)** told that we are allowed to use them (for example, in order to provide some nontrivial examples).

2.1.2 Proofs of Well-Known High School Rules

You are going to learn the language of proofs in the same way that a child learns to talk; by observing others prove things and imitating them, and eventually you will get the hang of it.

We begin by proving the familiar transitive law.

Transitive law

Theorem 2.1 *If $a < b$ and $b < c$, then $a < c$.*

Proof Of course, a picture shows that this theorem is "obvious":

$$a < b \text{ and } b < c \implies a < c$$

To give a rigorous proof, suppose $a < b$ and $b < c$. Then by definition of less than (recall the inequality law **(O2)** on p. 31), there are natural numbers d and e such that $b = a + d$ and $c = b + e$. Hence, by the associative law,

$$c = b + e = (a + d) + e = a + (d + e).$$

Thus, $c = a + f$, where $f = d + e \in \mathbb{N}$, so $a < c$ by definition of "less than." ∎

---

[2]Taking the axioms of set theory on faith, which we are doing even though we haven't listed many of them(!), we can define the natural numbers as certain sets; see [98, Section 11] or [149].

Before moving on, we briefly analyze this theorem in view of what we learned in Section 1.2 on p. 15. The **hypotheses** or **assumptions** of this theorem are that a, b, and c are natural numbers with $a < b$ and $b < c$, and the **conclusion** is that $a < c$. Note that the fact that a, b, and c are natural numbers and that natural numbers are assumed to satisfy all their arithmetic and order properties were left unwritten in the statement of the proposition, since these assumptions were understood within the context of this section. The "if–then" wording means: If the assumptions are true, then the conclusion is also true, or given that the assumptions are true, the conclusion follows. We can also reword Theorem 2.1 as follows:

$$a < b \text{ and } b < c \text{ implies (also written } \implies) a < c;$$

that is, the truth of the assumptions implies the truth of the conclusion. We can also state this theorem as follows:

$$a < b \text{ and } b < c \text{ only if } a < c,$$

that is, the hypotheses $a < b$ and $b < c$ hold only if it follows that $a < c$; stated another way,

$$a < c \text{ if } a < b \text{ and } b < c;$$

that is, the conclusion $a < c$ is true if, or given that, the hypotheses $a < b$ and $b < c$ are true. The kind of proof used in Theorem 2.1 is called a **direct proof**, where we take the hypotheses $a < b$ and $b < c$ as true and prove that the conclusion $a < c$ is true. We shall see other types of proofs later. We next give another easy and direct proof of the so-called "FOIL law" of multiplication. However, before proving this result, we note that the distributive law (**D**) holds from the right:

$$(a + b) \cdot c = ac + bc.$$

Indeed,

$$
\begin{aligned}
(a + b) \cdot c &= c \cdot (a + b) && \text{commutative law} \\
&= (c \cdot a) + (c \cdot b) && \text{distributive law} \\
&= (a \cdot c) + (b \cdot c) && \text{commutative law.}
\end{aligned}
$$

FOIL law

Theorem 2.2 *For any natural numbers a, b, c, d, we have*

$$(a + b) \cdot (c + d) = ac + ad + bc + bd, \quad (\textbf{\textit{First}} + \textbf{\textit{Outside}} + \textbf{\textit{Inside}} + \textbf{\textit{Last}}).$$

Proof We simply compute:

$$
\begin{aligned}
(a+b) \cdot (c+d) &= (a+b) \cdot c + (a+b) \cdot d \quad \text{distributive law} \\
&= (ac+bc) + (ad+bd) \quad \text{distributive law (from right)} \\
&= ac + (bc + (ad+bd)) \quad \text{associative law} \\
&= ac + ((bc+ad) + bd) \quad \text{associative law} \\
&= ac + ((ad+bc) + bd) \quad \text{commutative law} \\
&= ac + ad + bc + bd,
\end{aligned}
$$

where at the last step we dropped parentheses, as we know we can in sums of more than two numbers (consequence of the associative law).   ∎

We now prove the familiar cancellation properties of high school algebra.

Theorem 2.3 *Given any natural numbers a, b, c, we have*

$$a + c = b + c \quad \text{if and only if} \quad a = b.$$

In particular, given $a + c = b + c$, we can "cancel" c, obtaining $a = b$.

Proof Suppose that $a = b$. Then because a and b are just different letters for the same natural number, we have $a + c = b + c$.

We now have to prove that if $a + c = b + c$, then $a = b$. To prove this, we use a **proof by contraposition**. This is how it works. We need to prove that if the assumption "$P : a + c = b + c$" is true, then the conclusion "$Q : a = b$" is also true. Instead, we shall prove the logically equivalent **contrapositive** statement: If the conclusion Q is false, then the assumption P must also false. The statement that Q is false is just that $a \neq b$, and the statement that P is false is just that $a + c \neq b + c$. Thus, we must prove

$$\text{if } a \neq b, \text{ then } a + c \neq b + c.$$

To this end, assume that $a \neq b$; then either $a < b$ or $b < a$. Because the notation is entirely symmetric between a and b, we may assume that $a < b$. Then by definition of less than, we have $b = a + d$ for some natural number d. Hence, by the associative and commutative laws,

$$b + c = (a+d) + c = a + (d+c) = a + (c+d) = (a+c) + d.$$

Thus, by definition of less than, $a + c < b + c$, so $a + c \neq b + c$.   ∎

There is a multiplicative cancellation as well; see Problem 5b. Other examples using the fundamental properties (**A**), (**M**), (**D**), and (**O**) of the natural numbers are found in the exercises. We now concentrate on the induction property (**I**).

2.1.3 *Induction*

We all know that every natural number is greater than or equal to 1. Here is a proof!

Theorem 2.4 *Every natural number is greater than or equal to one.*

Proof Rewording this as an "if–then" statement, we need to prove that if n is a natural number, then $n \geq 1$. To prove this, let $M = \{n \in \mathbb{N}; \, n \geq 1\}$, the collection all natural numbers greater than or equal to one. Then M contains 1. If a natural number n belongs to M, then by definition of M, $n \geq 1$. This means that $n = 1$ or $n > 1$. In the first case, $n + 1 = 1 + 1$, so by definition of less than, $1 < n + 1$. In the second case, $n > 1$ means that $n = 1 + m$ for some $m \in \mathbb{N}$, so $n + 1 = (1 + m) + 1 = 1 + (m + 1)$. Again by definition of less than, $1 < n + 1$. In either case, $n + 1$ also belongs to M. Thus by induction, $M = \mathbb{N}$. ∎

Now we prove the Archimedean ordering property of the natural numbers.

Archimedean ordering of $\mathbb{N}$

Theorem 2.5 *Given any natural numbers a and b, there is a natural number n such that $b < a \cdot n$.*

Proof Let $a, b \in \mathbb{N}$; we need to *produce* an $n \in \mathbb{N}$ such that $b < a \cdot n$. By the previous theorem, either $a = 1$ or $a > 1$. If $a = 1$, then we set $n = b + 1$, in which case $b < b + 1 = 1 \cdot n$. If $1 < a$, then we can write $a = 1 + c$ for some natural number c. In this case, let $n = b$. Then,

$$a \cdot n = (1 + c) \cdot b = b + c \cdot b > b.$$ ∎

The following theorem contains an important property of the natural numbers. Its proof is an example of a **proof by contradiction** or **reductio ad absurdum**, whereby we start with the tentative assumption that the conclusion is false and then proceed with our argument until we eventually get a logical absurdity. Thus, the conclusion must have been true in the first place.

Well-ordering (principle) of $\mathbb{N}$

Theorem 2.6 *Every nonempty set of natural numbers has a smallest element, that is, an element less than or equal to every other member of the set.*

Proof Let A be a nonempty set of natural numbers; we need to show that A contains a smallest element. Well, either A has this property or not. Suppose, for the sake of (hopefully!) obtaining a contradiction, that A does not have a smallest element. From this assumption we shall derive a nonsense statement. Let $M = \{n \in \mathbb{N};$

$n < a$ for all $a \in A$}. Note that since a natural number is never less than itself, M does not contain any element of A. In particular, since A is nonempty, $M \neq \mathbb{N}$ (since M is missing elements of A). However, we shall prove by induction that $M = \mathbb{N}$. This, of course, will give us a contradiction.

To arrive at our contradiction, we first show that M contains 1. By Theorem 2.4, we know that 1 is less than or equal to every natural number; in particular, 1 is less than or equal to every element of A. Hence, if 1 is in A, then 1 would be the smallest element of A. However, we are assuming that A does not have a smallest element, so 1 cannot be in A. Hence, 1 is less than every element of A, so M contains 1.

Suppose that M contains a natural number n; we shall prove that M contains $n + 1$, that is, $n + 1$ is less than every element of A. Let a be any element of A and suppose that $a \neq n + 1$. Since $n < a$ (as $n \in M$), we can write $a = n + c$ for some natural number c. Note that $c \neq 1$, since by assumption, $a \neq n + 1$. Thus (by Theorem 2.4) $c > 1$, and so we can write $c = 1 + d$ for some natural number d. Hence,

$$a = n + c = n + 1 + d,$$

which shows that $n + 1 < a$. It follows that if $n + 1$ belonged to A, then $n + 1$ would be the smallest element of A, which we know cannot exist. Hence, $n + 1 \notin A$. In particular, $n + 1 < a$ for every element $a \in A$. Thus, $n + 1 \in M$, so by induction, $M = \mathbb{N}$, and we arrive at our desired contradiction. ∎

Finally, we remark that the symbol 2 denotes, by definition, the natural number $1 + 1$. Since $2 = 1 + 1$, $1 < 2$ by definition of less than. The symbol 3 denotes the natural number $2 + 1 = 1 + 1 + 1$. By definition of less than, $2 < 3$. Similarly, 4 is the number $3 + 1$, and so forth. Continuing in this manner, we can assign the usual symbols to the natural numbers that we are accustomed to in "everyday life"; see Section 2.5 starting on p. 69 for more details. In Problem 4 we see that there is no natural number between n and $n + 1$, so the sequence of symbols defined will cover all possible natural numbers.

All the letters in the following exercises represent natural numbers. In these exercises, you should use only the axioms and properties of the natural numbers established in this section. Remember that if you can't see how to prove something after some effort, take a break (e.g., take a bus ride somewhere) and come back to the problem later.[3]

▶ **Exercises 2.1**

1. Prove that every natural number greater than 1 can be written in the form $m + 1$, where m is a natural number.
2. Are there natural numbers a and b such that $a = a + b$? What logical inconsistency happens if such an equation holds?

---

[3] "I entered an omnibus to go to some place or other. At that moment when I put my foot on the step the idea came to me, without anything in my former thoughts seeming to have paved the way for it, that the transformations I had used to define the Fuchsian functions were identical with non-Euclidean geometry." Henri Poincaré (1854–1912).

3. Prove the following statements.

   (a) If $n^2 = 1$ (that is, $n \cdot n = 1$), then $n = 1$. Suggestion: Contrapositive.
   (b) There does not exist a natural number n such that $2n = 1$.
   (c) There does not exist a natural number n such that $2n = 3$.

4. Prove the following statements.

   (a) If $n \in \mathbb{N}$, then there is no $m \in \mathbb{N}$ such that $n < m < n + 1$.
   (b) If $n \in \mathbb{N}$, then there is a unique $m \in \mathbb{N}$ satisfying $n < m < n + 2$; in fact, prove that the only such natural number is $m = n + 1$. (That is, prove that $n + 1$ satisfies the inequality, and if m also satisfies the inequality, then $m = n + 1$.)

5. Prove the following statements for natural numbers a, b, c, d.

   (a) $(a + b)^2 = a^2 + 2ab + b^2$, where a^2 means $a \cdot a$ and b^2 means $b \cdot b$.
   (b) $a = b$ if and only if $a \cdot c = b \cdot c$. As a corollary, if $a \cdot c = c$, prove that $a = 1$. In particular, 1 is the only multiplicative identity.
   (c) $a < b$ if and only if $a + c < b + c$.
   (d) $a < b$ if and only if $a \cdot c < b \cdot c$.
   (e) If $a < b$ and $c < d$, then $a \cdot c < b \cdot d$.

6. Let A be a finite collection of natural numbers. Prove that A has a largest element, that is, A contains a number n such that $n \geq a$ for every element a in A.

7. Many books replace the induction axiom with the well-ordering principle. For this problem, let us assume the well-ordering principle instead of the induction axiom. In (ii), we *prove* the induction axiom.

   (i) Prove Theorem 2.4 using well-ordering. Suggestion: By well-ordering, $\mathbb{N}$ has a least element; call it n. We need to prove that $n \geq 1$. Assume that $n < 1$ and find another natural number less than n to derive a contradiction.
   (ii) Prove the induction property.

2.2 The Principle of Mathematical Induction

In this section we master the principle of mathematical induction by studying many applications of its use.

2.2.1 *Principle of Mathematical Induction*

The induction axiom of the natural numbers is the basis for the **principle of mathematical induction**, which goes as follows. Suppose for each $n \in \mathbb{N}$ we have a corresponding statement,

$$P_1, P_2, P_3, P_4, P_5, \ldots,$$

and suppose that **(1)** P_1 is true and **(2)** if n is a natural number and the statement P_n happens to be valid, then the statement P_{n+1} is also valid. Then it must be that every statement $P_1, P_2, P_3, \ldots$ is true. To see why every statement P_n is true, let M be the collection of all natural numbers n such that P_n is true. Then by **(1)**, M contains 1, and by **(2)**, if M contains a natural number n, then it contains $n + 1$. By the induction axiom, M must be all of $\mathbb{N}$; that is, P_n is true for every n. Induction is like dominoes: Line up infinitely many dominos in a row and knock down the first domino (that is, P_1 is true), and if the nth domino knocks down the $(n + 1)$st domino (that is, $P_n \implies P_{n+1}$), then *every* domino eventually gets knocked down (that is, all the statements $P_1, P_2, P_3, \ldots$ are true). Here's a picture to help visualize this concept (Fig. 2.1).

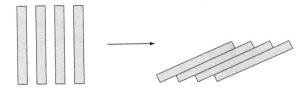

Fig. 2.1 Induction is like dominoes

We now illustrate this principle through some famous examples.

Remark: In order to present examples that have applicability in the sequel, we have to go outside the realm of natural numbers and assume familiarity with integers, real numbers, and complex numbers, and we shall use their properties freely. Integers are discussed in Section 2.3, real numbers in Sections 2.6 and 2.7, and complex numbers in Section 2.10.

2.2.2 Inductive Definitions: Powers and Sums

We of course know what 7^3 is, namely $7 \cdot 7 \cdot 7$. In general, by a^n, where a is a complex number called the **base** and n is a positive integer called the **exponent**, we mean

$$a^n = \underbrace{a \cdot a \cdots a}_{n \text{ times}}.$$

It's very common to use induction as follows to make powers more precise.[4]

Example 2.1 Let P_n denote the statement "the power a^n is defined." First, $a^1 := a$ defines a^1. (Recall that ":=" means "equals by definition.") Second, assuming that

---

[4] Although common, the inductive definitions for powers and for summation (see Example 2.3) are not perfectly rigorous, since we should be invoking *the recursion theorem*; see [111] or [149] for more details.

a^n has been defined, $a^{n+1} := a^n \cdot a$ defines a^{n+1}. Thus, the statement P_{n+1} holds, so by induction, a^n is defined for every natural number n.

Example 2.2 Using induction, we prove that for any natural numbers m and n, we have

$$a^{m+n} = a^m \cdot a^n. \tag{2.1}$$

Indeed, let us fix the natural number m and let P_n be the statement "Equation (2.1) holds for the natural number n." By definition of a^{m+1}, we have

$$a^{m+1} = a^m \cdot a = a^m \cdot a^1.$$

Thus, (2.1) holds for $n = 1$. Assume that (2.1) holds for a natural number n. Then by definition of powers and our induction hypothesis,

$$a^{m+(n+1)} = a^{(m+n)+1} = a^{m+n} \cdot a = a^m \cdot a^n \cdot a = a^m \cdot a^{n+1},$$

which is exactly the statement P_{n+1}. If $a \neq 0$ and we also define $a^0 := 1$, then as the reader can readily check, (2.1) continues to hold even if m or n is zero.

In elementary calculus, we were introduced to the summation notation. Let $a_0, a_1, a_2, a_3, \ldots$ be any list of complex numbers. For every natural number n, we define $\sum_{k=0}^{n} a_k$ as the sum of the numbers $a_0, \ldots, a_n$:

$$\sum_{k=0}^{n} a_k = a_0 + a_1 + \cdots + a_n.$$

By the way, in 1755 Euler introduced the sigma notation $\sum$ for summation [183].

Example 2.3 We also can define summation using induction. We define $\sum_{k=0}^{0} a_k = a_0$. For a natural number n, let P_n represent the statement "$\sum_{k=0}^{n} a_k$ is defined." We define

$$\sum_{k=0}^{1} a_k = a_0 + a_1.$$

Suppose that P_n holds for $n \in \mathbb{N}$; that is, $\sum_{k=0}^{n} a_k$ is defined. Then we define

$$\sum_{k=0}^{n+1} a_k = \left(\sum_{k=0}^{n} a_k \right) + a_{n+1}.$$

Thus, P_{n+1} holds. We conclude that the sum $\sum_{k=0}^{n} a_k$ is defined for $n = 0$ and for every natural number n.

2.2.3 Classic Examples: Arithmetic and Geometric Progressions

Example 2.4 We shall prove that for every natural number n, the sum of the first n natural numbers equals $n(n + 1)/2$; that is,

$$1 + 2 + \cdots + n = \frac{n(n + 1)}{2}. \tag{2.2}$$

Here, P_n represents the statement "Equation (2.2) holds." Certainly, $1 = \dfrac{1(1 + 1)}{2}$. Thus, our statement is true for $n = 1$. Suppose our statement holds for some n. Then adding $n + 1$ to both sides of (2.2), we obtain

$$1 + 2 + \cdots + n + (n + 1) = \frac{n(n + 1)}{2} + (n + 1)$$
$$= \frac{n(n + 1) + 2(n + 1)}{2} = \frac{(n + 1)(n + 1 + 1)}{2},$$

which is exactly the statement P_{n+1}. Hence, by the principle of mathematical induction, every single statement P_n is true.

The most famous story involving the formula (2.2) is that of a ten-year-old Carl Friedrich Gauss (1777–1855) [35, p. 497]:

> One day, in order to keep the class occupied, the teacher had the students add up all the numbers from one to a hundred, with instructions that each should place his slate on a table as soon as he had completed the task. Almost immediately Carl placed his slate on the table, saying, "There it is." The teacher looked at him scornfully while the others worked diligently. When the instructor finally looked at the results, the slate of Gauss was the only one to have the correct answer, 5050, with no further calculation. The ten-year-old boy evidently had computed mentally the sum of the arithmetic progression $1 + 2 + 3 + \cdots + 99 + 100$, presumably through the formula $m(m + 1)/2$.

We remark that the high school way to prove P_n is to write the sum of the first n integers forward and backwards:

$$S_n = 1 + 2 + \cdots + (n - 1) + n$$

and

$$S_n = n + (n - 1) + \cdots + 2 + 1.$$

Notice that the sum of each column is just $n + 1$. Since there are n columns, adding these two expressions, we obtain $2S_n = n(n + 1)$, which implies our result.

What if we sum just the odd integers? We get (proof left to you!)

$$1 + 3 + 5 + \cdots + (2n - 1) = n^2.$$

Do you see why the following picture makes this formula "obvious"?

Example 2.5 We now consider the sum of a geometric progression. Let $a \neq 1$ be any complex number. We prove that for every natural number n,

$$1 + a + a^2 + \cdots + a^n = \frac{1 - a^{n+1}}{1 - a}. \tag{2.3}$$

The sequence $1, a, a^2, a^3, \ldots, a^n$ makes up a **geometric progression**. Observe that

$$\frac{1 - a^2}{1 - a} = \frac{(1 + a)(1 - a)}{1 - a} = 1 + a,$$

so our assertion holds for $n = 1$. Suppose that the sum (2.3) holds for some n. Then adding a^{n+1} to both sides of (2.3), we obtain

$$1 + a + a^2 + \cdots + a^n + a^{n+1} = \frac{1 - a^{n+1}}{1 - a} + a^{n+1}$$
$$= \frac{1 - a^{n+1} + a^{n+1} - a^{n+2}}{1 - a} = \frac{1 - a^{n+2}}{1 - a},$$

which is exactly the Eq. (2.3) for $n + 1$. This completes the proof for the sum of a geometric progression.

The high school way to establish the sum of a geometric progression is to multiply

$$G_n = 1 + a + a^2 + \cdots + a^n$$

by a,

$$a\, G_n = a + a^2 + a^3 + \cdots + a^{n+1},$$

and then to subtract this equation from the preceding one and cancel like terms:

$$(1 - a)G_n = G_n - a\, G_n = (1 + a + \cdots + a^n) - (a + \cdots + a^{n+1}) = 1 - a^{n+1}.$$

Dividing by $1 - a$ proves (2.3). Splitting the fraction at the end of (2.3) and solving for $1/(1 - a)$, we obtain the following version of the geometric progression sum:

$$\frac{1}{1 - a} = 1 + a + a^2 + \cdots + a^n + \frac{a^{n+1}}{1 - a}. \tag{2.4}$$

2.2.4   *More Sophisticated Examples*

Here's a famous inequality due to Jacob Bernoulli (1654–1705) that we'll have to use on many occasions.

Bernoulli's inequality

> **Theorem 2.7** *For every real number $a > -1$ and every natural number n,*
>
> $$(1+a)^n \begin{cases} = 1 + na & \text{if } n = 1 \text{ or } a = 0, \\ > 1 + na & \text{if } n > 1 \text{ and } a \neq 0, \end{cases} \qquad \textbf{\textit{Bernoulli' inequality.}}$$

Proof If $a = 0$, then Bernoulli's inequality certainly holds (both sides equal 1), so we'll assume that $a \neq 0$. If $n = 1$, then Bernoulli's inequality is just $1 + a = 1 + a$, which is true. Suppose that Bernoulli's inequality holds for a number n. Then $(1+a)^n \geq (1 + na)$, where if $n = 1$, this is an equality, and if $n > 1$, this is a strict inequality. Multiplying Bernoulli's inequality by $1 + a > 0$, we obtain

$$(1+a)^{n+1} \geq (1+a)(1+na) = 1 + na + a + na^2.$$

Since $n\,a^2$ is positive, the expression on the right is greater than

$$1 + na + a = 1 + (n+1)a.$$

Combining this equation with the previous inequality proves Bernoulli's inequality for $n + 1$. By induction, Bernoulli's inequality holds for every $n \in \mathbb{N}$. ∎

If n is a natural number, the symbol $n!$ (read "n **factorial**") represents the product of the first n natural numbers. Thus,

$$n! := 1 \cdot 2 \cdot 3 \cdots (n-1) \cdot n.$$

(Of course, we can also define $n!$ using induction.) It is convenient to define $0! = 1$ so that certain formulas continue to hold for $n = 0$. Thus, $n!$ is defined for all nonnegative integers n, that is, for $n = 0, 1, 2, \ldots$. Given nonnegative integers n and k with $k \leq n$, we define the **binomial coefficient** $\binom{n}{k}$ by

$$\binom{n}{k} = \frac{n!}{k!(n-k)!}.$$

For example, for every nonnegative integer n,

$$\binom{n}{0} = \frac{n!}{0!(n-0)!} = \frac{n!}{n!} = 1 \quad \text{and} \quad \binom{n}{n} = \frac{n!}{n!(n-n)!} = \frac{n!}{n!} = 1.$$

Lemma 2.8 *For all nonnegative integers k, n with $1 \le k \le n$, we have*

$$\binom{n}{k-1} + \binom{n}{k} = \binom{n+1}{k}. \tag{2.5}$$

Proof The proof is but an algebra computation:

$$\begin{aligned}
\binom{n}{k-1} + \binom{n}{k} &= \frac{n!}{(k-1)!\,(n-k+1)!} + \frac{n!}{k!\,(n-k)!} \\
&= \frac{n!\,k}{k!\,(n-k+1)!} + \frac{n!\,(n-k+1)}{k!\,(n-k+1)!} \\
&= \frac{n!\,(n+1)}{k!\,(n+1-k)!} = \binom{n+1}{k}.
\end{aligned}$$

■

Using (2.5), we can build Blaise Pascal's (1623–1662) triangle (Fig. 2.2).

$$\begin{array}{ccccccccccc}
 & & & & & 1 & & & & & \\
 & & & & 1 & & 1 & & & & \\
 & & & 1 & & 2 & & 1 & & & \\
 & & 1 & & 3 & & 3 & & 1 & & \\
 & 1 & & 4 & & 6 & & 4 & & 1 & \\
1 & & 5 & & 10 & & 10 & & 5 & & 1
\end{array}$$

Fig. 2.2 First six rows of **Pascal's triangle**: Adding two adjacent entries in the nth row gives the entry below in the $(n+1)$st row

Problem 11 contains a generalization of the following important theorem.

Binomial theorem

Theorem 2.9 *For all complex numbers a and b, and $n \in \mathbb{N}$, we have*

$$(a+b)^n = \sum_{k=0}^{n} \binom{n}{k} a^k b^{n-k}, \quad \textit{binomial formula}.$$

Proof Here's the binomial formula written out in detail:

$$(a+b)^n = \binom{n}{0} b^n + \binom{n}{1} a\, b^{n-1} + \binom{n}{2} a^2\, b^{n-2} + \cdots + \binom{n}{n-1} a^{n-1} b + \binom{n}{n} a^n. \quad (2.6)$$

If $n = 1$, the right-hand side is

$$\binom{1}{0} b^1 + \binom{1}{1} a^1 = b + a = (a+b)^1,$$

so the $n = 1$ case holds. Suppose that (2.6) holds for a natural number n; we will prove that it holds for n replaced by $n + 1$. To do so, multiply (2.6) by a,

$$a(a+b)^n = \binom{n}{0} a\, b^n + \binom{n}{1} a^2\, b^{n-1} + \binom{n}{2} a^3\, b^{n-2} + \cdots + \binom{n}{n-1} a^n b + \binom{n}{n} a^{n+1},$$

and then by b:

$$b(a+b)^n = \binom{n}{0} b^{n+1} + \binom{n}{1} a\, b^n + \binom{n}{2} a^2\, b^{n-1} + \cdots + \binom{n}{n-1} a^{n-1} b^2 + \binom{n}{n} a^n b.$$

Now add $a(a+b)^n$ and $b(a+b)^n$, and combine like terms such as ab^n, $a^2 b^{n-1}$, and so forth; using our previous lemma, we obtain

$$(a+b)^{n+1} = \binom{n}{0} b^{n+1} + \binom{n+1}{1} a\, b^n + \binom{n+1}{2} a^2\, b^{n-1} \quad (2.7)$$

$$+ \cdots + \binom{n+1}{n-1} a^{n-1} b^2 + \binom{n+1}{n} a^n b + \binom{n}{n} a^{n+1}.$$

Since $\binom{n}{0} = \binom{n+1}{0}$ and $\binom{n}{n} = \binom{n+1}{n+1}$ (these just equal 1), we see that (2.7) is exactly (2.6) with n replaced by $n + 1$. This proves the binomial formula. ∎

2.2.5 Strong Form of Induction

Sometimes it is necessary to use the following "strong" induction using a "stronger" (albeit equivalent) hypothesis. For each natural number n, let P_n be a statement. Suppose that **(1)** P_1 is true and **(2)** if n is a natural number and if each statement P_m is true for every $m \le n$, then the statement P_{n+1} is also true. Then every single statement $P_1, P_2, P_3, \ldots$ is true. To see this, let M be all the natural numbers such that P_n is *not* true. We shall prove that M must be empty, which shows that P_n is true for every n. Indeed, suppose that M is not empty. Then by well-ordering, M contains a least element, say n. Since P_1 is true, M does not contain 1, so $n > 1$. Since $1, 2, \ldots, n-1$ are not in M (because n is the least element of M), the statements $P_1, P_2, \ldots, P_{n-1}$ must be true. Hence, by Property **(2)** of the statements, P_n must also be true. This shows that M does not contain n, which contradicts the assumption that n is in M. Thus, M must be empty. Problems 6, 9, and 10 contain exercises in which strong induction is useful.

As already stated, to illustrate nontrivial induction examples, in the exercises you may freely use common properties of integers, real, and complex numbers.

▶ **Exercises 2.2**

1. Consider the statement $1 + 2 + 3 + \cdots + n = (2n + 1)^2/8$. Prove that P_n implies P_{n+1}. However, is the statement true for all n?

2. Using induction, prove that for all complex numbers a and b and for all natural numbers m and n, we have $(ab)^n = a^n \cdot b^n$ and also $(a^m)^n = a^{mn}$. If a and b are nonzero, prove that these equations hold even if $m = 0$ or $n = 0$.

3. Prove the following (quite pretty) formulas/statements via induction:

   (a)
   $$\frac{1}{1 \cdot 2} + \frac{1}{2 \cdot 3} + \cdots + \frac{1}{n(n+1)} = \frac{n}{n+1}.$$

   (b)
   $$1^2 + 2^2 + \cdots + n^2 = \frac{n(n+1)(2n+1)}{6}.$$

   (c)
   $$1^3 + 2^3 + \cdots + n^3 = (1 + 2 + \cdots + n)^2 = \left(\frac{n(n+1)}{2}\right)^2.$$

   (d)
   $$\frac{1}{2} + \frac{2}{2^2} + \frac{3}{2^3} + \cdots + \frac{n}{2^n} = 2 - \frac{n+2}{2^n}.$$

   (e) For $a \neq 1$,
   $$(1 + a)(1 + a^2)(1 + a^4) \cdots (1 + a^{2^n}) = \frac{1 - a^{2^{n+1}}}{1 - a}.$$

   (f) $n^3 - n$ is always divisible by 3.

   (g) Every natural number n is either even or odd. Here, n is even means that $n = 2m$ for some $m \in \mathbb{N}$, and n odd means that $n = 1$ or $n = 2m + 1$ for some $m \in \mathbb{N}$.

   (h) $n < 2^n$ for all $n \in \mathbb{N}$. (Can you also prove this using Bernoulli's inequality?)

   (i) Using the identity (2.5), called **Pascal's rule**, prove that $\binom{n}{k}$ is a natural number for all $n, k \in \mathbb{N}$ with $1 \leq k \leq n$. (P_n is the statement "$\binom{n}{k} \in \mathbb{N}$ for all $1 \leq k \leq n$.")

4. In this problem we prove some nifty binomial formulas. Prove that

   $$(a) \sum_{k=0}^{n} \binom{n}{k} = 2^n, \quad (b) \sum_{k=0}^{n} (-1)^k \binom{n}{k} = 0,$$

   $$(c) \sum_{k \text{ odd}} \binom{n}{k} = 2^{n-1}, \quad (d) \sum_{k \text{ even}} \binom{n}{k} = 2^{n-1},$$

where $k = 1, 3, 5, \ldots$ and $k \le n$ in (c), and $k = 0, 2, 4, \ldots$ and $k \le n$ in (d).

5. (**Towers of Hanoi**) Induction can be used to analyze games! (See Problem 6 on p. 56 for the game of Nim.) For instance, the *towers of Hanoi* starts with three pegs and n disks of different sizes placed on one peg, with the biggest disk on the bottom and with the sizes decreasing to the top as shown here:

A move is made by taking the top disk off a stack and putting it on another peg so that there is no smaller disk below it. The object of the game is to transfer all the disks to another peg. Prove that the puzzle can be solved in $2^n - 1$ moves and that it cannot be solved in fewer than $2^n - 1$ moves.

6. (**The coin game**) Two people have n coins each, and they put them on a table, in separate piles. Then they take turns removing their own coins; they may take as many as they wish, but they must take at least one. The person removing the last coin(s) wins. Using strong induction, prove that the second person has a "foolproof winning strategy." More explicitly, prove that for each $n \in \mathbb{N}$, there is a strategy such that the second person will win the game with n coins each following that strategy.

7. We now prove the **arithmetic–geometric mean inequality** (AGMI): For all nonnegative (that is, ≥ 0) real numbers $a_1, \ldots, a_n$, we have

$$(a_1 \cdots a_n)^{1/n} \le \frac{a_1 + \cdots + a_n}{n}, \quad \text{or equivalently,} \quad a_1 \cdots a_n \le \left(\frac{a_1 + \cdots + a_n}{n} \right)^n.$$

The product $(a_1 \cdots a_n)^{1/n}$ is the **geometric mean** and the sum $\frac{a_1 + \cdots + a_n}{n}$ is the **arithmetic mean**, of the numbers $a_1, \ldots, a_n$.

(i) Show that $\sqrt{a_1 a_2} \le \frac{a_1 + a_2}{2}$. Suggestion: Expand $(\sqrt{a_1} - \sqrt{a_2})^2 \ge 0$.

(ii) By induction, show that the AGMI holds for 2^n terms for every natural number n.

(iii) We now prove the AGMI for n terms where n is not necessarily a power of 2. Let $a = (a_1 + \cdots + a_n)/n$. By Problem 3h, we know that $2^n - n$ is a natural number. Apply the AGMI to the 2^n terms $a_1, \ldots, a_n, a, a, \ldots, a$, where there are $2^n - n$ occurrences of a in this list, to derive the AGMI in general.

8. (Cf. [169]) In this problem we give another proof of the AGMI. The AGMI holds for one term, so assume that it holds for n terms; we shall prove the AGMI for $n + 1$ terms.

(i) Prove that if the AGMI holds when the $n + 1$ nonnegative real numbers $a_1, \ldots, a_{n+1}$ satisfy $a_1 \cdot a_2 \cdot a_3 \cdots a_{n+1} = 1$, then the AGMI holds for any $n + 1$ nonnegative real numbers.

(ii) By (i), we just have to verify that the AGMI holds when $a_1 \cdots a_{n+1} = 1$. Using the induction hypothesis, prove that $a_1 + \cdots + a_n + a_{n+1} \geq n(a_{n+1})^{-1/n} + a_{n+1}$.

(iii) Prove that $a_1 + \cdots + a_{n+1} \geq n + 1$, which is the AGMI for $n + 1$ terms, by proving the following lemma: For every $x > 0$, we have $nx^{-1/n} + x \geq n + 1$. Suggestion: In Bernoulli's inequality, replace the n there by $n + 1$ and let $a = x^{1/n} - 1$.

9. (**Fibonacci sequence**) The **Fibonacci sequence** is the sequence $F_0, F_1, F_2, F_3, \ldots$ defined recursively by

$$\boxed{F_0 = 0, \quad F_1 = 1, \quad F_n = F_{n-1} + F_{n-2} \text{ for all } n \geq 2.}$$

Using strong induction, prove that for every natural number n,

$$F_n = \frac{1}{\sqrt{5}} \left[\Phi^n - (-\Phi)^{-n} \right], \text{ where } \Phi = \frac{1 + \sqrt{5}}{2} \ (\Phi \text{ is called the } \textbf{golden ratio}).$$

Suggestion: Note that $\Phi^2 = \Phi + 1$ and hence $-\Phi^{-1} = 1 - \Phi = (1 - \sqrt{5})/2$.

10. (**Pascal's method**) Using a method due to Pascal, we generalize our formula (2.2) for the sum of the first n integers to sums of powers. See [17] for more on Pascal's method. For all natural numbers k, n, put $\sigma_k(n) := 1^k + 2^k + \cdots + n^k$ and set $\sigma_0(n) := n$.

(i) For all $n, k \in \mathbb{N}$, prove that

$$(n + 1)^{k+1} - 1 = \sum_{\ell=0}^{k} \binom{k + 1}{\ell} \sigma_\ell(n).$$

Suggestion: The left-hand side can be written as $\sum_{m=1}^{n} \left((m + 1)^{k+1} - m^{k+1} \right)$. Use the binomial theorem on $(m + 1)^{k+1}$.

(ii) Using the strong form of induction on k, prove that for all $k \in \mathbb{N}$, there are rational numbers $a_{k1}, \ldots, a_{kk} \in \mathbb{Q}$ such that

$$\sigma_k(n) = \frac{1}{k+1} n^{k+1} + a_{kk}n^k + \cdots + a_{k2}n^2 + a_{k1}n \text{ for all } n \in \mathbb{N}. \quad \textbf{(Pascal's formula)}$$

(iii) (Cf. [133]) Using the fact that $\sigma_3(n) = \frac{1}{4}n^4 + a_{33}n^3 + a_{32}n^2 + a_{31}n$, find the coefficients a_{31}, a_{32}, a_{33}. Suggestion: Consider the difference $\sigma_3(n) - \sigma_3(n - 1)$. (Using a similar method, can you find the coefficients in the sum for $\sigma_4(n)$?)

11. **(The multinomial theorem)** A **multi-index** is an n-tuple of nonnegative integers and is usually denoted by Greek letters, for instance $\alpha = (\alpha_1, \ldots, \alpha_n)$, where each α_k can be any of $0, 1, 2, \ldots$. We define $|\alpha| = \alpha_1 + \cdots + \alpha_n$ and $\alpha! = \alpha_1! \cdot \alpha_2! \cdots \alpha_n!$. Prove that for every natural number n, all complex numbers $a_1, \ldots, a_n$, and every natural number k, we have

$$(a_1 + \cdots + a_n)^k = \sum_{|\alpha|=k} \frac{k!}{\alpha!} a_1^{\alpha_1} \cdots a_n^{\alpha_n}.$$

Here, the summation is over all multi-indices α with $|\alpha| = k$. Suggestion: For the induction step, write $a_1 + \cdots + a_{n+1} = a + a_{n+1}$, where $a = a_1 + \cdots + a_n$, and use the binomial formula on $(a + a_{n+1})^k$.

2.3 The Integers

Have you ever wondered what it would be like in a world where the temperature was never below zero degrees Celsius? How boring it would be to never see snow! The natural numbers $1, 2, 3, \ldots$ are closed under addition and multiplication, which are essential for counting purposes used in everyday life. However, the natural numbers do not have negatives, which is an inconvenience mathematically. In particular, $\mathbb{N}$ is not closed under subtraction. For example, the equation

$$x + 7 = 4$$

does not have any solution x in the natural numbers. We can either accept that such an equation does not have solutions[5] or we can describe a number system in which such an equation does have solutions. We shall go the latter route, and in this section, we study the integers, which are closed under subtraction and have negatives.

2.3.1 Axioms for Integer Numbers

Incorporating zero and the negatives of the natural numbers,

$$0, -1, -2, -3, -4, \ldots,$$

---

[5]"The imaginary expression $\sqrt{(-a)}$ and the negative expression $-b$ have this resemblance, that either of them occurring as the solution of a problem indicates some inconsistency or absurdity. As far as real meaning is concerned, both are imaginary, since $0 - a$ is as inconceivable as $\sqrt{(-a)}$." Augustus De Morgan (1806–1871). [Nowadays we don't share De Morgan's views!]

into the natural numbers forms the **integers**:

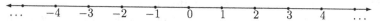

The set of integers is denoted by $\mathbb{Z}$. The natural numbers are also referred to as the **positive integers**, their negatives the **negative integers**, while the numbers $0, 1, 2, \ldots,$ the natural numbers plus zero, are called the **nonnegative integers** or **whole numbers**, and finally, $0, -1, -2, \ldots$ are the **nonpositive integers**. The following arithmetic properties of addition and multiplication of integers, like those for natural numbers, are familiar (in the following, a, b, c denote arbitrary integers):

Addition satisfies

(A1) $a + b = b + a$ (commutative law);
(A2) $(a + b) + c = a + (b + c)$ (associative law);
(A3) there is an integer denoted by 0 "zero" such that

$$a + 0 = a = 0 + a \quad \text{(existence of additive identity)};$$

(A4) for each integer a there is an integer denoted by the symbol $-a$ such that[6]

$$a + (-a) = 0 \quad \text{and} \quad (-a) + a = 0 \quad \text{(existence of additive inverse)}.$$

Multiplication satisfies

(M1) $a \cdot b = b \cdot a$ (commutative law);
(M2) $(a \cdot b) \cdot c = a \cdot (b \cdot c)$ (associative law);
(M3) there is an integer denoted by 1 "one," different from 0, such that

$$1 \cdot a = a = a \cdot 1 \quad \text{(existence of multiplicative identity)}.$$

As with the natural numbers, the $\cdot$ is sometimes dropped and the associative laws imply that expressions involving integers such as $a + b + c$ and abc make sense without the use of parentheses. Addition and multiplication are related by

(D) $a \cdot (b + c) = (a \cdot b) + (a \cdot c)$ (distributive law).

Of these arithmetic properties, the only additional properties listed that were not listed for natural numbers are **(A3)** and **(A4)**. **Subtraction** is the operation

$$a + (-b) = (-b) + a \quad \text{and is denoted by} \quad a - b,$$

so subtraction is, by definition, really just "adding negatives." A set together with the operations of addition and multiplication that satisfy properties **(A1)**–**(A4)**, **(M2)**, **(M3)**, and **(D)** is called a **ring**; essentially, a ring is just a set of objects closed

---

[6] At this moment, there could possibly be another integer, say $b \neq -a$, such that $a + b = 0$, but in Theorem 2.10 we prove that if such a b exists, then $b = -a$; so, additive inverses are unique.

under addition, multiplication, and subtraction. If the multiplication operation satisfies (**M1**), then the set is called a **commutative ring**.

The natural numbers, or positive integers, which we denote either by $\mathbb{N}$ or by $\mathbb{Z}^+$, is closed under addition and multiplication, and satisfies the **trichotomy** law, namely, for every integer a, exactly one of the following "positivity" properties holds:

(**P**) a is a positive integer, $a = 0$, or $-a$ is a positive integer.

Stated another way, property (**P**) means that $\mathbb{Z}$ is a union of disjoint sets,

$$\mathbb{Z} = \mathbb{Z}^+ \cup \{0\} \cup -\mathbb{Z}^+,$$

where $-\mathbb{Z}^+$ consists of all integers of the form $-a$, where $a \in \mathbb{Z}^+$.

Everyday experience convinces us that there is a set of integers satisfying properties (**A**), (**M**), (**D**), and (**P**). In fact, it's not difficult to *construct* the integers from the natural numbers (see, for instance, [149]), but for expediency, we'll just *assume* the existence of a set $\mathbb{Z}$ satisfying properties (**A**), (**M**), (**D**), and (**P**). From the properties listed above, we shall derive some well-known properties of the integers that you have known since grade school.

2.3.2 Proofs of Well-Known High School Rules

Since the integers satisfy the same arithmetic properties as the natural numbers, the same proofs as in Section 2.1 prove that the distributive law (**D**) holds from the right and the FOIL law holds. Also, the *cancellation property* in Theorem 2.3 holds: Given integers a, b, c,

$$a = b \text{ if and only if } a + c = b + c.$$

However, now this statement is easily proved using the fact that the integers have additive inverses. We prove only the "if" part: If $a + c = b + c$, then adding $-c$ to both sides of this equation, we obtain

$$(a + c) + (-c) = (b + c) + (-c) \implies a + (c + (-c)) = b + (c + (-c))$$
$$\implies a + 0 = b + 0,$$

or $a = b$. Comparing this proof with that of Theorem 2.3 shows the usefulness of having additive inverses.

We now show that we can always solve equations such as the one given at the beginning of this section. Moreover, we prove that there is only one additive identity.

Uniqueness of additive identities and inverses

Theorem 2.10 *For $a, b, x \in \mathbb{Z}$:*

(1) The equation

$$x + a = b \quad \text{holds if and only if} \quad x = b - a.$$

In particular, the only x that satisfies the equation $x + a = a$ is $x = 0$. Thus, there is only one additive identity.

(2) The only x that satisfies the equation

$$x + a = 0$$

is $x = -a$. Thus, each integer has exactly one additive inverse.

(3) Finally, $0 \cdot a = 0$. (zero $\times$ anything is zero).

Proof We have

$$x + a = b \iff (x + a) + (-a) = b + (-a) \iff x + (a + (-a)) = b - a$$
$$\iff x + 0 = b - a \iff x = b - a.$$

For the first $\iff$ we used cancellation. This proves *(1)*, and taking $b = 0$ in *(1)* implies *(2)*.

Since $0 = 0 + 0$, we have

$$0 \cdot a = (0 + 0) \cdot a = 0 \cdot a + 0 \cdot a.$$

Canceling $0 \cdot a$ (that is, adding $-(0 \cdot a)$ to both sides) proves *(3)*. ∎

By commutativity, $a + x = b$ if and only if $x = b - a$. Similarly, $a + x = a$ if and only if $x = -a$, and $a + x = 0$ if and only if $x = -a$.

Some of the most confusing rules of elementary mathematics are the rules involving negatives. These rules can be explained using accounting (getting/owing money), balloons attached to baskets containing weights, walking (or driving) forward and backward, behavior (pleasant being positive and grumpy being negative), and many others. I like to use the number line. One interpretation of subtraction $a - b$ is the "displacement from b to a". Here are some examples:

$$\dots\ -4\quad -3\quad -2\quad -1\quad 0\quad 1\ \dots \qquad\qquad \dots\ -4\quad -3\quad -2\quad -1\quad 0\quad 1\ \dots$$

$-4 - 1 =$ displacement from 1 to -4 $\qquad\qquad 1 - (-4) =$ displacement from -4 to 1
$\qquad\quad = -5 \qquad\qquad\qquad\qquad\qquad\qquad\qquad\qquad\ = 5$

Here are some more:

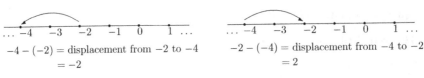

$-4-(-2)=$ displacement from -2 to -4 $-2-(-4)=$ displacement from -4 to -2

$= -2$ $= 2$

In the three cases in which we had $a - (-b)$, the answer is the same as $a + b$, whence the rule "subtracting a negative is adding a positive." What about multiplying negatives? Here's one way to understand multiplication conceptually. Let a, b be integers with $a \geq 0$. Then one common interpretation of $a \cdot b$ is the number obtained by "scaling b by the factor a." If $a > 1$, this scaling will stretch b; if $0 \leq a < 1$, it will shrink b instead. Since negatives are synonymous with "opposites," we can interpret $(-a) \cdot b$ as the number obtained by "scaling b, in its opposite direction, by the factor a." Here are some examples using this interpretation of multiplication:

In the first case, we stretch 3 by 2, and in the second, we stretch 3 by 2 in the opposite direction. We can stretch -3 using the same ideas:

From these pictures we can see in particular that $(-a) \cdot (-b) = a \cdot b$, at least for the example presented. If you were like me and didn't quite understand the rules $a - (-b) = a + b$ and $(-a) \cdot (-b) = a \cdot b$ back in grade school, it would be satisfying to know that we can in fact *prove* them:

Rules of sign

> **Theorem 2.11** *The following "rules of signs" hold:*
>
> *(1)* $-(-a) = a.$
> *(2)* $a \cdot (-1) = -a = (-1) \cdot a.$
> *(3)* $(-1) \cdot (-1) = 1.$
> *(4)* $(-a) \cdot (-b) = ab.$
> *(5)* $(-a) \cdot b = -(ab) = a \cdot (-b).$
> *(6)* $-(a+b) = (-a) + (-b).$ *In particular,* $-(a - b) = b - a.$
> *(7)* $a - (-b) = a + b.$

Proof We prove *(1)*–*(3)*, leaving *(4)*–*(7)* for you in Problem 1. To prove *(1)*, note that since $a + (-a) = 0$, by uniqueness of additive inverses proved in the previous theorem, the additive inverse of $-a$ is a, that is, $-(-a) = a$.

To prove *(2)*, observe that

$$a + a \cdot (-1) = a \cdot 1 + a \cdot (-1) = a \cdot (1 + (-1)) = a \cdot 0 = 0,$$

so by uniqueness of additive inverses, we have $-a = a \cdot (-1)$. By commutativity, $-a = (-1) \cdot a$ also holds.

By *(1)*, *(2)*, we get *(3)*: $(-1) \cdot (-1) = -(-1) = 1$. ∎

Everyone knows that $-0 = 0$. This fact follows from the formula $0 + 0 = 0$ (so the additive inverse of 0 is 0), or as an easy application of *(2)*: $-0 = 0 \cdot (-1) = 0$, since zero times anything is zero. We can also distribute over subtraction:

$$a(b - c) = a(b + (-c)) = ab + a(-c)$$
$$= ab - ac \quad \text{(by (5) of Theorem 2.11)}.$$

Trichotomy: Using the positivity assumption (**P**), we can order the integers in much the same way as the natural numbers are ordered. Given integers a and b, exactly one of the following holds:

(O1) $a = b$, that is, $b - a = 0$;
(O2) $a < b$, which means that $b - a$ is a positive integer;
(O3) $b < a$, which means that $a - b$ is a positive integer.

By our previous theorem, $-(b - a) = a - b$, so (**O3**) is just that $-(b - a)$ is a natural number; thus, (**O1**), (**O2**), and (**O3**) refer to $b - a$ being 0, positive, or negative.

Just as for natural numbers, we can define $\leq$, $>$, and $\geq$. For example, $0 < b$, or $b > 0$, means that $b - 0 = b$ is a positive integer. Thus, an integer b is greater than 0 is synonymous with b is a positive integer. (Of course, this agrees with our English usage of $b > 0$ to mean b is positive!) Similarly, $b < 0$ means that $0 - b = -b$ is a positive integer. As with the natural numbers, we have the transitive law: If $a < b$ and $b < c$, then $a < c$, and we also have the Archimedean ordering of $\mathbb{Z}$: Given a natural number a and integer b, there is a natural number n such that $b < a \cdot n$. To see this last property, note that if $b \leq 0$, then any natural number n works; if $b > 0$, then b is a natural number, and the Archimedean ordering of $\mathbb{N}$ applies to show the existence of n. We now prove some of the familiar inequality rules.

Inequality rules

Theorem 2.12 *The following inequality rules hold:*

(1) If $a < b$ and $c \leq d$, then $a + c < b + d$.
(2) If $a < b$ and $c > 0$, then $a \cdot c < b \cdot c$ (positive multipliers preserve inequalities).
(3) If $a < b$ and $c < 0$, then $a \cdot c > b \cdot c$ (negative multipliers switch inequalities).
(4) If $a > 0$ and $b > 0$, then $ab > 0$ (positive × positive is positive).
(5) If $a > 0$ and $b < 0$ (or vice versa), then $ab < 0$ (positive × negative is negative).
(6) If $a < 0$ and $b < 0$, then $ab > 0$. (negative × negative is positive)

Proof We prove *(1)–(3)* and leave *(4)–(6)* for you in Problem 1.

To prove *(1)*, we use associativity and commutativity to write

$$(b+d) - (a+c) = (b-a) + (d-c).$$

Since $a < b$, by definition of less than, $b - a$ is a natural number, and since $c \leq d$, $d - c$ is either zero (if $c = d$) or a natural number. Hence, $(b - a) + (d - c)$ is either adding two natural numbers or a natural number and zero; in either case, the result is a natural number. Thus, $a + c < b + d$.

If $a < b$ and $c > 0$, then (distributing over subtraction)

$$b \cdot c - a \cdot c = (b-a) \cdot c.$$

The number c is a natural number, and since $a < b$, the integer $b - a$ is a natural number, so their product $(b - a) \cdot c$ is also a natural number. Thus, $a \cdot c < b \cdot c$.

If $a < b$ and $c < 0$, then by our rules of sign,

$$a \cdot c - b \cdot c = (a-b) \cdot c = -(a-b) \cdot (-c) = (b-a) \cdot (-c).$$

Since $c < 0$, the integer $-c$ is a natural number, and since $a < b$, the integer $b - a$ is a natural number, so their product $(b - a) \cdot (-c)$ is also a natural number. Thus, $a \cdot c > b \cdot c$. ∎

We now prove that zero and one have the familiar properties that you know.

Zero, cancellation, and one

> **Theorem 2.13** *For integers a, b, c, we have the following:*
>
> *(1) If $a \cdot b = 0$, then $a = 0$ or $b = 0$.*
> *(2) If $a \cdot b = a \cdot c$, where $a \neq 0$, then $b = c$.*
> *(3) If $a \cdot b = a$, then $b = 1$, so 1 is the only multiplicative identity.*

Proof We give two proofs of *(1)*. Although **Proof I** is acceptable, **Proof II** is much preferred, because **Proof I** in fact boils down to a contrapositive statement, which **Proof II** goes to directly.

Proof I: Assume that $ab = 0$. We shall prove that $a = 0$ or $b = 0$. Now either $a = 0$ or $a \neq 0$. If $a = 0$, then we are done, so assume that $a \neq 0$. We need to prove that $b = 0$. Well, either $b = 0$ or $b \neq 0$. However, it cannot be true that $b \neq 0$, for according to the properties *(4)–(6)* of our rules for inequalities,

$$\text{if } a \neq 0 \text{ and } b \neq 0, \text{ then } a \cdot b \neq 0. \tag{2.8}$$

But $ab = 0$, so $b \neq 0$ cannot be true. This contradiction shows that $b = 0$.

Proof II: Our second proof of *(1)* is a **proof by contraposition**, which is essentially what we did in **Proof I** without stating it! Recall, as already explained in the proof of Theorem 2.3 on p. 396, that the technique of a proof by contraposition is that in order to prove the statement "if $a \cdot b = 0$, then $a = 0$ or $b = 0$," we can instead try to prove the contrapositive statement:

if $a \neq 0$ and $b \neq 0$, then $a \cdot b \neq 0$.

However, as explained above (2.8), the truth of this statement follows from our inequality rules. This gives another (better) proof of *(1)*.

To prove *(2)*, assume that $a \cdot b = a \cdot c$, where $a \neq 0$. Then,

$$0 = a \cdot b - a \cdot c = a \cdot (b - c).$$

By *(1)*, either $a = 0$ or $b - c = 0$. We are given that $a \neq 0$, so we must have $b - c = 0$, or adding c to both sides, $b = c$. *(3)* follows from *(2)* when $c = 1$. ∎

Property *(1)* of this theorem is the basis for solving quadratic equations in high school. For example, let us solve $x^2 - x - 6 = 0$. We first "factor"; that is, we observe that

$$(x - 3)(x + 2) = x^2 - x - 6 = 0.$$

By property *(1)*, we know that $x - 3 = 0$ or $x + 2 = 0$. Thus, $x = 3$ or $x = -2$.

2.3.3 Absolute Value

Given an integer a, we know that either $a = 0$, a is a positive integer, or $-a$ is a positive integer. The **absolute value** of the integer a is denoted by $|a|$ and is defined to be the "nonnegative part of a":

$$|a| = \begin{cases} a & \text{if } a \geq 0, \\ -a & \text{if } a < 0. \end{cases}$$

Thus, for instance, $|5| = 5$, while $|-2| = -(-2) = 2$. In the following theorem, we prove some (what should be) familiar rules of absolute value. To prove statements about absolute values, it's convenient to **prove by cases**.

Absolute value rules

Theorem 2.14 *For $a, b, x \in \mathbb{Z}$:*

(1) $|a| = 0$ *if and only if $a = 0$.*
(2) $|ab| = |a|\,|b|$.
(3) $|a| = |-a|$.
(4) *For $x \geq 0$, $|a| \leq x$ if and only if $-x \leq a \leq x$.*
(5) $-|a| \leq a \leq |a|$.
(6) $|a + b| \leq |a| + |b|$ *(triangle inequality)*.

Proof If $a = 0$, then by definition, $|0| = 0$. Conversely, we need to prove that if $|a| = 0$, then $a = 0$. We prove the contrapositive: If $a \neq 0$, then $|a| \neq 0$. But this is clear, because $|a|$ equals a or $-a$, both of which are not zero if a is not zero. This proves *(1)*.

To prove *(2)*, we consider four cases: $a \geq 0$ and $b \geq 0$, $a < 0$ and $b \geq 0$, $a \geq 0$ and $b < 0$, and lastly, $a < 0$ and $b < 0$. We shall freely use the *rules of sign* and *inequality rules*. If $a \geq 0$ and $b \geq 0$, then $ab \geq 0$, so $|ab| = ab = |a| \cdot |b|$. If $a < 0$ and $b \geq 0$, then $ab \leq 0$, so $|ab| = -ab = (-a) \cdot b = |a| \cdot |b|$. The case that $a \geq 0$ and $b < 0$ is handled similarly. Lastly, if $a < 0$ and $b < 0$, then $ab > 0$, so $|ab| = ab = (-a) \cdot (-b) = |a| \cdot |b|$.

By *(2)*, we have $|-a| = |(-1) \cdot a| = |-1| \cdot |a| = 1 \cdot |a| = |a|$, which proves *(3)*.

To prove *(4)*, let $x \geq 0$, and suppose $|a| \leq x$. Since $x \geq 0$, we have $-x \leq 0$, so we can write $|a| \leq x$ as $-x \leq |a| \leq x$. Thus, we must prove $-x \leq |a| \leq x$ if and only if $-x \leq a \leq x$. Consider two cases: $a \geq 0$, $a < 0$. In the first case, $-x \leq |a| \leq x$ is equivalent to $-x \leq a \leq x$, which proves *(4)* when $a \geq 0$. In the second case, $a < 0$, we have $-x \leq |a| \leq x$ is equivalent to $-x \leq -a \leq x$. Multiplying through by -1, we get the equivalent statement $x \geq a \geq -x$, or $-x \leq a \leq x$ (equivalent, since multiplying the last inequality by -1 gives the former inequality $-x \leq -a \leq x$). This prove *(4)* in the case $a < 0$.

Property *(5)* follows from *(4)* with $x = |a|$.

Finally, we prove the triangle inequality. From *(5)* we have $-|a| \leq a \leq |a|$ and $-|b| \leq b \leq |b|$. Adding these inequalities gives

$$-\big(|a| + |b|\big) \leq a + b \leq |a| + |b|.$$

Applying *(4)* gives the triangle inequality.   ∎

▶ Exercises 2.3

1. Finish the proofs in the *rules of signs* and *inequality rules* theorems.
2. For integers a, b, prove the inequalities

$$\big|\,|a| - |b|\,\big| \leq |a \pm b| \leq |a| + |b|.$$

3. Let b be an integer. Prove that the only integer a satisfying

$$b - 1 < a < b + 1$$

   is the integer $a = b$.
4. Let $n \in \mathbb{N}$. Assume properties of powers from Example 2.2 on p. 16.

   (a) Let a, b be *nonnegative* integers. Using a proof by contraposition, prove that if $a^n = b^n$, then $a = b$.

(b) We now consider the situation that a, b could be negative. So, let a, b be arbitrary integers. Suppose that $n = 2m$ for some positive integer m. Prove that if $a^n = b^n$, then $a = \pm b$.

(c) Again let a, b be arbitrary integers. Suppose that $n = 2m - 1$ for some natural number m. Prove the statement if $a^n = b^n$, then $a = b$, using a proof by cases. Here the cases consist of a, b both nonnegative, both negative, and when one is nonnegative and the other negative (in this last case, show that $a^n = b^n$ actually can never hold, so for this last case, the statement is superfluous).

5. In this problem we prove an integer version of induction. Let k be any integer (positive, negative, or zero) and suppose that we are given a list of statements

$$P_k, P_{k+1}, P_{k+2}, \ldots,$$

and suppose that **(1)** P_k is true and **(2)** if n is an integer with $n \geq k$ and the statement P_n happens to be valid, then the statement P_{n+1} is also valid. Prove that every single statement $P_k, P_{k+1}, P_{k+2}, \ldots$ is true.

6. **(Game of Nim)** Here's a fascinating example using strong induction and proof by cases; see the coin game in Problem 6 on p. 46 for a related game. Suppose that n stones are thrown on the ground. Two players take turns removing one, two, or three stones each. The last one to remove a stone loses. Let P_n be the statement that the player starting first has a foolproof winning strategy if n is of the form $n = 4k, 4k + 2$, or $4k + 3$ for some $k = 0, 1, 2, \ldots$, and the player starting second has a foolproof winning strategy if $n = 4k + 1$ for some $k = 0, 1, 2, \ldots$. In this problem we prove that P_n is true for all $n \in \mathbb{N}$.[7]

(i) Prove that P_1 is true. Assume that $P_1, \ldots, P_n$ hold. To prove that P_{n+1} holds, we prove by cases. The integer $n + 1$ can be of four types: $n + 1 = 4k$, $n + 1 = 4k + 1$, $n + 1 = 4k + 2$, or $n + 1 = 4k + 3$.

(ii) Case 1: $n + 1 = 4k$. The first player can remove one, two, or three stones; in particular, he can remove three stones (leaving $4k - 3 = 4(k - 1) + 1$ stones). Prove that the first person wins.

(iii) Case 2: $n + 1 = 4k + 1$. Prove that the second player will win regardless of whether the first person takes one, two, or three stones (leaving $4k$, $4(k - 1) + 3$, and $4(k - 1) + 2$ stones, respectively).

(iv) Case 3, Case 4: $n + 1 = 4k + 2$ or $n + 1 = 4k + 3$. Prove that the first player has a winning strategy in the cases that $n + 1 = 4k + 2$ and $n + 1 = 4k + 3$. Suggestion: Make the first player remove one and two stones, respectively.

---

[7] Here, we are assuming that every natural number can be written in the form $4k, 4k + 1, 4k + 2$, or $4k + 3$; you can either assume this fact (which is not difficult to prove) or wait until the division algorithm, Theorem 2.16 on p. 59, from which this fact follows immediately.

2.4 Primes and the Fundamental Theorem of Arithmetic

It is not always true that given two integers a and b, there is an integer q (for "quotient") such that

$$b = a\,q.$$

For instance, $2 = 4q$ can never hold for any integer q, nor can $17 = 2q$. This, of course, is exactly the reason rational numbers are needed! (Rational numbers will come up in Section 2.6.) The existence or nonexistence of such quotients opens up an incredible wealth of topics concerning prime numbers in number theory.[8]

2.4.1 Divisibility

If a and b are integers, with a *nonzero*, we say that a **divides** b and write $a|b$, if there is an integer q such that

$$b = a\,q.$$

We also say that b is **divisible** by a, or b is a **multiple** of a. We call a a **divisor** or **factor** of b, and q the **quotient** (of b divided by a).

Example 2.6 Thus, for example $4|(-16)$ with quotient -4, because $-16 = 4 \cdot (-4)$, and $(-2)|(-6)$ with quotient 3, because $-6 = (-2) \cdot 3$.

Note that we adopt the convention that *divisors are by definition nonzero*. To see why, note that for an arbitrary integer q, we have

$$0 = 0 \cdot q,$$

so if 0 were allowed to be a divisor, then 0 divided by itself would not have a unique quotient! However, nonzero divisors give rise to unique quotients. Indeed, if $a \neq 0$ and if we have $b = aq$ and $b = aq'$, then by cancellation (Theorem 2.13 on p. 375),

$$aq = aq' \implies q = q',$$

so quotients are unique. Because speaking of *the* quotient rather than *a* quotient is desirable in discussing divisibility, we define divisors to be nonzero. Thus comes the high school phrase "You can never divide by 0!" Here are some important properties of division.

---

[8]"Mathematicians have tried in vain to this day to discover some order in the sequence of prime numbers, and we have reason to believe that it is a mystery into which the human mind will never penetrate." Leonhard Euler (1707–1783) [225].

Divisibility rules

Theorem 2.15 *The following divisibility rules hold:*

(1) If $a|b$ and b is positive, then $|a| \leq b$.
(2) If $a|b$, then $a|bc$ for every integer c.
(3) If $a|b$ and $b|c$, then $a|c$.
(4) If $a|b$ and $a|c$, then $a|(bx + cy)$ for all integers x and y.

Proof Assume that $a|b$ and $b > 0$. Since $a|b$, we have $b = aq$ for an integer q. Assume momentarily that $a > 0$. By our inequality rules (Theorem 2.12 on p. 54), we know that "positive $\times$ negative is negative," so q cannot be negative. Also, q can't be zero, because $b \neq 0$. Therefore $q > 0$. By our rules for inequalities,

$$a = a \cdot 1 \leq a \cdot q = b \implies |a| \leq b.$$

Assume now that $a < 0$. Then $b = aq = (-a)(-q)$. Since $(-a) > 0$, by our proof for positive divisors that we just did, we have $(-a) \leq b$, that is, $|a| \leq b$.

We now prove *(2)*. If $a|b$, then $b = aq$ for some integer q. Hence,

$$bc = (aq)c = a(qc),$$

so $a|bc$.

To prove *(3)*, suppose that $a|b$ and $b|c$. Then $b = aq$ and $c = bq'$ for some integers q and q'. Hence,

$$c = bq' = (aq)q' = a(qq'),$$

so $a|c$.

Finally, assume that $a|b$ and $a|c$. Then $b = aq$ and $c = aq'$ for integers q and q'. Hence, for all integers x and y,

$$bx + cy = (aq)x + (aq')y = a(qx + q'y),$$

so $a|(bx + cy)$. ∎

2.4.2 The Division Algorithm

Although we cannot always divide one integer into another, we can always do so up to remainders.

Example 2.7 For example, although 2 does not divide 7, we can write

$$7 = 3 \cdot 2 + 1.$$

Another example is that although -3 does not divide -13, we do have

$$-13 = 5 \cdot (-3) + 2.$$

In general, if a and b are integers and

$$b = qa + r, \quad \text{where } 0 \le r < |a|,$$

then we call q the **quotient** (of b divided by a) and r the **remainder**. Such numbers always exist, as we now prove.

The division algorithm

Theorem 2.16 *Given any integers a and b with $a \ne 0$, there are unique integers q and r such that*

$$\boxed{b = qa + r \quad \text{with} \ \ 0 \le r < |a|.}$$

Moreover, if a and b are both positive, then q is nonnegative. Furthermore, a divides b if and only if $r = 0$.

Proof Assume for the moment that $a > 0$. Consider the list of integers

$$\ldots, 1+b-3a, \ 1+b-2a, \ 1+b-a, \ 1+b, \ 1+b+a, \ 1+b+2a, \ 1+b+3a, \ldots \tag{2.9}$$

extending indefinitely in both directions. Notice that since $a > 0$, for every integer n,

$$1+b+na < 1+b+(n+1)a,$$

so the integers in the list (2.9) are increasing. Moreover, by the Archimedean ordering of the integers, there is a natural number n such that $-1-b < an$, or $1+b+an > 0$. In particular, $1+b+ak > 0$ for $k \ge n$. Thus, far enough to the right in the list (2.9), all the integers are positive. Let A be the set of all natural numbers appearing in the list (2.9). By the well-ordering principle (Theorem 2.6 on p. 137), this set of natural numbers has a least element, let us call it $1+b+ma$, where m is an integer. This integer satisfies

$$1+b+(m-1)a < 1 \le 1+b+ma, \tag{2.10}$$

for if $1+b+(m-1)a \ge 1$, then $1+b+(m-1)a$ would be an element of A smaller than $1+b+ma$. Put $q = -m$ and $r = b+ma = b-qa$. Then $b = qa+r$ by construction, and substituting q and r into (2.10), we obtain

$$1+r-a < 1 \le 1+r.$$

Subtracting 1 from everything, we see that

$$r - a < 0 \le r.$$

Thus, $0 \le r$ and $r - a < 0$ (that is, $r < a$). Thus, we have found integers q and r such that $b = qa + r$ with $0 \le r < a$. Observe from (2.10) that if b is positive, then m can't be positive (otherwise, $1 + b + (m - 1)a \ge 1$, violating (2.10)). Thus, q is nonnegative if both a and b are positive. Assume now that $a < 0$. Then $-a > 0$, so by what we just did, there are integers s and r with $b = s(-a) + r$ and $0 \le r < -a$; that is, $b = qa + r$, where $q = -s$ and $0 \le r < |a|$.

We now prove uniqueness. Assume that we also have $b = q'a + r'$ with $0 \le r' < |a|$. We first prove that $r = r'$. Indeed, by symmetry in the primed and unprimed letters, we may assume that $r \le r'$. Then $0 < r' - r \le r' < |a|$. Moreover,

$$q'a + r' = qa + r \quad \Longrightarrow \quad (q' - q)a = r' - r.$$

This shows that $a|(r' - r)$. Now if $r < r'$, then $r' - r$ is positive, so by Property (1) of Theorem 2.15 we would have $|a| \le r' - r$. However, we have already stated that $r' - r < |a|$, so we must have $r = r'$. Then the equation $(q' - q)a = r' - r$ reads $(q' - q)a = 0$. Since $a \ne 0$, we must have $q' - q = 0$, or $q = q'$. Our proof of uniqueness is thus complete.

Finally, we prove that $a|b$ if and only if $r = 0$. If $a|b$, then $b = ac = ac + 0$ for some integer c. By uniqueness already established, we have $q = c$ and $r = 0$. Conversely, if $r = 0$, then $b = aq$, so $a|b$ by definition of divisibility. ∎

An integer n is **even** if we can write $n = 2m$ for some integer m, and **odd** if we can write $n = 2m + 1$ for some integer m.

Example 2.8 For instance, $0 = 2 \cdot 0$, so 0 is even; $1 = 2 \cdot 0 + 1$, so 1 is odd; and -1 is odd, since $-1 = 2 \cdot (-1) + 1$.

Using the division algorithm, we can easily prove that every integer is either even or odd. Indeed, dividing n by 2, the division algorithm implies that $n = 2m + k$, where $0 \le k < 2$, that is, where k is either 0 or 1. This shows that n is either even (if $k = 0$) or odd (if $k = 1$).

An important application of the division algorithm is to the so-called Euclidean algorithm for finding greatest common divisors; see Problem 4.

2.4.3 Prime Numbers

Consider the number 12. This number has six positive factors, or divisors, 1, 2, 3, 4, 6, and 12. The number 21 has four positive factors, 1, 3, 7, and 21. The number 1 has only one positive divisor, 1. However, as the reader can check, 17 has exactly two positive factors, 1 and 17. Similarly, 5 has exactly two positive factors, 1 and 5. Numbers such as 5 and 17 are given a special name: A natural number that has exactly two positive factors is called a **prime** number.[9] Another way to say this is that a prime number is a natural number with exactly two factors, itself and 1. (Thus, 1 is not prime.) A list of the first ten primes is

$$2, 3, 5, 7, 11, 13, 17, 19, 23, 29.$$

An integer greater than 1 that is not prime is called a **composite** number. Notice that

$$12 = 2 \times 6 = 2 \times 2 \times 3, \quad 21 = 3 \times 7, \quad 17 = 17, \quad 5 = 5.$$

In each of these circumstances, we have **factored**, or expressed as a product, each number into a product of its prime factors. Here, by *convention*, we consider a prime number as factored.

Lemma 2.17 *Every natural number other than 1 can be factored into primes.*

Proof We shall prove that for every natural number $m = 1, 2, 3, \ldots$, the number $m + 1$ can be factored into primes; which is to say, every natural number $n = 2, 3, 4, \ldots$ can be factored into primes. We prove this lemma using strong induction. By our convention, $n = 2$ is already in factored form. Assume that our theorem holds for all natural numbers $2, 3, 4, \ldots, n$; we shall prove that our theorem holds for the natural number $n + 1$. Now, $n + 1$ is either prime or composite. If it is prime, then it is already in factored form. If it is composite, then $n + 1 = pq$, where p and q are natural numbers greater than 1. By Theorem 2.15, both p and q are less than $n + 1$. By the induction hypothesis, p and q can be factored into primes. It follows that $n + 1 = pq$ can also be factored into primes.   ∎

One of the first questions that one may ask is how many primes there are. This was answered by Euclid of Alexandria (c. 325 B.C.–c. 265 B.C.): There are infinity many. The following proof is the original due to Euclid and is the classic "proof by contradiction."

Euclid's theorem

Theorem 2.18 *There are infinitely many primes.*

Proof We start with the tentative assumption that the theorem is false. Thus, we assume that there are finitely many primes. There being finitely many, we can list them:

$$p_1, \ p_2, \ \ldots, p_n.$$

---

[9]"I hope you will agree that there is no apparent reason why one number is prime and another not. To the contrary, upon looking at these numbers one has the feeling of being in the presence of one of the inexplicable secrets of creation." Don Zagier [271, p. 8].

Consider the number

$$(p_1 p_2 p_3 \cdots p_n) + 1.$$

This number is either prime or composite. It is greater than all the primes $p_1, \ldots, p_n$, so this number can't equal any $p_1, \ldots, p_n$. We conclude that n must be composite, so

$$p_1 p_2 p_3 \cdots p_n + 1 = ab, \tag{2.11}$$

for some natural numbers a and b. By our lemma, both a and b can be expressed as a product involving some of $p_1, \ldots, p_n$, which implies that ab also has such an expression. In particular, being a product of some of the $p_1, \ldots, p_n$, the right-hand side of (2.11) is divisible by at least one of the prime numbers $p_1, \ldots, p_n$. However, the left-hand side is certainly not divisible by any such prime, because if we divide the left-hand side by any one of the primes $p_1, \ldots, p_n$, we always get the remainder 1! This contradiction shows that our original assumption that the theorem is false must have been incorrect; hence there must be infinitely many primes. ∎

2.4.4 Fundamental Theorem of Arithmetic

Consider the integer 120, which can be factored as follows:

$$120 = 2 \times 2 \times 2 \times 3 \times 5.$$

A little verification shows that it is impossible to factor 120 into any primes other than the ones displayed. Of course, the order can be different, e.g.,

$$120 = 3 \times 2 \times 2 \times 5 \times 2.$$

It is of fundamental importance in mathematics that every natural number can be factored into a product of primes in only one way, apart from the order.

Fundamental theorem of arithmetic

Theorem 2.19 *Every natural number other than 1 can be factored into primes in only one way, except for the order of the factors.*

Proof For sake of contradiction, let us suppose that there are numbers that can be factored in more that one way. By the well-ordering principle, there is a smallest such natural number a. Thus, we can write a as a product of primes in two ways:

$$a = p_1 p_2 \cdots p_m = q_1 q_2 \cdots q_n.$$

Note that both m and n are greater than 1, for a single prime number has one prime factorization. We shall obtain a contradiction by showing there is a smaller natural number that has two factorizations. First, we observe that none of the primes p_j on the left equals any of the primes q_k on the right. Indeed, if, for example, $p_1 = q_1$, then by cancellation, we could divide them out, obtaining the natural number

$$p_2 p_3 \cdots p_m = q_2 q_3 \cdots q_n.$$

This number is smaller than a and the two sides must represent two distinct prime factorizations, for if these prime factorizations were the same apart from the orderings, then (since $p_1 = q_1$) the factorizations for a would also be the same apart from orderings. Since a is the smallest such number with more than one factorization, we conclude that none of the primes p_j equals a prime q_k. In particular, $p_1 \neq q_1$. By symmetry we may assume that $p_1 < q_1$. Now consider the natural number

$$
\begin{aligned}
b &= (q_1 - p_1) q_2 q_3 \cdots q_n \qquad\qquad\qquad\qquad (2.12)\\
&= q_1 q_2 \cdots q_n - p_1 q_2 \cdots q_n \\
&= p_1 p_2 \cdots p_m - p_1 q_2 \cdots q_n \quad (\text{since } p_1 \cdots p_m = a = q_1 \cdots q_m) \\
&= p_1 (p_2 p_3 \cdots p_m - q_2 q_3 \cdots q_n). \qquad\qquad\qquad (2.13)
\end{aligned}
$$

Since $0 < q_1 - p_1 < q_1$, the number b is less than a, so b can be factored in only one way apart from orderings. Observe that the number $q_1 - p_1$ cannot have p_1 as a factor, for if p_1 divides $q_1 - p_1$, then p_1 also divides $(q_1 - p_1) + p_1 = q_1$, which is impossible, because q_1 is prime. Thus, factoring $q_1 - p_1$ into its prime factors, none of which is p_1, the expression (2.12) and the fact that $p_1 \neq q_k$ for every k shows that b does not contain the factor p_1 in its factorization. On the other hand, factoring $p_2 p_3 \cdots p_m - q_2 q_3 \cdots q_n$ into its prime factors, the expression (2.13) clearly shows that p_1 is in the prime factorization of b. This contradiction ends the proof. ∎

Another popular way to prove the fundamental theorem of arithmetic uses the concept of the **greatest common divisor**; see Problem 5 for this proof.

In our first exercise, recall that the notation $n!$ (read "n factorial") for $n \in \mathbb{N}$ denotes the product of the first n positive integers: $n! = 1 \cdot 2 \cdot 3 \cdots n$.

▶ **Exercises 2.4**

1. A natural question is: How sparse are the primes? Prove that there are arbitrarily large gaps in the list of primes in the following sense: Given a positive integer k, there are k consecutive composite integers. Suggestion: Consider the integers

$$(k+1)! + 2, \ (k+1)! + 3, \ldots, (k+1)! + k, \ (k+1)! + k + 1.$$

2. Using the fundamental theorem of arithmetic, prove that if a prime p divides ab, where $a, b \in \mathbb{N}$, then p divides a or p divides b. Is this statement true if $p > 1$ is not prime?

3. Prove Lemma 2.17, that every natural number other than 1 can be factored into primes, using the well-ordering principle instead of induction.

4. (**The Euclidean algorithm**) Let a and b be two integers, both not zero. Consider the set of all positive integers that divide both a and b. This set is nonempty (it contains 1) and is finite (since integers larger than $|a|$ and $|b|$ cannot divide both a and b). This set therefore has a largest element (Problem 6 on p. 36), which we denote by (a, b) and call the **greatest common divisor** (GCD) of a and b. In this problem we find the GCD using the Euclidean algorithm.

   (i) Show that $(\pm a, b) = (a, \pm b) = (a, b)$ and $(0, b) = |b|$. Because of these equalities, we henceforth assume that a and b are positive.

   (ii) By the division algorithm, we know that there are unique nonnegative integers q_0 and r_0 such that $b = q_0 a + r_0$ with $0 \le r_0 < a$. Show that $(a, b) = (a, r_0)$.

   (iii) By successive divisions by remainders, we can write

$$b = q_0 \cdot a + r_0, \quad a = q_1 \cdot r_0 + r_1, \quad r_0 = q_2 \cdot r_1 + r_2,$$
$$r_1 = q_3 \cdot r_2 + r_3, \quad \cdots \quad r_{j-1} = q_j \cdot r_j + r_{j+1}, \quad \cdots, \tag{2.14}$$

   where the process is continued only as long as we don't get a zero remainder. Show that $a > r_0 > r_1 > r_2 > \cdots$, and using this fact, explain why we must eventually get a zero remainder.

   (iv) Let $r_{n+1} = 0$ be the first zero remainder. Show that $r_n = (a, b)$. Thus, the last positive remainder in the sequence (2.14) equals (a, b). This process for finding the GCD is called the **Euclidean algorithm**.

   (v) Using the Euclidean algorithm, find $(77, 187)$ and $(193, 245)$.

5. Working backward through the Eq. (2.14), show that for every two integers a, b, we have

$$(a, b) = r_n = k a + \ell b,$$

for some integers k and ℓ. Using this fact concerning the GCD, we shall give another proof of the fundamental theorem of arithmetic.

   (i) Prove that if a prime p divides a product ab, then p divides a or p divides b. (Problem 2 does not apply here, because in that problem we used the fundamental theorem of arithmetic, but now we are going to prove this fundamental theorem.) Suggestion: Either p divides a or it doesn't; if it does, we're done, and if not, then the GCD of p and a is 1. It follows that $1 = k p + \ell a$, for some integers k, ℓ. Multiply this equation by b.

   (ii) Using induction, prove that if a prime p divides a product $a_1 \cdots a_n$, then p divides some a_i.

   (iii) Using (ii), prove the fundamental theorem of arithmetic.

6. (**Modular arithmetic**) Given $n \in \mathbb{N}$, we say that $x, y \in \mathbb{Z}$ are **congruent modulo** n, written $x \equiv y \pmod{n}$, if $x - y$ is divisible by n. For $a, b, x, y, u, v \in \mathbb{Z}$, prove the following:

(a) $x \equiv y \pmod{n}$ if and only if x and y have the same remainder when divided by n.

(b) $x \equiv y \pmod{n}$, $y \equiv x \pmod{n}$, $x - y \equiv 0 \pmod{n}$ are equivalent statements.

(c) If $x \equiv y \pmod{n}$ and $y \equiv z \pmod{n}$, then $x \equiv z \pmod{n}$.

(d) If $x \equiv y \pmod{n}$ and $u \equiv v \pmod{n}$, then $ax + by \equiv au + bv \pmod{n}$.

(e) If $x \equiv y \pmod{n}$ and $u \equiv v \pmod{n}$, then $xu \equiv yv \pmod{n}$.

(f) Finally, prove that if $x \equiv y \pmod{n}$ and $m | n$, where $m \in \mathbb{N}$, then $x \equiv y \pmod{m}$.

7. (**Fermat's little theorem**) We assume the basics of modular arithmetic from Problem 6. Let p be prime. In this problem we prove that for every $x \in \mathbb{Z}$, we have $x^p \equiv x \pmod{p}$. This result is due to Pierre de Fermat (1601–1665).

(i) For all $k, n \in \mathbb{N}$ with $1 \le k \le n$, we assume that $n!$ is divisible by $k!(n - k)!$; the quotient is denoted by $\binom{n}{k}$. (This divisibility fact was an exercise in Problem 3i on p. 46.) Prove that for $1 \le k \le p$, $\binom{p}{k}$ is divisible by p.

(ii) Using (i) and the binomial theorem, prove that for all $x, y \in \mathbb{Z}$, $(x + y)^p \equiv x^p + y^p \pmod{p}$.

(iii) Using (ii) and induction, prove that $x^p \equiv x \pmod{p}$ for all $x \in \mathbb{N}$. Conclude that $x^p \equiv x \pmod{p}$ for all $x \in \mathbb{Z}$.

8. (**Pythagorean triples**) A **Pythagorean triple** consists of three natural numbers (x, y, z) such that $x^2 + y^2 = z^2$. For example, $(3, 4, 5)$ and $(6, 8, 10)$ are such triples. A Pythagorean triple is called **primitive** if x, y, z are **relatively prime**, or **coprime**, which means that x, y, z have no common prime factors. For instance, $(3, 4, 5)$ is primitive, while $(6, 8, 10)$ is not. In this problem we prove

$$(x, y, z) \text{ is primitive} \iff \begin{cases} x = 2mn, \ y = m^2 - n^2, \ z = m^2 + n^2, \ \text{or,} \\ x = m^2 - n^2, \ y = 2mn, \ z = m^2 + n^2, \end{cases}$$

where m, n are coprime, $m > n$, and m, n are of opposite parity; that is, one of m, n is even and the other is odd.

(i) Prove the "$\Longleftarrow$" implication. Henceforth, let (x, y, z) be a primitive Pythagorean triple.

(ii) Prove that x and y cannot both be even.

(iii) Show that x and y cannot both be odd.

(iv) Therefore, one of x, y is even and the other is odd; let us choose x as even and y as odd. (The other way around is handled similarly.) Show that z is odd and conclude that $z + y$ and $z - y$ are both divisible by 2. We shall denote the respective quotients by $u = \frac{1}{2}(z + y)$ and $v = \frac{1}{2}(z - y)$, which are both natural numbers.

(v) Show that $y = u - v$ and $z = u + v$ and then $x^2 = 4uv$. Conclude that uv is a perfect square (that is, $uv = k^2$ for some $k \in \mathbb{N}$).

(vi) Prove that u and v must be coprime, and from this fact and the fact that uv is a perfect square, conclude that u and v each must be a perfect square; say $u = m^2$ and $v = n^2$ for some $m, n \in \mathbb{N}$. Finally, prove the desired result.

9. **(Pythagorean triples, again)** If you like primitive Pythagorean triples, here's another problem: Prove that if m, n are coprime, $m > n$, and m, n are of the *same* parity (either both even or both odd), then both $m^2 - n^2$ and $m^2 + n^2$ are divisible by 2, and

$$(x, y, z) \text{ is primitive}, \quad \text{where } x = mn , \ y = \frac{m^2 - n^2}{2} , \ z = \frac{m^2 + n^2}{2}.$$

Here, the notations $y = \frac{m^2 - n^2}{2}$ and $z = \frac{m^2 + n^2}{2}$ refer to the quotients of $m^2 - n^2$ and $m^2 + n^2$ when divided by 2. Combining this with the previous problem, we see that given coprime natural numbers $m > n$, we have

$$(x, y, z) \text{ is primitive}, \quad \text{where } \begin{cases} x = 2mn , \ y = m^2 - n^2 , \ z = m^2 + n^2, \quad \text{or} \\ x = mn , \ y = \frac{m^2 - n^2}{2} , \ z = \frac{m^2 + n^2}{2}, \end{cases}$$

according as m and n have opposite or the same parity.

10. **(Mersenne primes)** A number of the form $M_n = 2^n - 1$ is called a **Mersenne number**, named after Marin Mersenne (1588–1648). If M_n is prime, it's called a **Mersenne prime**. For instance, $M_2 = 2^2 - 1 = 3$ is prime, $M_3 = 2^3 - 1 = 7$ is prime, but $M_4 = 2^4 - 1 = 15$ is not prime. However, $M_5 = 2^5 - 1 = 31$ is prime again. It it not known whether there exist infinitely many Mersenne primes. Prove that if M_n is prime, then n is prime. (The converse if false; for instance, M_{23} is composite.) Suggestion: Prove the contrapositive. Also, the polynomial identity $x^k - 1 = (x - 1)(x^{k-1} + x^{k-2} + \cdots + x + 1)$ might be helpful.

11. **(Perfect numbers)** A number $n \in \mathbb{N}$ is said to be **perfect** if it is the sum of its proper (that is, smaller than itself) divisors. For example, $6 = 1 + 2 + 3$ and $28 = 1 + 2 + 4 + 7 + 14$ are perfect. It's not known whether there exist any odd perfect numbers! In this problem, we prove that perfect numbers are related to Mersenne primes as follows:

$$n \text{ is even and perfect} \iff n = 2^m (2^{m+1} - 1) \text{ where } m \in \mathbb{N} \text{ with } 2^{m+1} - 1 \text{ prime.}$$

For instance, when $m = 1, 2^{1+1} - 1 = 3$ is prime, so $2^1 (2^{1+1} - 1) = 6$ is perfect. Similarly, we get 28 when $m = 2$, and the next perfect number is 496 when $m = 4$. (Note that when $m = 3, 2^{m+1} - 1 = 15$ is not prime.)

(i) Prove that if $n = 2^m (2^{m+1} - 1)$, where $m \in \mathbb{N}$ and $2^{m+1} - 1$ is prime, then n is perfect. Suggestion: The proper divisors of n are $1, 2, \ldots, 2^m, q, 2q, \ldots,$ $2^{m-1}q$, where $q = 2^{m+1} - 1$.

(ii) To prove the converse, we proceed systematically as follows. First prove that if $m, n \in \mathbb{N}$, then d is a divisor of $m \cdot n$ if and only if $d = d_1 \cdot d_2$, where d_1 and d_2 are divisors of m and n, respectively. Suggestion: Write $m = p_1^{m_1} \cdots p_k^{m_k}$ and $n = q_1^{n_1} \cdots q_\ell^{n_\ell}$ as products of prime factors. Observe that a divisor of $m \cdot n$ is just a number of the form $p_1^{i_1} \cdots p_k^{i_k} q_1^{j_1} \cdots q_\ell^{j_\ell}$, where $0 \le i_r \le m_r$ and $0 \le j_r \le n_r$.

(iii) For $n \in \mathbb{N}$, define $\sigma(n)$ as the sum of all the divisors of n *including* n itself. Using (ii), prove that if $m, n \in \mathbb{N}$, then $\sigma(m \cdot n) = \sigma(m) \cdot \sigma(n)$.

(iv) Let n be even and perfect and write $n = 2^m q$, where $m \in \mathbb{N}$ and q is odd. By (iii), $\sigma(n) = \sigma(2^m)\sigma(q)$. Working out both sides of $\sigma(n) = \sigma(2^m)\sigma(q)$, prove that

$$\sigma(q) = q + \frac{q}{2^{m+1} - 1}. \tag{2.15}$$

Suggestion: Since n is perfect, prove that $\sigma(n) = 2n$, and by the definition of σ, prove that $\sigma(2^m) = 2^{m+1} - 1$.

(v) From (2.15) and the fact that $\sigma(q) \in \mathbb{N}$, show that $q = k(2^{m+1} - 1)$, where $k \in \mathbb{N}$. From (2.15) (that $\sigma(q) = q + k$), prove that $k = 1$. Finally, conclude that $n = 2^m(2^{m+1} - 1)$, where $q = 2^{m+1} - 1$ is prime.

12. In this exercise, we show how to factor factorials (cf. [85]). First, note that given $n > 1$, the prime factors of $n!$ are exactly those primes less than or equal to n. Thus, to factor $n!$, for each prime $p \le n$, we need to know the greatest power of p that divides $n!$. To find this greatest power, proceed as follows.

(i) For $n, m \in \mathbb{N}$, we denote by $\lfloor n/m \rfloor$ the quotient when n is divided by m. If $1 < m \le n$, prove that there is a $k_0 \in \mathbb{N}$ such that for all $k \ge k_0$, $\lfloor n/m^k \rfloor = 0$.

(ii) Given $n > 1$ and a prime $p \le n$, we shall prove that the greatest power of p that divides $n!$ is

$$e_p(n) := \sum_{k \ge 1} \left\lfloor \frac{n}{p^k} \right\rfloor,$$

where we sum only over those k's such that $\lfloor n/p^k \rfloor > 0$. (The sum is a finite sum by (i).) To this end, first show that

$$\left\lfloor \frac{n+1}{p^k} \right\rfloor - \left\lfloor \frac{n}{p^k} \right\rfloor = \begin{cases} 1 & \text{if } p^k \mid (n+1) \\ 0 & \text{if } p^k \nmid (n+1). \end{cases}$$

(iii) For all $n > 1$ and prime $p \le n$, prove that the difference $e_p(n+1) - e_p(n)$ is the greatest integer j such that p^j divides $n + 1$.

(iv) Now prove that $e_p(n)$ is the greatest power of p that divides $n!$ by induction on n. Suggestion: For the induction step, write $(n + 1)! = (n + 1) n!$.

(v) Find $e_2, e_3, e_5, e_7, e_{11}$ for $n = 12$ and then factor $12!$ into primes.

2.5 Decimal Representations of Integers

Since grade school, we have represented numbers in *base* 10. In this section we explore the use of arbitrary bases.

2.5.1 Decimal Representations of Integers

We need to make a careful distinction between integers and the symbols used to represent them.

Example 2.9 In our everyday notation, we have

$$2 = 1 + 1, \quad 3 = 2 + 1, \quad 4 = 3 + 1, \quad \ldots,$$

where 1 is our symbol for the multiplicative unit.

Example 2.10 The Romans used the symbol I in place of 1, and for the other numbers,

$$II = I + I, \quad III = II + I, \quad IV = III + I.$$

If you want to be proficient in using Roman numerals, see [223].

Example 2.11 We could be creative and make up our own symbols for integers: e.g.,

$$i = 1 + 1, \quad \text{like} = i + 1, \quad \text{math} = \text{like} + 1, \ldots .$$

As you may imagine, it would be very inconvenient to make up a different symbol for every single number! For this reason, we write numbers with respect to "bases." For instance, undoubtedly because we have ten fingers, the base ten, or **decimal**, system is the most widespread system for making symbols for the integers. In this system, we use the symbols $0, 1, 2, \ldots, 9$, called **digits**, for zero and the first nine positive integers, to write any integer using the number ten expressed by the symbol $10 := 9 + 1$ as the "base" with which to express numbers.

Example 2.12 Consider the *symbol* 12. This symbol represents the *number* twelve, which is the number

$$1 \cdot 10 + 2,$$

where 1 is now used in its dual role as a multiplicative unit.

Example 2.13 The symbol 4321 represents the number a given in words by four thousand three hundred twenty-one:

$$a = 4000 + 300 + 20 + 1 = 4 \cdot 10^3 + 3 \cdot 10^2 + 2 \cdot 10 + 1.$$

Note that the digits 1, 2, 3, 4 in the symbol 4321 are exactly the remainders produced after successive divisions of a and its quotients by 10. For example,

$$a = 432 \cdot 10 + 1 \quad \text{(remainder 1)}.$$

Now divide the quotient 432 by 10:

$$432 = 43 \cdot 10 + 2 \quad \text{(remainder 2)}.$$

Continuing dividing the quotients by 10, we get

$$43 = 4 \cdot 10 + 3, \quad \text{(remainder 3)}, \quad \text{and finally,} \quad 4 = 0 \cdot 10 + 4, \quad \text{(remainder 4)}.$$

We shall use this technique of successive divisions in the proof of Theorem 2.20 below. In general, the *symbol* $a = a_n a_{n-1} \ldots a_1 a_0$ represents the *number*

$$a = a_n \cdot 10^n + a_{n-1} \cdot 10^{n-1} + \cdots + a_1 \cdot 10 + a_0 \quad \text{(in base 10)}.$$

As with our previous example, the digits $a_0, a_1, \ldots, a_n$ are exactly the remainders produced after successive divisions of a and the resulting quotients by 10.

2.5.2   *Other Common Bases*

We now consider other bases; for instance, the base 7, or **septimal**, system. Here, we use the symbols 0, 1, 2, 3, 4, 5, 6, 7 to represent zero and the first seven natural numbers and the symbols $0, 1, \ldots, 6$ are the **digits** in base 7. Then we write an integer a as $a_n a_{n-1} \ldots a_1 a_0$ in base 7 if

$$a = a_n \cdot 7^n + a_{n-1} \cdot 7^{n-1} + \cdots + a_1 \cdot 7 + a_0.$$

Example 2.14 For instance, the number with symbol 10 in base 7 is really the number seven itself, since
$$10 \ (\text{base 7}) = 1 \cdot 7 + 0.$$

Example 2.15 The number one hundred one has the symbol 203 in the septimal system because
$$203 \ (\text{base 7}) = 2 \cdot 7^2 + 0 \cdot 7 + 3,$$

and in our familiar base 10 or decimal notation, the number on the right is just $2 \cdot 49 + 3 = 98 + 3 = 101$.

A base usually associated with computing is base 2, or the **binary** or **dyadic** system. In this case, we write numbers using only the digits 0 and 1. Thus, an integer a is written as $a_n a_{n-1} \ldots a_1 a_0$ in base 2 if

$$a = a_n \cdot 2^n + a_{n-1} \cdot 2^{n-1} + \cdots + a_1 \cdot 2 + a_0.$$

Example 2.16   For instance, the symbol 10101 in the binary system represents the number

$$10101 \ (\text{base 2}) = 1 \cdot 2^4 + 0 \cdot 2^3 + 1 \cdot 2^2 + 0 \cdot 2^1 + 1.$$

In familiar base 10 or decimal notation, the number on the right is $16 + 4 + 2 + 1 = 21$.

Example 2.17   The symbol 10 in base 2 is really the number 2 itself, since

$$10 \ (\text{base 2}) = 1 \cdot 2 + 0.$$

Not only are binary numbers useful for computing, they can help you win the *Game of Nim*; see [216]. (See also Problem 6 on p. 56.) Other common bases include eight (the **octal system**) and sixteen (the **hexadecimal system**).

We remark that one can develop addition and multiplication tables with respect to any base (the binary tables are really easy); see, for instance, [57, p. 7]. Once a base is fixed, we shall not make a distinction between a number and its representation in the chosen base. In particular, throughout this book we always use base 10 and write numbers with respect to this base unless stated otherwise.

2.5.3   Arbitrary Base Expansions of Integers

We now show that a number can be written with respect to any base. Fix a natural number $b > 1$, called a **base**. Let a be a natural number and suppose that it can be written as a sum of the form

$$a = a_n \cdot b^n + a_{n-1} \cdot b^{n-1} + \cdots + a_1 \cdot b + a_0,$$

where $0 \le a_k < b$ and $a_n \ne 0$. Then the symbol $a_n a_{n-1} \ldots a_1 a_0$ is called the b-**adic representation** of a. A couple of questions arise: First, does every natural number have such a representation, and second, if a representation exists, is it unique? The answer to both questions is yes.

In the following proof, we shall use the following useful "telescoping" sum several times:

$$\sum_{k=0}^{n}(b-1)b^k = \sum_{k=0}^{n}(b^{k+1}-b^k)$$
$$= (b^1 - b^0) + (b^2 - b^1) + (b^3 - b^2) + \cdots + (b^{n+1} - b^n)$$
$$= b^{n+1} - b^0 = b^{n+1} - 1.$$

Theorem 2.20 *Every natural number has a unique b-adic representation.*

Proof We first prove existence, and then uniqueness.

Step 1: We first prove existence using the technique of successive divisions we talked about before. Using the division algorithm, we form the remainders produced after successive divisions of a and its quotients by b:

$$a = q_0 \cdot b + a_0 \text{ (remainder } a_0), \quad q_0 = q_1 \cdot b + a_1, \text{ (remainder } a_1),$$
$$q_1 = q_2 \cdot b + a_2, \text{ (remainder } a_2), \ldots, \quad q_{j-1} = q_j \cdot b + a_j, \text{ (remainder } a_j), \ldots$$
$$(2.16)$$

and so forth. By the division algorithm, we have $q_j \geq 0$ and $0 \leq a_j < b$. Moreover, since $b > 1$ (that is, $b \geq 2$), from the equation $a = q_0 \cdot b + a_0$ it is evident that as long as the quotient q_0 is positive, we have

$$a = q_0 \cdot b + a_0 > q_0 \cdot 1 + 0 = q_0,$$

and in general, as long as the quotient q_j is positive, we have

$$q_{j-1} = q_j \cdot b + a_j > q_j.$$

These inequalities imply that $a > q_0 > q_1 > q_2 > \cdots \geq 0$, where the strict inequality $>$ holds as long as the quotients remain positive. Since there are only a numbers from 0 to $a - 1$, at some point the quotients must eventually reach zero. Let us say that $q_n = 0$ is the first time the quotients hit zero. If $n = 0$, then we have

$$a = 0 \cdot b + a_0,$$

so a has the b-adic representation a_0. Suppose that $n > 0$. Then the sequence (2.16) stops at $j = n$. Combining the first and second equations in (2.16), we get

$$a = q_0 \cdot b + a_0 = (q_1 \cdot b + a_1)b + a_0 = q_1 \cdot b^2 + a_1 b + a_0.$$

Combining this equation with the third equation in (2.16), we get

$$a = (q_2 \cdot b + a_2) \cdot b^2 + a_1 b + a_0 = q_2 \cdot b^3 + a_2 \cdot b^2 + a_1 b + a_0.$$

Continuing this process (slang for "by use of induction"), we eventually arrive at

$$a = (0 \cdot b + a_n) \cdot b^n + a_{n-1} \cdot b^{n-1} + \cdots + a_1 \cdot b + a_0$$
$$= a_n \cdot b^n + a_{n-1} \cdot b^{n-1} + \cdots + a_1 \cdot b + a_0.$$

This shows the existence of a b-adic representation of a.

Step 2: We now show that this representation is unique. Suppose that a has another such representation:

$$a = \sum_{k=0}^{n} a_k b^k = \sum_{k=0}^{m} c_k b^k, \tag{2.17}$$

where $0 \leq c_k < b$ and $c_m \neq 0$. We first prove that $n = m$. Indeed, hoping to get a contradiction, let's suppose that $n \neq m$, say $n < m$. Then,

$$a = \sum_{k=0}^{n} a_k b^k \leq \sum_{k=0}^{n} (b-1) b^k = b^{n+1} - 1.$$

Since $n < m \implies n + 1 \leq m$, it follows that $b^{n+1} \leq b^m$, so

$$a \leq b^m - 1 < b^m \leq c_m \cdot b^m \leq \sum_{k=0}^{m} c_k b^k = a \implies a < a.$$

This contradiction shows that $n = m$. Now let us assume that some digits in the expressions for a differ; let p be the largest integer such that a_p differs from the corresponding c_p, say $a_p < c_p$. Since $a_p < c_p$, we have $a_p - c_p \leq -1$. Now subtracting the two expressions for a in (2.17), we obtain

$$0 = a - a = \sum_{k=0}^{p} (a_k - c_k) b^k = \sum_{k=0}^{p-1} (a_k - c_k) b^k + (a_p - c_p) b^p$$

$$\leq \sum_{k=0}^{p-1} (b-1) b^k + (-b^p) = (b^p - 1) - b^p = -1,$$

a contradiction. Thus, the two representations of a must be equal. ∎

Lastly, we remark that if a is negative, then $-a$ is positive, so $-a$ has a b-adic representation. The negative of this representation is by definition the b-adic representation of a.

▶ **Exercises 2.5**

1. In this exercise we consider the base twelve or **duodecimal system**. For this system we need two more digit symbols. Let α denote ten and β denote eleven. Then the digits for the duodecimal system are $0, 1, 2, \ldots, 9, \alpha, \beta$.

(a) In the duodecimal system, what is the symbol for twelve, twenty-two, twenty-three, one hundred thirty-one?

(b) What numbers do the following symbols represent? $\alpha\alpha\alpha$, 12, and $2\beta\beta1$.

In the following "divisibility" exercises, we shall establish the validity of grade school divisibility "tricks"; cf. [121].

2. Let $a = a_n a_{n-1} \ldots a_0$ be the decimal (=base 10) representation of a natural number a. Let us first consider divisibility by 2, 5, 10.

   (a) Prove that a is divisible by 10 if and only if $a_0 = 0$.
   (b) Prove that a is divisible by 2 if and only if a_0 is even.
   (c) Prove that a is divisible by 5 if and only if $a_0 = 0$ or $a_0 = 5$.

3. We now consider 4 and 8.

   (a) Prove that a is divisible by 4 if and only if the number $a_1 a_0$ (written in decimal notation) is divisible by 4.
   (b) Prove that a is divisible by 8 if and only if $a_2 a_1 a_0$ is divisible by 8.

4. We consider divisibility by 3, 6, 9. Suggestion: Before considering these tests, prove that $10^k - 1$ is divisible by 9 for every nonnegative integer k.

   (a) Prove that a is divisible by 3 if and only if the sum of the digits (that is, $a_n + \cdots + a_1 + a_0$) is divisible by 3.
   (b) Prove that a is divisible by 6 if and only if a is even and the sum of the digits is divisible by 3.
   (c) Prove that a is divisible by 9 if and only if the sum of the digits is divisible by 9.

5. Prove that a is divisible by 7 if and only if the alternating sum

$$a_2 a_1 a_0 - a_5 a_4 a_3 + a_8 a_7 a_6 - a_{11} a_{10} a_9 + \cdots = \sum_{k=0,1,2,\ldots} (-1)^k a_{3k+2} a_{3k+1} a_{3k}$$

is divisible by 7. Here, $a_{3k+2} a_{3k+1} a_{3k}$ is the number $a_{3k+2} 10^2 + a_{3k+1} 10^1 + a_{3k}$. Suggestion: First prove that $10^{3k} + 1$, where k is odd, and $10^{3k} - 1$, where k is even, are each divisible by 7. Which of the two numbers 57,092 and 49,058 is divisible by 7? An analogous trick works for divisibility by 13; can you state it and then prove it?

6. Here's another divisibility by 7 result. Prove that a is divisible by 7 if and only if

$$(a_n a_{n-1} \ldots a_1) - 2a_0$$

is divisible by 7. Here, $a_n a_{n-1} \ldots a_1$ is the number $a_n 10^{n-1} + a_{n-1} 10^{n-2} + \cdots + a_2 10^1 + a_1$. Suggestion: First prove that if $m, n \in \mathbb{Z}$, then $10m + n$ is divisible by 7 if and only if $m - 2n$ is divisible by 7. To prove this claim, consider $10m + n + 4(m - 2n)$. Next, prove that $a = 10m + n$, where $m = a_n a_{n-1} \ldots a_1$ and $n = a_0$, and then use the claim to prove the result.

7. Prove that a is divisible by 11 if and only if the difference between the sums of the even and odd digits,

$$(a_0 + a_2 + a_4 + \cdots) - (a_1 + a_3 + a_5 + \cdots) = \sum_{k=0}^{n} (-1)^k a_k,$$

is divisible by 11. Suggestion: First prove that $10^{2k} - 1$ and $10^{2k+1} + 1$ are each divisible by 11 for every nonnegative integer k.

8. Using modular arithmetic from Problem 6 on p. 66, we can give alternative derivations of the above "tricks." Take, for example, the "9 trick" and the "11 trick."

(a) Show that $10^k \equiv 1 \pmod 9$ for every $k = 0, 1, 2, \ldots$. Using this fact, prove that a is divisible by 9 if and only if the sum of the digits of a is divisible by 9.

(b) Show that $10^k \equiv (-1)^k \pmod{11}$ for every $k = 0, 1, 2, \ldots$. Using this fact, prove that a is divisible by 11 if and only if the difference between the sums of the even and odd digits of a is divisible by 11.

2.6 Real Numbers: Rational and "Mostly" Irrational

Imagine a world in which you couldn't give half a cookie to your friend or where you couldn't buy a quarter pound of cheese at the grocery store; this is a world without rational numbers. In this section we discuss rational and real numbers, and we shall discover, as the Greeks did 2500 years ago, that the rational numbers are not sufficient for the purposes of geometry. Irrational numbers make up the missing lengths. We shall discover in the next few sections that there are vastly, immensely, fantastically (any other synonyms I missed?) "more" irrational numbers than rational numbers.

2.6.1 The Real and Rational Numbers

The set of real numbers is denoted by $\mathbb{R}$. The reader is certainly familiar with the following arithmetic properties of real numbers (in what follows, a, b, c denote real numbers):

Addition satisfies

(A1) $a + b = b + a$ (commutative law);
(A2) $(a + b) + c = a + (b + c)$ (associative law);
(A3) there is a real number denoted by 0 "zero" such that

$$a + 0 = a = 0 + a \quad \text{(existence of additive identity)};$$

(A4) for each a, there is a real number denoted by the symbol $-a$ such that

$$a + (-a) = 0 \quad \text{and} \quad (-a) + a = 0 \quad \text{(existence of additive inverse)}.$$

Multiplication satisfies

(M1) $a \cdot b = b \cdot a$ (commutative law);
(M2) $(a \cdot b) \cdot c = a \cdot (b \cdot c)$ (associative law);
(M3) there is a real number denoted by 1, "one," different from 0, such that

$$1 \cdot a = a = a \cdot 1 \quad \text{(existence of multiplicative identity)};$$

(M4) for $a \neq 0$ there is a real number, denoted by the symbol a^{-1}, such that

$$a \cdot a^{-1} = 1 \quad \text{and} \quad a^{-1} \cdot a = 1 \quad \text{(existence of multiplicative inverse)}.$$

As with the integers, the $\cdot$ is sometimes dropped, and the associative laws imply that expressions such as $a + b + c$ and abc make sense without using parentheses. Addition and multiplication are related by

(D) $a \cdot (b + c) = (a \cdot b) + (a \cdot c)$ (distributive law).

Of these arithmetic properties, the only additional property listed that was not listed for the integers is (M4), the existence of a multiplicative inverse for each nonzero real number. We denote

$$a + (-b) = (-b) + a \quad \text{by} \quad b - a$$

and
$$a \cdot b^{-1} = b^{-1} \cdot a \quad \text{by} \quad a/b \text{ or } \frac{a}{b}.$$

The positive real numbers, denoted by $\mathbb{R}^+$, is closed under addition, multiplication, and **trichotomy** law holds: Given a real number a, exactly one of the following "positivity" properties holds:

(P) a is a positive real number, $a = 0$, or $-a$ is a positive real number.

A set together with operations of addition and multiplication that satisfies properties (A1)–(A4), (M1)–(M4), and (D) is called a **field**. If in addition, the set has a "positive set" closed under addition and multiplication satisfying (P), then the set is called an **ordered field**.

A **rational number** is a real number that can be written in the form a/b, where a and b are integers with $b \neq 0$, and the set of all such numbers is denoted by $\mathbb{Q}$. Both the real numbers and the rational numbers are ordered fields. Now what is

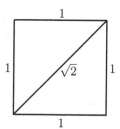

Fig. 2.3 The Greeks' discovery of irrational numbers

the difference between the real and rational numbers? The difference was discovered more than 2500 years ago by the Greeks, who found out that the length of the diagonal of a unit square is not a rational number (see Fig. 2.3 and Proof in Theorem 2.24). Because this length is not a rational number, we call $\sqrt{2}$ **irrational**.[10] Thus, there are "gaps" in the rational numbers. Now it turns out that *every* length is a real number. This fact is known as the completeness axiom of the real numbers. Thus, the real numbers have no "gaps." To finish up the list of axioms for the real numbers, we state this completeness axiom now but we leave the terms in the axiom undefined until Section 2.7 (so don't worry if some of these words seem foreign).

(C) (**Completeness axiom of the real numbers**) Every nonempty set of real numbers that is bounded above has a supremum, that is, a least upper bound, in the set of real numbers.

In the optional Section 2.8, we *construct* the set of real numbers $\mathbb{R}$ from the rationals.[11] In particular, we have $\mathbb{N} \subseteq \mathbb{R}^+$, and $\mathbb{R}$ has all the arithmetic, positivity, and completeness properties listed above. Until we get to Section 2.8 we shall *assume* this construction in order to prove some interesting results.

2.6.2 Proofs of Well-Known High School Rules

Since the real numbers satisfy the same (and even more) arithmetic properties as the natural and integer numbers, the same proofs as in Sections 2.1 and 2.3 prove the uniqueness of additive identities and inverses, rules of sign, properties of zero and one (in particular, the uniqueness of the multiplicative identity), etc.

Also, the real numbers are ordered in the same way as the integers. Given real numbers a and b, exactly one of the following holds:

---

[10]"The idea of the continuum seems simple to us. We have somehow lost sight of the difficulties it implies ...We are told such a number as the square root of 2 worried Pythagoras and his school almost to exhaustion. Being used to such queer numbers from early childhood, we must be careful not to form a low idea of the mathematical intuition of these ancient sages; their worry was highly credible." Erwin Schrödinger (1887–1961).

[11]See [149] for a construction of the rationals from the integers.

(O1) $a = b$;
(O2) $a < b$, which means that $b - a$ is a positive real number;
(O3) $b < a$, which means that $a - b$ is a positive real number.

Just as for integers, we can define $\leq$, $>$, and $\geq$, and **(O3)** can be stated as $-(b - a)$ is a positive real number. One can define the absolute value of a real number in the exact same way as it is defined for integers. Since the real numbers satisfy the same order properties as the integers, the same proofs as in Section 2.3 prove the inequality rules, absolute value rules, etc. for real numbers. Using the inequality rules, we can prove the following well-known fact from high school:

$$\text{if } a > 0, \text{ then } a^{-1} > 0.$$

Indeed, by definition of a^{-1}, we have $a \cdot a^{-1} = 1$. Since $1 > 0$ (recall that $1 \in \mathbb{N} \subseteq \mathbb{R}^+$) and $a > 0$, we have positive $\times$ a^{-1} = positive; the only way this is possible is if $a^{-1} > 0$ by the inequality rules. Here are other high school facts that can be proved using the inequality rules:

$$\text{If } 0 < a < 1, \text{ then } a^{-1} > 1 \text{ and if } a > 1, \text{ then } a^{-1} < 1.$$

Indeed, if $a < 1$ with a positive, then multiplying by $a^{-1} > 0$, we obtain

$$a \cdot (a^{-1}) < 1 \cdot (a^{-1}) \quad \Longrightarrow \quad 1 < a^{-1}.$$

Similarly, if $1 < a$, then multiplying through by $a^{-1} > 0$, we get $a^{-1} < 1$.
   Here are some more high school facts.

Uniqueness of multiplicative inverse

> **Theorem 2.21** *If a and b are real numbers with $a \neq 0$, then $x \cdot a = b$ if and only if $x = ba^{-1} = b/a$. In particular, setting $b = 1$, the only x that satisfies the equation $x \cdot a = 1$ is $x = a^{-1}$. Thus, each real number has exactly one multiplicative inverse.*

Proof If $x = b \cdot a^{-1}$, then

$$x \cdot a = (ba^{-1}) \cdot a = b(a^{-1}a) = b \cdot 1 = b.$$

Conversely, if x satisfies $x \cdot a = b$, then

$$ba^{-1} = (x \cdot a)a^{-1} = x \cdot (a\,a^{-1}) = x \cdot 1 = x. \qquad \blacksquare$$

   Note that $x \cdot 0 = 0$ for every real number x. (This is the real number version of Theorem 2.13 on p. 53.) In particular, 0 has no multiplicative inverse (there is no

"0^{-1}" such that $0 \cdot 0^{-1} = 1$), whence the high school saying, "You can't divide by zero."

Before discussing the famous "fraction rules," recall from Section 2.3.2 beginning on p. 32 that multiplication $a \cdot b$ can be thought of as "scaling b by a" (reversing the direction if a is negative). Using this interpretation, let's find $(1/2) \cdot (3/4)$:

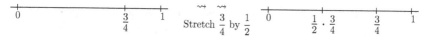

On the other hand, from Fig. 2.4 we notice geometrically that $(1/2) \cdot (3/4)$ coincides with $3/8$. Thus, $(1/2) \cdot (3/4) = 3/8$. Another way to interpret multiplication $a \cdot b$ is "a groups of b." (For instance, $3 \cdot 4$ is 3 groups of 4, or 12. I teach this interpretation to my children!) In our example, $(1/2) \cdot (3/4)$ is a "half-group of $3/4$," which is $3/8$ by the preceding figure. Thus, $(1/2) \cdot (3/4) = 3/8$ using either the scaling or grouping interpretation of multiplication. Of course, we could have arrived at the same answer a lot faster by simply following the rule of multiplying the numerators and denominators! We shall prove this rule, and many other "fraction rules" (even for fractions of real numbers), in the following theorem.

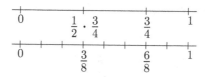

Fig. 2.4 *Top* Stretching $3/4$ by $1/2$ is the same as halving $3/4$, which is equivalent to making a "half-group of $3/4$." *Bottom* We divide the interval $[0, 1]$ into eight equal parts. We see that $(1/2) \cdot (3/4) = 3/8$

Fraction rules

Theorem 2.22 *For $a, b, c, d \in \mathbb{R}$, the following fraction rules hold (all denominators are assumed to be nonzero):*

$$(1)\ \frac{a}{a} = 1,\ \frac{a}{1} = a, \qquad (2)\ \frac{a}{-b} = -\frac{a}{b};$$

$$(3)\ \frac{a}{b} \cdot \frac{c}{d} = \frac{ac}{bd}, \qquad (4)\ \frac{a}{b} = \frac{ac}{bc};$$

$$(5)\ \frac{1}{a/b} = \frac{b}{a}, \qquad (6)\ \frac{a/b}{c/d} = \frac{a}{b} \cdot \frac{d}{c} = \frac{ad}{bc}, \qquad (7)\ \frac{a}{b} \pm \frac{c}{d} = \frac{ad \pm bc}{bd}.$$

Proof The proofs of these rules are really very elementary, so we prove only *(1)–(3)* and leave *(4)–(7)* to you in Problem 1.

We have $a/a = a \cdot a^{-1} = 1$, and since $1 \cdot 1 = 1$, by uniqueness of the multiplicative inverses we have $1^{-1} = 1$ and therefore $a/1 = a \cdot 1^{-1} = a \cdot 1 = a$.

To prove *(2)*, note that by our rules of sign,

$$(-b) \cdot (-b^{-1}) = b \cdot b^{-1} = 1,$$

and therefore by uniqueness of multiplicative inverses, we must have $(-b)^{-1} = -b^{-1}$. Thus, $a/(-b) := a \cdot (-b)^{-1} = a \cdot -b^{-1} = -a \cdot b^{-1} = -a/b$.

To prove *(3)*, observe that $b \cdot d \cdot b^{-1} \cdot d^{-1} = (bb^{-1}) \cdot (dd^{-1}) = 1 \cdot 1 = 1$, so by uniqueness of multiplicative inverses, $(bd)^{-1} = b^{-1}d^{-1}$. Thus,

$$\frac{a}{b} \cdot \frac{c}{d} = a \cdot b^{-1} \cdot c \cdot d^{-1} = a \cdot c \cdot b^{-1} \cdot d^{-1} = ac \cdot (bd)^{-1} = \frac{ac}{bd}. \qquad \blacksquare$$

We already know what a^n means for $n = 0, 1, 2, \ldots$ For negative integers, we define powers by

$$\boxed{a^{-n} := \frac{1}{a^n}, \quad a \neq 0, \ n = 1, 2, 3, \ldots}$$

Here are the familiar power rules.

Power rules

> **Theorem 2.23** *For $a, b \in \mathbb{R}$ and for integers m, n,*
>
> $$a^m \cdot a^n = a^{m+n}; \quad a^m \cdot b^m = (ab)^m; \quad \left(a^m\right)^n = a^{mn},$$
>
> *provided that the individual powers are defined (e.g., a and b are nonzero if an exponent is negative). If n is a natural number and $a, b \geq 0$, then*
>
> $$a < b \quad \text{if and only if} \quad a^n < b^n.$$
>
> *In particular, $a \neq b$ (both a, b nonnegative) if and only if $a^n \neq b^n$.*

Proof We leave the proof of the first three rules to the reader, since we already dealt with proving such rules in the problems of Section 2.2 on p. 86. Consider the last rule. Let $n \in \mathbb{N}$ and let $a, b \geq 0$ be not both zero (if both are zero, then $a = b$ and $a^n = b^n$, and there is nothing to prove). Observe that

$$(b - a) \cdot c = b^n - a^n, \quad \text{where } c = b^{n-1} + b^{n-2} a + \cdots + b a^{n-2} + a^{n-1}, \quad (2.18)$$

which is verified by multiplying out:

$$(b - a) (b^{n-1} + b^{n-2} a + \cdots + a^{n-1}) = (b^n + b^{n-2} a^2 + b^{n-1} a + \cdots + b a^{n-1})$$
$$- (b^{n-1} a + b^{n-2} a^2 + \cdots + b a^{n-1} + a^n) = b^n - a^n.$$

The formula for c (and the fact that $a, b \geq 0$ are not both zero) shows that $c > 0$. Therefore, the equation $(b - a) \cdot c = b^n - a^n$ shows that $(b - a) > 0$ if and only if $b^n - a^n > 0$. Therefore, $a < b$ if and only if $a^n < b^n$. ∎

If n is a natural number, then the nth **root** of a real number a is a real number b such that $b^n = a$, if such a b exists. For $n = 2$, we call b a **square root**, and if $n = 3$, a **cube root**. If $a \geq 0$, then according to the last power rule in Theorem 2.23, a can have at most one nonnegative nth root. In Theorem 2.32 on p. 93, we prove that *every* nonnegative real number has a unique nonnegative nth root.

Root notation :   If $b \geq 0$ satisfies $b^n = a$, we write b as $\sqrt[n]{a}$ or $a^{1/n}$.

If $n = 2$, we always write $\sqrt{a}$ or $a^{1/2}$ instead of $\sqrt[2]{a}$.

Example 2.18 We see that -3 is a square root of 9, since $(-3)^2 = 9$. Also, 3 is a square root of 9, since $3^2 = 9$. Since 3 is nonnegative, $\sqrt{9} = 3$. Here's a puzzle: Which two real numbers are their own nonnegative square roots?

We now show that "most" real numbers occurring in the study of polynomials, trigonometry, and logarithms are not rational numbers, that is, ratios of integers. These examples will convince the reader that there are many "gaps" in the rational numbers and of the importance of irrational numbers to mathematics.

> **Remark**: For the rest of this section, we shall *assume* that $\sqrt[n]{a}$ exists for every $a \geq 0$ (see Theorem 2.32 on p. 93), and we shall assume basic facts concerning the trig and log functions (to be proved in Sections 4.9 and 4.7, respectively). We make these assumptions in order to present interesting examples that will convince you without a shadow of a doubt that irrational numbers are indispensable in mathematics.

2.6.3   Irrational Roots and the Rational Zeros Theorem

We begin by showing that $\sqrt{2}$ is not rational. Again, we shall prove that $\sqrt{2}$ exists, in Theorem 2.32 to come. Before proving irrationality, we establish some terminology. We say that a rational number a/b is in **lowest terms** if a and b do not have any common prime factor in their prime factorizations. By Property *(4)* of the fraction rules, we can always "cancel" common factors to put a rational number in lowest terms.

Irrationality of $\sqrt{2}$

Theorem 2.24 *The number $\sqrt{2}$ is not rational.*

Proof We provide three proofs. The first is a geometric proof, while the second is an algebraic version of the same proof! The third proof is the "standard" proof in this business. (See also Problems 6–10 and the articles [71, 103, 156], [233, p. 39] and [237] for more proofs.)

Proof I: This proof is not meant to be rigorous, but rather to motivate **Proof II** below geometrically. We assume common beliefs from high school geometry, in particular, those concerning similar triangles.

Suppose, by way of contradiction, that $\sqrt{2} = a/b$, where $a, b \in \mathbb{N}$. Then $a^2 = 2b^2 = b^2 + b^2$, so by the Pythagorean theorem, the isosceles triangle with sides a, b, b is a right triangle, as shown here.

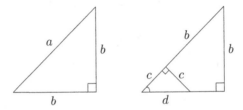

Fig. 2.5 In the *right picture*, we write a as $b + c$, and we draw a perpendicular, forming a new isosceles triangle with equal sides c, c and base d (The *smaller triangle* is isosceles because it is similar to the original. Indeed, the *smaller triangle*, just like the original one, has a 90° angle, and it shares an angle with the original triangle)

Hence, there is an isosceles right triangle whose lengths are (of course, positive) integers. By taking a smaller triangle if necessary, we may assume that a, b, b are the lengths of the smallest such triangle. We shall derive a contradiction by producing another isosceles right triangle with integer side lengths and a smaller hypotenuse. In fact, consider the triangle d, c, c drawn in Fig. 2.5. Note that $a = b + c$, so $c = a - b \in \mathbb{Z}$. To see that $d \in \mathbb{Z}$, observe that since the ratio of corresponding sides of similar triangles are in proportion, we have

$$\frac{d}{c} = \frac{a}{b} \implies d = \frac{a}{b} \cdot c = \frac{a}{b}(a - b) = \frac{a^2}{b} - a = 2b - a, \qquad (2.19)$$

where we used that $a^2 = b^2 + b^2 = 2b^2$. Therefore, $d = 2b - a \in \mathbb{Z}$ as well. Thus, we have indeed produced a smaller isosceles right triangle with integer side lengths.

Proof II: We now make **Proof I** rigorous. Suppose that $\sqrt{2} = a/b$ ($a, b \in \mathbb{N}$). By well-ordering, we may assume that a is the smallest positive numerator that $\sqrt{2}$ can have as a fraction; explicitly,

$$a = \text{least element of } \left\{ n \in \mathbb{N}; \ \sqrt{2} = \frac{n}{m} \text{ for some } m \in \mathbb{N} \right\}.$$

Motivated by (2.19), we *claim* that

$$\sqrt{2} = \frac{d}{c} \quad \text{where} \quad d = 2b - a, c = a - b \text{ are integers with } 0 < d < a. \quad (2.20)$$

Once we prove this claim, we contradict the minimality of a. Of course, the facts in (2.20) were derived from Fig. 2.5 geometrically, but now we actually prove these facts! First, to prove that $\sqrt{2} = d/c$, that is, $a/b = d/c$, observe that

$$\frac{a^2}{b^2} = 2 \implies \frac{a^2}{b} = 2b \implies \frac{a^2}{b} - a = 2b - a \implies \frac{a(a-b)}{b} = 2b - a.$$

Dividing by $a - b$ gives $a/b = d/c$, as required. To prove that $0 < d < a$, note that since $1 < 2 < 4$, that is, $1^2 < (\sqrt{2})^2 < 2^2$, by the (last statement of the) power rules in Theorem 2.23, we have $1 < \sqrt{2} < 2$, or $1 < a/b < 2$. Multiplying by b, we get $b < a < 2b$, which implies $2b - a > 0$ and $b < a$. Thus,

$$0 < 2b - a < 2a - a = a.$$

Hence, $0 < d < a$, and we get our a contradiction. The following proofs are variations on the fundamental theorem of arithmetic.

Proof III: We first establish the fact that if the square of an integer has a factor of 2, then the integer itself has a factor of 2. A quick way to prove this fact is using the fundamental theorem of arithmetic: For every integer m, the factors of m^2 are exactly the squares of the factors of m. Therefore, m^2 has a prime factor p if and only if m itself has the prime factor p. In particular, m^2 has the prime factor 2 if and only if m has the factor 2, which establishes our fact. A proof without using the fundamental theorem goes as follows. An integer m is either even or odd, that is, $m = 2n$ or $m = 2n + 1$, where n is the quotient of m when divided by 2. In the odd case, $m = 2n + 1$, we have

$$m^2 = (2n + 1)^2 = 4n^2 + 4n + 1 = 2(2n^2 + 2n) + 1,$$

so m^2 is odd. Thus, if m^2 is even, m itself must be even. Now suppose that $\sqrt{2}$ were a rational number, say

$$\sqrt{2} = \frac{a}{b},$$

where a/b is in lowest terms. Hence, $a = \sqrt{2}\, b$, and therefore $a^2 = 2b^2$. The number $2b^2 = a^2$ has a factor of 2, so a must have a factor of 2. Therefore, $a = 2c$ for some integer c. Now $a^2 = 2b^2$ implies

$$(2c)^2 = 2b^2 \implies 4c^2 = 2b^2 \implies 2c^2 = b^2.$$

The number $2c^2 = b^2$ has a factor of 2, so b must also have a factor of 2. Thus, we have shown that both a and b have a factor of 2. This contradicts the assumption that a and b have no common factors.

Proof IV: Suppose $\sqrt{2} = a/b$ is rational. Then

$$a^2 = 2b^2. \tag{2.21}$$

Since the number of prime factors doubles when a number is squared, the left side of (2.21) has an even number of prime factors. Since b^2 has an even number of prime factors and $2b^2$ has one more (an extra 2) than b^2, the right-hand side of (2.21) has an odd number of prime factors, a contradiction. We can also get a contradiction by noting that a^2 has an even number of factors of 2 (including the possibility zero if there are no such factors), while $2b^2$ has an odd number of factors of 2. ∎

The following theorem gives another method to prove the irrationality of $\sqrt{2}$ and also many other numbers. For a natural number n, recall that a (real-valued) **nth-degree polynomial** is a function $p(x) = a_n x^n + \cdots + a_1 x + a_0$, where $a_k \in \mathbb{R}$ for each k, and the **leading coefficient** a_n is not zero.

Rational zeros theorem

> **Theorem 2.25** *If a polynomial equation*
>
> $$c_n x^n + c_{n-1} x^{n-1} + \cdots + c_1 x + c_0 = 0, \quad c_n \neq 0,$$
>
> *where the c_k are integers, has a nonzero rational solution a/b, where a/b is in lowest terms, then a divides c_0 and b divides c_n.*

Proof Suppose that a/b is a rational solution of our equation with a/b in lowest terms. Since it is a solution, we have

$$c_n \left(\frac{a}{b}\right)^n + c_{n-1} \left(\frac{a}{b}\right)^{n-1} + \cdots + c_1 \left(\frac{a}{b}\right) + c_0 = 0.$$

Multiplying both sides by b^n, we obtain

$$c_n a^n + c_{n-1} a^{n-1} b + \cdots + c_1 a b^{n-1} + c_0 b^n = 0. \tag{2.22}$$

Bringing everything to the right except for $c_0 b^n$ and then factoring out an a, we obtain

$$
\begin{aligned}
c_0 b^n &= -c_n a^n - c_{n-1} a^{n-1} b - \cdots - c_1 a b^{n-1} \\
&= a(-c_n a^{n-1} - c_{n-1} a^{n-2} - \cdots - c_1 b^{n-1}).
\end{aligned}
$$

This formula shows that every prime factor of a occurs in the product $c_0 b^n$. By assumption, a and b have no common prime factors, and hence every prime factor of a must occur in c_0. This shows that a divides c_0.

To prove that b divides c_n, we rewrite (2.22) as

$$c_n a^n = -c_{n-1} a^{n-1} b - \cdots - c_1 a\, b^{n-1} - c_0 b^n$$
$$= b(-c_{n-1} a^{n-1} - \cdots - c_1 a\, b^{n-2} - c_0 b^{n-1}).$$

It follows that every prime factor of b occurs in $c_n a^n$. Since a and b have no common prime factors, every prime factor of b must occur in c_n. Thus, b divides c_n. ∎

Example 2.19 (**Irrationality of $\sqrt{2}$, Proof V**) Observe that $\sqrt{2}$ is a solution of the polynomial equation $x^2 - 2 = 0$. The rational zeros theorem implies that if the equation $x^2 - 2 = 0$ has a rational solution, say a/b in lowest terms, then a must divide $c_0 = -2$ and b must divide $c_2 = 1$. It follows that a can equal ± 1 or ± 2 and b can equal only ± 1. Therefore, if there are rational solutions of $x^2 - 2 = 0$, they must be $x = \pm 1$ or $x = \pm 2$. However,

$$(\pm 1)^2 - 2 = -1 \neq 0 \quad \text{and} \quad (\pm 2)^2 - 2 = 2 \neq 0,$$

so $x^2 - 2 = 0$ has no rational solutions. Therefore $\sqrt{2}$ is not rational.

A similar argument using the equation $x^n - a = 0$ proves the following.

> **Corollary 2.26** *If a and n are natural numbers and a is not the nth power of a natural number, then the nth root $\sqrt[n]{a}$ is irrational.*

2.6.4 Irrationality of Trigonometric Numbers

Let $0 < \theta < 90°$ be an angle whose measurement in degrees is rational. Following [155], we shall prove that $\cos\theta$ is irrational except when $\theta = 60°$, in which case

$$\cos 60° = \frac{1}{2}.$$

The proof of this result is based on the rational zeros theorem and Lemma 2.27 below. See Problem 5 for corresponding statements for sine and tangent. Of course, at this point, and only for purposes of illustration, we have to assume basic knowledge of the trigonometric functions. In Section 4.9 beginning on p. 323, we shall define these function rigorously and establish their usual properties.

Lemma 2.27 *For every natural number n, we can write* $2 \cos n\theta$ *as an nth-degree polynomial in* $2 \cos \theta$ *with integer coefficients and with leading coefficient one.*

Proof We need to prove that

$$2 \cos n\theta = (2 \cos \theta)^n + a_{n-1}(2 \cos \theta)^{n-1} + \cdots + a_1(2 \cos \theta) + a_0, \qquad (2.23)$$

where the coefficients $a_{n-1}, a_{n-2}, \ldots, a_0$ are integers. For $n = 1$, we can write $2 \cos \theta = (2 \cos \theta)^1 + 0$, so our proposition holds for $n = 1$. To prove our result in general, we use the strong form of induction. Assume that our proposition holds for $1, 2, \ldots, n$. Before proceeding to show that our lemma holds for $n + 1$, we shall prove the identity

$$2 \cos \big((n + 1)\theta\big) = \Big(2 \cos n\theta\Big)\Big(2 \cos \theta\Big) - 2 \cos \big((n - 1)\theta\big). \qquad (2.24)$$

To verify this identity, consider the well-known identities (cf. Theorem 4.35 on p. 325)

$$\cos(\alpha + \beta) = \cos \alpha \cos \beta - \sin \alpha \sin \beta$$
$$\cos(\alpha - \beta) = \cos \alpha \cos \beta + \sin \alpha \sin \beta.$$

Adding these equations, we obtain $\cos(\alpha + \beta) + \cos(\alpha - \beta) = 2 \cos \alpha \cos \beta$, or

$$\cos(\alpha + \beta) = 2 \cos \alpha \cos \beta - \cos(\alpha - \beta).$$

Setting $\alpha = n\theta$ and $\beta = \theta$, and then multiplying the result by 2, we get (2.24).

Now, since our lemma holds for $1, \ldots, n$, we see that in particular, $2 \cos n\theta$ and $2 \cos \big((n - 1)\theta\big)$ can be written as an nth-degree and an $(n - 1)$th-degree polynomial, respectively, in $2 \cos \theta$ with integer coefficients and with leading coefficient one. Substituting these polynomials into the right-hand side of the identity (2.24) shows that $2 \cos \big((n + 1)\theta\big)$ can be expressed as an $(n + 1)$th-degree polynomial in $2 \cos \theta$ with integer coefficients and with leading coefficient one. This proves our lemma. ∎

We are now ready to prove our main result.

Theorem 2.28 *Let* $0 < \theta < 90°$ *be an angle whose measurement in degrees is rational. Then* $\cos \theta$ *is rational if and only if* $\theta = 60°$.

Proof If $\theta = 60°$, then we know that $\cos \theta = 1/2$, which is rational. Assume now that θ is rational, say $\theta = a/b$, where a and b are natural numbers. Then choosing

$n = b \cdot 360°$, we have $n\theta = (b \cdot 360°) \cdot (a/b) = a \cdot 360°$. Thus, $n\theta$ is a multiple of $360°$, so $\cos n\theta = 1$. Substituting $n\theta$ into the Eq. (2.23), we obtain

$$(2\cos\theta)^n + a_{n-1}(2\cos\theta)^{n-1} + \cdots + a_1(2\cos\theta) + a_0 - 2 = 0,$$

where the coefficients are integers. Hence, $2\cos\theta$ is a solution of the equation

$$x^n + a_{n-1}x^{n-1} + \cdots + a_1 x + a_0 - 2 = 0.$$

By the rational zeros theorem, it follows that every rational solution of this equation must be an integer. So, if $2\cos\theta$ is rational, then it must be an integer. Since $0 < \theta < 90°$ and cosine is strictly between 0 and 1 for these θ's, the only integer that $2\cos\theta$ can equal is 1. Thus, $2\cos\theta = 1$ or $\cos\theta = 1/2$, and so θ must be $60°$. ∎

2.6.5 Irrationality of Logarithmic Numbers

Recall that the **(common) logarithm** to the base 10 of a real number a is defined to be the unique number x such that

$$10^x = a.$$

In Section 4.7, starting on p. 300, we define logarithms rigorously, but for now, in order to demonstrate another interesting example of irrational numbers, we shall assume familiarity with such logarithms from high school. We also assume basic facts concerning powers that we'll prove in the next section.

> **Theorem 2.29** Let $r > 0$ be a rational number. Then $\log_{10} r$ is rational if and only if $r = 10^n$, where n is an integer, in which case
>
> $$\log_{10} r = n.$$

Proof If $r = 10^n$, where $n \in \mathbb{Z}$, then $\log_{10} r = n$, so $\log_{10} r$ is rational. Assume now that $\log_{10} r$ is rational; we'll show that $r = 10^n$ for some $n \in \mathbb{Z}$. We may assume that $r > 1$, because if $r = 1$, then $r = 10^0$, and we're done, and if $r < 1$, then $r^{-1} > 1$, and $\log_{10} r^{-1} = -\log_{10} r$ is rational, so we can get the $r < 1$ result from the $r > 1$ result. We henceforth assume that $r > 1$. Let $r = a/b$, where a and b are natural numbers with no common factors. Assume that $\log_{10} r = c/d$, where c and d are natural numbers with no common factors. Then $r = 10^{c/d}$, which implies that $r^d = 10^c$, or after setting $r = a/b$, we get $(a/b)^d = 10^c$, or after some algebra, we have

$$a^d = 2^c \cdot 5^c \cdot b^d. \tag{2.25}$$

By assumption, a and b do not have any common prime factors. Hence, expressing a and b in their prime factorizations in (2.25) and using the fundamental theorem of arithmetic, we see that all the prime factors of b would also be prime factors of a. We conclude that b can have no prime factors, that is, $b = 1$, and a can have only the prime factors 2 and 5. Thus,

$$a = 2^m \cdot 5^n \quad \text{and} \quad b = 1$$

for some nonnegative integers m and n. Now according to (2.25),

$$2^{md} \cdot 5^{nd} = 2^c \cdot 5^c.$$

Again by the fundamental theorem of arithmetic, we must have $md = c$ and $nd = c$. However, c and d have no common factors, so $d = 1$, and therefore $m = c = n$. This, together with the fact that $b = 1$, proves that

$$r = \frac{a}{b} = \frac{a}{1} = 2^m \cdot 5^n = 2^n \cdot 5^n = 10^n.$$

∎

In the following exercises, assume that square roots and cube roots exist for nonnegative real numbers; these facts will be proved in the next section.

▶ **Exercises 2.6**

1. Prove properties *(4)–(7)* in the "fraction rules" theorem.
2. Let a be a positive real number and let m, n be nonnegative integers with $m < n$. If $0 < a < 1$, prove that $a^n < a^m$, and if $a > 1$, prove that $a^m < a^n$.
3. Let α be an irrational number. Prove that $-\alpha$ and α^{-1} are irrational. If r is a nonzero rational number, prove that the addition, subtraction, multiplication, and division of α and r are again irrational. As an application of this result, deduce that

$$-\sqrt{2}, \qquad \frac{1}{\sqrt{2}}, \qquad \sqrt{2}+1, \qquad 4-\sqrt{2}, \qquad 3\sqrt{2}, \qquad \frac{\sqrt{2}}{10}, \qquad \frac{7}{\sqrt{2}}$$

   are each irrational.
4. In this problem we prove that various numbers are irrational.

   (a) Prove that $\sqrt{6}$ is irrational using **Proof III** in Theorem 2.24. From the fact that $\sqrt{6}$ is irrational, and without using any irrationality facts concerning $\sqrt{2}$ and $\sqrt{3}$, prove that $\sqrt{2} + \sqrt{3}$ is irrational. Suggestion: Consider $(\sqrt{2} + \sqrt{3})^2$.
   (b) Use the rational zeros theorem to give another proof that $\sqrt{2} + \sqrt{3}$ is irrational. Suggestion: Let $x = \sqrt{2} + \sqrt{3}$, and then show that $x^4 - 10x^2 + 1 = 0$.

(c) Using the rational zeros theorem, prove that $(2\sqrt[3]{6}+7)/3$ is irrational and $\sqrt[3]{2}-\sqrt{3}$ is irrational. (If $x = \sqrt[3]{2} - \sqrt{3}$, you should end up with a sixth-degree polynomial equation for x, to which you can apply the rational zeros theorem.)

5. In this problem we look at irrational values of sine and tangent. Let $0 < \theta < 90°$ be an angle whose measurement in degrees is rational. You may assume *any* standard facts about the trigonometric functions and their identities.

   (a) Prove that $\sin\theta$ is rational if and only if $\theta = 30°$, in which case $\sin 30° = 1/2$. Suggestion: Do *not* try to imitate the proof of Theorem 2.28. Instead, use a trig identity to write sine in terms of cosine.
   (b) Prove that $\tan\theta$ is rational if and only if $\theta = 45°$, in which case $\tan\theta = 1$. Suggestion: Use the identity $\cos 2\theta = \frac{1-\tan^2\theta}{1+\tan^2\theta}$.

6. **(Irrationality of $\sqrt{2}$, Proof VI)** This problem is a variation on **Proof II**, the algebraic Pythagorean proof of Theorem 2.24. Suppose, hoping to get a contradiction, that $\sqrt{2} = a/b$ with $a, b \in \mathbb{N}$ and where, by well-ordering, we may assume that b is the smallest positive denominator for $\sqrt{2}$. Derive a contradiction.

7. **(Irrationality of $\sqrt{2}$, Proof VII)** Prove that if x and y are integers with no common factors and x/y is an integer, then $y = 1$. Now prove that $\sqrt{2}$ is irrational.

8. (Cf. [42]) **(Irrationality of $\sqrt{2}$, Proof VIII)** Assume that $\sqrt{2}$ is rational.

   (i) Show that there is a smallest natural number n such that $n\sqrt{2}$ is an integer.
   (ii) Show that $m := n\sqrt{2} - n = n(\sqrt{2} - 1)$ is a natural number smaller than n.
   (iii) Finally, show that $m\sqrt{2}$ is an integer to get a contradiction.

9. (Cf. [103]) **(Irrationality of $\sqrt{2}$, Proof IX)** Here's a base 10 proof.

   (i) Show that if $x \in \mathbb{N}$ is expressed in base 10, then the ones digit of x^2 is one of $0, 1, 4, 5, 6, 9$.
   (ii) Suppose now that $\sqrt{2} = a/b$, where $a, b \in \mathbb{N}$, is written in lowest terms. Using that $a^2 = 2b^2$, prove that a and b each have a common factor of 5, a contradiction.

10. **(Irrationality of $\sqrt{2}$, Proof X)** Here's a proof due to Marcin Mazur [161].

    (i) Show that $\sqrt{2} = \frac{-4\sqrt{2}+6}{3\sqrt{2}-4}$.
    (ii) Now suppose that $\sqrt{2} = a/b$ $(a, b \in \mathbb{N})$, where a is the smallest positive numerator that $\sqrt{2}$ can have as a fraction. Using the formula in (i), derive a contradiction.

2.7 The Completeness Axiom of $\mathbb{R}$ and Its Consequences

The completeness axiom of the real numbers essentially states that the real numbers have no "gaps." As discovered in the previous section, this property is quite in contrast to the nature of the rational numbers, which have many "gaps." In this section we discuss the completeness axiom and its consequences. Another consequence of the completeness axiom is that the real numbers are uncountable, while the rationals are countable, but we leave that breathtaking subject for Section 2.11.

2.7.1 The Completeness Axiom

Before discussing the completeness axiom, we need to talk about lower and upper bounds of sets.

A set $A \subseteq \mathbb{R}$ is said to be **bounded above** if there is a real number b greater than or equal to every number in A in the sense that for each a in A, we have $a \leq b$. Every such number b, if such exists, is called an **upper bound** for A. Here's an example of a set and two upper bounds, b_1 and b_2:

Suppose that b is an upper bound for A. Then b is called the **least upper bound**, or **supremum**, for A if b is just that, the least of all upper bounds for A, in the sense that it is less than every other upper bound for A. This supremum, if it exists, is denoted by sup A. The supremum in the previous picture is the number b_1. We shall use both terminologies "least upper bound" and "supremum" interchangeably.

Example 2.20 Consider the interval $I = [0, 1)$:

This interval is bounded above by, for instance, 1, 3/2, 22/7, 10, 1000. In fact, every upper bound for I is just a real number greater than or equal to 1. The least upper bound is 1, since 1 is the smallest upper bound. Note that $1 \notin I$.

Example 2.21 Now let $J = (0, 1]$:

This set is also bounded above, and every upper bound for J is, as before, just a real number greater than or equal to 1. The least upper bound is 1. In this case, $1 \in J$.

These examples show that the supremum of a set, if it exists, may or may not belong to the set. Some sets do not have supremums.

Example 2.22 $\mathbb{Z}$ is not bounded above (Lemma 2.35 on p. 95), nor is $(0, \infty)$.

We now summarize. Let $A \subseteq \mathbb{R}$ be bounded above. Then that a number b is the least upper bound, or supremum, for A means two things concerning b:

> **(L1)** for all a in A, $a \leq b$;
> **(L2)** if c is an upper bound for A, then $b \leq c$.

(L1) just means that b is an upper bound for A, and **(L2)** means that b is the least, or lowest, upper bound for A. Instead of **(L2)**, it is convenient to substitute its *contrapositive*:

> **(L2′)** if $c < b$, then for some a in A we have $c < a$.

This just says that every number c smaller than b is *not* an upper bound for A, which is to say that there is no upper bound for A that is smaller than b.

We can also talk about lower bounds. A set $A \subseteq \mathbb{R}$ is said to be **bounded below** if there is a real number b less than or equal to every number in A in the sense that for each a in A we have $b \leq a$. Every such number b, if such exists, is called a **lower bound** for A. If b is a lower bound for A, then b is called the **greatest lower bound**, or **infimum**, for A if b is just that, the greatest lower bound for A, in the sense that it is greater than every other lower bound for A. This infimum, if it exists, is denoted by inf A. We shall use both terminologies "greatest lower bound" and "infimum" interchangeably.

Example 2.23 The sets $I = [0, 1)$ and $J = (0, 1]$ are both bounded below (by, e.g., $0, -1/2, -1, -1000$), and in both cases the greatest lower bound is 0.

Thus, the infimum of a set, if it exists, may or may not belong to the set.

Example 2.24 The sets $\mathbb{Z}$ (see Lemma 2.35 on p. 96) and $(-\infty, 0)$ are not bounded below.

We now summarize. Let $A \subseteq \mathbb{R}$ be bounded below. Then that a number b is the greatest lower bound, or infimum, for A means two things concerning b:

> **(G1)** for all a in A, $b \leq a$;
> **(G2)** if c is a lower bound for A, then $c \leq b$.

(G1) says that b is a lower bound for A, and **(G2)** says that b is the greatest lower bound for A. Instead of **(G2)**, it is convenient to substitute its contrapositive:

> **(G2′)** if $b < c$, then for some a in A we have $a < c$.

This just says that every number c greater than b is not a lower bound for A, which is to say that there is no lower bound for A that is greater than b.

In the examples given so far (e.g., the intervals I and J), we have shown that if a set has an upper bound, then it has a least upper bound. This is a general phenomenon, called the **completeness axiom** of the real numbers:

(C) (**Completeness axiom of the real numbers**) Every nonempty set of real numbers that is bounded above has a supremum, that is, a least upper bound.

In Section 2.8, when we construct $\mathbb{R}$, we'll see that this completeness property follows immediately from the method of construction (see Theorem 2.41 on p. 108). Using the following lemma, we can prove the corresponding statement for infimums.

Lemma 2.30 *If A is nonempty and bounded below, then its reflected set $-A :=$ $\{-a\,;\ a \in A\}$ is nonempty and bounded above. Moreover,* $\inf A = -\sup(-A)$ *in the sense that* $\inf A$ *exists and this formula for* $\inf A$ *holds.*

Proof If A is nonempty and bounded below, there is a real number c such that $c \leq a$ for all a in A. Therefore, $-a \leq -c$ for all a in A, and hence the set $-A$ is bounded above. By the completeness axiom, $-A$ has a least upper bound, which we denote by b. Our lemma is finished once we show that $-b$ is the greatest lower bound for A. To see this, we know that $-a \leq b$ for all a in A, and so $-b \leq a$ for all a in A. Thus, $-b$ is a lower bound for A. Suppose that $b' \leq a$ for all a in A. Then $-a \leq -b'$ for all a in A, and so $b \leq -b'$, since b is the least upper bound for $-A$. Thus, $b' \leq -b$, and hence $-b$ is indeed the greatest lower bound for A. ∎

This lemma immediately gives the following theorem.

Theorem 2.31 *Every nonempty set of real numbers that is bounded below has an infimum, that is, greatest lower bound.*

The completeness axiom is powerful because it "produces out of thin air," so to speak, real numbers with certain properties. This ability to "produce" numbers has some profound consequences, as we now intend to demonstrate!

2.7.2 Existence of nth Roots

As a first consequence of the completeness axiom, we show that every nonnegative real number has a unique nonnegative nth root, where $n \in \mathbb{N}$. To see how the completeness axiom can "produce" roots, consider the existence of $\sqrt{2}$. How do we prove there is a real number $b \geq 0$ such that $b^2 = 2$? All we do is consider the set

$$A = \{x \in \mathbb{R}\,;\ x^2 < 2 \text{ or } x \leq 0\}.$$

In high school you would rewrite A as the interval $(-\infty, \sqrt{2})$, as seen here:

$\{x \in \mathbb{R}\,;\ x < \sqrt{2}\} = (-\infty, \sqrt{2})$ — the set of $x \in \mathbb{R}$ such that $x < \sqrt{2}$

However, we purposely wrote A as those x with $x^2 < 2$ or $x \leq 0$ because at this point, we don't know whether $\sqrt{2}$ exists, so here's A as written above:

the set of $x \in \mathbb{R}$ such that $x^2 < 2$ or $x \le 0$

This set looks as though it is "cutting" $\mathbb{R}$ into two halves at the point $\sqrt{2}$, which is in fact one inspiration for the construction method of $\mathbb{R}$ detailed in Section 2.8. Although the pictures above are not fully accurate, since we haven't proven that $\sqrt{2}$ exists yet, we observe that if $\sqrt{2}$ did exist, then it would be the supremum of the set A. We now use this observation to *prove* that $\sqrt{2}$ exists! Indeed, since A is nonempty, all we have to do is to show that A is bounded above. Axiom (**C**) then kicks in and says that $b := \sup A$ exists. We now can try to prove that $b^2 = 2$, which will prove the existence of $\sqrt{2}$. In Theorem 2.32, we will use this type of argument to show that nth roots exist. To prove the theorem, we need the fact that if $\xi, \eta > 0$, then for every natural number $k > 1$,

$$1 < \xi \implies \xi < \xi^k \quad \text{and} \quad \eta < 1 \implies \eta^k < \eta.$$

These properties follow from the power rules in Theorem 2.23 on p. 80. For example, $1 < \xi$ implies $1 = 1^{k-1} < \xi^{k-1}$; then multiplying by ξ, we get $\xi < \xi^k$.

Existence/uniqueness of nth roots

Theorem 2.32 *Every nonnegative real number has a unique nonnegative nth root.*

Proof First of all, uniqueness follows from the last power rule in Theorem 2.23. Note that the nth root of zero exists and equals zero, and certainly 1th roots always exist. So let $a > 0$ and $n \ge 2$; we shall prove that $\sqrt[n]{a}$ exists.

Step 1: As in the $\sqrt{2}$ case outlined above, we define the tentative $\sqrt[n]{a}$ as a supremum. Let A be the set of real numbers x such that $x^n < a$ or $x \le 0$. Here are pictures of A for $n = 2, \ldots, 5$, noting that we have to prove that the radicals exist:

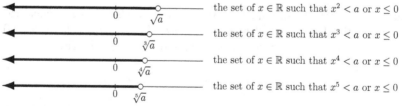

the set of $x \in \mathbb{R}$ such that $x^2 < a$ or $x \le 0$

the set of $x \in \mathbb{R}$ such that $x^3 < a$ or $x \le 0$

the set of $x \in \mathbb{R}$ such that $x^4 < a$ or $x \le 0$

the set of $x \in \mathbb{R}$ such that $x^5 < a$ or $x \le 0$

Since A contains 0, A is nonempty. We claim that A is bounded above by $a + 1$. To see this, let $x \in A$. If $x \le 0$, then of course $x \le a + 1$. Suppose now that $x > 0$, and observe that

$$x^n < a < a + 1 < (a+1)^n, \quad \text{which implies} \quad x^n < (a+1)^n.$$

The power rules theorem on p. 79 then implies that $x < a + 1$. Therefore, A is bounded above by $a + 1$. Being nonempty and bounded above, by the axiom of completeness A has a least upper bound, which we denote by $b \ge 0$. We shall prove that $b^n = a$, which proves our theorem. Since either $b^n = a$, $b^n < a$, or $b^n > a$, we shall prove that the latter two cases cannot occur.

Step 2: Suppose that $b^n < a$. In particular, b belongs to A. To get a contradiction, we increase b just a little to find a number b_1 in A with $b < b_1$ such that $b_1^n < a$ still holds, which contradicts that b was supposedly an upper bound for A. To find such a b_1, observe that—see the remark before this theorem—for every $0 < \varepsilon < 1$, we have $\varepsilon^m < \varepsilon$ for every natural number $m > 1$, so by the binomial theorem on p. 43,

$$(b + \varepsilon)^n = b^n + nb^{n-1}\varepsilon + \frac{n(n-1)}{2}b^{n-2}\varepsilon^2 + \cdots + \varepsilon^n$$

$$< b^n + nb^{n-1}\varepsilon + \frac{n(n-1)}{2}b^{n-2}\varepsilon + \cdots + \varepsilon$$

$$= b^n + \varepsilon c,$$

where c is a positive number. Since $b^n < a$, we have $(a - b^n)/c > 0$. Now let ε equal $(a - b^n)/c$ or $1/2$, whichever is smaller (or equal to $1/2$ if $(a - b^n)/c = 1/2$). Then $0 < \varepsilon < 1$ and $\varepsilon \leq (a - b^n)/c$, so

$$(b + \varepsilon)^n < b^n + \varepsilon c \leq b^n + \frac{a - b^n}{c} \cdot c = a.$$

Thus, $b_1 := b + \varepsilon$ also belongs to A, which gives the desired contradiction.

Step 3: Now suppose that $b^n > a$. To get a contradiction, we now decrease b just a little to find a number $b_2 < b$ that's still an upper bound for A, which contradicts that b was supposedly the least upper bound for A. To find such a b_2, observe that given any real number ε with $0 < \varepsilon < b$, we have $0 < \varepsilon b^{-1} < 1$, so by Bernoulli's inequality (Theorem 2.7 on p. 42),

$$(b - \varepsilon)^n = b^n\left(1 - \varepsilon b^{-1}\right)^n > b^n\left(1 - n\varepsilon b^{-1}\right) = b^n - \varepsilon c,$$

where $c = nb^{n-1} > 0$. Since $a < b^n$, we have $(b^n - a)/c > 0$. Let ε equal $(b^n - a)/c$ or $b/2$, whichever is smaller (or equal to $b/2$ if $(b^n - a)/c = b/2$). Then $0 < \varepsilon < b$ and $\varepsilon \leq (b^n - a)/c$, which implies that $-\varepsilon c \geq -(b^n - a)$. Therefore,

$$(b - \varepsilon)^n > b^n - \varepsilon c \geq b^n - (b^n - a) = a.$$

It follows that $b_2 := b - \varepsilon$ is an upper bound for A (can you prove this?), which gives the desired contradiction. ∎

In particular, $\sqrt{2}$ exists and, as we already know, is an irrational number. Here are proofs of the familiar root rules memorized from high school.

Root rules

> **Theorem 2.33** *For every nonnegative real numbers a and b and natural number n, we have*
>
> $$\sqrt[n]{ab} = \sqrt[n]{a}\,\sqrt[n]{b}, \quad \sqrt[m]{\sqrt[n]{a}} = \sqrt[mn]{a}, \quad \text{and} \quad a < b \iff \sqrt[n]{a} < \sqrt[n]{b}.$$

Proof Let $x = \sqrt[n]{a}$ and $y = \sqrt[n]{b}$. Then, $x^n = a$ and $y^n = b$, so by the power rules theorem on p. 79,

$$(xy)^n = x^n\, y^n = ab.$$

By uniqueness of nth roots, we must have $xy = \sqrt[n]{ab}$. This proves the first identity. The second identity is proved similarly. Finally, again by the power rules theorem, we have $\sqrt[n]{a} < \sqrt[n]{b} \iff \left(\sqrt[n]{a}\right)^n < \left(\sqrt[n]{b}\right)^n \iff a < b$, which proves the last statement of our theorem. ∎

Using the notation $a^{1/n}$ instead of $\sqrt[n]{a}$, the root rules look like

$$(ab)^{\frac{1}{n}} = a^{\frac{1}{n}} b^{\frac{1}{n}}, \quad \left(a^{\frac{1}{n}}\right)^{\frac{1}{m}} = a^{\frac{1}{mn}}, \quad \text{and} \quad a < b \iff a^{\frac{1}{n}} < b^{\frac{1}{n}}.$$

Given $a \in \mathbb{R}$ with $a \geq 0$ and $r = m/n$, where $m \in \mathbb{Z}$ and $n \in \mathbb{N}$, we define

$$a^r := \left(a^{1/n}\right)^m, \tag{2.26}$$

provided that $a \neq 0$ when $m \leq 0$. One can check that the right-hand side is defined independently of the representation of r; that is, if $r = p/q = m/n$ for some other $p \in \mathbb{Z}$ and $q \in \mathbb{N}$, then $(a^{1/q})^p = (a^{1/n})^m$. Combining the power rules theorem for integer powers and the root rules theorem above, we get the following theorem.

Power rules for rational powers

> **Theorem 2.34** *For $a, b \in \mathbb{R}$ with $a, b \geq 0$, and $r, s \in \mathbb{Q}$, provided that the individual powers are defined, we have*
>
> $$a^r \cdot a^s = a^{r+s}, \quad a^r \cdot b^r = (ab)^r, \quad \left(a^r\right)^s = a^{rs},$$
>
> *and if r is nonnegative, then*
>
> $$a < b \iff a^r < b^r.$$

We shall define a^x for every nonnegative real number a and real number x in Section 4.7 and prove a similar theorem (see Theorem 4.33 on p. 305); see also Problem 9 for another way to define a^x.

2.7.3 The Archimedean Property and Its Consequences

Another consequence of the completeness axiom is the following "obvious" fact.

> **Lemma 2.35** *The set $\mathbb{N}$ is not bounded above, and the set $\mathbb{Z}$ is bounded neither above nor below.*

Proof We prove the claim for $\mathbb{N}$, leaving the claim for $\mathbb{Z}$ to you. Assume, to get a contradiction, that $\mathbb{N}$ *is* bounded above. Then $\mathbb{N}$ must have a least upper bound, say b. Since $b - 1$ is smaller than the least upper bound b, there must be a natural number m such that $b - 1 < m$, which implies that $b < m + 1$. However, $m + 1$ is a natural number, so b cannot be an upper bound for $\mathbb{N}$, a contradiction. ■

This lemma yields many useful results.

The $1/n$-principle

> **Theorem 2.36** *Given a real number $x > 0$, there is a natural number n such that $\frac{1}{n} < x$.*

Proof Indeed, since $\mathbb{N}$ is not bounded above, $\frac{1}{x}$ is not an upper bound, so there is $n \in N$ such that $\frac{1}{x} < n$. This implies that $\frac{1}{n} < x$, and we're done. ■

Here's an example showing the $1/n$-principle in action.

Example 2.25 Consider the set

$$A = \left\{ 1 - \frac{3}{n} \, ; \, n = 1, 2, 3, \dots \right\}$$

There are infinitely many points "accumulating" near 1, but we were able to draw only a handful of points. (Note that 1 is not in the set A, which is why we drew an open circle at 1.) Based on this picture, we conjecture that $\sup A = 1$ and $\inf A = -2$. To show that $\sup A = 1$, we need to prove two things: that 1 is an upper bound for A and that 1 is the least of all upper bounds for A. First, we show that 1 is an upper bound. To see this, observe that for every $n \in \mathbb{N}$,

$$\frac{3}{n} \geq 0 \quad \Longrightarrow \quad -\frac{3}{n} \leq 0 \quad \Longrightarrow \quad 1 - \frac{3}{n} \leq 1.$$

Thus, for all $a \in A$, $a \leq 1$, so 1 is indeed an upper bound for A. Second, we must show that 1 is the least of all upper bounds. So assume that $c < 1$; we'll show that c cannot be an upper bound by showing that there is $a \in A$ such that $c < a$, that is, there is $n \in \mathbb{N}$ such that $c < 1 - 3/n$. Observe that

$$c < 1 - \frac{3}{n} \quad \Longleftrightarrow \quad \frac{3}{n} < 1 - c \quad \Longleftrightarrow \quad \frac{1}{n} < \frac{1-c}{3}. \tag{2.27}$$

Since $c < 1$, we have $(1 - c)/3 > 0$, so by the $1/n$-principle, there exists $n \in \mathbb{N}$ such that $1/n < (1 - c)/3$. Hence, by (2.27), there is $n \in \mathbb{N}$ such that $c < 1 - 3/n$. This shows that c is not an upper bound for A.

To show that $\inf A = -2$, we need to prove two things: That -2 is a lower bound and that -2 is the greatest of all lower bounds. First, to prove that -2 is a lower bound, observe that for every $n \in \mathbb{N}$,

$$\frac{3}{n} \leq 3 \quad \Longrightarrow \quad -3 \leq -\frac{3}{n} \quad \Longrightarrow \quad -2 \leq 1 - \frac{3}{n}.$$

Thus, for all $a \in A$, $-2 \leq a$, so -2 is indeed a lower bound for A. Second, to see that -2 is the greatest of all lower bounds, assume that $-2 < c$; we'll show that c cannot be a lower bound by showing there is $a \in A$ such that $a < c$; that is, there is $n \in \mathbb{N}$ such that $1 - 3/n < c$. In fact, simply take $n = 1$. Then $1 - 3/n = 1 - 3 = -2 < c$. This shows that c is not a lower bound for A.

Here's another useful consequence of the fact that $\mathbb{N}$ is not bounded above.[12]

Archimedean property

> **Theorem 2.37** *Given a real number $y > 0$ and a real number x, there is a unique integer n such that*
>
> $$ny \leq x < (n + 1)y.$$
>
> *In particular (set $y = 1$), for every real number x, there is a unique integer n such that $n \leq x < n + 1$, which is obvious if we view the real numbers as a line.*

Proof The inequality $ny \leq x < (n + 1)y$ is, after division by y, equivalent to

$$n \leq z < n + 1, \quad \text{where } z = \frac{x}{y}.$$

We shall work with this inequality instead of the original. If z is an integer, then $n = z$ is the unique integer satisfying $n \leq z < n + 1$. We henceforth assume that z is not an integer. Consider the set $A = \{m \in \mathbb{N};\ |z| < m\}$. This set is not empty, since $\mathbb{N}$ is not bounded above, so by the well-ordering of $\mathbb{N}$, A contains a least element, say $\ell \in \mathbb{N}$. Then $|z| < \ell$ (because $\ell \in A$). We claim that $\ell - 1 < |z|$. Indeed, if we had $|z| < \ell - 1$ (we cannot have $|z| = \ell - 1$, since z is not an integer), then we would have $\ell - 1 \in A$, which contradicts (since $\ell - 1 < \ell$) that ℓ is the least element of A. Thus, $\ell - 1 < |z| < \ell$. Now if $z > 0$, we get $n < z < n + 1$, where $n = \ell - 1$. If $z < 0$, then we obtain $\ell - 1 < -z < \ell$, or after multiplying through by -1, we see that $n < z < n + 1$, where $n = -\ell$. This proves existence.

---

[12]The "Archimedean property" might equally well be called the "Eudoxan property" after Eudoxus of Cnidus (408 B.C.–355 B.C.); see [181] and [129, p. 7].

To prove uniqueness, assume that $n \le z < n+1$ and $m \le z < m+1$ for $m, n \in \mathbb{Z}$. These inequalities imply that $n \le z < m+1$, so $n < m+1$. Similarly, we have $m < n+1$. Thus, $n < m+1 < (n+1)+1 = n+2$, or $0 < m-n+1 < 2$. This implies that $m-n+1 = 1$, or $m = n$. ∎

We remark that some authors replace the integer n by the integer $n-1$ in the Archimedean property so that it reads: Given a real number $y > 0$ and a real number x, there is a unique integer n such that $(n-1)y \le x < ny$. We'll use this formulation of the Archimedean property in the proof of Theorem 2.38 below.

Using the Archimedean property, we can define the **greatest integer function** as follows. Given $a \in \mathbb{R}$, we define $\lfloor a \rfloor$ as the greatest integer less than or equal to a, that is, $\lfloor a \rfloor$ is the unique integer n satisfying the inequalities $n \le a < n+1$. The greatest integer function looks like a staircase, as seen in Fig. 2.6. This function will come up at various times in the sequel. We now prove an important fact concerning the rational and irrational numbers.

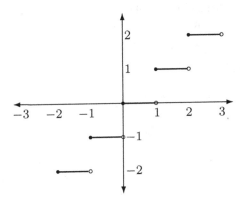

Fig. 2.6 The greatest integer function

Density of the (ir)rationals

Theorem 2.38 *Between every two real numbers there is a rational number and an irrational number.*

Proof Let $x < y$. We first prove that there is a rational number between x and y. Indeed, $y - x > 0$, so by the $1/n$-principle, there is a natural number n such that $1/n < y - x$. Then by the Archimedean principle, there is an integer m such that $m - 1 \le nx < m$. The right half of the inequality, namely $nx < m$, implies that $x < m/n$. The other half, $m - 1 \le nx$, implies (recalling that $1/n < y - x$)

$$\frac{m}{n} - \frac{1}{n} \le x \quad \Longrightarrow \quad \frac{m}{n} \le \frac{1}{n} + x \quad \Longrightarrow \quad \frac{m}{n} < (y - x) + x = y.$$

Thus, $x < m/n < y$.

To prove that between x and y there is an irrational number, note that $x - \sqrt{2} < y - \sqrt{2}$, so by what we just proved above, there is a rational number r such that $x - \sqrt{2} < r < y - \sqrt{2}$. Adding $\sqrt{2}$, we obtain

$$x < \xi < y, \quad \text{where } \xi = r + \sqrt{2}.$$

Note that ξ is irrational, for if it were rational, then $\sqrt{2} = \xi - r$ would also be rational, which we know is false. This completes our proof. ∎

2.7.4 The Nested Intervals Property

A sequence of sets $\{A_n\}$ is said to be **nested** if

$$A_1 \supseteq A_2 \supseteq A_3 \supseteq \cdots \supseteq A_n \supseteq A_{n+1} \supseteq \cdots ,$$

that is, if $A_k \supseteq A_{k+1}$ for each k; here's a picture to keep in mind:

Example 2.26 If $A_n = \left(0, \frac{1}{n}\right)$, then $\{A_n\}$ is a nested sequence:

Note that

$$\bigcap_{n=1}^{\infty} A_n = \bigcap_{n=1}^{\infty} \left(0, \frac{1}{n}\right) = \varnothing.$$

Indeed, if $x \in \bigcap A_n$, which means that $x \in \left(0, \frac{1}{n}\right)$ for every $n \in \mathbb{N}$, then $0 < x < 1/n$ for all $n \in \mathbb{N}$. However, by Theorem 2.36, there is an n such that $0 < 1/n < x$. This shows that $x \notin (0, 1/n)$, contradicting that $x \in \left(0, \frac{1}{n}\right)$ for every $n \in \mathbb{N}$. Therefore, $\bigcap A_n$ must be empty.

Example 2.27 Now on the other hand, if $A_n = \left[0, \frac{1}{n}\right]$, then $\{A_n\}$ is a nested sequence, but in that case,

$$\bigcap_{n=1}^{\infty} A_n = \bigcap_{n=1}^{\infty} \left[0, \frac{1}{n}\right] = \{0\} \neq \varnothing.$$

The difference between the first example and the second is that the second example is a nested sequence of closed and bounded intervals. Here, bounded means bounded above and below. It is a general fact that the intersection of a nested sequence of nonempty closed and bounded intervals is nonempty. This is the content of the nested intervals theorem.

Nested intervals theorem

> **Theorem 2.39** *The intersection of a nested sequence of nonempty closed and bounded intervals in $\mathbb{R}$ is nonempty.*

Proof Let $\{I_n = [a_n, b_n]\}$ be a nested sequence of nonempty closed and bounded intervals. Being nested, we in particular have $I_2 = [a_2, b_2] \subseteq [a_1, b_1] = I_1$, and so $a_1 \le a_2 \le b_2 \le b_1$. Since $I_3 = [a_3, b_3] \subseteq [a_2, b_2]$, we have $a_2 \le a_3 \le b_3 \le b_2$, and so $a_1 \le a_2 \le a_3 \le b_3 \le b_2 \le b_1$. In general, we see that for every n,

$$a_1 \le a_2 \le a_3 \le \cdots \le a_n \le b_n \le \cdots \le b_3 \le b_2 \le b_1.$$

More visually, we have

Let $a = \sup\{a_k \; ; \; k = 1, 2, \ldots\}$, which exists, since all the a_k are bounded above (by each of the b_n). By definition of supremum, $a_n \le a$ for each n. Also, since every b_n is an upper bound for the set $\{a_k \; ; \; k = 1, 2, \ldots\}$, by definition of supremum, $a \le b_n$ for each n. Thus, $a \in I_n$ for each n, and our proof is complete. ∎

Example 2.28 The "bounded" assumption cannot be dropped, for if $A_n = [n, \infty)$, then $\{A_n\}$ is a nested sequence of closed intervals, but

$$\bigcap_{n=1}^{\infty} A_n = \varnothing.$$

We end this section with a discussion of maximums and minimums. Given a set A of real numbers, a number a is called the **maximum** of A if $a \in A$ and $a = \sup A$, in which case we write $a = \max A$. Similarly, a is called the **minimum** of A if $a \in A$ and $a = \inf A$, in which case we write $a = \min A$. For instance, $1 = \max(0, 1]$, but $(0, 1)$ has no maximum, only a supremum, which is also 1. In Problem 4, we prove that every finite set has a maximum.

▶ **Exercises 2.7**

1. What are the supremums and infimums of the following sets? Give careful proofs of your answers. The "$1/n$-principle" might be helpful in some of your proofs.

   (a) $A = \{1 + \frac{5}{n} \; ; \; n = 1, 2, 3, \ldots\}$ (b) $B = \{3 - \frac{8}{n^3} \; ; \; n = 1, 2, 3, \ldots\}$
   (c) $C = \{1 + (-1)^n \frac{1}{n} \; ; \; n = 1, 2, 3, \ldots\}$ (d) $D = \{(-1)^n + \frac{1}{n} \; ; \; n = 1, 2, 3, \ldots\}$
   (e) $E = \left\{\sum_{k=1}^{n} \frac{1}{2^k} \; ; \; n = 1, 2, \ldots\right\}$ (f) $F = \left\{(-1)^n + \frac{(-1)^{n+1}}{n} \; ; \; n = 1, 2, 3, \ldots\right\}$.

2. Are the following sets bounded above? Are they bounded below? If the supremum or infimum exists, find it and prove your answer.

   (a) $A = \left\{1 + n^{(-1)^n} \; ; \; n = 1, 2, 3, \ldots\right\}$, (b) $B = \left\{2^{n(-1)^n} \; ; \; n = 1, 2, 3, \ldots\right\}$.

3. (Various properties of supremums/infimums)

   (a) If $A \subseteq \mathbb{R}$ is bounded above and A contains one of its upper bounds, prove that this upper bound is in fact the supremum of A.

   (b) Let $A \subseteq \mathbb{R}$ be a nonempty bounded set. For $x, y \in \mathbb{R}$, define a new set $xA + y$ by $xA + y := \{xa + y \, ; \, a \in A\}$. Consider the case $y = 0$. Prove that

   $$x > 0 \implies \inf(xA) = x \inf A, \quad \sup(xA) = x \sup A;$$

   $$x < 0 \implies \inf(xA) = x \sup A, \quad \inf(xA) = x \sup A.$$

   (c) With $x = 1$, prove that $\inf(A + y) = \inf(A) + y$ and $\sup(A + y) = \sup(A) + y$.

   (d) What are the formulas for $\inf(xA + y)$ and $\sup(xA + y)$?

   (e) If $A \subseteq B$ and B is bounded, prove that $\sup A \leq \sup B$ and $\inf B \leq \inf A$.

4. In this problem we prove some facts concerning maximums and minimums.

   (a) Let $A \subseteq \mathbb{R}$ be nonempty. An element $a \in A$ is called the **maximum**, respectively **minimum**, element of A if $a \geq x$, respectively $a \leq x$, for all $x \in A$. Prove that A has a maximum (respectively minimum) if and only if $\sup A$ exists and $\sup A \in A$ (respectively $\inf A$ exists and $\inf A \in A$).

   (b) Let $A \subseteq \mathbb{R}$ and suppose that A has a maximum, say $a = \max A$. Given $b \in \mathbb{R}$, prove that $A \cup \{b\}$ also has a maximum, and $\max(A \cup \{b\}) = \max\{a, b\}$.

   (c) Prove that a nonempty finite set of real numbers has a maximum and minimum, where by finite we mean a set of the form $\{a_1, a_2, \ldots, a_n\}$, where $a_1, \ldots, a_n \in \mathbb{R}$.

5. If $A \subseteq \mathbb{R}^+$ is nonempty and closed under addition, prove that A is not bounded above. (As a corollary, we get another proof that $\mathbb{N}$ is not bounded above.) If $A \subseteq (1, \infty)$ is nonempty and closed under multiplication, prove that A is not bounded above.

6. Using the Archimedean property, prove that if $a, b \in \mathbb{R}$ and $b - a > 1$, then there is $n \in \mathbb{Z}$ such that $a < n < b$. Using this result, can you give another proof that between every two real numbers there is a rational number?

7. If $a \in \mathbb{R}$, prove that $|a| = \sqrt{a^2}$.

8. Here are some more power rules for you to prove. Let $p, q \in \mathbb{Q}$.

   (a) If $p < q$ and $a > 1$, then $a^p < a^q$.

   (b) If $p < q$ and $0 < a < 1$, then $a^q < a^p$.

   (c) Let $a > 0$ and let $p < q$. Prove that $a > 1$ if and only if $a^p < a^q$.

9. (**Real numbers to real powers**) We define $0^x := 0$ for all real $x > 0$; otherwise, 0^x is undefined. We now define a^x for $a > 0$ and $x \in \mathbb{R}$. First, assume that $a \geq 1$ and $x \geq 0$.

(a) Prove that $A = \{a^r ; 0 \leq r \leq x, r \in \mathbb{Q}\}$ is bounded above, where a^r is defined in (2.26). Define $a^x := \sup A$. Prove that if $x \in \mathbb{Q}$, then this definition of a^x agrees with the definition (2.26).

(b) For $a, b, x, y \in \mathbb{R}$ with $a, b \geq 1$ and $x, y \geq 0$, prove that

$$a^x \cdot a^y = a^{x+y}; \quad a^x \cdot b^x = (ab)^x; \quad (a^x)^y = a^{xy}. \qquad (2.28)$$

(In the equality $(a^x)^y = a^{xy}$, you should first show that $a^x \geq 1$, so $(a^x)^y$ is defined.)

(c) If $0 < a < 1$ and $x \geq 0$, define $a^x := 1/(1/a)^x$; note that $1/a > 1$, so $(1/a)^x$ is defined. Finally, if $a > 0$ and $x < 0$, define $a^x := (1/a)^{-x}$; note that $-x > 0$, so $(1/a)^{-x}$ is defined. Prove (2.28) for all $a, b, x, y \in \mathbb{R}$ with $a, b > 0$ and $x, y \in \mathbb{R}$.

10. Let $p(x) = ax^2 + bx + c$ be a quadratic polynomial with real coefficients and with $a \neq 0$. Prove that $p(x)$ has a real root (that is, an $x \in \mathbb{R}$ with $p(x) = 0$) if and only if $b^2 - 4ac \geq 0$, in which case the root(s) are given by the quadratic formula:

$$x = \frac{-b \pm \sqrt{b^2 - 4ac}}{2a}.$$

11. Let $\{I_n = [a_n, b_n]\}$ be a nested sequence of nonempty closed and bounded intervals and put $A = \{a_n ; n \in \mathbb{N}\}$ and $B = \{b_n ; n \in \mathbb{N}\}$. Show that $\sup A$ and $\inf B$ exist and $\bigcap I_n = [\sup A, \inf B]$.

12. In this problem we give a characterization of the completeness axiom (**C**) of $\mathbb{R}$ in terms of intervals as explained in [50]. A subset A of $\mathbb{R}$ is **convex** if given x and y in A and $t \in \mathbb{R}$ with $x < t < y$, we have $t \in A$.

(a) Prove that all convex subsets of $\mathbb{R}$ are intervals.

(b) Let us pretend that instead of assuming axiom (**C**) for $\mathbb{R}$, we take as an axiom that all convex subsets of $\mathbb{R}$ are intervals. Prove the completeness property (**C**) of $\mathbb{R}$. Suggestion: Let I be the set of all upper bounds of a nonempty set A that is bounded above. Show that I is convex.

This problem shows that the completeness axiom is equivalent to the statement that all convex sets are intervals.

2.8 ★ Construction of the Real Numbers via Dedekind Cuts

We take as our starting point the existence of the ordered field $\mathbb{Q}$ of rational numbers.[13] Then following Richard Dedekind's (1831–1916) book *Continuity and Irrational Numbers* [59], we construct the real numbers as certain *sets* of rational numbers called

---

[13] The rationals are not difficult to construct from the integers; see, for instance, [149].

"cuts." This construction, like all our set theory, is from the "*very* naive" viewpoint. We shall focus on understanding conceptually how the reals and its operations arise geometrically from the rationals, leaving many of the tedious details to the reader. You can also check out [149], which has a complete account of all the details. In Edmund Landau's (1877–1938) famous book [137, p. xi] in which he constructs the real numbers, he wrote (dated 1929)

> I hope that I have written this book, after a preparation stretching over decades, in such a way that a normal student can read it in two days. And then (since he already knows the formal rules from school) he may forget its contents, with the exception of the axiom of induction and of Dedekind's fundamental theorem.

Although one could read Landau's book in two days, few beginners could fully comprehend his book in two days. To master the ideas presented here, we recommend this section to be read slowly, taking more than two days!

2.8.1 Construction of the Reals from the Rationals

We are given the rational numbers; the usual picture is that the rationals are laid out in a line but that there are lots of tiny gaps at the irrationals (Fig. 2.7).

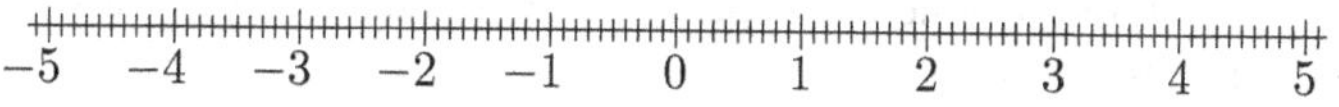

Fig. 2.7 Here, we draw the rationals as the small vertical sticks. For clarity of our pictures, we will henceforth leave out the sticks

The goal of this section is to fill in the holes. In order to gain some insight, let us assume for the moment that we are given the real numbers. We ask:

How can we uniquely describe a real number using rational numbers?

To answer this question, recall that we used a left-infinite open interval in our proof on p. 93 that $\sqrt{2}$ exists; there we used an interval of real numbers, but we could have used an interval of rational numbers (Fig. 2.8).

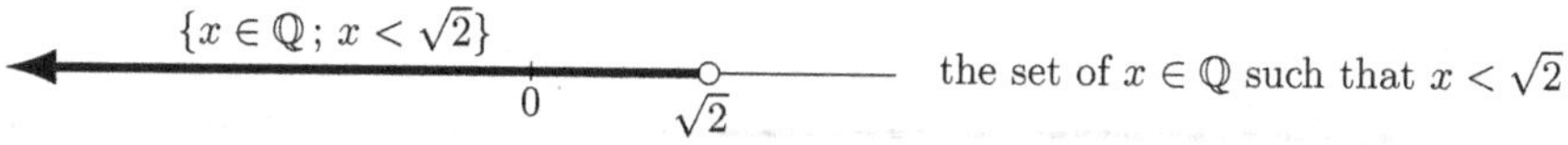

the set of $x \in \mathbb{Q}$ such that $x < \sqrt{2}$

Fig. 2.8 The number $\sqrt{2}$ is the supremum of the set shown

Actually, at that point we did not know that $\sqrt{2}$ existed, so we should write the interval in a way that does not mention $\sqrt{2}$:

$\{x \in \mathbb{Q}\,;\, x^2 < 2 \text{ or } x \le 0\}$

the set of $x \in \mathbb{Q}$ such that $x^2 < 2$ or $x \le 0$

The idea here is that $\sqrt{2}$ "cuts" $\mathbb{Q}$ into two halves and is the supremum (respectively infimum) of the left (respectively right) half. More generally, for every $a \in \mathbb{R}$ (still temporarily assuming that $\mathbb{R}$ is given), the intuition is that a "cuts" the rational numbers into left and right halves (Fig. 2.9).

$\{x \in \mathbb{Q}\,;\, x < a\}$

Fig. 2.9 The real number a "cuts" the rationals into left and right halves. We will focus on the left side of the "cut"

Then a is uniquely determined as the supremum of the set $\{x \in \mathbb{Q}\,;\, x < a\}$. Thus, we have a one-to-one correspondence:

real numbers $\longleftrightarrow$ left-infinite open intervals of rationals.

For simplicity, we shall call a "left-infinite open interval of rationals" a **cut** (or **Dedekind cut**). Conclusion: Assuming that real numbers are given, we see that the real numbers are in one-to-one correspondence with cuts! Now comes Dedekind's stroke of genius [59].

> **Idea** : Since real numbers are in one-to-one correspondence with cuts . . . why don't we just *define* real numbers *to be* cuts!

This indeed is what we shall do! Before moving on, let's define cuts in a precise way. A **cut** is a nonempty set $a \subseteq \mathbb{Q}$ that is (**1**) bounded above; (**2**) left-infinite; and (**3**) has no maximum element. Explicitly, a nonempty set $a \subseteq \mathbb{Q}$ is a **cut** if:

(**1**) There is $y \in \mathbb{Q}$ such that for all $x \in a$, $x \le y$.
(**2**) If $x \in a$, then for all $r \in \mathbb{Q}$ with $r < x$, we have $r \in a$.
(**3**) If $x \in a$, then there is $r \in a$ such that $x < r$.

You should picture a cut as in Fig. 2.10. We no longer have to *assume* that $\mathbb{R}$ exists, for now we can *define* it. We come to the main definition of this section:

a

Fig. 2.10 A cut is a left-infinite open interval of rationals. We will see later (Theorem 2.42) that the hole you see *is* the cut a!

> The set of **real numbers** $\mathbb{R}$ is, by definition, the collection of cuts. A **real number** is a cut. Thus, $\mathbb{R} := \{a \, ; \, a \text{ is a cut}\}$.

Thus, *real number* and *cut* are synonyms.

2.8.2 Examples of Cuts

Example 2.29 Here are some cuts representing rational numbers (Fig. 2.11).

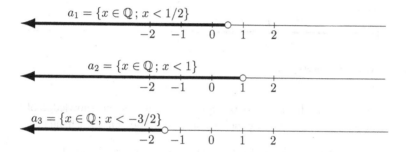

Fig. 2.11 The rational numbers $1/2$, 1, and $-3/2$ are the cuts a_1, a_2, a_3

There are also cuts representing irrational numbers; here again is our motivating example $\sqrt{2}$ (Fig. 2.12).

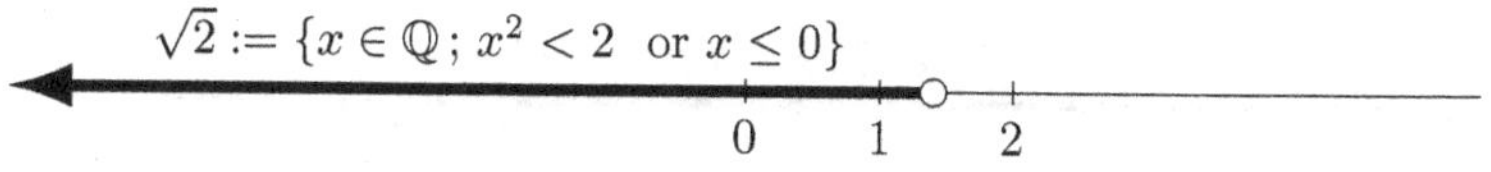

$$\sqrt{2} := \{x \in \mathbb{Q} \, ; \, x^2 < 2 \text{ or } x \leq 0\}$$

Fig. 2.12 $\sqrt{2}$ is defined as a cut

Although it's obvious that $\sqrt{2}$ is a cut, we do have to prove it. See Problem 2.

In general, given a *rational* number a, we define the *real* number a^* as follows:

> **Rational numbers define real numbers** : $a^* := \{x \in \mathbb{Q} \, ; \, x < a\}$.

In this way, we can think of $\mathbb{Q}$ as a subset of $\mathbb{R}$. Let's prove that a^* is a cut. Note that a^* is bounded above by $a + 1$ and by definition it's left-infinite. To see that a^* has no maximum element, let $x \in a^*$. Then $x < a$, so their average $r := (x + a)/2$ is a rational number, and it's easy to check that $r \in a^*$ and $x < r$. Thus, a^* is a cut.

Particularly important real numbers include

$$0^* := \{x \in \mathbb{Q}; \ x < 0\} \quad \text{and} \quad 1^* := \{x \in \mathbb{Q}; \ x < 1\}.$$

See Fig. 2.13 for pictures of 0^* and 1^*. As explained in Problem 12, we can treat a *rational* number a and its corresponding *real* number a^* as identical. We use the star $*$ during the construction of $\mathbb{R}$ for purposes of clarity; after the construction, we shall never write a^* for a rational a again!

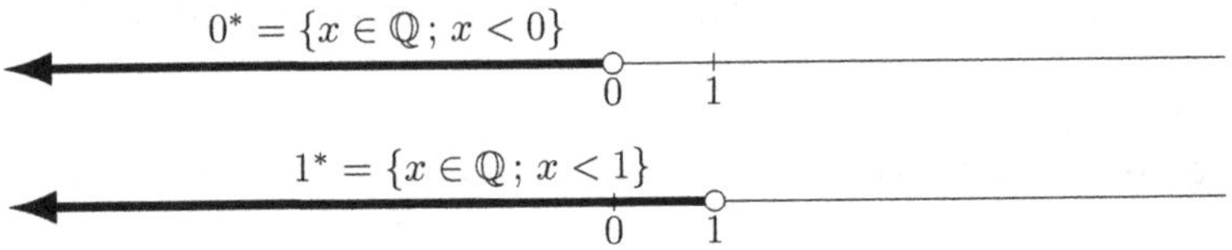

Fig. 2.13 The real numbers 0^* and 1^*

In order to present the next two examples, we shall assume knowledge of limits from elementary calculus, leaving their rigorous development for Chapter 3.

Example 2.30 On p. 226 we will define the real number e as the limit

$$e = \lim_{n \to \infty} \left(1 + \frac{1}{n}\right)^n.$$

In decimal notation to be covered in Section 3.8 on p. 226, we have $e = 2.71828\ldots$ Here's a cut that one might believe defines e.

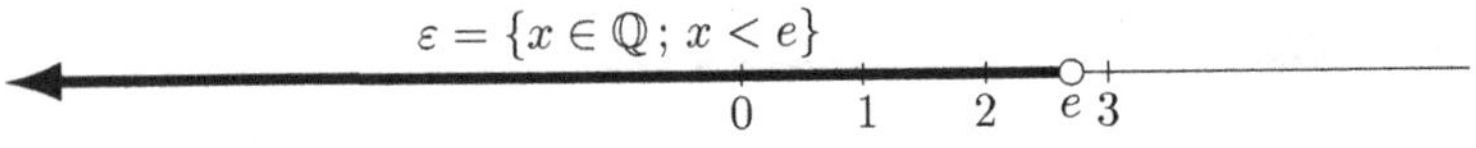

Fig. 2.14 Is $\varepsilon = \{x \in \mathbb{Q}; \ x < e\}$ a valid cut for e?

Unfortunately, this cut is *cheating*, since e is known to be irrational, which we'll prove in Theorem 3.31 on p. 222; hence the cut ε in Fig. 2.14 is not entirely written in terms of rational numbers. We can fix this issue as follows. For each n, let

$$e_n = \left(1 + \frac{1}{n}\right)^n.$$

We have $e_1 = (1 + 1)^1 = 2$, $e_2 = (1 + 1/2)^2 = 2\frac{1}{4}$ and so forth. The sequence $e_1, e_2, e_3, \ldots$ increases upward toward e as shown here:

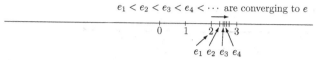

Now for each n, each e_n is a rational number, so the following cut make sense:

$$e_n^* = \{x \in \mathbb{Q} \, ; \, x < e_n\}.$$

We can now define e as the cut (see Problem 3)

$$e := \bigcup_{n=1}^{\infty} e_n^*.$$

Here's a picture of e (Fig. 2.15).

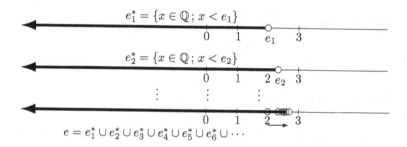

Fig. 2.15 The real number e is a union of cuts

Example 2.31 In Theorem 5.3 on p. 389 we will express the real number $\pi = 3.14159\ldots$ as the limit

$$\pi = \lim_{n \to \infty} p_n \, ,$$

where for each n, we have

$$p_n = 2 \cdot \frac{1}{1} \cdot \frac{2}{1} \cdot \frac{2}{3} \cdot \frac{4}{3} \cdot \frac{4}{5} \cdot \frac{6}{5} \cdot \frac{6}{7} \cdot \frac{8}{7} \cdots \frac{2n}{2n-1} \cdot \frac{2n}{2n+1}.$$

We have $p_1 < p_2 < p_3 < p_4 < \cdots$ and these numbers increase to π. For each n, define the cut

$$p_n^* = \{x \in \mathbb{Q} \, ; \, x < p_n\}.$$

Then we can define π as the cut

$$\pi := \bigcup_{n=1}^{\infty} p_n^* \, ,$$

as shown here:

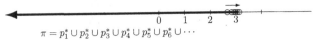

$$\pi = p_1^* \cup p_2^* \cup p_3^* \cup p_4^* \cup p_5^* \cup p_6^* \cup \cdots$$

2.8.3 Order and Completeness Are Easy!

Consider rational numbers a and b and their corresponding cuts $a^* = \{x \in \mathbb{Q} \, ; \, x < a\}$ and $b^* = \{x \in \mathbb{Q} \, ; \, x < b\}$. Here's a picture (Fig. 2.16).

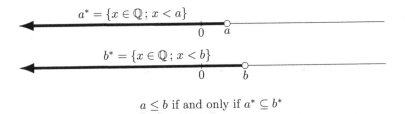

$$a \leq b \text{ if and only if } a^* \subseteq b^*$$

Fig. 2.16 Inequalities are equivalent to subset inclusion!

This picture suggests the following lemma (we leave its proof for Problem 4).

Lemma 2.40 *For rational numbers a, b, we have $a \leq b$ if and only if their cuts satisfy $a^* \subseteq b^*$, and $a = b$ if and only if $a^* = b^*$.*

Due to this lemma, for every two *real* numbers a, b (which, remember, are just cuts), we declare the following definition of inequality.

Definition of inequality :   We define $a \leq b$ if $a \subseteq b$.

Of course,

$$b \geq a \text{ is an equivalent way to write } a \leq b,$$

and

$$\text{we write } a < b \text{ if } a \leq b \text{ and } a \neq b.$$

This definition of inequality behaves exactly as you think. For example, for all real numbers a, b, c, we have **transitivity**,

$$a \leq b \text{ and } b \leq c \implies a \leq c,$$

and **trichotomy**: Exactly one of the following holds,

$$a = b, \quad a < b, \quad \text{or} \quad a > b.$$

You will prove these results in Problem 5.

Example 2.32 For a and b *rational* numbers, observe that

$$a \leq b \quad \Longleftrightarrow \quad a^* \leq b^*;$$

Here, the left-hand "$\leq$" is an inequality in $\mathbb{Q}$, while the right-hand "$\leq$" is an inequality in $\mathbb{R}$. Indeed, Lemma 2.40 says that $a \leq b$ if and only if $a^* \subseteq b^*$, which by definition of "$\leq$" in $\mathbb{R}$, means that $a^* \leq b^*$. Thus, the ordering in $\mathbb{Q}$ is consistent with the ordering of rational numbers in $\mathbb{R}$.

Now that we have a relation "$\leq$" we can define upper bounds, least upper bounds, etc., exactly as we discussed throughout Section 2.7.1 starting on p. 90. What I like about Dedekind's description of $\mathbb{R}$ is that the completeness axiom is immediate.

The completeness axiom holds!

> **Theorem 2.41** *Every bounded nonempty subset of $\mathbb{R}$ has a supremum, that is, a least upper bound.*

Proof Before we get to the proof, consider as an example, to gain insight, the set $A \subseteq \mathbb{R}$, consisting of five points, shown in Fig. 2.17. By definition of real numbers, the set A is a collection of cuts. So, each point in Fig. 2.17 is really a cut. Thus, let us picture the five points in Fig. 2.17 as the cuts in Fig. 2.18. It's clear that the "biggest" cut—the supremum of A—is obtained by taking the union of each of the cuts. This idea in fact always works: Given a subset of $\mathbb{R}$ bounded above, if we take the union of its elements, we get the supremum of the set! Fig. 2.19 reiterates this idea. And now back to the proof of our theorem. Let $A \subseteq \mathbb{R}$ be nonempty and bounded above; we must prove that A has a supremum. As proved in Problem 3, the following is a cut:

$$A$$

Fig. 2.17 If we think of $\mathbb{R}$ as a line, the subset A of $\mathbb{R}$ is a collection of points on the line. We want to find sup A

$$b := \bigcup_{a \in A} a.$$

We claim that b is the supremum of A. First, if $a \in A$, then since b is the union of all such a's, we certainly have $a \subseteq b$, which is to say, $a \leq b$. Thus, b is an upper bound.

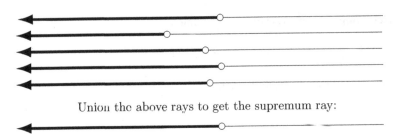

Union the above rays to get the supremum ray:

Fig. 2.18 The points in A are really cuts of $\mathbb{Q}$. To find the "biggest" cut, all we do is take their union!

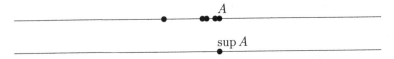

Fig. 2.19 The supremum of a set bounded above is obtained by taking the union of its points (the points are sets, so we can take their union)

We now prove that b is the least of all upper bounds, so suppose c is another upper bound. This means that for all $a \in A$, we have $a \le c$, that is, $a \subseteq c$. It follows that the union of all the a's is also a subset of c:

$$b := \bigcup_{a \in A} a \subseteq c.$$

Thus, $b \le c$, so b is the least of all upper bounds.   ∎

We know that a rational number a gives the real number

$$a^* = \{x \in \mathbb{Q}; \ x < a\}.$$

Since $x < a$ is equivalent to the corresponding real numbers satisfying $x^* < a^*$ (see our discussion in Example 2.32), we can write

$$a^* = \{x \in \mathbb{Q}; \ x^* < a^*\} \quad \text{for a } rational \text{ number } a.$$

Now if a is a *real* number, we will prove an analogous formula (see Theorem 2.42 below):

$$\boxed{a = \{x \in \mathbb{Q}; \ x^* < a\}.}$$

Note that on the right-hand side, x^* refers to the real number $x^* = \{r \in \mathbb{Q}; \ r < x\}$, and the inequality "$x^* < a$" is meant in the sense of real numbers, that is,

$$\textbf{(1)} \ \{r \in \mathbb{Q}; \ r < x\} \subseteq a \quad \text{and} \quad \textbf{(2)} \ \{r \in \mathbb{Q}; \ r < x\} \ne a. \tag{2.29}$$

Thus, we can picture arbitrary cuts as left-infinite open intervals of rational numbers starting from a (Fig. 2.20).

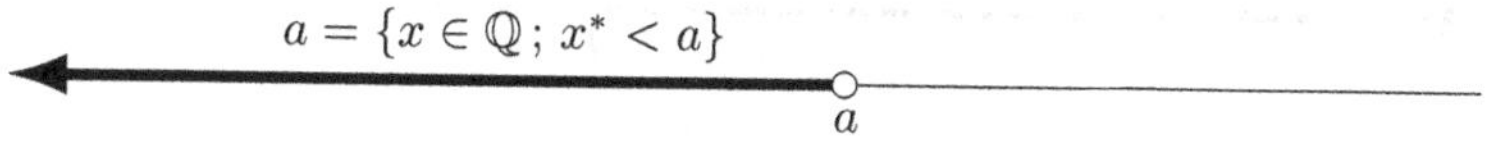

$$a = \{x \in \mathbb{Q}; \, x^* < a\}$$

Fig. 2.20 The cut (real number) a equals $\{x \in \mathbb{Q}; \, x^* < a\}$

Theorem 2.42 *For every real number a, we have*

$$a = \{x \in \mathbb{Q}; \, x^* < a\};$$

that is, for a rational number x, we have $x \in a \iff x^* < a$.

Proof Let $\alpha = \{x \in \mathbb{Q}; \, x^* < a\}$; we will show that $a = \alpha$. Since both a and α are subsets of $\mathbb{Q}$, we need to show that $a \subseteq \alpha$ and $\alpha \subseteq a$.

Proof that $a \subseteq \alpha$: Let $x \in a$. To prove that $x \in \alpha$, we need to show that $x^* < a$, which means that **(1)** and **(2)** in (2.29) hold. To prove **(1)**, let $r \in \mathbb{Q}$ with $r < x$. Since $x \in a$ and a is left-infinite (by definition of a cut), we have $r \in a$. This proves **(1)**. Also, since $x \in a$ and a has no maximum element, there is a rational number $s \in a$ with $x < s$. Therefore, $s \in a$ and s is not in $\{r \in \mathbb{Q}; \, r < x\}$. This proves **(2)**.

Proof that $\alpha \subseteq a$: Let $x \in \alpha$; we will show that $x \in a$. By definition of α, $x^* < a$, which means that **(1)** and **(2)** in (2.29) hold. In particular, **(2)** implies that there is a rational number $y \in a$ such that $y \geq x$. If $y = x$, then $x \in a$, and if $y > x$, then since a is left-infinite (by definition of a cut), we also have $x \in a$. ■

2.8.4 Operations on the Reals

We now show that the set $\mathbb{R}$ we've defined satisfies all the ordered field axioms discussed on p. 75. We first define addition. How do we define the sum of cuts? Consider first *rational* numbers a, b and their corresponding cuts a^* and b^* shown in the first two pictures here.

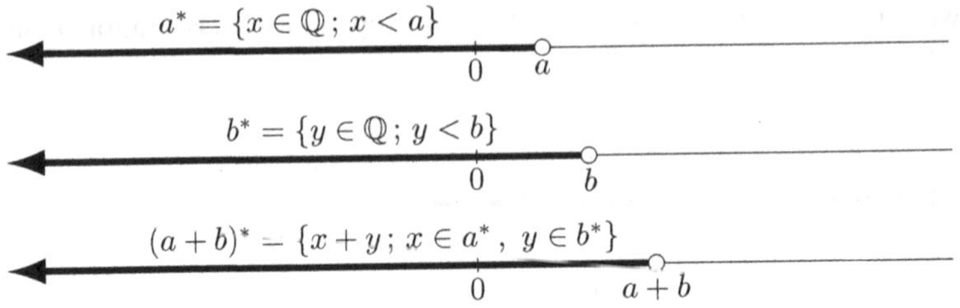

Fig. 2.21 One can show that $(a+b)^* := \{z \in \mathbb{Q};\ z < a+b\}$ equals the set $\{x+y;\ x \in a^*,\ y \in b^*\}$

It's not difficult to prove that the following equality is true regarding the cut corresponding to the rational number $a+b$ (see the last picture in Fig. 2.21):

$$(a+b)^* = \{x+y;\ x \in a^*,\ y \in b^*\}.$$

Inspired by this observation, we make the following definition for all *real* numbers $a, b \in \mathbb{R}$:

> **Definition of addition :** $a + b := \{x+y;\ x \in a,\ y \in b\}.$

Of course, we should check that the right-hand side is indeed a cut.

Sum of cuts is a cut

> **Lemma 2.43** *For real numbers a, b, the sum $a + b$ is also a real number.*

Proof To show that $a + b$ is a cut, we need to show that it's a subset of $\mathbb{Q}$, that is, that it is

(**1**) bounded above; (**2**) left-infinite; (**3**) has no maximum element.

Proof of (**1**): By the definition of cut, there are rational numbers c_1 and c_2 such that for all $x \in a$ and $y \in b$, we have $x \le c_1$ and $y \le c_2$. Hence, for all $x \in a$ and $y \in b$, $x + y \le c$, where $c = c_1 + c_2$. Thus, $a + b$ is bounded above by the rational number c.

Proof of (**2**): Let $p \in a + b$, say $p = x + y$, where $x \in a$ and $y \in b$, and let $r < p$; we must show that $r \in a + b$. However, observe that since $p = x + y$ and $r < p$, we have $r < x + y$. Hence, $r - y < x$, and therefore, since a is a cut, we know that $r - y \in a$. Since we already know that $y \in b$, it follows that

$$r = (r - y) + y \in a + b.$$

Thus, $a + b$ is left-infinite.

Proof of (**3**): Again let $p \in a + b$, say $p = x + y$, where $x \in a$ and $y \in b$; we must show that there is $r \in a + b$ with $p < r$. However, since a is a cut, there is an element $s \in a$ with $x < s$. Therefore, if we let $r := s + y$, then $r \in a + b$ and

$$p = x + y < s + y = r.$$

This shows that $a + b$ has no maximum element. ∎

We now consider additive inverses. Given a *rational* number a, its additive inverse $-a$ defines the cut

$$(-a)^* = \{x \in \mathbb{Q}; \ x < -a\} = \{x \in \mathbb{Q}; \ -x > a\},$$

where we used that $x < -a$ is equivalent to $-x > a$. Since $-x > a$ is also equivalent to the real number inequality $(-x)^* > a^*$, we can write

$$(-a)^* = \{x \in \mathbb{Q}; \ a^* < (-x)^*\}.$$

Here's a picture (Fig. 2.22).

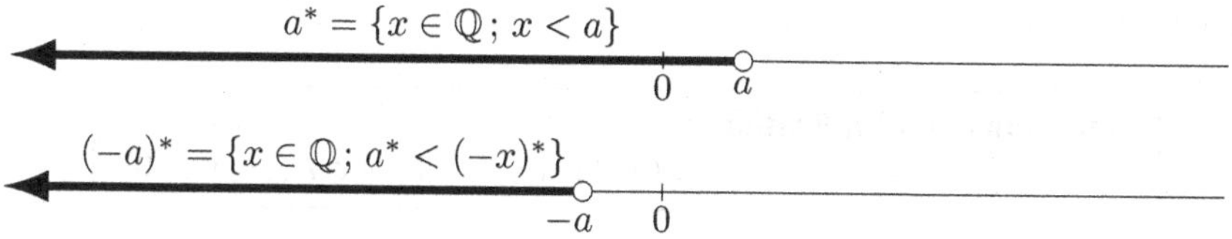

Fig. 2.22 For $x \in \mathbb{Q}$, we have $x < -a$ if and only if $a^* < (-x)^*$

Generalizing to real numbers now, we are led to make the following definition of the additive inverse of a *real* number a.

Definition of additive inverse : $-a := \{x \in \mathbb{Q}; \ a < (-x)^*\}.$

In Problem 9 you will prove that $-a$ is indeed a real number. As usual, subtraction is defined, for all real numbers $a, b \in \mathbb{R}$, by

$$a - b := a + (-b).$$

We define the set of *positive* real numbers (or positive cuts) as

$$\mathbb{R}^+ := \{a \in \mathbb{R}; \ a > 0^*\} \subseteq \mathbb{R}.$$

To define multiplication, again consider two *rational* numbers a, b and assume that a and b are positive. Then it's not difficult to prove that the following equality is true regarding the cut corresponding to the rational number ab (see Fig. 2.23):

$$(ab)^* = \{x \in \mathbb{Q} ; \ x \le 0\} \cup \{xy ; \ x \in a^*, \ y \in b^*, \ x, y > 0\}.$$

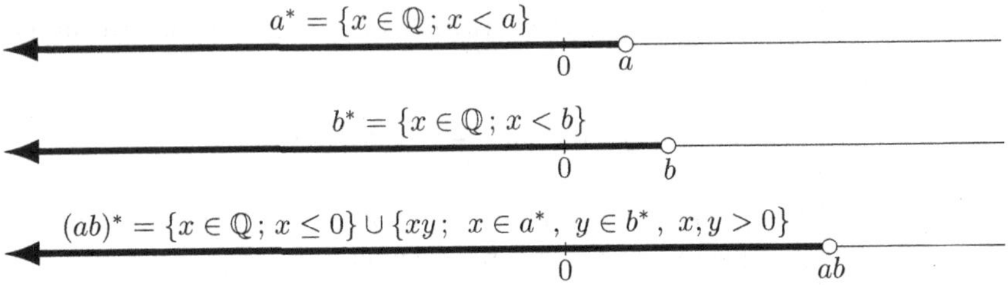

$$a^* = \{x \in \mathbb{Q} ; x < a\}$$

$$b^* = \{x \in \mathbb{Q} ; x < b\}$$

$$(ab)^* = \{x \in \mathbb{Q} ; x \le 0\} \cup \{xy ; \ x \in a^*, \ y \in b^*, \ x, y > 0\}$$

Fig. 2.23 One can show that $(ab)^* := \{z \in \mathbb{Q} ; \ z < ab\}$, equals the set $\{x \in \mathbb{Q} ; \ x \le 0\} \cup \{xy ; \ x \in a^*, \ y \in b^*, \ x, y > 0\}$

Inspired by this observation, we make the following definition for all positive *real* numbers $a, b \in \mathbb{R}^+$.

Definition of multiplication :
$$a \cdot b := \{x \in \mathbb{Q} ; \ x \le 0\} \cup \{xy ; \ x \in a, \ y \in b, \ x, y > 0\}.$$

In Problem 11 you will prove that $a \cdot b$ is a cut. You will also define multiplication of real numbers that may not both be positive.

Finally, we consider multiplicative inverses. If a is a positive *rational* number, observe that (see Fig. 2.24)

$$(a^{-1})^* := \{x \in \mathbb{Q} ; \ x < a^{-1}\} = \{x \in \mathbb{Q} ; \ x \le 0\} \cup \{x \in \mathbb{Q} ; \ x < a^{-1}, \ x > 0\}$$
$$= \{x \in \mathbb{Q} ; \ x \le 0\} \cup \{x \in \mathbb{Q} ; \ a < x^{-1}, \ x > 0\}$$
$$= \{x \in \mathbb{Q} ; \ x \le 0\} \cup \{x \in \mathbb{Q} ; \ a^* < (x^{-1})^*, \ x > 0\}.$$

Again, it's always good to have a corresponding picture.

Generalizing to real numbers, we make the following declaration for $a \in \mathbb{R}^+$.

Definition of multiplicative inverse :
$$a^{-1} := \{x \in \mathbb{Q} ; \ x \le 0\} \cup \{x \in \mathbb{Q} ; \ a < (x^{-1})^*, \ x > 0\}.$$

In Problem 13 you will prove that a^{-1} is indeed a real number. If a is a negative real number, meaning $a < 0^*$, we define

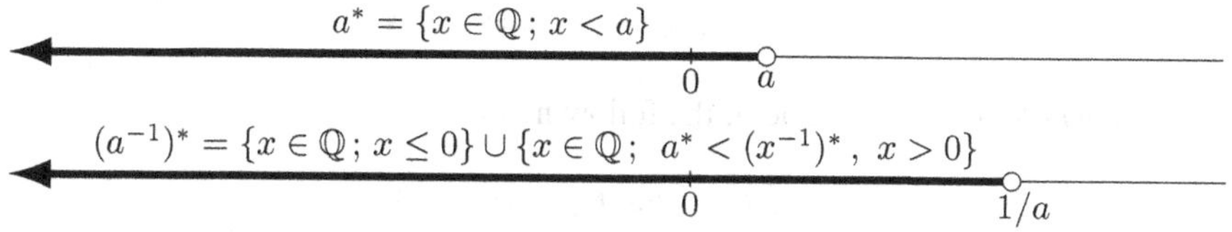

Fig. 2.24 Multiplicative inverse of a positive rational number. In this picture we assume $0 < a < 1$, so $1/a$ is larger than a

$$a^{-1} := (-a)^{-1},$$

where $-a$ is the additive inverse of a. Then division is defined, for all real numbers $a, b \in \mathbb{R}$ with $b \neq 0^*$, by

$$\frac{a}{b} := a \cdot b^{-1}.$$

We now come to the main result of this section.

The reals $\mathbb{R}$

Theorem 2.44 *The set of real numbers $\mathbb{R}$ is a complete ordered field; that is, $\mathbb{R}$ satisfies axioms (A), (M), (D), (P), and (D) listed on pp. 75–79.*

We leave some parts of this proof to the problems, and those we don't ask you to do, please feel free to try them! Please see [149] for a complete account. Finally, we remark that as shown in Problem 12, for all intents and purposes we can treat a *rational* number a and its corresponding real number a^* as identical. For this reason, we can regard $\mathbb{Q}$ as a subset of $\mathbb{R}$, and after this section, we shall never again write a^* for a.

▶ **Exercises 2.8**

1. Judging from the front cover, you can see that my favorite formula is Euler's sum for $\pi^2/6$. Given that $\pi^2/6 = \frac{1}{1^2} + \frac{1}{2^2} + \frac{1}{3^2} + \frac{1}{4^2} + \cdots$, express $\pi^2/6$ as a Dedekind cut.

2. **(Roots of rational numbers)** Let $n \in \mathbb{N}$ and let a be a positive rational number. Define

$$\sqrt[n]{a} := \{x \in \mathbb{Q}; \ x \leq 0\} \cup \{x \in \mathbb{Q}; \ x^n < a, \ x > 0\}.$$

Prove that $\sqrt[n]{a}$ is a cut. Suggestion: To prove that $\sqrt[n]{a}$ has no maximum, review the argument in **Step 2** of Theorem 2.32 on p. 93.

3. **(Unions of cuts)** Let A be a collection of cuts bounded above. Thus, there is a cut c such that $a \leq c$ (that is, $a \subseteq c$) for all $a \in A$. Prove that $\bigcup_{a \in A} a$ is also a cut.

4. Prove Lemma 2.40.

5. For all real numbers a, b, c, prove *transitivity*:

$$a \leq b \quad \text{and} \quad b \leq c \quad \Longrightarrow \quad a \leq c,$$

and *trichotomy*: Exactly one of the following holds,

$$a = b, \quad a < b, \quad \text{or} \quad a > b.$$

Suggestion: For trichotomy, first prove that two conditions cannot be simultaneously satisfied, so we just have to prove that at least one condition is satisfied. Pick two conditions and assume that they do not hold, and then try to prove that the third does hold.

6. Let $r \in \mathbb{Q}^+$ and let $a \in \mathbb{R}$. Prove that $a + (-r)^* < a$. Suggestion: If not, prove that for all $x \in a$, we have $x + r \in a$. Conclude by induction that $x + nr \in a$ for all $n \in \mathbb{N}$.

7. In this problem we study a couple of different viewpoints on real numbers. Let $a \in \mathbb{R}$.

   (i) Let $b = \bigcup_{x \in a} x^*$; that is, b is the union of the points in a, viewed as real numbers. Prove that $a = b$.

   (ii) Let $A = \{x^* ; \ x \in a\}$; that is, A is the collection of all the rational numbers in a viewed as real numbers. Prove that $\sup A = a$.

8. Let $r \in \mathbb{Q}^+$ and let $a \in \mathbb{R}$. Prove that there is a rational number y such that

$$y^* < a < (y + r)^*.$$

Note that $y^* < a$ just means that $y \in a$. Suggestion: Use that $a + (-r/2)^* < a$ from Problem 6, so $a + (-r/2)^*$ cannot be an upper bound for the set A in Problem 7.

9. (**Additive inverses of cuts**) In this problem we study additive inverses.

   (a) For every real number a, prove that

$$-a := \{x \in \mathbb{Q} ; \ a < (-x)^*\}$$

   is a real number. Suggestion: To prove that $-a^*$ is bounded above, pick an arbitrary rational number $y \in a$ and prove that $-y$ is an upper bound for $-a$.

   (b) If $a > 0^*$, prove that $-a < 0^*$. If $a < 0^*$, prove that $-a > 0^*$.

   (c) If $a \in \mathbb{R}$, prove that exactly one of the following holds: $a = 0^*$, $a \in \mathbb{R}^+$, $-a \in \mathbb{R}^+$.

10. (**Addition of cuts**)

   (a) Prove that addition of real numbers is commutative and associative.

   (b) If $a \in \mathbb{R}$, prove that $a + 0^* = a$.

Suggestion: Recalling that $a + 0^*$ and a are sets, you need to show that $a + 0^* \subseteq a$ and $a \subseteq a + 0^*$. To prove the latter, let $x \in a$ and pick, using that a has no maximum element, $r \in a$ with $x < r$. Consider $x = r + (x - r)$.

(c) If $a \in \mathbb{R}$, prove that $a + (-a) = 0^*$.

Suggestion: You need to prove that $a + (-a) \subseteq 0^*$ and $0^* \subseteq a + (-a)$. To prove the latter, let $x \in 0^*$. Then $-x \in \mathbb{Q}^+$. Apply Problem 8 with $r = -x$.

(d) For $a, b \in \mathbb{R}$, prove that $a < b \iff b - a > 0^*$.

(e) For $a, b \in \mathbb{R}$, prove that $a < b \iff b = a + c$, where $c > 0^*$.

11. **(Multiplication of cuts)**

(a) For positive real numbers a, b, prove that

$$a \cdot b := \{x \in \mathbb{Q}; \ x \le 0\} \cup \{xy; \ x \in a, \ y \in b, \ x, y > 0\}$$

is a real number.

(b) For positive real numbers, prove that multiplication is commutative and associative.

(c) For positive real numbers, prove the distributive property.

(d) For all positive real numbers a, prove that $a \cdot 1^* = a$.

(e) Let $\sqrt{2} = \{x \in \mathbb{Q}; \ x \le 0\} \cup \{x \in \mathbb{Q}; \ x^2 < 2, \ x > 0\}$. Using the definition of multiplication of cuts, prove that

$$\sqrt{2} \cdot \sqrt{2} = 2^*, \quad \text{where } 2^* = \{x \in \mathbb{Q}; \ x < 2\}.$$

Remark: For real numbers a, b, at least one of which is not positive, we define

$$a \cdot b := \begin{cases} 0 & \text{if either } a = 0^* \text{ or } b = 0^*, \\ -\big((-a) \cdot b\big) & \text{if } a < 0^* \text{ and } b > 0^*, \\ -\big(a \cdot (-b)\big) & \text{if } a > 0^* \text{ and } b < 0^*, \\ (-a) \cdot (-b) & \text{if } a < 0^* \text{ and } b < 0^*. \end{cases}$$

Based on this case-by-case definition, proving multiplication results by considering every case can get a bit tedious!

12. Define

$$f : \mathbb{Q} \to \mathbb{R} \quad \text{by} \quad f(a) = a^* \quad \text{for all } a \in \mathbb{Q}, \text{ where } a^* = \{x \in \mathbb{Q}; \ x < a\}.$$

(i) Prove that f is an injection. This shows that the image, or range, of f is a replica of $\mathbb{Q}$. We next prove that f replicates the ordered field properties of $\mathbb{Q}$.

(ii) Prove that for all $a, b \in \mathbb{Q}$, $f(a + b) = f(a) + f(b)$.

(iii) Prove that for all $a, b \in \mathbb{Q}$, $f(a \cdot b) = f(a) \cdot f(b)$. See the remark in the previous problem for how to define $f(a) \cdot f(b)$ in case $f(a)$ or $f(b)$ is nonpositive.

(iv) Prove that $a \in \mathbb{Q}^+ \iff f(a) \in \mathbb{R}^+$. We conclude that the replica of $\mathbb{Q}$ in $\mathbb{R}$ behaves, both algebraically and with respect to order, exactly like $\mathbb{Q}$. For this reason, we *identify* $\mathbb{Q}$ and the subset of real numbers $\{a^*\,;\ a \in \mathbb{Q}\}$.

13. (**Multiplicative inverses of cuts**)

(a) For every positive real number a, prove that

$$a^{-1} := \{x \in \mathbb{Q}\,;\ x \le 0\} \cup \{x \in \mathbb{Q}\,;\ a < (x^{-1})^*,\ x > 0\}$$

is a real number. (For a negative, we define $a^{-1} := -(-a)^{-1}$.) Suggestion: To prove that a^{-1} is bounded above, pick an arbitrary rational number $y \in a$ with $y > 0$ and prove that y^{-1} is an upper bound for a^{-1}.

(b) Prove that $a \cdot a^{-1} = 1^*$ as follows. First prove that $a \cdot a^{-1} \le 1^*$. To prove $1^* \le a \cdot a^{-1}$ is more complicated. Let $z \in \mathbb{Q}$ with $0 < z < 1$. We need to show that $z \in a \cdot a^{-1}$; to do so, proceed as follows.

(i) Let $z_1 \in \mathbb{Q}$ with $z < z_1 < 1$. Using an inequality property proved in Problem 10 and the distributive property, prove that $a \cdot z_1^* < a$.

(ii) Pick $x \in \mathbb{Q}$ with $a \cdot z_1^* < x^* < a$ (why does such an x exist?) and let $y = x^{-1}z$. Prove that $y \in a^{-1}$ and conclude that $z \in a \cdot a^{-1}$.

2.9   *m*-Dimensional Euclidean Space

The plane $\mathbb{R}^2$ is said to be two-dimensional, because to locate a point in the plane requires two real numbers, its ordered pair of coordinates. Similarly, we are all familiar with $\mathbb{R}^3$, which is said to be three-dimensional because to represent any point in space we need an ordered triple of real numbers. In this section we generalize these considerations to m-dimensional space $\mathbb{R}^m$ and study its properties.

2.9.1   *The Vector Space Structure of* $\mathbb{R}^m$

Recall that the set $\mathbb{R}^m$ is just the product $\mathbb{R}^m := \mathbb{R} \times \cdots \times \mathbb{R}$ (m copies of $\mathbb{R}$), or explicitly, the set of all m-tuples of real numbers,

$$\mathbb{R}^m := \{(x_1, \ldots, x_m)\,;\ x_1, \ldots, x_m \in \mathbb{R}\}.$$

We call elements of $\mathbb{R}^m$ **vectors** (or **points**), and we use the notation 0 for the m-tuple of zeros $(0, \ldots, 0)$ (m zeros); it will always be clear from context whether 0 refers to the real number zero or the m-tuple of zeros. In elementary calculus, we usually focus on the cases $m = 1$, $m = 2$, and $m = 3$, as shown in Fig. 2.25. Given $x = (x_1, \ldots, x_m)$ and $y = (y_1, \ldots, y_m)$ in $\mathbb{R}^m$ and a real number a, we define

$$x + y = (x_1 + y_1, \ldots, x_m + y_m) \quad \text{and} \quad ax = (ax_1, \ldots, ax_m).$$

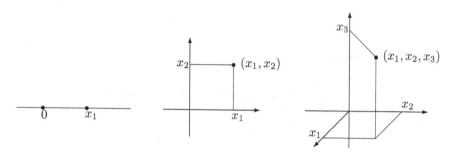

Fig. 2.25 $\mathbb{R}$, $\mathbb{R}^2$, and $\mathbb{R}^3$. The zero vector is the origin 0, $(0, 0)$, and $(0, 0, 0)$, respectively

We also define

$$-x = (-x_1, \ldots, -x_m).$$

With these definitions, observe that

$$x + y = (x_1 + y_1, \ldots, x_m + y_m) = (y_1 + x_1, \ldots, y_m + x_m) = y + x$$

and

$$x + 0 = (x_1 + 0, \ldots, x_m + 0) = (x_1, \ldots, x_m) = x,$$

and similarly, $0 + x = x$. These computations prove properties (A1) and (A3) below, and you can check that the following further properties of addition are satisfied: Addition satisfies

(A1) $x + y = y + x$ (commutative law);
(A2) $(x + y) + z = x + (y + z)$ (associative law);
(A3) there is an element 0 such that $x + 0 = x = 0 + x$ (additive identity);
(A4) for each x, there is $-x$ such that

$$x + (-x) = 0 \quad \text{and} \quad (-x) + x = 0 \quad \text{(additive inverse)}.$$

Of course, we usually write $x + (-y)$ as $x - y$.

Multiplication by real numbers satisfies

(M1) $1 \cdot x = x$ (multiplicative identity);
(M2) $(ab)x = a(bx)$ (associative law);

and finally, addition and multiplication are related by

(D) $a(x + y) = ax + ay$ and $(a + b)x = ax + bx$ (distributive law).

Every set with an operation of "+" and an operation of multiplication by real numbers that satisfies properties **(A1)**–**(A4)**, **(M1)**–**(M2)**, and **(D)**, is called a **real vector space**. The elements of the set are then called **vectors**. If the scalars $a, b, 1$ in **(M1)**–**(M2)** and **(D)** are elements of a field $\mathbb{F}$, then we say that the vector space is an $\mathbb{F}$ vector space or a vector space **over** $\mathbb{F}$. In particular, $\mathbb{R}^m$ is a real vector space.

2.9.2 Inner Products

We now review *inner products*, also called dot products in elementary calculus. We all probably know that given any two vectors $x = (x_1, x_2, x_3)$ and $y = (y_1, y_2, y_3)$ in $\mathbb{R}^3$, the dot product $x \cdot y$ is the number

$$x \cdot y = x_1 y_1 + x_2 y_2 + x_3 y_3.$$

We generalize this to $\mathbb{R}^m$ as follows: If $x = (x_1, \ldots, x_m)$ and $y = (y_1, \ldots, y_m)$, then we define the **inner product** (also called the **dot product** or **scalar product**) $\langle x, y \rangle$ as the real number

$$\langle x, y \rangle = x_1 y_1 + x_2 y_2 + \cdots + x_m y_m = \sum_{j=1}^{m} x_j y_j.$$

It is also common to denote $\langle x, y \rangle$ by $x \cdot y$ or (x, y), but we prefer the angle bracket notation $\langle x, y \rangle$, which is popular in physics, because the dot "$\cdot$" can be confused with multiplication, and the parentheses "$(\ ,\ )$" can be confused with ordered pair.

In the following theorem we summarize some of the main properties of $\langle \cdot, \cdot \rangle$.

Theorem 2.45 *For all vectors x, y, z in $\mathbb{R}^m$ and every real number a,*

(i) $\langle x, x \rangle \geq 0$ and $\langle x, x \rangle = 0$ if and only if $x = 0$.
(ii) $\langle x + y, z \rangle = \langle x, z \rangle + \langle y, z \rangle$ and $\langle x, y + z \rangle = \langle x, y \rangle + \langle x, z \rangle$.
(iii) $\langle a\, x, y \rangle = a \langle x, y \rangle$ and $\langle x, a\, y \rangle = a \langle x, y \rangle$.
(iv) $\langle x, y \rangle = \langle y, x \rangle$.

Proof To prove *(i)*, just note that

$$\langle x, x \rangle = x_1^2 + x_2^2 + \cdots + x_m^2$$

and $x_j^2 \geq 0$ for each j. If $\langle x, x \rangle = 0$, then since the only way a sum of nonnegative numbers can be zero is that each number is zero, we must have $x_j^2 = 0$ for each j.

Hence, every x_j is equal to zero, and therefore, $x = (x_1, \ldots, x_m) = 0$. Conversely, if $x = 0$, that is, $x_1 = 0, \ldots, x_m = 0$, then of course $\langle x, x \rangle = 0$ too.

To prove *(ii)*, we just compute:

$$\langle x + y, z \rangle = \sum_{j=1}^{m} (x_j + y_j) z_j = \sum_{j=1}^{m} (x_j z_j + y_j z_j) = \langle x, z \rangle + \langle y, z \rangle.$$

The other identity $\langle x, y + z \rangle = \langle x, y \rangle + \langle x, z \rangle$ is proved similarly. The proofs of *(iii)* and *(iv)* are also simple computations, so we leave their proofs to the reader. ∎

We remark that every real vector space V with an operation that assigns to every two vectors x and y in V a real number $\langle x, y \rangle$ satisfying properties *(i)*–*(iv)* of Theorem 2.45 is called a **real inner product space**, and the operation $\langle \cdot, \cdot \rangle$ is called an **inner product** on V. In particular, $\mathbb{R}^m$ is a real inner product space.

2.9.3 The Norm in $\mathbb{R}^m$

Recall that the length of a vector $x = (x_1, x_2)$ in $\mathbb{R}^2$, which we denote by $|x|$, is just the distance of the point x from the origin, as shown in the picture:

We generalize this concept as follows. The **length** or **norm** of a vector $x = (x_1, \ldots, x_m)$ in $\mathbb{R}^m$ is by definition the nonnegative real number

$$\boxed{|x| = \sqrt{x_1^2 + \cdots + x_m^2} = \sqrt{\langle x, x \rangle}.}$$

We interpret the norm $|x|$ as the length of the vector x, or the distance of x from the origin 0. In particular, the squared length $|x|^2$ of the vector x is given by

$$|x|^2 = \langle x, x \rangle.$$

Alert: For $m > 1$, $|x|$ does not mean absolute value of a real *number* x; it means norm of a *vector* x. However, if $m = 1$, then "norm" and "absolute value" are the same, because for $x = x_1 \in \mathbb{R}^1 = \mathbb{R}$, the above definition of norm is $\sqrt{x_1^2}$, which is exactly the absolute value of x_1 according to Problem 7 on p. 102.

The following inequality relates the norm and the inner product. It is commonly called the **Schwarz inequality** or **Cauchy–Schwarz inequality** after Hermann Schwarz (1843–1921), who proved it for integrals in 1885, and Augustin Cauchy (1789–1857), who proved it for sums in 1821. However (see [101] for the history), it should be called the **Cauchy–Bunyakovsky–Schwarz inequality**, because Viktor

Bunyakovsky (1804–1889), a student of Cauchy, stated the inequality some 25 years before Schwarz. (Note: There is no "t" before the "z" in "Schwarz." The German mathematician Hermann Schwarz is not to be confused with the French mathematician Laurent-Moïse Schwartz (1915–2002).)

CBS inequality

Theorem 2.46 *For all vectors x, y in $\mathbb{R}^m$, we have*

$$|\langle x, y \rangle| \le |x|\,|y| \qquad \textbf{\textit{CBS inequality}}.$$

Proof If $y = 0$, then both sides of the CBS inequality are zero, so we henceforth assume that $y \ne 0$. Taking the squared length of the vector $x - \frac{\langle x, y \rangle}{|y|^2} y$, we get

$$
\begin{aligned}
0 \le \left| x - \frac{\langle x, y \rangle}{|y|^2} y \right|^2 &= \left\langle x - \frac{\langle x, y \rangle}{|y|^2} y, x - \frac{\langle x, y \rangle}{|y|^2} y \right\rangle \\
&= \langle x, x \rangle - \frac{\langle x, y \rangle}{|y|^2} \langle x, y \rangle - \frac{\langle x, y \rangle}{|y|^2} \langle y, x \rangle + \frac{\langle x, y \rangle \langle x, y \rangle}{|y|^4} \langle y, y \rangle \\
&= |x|^2 - \frac{\langle x, y \rangle^2}{|y|^2} - \frac{\langle x, y \rangle^2}{|y|^2} + \frac{\langle x, y \rangle^2}{|y|^2}.
\end{aligned}
$$

Canceling the last two terms, we see that

$$
0 \le |x|^2 - \frac{\langle x, y \rangle^2}{|y|^2}, \quad \text{or} \quad \langle x, y \rangle^2 \le |x|^2 |y|^2.
$$

Taking square roots proves the CBS inequality. As a side remark, the vector $x - \frac{\langle x, y \rangle}{|y|^2} y$ whose squared length we took didn't come out of a hat! You might recall (or you can look it up) from your "multivariable calculus" or "vector calculus" course that the projection of x onto y and the projection of x onto the orthogonal complement of y are given by $\frac{\langle x, y \rangle}{|y|^2} y$ and $x - \frac{\langle x, y \rangle}{|y|^2} y$, respectively, as seen here:

Thus, all we did above was take the squared length of the projection of x onto the orthogonal complement of y. ∎

In the following theorem we list some of the main properties of the norm $|\cdot|$.

> **Theorem 2.47** *For all vectors x, y in* $\mathbb{R}^m$ *and every real number a,*
>
> *(i)* $|x| \geq 0$ *and* $|x| = 0$ *if and only if* $x = 0$.
> *(ii)* $|a\,x| = |a|\,|x|$.
> *(iii)* $|x + y| \leq |x| + |y|$ *(**triangle inequality**)*.
> *(iv)* $\big||x| - |y|\big| \leq |x \pm y| \leq |x| + |y|$.

Proof Property *(i)* of Theorem 2.45 implies *(i)*. To prove *(ii)*, observe that

$$|ax| = \sqrt{\langle ax, ax \rangle} = \sqrt{a^2 \langle x, x \rangle} = |a| \sqrt{\langle x, x \rangle} = |a|\,|x|,$$

and therefore $|ax| = |a|\,|x|$. To prove the triangle inequality, we use the CBS inequality to get

$$
\begin{aligned}
|x + y|^2 = \langle x + y, x + y \rangle &= |x|^2 + \langle x, y \rangle + \langle y, x \rangle + |y|^2 \\
&= |x|^2 + 2\langle x, y \rangle + |y|^2 \\
&\leq |x|^2 + 2|x|\,|y| + |y|^2,
\end{aligned}
$$

where we used that $\langle x, y \rangle \leq |\langle x, y \rangle| \leq |x|\,|y|$, which follows from the CBS inequality. Thus,

$$|x + y|^2 \leq (|x| + |y|)^2.$$

Taking the square root of both sides proves the triangle inequality.

The second half of *(iv)* follows from the triangle inequality:

$$|x \pm y| = |x + (\pm 1)y| \leq |x| + |(\pm 1)y| = |x| + |y|.$$

To prove the first half, $\big||x| - |y|\big| \leq |x \pm y|$, we use the triangle inequality to get

$$|x| - |y| = |(x - y) + y| - |y| \leq |x - y| + |y| - |y| = |x - y|$$
$$\implies \quad |x| - |y| \leq |x - y|. \quad (2.30)$$

Switching the letters x and y in (2.30), we get $|y| - |x| \leq |y - x|$, or equivalently, $-(|x| - |y|) \leq |x - y|$. Combining this with (2.30), we see that

$$|x| - |y| \leq |x - y| \quad \text{and} \quad -(|x| - |y|) \leq |x - y|.$$

By the definition of absolute value, it follows that $\big||x| - |y|\big| \leq |x - y|$. Replacing y with $-y$ and using that $|-y| = |y|$, we get $\big||x| - |y|\big| \leq |x + y|$. This finishes the proof of *(iv)*. ∎

We remark that every real vector space V with an operation that assigns to every vector x in V a nonnegative real number $|x|$ such that $|\cdot|$ satisfies properties *(i)–(iii)* of Theorem 2.47 is called a **real normed space**, and the operation $|\cdot|$ is called a **norm** on V. In particular, $\mathbb{R}^m$ is a real normed space. Problem 7 in the exercises explores a different norm on $\mathbb{R}^m$.

In analogy with the distance between two real numbers, we define the **distance** between two vectors x and y in $\mathbb{R}^m$ to be the number

$$|x - y|.$$

In particular, the triangle inequality implies that given any other vector z, we have

$$|x - y| = |(x - z) + (z - y)| \leq |x - z| + |z - y|,$$

that is,

$$|x - y| \leq |x - z| + |z - y|. \tag{2.31}$$

If $m = 2$ and we plot x, y and z in the plane and then draw the triangle with vertices at x, y, z, then (2.31) says that the distance between the two points x and y is shorter than the distance traversed by going from x to z and then from z to y; see Fig. 2.26.

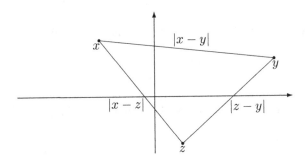

Fig. 2.26 Why the triangle inequality (2.31) is obvious

Finally, we remark that the norm $|\cdot|$ on $\mathbb{R}^m$ is sometimes called the **ball norm** on $\mathbb{R}^m$ for the following reason. Let $r > 0$, take for example $m = 3$, and let $x = (x_1, x_2, x_3) \in \mathbb{R}^3$. Then $|x| < r$ means that $\sqrt{x_1^2 + x_2^2 + x_3^2} < r$, or squaring both sides, we get

$$x_1^2 + x_2^2 + x_3^2 < r^2,$$

which simply says that x is inside the ball of radius r. So if $c \in \mathbb{R}^3$, then $|x - c| < r$ just means that

$$(x_1 - c_1)^2 + (x_2 - c_2)^2 + (x_3 - c_3)^2 < r^2,$$

which is to say that x is inside the ball of radius r that is centered at the point $c = (c_1, c_2, c_3)$:

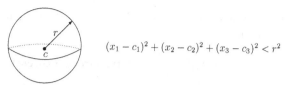

$$(x_1 - c_1)^2 + (x_2 - c_2)^2 + (x_3 - c_3)^2 < r^2$$

Generalizing this notion to m-dimensional space, given c in $\mathbb{R}^m$, we call the set of all x such that $|x - c| < r$, or after squaring both sides,

$$(x_1 - c_1)^2 + (x_2 - c_2)^2 + \cdots + (x_m - c_m)^2 < r^2,$$

the **open ball** of radius r centered at c. We denote this set by B_r, or $B_r(c)$ to emphasize that the center of the ball is c. In set notation, we have

$$\boxed{B_r(c) := \{x \in \mathbb{R}^m ; |x - c| < r\}.} \tag{2.32}$$

The set of x with $<$ replaced by $\leq$ is called the **closed ball** of radius r centered at c and is denoted by $\overline{B}_r$ or $\overline{B}_r(c)$,

$$\boxed{\overline{B}_r(c) := \{x \in \mathbb{R}^m ; |x - c| \leq r\}.}$$

If $m = 1$, then the ball concept reduces to intervals in $\mathbb{R}^1 = \mathbb{R}$:

$$x \in B_r(c) \iff |x - c| < r \iff -r < x - c < r$$
$$\iff c - r < x < c + r \iff x \in (c - r, c + r).$$

Thus,

$$\boxed{\text{for } m = 1, \text{ we have } B_r(c) = (c - r, c + r).}$$

Similarly, for $m = 1$, $\overline{B}_r(c)$ is just the closed interval $[c - r, c + r]$.

We end this section with the following very important inequality, which you should memorize:

$$\boxed{\text{For all } a, b \in \mathbb{R}, \quad ab \leq \frac{1}{2}(a^2 + b^2).} \tag{2.33}$$

To prove this inequality, just multiply out $(a - b)^2 \geq 0$. This inequality is used to give another proof of the CBS inequality in Problem 2, but it also has a ton of uses that you'll find as you do the problems in this book.

▶ **Exercises 2.9**

1. Let $x, y \in \mathbb{R}^m$. Prove that

$$|x + y|^2 + |x - y|^2 = 2|x|^2 + 2|y|^2 \quad \textbf{(parallelogram law)}.$$

Vectors x and y are said to be **orthogonal** if $\langle x, y \rangle = 0$. Prove that x and y are orthogonal if and only if

$$|x + y|^2 = |x|^2 + |y|^2 \quad \textbf{(Pythagorean theorem)}.$$

2. **(CBS inequality, Proof II)** Here's another way to prove the CBS inequality.

   (i) Let $x, y \in \mathbb{R}^m$ with $|x| = 1$ and $|y| = 1$. Using (2.33), prove that $|\langle x, y \rangle| \le 1$.
   (ii) Now let x and y be arbitrary nonzero vectors of $\mathbb{R}^m$. Applying (b) to the vectors $x/|x|$ and $y/|y|$, derive the CBS inequality.

3. **(CBS inequality, Proof III)** Here's an "algebraic" proof. Let $x, y \in \mathbb{R}^m$ with $y \neq 0$ and let $p(t) = |x + ty|^2$ for $t \in \mathbb{R}$. Note that $p(t) \ge 0$ for all t.

   (i) Show that $p(t)$ can be written in the form $p(t) = a t^2 + 2bt + c$, where a, b, c are real numbers with $a \neq 0$.
   (ii) Using the fact that $p(t) \ge 0$ for all t, prove the CBS inequality. Suggestion: Write $p(t) = a(t + b/a)^2 + (c - b^2/a)$.

4. Prove that for all vectors x and y in $\mathbb{R}^m$, we have

$$2|x|^2|y|^2 - 2\left(\sum_{n=1}^m x_n y_n\right)^2 = \sum_{k=1}^m \sum_{\ell=1}^m (x_k y_\ell - x_\ell y_k)^2 \quad \textbf{(Lagrange identity)},$$

after Joseph-Louis Lagrange (1736–1813). Suggestion: Observe that

$$2|x|^2|y|^2 - 2\left(\sum_{n=1}^m x_n y_n\right)^2 = 2\left(\sum_{k=1}^m x_k^2\right)\left(\sum_{\ell=1}^m y_\ell^2\right) - 2\left(\sum_{k=1}^m x_k y_k\right)\left(\sum_{\ell=1}^m x_\ell y_\ell\right)$$

$$= 2\sum_{k=1}^m \sum_{\ell=1}^m x_k^2 y_\ell^2 - 2\sum_{k=1}^m \sum_{\ell=1}^m x_k y_k x_\ell y_\ell.$$

Prove that $\sum_{k=1}^m \sum_{\ell=1}^m (x_k y_\ell - x_\ell y_k)^2$, when expanded out, has the same form.

5. **(CBS inequality, Proof IV, and equality in CBS)**

   (a) Prove the CBS inequality from Lagrange's identity.
   (b) Using Lagrange's identity, prove that equality holds in the CBS inequality (that is, $|\langle x, y \rangle| = |x| |y|$) if and only if x and y are collinear, which is to say that $x = 0$ or $y = cx$ for some $c \in \mathbb{R}$.
   (c) Instead of using Lagrange's identity, go back through the proof of Theorem 2.46 and from that argument, prove that equality holds in the CBS inequality if and only if x and y are collinear.
   (d) Now show that equality holds in the triangle inequality (that is, $|x + y| = |x| + |y|$) if and only if $x = 0$ or $y = cx$ for some $c \ge 0$.

6. (**Laws of trigonometry**) *In this problem we assume knowledge of the trigonometric functions; see Section 4.9 for a rigorous development of these functions.* By the CBS inequality, given any two nonzero vectors x, $y \in \mathbb{R}^m$, we have $\frac{|\langle x, y \rangle|}{|x| |y|} \leq 1$. In particular, there is a unique angle $\theta \in [0, \pi]$ such that $\cos \theta = \frac{\langle x, y \rangle}{|x| |y|}$. The number θ is by definition the **angle** between the vectors x, y.

(a) Consider the triangle seen here:

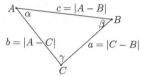

Prove the following three **laws of cosines**:

$$a^2 = b^2 + c^2 - 2bc \cos \alpha, \quad b^2 = a^2 + c^2 - 2ac \cos \beta, \quad c^2 = a^2 + b^2 - 2ab \cos \gamma.$$

Suggestion: To prove the last equality, observe that $c^2 = |A - B|^2 = |x - y|^2$, where $x = A - C$, $y = B - C$. Compute the dot product $|x - y|^2 = \langle x - y, x - y \rangle$.

(b) Using that $\sin^2 \alpha = 1 - \cos^2 \alpha$ and that $a^2 = b^2 + c^2 - 2bc \cos \alpha$, prove that

$$\frac{\sin^2 \alpha}{a^2} = 4 \frac{s(s - a)(s - b)(s - c)}{a^2 b^2 c^2}, \tag{2.34}$$

where $s := (a + b + c)/2$ is called the **semiperimeter**. From (2.34), conclude that

$$\frac{\sin \alpha}{a} = \frac{\sin \beta}{b} = \frac{\sin \gamma}{c} \quad \text{(law of sines)}.$$

(c) Assume the following formula: Area of a triangle $= \frac{1}{2}$ base $\times$ height. Use this formula together with (2.34) to prove that the area of the triangle above is given by

$$\text{Area} = \sqrt{s(s - a)(s - b)(s - c)} \quad \text{(\textbf{Heron's formula})},$$

a formula named after Heron of Alexandria (c. 10–c. 75).

7. For $x = (x_1, x_2, \ldots, x_m)$ in $\mathbb{R}^m$, define

$$\|x\|_\infty = \max \{|x_1|, |x_2|, \ldots, |x_m|\},$$

called the **sup (or supremum) norm**. Of course, we we need to show that this is a norm.

(a) Show that $\| \cdot \|_\infty$ defines a norm on $\mathbb{R}^m$, that is, $\| \cdot \|_\infty$ satisfies properties (i)–(iii) of Theorem 2.47.

(b) In $\mathbb{R}^2$, what is the set of all points $x = (x_1, x_2)$ such that $\|x\|_\infty \leq 1$? Draw a picture of this set. Do you see why $\|\cdot\|_\infty$ is often called the **box norm**?

(c) Show that for every x in $\mathbb{R}^m$, $\|x\|_\infty \leq |x| \leq \sqrt{m}\,\|x\|_\infty$, where $|x|$ denotes the usual *ball norm* of x.

(d) Let $r > 0$ and let $\overline{B}_r$ denote the closed ball in $\mathbb{R}^m$ of radius r centered at the origin (the set of x in $\mathbb{R}^m$ such that $|x| \leq r$). Let $\overline{\text{Box}}_r$ denote the closed ball in $\mathbb{R}^m$ of radius r in the sup norm centered at the origin (the set of x in $\mathbb{R}^m$ such that $\|x\|_\infty \leq r$). Prove that $\overline{B}_1 \subseteq \overline{\text{Box}}_1 \subseteq \overline{B}_{\sqrt{m}}$. When $m = 2$, give a "proof by picture" of these set inequalities by drawing the three sets $\overline{B}_1$, $\overline{\text{Box}}_1$, and $\overline{B}_{\sqrt{2}}$.

2.10   The Complex Number System

Imagine a world in which we could not solve the equation $x^2 - 2 = 0$. This is an exclusively rational numbers world. Such a world is a world in which the length of the diagonal of a unit square would not make sense; a very poor world indeed! Imagine now a world in which every time we tried to solve a quadratic equation such as $x^2 + 1 = 0$, we got "stuck" and could not proceed further. The complex number system (introducing so-called "imaginary numbers") alleviates this inconvenience to mathematics and in fact also to science, since it turns out that complex numbers are *necessary* for describing nature.[14]

Remark: This book is considered a *real* analysis book, so you may wonder why *complex* numbers are here. The answer is that real analysis is not just about real numbers! Real analysis is more the study of limit processes, including sequences, series, continuity, and so forth, which traditionally were developed for analyzing real numbers and real-valued functions of a real variable; hence the name "real analysis." These "real variable techniques" apply not only to real numbers, but also to complex numbers and even vectors, so these topics fit naturally in a real analysis course. "Complex analysis," on the other hand, is not just about complex numbers! It more refers to the notions surrounding the calculus of complex-valued functions of a complex variable, in particular to complex-differentiable (so-called "holomorphic") functions.

---

[14] "Furthermore, the use of complex numbers is in this case not a calculational trick of applied mathematics but comes close to being a necessity in the formulation of the laws of quantum mechanics …It is difficult to avoid the impression that a miracle confronts us here." Nobel laureate Eugene Wigner (1902–1995) responding to the "miraculous" appearance of complex numbers in the formulation of quantum mechanics. [173, p. 208], [262, 263].

2.10.1   Definition of Complex Numbers

In high school we learned that a complex, or "imaginary," number is a "number of the form $a + b\,i$," where a and b are real numbers and i is the "imaginary unit" satisfying $i^2 = -1$. We even manipulated complex numbers assuming the usual laws of arithmetic, such as addition,

$$(a + b\,i) + (c + d\,i) = (a + c) + (b + d)\,i, \tag{2.35}$$

and multiplication,

$$\begin{aligned} (a + b\,i) \cdot (c + d\,i) &= ac + ad\,i + bc\,i + bd\,i^2 \\ &= ac + ad\,i + bc\,i - bd \\ &= (ac - bd) + (ad + bc)\,i. \end{aligned} \tag{2.36}$$

We also found that if $a + b\,i \neq 0$, then

$$\frac{1}{a + b\,i} = \frac{a - b\,i}{(a + b\,i)(a - b\,i)} = \frac{a - b\,i}{a^2 + b^2}. \tag{2.37}$$

If complex numbers are "imaginary," we certainly manipulated them as if they really existed! We even drew them in the "complex plane" as if they really existed:

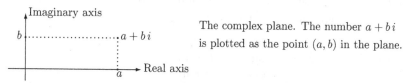

The complex plane. The number $a + bi$ is plotted as the point (a, b) in the plane.

Of course, complex numbers are not imaginary at all, and we shall define them now. The idea is that the so-called "complex plane" looks the same as $\mathbb{R}^2$, the real two-dimensional plane, which is already defined! So why not *define* complex numbers as elements of $\mathbb{R}^2$? With this in mind, we define the **complex number system** $\mathbb{C}$ as the set $\mathbb{R}^2$ together with the following arithmetic rules. Let $z, w \in \mathbb{C}$, which means $z = (a, b)$ and $w = (c, d)$ for real numbers a, b, c, d. Since z and w are vectors in $\mathbb{R}^2$, from Section 2.9 we already know how to add z and w:

$$\boxed{z + w = (a + c, b + d),}$$

which is the ordered pair version of the identity (2.35). The new ingredient is multiplication: Led by (2.36), we define

$$\boxed{z \cdot w = (ac - bd,\ ad + bc)\,;} \tag{2.38}$$

we sometimes drop the dot and just write zw. We define $-z = (-a, -b)$, and we write 0 for $(0, 0)$. Finally, if $z = (a, b) \neq 0$ (that is, $a \neq 0$ and $b \neq 0$), then led by (2.37), we define

$$\boxed{z^{-1} = \left(\frac{a}{a^2 + b^2}, \frac{-b}{a^2 + b^2} \right).}$$
(2.39)

We can then define division as

$$\frac{w}{z} := w \cdot z^{-1}.$$

In summary, $\mathbb{C}$ as a set is just the real vector space $\mathbb{R}^2$ with the extra operation of multiplication (and division by nonzero elements).

> **Theorem 2.48** *The set of complex numbers is a field with $(0, 0)$ (denoted henceforth by 0) and $(1, 0)$ (denoted henceforth by 1) the additive and multiplicative identities, respectively.*

Proof If $z, w, u \in \mathbb{C}$, then we need to show that addition satisfies

   (A1) $z + w = w + z$ (commutative law);
   (A2) $(z + w) + u = z + (w + u)$ (associative law);
   (A3) $z + 0 = z = 0 + z$ (additive identity);
   (A4) $z + (-z) = 0$ and $(-z) + z = 0$ (additive inverse).

We also need to show that multiplication satisfies

   (M1) $z \cdot w = w \cdot z$ (commutative law);
   (M2) $(z \cdot w) \cdot u = z \cdot (w \cdot u)$ (associative law);
   (M3) $1 \cdot z = z = z \cdot 1$ (multiplicative identity);
   (M4) for $z \neq 0$, we have $z \cdot z^{-1} = 1$ and $z^{-1} \cdot z = 1$ (multiplicative inverse).

Finally, we need to prove that multiplication and addition are related by

(D) $z \cdot (w + u) = (z \cdot w) + (z \cdot u)$ (distributive law).

The proofs of all these properties are very easy and merely involve using the definition of addition and multiplication, so we leave all the proofs to the reader, except for **(M4)**. Here, by definition of multiplication,

$$\begin{aligned}
z \cdot z^{-1} &= (a, b) \cdot \left(\frac{a}{a^2 + b^2}, \frac{-b}{a^2 + b^2} \right) \\
&= \left(a \cdot \frac{a}{a^2 + b^2} - b \cdot \frac{-b}{a^2 + b^2}, \ a \cdot \frac{-b}{a^2 + b^2} + b \cdot \frac{a}{a^2 + b^2} \right) \\
&= \left(\frac{a^2}{a^2 + b^2} + \frac{b^2}{a^2 + b^2}, 0 \right) = (1, 0) = 1.
\end{aligned}$$

Similarly, $z^{-1} \cdot z = 1$, and **(M4)** is proven.   ∎

2.10.2 Reminiscences of High School

We now explain how to make certain identifications so that our definition of complex numbers looks exactly like what we learned in high school. First, we consider $\mathbb{R}$ a subset of $\mathbb{C}$ by the *identification* of a real number a with the ordered pair $(a, 0)$; in other words, for the sake of notational convenience, we do not make a distinction between the *complex* number $(a, 0)$ and the *real* number a. Observe that by definition of addition of complex numbers,

$$(a, 0) + (b, 0) = (a + b, 0).$$

Under our identification, the left-hand side is "$a + b$," and the right-hand side is also "$a + b$." By definition (2.38) of complex multiplication, we have

$$(a, 0) \cdot (b, 0) = (a \cdot b - 0 \cdot 0, \ a \cdot 0 + 0 \cdot b) = (ab, 0),$$

which is to say, "$a \cdot b = ab$" under our identification. Thus, our identification of $\mathbb{R}$ preserves the arithmetic operations of $\mathbb{R}$. We shall henceforth consider $\mathbb{R}$ a subset of $\mathbb{C}$ and write complex numbers of the form $(a, 0)$ as just a.

We define the complex number i, notation introduced in 1777 by Euler [183], as the complex number

$$\boxed{i := (0, 1).}$$

There is certainly nothing "imaginary" about the ordered pair $(0, 1)$! If seeing is believing, here's a picture of this "imaginary" i:

Using the definition (2.38) of complex multiplication, we have

$$i^2 = i \cdot i = (0, 1) \cdot (0, 1)$$
$$= (0 \cdot 0 - 1 \cdot 1, \ 0 \cdot 1 + 1 \cdot 0) = (-1, 0),$$

which is to say that $i^2 = -1$ under our identification of $(-1, 0)$ with -1. Thus, the complex number $i = (0, 1)$ is the "imaginary unit" from years past; however, our definition of i avoids the mysterious obscurity usually associated with it in high school.[15] Let $z = (a, b)$ be a complex number. Then by the definition of complex

---

[15]"That this subject [imaginary numbers] has hitherto been surrounded by mysterious obscurity, is to be attributed largely to an ill adapted notation. If, for example, $+1$, -1, and the square root of -1 had been called direct, inverse and lateral units, instead of positive, negative and imaginary (or even impossible), such an obscurity would have been out of the question." Carl Friedrich Gauss (1777–1855).

addition and multiplication, the definition of i, and our identification of $\mathbb{R}$ as a subset of $\mathbb{C}$, we see that

$$a + bi = a + (b, 0) \cdot (0, 1) = a + (b \cdot 0 - 0 \cdot 1, b \cdot 1 + 0 \cdot 0)$$
$$= (a, 0) + (0, b)$$
$$= (a, b) = z.$$

Thus, $z = a + bi$. By commutativity, we also have $z = a + ib$. In conclusion, we can write complex numbers just as we did in high school! However, now we know that i is not "imaginary" but is genuinely defined (as $(0, 1)$). We call a the **real part** of z, and b the **imaginary part** of z, and we denote them by $a = \operatorname{Re} z$ and $b = \operatorname{Im} z$, so that

$$\boxed{z = \operatorname{Re} z + i \operatorname{Im} z.}$$

From this point on, we shall typically use the notation $z = a + bi = a + ib$ instead of $z = (a, b)$ for complex numbers.

Before moving on, we remark that the definition (2.38) of multiplication, however weird it may look, is very concrete viewed geometrically. Consider, for instance, multiplication by i. If $z = a + ib$, then

$$i \cdot z = i \cdot (a + ib) = -b + ia.$$

Here's a picture of $i \cdot z$:

   $i \cdot z = -b + ia$ has horizontal component $-b$
                        and vertical component a

Notice that $i \cdot z$ is just $z = a + ib$ rotated counterclockwise by $90°$. (Think about rotating the triangle that z forms by $90°$.) Thus, *multiplication by i acts as rotation by $90°$ counterclockwise*! In particular, the "mysterious" identity $i \cdot i = -1$ just says that if we rotate the north-pointing unit vector $i = (0, 1)$ by $90°$ counterclockwise, we get the west-pointing unit vector -1 (that is, $(-1, 0)$):

Multiplication by general complex numbers acts by rotation together with a scaling; see p. 336. See [147, Chapter 2] for more on the geometry of complex numbers.

2.10.3 Absolute Values and Complex Conjugates

We define the **absolute value** (or **length**, or **modulus**) of a complex number $z = (a, b)$ as the usual length (or norm) of (a, b):

$$\boxed{|z| := |(a, b)| = \sqrt{a^2 + b^2}.}$$

Thus, $|z|^2 = a^2 + b^2$. We define the **complex conjugate** of $z = (a, b)$ as the complex number $\overline{z} = (a, -b)$, that is, $\overline{z} = a - bi$. Note that if $|z| \neq 0$, then according to the definition (2.39) of z^{-1}, we have

$$z^{-1} = \frac{\overline{z}}{|z|^2},$$

so the inverse of a complex number can be expressed in terms of the complex conjugate and absolute value. In the next theorem we list other properties of the complex conjugate.

Theorem 2.49 *If z and w are complex numbers, then*

(1) $\overline{\overline{z}} = z$;
(2) $\overline{z + w} = \overline{z} + \overline{w}$ *and* $\overline{zw} = \overline{z} \cdot \overline{w}$;
(3) $z\overline{z} = |z|^2$;
(4) $z + \overline{z} = 2\operatorname{Re} z$ *and* $z - \overline{z} = 2i \operatorname{Im} z$.

Proof The proofs of all these properties are very easy and merely involve using the definition of complex conjugation, so we'll prove only *(3)*, leaving the rest for Problem 3. We have

$$z\overline{z} = (a + bi)(a - bi) = a^2 + a(-bi) + (bi)a + (bi)(-bi)$$
$$= a^2 - abi + abi - b^2 \cdot i^2 = a^2 - b^2 \cdot (-1) = a^2 + b^2 = |z|^2. \quad \blacksquare$$

In the final theorem of this section we list various properties of absolute value. Note that Properties *(1)* and *(5)* follow from the properties of the norm on $\mathbb{R}^2$. The remaining properties are left for Problem 3.

Theorem 2.50 *For all complex numbers z, w, we have*

(1) $|z| \geq 0$ *and* $|z| = 0$ *if and only if* $z = 0$;
(2) $|\bar{z}| = |z|$;
(3) $|\operatorname{Re} z| \leq |z|$;
(4) $|z\,w| = |z|\,|w|$;
(5) $|z + w| \leq |z| + |w|$ *(**triangle inequality**).*

An induction argument shows that for every collection of n complex numbers $z_1, \ldots, z_n$,

$$|z_1\, z_2 \cdots z_n| = |z_1|\,|z_2| \cdots |z_n|.$$

In particular, setting $z_1 = z_2 = \cdots = z_n = z$, we see that $|z^n| = |z|^n$.

▶ **Exercises 2.10**

1. Show that $z \in \mathbb{C}$ is a real number if and only if $\bar{z} = z$.
2. If w is a complex root of a polynomial $p(z) = z^n + a_{n-1}\, z^{n-1} + \cdots + a_1 z + a_0$ with real coefficients (that is, $p(w) = 0$ and each a_k is real), prove that $\bar{w}$ is also a root.
3. Prove properties *(1)*, *(2)*, and *(4)* of Theorem 2.49 and properties *(2)*, *(3)*, and *(4)* of Theorem 2.50.
4. If $z \in \mathbb{C}$, prove that there exist a nonnegative real number r and a complex number ω with $|\omega| = 1$ such that $z = r\,\omega$. If z is nonzero, show that r and ω are uniquely determined by z, that is, if $z = r'\,\omega'$, where $r' \geq 0$ and $|\omega'| = 1$, then $r' = r$ and $\omega' = \omega$. The decomposition $z = r\,\omega$ is called the **polar decomposition** of z. (On p. 335 in Section 4.9 we relate the polar decomposition to the trigonometric functions.)

2.11 Cardinality and "Most" Real Numbers Are Irrational

In Section 2.6, we saw that in a sense (concerning roots, trig functions, logarithms—objects of practical interest), there appear to be vastly more irrational numbers than rationals. This suggests the following question[16] How much more? In this section we discuss Georg Cantor's (1845–1918) discovery that the rational numbers have, in some sense, the same number of elements as the natural numbers do! It turns out that the irrational numbers (and the entire set of real numbers) have many more elements and are impossible to count; thus they are said to be uncountable. We also discuss algebraic and transcendental numbers and their countability properties.

---

[16]"In mathematics the art of proposing a question must be held of higher value than solving it." (A thesis defended at Cantor's doctoral examination.) Georg Cantor (1845–1918).

2.11.1 Cardinality

Cardinality is a mathematical concept that extends the idea of "number of elements" from finite sets to infinite ones. Formally, two sets A and B are said to have the same **cardinality** if there is a bijection between the two sets. Of course, if $f : A \longrightarrow B$ is a bijection, then $g = f^{-1} : B \longrightarrow A$ is a bijection, so the notion of cardinality does not depend on "which way the bijection goes." We think of A and B as having the same number of elements, since the bijection sets up a one-to-one correspondence between elements of the two sets. A set A is said to be **finite** if it is empty, in which case we say that A has zero elements, or if it has the same cardinality as a set of the form

$$\mathbb{N}_n := \{1, 2, \ldots, n\}$$

for some natural number n, in which case we say that A has n **elements**. If A is not finite, it is said to be **infinite**.[17] A set is called **countable** if it has the same cardinality as a finite set or the set of natural numbers. To distinguish between finite and infinite countable sets, we call a set **countably infinite** if it has the cardinality of the natural numbers. Finally, a set is **uncountable** if it is not countable; such a set is not finite and does not have the cardinality of $\mathbb{N}$. See Fig. 2.27 for relationships between finite and infinite sets. If A is countable and f is a bijection from either $\mathbb{N}$ or some $\mathbb{N}_n$ onto A, then A can be expressed as a list:

$$A = \{a_1, a_2, a_3, \ldots\},$$

where $a_1 = f(1), a_2 = f(2), \ldots$ Thus, countable sets are sets whose elements can be sequenced.

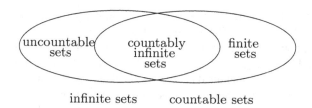

Fig. 2.27 Infinite sets are uncountable or countably infinite, and countable sets are countably infinite or finite. Infinite sets and countable sets intersect in the countably infinite sets

Example 2.33 $\mathbb{Z}$ is countably infinite, since the function $f : \mathbb{Z} \longrightarrow \mathbb{N}$ defined by

$$f(n) = \begin{cases} 2n & \text{if } n > 0 \\ 2|n| + 1 & \text{if } n \leq 0, \end{cases}$$

---

[17]"Even in the realm of things which do not claim actuality, and do not even claim possibility, there exist beyond dispute sets which are infinite." Bernard Bolzano (1781–1848).

is a bijection of $\mathbb{Z}$ onto $\mathbb{N}$. Notice that f labels the integers with the natural numbers in a back-and-forth fashion, as seen here:

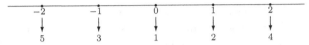

If two sets A and B have the same cardinality, we write $\text{card}(A) = \text{card}(B)$. One can check that if $\text{card}(A) = \text{card}(B)$ and $\text{card}(B) = \text{card}(C)$, then $\text{card}(A) = \text{card}(C)$. Thus, cardinality satisfies a "transitive law."

Example 2.34 As a consequence of this "transitive law," any two countably infinite sets have the same cardinality.

It is "obvious" that a set cannot have both n elements and m elements where $n \neq m$, but this still needs proof! The proof is based on the **pigeonhole principle** (Fig. 2.28).

More pigeons than holes!

Fig. 2.28 If $m > n$ and m pigeons are put into n pigeonholes, then at least two pigeons must be put into the same pigeonhole

Pigeonhole principle

Theorem 2.51 *For $m > n$, there is no injection from $\mathbb{N}_m$ to $\mathbb{N}_n$.*

Proof We proceed by induction on n. Let $m > 1$ and $f : \mathbb{N}_m \longrightarrow \{1\}$ be a function. Then $f(m) = f(1) = 1$, so f is not an injection.

Assume that our theorem is true for n; we shall prove that it's true for $n + 1$. Let $m > n + 1$ and let $f : \mathbb{N}_m \longrightarrow \mathbb{N}_{n+1}$. We shall prove that f is not an injection. First of all, if the range of f is contained in $\mathbb{N}_n \subseteq \mathbb{N}_{n+1}$, then we can consider f a function into $\mathbb{N}_n$, and hence by the induction hypothesis, f is not an injection. So assume that $f(a) = n + 1$ for some $a \in \mathbb{N}_m$. If there is another element of $\mathbb{N}_m$ whose image is $n + 1$, then f is not injection, so we may furthermore assume that a is the only element of $\mathbb{N}_m$ whose image is $n + 1$. Then $f(k) \in \mathbb{N}_n$ for $k \neq a$, so we can define a function $g : \mathbb{N}_{m-1} \longrightarrow \mathbb{N}_n$ by "skipping" $f(a) = n + 1$:

$$g(1) := f(1), \ g(2) := f(2), \ldots, g(a - 1) := f(a - 1), \ g(a) := f(a + 1),$$
$$g(a + 1) := f(a + 2), \ldots, g(m - 1) := f(m).$$

Since $m > n + 1$, we have $m - 1 > n$, so by the induction hypothesis, g is not an injection. The definition of g shows that $f : \mathbb{N}_m \longrightarrow \mathbb{N}_{n+1}$ cannot be an injection either, which completes the proof of our theorem. ∎

We now prove that the number of elements of a finite set is unique. We also prove the "obvious" fact that an countably infinite set is not finite.

Theorem 2.52 *The number of elements of a finite set is unique, and countably infinite sets are not finite.*

Proof Suppose, to get a contradiction, that for some finite set A, there are bijections $f : A \longrightarrow \mathbb{N}_n$ and $g : A \longrightarrow \mathbb{N}_m$ for some natural numbers $m > n$. (We may omit the case that A is empty.) Then

$$f \circ g^{-1} : \mathbb{N}_m \longrightarrow \mathbb{N}_n$$

is a bijection, in particular an injection, contradicting the pigeonhole principle.

Now suppose that for some set A, there are bijections $f : A \longrightarrow \mathbb{N}_n$ and $g : A \longrightarrow \mathbb{N}$. Then,

$$f \circ g^{-1} : \mathbb{N} \longrightarrow \mathbb{N}_n$$

is a bijection, so an injection, and so in particular, its restriction to $\mathbb{N}_{n+1} \subseteq \mathbb{N}$ is an injection. This again is impossible by the pigeonhole principle. ∎

2.11.2 Basic Results on Countability

The following is intuitively obvious.

Lemma 2.53 *A subset of a countable set is countable.*

Proof Let A be a nonempty subset of a countable set B, where for definiteness we assume that B is countably infinite. (The finite case is easy.) Let $f : \mathbb{N} \longrightarrow B$ be a bijection. Using the well-ordering principle (see p. 35), we can define

$$n_1 = \text{smallest element of } \{n \in \mathbb{N}; \ f(n) \in A\}.$$

If $A \neq \{f(n_1)\}$, then via well-ordering, we can define

$$n_2 = \text{smallest element of } \{n \in \mathbb{N} \backslash \{n_1\}; \ f(n) \in A \}.$$

Note that $n_1 < n_2$ (why?). If $A \neq \{f(n_1), f(n_2)\}$, then we can define

$$n_3 = \text{smallest element of } \{n \in \mathbb{N} \backslash \{n_1, n_2\}; \ f(n) \in A \}.$$

Then $n_1 < n_2 < n_3$. We can continue this process by induction, defining n_{k+1} as the smallest element in the set $\{n \in \mathbb{N} \backslash \{n_1, \ldots, n_k\}; \ f(n) \in A\}$ as long as $A \neq \{f(n_1), \ldots, f(n_k)\}$.

There are two possibilities: the above process terminates, or it continues indefinitely. If the process terminates, let n_m be the last natural number that can be defined in this process. Then $A = \{f(n_1), \ldots, f(n_m)\}$, a finite set. Suppose now that the above process continues indefinitely. Then the above recursive procedure produces an infinite sequence of increasing natural numbers $n_1 < n_2 < n_3 < n_4 < \cdots$. One can check (for instance, by induction) that $k < n_{k+1}$ for all k. We claim that the map $h : \mathbb{N} \longrightarrow A$ defined by $h(k) := f(n_k)$ is a bijection, which shows that A is countably infinite. It is certainly injective, because f is. To see that h is surjective, let $a \in A$. Then, because f is surjective, there is $\ell \in \mathbb{N}$ such that $f(\ell) = a$. We claim that $\ell \in \{n_1, \ldots, n_\ell\}$. Indeed, if not, then $\ell \in \{n \in \mathbb{N} \backslash \{n_1, n_2, \ldots n_\ell\}; \ f(n) \in A\}$, so by definition of $n_{\ell+1}$,

$$n_{\ell+1} = \text{smallest element of } \{n \in \mathbb{N} \backslash \{n_1, n_2, \ldots n_\ell\}; \ f(n) \in A \} \leq \ell.$$

However, this contradicts the fact that $k < n_{k+1}$ for all k. Hence, $\ell = n_j$ for some j, so $h(j) = f(n_j) = f(\ell) = a$. This proves that h is surjective. ∎

Theorem 2.54 *A finite product of countable sets is countable, and a countable union of countable sets is countable.*

Proof We consider the product of only two countably infinite sets (the other cases are left to the reader). The countability of the product of more than two countable sets can be handled by induction. If A and B are countably infinite, then it follows that $\text{card}(A \times B) = \text{card}(\mathbb{N} \times \mathbb{N})$, so it suffices to show that $\text{card}(\mathbb{N} \times \mathbb{N}) = \text{card}(\mathbb{N})$. Let $C \subseteq \mathbb{N}$ consist of all natural numbers of the form $2^n \, 3^m$, where $n, m \in \mathbb{N}$. Being an infinite subset of $\mathbb{N}$, it follows, by our lemma, that C is countably infinite. Consider the function $f : \mathbb{N} \times \mathbb{N} \longrightarrow C$ defined by

$$f(n, m) = 2^n \, 3^m.$$

By unique factorization, f is one-to-one (hence a bijection), so $\mathbb{N} \times \mathbb{N}$, like C, is countably infinite. See Problem 1 for other proofs that $\mathbb{N} \times \mathbb{N}$ is countably infinite.

Consider a set $A = A_1 \cup A_2 \cup A_3 \cup \ldots$, that is, a countable union of countable sets $A_1, A_2, A_3, \ldots$; we shall prove that A is countable. Since each A_n is countable, we can list the (distinct) elements of A_n:

$$A_n = \{a_{n1}, a_{n2}, a_{n3}, \ldots\}.$$

We define a function $g : A \to \mathbb{N} \times \mathbb{N}$ as follows. Given $a \in A = A_1 \cup A_2 \cup A_3 \cup \ldots$, let n be the *smallest* natural number such that $a \in A_n$; then $a = a_{nm}$ for a unique

m, and we define $g(a) = (n, m)$. This defines $g : A \longrightarrow \mathbb{N} \times \mathbb{N}$, which is one-to-one, as the reader can check. So, A has the same cardinality as the subset $g(A)$ of the countable set $\mathbb{N} \times \mathbb{N}$. Since subsets of countable sets are countable, it follows that A has the same cardinality as a countable set, so A is countable.   ∎

Example 2.35 (Cf. Example 2.33) As an easy application of this theorem, we observe that $\mathbb{Z} = \mathbb{N} \cup \{0\} \cup (-\mathbb{N})$, and since each set on the right is countable, their union $\mathbb{Z}$ is also countable.

We leave the proof of the following lemma as Problem 3.

Lemma 2.55 *Every infinite set has a countably infinite subset.*

The following theorem says that countable sets "don't count" in the sense that they do not add anything to cardinalities of infinite sets.

Countable sets don't count

Theorem 2.56 *For every infinite set A and countable set B, the sets A and $A \cup B$ have the same cardinality.*

Proof Let A be an infinite set and let B be countable. Here's a picture to think of:

The sets A and B are drawn as ellipses intersecting in a set that we'll call D. We can write $A = C \cup D$, where C and $D \subseteq B$ are disjoint (explicitly, $C = A \backslash B$ and $D = A \cap B$). Since a subset of a countable set is countable, D is countable. We find a bijection between A and $A \cup B$ in two cases.

Case 1: C is finite. Since $A = C \cup D$ by assumption is infinite, and D is countable, it follows that D must be countably infinite. Since a union of countable sets is countable, A is countably infinite. Therefore, $A \cup B$ is countably infinite as well. Thus, there is a bijection between A and $A \cup B$.

Case 2: C is infinite. By our lemma, there is a subset $E \subseteq C$ that is countably infinite. Let $F = C \backslash E$; here's a picture to think of:

We have
$$A = E \cup F \cup D \quad \text{and} \quad A \cup B = E \cup F \cup B.$$

Since unions of countable sets are countable, the sets $E \cup D$ and $E \cup B$ are countable; they are both countably infinite, since E is infinite. Thus, there is a bijection between

$E \cup D$ and $E \cup B$. The identity map is a bijection from F onto F, so there is a bijection between A and $A \cup B$. This completes the proof.   ■

2.11.3 Real, Rational, and Irrational Numbers

Sets that have the same cardinality as $\mathbb{N}$, that is, countably infinite sets, are said to have cardinality $\aleph_0$ "ahh-lef null"; for example, we proved that $\mathbb{Z}$ has cardinality $\aleph_0$. It may be surprising to learn that the rationals also have cardinality $\aleph_0$.

Theorem 2.57 $\mathbb{Q}$ *has cardinality* $\aleph_0$.

Proof Let $A = \{(m, n) \in \mathbb{Z} \times \mathbb{N}; \ m \text{ and } n \text{ have no common factors}\}$. Since a product of countable sets is countable, $\mathbb{Z} \times \mathbb{N}$ is countable, and since A is a subset of a countable set, A is countable. Moreover, A is infinite, since, for example, all numbers of the form $(m, 1)$ belong to A, where $m \in \mathbb{Z}$. Thus, A has cardinality $\aleph_0$. Define $f : A \longrightarrow \mathbb{Q}$ by

$$f(m, n) = \frac{m}{n}, \quad \text{for } (m, n) \in A.$$

One can show that f is a bijection, so $\operatorname{card}(\mathbb{Q}) = \operatorname{card}(A) = \operatorname{card}(\mathbb{N})$.   ■

Sets that have the same cardinality as $\mathbb{R}$ are said to have cardinality $\mathfrak{c}$, where "$\mathfrak{c}$" is for "continuum" (since $\mathbb{R}$ looks like a continuous line). We call an interval **nontrivial** if it neither is empty nor consists of a single point.

Theorem 2.58 *Every nontrivial interval has cardinality* $\mathfrak{c}$, *and the set of irrational numbers in a nontrivial interval also has cardinality* $\mathfrak{c}$.

Proof There are many types of intervals, so to make the proof short, we shall focus on intervals of the form (a, b) and $[a, b]$, where a and b are real, $a < b$, leaving the other cases for your enjoyment.

Case 1: Consider (a, b). Define $f : (-1, 1) \longrightarrow (a, b)$ by

$$f(x) = \frac{b - a}{2}(x + 1) + a.$$

Then f is a bijection. The function $g : \mathbb{R} \longrightarrow (-1, 1)$ defined by

$$g(x) = \frac{x}{\sqrt{1 + x^2}}$$

is also a bijection. (The inverse is $g^{-1} : (-1, 1) \longrightarrow \mathbb{R}$ given by $g^{-1}(y) = y/\sqrt{1 - y^2}$.) Thus, $f \circ g : \mathbb{R} \longrightarrow (a, b)$ is a bijection. Hence, $\text{card}(a, b) = \mathfrak{c}$. Let A and B be the collections of irrational and rational numbers in (a, b), respectively. Since subsets of countable sets are countable, B is countable, so, since "countable sets don't count" (Theorem 2.56), A and $A \cup B = (a, b)$ have the same cardinality.

Case 2: Consider $[a, b]$. Since $[a, b] = (a, b) \cup \{a, b\}$ and $\{a, b\}$ is countable, and "countable sets don't count," it follows that $\text{card}[a, b] = \text{card}(a, b) = \mathfrak{c}$. If A' is the collection of irrational numbers in $[a, b]$ and B' is the collection of rationals in $[a, b]$, then B' is countable, being a subset of a countable set. Since "countable sets don't count," A' and $A' \cup B' = [a, b]$ have the same cardinality.   ∎

An obvious question is whether $\mathfrak{c} = \aleph_0$. In other words, is $\mathbb{R}$ countably infinite? Cantor in fact proved that $\mathbb{R}$ is uncountable, so $\mathbb{R}$ has so many elements that we can't sequentially list them all. The following proof is close to, but not exactly, Cantor's original proof (see [94] for a nice exposition on his original proof.) He gave another proof of this result, which we present in Section 3.8 on p. 233.

Cantor's first proof

Theorem 2.59 $\mathfrak{c} \neq \aleph_0$.

Proof Suppose, for the sake of contradiction, that we can list the reals: $\mathbb{R} = \{c_1, c_2, \dots\}$. Take any interval $I_1 = [a_1, b_1]$, where $a_1 < b_1$, that does not contain c_1. Now let $I_2 = [a_2, b_2] \subseteq I_1$, where $a_2 < b_2$, be an interval that does not contain c_2 (Fig. 2.29).

Fig. 2.29 Divide the interval I_1 into thirds. At least one of the three subintervals does not contain c_2; call that subinterval $[a_2, b_2]$

By induction, we construct a sequence of nested closed and bounded intervals $I_n = [a_n, b_n]$ that do not contain c_n. By the nested intervals theorem on p. 100, there is a point c in every I_n. By construction, I_n does not contain c_n, so c cannot equal any c_n, which contradicts that $\{c_1, c_2, \dots\}$ is a list of all the real numbers.   ∎

In particular, by Theorem 2.58, every nontrivial interval and the set of irrational numbers in every nontrivial interval are uncountable.

So far, all infinite subsets of $\mathbb{R}$ we've looked at ($\mathbb{Z}$, $\mathbb{Q}$, nontrivial intervals, irrationals, etc.) have cardinality $\aleph_0$ or $\mathfrak{c}$. Are there infinite subsets of $\mathbb{R}$ with other cardinalities? Cantor conjectured that the answer is no, which became known as the "continuum hypothesis": Every infinite subset of $\mathbb{R}$ must have cardinality $\aleph_0$ or $\mathfrak{c}$. Cantor could not prove his conjecture, and in fact, many years later, Kurt Gödel (1906–1978) and Paul Cohen (1934–2007) proved that the continuum hypothesis is undecidable—cannot be proved or disproved—in the standard axioms of mathematics. See [58, 80] for more details on this fascinating story.

2.11.4   Roots of Polynomials

We already know that the real numbers can be classified into two disjoint sets, the rationals and irrationals. There is another important classification, into algebraic and transcendental numbers. These numbers have to do with roots of polynomials, so we begin by discussing some elementary properties of polynomials. With an eye toward later applications, we shall consider complex polynomials in this subsection, noting that Lemma 2.60 and Theorem 2.61 below are also valid for real polynomials (polynomials with real coefficients).

Let $n \geq 1$ and let

$$p(z) = a_n z^n + a_{n-1} z^{n-1} + \cdots + a_2 z^2 + a_1 z + a_0, \quad a_n \neq 0, \tag{2.40}$$

be an nth-degree polynomial with *complex* coefficients (meaning that each a_k is in $\mathbb{C}$).

Lemma 2.60   *For every $z, a \in \mathbb{C}$, we can write*

$$p(z) - p(a) = (z - a)\, q(z),$$

where $q(z)$ is a polynomial of degree $n - 1$.

Proof First of all, observe that given a polynomial $f(z)$ and a complex number b, the "shifted" function $f(z + b)$ is also a polynomial in z of the same degree as f; this can be easily proved using the formula (2.40) for a polynomial. In particular, $P(z) = p(z + a) - p(a)$ is a polynomial of degree n and hence can written in the form

$$P(z) = b_n z^n + b_{n-1} z^{n-1} + \cdots + b_2 z^2 + b_1 z + b_0,$$

where $b_n \neq 0$ (in fact, $b_n = a_n$, but this isn't needed). Notice that $P(0) = p(a) - p(a) = 0$. It follows that $b_0 = 0$, so

$$P(z) = z\, Q(z) \quad , \quad \text{where} \quad Q(z) = b_n z^{n-1} + b_{n-1} z^{n-2} + \cdots + b_2 z + b_1$$

is a polynomial of degree $n - 1$. Replacing z with $z - a$, we obtain

$$p(z) - p(a) = P(z - a) = (z - a)\, q(z),$$

where $q(z) = Q(z - a)$ is a polynomial of degree $n - 1$.   ∎

Suppose that $a \in \mathbb{C}$ is a **root** of $p(z)$, which means that $p(a) = 0$. Then according to our lemma, we can write $p(z) = (z - a)q(z)$, where q is a polynomial of degree $n - 1$. If $q(a) = 0$, then again by our lemma, we can write $q(z) = (z - a)r(z)$, where

$r(z)$ is a polynomial of degree $n - 2$. Thus, $p(z) = (z - a)^2 r(z)$. Continuing this process, which must stop by at least the nth step (because the degree of a polynomial cannot be negative), we can write

$$p(z) = (z - a)^k s(z),$$

where $s(z)$ is a polynomial of degree $n - k$ and $s(a) \neq 0$. Because of the factor $(z - a)^k$, we say that a is a root of $p(z)$ of **multiplicity** k.

Theorem 2.61 *Every nth-degree complex polynomial (see the expression (2.40)) has at most n complex roots counting multiplicities.*

Proof The proof is by induction. Certainly this theorem holds for polynomials of degree 1 (if $p(z) = a_1 z + a_0$ with $a_1 \neq 0$, then $p(z) = 0$ if and only if $z = -a_0/a_1$). Suppose that this theorem holds for polynomials of degree n. Let p be a polynomial of degree $n + 1$. If p has no roots, then this theorem holds for p, so suppose that p has a root, call it a. Then by our lemma, we can write

$$p(z) = (z - a)q(z),$$

where q is a polynomial of degree n. Note that the roots of p are $z = a$ and the roots of q. Since by induction, q has at most n roots counting multiplicities, it follows that the polynomial p has at most $n + 1$ such roots. ∎

As a consequence of the fundamental theorem of algebra, see p. 344 in Section 4.10, every polynomial of degree n has exactly n complex roots counting multiplicities.

2.11.5 Uncountability of Transcendental Numbers

We now return to real numbers. We already know that a rational number is a real number that can be written as a ratio of integers, and a number is irrational, by definition, if it is not rational. An important class of numbers that generalizes rational numbers is called the algebraic numbers. To motivate this generalization, let $r = a/b$, where $a, b \in \mathbb{Z}$ with $b \neq 0$, be a rational number. Then r is a root of the polynomial equation

$$bx - a = 0.$$

Therefore, every rational number is the root of a (linear, or degree-1) polynomial with integer coefficients. In general, an **algebraic number** is a real number that is a root of some polynomial with *integer* coefficients. A real number is called **transcendental** if it is not algebraic. (These numbers are transcendental because, as remarked by

Euler, they surpass, that is "transcend," the powers of ordinary algebra to solve for them [67].) As demonstrated above, we already know that every rational number is algebraic. But there are many more algebraic numbers.

Example 2.36 The numbers $\sqrt{2}$ and $\sqrt[3]{5}$ are both algebraic, being roots of the polynomials

$$x^2 - 2 \quad \text{and} \quad x^3 - 5,$$

respectively. On the other hand, the numbers e and π are examples of transcendental numbers; for proofs see [146, 174, 175].

The numbers $\sqrt{2}$ and $\sqrt[3]{5}$ are irrational, so there are irrational numbers that are algebraic. Thus, the algebraic numbers include all rational numbers and many irrational numbers as well, namely those irrational numbers that are roots of polynomials with integer coefficients. The following (seemingly counterintuitive) result was discovered by Cantor [94] (we leave the proof to Problem 4 below).

Cardinality of algebraics and transcendentals

Theorem 2.62 *The set of all algebraic numbers is countable, and the set of all transcendental numbers is uncountable.*

▶ **Exercises 2.11**

1. Here are some countability proofs.

(a) Prove that the set of prime numbers is countably infinite.

(b) Let $\mathbb{N}_0 = \{0, 1, 2, \dots\}$. Show that $\mathbb{N}_0$ is countably infinite. Define $f : \mathbb{N}_0 \times \mathbb{N}_0 \longrightarrow \mathbb{N}_0$ by $f(0, 0) = 0$ and for $(m, n) \neq (0, 0)$, define

$$f(m, n) = \left(1 + 2 + 3 + \cdots + (m + n)\right) + n = \frac{1}{2}(m + n)(m + n + 1) + n.$$

Prove that f is a bijection. Suggestion: Define $g : \mathbb{N}_0 \longrightarrow \mathbb{N}_0 \times \mathbb{N}_0$ as follows. If $k \in \mathbb{N}_0$, choose $\ell \in \mathbb{N}$ so that $\ell(\ell - 1)/2 \leq k < \ell(\ell + 1)/2$ and put

$$g(k) = \left(\frac{\ell(\ell + 1)}{2} - k - 1 \,,\; k - \frac{\ell(\ell - 1)}{2}\right).$$

Prove that f and g are inverse functions. **Remark:** The function f counts $\mathbb{N}_0 \times \mathbb{N}_0$, as shown here:

$$
\begin{array}{llll}
\vdots & \vdots & \vdots & \\
(0,2) & (1,2) & (2,2) & \cdots \\
(0,1) & (1,1) & (2,1) & \cdots \\
(0,0) & (1,0) & (2,0) & \cdots
\end{array}
$$

    (c) Write $\mathbb{Q}$ as a countable union of countable sets, giving thereby another proof that the rational numbers are countable.

    (d) Prove that $f : \mathbb{N} \times \mathbb{N} \longrightarrow \mathbb{N}$ defined by $f(m, n) = 2^{m-1}(2n - 1)$ is a bijection; this gives another proof that $\mathbb{N} \times \mathbb{N}$ is countable.

2. Here are some formulas for polynomials in terms of roots.

    (a) If $c_1, \ldots, c_k$ are roots of a polynomial $p(z)$ of degree n (with each root repeated according to multiplicity), prove that $p(z) = (z - c_1)(z - c_2) \cdots (z - c_k) q(z)$, for some polynomial $q(z)$ of degree $n - k$.

    (b) If $k = n$, prove that $p(z) = a_n(z - c_1)(z - c_2) \cdots (z - c_n)$, where a_n is the coefficient of z^n in the formula (2.40) for $p(z)$.

3. If A is an infinite set, prove that it has a countably infinite subset. Suggestion: Define $f : \mathbb{N} \to A$ as follows. First pick a point a_1 from A and define $f(1) = a_1$. Assume that $f(1), \ldots, f(n)$ have been defined, then pick a point a_{n+1} from $A \setminus \{f(1), \ldots, f(n)\}$, then define $f(n + 1) = a_{n+1}$. The desired countably infinite subset of A is the range $f(\mathbb{N})$.

4. In this problem we prove Theorem 2.62.

    (i) For a nonconstant polynomial with integer coefficients $a_n x^n + a_{n-1} x^{n-1} + \cdots + a_1 x + a_0$ (nonconstant meaning $n \geq 1$ and $a_n \neq 0$), we define its **index** as the natural number

$$n + |a_n| + |a_{n-1}| + |a_{n-2}| + \cdots + |a_2| + |a_1| + |a_0|.$$

    Given $k \in \mathbb{N}, k \geq 2$, prove there are at most finitely many nonconstant polynomials with integer coefficients having index k.

    (ii) For $k \in \mathbb{N}, k \geq 2$, let A_k be the set of all roots (algebraic numbers) of nonconstant polynomials with integer coefficients of index k. Prove that A_k is a finite set and the set of all algebraic numbers is the union $\bigcup_{k=2}^{\infty} A_k$.

    (iii) Complete the proof of the theorem.

5. Let X be a set and denote the set of all functions from X to $\{0, 1\}$ by $\{0, 1\}^X$. Define a map from the power set of X to $\{0, 1\}^X$ by

$$f : \mathscr{P}(X) \longrightarrow \{0, 1\}^X, \qquad X \supseteq A \longmapsto f(A) = \chi_A,$$

where χ_A is the characteristic function of A. Prove that f is a bijection. Conclude that $\mathscr{P}(X)$ has the same cardinality as $\{0, 1\}^X$.

6. Suppose that $\mathrm{card}(X) = n \in \mathbb{N}$. Prove that $\mathrm{card}(\mathscr{P}(X)) = 2^n$. Suggestion: There are many proofs you can come up with; here's one using the previous prob-

lem. Assuming that $X = \{0, 1, \ldots, n - 1\}$, which we may (why?), we just have to prove that $\mathrm{card}(\{0, 1\}^X) = 2^n$. To prove this, define $F : \{0, 1\}^X \longrightarrow \{0, 1, 2, \ldots, 2^n - 1\}$ as follows: If $f : X \longrightarrow \{0, 1\}$ is a function, then denoting $f(k)$ by a_k, define

$$F(f) = a_{n-1} 2^{n-1} + a_{n-2} 2^{n-2} + \cdots + a_1 2^1 + a_0.$$

Prove that F is a bijection. Suggestion: Review some material about binary representations from Section 2.5.

7. (**Cantor's theorem**) This theorem is simple to prove yet profound in nature.

   (a) Prove that there can never be a surjection of a set A onto its power set $\mathscr{P}(A)$. (This is called Cantor's theorem.) In particular, $\mathrm{card}(A) \neq \mathrm{card}(\mathscr{P}(A))$. Suggestion: Suppose not and let f be such a surjection. Consider the set

   $$B = \{a \in A \,;\, a \notin f(a)\} \subseteq A.$$

   Derive a contradiction from the assumption that f is surjective. Cantor's theorem shows that by taking power sets, one can always get bigger and bigger sets.

   (b) Prove that the set of all subsets of $\mathbb{N}$ is uncountable.

   (c) From Cantor's theorem and Problem 5, prove that the set of all sequences of 0's and 1's is uncountable. Here, a **sequence** is just function from $\mathbb{N}$ to $\{0, 1\}$, which can also be thought of as a list $(a_1, a_2, a_3, a_4, \ldots)$ where each a_k is either 0 or 1.

8. (**Vredenduin's paradox** [252]) Here is another paradox related to Russell's paradox. Assume that $A = \{\{a\} \,;\, a \text{ is a set}\}$ is a set. Let $B \subseteq A$ be the subset consisting of all sets of the form $\{a\}$, where $a \in \mathscr{P}(A)$. Define

   $$g : \mathscr{P}(A) \longrightarrow B \quad \text{by} \quad g(V) = \{V\} \text{ for all } V \in \mathscr{P}(A).$$

   Show that g is a bijection and then derive a contradiction to Cantor's theorem. This shows that A is not a set.

9. We define a **statement** as a finite string of symbols, say found in a word processing program (we regard a space as a symbol and we assume that there are finitely many symbols). For example, *The formula $e^{i\pi} + 1 = 0$ is beautiful!* is a statement.

   (a) Let A be the set of all statements. What's the cardinality of A?

   (b) Is the set of all possible mathematical proofs countable? Why?

Chapter 3
Infinite Sequences of Real and Complex Numbers

Notable enough, however, are the controversies over the series $1 - 1 + 1 - 1 + 1 - \cdots$,
whose sum was given by Leibniz as 1/2, although others disagree. ... Understanding of this
question is to be sought in the word "sum"; this idea, if thus conceived—namely, the sum of
a series is said to be that quantity to which it is brought closer as more terms of the series
are taken—has relevance only for convergent series, and we should in general give up the
idea of sum for divergent series.
Leonhard Euler (1707–1783).

Analysis is often described as the study of *infinite processes*, of which the study of
sequences and series forms the backbone. It is dealing with the concept of *infinite* in
infinite processes that makes analysis technically challenging. In fact, the subject of
sequences is when real analysis becomes "really hard."

Let us consider the following infinite series that Euler mentioned:

$$s = 1 - 1 + 1 - 1 + 1 - 1 + 1 - 1 + \cdots.$$

Let's manipulate this infinite series without being too careful. First, we notice that

$$s = (1 - 1) + (1 - 1) + (1 - 1) + \cdots = 0 + 0 + 0 + \cdots = 0,$$

so $s = 0$. On the other hand,

$$s = 1 - (1 - 1) - (1 - 1) - (1 - 1) - \cdots = 1 - 0 - 0 - 0 - \cdots = 1,$$

so $s = 1$. Finally, we can get Leibniz's value of $1/2$ as follows:

$$
\begin{aligned}
2s = 2 - 2 + 2 - 2 + \cdots &= 1 + 1 - 1 - 1 + 1 + 1 - 1 - 1 + \cdots \\
&= 1 + (1 - 1) - (1 - 1) + (1 - 1) - (1 - 1) + \cdots \\
&= 1 + 0 - 0 + 0 - 0 + \cdots = 1,
\end{aligned}
$$

© Paul Loya 2017
P. Loya, *Amazing and Aesthetic Aspects of Analysis*,
https://doi.org/10.1007/978-1-4939-6795-7_3

so $s = 1/2$. This example shows us that we need to be careful in dealing with the infinite. In the pages that follow, we "tame the infinite" with rigorous definitions.

Another highlight of this chapter is our study of the number e (Euler's number), which you have seen countless times in calculus and which pops up everywhere including economics (compound interest), population growth, radioactive decay, and probability. We shall prove two of the most famous formulas for this number:

$$e = 1 + \frac{1}{1!} + \frac{1}{2!} + \frac{1}{3!} + \frac{1}{4!} + \cdots \qquad \text{and} \qquad e = \lim_{n \to \infty} \left(1 + \frac{1}{n}\right)^n.$$

See [55, 158] for more on this incredible and versatile number. Another number we'll look at is the golden ratio $\Phi = \frac{1+\sqrt{5}}{2}$, which has strikingly pretty formulas

$$\Phi = \sqrt{1 + \sqrt{1 + \sqrt{1 + \sqrt{1 + \cdots}}}} = 1 + \cfrac{1}{1 + \cfrac{1}{1 + \cfrac{1}{1 + \cfrac{1}{1 + \ddots}}}}.$$

In Section 3.1 we begin our study of infinite processes by learning about sequences and their limits; then in Section 3.2 we discuss the properties of sequences. Sections 3.3 and 3.4 are devoted to answering the question of when a given sequence converges; in these sections we'll also derive the above formulas for Φ. Next, in Section 3.5, we study infinite series, which really constitute a special case of the study of infinite sequences. The exponential function, called by many "the most important function in mathematics" [205, p. 1], is our subject of study in Section 3.7. This function is defined by

$$\exp(z) = \sum_{n=0}^{\infty} \frac{z^n}{n!}, \qquad z \in \mathbb{C}.$$

We shall derive a few of the exponential function's many properties, including its relationship to Euler's number e. As a bonus prize, in Section 3.7 we'll also prove that e is irrational, and we look at a useful (but little publicized) theorem called *Tannery's theorem*, which is a very handy result that we'll use a lot in subsequent sections. Finally, in Section 3.8 we see how real numbers can be represented as decimals (with respect to arbitrary bases), and we look at Cantor's famous "constructive" diagonal argument.

CHAPTER 3 OBJECTIVES: THE STUDENT WILL BE ABLE TO . . .

- Apply the rigorous ε-N definition of convergence for sequences and series.
- Decide when a sequence is monotone, Cauchy, or has a convergent subsequence (Bolzano–Weierstrass), and when a series converges (absolutely).

- Define the exponential function and the number e.
- Explain Cantor's diagonal argument.

3.1 Convergence and ε-N Arguments for Limits of Sequences

Undeniably, the most important concept in all of undergraduate analysis is the notion of convergence. Intuitively, if a sequence $\{a_n\}$ in $\mathbb{R}^m$ converges to an element a in $\mathbb{R}^m$, then a_n is "as close as we want" to a for every n "sufficiently large." In this section we make the terms in quotes rigorous, which introduces the first bona fide technical definition in this book: the ε-N definition of limit.

3.1.1 Definition of Convergence

A **sequence** in $\mathbb{R}^m$ can be thought of as a list

$$a_1, \ a_2, \ a_3, \ a_4, \ \ldots$$

of vectors, or points, a_n in $\mathbb{R}^m$. In the language of functions, a sequence is simply a function $f : \mathbb{N} \longrightarrow \mathbb{R}^m$, where we denote $f(n)$ by a_n. Usually a sequence is denoted by $\{a_n\}$ or by $\{a_n\}_{n=1}^{\infty}$. Of course, we are not restricted to $n \geq 1$, and we could just as well start at any integer, e.g., $\{a_n\}_{n=-5}^{\infty}$. For convenience, in most of our proofs we shall work with sequences starting at $n = 1$, although all the results we shall discuss work for sequences starting with any index.

Example 3.1 Some examples of real sequences (that is, sequences in $\mathbb{R}^1 = \mathbb{R}$) include[1]

$$3, \ 3.1, \ 3.14, \ 3.141, \ 3.1415, \ \ldots$$

and

$$1, \ \frac{1}{2}, \ \frac{1}{3}, \ \frac{1}{4}, \ \frac{1}{5}, \ \frac{1}{6}, \ \ldots, a_n = \frac{1}{n}, \ldots \ .$$

We are interested mostly in real and complex sequences. Here, by a complex sequence we simply mean a sequence in $\mathbb{R}^2$, where we are free to use the notation i for $(0, 1)$ and the field properties of complex numbers.

Example 3.2 The following sequence is a complex sequence:

$$i, \ i^2 = -1, \ i^3 = -i, \ i^4 = 1, \ \ldots, a_n = i^n, \ldots \ .$$

---

[1] We'll talk about decimal expansions of real numbers in Section 3.8 and π in Chapter 4.

This sequence goes counterclockwise around the unit circle:

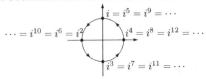

Although we shall focus on real and complex sequences in this book, later on you might deal with topology and calculus in $\mathbb{R}^m$ (as in, for instance, [146]), so for your later psychological health we might as well get used to working with $\mathbb{R}^m$ instead of $\mathbb{R}^1$.

We now try to motivate a precise definition of convergence. To begin, what does "converge" ("approach" or "getting closer and closer") mean in everyday life? For instance, what is a mathematical way to describe an airplane converging to an airport, as seen here:

One description of "convergence" is that given any radius of the airport, from some point on the airplane will be within that radius of the airport. Here's a picture:

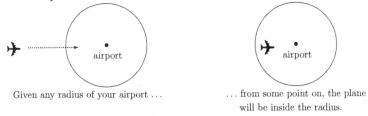

Given any radius of your airport ...   ... from some point on, the plane
                                                        will be inside the radius.

For example, eventually the airplane will be within 1 km of the airport, within 0.5 km of the airport, within 0.1 km of the airport, and so forth. No matter what radius of the airport you choose, eventually the airplane will be within that radius.

This idea works to describe convergence of sequences! Thus, without being too precise, a sequence $\{a_n\}$ in $\mathbb{R}^m$ converges to an element L in $\mathbb{R}^m$ if for any radius of the target L, from some point on, the a_n are within that radius of L, as shown in Fig. 3.1. Following tradition, we use the Greek symbol epsilon ε to denote a given radius. We now make convergence precise. First, when we say that the a_n "are within ε of L," we simply mean the distance between L and a_n is less than ε:

Fig. 3.1 Given any radius $\varepsilon > 0$ of L, from some point on, the a_n are within the radius ε of L

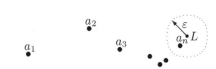

$$|a_n - L| < \varepsilon.$$

Second, when we say that some property holds "from some point on," we mean that the property holds for all $n > N$ for some (usually large) real number N. To summarize, for a_n to converge to L, we mean given any $\varepsilon > 0$ (a "radius of L"), there is a real number N such that

$$n > N \text{ ("from some point on")} \implies |a_n - L| < \varepsilon \text{ ("}a_n \text{ is within } \varepsilon \text{ of } L\text{").}$$

Note: "from some point on" is often reworded as "for n sufficiently large." We now conclude our findings with a precise definition: A sequence $\{a_n\}$ in $\mathbb{R}^m$ is said to **converge** (or **tend**) to an element L in $\mathbb{R}^m$ if for every $\varepsilon > 0$, there is an $N \in \mathbb{R}$ such that for all $n > N$, $|a_n - L| < \varepsilon$. Because this definition is so important, we display it:

A sequence $\{a_n\}$ in $\mathbb{R}^m$ converges to $L \in \mathbb{R}^m$ if for every $\varepsilon > 0$ there is an N such that
$$n > N \implies |a_n - L| < \varepsilon.$$

We call $\{a_n\}$ a **convergent sequence**, L the **limit**[2] of $\{a_n\}$, and we usually denote the fact that $\{a_n\}$ converges to L in one of four ways:

$$a_n \to L, \qquad a_n \to L \text{ as } n \to \infty, \qquad \lim a_n = L, \qquad \lim_{n \to \infty} a_n = L.$$

If a sequence does not converge (to any element of $\mathbb{R}^m$), we say that it **diverges**.

We can also state the definition of convergence in terms of open balls. Observe that $|a_n - L| < \varepsilon$ is just saying that $a_n \in B_\varepsilon(L)$, the open ball of radius ε centered at L (see formula (2.32) on p. 125). Therefore, $a_n \to L$ in $\mathbb{R}^m$ if

---

for every $\varepsilon > 0$, there is an $N \in \mathbb{R}$ such that $n > N \implies a_n \in B_\varepsilon(L)$.

This open ball form of limit occurs in many subjects, such as metric spaces in topology. In the case $m = 1$, the ε-ball around L is just the interval $(L - \varepsilon, L + \varepsilon)$ (see our discussion on p. 125 at the end of Section 2.9). Therefore,

---

A real sequence $\{a_n\}$ converges to $L \in \mathbb{R}$ if and only if, for every $\varepsilon > 0$ there is an N such that
$$n > N \implies L - \varepsilon < a_n < L + \varepsilon.$$

Figure 3.2 shows one way of thinking about limits of real sequences.

---

[2] "One magnitude is said to be the limit of another magnitude when the second may approach the first within any given magnitude, however small, though the second may never exceed the magnitude it approaches." Jean d'Alembert (1717–1783). The article on "Limite" in the Encyclopédie 1754.

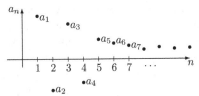

Fig. 3.2 Plot the points a_n on the real line. Then $a_n \to L$ if and only if for every $\varepsilon > 0$, if we draw the interval $(L - \varepsilon, L + \varepsilon)$ about L, then "from some point on," the a_n are inside the interval

Another common way of thinking about limits of real sequences is the *graph method*, in which we first plot the sequence in the xy-plane by plotting $x = n$ versus $y = a_n$:

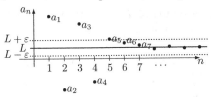

Then, $a_n \to L$ if and only if for every $\varepsilon > 0$, if we draw the interval $(L - \varepsilon, L + \varepsilon)$ about L on the y-axis, then "from some point on," the a_n are inside the interval:

3.1.2 Standard Examples of ε-N Arguments

We now give some standard examples of using our ε-N definition of limit.

Example 3.3 We shall prove that the sequence $\{1/2, 1/3, 1/4, \dots\}$ converges to zero:

$$\lim \frac{1}{n+1} = 0.$$

This is made "obvious" by plotting the points in this sequence:

In general, every sequence $\{a_n\}$ that converges to zero is called a **null sequence**. Thus, we claim that $\{1/(n + 1)\}$ is a null sequence. Let $\varepsilon > 0$ be a positive real number. Our goal: To prove that there exists a real number N such that

$$n > N \quad \Longrightarrow \quad \left| \frac{1}{n+1} - 0 \right| = \frac{1}{n+1} < \varepsilon. \tag{3.1}$$

To find such a number N, we can proceed in many ways. Here are two.

(I) Our first method is direct: We observe that[3]

$$\frac{1}{n+1} < \varepsilon \quad \Longleftrightarrow \quad \frac{1}{\varepsilon} < n+1 \quad \Longleftrightarrow \quad \frac{1}{\varepsilon} - 1 < n. \qquad (3.2)$$

For this reason, let us choose N to be the real number $N = 1/\varepsilon - 1$. Let $n > N$, that is, $N < n$, or using the definition of N, $1/\varepsilon - 1 < n$. Then by (3.2), we have $1/(n+1) < \varepsilon$. In summary, for $n > N$, we have proved that $1/(n+1) < \varepsilon$. This proves (3.1) with $N = 1/\varepsilon - 1$. Thus, by the definition of convergence, $1/(n+1) \to 0$.

(II) Another technique is to try to simplify $1/(n+1)$ so that it's easy to work with. Since $n < n+1$, we have $1/(n+1) < 1/n$, so

$$\text{for every } n \in \mathbb{N}, \quad \frac{1}{n+1} < \frac{1}{n}. \qquad (3.3)$$

Thus, if we can make $1/n < \varepsilon$, we get $1/(n+1) < \varepsilon$ for free. Now, it's easy to make $1/n < \varepsilon$, for

$$\frac{1}{n} < \varepsilon \quad \Longleftrightarrow \quad \frac{1}{\varepsilon} < n.$$

With this preliminary work done, let us now choose $N = 1/\varepsilon$. Let $n > N$, which is to say, $1/\varepsilon < n$. Then we certainly have $1/n < \varepsilon$, and hence by (3.3), we know that $1/(n+1) < \varepsilon$, too. In summary, for $n > N$, we have proved that $1/(n+1) < \varepsilon$. This proves (3.1) with $N = 1/\varepsilon$.

Note that in **(I)** and **(II)**, we found different N's (namely $N = 1/\varepsilon - 1$ in **(I)** and $N = 1/\varepsilon$ in **(II)**), but this doesn't matter, because to prove (3.1), we just need to show that such an N *exists*; there is never a unique N that works, since if one N works, every larger N will also work. We remark that a similar argument shows that the sequence $\{1/n\}$ is also a null sequence: $\lim \frac{1}{n} = 0$.

Here is a recipe for attacking most limit problems:

Four-step recipe on how to prove $a_n \to L$:
(1) Always start your proof with "Let $\varepsilon > 0$."
(2) "Massage" (make simpler by manipulating) $|a_n - L|$ so that

$$|a_n - L| \le \alpha_n$$

where α_n is a simple expression.
(3) Show that there is N such that $n > N \implies \alpha_n < \varepsilon$.
(4) Conclude that $n > N \implies |a_n - L| < \varepsilon$.

---

[3]Note that we are *not* claiming that $1/(n+1) < \varepsilon$ is true for every n. We are just writing down the *statement* $1/(n+1) < \varepsilon$ and statements equivalent to it. The point is to discover an equivalent statement of the form $n >$ some real number. This real number is an N making (3.1) true.

In Part (**I**) of Example 3.3 above, α_n was $1/(n+1)$ (which equaled $|a_n - L|$ in the example), while in Part (**II**), we chose $\alpha_n = 1/n$ (which is a little simpler than $|a_n - L|$). In general, there are many different α_n that work. Here's a harder example to consider, in which we find several α_n.

Example 3.4 Let's prove that[4]

$$\lim \frac{2n^2 - n}{n^2 - 9} = 2.$$

For the sequence $a_n = (2n^2 - n)/(n^2 - 9)$, we take the indices to be $n = 4, 5, 6, \ldots$ (since for $n = 3$, the quotient is undefined). Here's a picture of this sequence (Fig. 3.3):

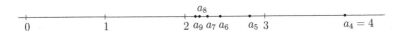

Fig. 3.3 We have $a_4 = 4$, $a_5 \approx 2.81$, $a_6 \approx 2.44$, $a_7 = 2.275$, $a_8 \approx 2.18$ and $a_9 = 2.125$. The points a_n seem to be getting closer to 2

Let $\varepsilon > 0$ be given. We want to prove that there exists a real number N such that the following statement holds:

$$n > N \quad \Longrightarrow \quad \left| \frac{2n^2 - n}{n^2 - 9} - 2 \right| < \varepsilon. \tag{3.4}$$

We now "massage" (simplify) the absolute value on the right as much as we can. For instance, we first can combine fractions:

$$\left| \frac{2n^2 - n}{n^2 - 9} - 2 \right| = \left| \frac{2n^2 - n}{n^2 - 9} - \frac{2n^2 - 18}{n^2 - 9} \right| = \left| \frac{18 - n}{n^2 - 9} \right|. \tag{3.5}$$

Before we do any more massaging, we note the following simple

---
Fraction fact: A fraction a/b of positive numbers can be made larger by increasing the numerator a or decreasing the denominator b (while still keeping the denominator positive).

Now going back to (3.5), we make the numerator of (3.5) *larger* using the triangle inequality: for $n = 4, 5, \ldots$, we have

---
[4]How did we know that the limit is 2? The trick to finding the limit of the quotient of polynomials in n of the same degree is to divide the leading coefficients. Here, the leading coefficients of $2n^2 - n$ and $n^2 - 9$ are 2 and 1, respectively, so the limit is $2/1 = 2$.

$$\left| \frac{18-n}{n^2-9} \right| \leq \frac{18+n}{n^2-9}.$$

Second, just for icing on the cake, let us make the top of the right-hand fraction a little simpler by observing that $18 \leq 18n$, so we conclude that

$$\frac{18+n}{n^2-9} \leq \frac{18n+n}{n^2-9} = \frac{19n}{n^2-9}.$$

In conclusion, we have "massaged" our expression to the following inequality:

$$\left| \frac{2n^2-n}{n^2-9} - 2 \right| \leq \frac{19n}{n^2-9}.$$

We now work with the denominator n^2-9 and make it *smaller*; this will make $19n/(n^2-9)$ bigger and hopefully simpler, and then we can make the resulting simple expression $< \varepsilon$ as desired. Here are three slightly different ways to do so.

(I) For our first method to make n^2-9 smaller, instead of subtracting 9 from n^2, we subtract the bigger number $9n$ from n^2. It follows that

$$\text{for } n > 9, \qquad \frac{19n}{n^2-9} \leq \frac{19n}{n^2-9n} = \frac{19n}{n(n-9)} = \frac{19}{n-9}.$$

We put $n > 9$ to ensure that n^2-9 is positive. Thus,

$$\text{for } n > 9, \qquad \left| \frac{2n^2-n}{n^2-9} - 2 \right| \leq \frac{19}{n-9}, \qquad (3.6)$$

and the right-hand side is simple! Now, for $n > 9$,

$$\frac{19}{n-9} < \varepsilon \iff \frac{19}{\varepsilon} < n-9 \iff 9+\frac{19}{\varepsilon} < n. \qquad (3.7)$$

Thus, let us pick $N = 9 + 19/\varepsilon$. We'll prove that this N works for (3.4). Let $n > N$, or the same thing, $n > 9 + 19/\varepsilon$ (which implies, in particular, that $n > 9$). Then,

$$\left| \frac{2n^2-n}{n^2-9} - 2 \right| \overset{\text{by (3.6)}}{\leq} \frac{19}{n-9} \overset{\text{by (3.7)}}{<} \varepsilon,$$

and we're done.

(II) For our second method to make n^2-9 smaller, we factor:

$$n^2-9 = (n+3)(n-3) \geq n(n-3).$$

Hence,

$$\frac{19n}{n^2 - 9} \leq \frac{19n}{n(n-3)} = \frac{19}{n-3},$$

and we get

$$\left| \frac{2n^2 - n}{n^2 - 9} - 2 \right| \leq \frac{19}{n-3}. \tag{3.8}$$

Now,

$$\frac{19}{n-3} < \varepsilon \iff 3 + \frac{19}{\varepsilon} < n. \tag{3.9}$$

Thus, let us pick $N = 3 + 19/\varepsilon$. We'll prove that this N works for (3.4). Let $n > N$, or the same thing, $n > 3 + 19/\varepsilon$. Then,

$$\left| \frac{2n^2 - n}{n^2 - 9} - 2 \right| \overset{\text{by (3.8)}}{\leq} \frac{19}{n-3} \overset{\text{by (3.9)}}{<} \varepsilon,$$

and we're done.

(III) For our third method to make $n^2 - 9$ smaller, observe that $n^2/2 > 9$ for $n > 4$. Thus, instead of subtracting 9 from n^2, for $n > 4$ we subtract the bigger number $n^2/2$ from n^2. It follows that

$$\text{for } n > 4, \quad \frac{19n}{n^2 - 9} \leq \frac{19n}{n^2 - n^2/2} = \frac{19n}{n^2/2} = \frac{38}{n}.$$

Thus,

$$\text{for } n > 4, \quad \left| \frac{2n^2 - n}{n^2 - 9} - 2 \right| \leq \frac{38}{n}. \tag{3.10}$$

Since

$$\frac{38}{n} < \varepsilon \iff \frac{38}{\varepsilon} < n, \tag{3.11}$$

let us pick $N = \max\{4, 38/\varepsilon\}$. If $n > N$, then both $n > 4$ and $n > 38/\varepsilon$ hold, so

$$\left| \frac{2n^2 - n}{n^2 - 9} - 2 \right| \overset{\text{by (3.10)}}{\leq} \frac{38}{n} \overset{\text{by (3.11)}}{<} \varepsilon.$$

3.1.3   Sophisticated Examples of ε-N Arguments

We now give some very famous classical examples of ε-N arguments.

Example 3.5 Let a be a complex number with $|a| < 1$ and consider the sequence $a, a^2, a^3, \ldots$ (so that $a_n = a^n$ for each n). We shall prove that $\{a_n\}$ is a null sequence, that is,

$$\boxed{\lim a^n = 0.}$$

Let $\varepsilon > 0$ be a positive real number. We need to prove that there is a real number N such that the following statement holds:

$$n > N \quad \Longrightarrow \quad \left|a^n - 0\right| = |a|^n < \varepsilon. \tag{3.12}$$

If $a = 0$, then any N would do, so we might as well assume that $a \neq 0$. In this case,[5] since the real number $|a|$ is less than 1, we can write $|a| = \frac{1}{1+b}$, where $b > 0$; in fact, we can simply take $b = -1 + 1/|a|$. (Since $|a| < 1$, we have $1/|a| > 1$, so $b > 0$.) Therefore,

$$|a|^n = \frac{1}{(1+b)^n}, \quad \text{where } b > 0.$$

We now try to make $(1 + b)^n$ *smaller*. To do so, we use Bernoulli's inequality (Theorem 2.7 on p. 42) to write

$$(1+b)^n \geq 1 + nb \geq nb \quad \Longrightarrow \quad \frac{1}{(1+b)^n} \leq \frac{1}{nb}.$$

Hence,

$$|a|^n \leq \frac{1}{nb}. \tag{3.13}$$

Thus, we can satisfy (3.12) by making $1/(nb) < \varepsilon$ instead. Now,

$$\frac{1}{nb} < \varepsilon \quad \Longleftrightarrow \quad \frac{1}{b\varepsilon} < n. \tag{3.14}$$

For this reason, let us pick $N = 1/(b\varepsilon)$. Let $n > N$ (that is, $1/(b\varepsilon) < n$). Then,

$$|a|^n \overset{\text{by (3.13)}}{\leq} \frac{1}{nb} \overset{\text{by (3.14)}}{<} \varepsilon.$$

This proves (3.12) and thus, by the definition of convergence, $a^n \to 0$.

Example 3.6 For our next example, let $a > 0$ be a positive real number and consider the sequence $a, a^{1/2}, a^{1/3}, \ldots$ (so that $a_n = a^{1/n}$ for each n). We shall prove that $a_n \to 1$, that is,

$$\boxed{\lim a^{1/n} = 1.}$$

---

[5] You might be tempted to use logarithms on (3.12) to say that $|a|^n < \varepsilon$ if and only if $n \log |a| < \log \varepsilon$, or $n > \log \varepsilon / \log |a|$ (noting that $\log |a| < 0$, since $|a| < 1$). However, we have not yet developed the theory of logarithms! We will define logarithms on p. 300 in Section 4.7.

If $a = 1$, then the sequence $a^{1/n}$ is just the constant sequence $1, 1, 1, 1, \ldots$, which certainly converges to 1 (can you prove this?). We consider two cases.

(I) Suppose that $a > 1$; we shall consider the case $0 < a < 1$ afterward. Let $\varepsilon > 0$ be a positive real number. We need to prove that there is a real number N such that

$$n > N \implies \left| a^{1/n} - 1 \right| < \varepsilon. \tag{3.15}$$

By our familiar root rules (Theorem 2.33 on p. 95), we know that $a^{1/n} > 1^{1/n} = 1$ and therefore $b_n := a^{1/n} - 1 > 0$. By Bernoulli's inequality (Theorem 2.7 on p. 42), it follows that

$$a = \left(a^{1/n} \right)^n = (1 + b_n)^n \geq 1 + nb_n \geq nb_n \implies b_n \leq \frac{a}{n}.$$

Hence,

$$\left| a^{1/n} - 1 \right| = |b_n| \leq \frac{a}{n}. \tag{3.16}$$

Thus, we can satisfy (3.15) by making $a/n < \varepsilon$ instead. Now,

$$\frac{a}{n} < \varepsilon \iff \frac{a}{\varepsilon} < n. \tag{3.17}$$

For this reason, let us pick $N = a/\varepsilon$. Let $n > N$ (that is, $a/\varepsilon < n$). Then,

$$\left| a^{1/n} - 1 \right| \overset{\text{by (3.16)}}{\leq} \frac{a}{n} \overset{\text{by (3.17)}}{<} \varepsilon.$$

So, by definition of convergence, $a^{1/n} \to 1$.

(II) Now consider the case $0 < a < 1$. Let $\varepsilon > 0$ be a positive real number. We need to prove that there is a real number N such that

$$n > N \implies \left| a^{1/n} - 1 \right| < \varepsilon.$$

Since $0 < a < 1$, we have $1/a > 1$, so by Case (I), we know that $1/a^{1/n} = (1/a)^{1/n} \to 1$. Thus, there is a real number N such that

$$n > N \implies \left| \frac{1}{a^{1/n}} - 1 \right| < \varepsilon.$$

Multiplying both sides of the right-hand inequality by the positive real number $a^{1/n}$, we get $n > N \implies \left| 1 - a^{1/n} \right| < a^{1/n}\varepsilon$. Since $0 < a < 1$, by our root rules, $a^{1/n} < 1^{1/n} = 1$, so $a^{1/n}\varepsilon < 1 \cdot \varepsilon = \varepsilon$. Hence,

$$n > N \implies \left| a^{1/n} - 1 \right| < \varepsilon,$$

which shows that $a^{1/n} \to 1$, as we wished to show.

Example 3.7 We come to our last example, which may seem surprising at first. Consider the sequence $a_n = n^{1/n}$. We already know that if $a > 0$ is a fixed real number, then $a^{1/n} \to 1$. In our present case, $a_n = n^{1/n}$, so the "a" is increasing with n, and it is not at all obvious what $n^{1/n}$ converges to, if anything! However, we shall prove that

$$\boxed{\lim n^{1/n} = 1.}$$

For $n > 1$, by our root rules we know that $n^{1/n} > 1^{1/n} = 1$, so for $n > 1$, we conclude that $b_n := n^{1/n} - 1 > 0$. By the binomial theorem (Theorem 2.9 on p. 43), we have

$$n = (n^{1/n})^n = (1 + b_n)^n = 1 + \binom{n}{1} b_n + \binom{n}{2} b_n^2 + \cdots + \binom{n}{n} b_n^n.$$

Since $b_n > 0$, all the terms on the right-hand side are positive, so dropping all the terms except the third term on the right, we see that for $n > 1$,

$$n > \binom{n}{2} b_n^2 = \frac{n!}{2!\,(n-2)!} b_n^2 = \frac{n(n-1)}{2} b_n^2.$$

Canceling the n's from both sides, we obtain for $n > 1$,

$$b_n^2 < \frac{2}{n-1} \implies b_n < \frac{\sqrt{2}}{\sqrt{n-1}}.$$

Hence, for $n > 1$,

$$\left| n^{1/n} - 1 \right| = |b_n| < \frac{\sqrt{2}}{\sqrt{n-1}}. \tag{3.18}$$

Let $\varepsilon > 0$ be given. Then

$$\frac{\sqrt{2}}{\sqrt{n-1}} < \varepsilon \iff 1 + \frac{2}{\varepsilon^2} < n. \tag{3.19}$$

For this reason, let us pick $N = 1 + 2/\varepsilon^2$. Let $n > N$ (that is, $1 + 2/\varepsilon^2 < n$). Then,

$$\left| n^{1/n} - 1 \right| \overset{\text{by (3.18)}}{\leq} \frac{\sqrt{2}}{\sqrt{n-1}} \overset{\text{by (3.19)}}{<} \varepsilon.$$

Thus, by definition of convergence, $n^{1/n} \to 1$.

▶ **Exercises 3.1**

1. Using the ε-N definition of limit, prove that

(a) $\lim \dfrac{(-1)^n}{n} = 0$, (b) $\lim \left(2 + \dfrac{3}{n}\right) = 2$, (c) $\lim \dfrac{n}{n-1} = 1$, (d) $\lim \dfrac{(-1)^n}{\sqrt{n}-1} = 0$.

2. Using the ε-N definition of limit, prove that

$$(a) \lim \frac{5n^2+2}{n^3-3n+1} = 0, \quad (b) \lim \frac{n^2-\sqrt{n}}{3n^2-2} = \frac{1}{3}, \quad (c) \lim \left[\sqrt{n^2+n}-n\right] = \frac{1}{2}.$$

3. Here is another method to prove that $a^{1/n} \to 1$.

   (i) Note that for every b, we have $b^n - 1 = (b-1)(b^{n-1} + b^{n-2} + \cdots + b + 1)$. Using this formula, prove that if $a \geq 1$, then $a - 1 \geq n(a^{1/n} - 1)$. Suggestion: Let $b = a^{1/n}$.
   (ii) Now prove that for every $a > 0$, $a^{1/n} \to 1$. (Do the case $a \geq 1$ first, then consider the case $0 < a < 1$.)
   (iii) More generally, if $\{r_n\}$ is a sequence of rational numbers with $r_n \to 0$, prove that for every $a > 0$, $a^{r_n} \to 1$. Suggestion: Assume that $a \geq 1$. Then try to use (i) to prove that if $0 < r < 1$ is rational, then $a^r - 1 \leq C\,r$, where $C = a(a-1)$. Next prove that if $0 < r < 1$ is rational, then $|a^{-r} - 1| \leq C\,r$ as well. Conclude that if $0 \leq |r| < 1$, then $|a^r - 1| \leq C\,|r|$. Now prove that for every $a \geq 1$, $a^{r_n} \to 1$. Lastly, do the case $0 < a < 1$.

4. Let a be a complex number with $|a| < 1$. We already know that $a^n \to 0$. In this problem we prove the somewhat surprising fact that $na^n \to 0$. Although n grows very large, this limit shows that a^n must go to zero faster than n grows.

   (i) As in Example 3.5, we can write $|a| = 1/(1+b)$, where $b > 0$. Using the binomial theorem, show that for $n > 1$, $|a|^n < \dfrac{1}{\binom{n}{2} b^2}$.
   (ii) Show that $n|a|^n < \dfrac{2}{(n-1) b^2}$.
   (iii) Now prove that $na^n \to 0$.

5. Here's an even more surprising fact. Let a be a complex number with $|a| < 1$. Prove that given a natural number $k > 0$, we have $n^k a^n \to 0$. Suggestion: Let $\alpha := |a|^{1/k} < 1$ and use the fact that $n\alpha^n \to 0$ by the previous problem.

6. If $\{a_n\}$ is a sequence of nonnegative real numbers and $a_n \to L$, prove the following:

   (i) $L \geq 0$.
   (ii) $\sqrt{a_n} \to \sqrt{L}$. (You need to consider two cases, $L = 0$ and $L > 0$.)

7. Let $\{a_n\}$ be a sequence in $\mathbb{R}^m$ and let $L \in \mathbb{R}^m$. Form the negation of the definition that $a_n \to L$, thus giving a statement that $a_n \nrightarrow L$ (the sequence $\{a_n\}$ does not tend to L). Using your negation, prove that the sequence $\{(-1)^n\}$ diverges, that is, does not converge to any real number. In the next section we shall find an easy way to verify that a sequence diverges using the notion of subsequences.

8. (**Infinite products**—see Chapter 7 for more on this amazing topic!) In this problem we investigate the **infinite product**

$$\frac{2^2}{1\cdot 3}\cdot\frac{3^2}{2\cdot 4}\cdot\frac{4^2}{3\cdot 5}\cdot\frac{5^2}{4\cdot 6}\cdot\frac{6^2}{5\cdot 7}\cdot\frac{7^2}{6\cdot 8}\cdots . \tag{3.20}$$

We interpret this infinite product as the limit of the **partial products**

$$a_1 = \frac{2^2}{1\cdot 3}, \quad a_2 = \frac{2^2}{1\cdot 3}\cdot\frac{3^2}{2\cdot 4}, \quad a_3 = \frac{2^2}{1\cdot 3}\cdot\frac{3^2}{2\cdot 4}\cdot\frac{4^2}{3\cdot 5},\cdots .$$

In other words, for each $n \in \mathbb{N}$, we define $a_n := \frac{2^2}{1\cdot 3}\cdot\frac{3^2}{2\cdot 4}\cdots\frac{(n+1)^2}{n\cdot(n+2)}$. We prove that the sequence $\{a_n\}$ converges as follows.

(i) Prove that $a_n = \frac{2(n+1)}{n+2}$.
(ii) Now prove that $a_n \to 2$. We often write the infinite product (3.20) using $\prod$ notation and we express the limit $\lim a_n = 2$ as

$$\frac{2^2}{1\cdot 3}\cdot\frac{3^2}{2\cdot 4}\cdot\frac{4^2}{3\cdot 5}\cdot\frac{5^2}{4\cdot 6}\cdots = 2 \quad \text{or} \quad \prod_{n=1}^{\infty}\frac{(n+1)^2}{n(n+2)} = 2.$$

3.2 A Potpourri of Limit Properties for Sequences

Now that we have a working knowledge of the ε-N definition of limit, we move to studying the properties of limits that will be used throughout the rest of the book. In particular, we learn the *algebra of limits*, which allows us to combine convergent sequences to form other convergent sequences. Finally, we discuss the notion of properly divergent sequences.

3.2.1 Basic Limit Theorems

Recall that a **null sequence** is a sequence converging to zero. The following theorem is our "four step recipe" on p. 153.

Limit recipe theorem

> **Theorem 3.1** *If $\{a_n\}$ is a sequence in $\mathbb{R}^m$, $L \in \mathbb{R}^m$, and $|a_n - L| \le \alpha_n$ for some null sequence $\{\alpha_n\}$ of nonnegative real numbers, then $a_n \to L$.*

Proof If $\varepsilon > 0$ is given, there is an N such that $n > N$ implies $|\alpha_n - 0| = \alpha_n < \varepsilon$. It follows that for $n > N$, $|a_n - L| < \varepsilon$, and we're done. ∎

That was easy! We now prove slightly harder results. We begin by showing that limits are unique, that is, convergent sequences cannot have two different limits.

Uniqueness of limits

Theorem 3.2 *Sequences can have at most one limit.*

Proof Let $\{a_n\}$ be a sequence in $\mathbb{R}^m$ and suppose that $a_n \to L$ and $a_n \to L'$. We shall prove that $L = L'$. To do so, assume (in order to get a contradiction) that $L \neq L'$ and let $\varepsilon = |L - L'|/2$; here's a picture:

$$\varepsilon = |L - L'|/2$$

Since $a_n \to L$, there is an N such that $|a_n - L| < \varepsilon$ for all $n > N$, and since $a_n \to L'$, there is N' such that $|a_n - L'| < \varepsilon$ for all $n > N'$. By the triangle inequality,

$$|L - L'| = |(L - a_n) + (a_n - L')| \leq |L - a_n| + |a_n - L'|.$$

In particular, for n greater than the larger of N and N', we see that

$$|L - L'| \leq |L - a_n| + |a_n - L'| < \varepsilon + \varepsilon = 2\varepsilon;$$

that is, $|L - L'| < |L - L'|$, an impossibility. Thus, $L = L'$. ∎

It is important that the convergence or divergence of a sequence depends only on the "tail" of the sequence, that is, on the terms of the sequence for large n. This fact is more or less obvious.

Example 3.8 Consider the sequence

$$-100, \ 100, \ 50, \ 1000, \ \frac{1}{2}, \ \frac{1}{4}, \ \frac{1}{8}, \ \frac{1}{16}, \ \frac{1}{32}, \ \frac{1}{64}, \ \frac{1}{128}, \ \dots, \ \frac{1}{2^k}, \ \dots.$$

This sequence converges to zero, and the first few terms don't change this fact.

Given a sequence $\{a_n\}$ in $\mathbb{R}^m$ and a nonnegative integer $k = 0, 1, 2, \dots$, we call the sequence $\{a_{k+1}, a_{k+2}, a_{k+3}, a_{k+4}, \dots\}$ a k-**tail** (of the sequence $\{a_n\}$). We'll leave the following proof to the reader.

Tails theorem for sequences

Theorem 3.3 *A sequence converges if and only if every tail converges, if and only if some tail converges.*

We now show that convergence in $\mathbb{R}^m$ can be reduced to convergence in $\mathbb{R}$, which is why real sequences are so important. Let $\{a_n\}$ be a sequence in $\mathbb{R}^m$. Since $a_n \in \mathbb{R}^m$, we can express a_n in terms of its m-tuple of components:

$$a_n = (a_{1n}, a_{2n}, \ldots, a_{mn}).$$

Notice that each coordinate, say the kth one a_{kn}, is a real number, and so $\{a_{kn}\} = \{a_{k1}, a_{k2}, a_{k3}, \ldots\}$ is a sequence in $\mathbb{R}$. Given $L = (L_1, L_2, \ldots, L_m) \in \mathbb{R}^m$, in the following theorem we prove that if $a_n \to L$, then for each $k = 1, \ldots, m$, $a_{kn} \to L_k$ as $n \to \infty$ as well. Conversely, we shall prove that if for each $k = 1, \ldots, m$, $a_{kn} \to L_k$ as $n \to \infty$, then $a_n \to L$ as well.

Component theorem

Theorem 3.4 *A sequence in $\mathbb{R}^m$ converges to $L \in \mathbb{R}^m$ if and only if each component sequence converges in $\mathbb{R}$ to the corresponding component of L.*

Proof Suppose first that $a_n \to L$. Fixing k, we shall prove that $a_{kn} \to L_k$. Let $\varepsilon > 0$. Since $a_n \to L$, there is an N such that for all $n > N$, $|a_n - L| < \varepsilon$. Hence, by definition of the norm on $\mathbb{R}^m$, for all $n > N$,

$$(a_{kn} - L_k)^2 \le (a_{1n} - L_1)^2 + (a_{2n} - L_2)^2 + \cdots + (a_{mn} - L_m)^2 = |a_n - L|^2 < \varepsilon^2.$$

Taking square roots of both sides shows that for all $n > N$, $|a_{kn} - L_k| < \varepsilon$, which shows that $a_{kn} \to L_k$.

Suppose now that for each $k = 1, \ldots, m$, $a_{kn} \to L_k$. Let $\varepsilon > 0$. Since $a_{kn} \to L_k$, there is an N_k such that for all $n > N_k$, $|a_{kn} - L_k| < \varepsilon/\sqrt{m}$. Let N be the largest of the numbers $N_1, N_2, \ldots, N_m$. Then for $n > N$, we have

$$|a_n - L|^2 = (a_{1n} - L_1)^2 + (a_{2n} - L_2)^2 + \cdots + (a_{mn} - L_m)^2$$
$$< \left(\frac{\varepsilon}{\sqrt{m}}\right)^2 + \left(\frac{\varepsilon}{\sqrt{m}}\right)^2 + \cdots + \left(\frac{\varepsilon}{\sqrt{m}}\right)^2 = \frac{\varepsilon^2}{m} + \frac{\varepsilon^2}{m} + \cdots + \frac{\varepsilon^2}{m} = \varepsilon^2.$$

Taking square roots of both sides shows that for all $n > N$, $|a_n - L| < \varepsilon$, which shows that $a_n \to L$. ∎

Example 3.9 Let us apply this theorem to $\mathbb{C}$ (which, recall, is just $\mathbb{R}^2$ with a special multiplication). Let $c_n = (a_n, b_n) = a_n + i b_n$ be a complex sequence (here we switch notation from a_n to c_n in the theorem and we let $c_{1n} = a_n$ and $c_{2n} = b_n$). Then it follows that $c_n \to c = a + ib$ if and only if $a_n \to a$ and $b_n \to b$. In other words, $c_n \to c$ if and only if the real and imaginary parts of c_n converge to the real and imaginary parts, respectively, of c. For instance, from Example 3.6 on p. 157 and Example 3.3 on p. 152, it follows that for every real $a > 0$, we have

$$\frac{1}{n+1} + i a^{1/n} \to 0 + i \cdot 1 = i.$$

A sequence $\{a_n\}$ in $\mathbb{R}^m$ (convergent or not) is said to be **bounded** if there is a constant C such that $|a_n| \le C$ for all n.

Example 3.10 The sequence $\{n\} = \{1, 2, 3, 4, 5, \ldots\}$ is not bounded, by the Archimedean property of $\mathbb{R}$ (see p. 96). Also if $a > 1$ is a real number, then the sequence $\{a^n\} = \{a^1, a^2, a^3, \ldots\}$ is not bounded. One way to see this uses Bernoulli's inequality on p. 41: We can write $a = 1 + r$, where $r > 0$, so by Bernoulli's inequality,

$$a^n = (1 + r)^n \geq 1 + nr > nr,$$

and nr can be made greater than any constant C by the Archimedean property of $\mathbb{R}$. Thus, $\{a^n\}$ cannot be bounded. On the other hand, the sequence $\{(-1)^n\}$ is bounded, since $|(-1)^n| \leq 1$ for all n.

We now prove the fundamental fact that *if a sequence converges, then it must be bounded.* The converse is false, as the example $\{(-1)^n\}$ shows.

Convergent sequences are bounded

Theorem 3.5 *Every convergent sequence is bounded.*

Proof If $a_n \to L$ in $\mathbb{R}^m$, then with $\varepsilon = 1$ in the definition of convergence, there is an N such that for all $n > N$, we have $|a_n - L| < 1$, which, by the triangle inequality, implies that

$$n > N \implies |a_n| = |(a_n - L) + L| \leq |a_n - L| + |L| < 1 + |L|. \qquad (3.21)$$

Let k be a natural number greater than N and let

$$C := \max \{|a_1|, |a_2|, \ldots, |a_{k-1}|, |a_k|, 1 + |L|\}.$$

Then $|a_n| \leq C$ for $n = 1, 2, \ldots, k$, and by (3.21), $|a_n| < C$ for $n > k$. Thus, $|a_n| \leq C$ for all n, and hence $\{a_n\}$ is bounded. ∎

Forming the contrapositive, we know that if a sequence is not bounded, then the sequence cannot converge. Therefore, this theorem can be used to prove that certain sequences *do not* converge.

Example 3.11 Each of sequences $\{n\}$, $\{1 + in^2\}$, $\{2^n + i/n\}$ is not bounded and therefore does not converge.

3.2.2 *Real Sequences and Preservation of Inequalities*

Real sequences have certain properties that general sequences in $\mathbb{R}^m$ and complex sequences do not have, namely those corresponding to the order properties of $\mathbb{R}$. The first theorem we discuss is the famous squeeze theorem, illustrated on the next page:

The squeeze theorem says that if a sequence $\{b_n\}$ is squeezed between two sequences $\{a_n\}$ and $\{c_n\}$ that converge to a common value, then $\{b_n\}$ also converges to that same value. To prove this result, we shall use the following formulation of convergence discussed in Section 3.1 on p. 149:

> A real sequence $\{a_n\}$ converges to $L \in \mathbb{R}$ if and only if for every $\varepsilon > 0$ there is an N such that
> $$n > N \implies L - \varepsilon < a_n < L + \varepsilon.$$

In the following theorem, the phrase "for n sufficiently large ..." means "there is an n_0 such that for $n > n_0 \ldots$"

Squeeze theorem

Theorem 3.6 *Let $\{a_n\}$, $\{b_n\}$, and $\{c_n\}$ be sequences in $\mathbb{R}$ with $\{a_n\}$ and $\{c_n\}$ convergent such that $\lim a_n = \lim c_n$ and for n sufficiently large, $a_n \leq b_n \leq c_n$. Then the sequence $\{b_n\}$ is also convergent, and*

$$\lim a_n = \lim b_n = \lim c_n.$$

Proof Let $L = \lim a_n = \lim c_n$ and let $\varepsilon > 0$. By the tails theorem, we may assume that $a_n \leq b_n \leq c_n$ for all n. Since $a_n \to L$, there is an N_1 such that for $n > N_1$, $L - \varepsilon < a_n < L + \varepsilon$, and since $c_n \to L$, there is an N_2 such that for $n > N_2$, $L - \varepsilon < c_n < L + \varepsilon$. Let N be the maximum of N_1 and N_2. Then for $n > N$,

$$L - \varepsilon < a_n \leq b_n \leq c_n < L + \varepsilon,$$

which implies that for $n > N$, $L - \varepsilon < b_n < L + \varepsilon$. Thus, $b_n \to L$. ∎

Example 3.12 Here's a neat sequence involving the squeeze theorem. Consider $\{b_n\}$, where

$$b_n = \frac{1}{(n+1)^2} + \frac{1}{(n+2)^2} + \frac{1}{(n+3)^2} + \cdots + \frac{1}{(2n)^2} = \sum_{k=1}^{n} \frac{1}{(n+k)^2}.$$

Observe that for $k = 1, 2, \ldots, n$, we have

$$0 \leq \frac{1}{(n+k)^2} \leq \frac{1}{(n+0)^2} = \frac{1}{n^2}.$$

Thus,

$$0 \le \sum_{k=1}^{n} \frac{1}{(n+k)^2} \le \sum_{k=1}^{n} \frac{1}{n^2} = \left(\sum_{k=1}^{n} 1\right) \cdot \frac{1}{n^2} = n \cdot \frac{1}{n^2}.$$

Therefore,

$$0 \le b_n \le \frac{1}{n}.$$

Since $a_n = 0 \to 0$ and $c_n = 1/n \to 0$, by the squeeze theorem, $b_n \to 0$ as well.

Example 3.13 (*Real numbers as limits of (ir)rational numbers*) We claim that given a $c \in \mathbb{R}$, there are sequences of rational numbers $\{r_n\}$ and irrational numbers $\{q_n\}$, both converging to c. Indeed, for each $n \in \mathbb{N}$ we have $c - \frac{1}{n} < c$, so by Theorem 2.38 on p. 98, there is a rational number r_n and irrational number q_n such that

$$c - \frac{1}{n} < r_n < c \quad \text{and} \quad c - \frac{1}{n} < q_n < c.$$

Since $c - \frac{1}{n} \to c$ and $c \to c$, by the squeeze theorem, we have $r_n \to c$ and $q_n \to c$.

The following theorem states that real sequences preserve inequalities.

Limits preserve inequalities

> **Theorem 3.7** *Let $\{a_n\}$ and $\{b_n\}$ converge in $\mathbb{R}$.*
>
> *(1) If $a_n \le b_n$ for n sufficiently large, then $\lim a_n \le \lim b_n$.*
> *(2) If $c \le a_n \le d$ for n sufficiently large, then $c \le \lim a_n \le d$.*

Proof We shall prove the *contrapositive* of *(1)*. To this end, assume that $a_n \to a :=$ $\lim a_n$ and $b_n \to b := \lim b_n$, where $a > b$. Let $\varepsilon = (a - b)/2$, which is positive by assumption. Since $a_n \to a$, there is an N_1 such that for $n > N_1, a - \varepsilon < a_n < a + \varepsilon$, and since $b_n \to b$, there is an N_2 such that for $n > N_2, b - \varepsilon < b_n < b + \varepsilon$. Let N be the larger of N_1 and N_2. Then for $n > N$, we have

$$a - \varepsilon < a_n < a + \varepsilon \quad \text{and} \quad b - \varepsilon < b_n < b + \varepsilon.$$

Hence, for $n > N$,

$$a_n - b_n > (a - \varepsilon) - (b + \varepsilon) = a - b - 2\varepsilon = 0;$$

that is, for $n > N, a_n > b_n$. This proves our first result.

(2) follows from *(1)* applied to the constant sequences $\{c, c, c, \dots\}$, which converges to c, and $\{d, d, d, \dots\}$, which converges to d:

$$c = \lim c_n \leq \lim a_n \leq \lim d_n = d.$$ ■

If $c < a_n < d$ for n sufficiently large, must it be true that $c < \lim a_n < d$ (with strict inequalities)? The answer is no. Can you give a counterexample?

3.2.3   Subsequences

For the rest of this section we focus on general sequences in $\mathbb{R}^m$ and not just $\mathbb{R}$. A subsequence is just a sequence formed by picking out certain (countably many) terms of a given sequence. More precisely, let $\{a_n\}$ be a sequence in $\mathbb{R}^m$. Let $\nu_1 < \nu_2 < \nu_3 < \cdots$ be a sequence of natural numbers that is increasing. Then the sequence $\{a_{\nu_n}\}$ given by

$$a_{\nu_1}, \ a_{\nu_2}, \ a_{\nu_3}, \ a_{\nu_4}, \ \ldots$$

is called a **subsequence** of $\{a_n\}$.

Example 3.14   Consider the sequence

$$\frac{1}{1}, \frac{1}{2}, \frac{1}{3}, \frac{1}{4}, \frac{1}{5}, \frac{1}{6}, \ \ldots, a_n = \frac{1}{n}, \ldots.$$

Choosing $2, 4, 6, \ldots, \nu_n = 2n, \ldots$, we get the subsequence

$$\frac{1}{2}, \frac{1}{4}, \frac{1}{6}, \ \ldots, a_{\nu_n} = \frac{1}{2n}, \ldots.$$

Example 3.15   As another example, choosing $1!, 2!, 3!, 4!, \ldots, \nu_n = n!, \ldots$, we get the subsequence

$$\frac{1}{1!}, \frac{1}{2!}, \frac{1}{3!}, \ \ldots, a_{\nu_n} = \frac{1}{n!}, \ldots.$$

Notice that both subsequences, $\{1/(2n)\}$ and $\{1/n!\}$ also converge to zero, the same limit as the original sequence $\{1/n\}$. This is a general fact: If a sequence converges, then every subsequence of it must converge to the same limit.

Subsequence convergence theorem

> **Theorem 3.8**   *Every subsequence of a convergent sequence converges to the same limit as the original sequence.*

Proof   Let $\{a_n\}$ be a sequence in $\mathbb{R}^m$ converging to $L \in \mathbb{R}^m$. Let $\{a_{\nu_n}\}$ be a subsequence and let $\varepsilon > 0$. Since $a_n \to L$, there is an N such that for all $n > N$, $|a_n - L| < \varepsilon$. Since $\nu_1 < \nu_2 < \nu_3 < \cdots$ is an increasing sequence of natural numbers, one can check (for instance, by induction) that $n \leq \nu_n$ for all n. Thus, for $n > N$, we have $\nu_n > N$, and hence for $n > N$, we have $|a_{\nu_n} - L| < \varepsilon$. This proves that $a_{\nu_n} \to L$ and completes the proof.   ∎

This theorem gives perhaps the easiest way to prove that a sequence *does not* converge.

Example 3.16 Consider the sequence

$$i, \; i^2 = -1, \; i^3 = -i, \; i^4 = 1, \; i^5 = i, \; i^6 = -1, \ldots, a_n = i^n, \ldots,$$

as seen here:

Choosing $1, 5, 9, 13, \ldots, \nu_n = 4n - 3, \ldots$, we get the subsequence

$$i, \; i, \; i, \; i, \; \ldots,$$

which converges to i. On the other hand, choosing $2, 6, 10, 14, \ldots, \nu_n = 4n - 2, \ldots$, we get the subsequence

$$-1, \; -1, \; -1, \; -1, \ldots,$$

which converges to -1. Since these two subsequences do not converge to the same limit, the original sequence $\{i^n\}$ cannot converge. Indeed, if $\{i^n\}$ did converge, then every subsequence of $\{i^n\}$ would have to converge to the same limit as $\{i^n\}$, but we have found subsequences that converge to different limits.

3.2.4 Algebra of Limits

Let $\{a_n\}$ and $\{b_n\}$ be sequences in $\mathbb{R}^m$. Given real numbers c, d, we can form the **linear combination sequence** $\{c\, a_n + d\, b_n\}$ (where the nth term of the sequence is $c\, a_n + d\, b_n$). As a special case, the sum of these sequences is the sequence $\{a_n + b_n\}$, and the difference is the sequence $\{a_n - b_n\}$, and choosing $d = 0$, the multiple of $\{a_n\}$ by c is the sequence $\{c\, a_n\}$. The **sequence of norms** of the sequence $\{a_n\}$ is the sequence of real numbers $\{|a_n|\}$. We can reinterpret the definition of limit in terms of norm sequences:

Example 3.17 (Null and norm sequences) If $\{a_n\}$ is a sequence in $\mathbb{R}^m$ and $L \in \mathbb{R}^m$, we claim that

$a_n \to L$ *if and only if the sequence of norms* $\{|a_n - L|\}$ *is a null sequence.*

Indeed, this statement is just a rewording of the definition of limit! To see this, note that if we put $b_n = |a_n - L|$, then the definition of $a_n \to L$ is

> Given $\varepsilon > 0$, there is an $N \in \mathbb{R}$ such that $n > N$ implies $b_n < \varepsilon$.

Since $b_n = |b_n - 0|$, we are just saying that $b_n \to 0$, that is, $\{|a_n - L|\}$ is a null sequence. In particular, the sequence $\{a_n\}$ is a null sequence if and only if $\{|a_n|\}$ is a null sequence.

The following example uses the renowned $\varepsilon/2$-**trick**,, which says that

$$\boxed{\text{We can make } \alpha + \beta < \varepsilon \text{ by taking } \alpha < \frac{\varepsilon}{2} \text{ and } \beta < \frac{\varepsilon}{2}.}$$

Example 3.18 The easiest case of the algebra of limits concerns null sequences:

Linear combinations of null sequences are again null.

To see this, let $\{a_n\}$ and $\{b_n\}$ be null sequences and let c, d be real numbers; we will show that $\{c\,a_n + d\,b_n\}$ is null. Let $\varepsilon > 0$. By the triangle inequality,

$$|c\,a_n + d\,b_n| \le |c|\,|a_n| + |d|\,|b_n|.$$

We use the $\varepsilon/2$-trick as follows. First, since $a_n \to 0$, there is an N_1 such that for all $n > N_1$, $|c|\,|a_n| < \varepsilon/2$. (If $|c| = 0$, any N_1 will work; if $|c| > 0$, then choose N_1 corresponding to the error $\varepsilon/(2|c|)$ in the definition of convergence for $a_n \to 0$.) Second, since $b_n \to 0$, there is an N_2 such that for all $n > N_2$, $|d|\,|b_n| < \varepsilon/2$. Now setting N as the larger of N_1 and N_2, it follows that for $n > N$,

$$|c\,a_n + d\,b_n| \le |c|\,|a_n - a| + |d|\,|b_n - b| < \frac{\varepsilon}{2} + \frac{\varepsilon}{2} = \varepsilon.$$

Therefore, the sequence $\{c\,a_n + d\,b_n\}$ is null, as claimed.

Using this example, we prove the following *algebra of limits* theorem.

Theorem 3.9 *Linear combinations and norms of convergent sequences converge to the corresponding linear combinations and norms of the limits.*

Proof We consider first linear combinations. If $a_n \to a$ and $b_n \to b$, we shall prove that $c\,a_n + d\,b_n \to c\,a + d\,b$. To see this, observe that by the triangle inequality,

$$\begin{aligned}
|c\,a_n + d\,b_n - (c\,a + d\,b)| &= |c\,(a_n - a) + d\,(b_n - b)| \\
&\le |c|\,|a_n - a| + |d|\,|b_n - b|. \quad (3.22)
\end{aligned}$$

By Example 3.17, we know that both $\{|a_n - a|\}$ and $\{|b_n - b|\}$ are null sequences, and since linear combinations of null sequences are null, it follows that the sequence in (3.22) to the right of $\leq$ is a null sequence. By the limit recipe theorem (Theorem 3.1 on p. 161), it follows that $c\,a_n + d\,b_n \to c\,a + d\,b$.

Assuming that $a_n \to a$, we show that $|a_n| \to |a|$. Indeed, as a consequence of the triangle inequality (see Property *(iv)* in Theorem 2.47 on p. 122), we have

$$\big|\,|a_n| - |a|\,\big| \leq |a_n - a|.$$

Since $\{|a_n - a|\}$ is null, by the limit recipe theorem, $|a_n| \to |a|$.   ∎

Let $\{a_n\}$ and $\{b_n\}$ be complex sequences. Given complex numbers c, d, the same proof detailed above shows that $c\,a_n + d\,b_n \to c\,a + d\,b$. However, since they are complex sequences, we can also multiply these sequences, term by term, defining the **product** sequence as the sequence $\{a_n b_n\}$ (where the nth term of the sequence is the product $a_n b_n$). Also, assuming that $b_n \neq 0$ for each n, we can divide the sequences, term by term, defining the **quotient** sequence as the sequence $\{a_n / b_n\}$.

> **Theorem 3.10** *Products of convergent complex sequences converge to the corresponding products of the limits. Quotients of convergent complex sequences, where the denominator sequence is a nonzero sequence converging to a nonzero limit, converge to the corresponding quotient of the limits.*

Proof Given $a_n \to a$ and $b_n \to b$, we shall prove that $a_n b_n \to a\,b$. To see this, observe that by the triangle inequality,

$$|a_n b_n - a\,b| = |a_n(b_n - b) + b(a_n - a)| \leq |a_n|\,|b_n - b| + |b|\,|a_n - a|.$$

Since convergent sequences are bounded (Theorem 3.5), there is a constant C such that $|a_n| \leq C$ for all n. Hence,

$$|a_n b_n - a\,b| \leq C\,|b_n - b| + |b|\,|a_n - a|.$$

The right-hand side is a null sequence, being a linear combination of null sequences. Thus by the limit recipe theorem, $a_n b_n \to a\,b$.

Now assume that $b_n \neq 0$ for each n and $b \neq 0$; we shall prove that $a_n / b_n \to a/b$. Since we can write this limit statement as a product: $a_n \cdot b_n^{-1} \to a \cdot b^{-1}$, all we have to do is show that $b_n^{-1} \to b^{-1}$. To see this, note that

$$|b_n^{-1} - b^{-1}| = |b_n\,b|^{-1}\,|b_n - b|.$$

Let N be chosen in accordance with the error $|b|/2$ in the definition of convergence for $b_n \to b$. Then for $n > N$,

$$|b| = |b - b_n + b_n| \le |b - b_n| + |b_n| < \frac{|b|}{2} + |b_n|.$$

Bringing $|b|/2$ to the left, for $n > N$ we have $|b|/2 < |b_n|$, or

$$|b_n|^{-1} < 2|b|^{-1}, \qquad n > N.$$

Hence, for $n > N$,

$$|b_n^{-1} - b^{-1}| = |b_n b|^{-1} |b_n - b| \le C |b_n - b|,$$

where $C = 2|b|^{-2}$. Since $\{C |b_n - b|\}$ is a null sequence, by the limit recipe theorem on p. 161, $b_n^{-1} \to b^{-1}$, and our proof is complete. ∎

These two *algebra of limits* theorems can be used to evaluate limits in an easy manner.

Example 3.19 Since $\lim \frac{1}{n} = 0$, by our product theorem (Theorem 3.10), we have

$$\lim \frac{1}{n^2} = \left(\lim \frac{1}{n} \right) \cdot \left(\lim \frac{1}{n} \right) = 0 \cdot 0 = 0.$$

Example 3.20 Since the constant sequence 1 converges to 1, by our linear combination theorem (Theorem 3.9), for every number a, we have

$$\lim \left(1 + \frac{a}{n^2} \right) = \lim 1 + a \cdot \lim \frac{1}{n^2} = 1 + a \cdot 0 = 1.$$

Example 3.21 Now dividing the top and bottom of $\frac{n^2+3}{n^2+7}$ by $1/n^2$ and using our theorem on quotients and the limit we just found in the previous example, we obtain

$$\lim \frac{n^2 + 3}{n^2 + 7} = \frac{\lim \left(1 + \frac{3}{n^2} \right)}{\lim \left(1 + \frac{7}{n^2} \right)} = \frac{1}{1} = 1.$$

3.2.5 *Properly Divergent Sequences*

In dealing with sequences of real numbers, inevitably infinities occur. For instance, we know that the sequence $\{n^2\}$ diverges, since it is unbounded. However, in elementary calculus, we would usually write $n^2 \to +\infty$ or $\lim n^2 = +\infty$, which suggests that this sequence converges to the number "infinity." We now make this notion precise.

A sequence $\{a_n\}$ of real numbers **diverges to** $+\infty$ if given any real number $M > 0$, there is a real number N such that for all $n > N$, $a_n > M$; here's a picture:

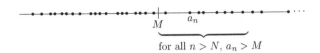

for all $n > N$, $a_n > M$

Thus, for every M, the a_n are eventually greater than M. If instead of plotting the a_n on the real line, we graph n versus a_n, then divergence to $+\infty$ looks like this:

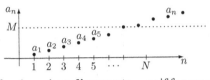

A sequence $\{a_n\}$ of real numbers **diverges to** $-\infty$ if for every real number $M < 0$, there is a real number N such that for all $n > N$, $a_n < M$. In the first case, we write $\lim a_n = +\infty$ or $a_n \to +\infty$ (sometimes we drop the "+" in front of ∞), and in the second case, we write $\lim a_n = -\infty$ or $a_n \to -\infty$. In either case, we say that $\{a_n\}$ is **properly divergent**. It is important to understand that the symbols $+\infty$ and $-\infty$ are simply notation and that they do not represent real numbers.[6] We now present some examples.

Example 3.22 Given a natural number k, we shall prove that $\lim n^k = +\infty$. To see this, let $M > 0$. Then we want to prove that there is an N such that for all $n > N$, $n^k > M$. To do so, observe that $n^k > M$ if and only if $n > M^{1/k}$. For this reason, we choose $N = M^{1/k}$. With this choice of N, for all $n > N$, we certainly have $n^k > M$, and hence $n^k \to +\infty$, as stated. Using a very similar argument, one can show that $-n^k \to -\infty$.

Example 3.23 In Example 3.10 on p. 164, we showed that given a real number $a > 1$, the sequence $\{a^n\}$ diverges to $+\infty$.

Because $\pm\infty$ are not real numbers, some of the limit theorems we have proved in this section are not valid when $\pm\infty$ are the limits, but many do hold under certain conditions. For example, if $a_n \to +\infty$ and $b_n \to +\infty$, then for all nonnegative real numbers c, d, at least one of which is positive, the reader can check that

$$c\, a_n + d\, b_n \to +\infty.$$

If c, d are nonpositive, with at least one of them negative, then $c\, a_n + d\, b_n \to -\infty$. If c and d have opposite signs, then there is no general result. For example, if $a_n = n$, $b_n = n^2$, and $c_n = n + (-1)^n$, then $a_n, b_n, c_n \to +\infty$, but

$$\lim(a_n - b_n) = -\infty, \quad \lim(b_n - c_n) = +\infty, \quad \text{and} \quad \lim(a_n - c_n) \text{ does not exist!}$$

---

[6]It turns out that $\pm\infty$ form part of a number system called the **extended real numbers**, which consists of the real numbers together with the symbols $+\infty = \infty$ and $-\infty$. One can define addition, multiplication, and order in this system, with certain exceptions (such as subtraction of infinities is not allowed). If you take measure theory, you will study this system.

We encourage the reader to think about which limit theorems extend to the case of infinite limits. For example, here is a **squeeze law**: If $a_n \le b_n$ for all n sufficiently large and $a_n \to +\infty$, then $b_n \to +\infty$ as well. Some more limit theorems for infinite limits are presented in the exercises (see e.g., Problem 10).

▶ **Exercises 3.2**

1. Evaluate the following limits using limits already proved in the text or exercises and invoking the "algebra of limits."

   (a) $\lim \dfrac{(-1)^n n}{n^2 + 5}$, (b) $\lim \dfrac{(-1)^n}{n + 10}$, (c) $\lim \dfrac{2^n}{3^n + 10}$, (d) $\lim \left(7 + \dfrac{3}{n}\right)^2$.

2. If $\{a_n\}$ is a sequence in $\mathbb{R}^m$ and $\lim |a_n| = 0$ (that is, $\{|a_n|\}$ is null), we know that $a_n \to 0$. It is important that zero is the limit in the hypothesis. Indeed, give an example of a sequence for which $\lim |a_n|$ exists and is nonzero, but $\{a_n\}$ diverges.

3. Why do the following sequences diverge?

   (a) $\{(-1)^n\}$, (b) $\left\{ a_n = \displaystyle\sum_{k=0}^{n} (-1)^k \right\}$, (c) $\{a_n = 2^{n(-1)^n}\}$, (d) $\{i^n + 1/n\}$.

4. Find the limits of each of the following sequences:

   (a) $a_n = \displaystyle\sum_{k=1}^{n} \dfrac{1}{\sqrt{n^2 + k}}$, (b) $b_n = \displaystyle\sum_{k=1}^{n} \dfrac{1}{\sqrt{n + k}}$, (c) $c_n = \dfrac{1}{n} \displaystyle\sum_{k=n}^{2n} \dfrac{1}{k}$.

5. (a) Let $a_1 \in \mathbb{R}$ and for $n \ge 1$, define $a_{n+1} = \dfrac{\text{sgn}(a_n) + 10(-1)^n}{\sqrt{n}}$. Here, $\text{sgn}(x) = 1$ if $x > 0$, $\text{sgn}(x) = 0$ if $x = 0$, and $\text{sgn}(x) = -1$ if $x < 0$. Find $\lim a_n$.

   (b) Let $a_1 \in [-1, 1]$, and for $n \ge 1$, define $a_{n+1} = \dfrac{a_n}{|a_n| + 1}$. Find $\lim a_n$. Suggestion: Can you prove that $-1/n \le a_n \le 1/n$ for all $n \in \mathbb{N}$?

6. If $\{a_n\}$ and $\{b_n\}$ are complex sequences with $\{a_n\}$ bounded and $b_n \to 0$, prove that $a_n b_n \to 0$. Why can't we use Theorem 3.10 in this situation?

7. (**The root test for sequences**) Let $\{a_n\}$ be a sequence of positive real numbers such that $R := \lim a_n^{1/n}$ exists. (That is, $\{a_n^{1/n}\}$ converges and we denote its limit by R.)

   (i) If $R < 1$, prove that $\lim a_n = 0$. Suggestion: Show that there is a real number r with $0 < r < 1$ such that $0 < a_n < r^n$ for all n sufficiently large, that is, that there is an N such that $0 < a_n < r^n$ for all $n > N$.

   (ii) If, however, $R > 1$, prove that $\{a_n\}$ is not bounded, and hence diverges.

   (iii) When $R = 1$, the test is inconclusive: Give an example of a convergent sequence and a divergent sequence, both of which satisfy $R = 1$.

8. (**The ratio test for sequences**) Let $\{a_n\}$ be a sequence of positive real numbers such that $R := \lim(a_{n+1}/a_n)$ exists.

(i) If $R < 1$, prove that $\lim a_n = 0$. Suggestion: Show that there are real numbers C, r with $C > 0$ and $0 < r < 1$ such that $0 < a_n < C r^n$ for all n sufficiently large.

(ii) If, however, $R > 1$, prove that $\{a_n\}$ is not bounded, and hence diverges.

(iii) When $R = 1$ the test is inconclusive: Give an example of a convergent sequence and a divergent sequence, both of which satisfy $R = 1$.

(iv) Given $a > 0$, find $\lim(a^n/n!)$.

9. Which of the following sequences are properly divergent? Prove your answers.

$$(a)\ \{\sqrt{n^2+1}\}, \quad (b)\ \{n(-1)^n\}, \quad (c)\ \left\{\frac{3^n - 10}{2^n}\right\}, \quad (d)\ \left\{\frac{n}{\sqrt{n+10}}\right\}.$$

10. Let $\{a_n\}$ and $\{b_n\}$ be sequences of real numbers with $\lim a_n = +\infty$ and $b_n \neq 0$ for n large and suppose that for some real number R, we have

$$\lim \frac{a_n}{b_n} = R.$$

(i) If $R > 0$, prove that $\lim b_n = +\infty$.

(ii) If $R < 0$, prove that $\lim b_n = -\infty$.

(iii) Can you draw any conclusions if $R = 0$?

3.3 The Monotone Criteria, the Bolzano–Weierstrass Theorem, and e

Up to now, we have proved the convergence of a sequence by first exhibiting an a priori limit of the sequence and then proving that the sequence converged to the exhibited value. For instance, we showed that the sequence $\{1/(n+1)\}$ converges by showing that it converges to 0. Can we still determine whether a sequence converges without producing an a priori limit value? The answer is yes, and there are two ways to do this. One is called the monotone criterion, and the other is the Cauchy criterion. We study the monotone criterion in this section and save Cauchy's criterion for the next. In this section we work strictly with sequences of *real numbers*.

3.3.1 Monotone Criterion

A **monotone sequence** $\{a_n\}$ of real numbers is a sequence that is either **nondecreasing**, $a_n \leq a_{n+1}$ for each n,

$$a_1 \leq a_2 \leq a_3 \leq \cdots ,$$

or **nonincreasing**, $a_n \geq a_{n+1}$ for each n,

$$a_1 \geq a_2 \geq a_3 \geq \cdots .$$

Here is a picture to keep in mind:

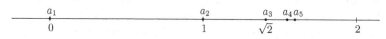

nondecreasing   nonincreasing

Example 3.24 Consider the sequence of real numbers $\{a_n\}$ defined inductively as follows:

$$a_1 = 0, \qquad a_{n+1} = \sqrt{1 + a_n}, \quad n \in \mathbb{N}. \qquad (3.23)$$

Thus,

$$a_1 = 0, \quad a_2 = 1, \quad a_3 = \sqrt{1 + \sqrt{1}}, \quad a_4 = \sqrt{1 + \sqrt{1 + \sqrt{1}}}, \ldots .$$

Observe that

$$a_{n+1} = \sqrt{1 + \sqrt{1 + \sqrt{1 + \sqrt{\cdots + \sqrt{1}}}}}, \qquad (3.24)$$

where there are n square roots here. To get a feeling for a sequence, we always recommend to plot a few points, which we do in Fig. 3.4. We claim that this sequence is nondecreasing: $0 = a_1 \leq a_2 \leq a_3 \leq \cdots$. To see this, we use induction to prove that $0 \leq a_n \leq a_{n+1}$ for each n. If $n = 1$, then $a_1 = 0 \leq 1 = a_2$. Assume that $0 \leq a_n \leq a_{n+1}$. Then $1 + a_n \leq 1 + a_{n+1}$, so using that square roots preserve inequalities (the root rules on p. 94), we see that

a_1		a_2	a_3 $a_4 a_5$	
0		1	$\sqrt{2}$	2

Fig. 3.4 We have $a_1 = 0$, $a_2 = 1$, $a_3 = \sqrt{2}$, $a_4 = \sqrt{1 + \sqrt{2}} \approx 1.554$, $a_5 \approx 1.598$, $a_6 = 1.612$, $a_7 \approx 1.616, \ldots$

$$a_{n+1} = \sqrt{1 + a_n} \leq \sqrt{1 + a_{n+1}} = a_{n+2}.$$

This establishes the induction step, so we conclude that our sequence $\{a_n\}$ is nondecreasing. We also claim that $\{a_n\}$ is bounded. Indeed, based on Fig. 3.4, we conjecture that $a_n \leq 2$ for each n. Again we proceed by induction. First, we have $a_1 = 0 \leq 2$. If $a_n \leq 2$, then by the definition of a_{n+1}, we have

$$a_{n+1} = \sqrt{1 + a_n} \leq \sqrt{1 + 2} \leq \sqrt{4} = 2,$$

which proves that $\{a_n\}$ is bounded by 2.

The monotone criterion in Theorem 3.11 states in particular that if a monotone sequence of real numbers is bounded, then it must converge. From a physical standpoint, this is obvious; see the car in Fig. 3.5. The same is true of a sequence of numbers, as in Fig. 3.6. In particular, this implies that our recursive sequence (3.23) in Example 3.24 converges.

Fig. 3.5 A car rolling forward (is monotone, nondecreasing) with a wall in front (is bounded) will come to rest either before or at the wall

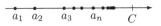

Fig. 3.6 If $\{a_n\}$ is nondecreasing and bounded (say by a constant C), like a car rolling toward a wall, it must "cluster" before, or at, C

Monotone criterion

> **Theorem 3.11** *A monotone sequence of real numbers converges if and only if the sequence is bounded.*

Proof We already know that if a sequence converges, then it must be bounded. So, let $\{a_n\}$ be a bounded monotone sequence; we must prove that it converges. If $\{a_n\}$ is nonincreasing, $a_1 \geq a_2 \geq a_3 \geq \cdots$, then the sequence $\{-a_n\}$ is nondecreasing: $-a_1 \leq -a_2 \leq \cdots$. Thus, if we prove that bounded *nondecreasing* sequences converge, then $\lim(-a_n)$ would exist. This would imply that $\lim a_n = -\lim(-a_n)$ exists too. So it remains to prove our theorem under the assumption that $\{a_n\}$ is *nondecreasing*: $a_1 \leq a_2 \leq \cdots$. Let L equal the supremum of the set $\{a_1, a_2, a_3, \dots\}$; this supremum exists because the sequence is bounded. Let $\varepsilon > 0$. Then $L - \varepsilon$ is smaller than L. Since L is the least upper bound of the set $\{a_1, a_2, a_3, \dots\}$ and $L - \varepsilon < L$, there must exist an N such that $L - \varepsilon < a_N \leq L$. Since the sequence is nondecreasing, for all $n > N$ we must also have $L - \varepsilon < a_n \leq L$. Since $L < L + \varepsilon$, we conclude that

$$n > N \implies L - \varepsilon < a_n < L + \varepsilon.$$

Hence, $\lim a_n = L$.   ∎

Example 3.25 The monotone criterion implies that our sequence (3.23) converges, to some real number L. Squaring both sides of a_{n+1}, we see that

$$a_{n+1}^2 = 1 + a_n.$$

The subsequence $\{a_{n+1}\}$ also converges to L by Theorem 3.8 on p. 167. Therefore, by the algebra of limits,

$$L^2 = \lim a_{n+1}^2 = \lim(1 + a_n) = 1 + L.$$

Solving for L, we get (using the quadratic formula),

$$L = \frac{1 \pm \sqrt{5}}{2}.$$

Since $0 = a_1 \leq a_2 \leq a_3 \leq \cdots \leq a_n \to L$, and limits preserve inequalities (Theorem 3.7 on p. 166), the limit L cannot be negative, so we conclude that

$$L = \frac{1 + \sqrt{5}}{2}.$$

This number is called the **golden ratio** and is denoted by Φ. In view of the expressions found in (3.24), we can interpret Φ as the infinite "continued square root":

$$\Phi = \frac{1 + \sqrt{5}}{2} = \sqrt{1 + \sqrt{1 + \sqrt{1 + \sqrt{1 + \sqrt{1 + \cdots}}}}}. \tag{3.25}$$

There are many stories about Φ; unfortunately, many of them are false, see [159].

Our next important theorem is the monotone subsequence theorem. It says that given any sequence of real numbers, whether or not it converges, you can always choose from it a monotone subsequence. Here's a picture (Fig. 3.7).

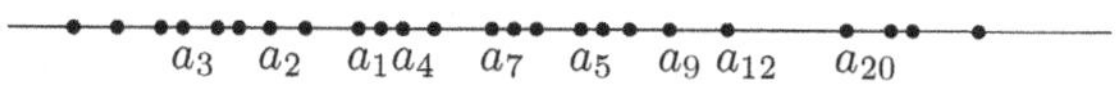

Fig. 3.7 The *dots* represents some points in a sequence $\{a_n\}$, and we label some of the points. The finitely many points $a_1, a_4, a_5, a_9, a_{12}, a_{20}$ are nondecreasing, and if there are infinitely many points in the sequence to the *right* of a_{20}, then we can continue to choose points in the sequence to get a nondecreasing subsequence

There are many nice proofs of the following theorem, such as those in [19, 170, 241]. In the proof we use, the notion of the **maximum** of a set A of real numbers, by which we mean a number a that satisfies $a \in A$ and $a = \sup A$, in which case we write $a = \max A$.

Monotone subsequence theorem

Theorem 3.12 *Every sequence of real numbers has a monotone subsequence.*

Proof Let $\{a_n\}$ be a sequence of real numbers. Then the statement "for every $n \in \mathbb{N}$, the maximum of the set $\{a_n, a_{n+1}, a_{n+2}, \dots\}$ exists" is either a true statement, or it's false, which means "there is an $m \in \mathbb{N}$ such that the maximum of the set $\{a_m, a_{m+1}, a_{m+2}, \dots\}$ does not exist."

Case 1: Suppose we are in the first case: for each n, $\{a_n, a_{n+1}, a_{n+2}, \dots\}$ has a greatest member. In particular, we can choose a_{ν_1} such that

$$a_{\nu_1} = \max\{a_1, a_2, \dots\}.$$

Now $\{a_{\nu_1+1}, a_{\nu_1+2}, \dots\}$ has a greatest member, so we can choose a_{ν_2} such that

$$a_{\nu_2} = \max\{a_{\nu_1+1}, a_{\nu_1+2}, \dots\}.$$

Since a_{ν_2} is obtained by taking the maximum of a smaller set of elements, we have $a_{\nu_2} \le a_{\nu_1}$. Let

$$a_{\nu_3} = \max\{a_{\nu_2+1}, a_{\nu_2+2}, \dots\}.$$

Since a_{ν_3} is obtained by taking the maximum of a smaller set of elements than the set defining a_{ν_2}, we have $a_{\nu_3} \le a_{\nu_2}$. Continuing by induction, we construct a monotone (nonincreasing) subsequence.

Case 2: Suppose that the maximum of the set $A = \{a_m, a_{m+1}, a_{m+2}, \dots\}$ does not exist, for some $m \ge 1$. Let $a_{\nu_1} = a_m$. Since A has no maximum, there is a $\nu_2 > m$ such that

$$a_m < a_{\nu_2},$$

for if there were no such a_{ν_2}, then a_m would be a maximum element of A, which we know is not possible. Since none of the elements $a_m, a_{m+1}, \dots, a_{\nu_2}$ is a maximum element of A, there must exist a $\nu_3 > \nu_2$ such that

$$a_{\nu_2} < a_{\nu_3},$$

for otherwise one of $a_m, \dots, a_{\nu_2}$ would be a maximum element of A. Similarly, since none of $a_m, \dots, a_{\nu_2}, \dots, a_{\nu_3}$ is a maximum element of A, there must exist a $\nu_4 > \nu_3$ such that

$$a_{\nu_3} < a_{\nu_4}.$$

Continuing by induction, we construct a monotone (nondecreasing) sequence $\{a_{\nu_k}\}$. ∎

3.3.2 The Bolzano–Weierstrass Theorem

The following theorem, named after Bernard Bolzano (1781–1848) and Karl Weierstrass (1815–1897), is one of the most important results in analysis and will be frequently employed in the sequel.

Bolzano–Weierstrass theorem for $\mathbb{R}$

> **Theorem 3.13** *Every bounded sequence in $\mathbb{R}$ has a convergent subsequence. In fact, if the sequence is contained in a closed interval I, then the limit of the convergent subsequence is also in I.*

Proof Let $\{a_n\}$ be a bounded sequence in $\mathbb{R}$. By the monotone subsequence theorem, this sequence has a monotone subsequence $\{a_{\nu_n}\}$, which of course is also bounded. By the monotone criterion (Theorem 3.11), this subsequence converges. Suppose that $\{a_n\}$ is contained in a closed interval $I = [a, b]$. Then $a \leq a_{\nu_n} \leq b$ for each n. Since limits preserve inequalities, the limit of the subsequence $\{a_{\nu_n}\}$ also lies in $[a, b]$. ∎

Using induction on m (we already did the $m = 1$ case), we leave the proof of the following generalization to you, if you're interested.

Bolzano–Weierstrass theorem for $\mathbb{R}^m$

> **Theorem 3.14** *Every bounded sequence in $\mathbb{R}^m$ has a convergent subsequence.*

Example 3.26 For many sequences it's easy to find convergent subsequences explicitly. For example, we've already looked at the complex, or $\mathbb{R}^2$, sequence

$$i, \ i^2 = -1, \ i^3 = -i, \ i^4 = 1, \ i^5 = i, \ i^6 = -1, \ldots, a_n = i^n, \ldots .$$

With $n = 1, 5, 9, 13, \ldots$ we get the convergent subsequence

$$i, \ i, \ i, \ i, \ \ldots .$$

There are many other convergent subsequences, all of which will converge to either $i, -1, -i$, or 1.

3.3.3 The Number e

We now define Euler's constant e by a method that has been around for ages; cf. [129, p. 82], [261]. Consider the two sequences whose terms are given by

$$a_n = \left(1 + \frac{1}{n}\right)^n = \left(\frac{n+1}{n}\right)^n \quad \text{and} \quad b_n = \left(1 + \frac{1}{n}\right)^{n+1} = \left(\frac{n+1}{n}\right)^{n+1},$$

where $n = 1, 2, \ldots$. We shall prove that the sequence $\{a_n\}$ is bounded above and is **strictly increasing**, which means that $a_n < a_{n+1}$ for all n. We'll also prove that $\{b_n\}$ is bounded below and **strictly decreasing**, which means that $b_n > b_{n+1}$ for all n. In particular, the limits $\lim a_n$ and $\lim b_n$ exist by the monotone criterion. Notice that

$$b_n = a_n \left(1 + \frac{1}{n}\right),$$

and $1 + 1/n \to 1$, so if sequences $\{a_n\}$ and $\{b_n\}$ converge, they must converge to the same limit. This limit is denoted by the letter e, introduced in 1727 by Euler perhaps because "e" is the first letter in "exponential" [35, p. 442], *not* because "e" is the first letter of his last name!

The proof that the sequences above are monotone follows from Bernoulli's inequality on p. 42. First, to see that $b_{n-1} > b_n$ for $n \geq 2$, observe that

$$\frac{b_{n-1}}{b_n} = \left(\frac{n}{n-1}\right)^n \left(\frac{n}{n+1}\right)^{n+1} = \left(\frac{n^2}{n^2-1}\right)^n \left(\frac{n}{n+1}\right)$$

$$= \left(1 + \frac{1}{n^2-1}\right)^n \left(\frac{n}{n+1}\right).$$

According to Bernoulli's inequality, we have

$$\left(1 + \frac{1}{n^2-1}\right)^n > 1 + \frac{n}{n^2-1} > 1 + \frac{n}{n^2} = \frac{n+1}{n},$$

which implies that

$$\frac{b_{n-1}}{b_n} > \frac{n+1}{n} \cdot \frac{n}{n+1} = 1 \quad \Longrightarrow \quad b_{n-1} > b_n.$$

This proves that $\{b_n\}$ is strictly decreasing. Certainly $b_n > 0$ for each n, so the sequence b_n is bounded below and hence converges.

To see that $a_{n-1} < a_n$ for $n \geq 2$, we proceed in a similar manner:

$$\frac{a_n}{a_{n-1}} = \left(\frac{n+1}{n}\right)^n \left(\frac{n-1}{n}\right)^{n-1} = \left(\frac{n^2-1}{n^2}\right)^n \left(\frac{n}{n-1}\right)$$

$$= \left(1 - \frac{1}{n^2}\right)^n \left(\frac{n}{n-1}\right).$$

Bernoulli's inequality for $n \geq 2$ implies that

$$\left(1 - \frac{1}{n^2}\right)^n > 1 - \frac{n}{n^2} = 1 - \frac{1}{n} = \frac{n-1}{n},$$

so

$$\frac{a_n}{a_{n-1}} > \frac{n-1}{n} \cdot \frac{n}{n-1} = 1 \implies a_{n-1} < a_n.$$

This shows that $\{a_n\}$ is strictly increasing. Finally, since $a_n < b_n \le b_1 = 4$, the sequence $\{a_n\}$ is bounded above.

In conclusion, we have proved that the limit

$$e := \lim_{n \to \infty}\left(1 + \frac{1}{n}\right)^n$$

exists, which equals by definition the number denoted by e. Moreover, since $a_n < b_n$ for all n, it follows that

$$\left(1 + \frac{1}{n}\right)^n < e < \left(1 + \frac{1}{n}\right)^{n+1}, \qquad \text{for all } n. \tag{3.26}$$

We shall need this inequality later when we discuss the Euler–Mascheroni constant. This inequality is also useful in studying a "weak form" of Stirling's formula, which we now describe. Recall that $0! = 1$ and that given a positive integer n, we define $n!$ (which we read "n factorial") as $n! = 1 \cdot 2 \cdot 3 \cdots n$. For n positive, observe that $n!$ is less than n^n, or equivalently, $\sqrt[n]{n!}/n < 1$. A natural question to ask is, How much less than one is the ratio $\sqrt[n]{n!}/n$? Using (3.26), in Problem 6 you will prove that

$$\lim \frac{\sqrt[n]{n!}}{n} = \frac{1}{e}, \qquad \text{("weak form" of Stirling's formula).} \tag{3.27}$$

▶ **Exercises 3.3**

1. (a) Show that the sequence defined inductively by $a_{n+1} = \frac{1}{3}(2a_n + 4)$ with $a_1 = 0$ is nondecreasing and bounded above by 4. Find the limit.

   (b) Let $\alpha \ge 1$ and let $a_1 = 1$. Show that the sequence defined inductively by $a_{n+1} = \sqrt{\alpha\, a_n}$ is nondecreasing and bounded above by α. Find the limit.

   (c) Let $a > 0$. Show that the sequence defined inductively by $a_{n+1} = a_n/(1 + 2a_n)$ with $a_1 = a$ is a bounded monotone sequence. Find the limit.

   (d) Let $a \ge 6$. Prove that the sequence defined inductively by $a_{n+1} = 5 + \sqrt{a_n - 5}$ with $a_1 = a$ is a bounded monotone sequence. Find the limit.

   (e) Let $b > 0$ and let $a_1 \ge b/2$. Show that the sequence defined inductively by $a_{n+1} = b - b^2/(4a_n)$ is a bounded monotone sequence. Find the limit. Suggestion: Pick, e.g., $b = 2$ and $a_1 = 2$ and calculate a few values of a_n to conjecture whether $\{a_n\}$ is, for general $b > 0$ and $a_1 \ge b/2$, nondecreasing or nonincreasing. Also conjecture bounds. Now prove your conjectures.

(f) Let $a \in \mathbb{R}$ with $1 \le a \le 3$. Show that the sequence defined inductively by $a_{n+1} = 1/(4 - a_n)$ with $a_1 = a$ is a bounded monotone sequence. Find the limit.

2. Show that the sequence with $a_n = \sum_{k=1}^{n} \frac{1}{n+k} = \frac{1}{n+1} + \frac{1}{n+2} + \cdots + \frac{1}{n+n}$ is a bounded monotone sequence and its limit L satisfies $1/2 \le L \le 1$. The limit of this sequence is not at all obvious (it equals $\log 2$; see Problem 8 on p. 312).

3. **(Computing square roots)** In this problem we give two different ways to express square roots in terms of sequences.

   (1) Let $a > 0$. Let a_1 be a positive number and define

   $$a_{n+1} = \frac{1}{2}\left(a_n + \frac{a}{a_n}\right), \qquad n \ge 1.$$

   (i) Show that $a_n > 0$ for all n and $a_{n+1}^2 - a \ge 0$ for all n.
   (ii) Show that $\{a_n\}$ is nonincreasing for $n \ge 2$.
   (iii) Conclude that $\{a_n\}$ converges and find its limit.

   (2) Let $a \ge 0$ and fix a real number $k > 0$ such that $\sqrt{a} \le k$. Show that the sequence defined inductively by $a_{n+1} = a_n + \frac{1}{2k}(a - a_n^2)$ with $a_1 = 0$ is nondecreasing and bounded above by $\sqrt{a}$. Prove that $a_n \to \sqrt{a}$. Suggestion: Assuming $a_n \le \sqrt{a}$, to prove that $a_{n+1} \le \sqrt{a}$, write $a_n = \sqrt{a} - b$, where $b \ge 0$.

4. (Cf. [268]) In this problem we analyze the constant e based on the arithmetic–geometric mean inequality (AGMI) (see Problem 7 on p. 46). Assume the AGMI, which states that given $n + 1$ nonnegative real numbers $x_1, \ldots, x_{n+1}$,

   $$x_1 \cdot x_2 \cdots x_{n+1} \le \left(\frac{x_1 + x_2 + \cdots + x_{n+1}}{n + 1}\right)^{n+1}.$$

   (i) Put $x_k = (1 + 1/n)$ for $k = 1, \ldots, n$ and $x_{n+1} = 1$ in the AGMI to prove that the sequence $a_n = (1 + \frac{1}{n})^n$ is nondecreasing.
   (ii) If $b_n = (1 + 1/n)^{n+1}$, then show that for $n \ge 2$,

   $$\frac{b_n}{b_{n-1}} = \left(1 - \frac{1}{n^2}\right)^n \left(1 + \frac{1}{n}\right) = \underbrace{\left(1 - \frac{1}{n^2}\right) \cdots \left(1 - \frac{1}{n^2}\right)}_{n \text{ times}} \left(1 + \frac{1}{n}\right).$$

   Applying the AGMI to the right-hand side, show that $b_n/b_{n-1} \le 1$, which shows that the sequence $\{b_n\}$ is nonincreasing.

   (iii) Conclude that both sequences $\{a_n\}$ and $\{b_n\}$ converge. Of course, just as in the text, we denote their common limit by e.

5. **(Continued roots)** For more on this subject, see [4], [109], [162, p. 775], [230], [115].

(1) Fix $k \in \mathbb{N}$ with $k \geq 2$ and fix $a > 0$. Show that the sequence defined inductively by $x_{n+1} = \sqrt[k]{a + x_n}$ with $x_1 = \sqrt[k]{a}$ is a bounded monotone sequence. Prove that the limit L is a root of the equation $x^k - x - a = 0$. The limit L can be thought of as the *continued root* (can you see why?)

$$L = \sqrt[k]{a + \sqrt[k]{a + \sqrt[k]{a + \sqrt[k]{a + \cdots}}}}.$$

(2) Let $\{a_n\}$ be a sequence of nonnegative real numbers. Define the sequence $\{\alpha_n\}$ by

$$\alpha_1 = \sqrt{a_1}, \quad \alpha_2 = \sqrt{a_1 + \sqrt{a_2}}, \quad \alpha_3 = \sqrt{a_1 + \sqrt{a_2 + \sqrt{a_3}}},$$

and so forth.[7] Prove that $\{\alpha_n\}$ converges if and only if there is a constant $M \geq 0$ such that $\sqrt[2^n]{a_n} \leq M$ for all n. Suggestion: To prove "only if," show that $\sqrt[2^n]{a_n} \leq \alpha_n$, and to prove "if," show that $\alpha_n \leq$

$$\sqrt{M^2 + \sqrt{M^{2^2} + \sqrt{\cdots + \sqrt{M^{2^n}}}}} = M b_n \quad \text{where } b_n = \sqrt{1 + \sqrt{1 + \sqrt{\cdots + \sqrt{1}}}}$$

(where there are n square roots) is found in (3.23); in particular, in Example 3.24 we showed that $b_n \leq 2$.

(3) Show that

$$\sqrt{1 + \sqrt{2 + \sqrt{3 + \sqrt{4 + \sqrt{5 + \cdots}}}}}$$

can be defined. This number is called **Kasner's number**, named after Edward Kasner (1878–1955), and is approximately $1.75793\ldots$.

6. In this problem we prove (3.27), the "weak form" of Stirling's formula.

(i) Prove that for each natural number n, $(n - 1)! \leq n^n e^{-n} e \leq n!$. Suggestion: Can you use induction and (3.26)? (You can also prove these inequalities using integrals as in [138, p. 219], but using (3.26) gives an "elementary" proof that is free of integration theory.)

(ii) Using (i), prove that for every natural number n,

$$\frac{e^{1/n}}{e} \leq \frac{\sqrt[n]{n!}}{n} \leq \frac{e^{1/n} n^{1/n}}{e}.$$

(iii) Now prove (3.27). Using (3.27), prove that

---

[7] For each $n \in \mathbb{N}$, define $f_n : [0, \infty) \to [0, \infty)$ by $f_n(x) = \sqrt{a_n + x}$. Then $\alpha_1 = f_1(0)$, $\alpha_2 = f_1(f_2(0))$, $\alpha_3 = f_1(f_2(f_3(0)))$, and in general, $\alpha_n := (f_1 \circ f_2 \circ \cdots \circ f_n)(0)$.

$$\lim \left(\frac{(3n)!}{n^{3n}} \right)^{1/n} = \frac{27}{e^3} \quad \text{and} \quad \lim \left(\frac{(3n)!}{n!\, n^{2n}} \right)^{1/n} = \frac{27}{e^2}.$$

3.4 Completeness, the Cauchy Criterion, and Contractive Sequences

The monotone criterion gives a criterion for convergence (in $\mathbb{R}$) of a monotone sequence of real numbers. Now what if the sequence is not monotone? The Cauchy criterion, originating with Bolzano, but then made into a formulated *criterion* by Cauchy [129, p. 87], gives a convergence criterion for general sequences of real numbers, and more generally, sequences of complex numbers and vectors.

3.4.1 Cauchy Sequences

Intuitively, a Cauchy sequence is a sequence whose points eventually cluster, clump up, or accumulate closer and closer to each other, as seen here (Fig. 3.8):

Fig. 3.8 A Cauchy sequence in $\mathbb{C} = \mathbb{R}^2$. The points get closer and closer to each other the farther you go in the sequence. That is, given any distance, any two points far enough down the sequence will be within that distance

Here's a precise definition: A sequence $\{a_n\}$ in $\mathbb{R}^m$ is said to be **Cauchy** if

for every $\varepsilon > 0$, there is $N \in \mathbb{R}$ such that $k, n > N \implies |a_k - a_n| < \varepsilon$.

Here, the ε tells "how close" you want the points to be to each other, and the N tells you how "far along" the sequence you have to go to have the points at least that close.

Example 3.27 The sequence of real numbers with $a_n = \frac{2n-1}{n-3}$ and $n \geq 4$ is Cauchy. To see this, let $\varepsilon > 0$. We need to prove there is a real number N such that

$$k, n > N \implies \left| \frac{2k-1}{k-3} - \frac{2n-1}{n-3} \right| < \varepsilon.$$

To see this, we "massage" the right-hand expression:

$$\left| \frac{2k-1}{k-3} - \frac{2n-1}{n-3} \right| = \left| \frac{(2k-1)(n-3) - (2n-1)(k-3)}{(k-3)(n-3)} \right|$$

$$= \left| \frac{5(n-k)}{(k-3)(n-3)} \right| \leq \left| \frac{5n}{(k-3)(n-3)} \right| + \left| \frac{5k}{(k-3)(n-3)} \right|.$$

Now observe that for $n \geq 4$, we have $\frac{n}{4} \geq 1$, so

$$n - 3 \geq n - \left(3 \cdot \frac{n}{4} \right) = n - \frac{3n}{4} = \frac{n}{4} \quad \Longrightarrow \quad \frac{1}{n-3} \leq \frac{4}{n}.$$

Thus, for $n, k \geq 4$, we have

$$\left| \frac{5n}{(k-3)(n-3)} \right| + \left| \frac{5k}{(k-3)(n-3)} \right| < 5n \cdot \frac{4}{k} \cdot \frac{4}{n} + 5k \cdot \frac{4}{k} \cdot \frac{4}{n} = \frac{80}{k} + \frac{80}{n}.$$

Hence,

$$\text{for } k, n \geq 4, \quad \left| \frac{2k-1}{k-3} - \frac{2n-1}{n-3} \right| < \frac{80}{k} + \frac{80}{n}.$$

Now to make the left-hand side less than ε, all we have to do is make the right-hand side less than ε, and we can do this by noticing that we can make

$$\frac{80}{k} + \frac{80}{n} < \varepsilon \quad \text{by making} \quad \frac{80}{k} < \frac{\varepsilon}{2} \quad \text{and} \quad \frac{80}{n} < \frac{\varepsilon}{2}.$$

These latter inequalities hold if and only if $k, n > 160/\varepsilon$. For this reason, let us pick N to be the larger of 3 and $160/\varepsilon$. Let $k, n > N$ (that is, $k, n \geq 4$ and $k, n > 160/\varepsilon$). Then,

$$\left| \frac{2k-1}{k-3} - \frac{2n-1}{n-3} \right| < \frac{80}{k} + \frac{80}{n} < \frac{\varepsilon}{2} + \frac{\varepsilon}{2} = \varepsilon.$$

This shows that the sequence $\{a_n\}$ is Cauchy. Notice that this sequence, $\{\frac{2n-1}{n-3}\}$, converges (to the number 2).

Example 3.28 Here's a more sophisticated example of a Cauchy sequence. Let $a_0 = 0$, $a_1 = 1$, and for $n \geq 2$, we let a_n be the arithmetic mean between the previous two terms:

$$a_n = \frac{a_{n-2} + a_{n-1}}{2}, \quad n \geq 2.$$

This sequence is certainly not monotone, as seen here (Fig. 3.9):

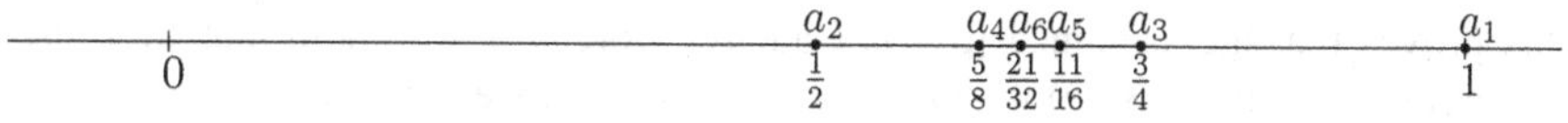

Fig. 3.9 The number a_n is always halfway between a_{n-1} and a_{n-2}

However, we shall prove that $\{a_n\}$ is Cauchy. To do so, we first prove by induction that

$$a_{n+1} - a_n = \left(-\frac{1}{2}\right)^n. \tag{3.28}$$

Since $a_0 = 0$ and $a_1 = 1$, this equation holds for $n = 0$. Assume that the equation holds for n. Then

$$a_{n+2} - a_{n+1} = \frac{a_n + a_{n+1}}{2} - a_{n+1} = \frac{1}{2}\left(a_n - a_{n+1}\right) = -\frac{1}{2}\left(-\frac{1}{2}\right)^n = \left(-\frac{1}{2}\right)^{n+1},$$

which proves the induction step. With (3.28) in hand, we show that the sequence $\{a_n\}$ is Cauchy. Let k, n be any natural numbers, where by symmetry, we assume that $k \leq n$ (otherwise, just switch k and n in what follows). We now form the *telescoping sum*

$$a_n = \left(a_n - a_{n-1}\right) + \left(a_{n-1} - a_{n-2}\right) + \left(a_{n-2} - a_{n-3}\right) + \cdots + \left(a_{k+1} - a_k\right) + a_k,$$

noticing that all the terms on the right after a_n cancel in pairs. Let $n = k + j$, where $j \geq 0$. Then in the telescoping sum, bringing a_k to the left-hand side, using (3.28) to rewrite the terms in parentheses, and using the sum of a geometric progression, Eq. (2.3) on p. 41, we obtain

$$a_n - a_k = r^{n-1} + r^{n-2} + r^{n-3} + \cdots + r^k \quad \left(\text{where we put } r = -\frac{1}{2}\right)$$

$$= r^k\left[1 + r + r^2 + \cdots + r^{j-1}\right]$$

$$= r^k \cdot \frac{1 - r^j}{1 - r}$$

$$= r^k \cdot \frac{2}{3} \cdot \left(1 - r^j\right), \tag{3.29}$$

as $1 - r = 3/2$. Since $\frac{2}{3} \cdot \left|1 - r^j\right| \leq \frac{2}{3} \cdot \left(1 + 1/2\right) = 1$, we conclude that

$$|a_k - a_n| \leq \frac{1}{2^k}, \qquad \text{for all } k, n \text{ with } k \leq n.$$

Now let $\varepsilon > 0$. Since $1/2 < 1$, we know that $1/2^k \to 0$ as $k \to \infty$ (see Example 3.5 on p. 156). Therefore, there is an N such that for all $k > N$, $1/2^k < \varepsilon$. In particular, for such k, it follows that

$$|a_k - a_n| < \varepsilon.$$

This proves that the sequence $\{a_n\}$ is Cauchy. Moreover, we claim that $\{a_n\}$ also converges. Indeed, by (3.29) with $k = 0$, so that $n = j$, we see that

$$a_n = \frac{2}{3} \cdot (1 - r^n), \quad \text{where } r = -\frac{1}{2}.$$

Since $|r| < 1$, we know that $r^n \to 0$. Hence,

$$\lim a_n = \frac{2}{3}.$$

We have thus far given two examples of Cauchy sequences, both of which also converge. In Theorem 3.16 below, we shall prove that every Cauchy sequence must converge. In real life this is "obvious"; for example, consider an airplane (Fig. 3.10):

Fig. 3.10 *Left* an airplane circling above making tighter and tighter circles (the Cauchy condition). *Right* you conclude that there must an airport nearby to which the airplane is converging

Similarly, if you see a mouse in your yard walking in tighter and tighter circles, you may infer that there is some food to which it is converging; or if you notice students coming from various directions clustering, you know that there is some event to which they are converging. The same is the case with $\mathbb{R}^m$.

3.4.2 Cauchy Criterion

The following two proofs use the $\varepsilon/2$-*trick*.

Lemma 3.15 *A Cauchy sequence in $\mathbb{R}^m$ that has a convergent subsequence is itself convergent (with the same limit as the subsequence).*

Proof Let $\{a_n\}$ be a Cauchy sequence and assume that $a_{\nu_n} \to L$ for some subsequence of $\{a_n\}$. We shall prove that $a_n \to L$. Let $\varepsilon > 0$. Since $\{a_n\}$ is Cauchy, there is an N such that

$$k, n > N \implies |a_k - a_n| < \frac{\varepsilon}{2}.$$

Since $a_{\nu_n} \to L$, there is a natural number $k \in \{\nu_1, \nu_2, \nu_3, \nu_4, \dots\}$ with $k > N$ such that

$$|a_k - L| < \frac{\varepsilon}{2}.$$

Now let $n > N$ be arbitrary. Then using the triangle inequality and the two inequalities displayed, we see that

$$|a_n - L| = |a_n - a_k + a_k - L| \le |a_n - a_k| + |a_k - L| < \frac{\varepsilon}{2} + \frac{\varepsilon}{2} = \varepsilon.$$

This proves that $a_n \to L$, and our proof is complete.   ■

Cauchy criterion

Theorem 3.16 *A sequence in $\mathbb{R}^m$ converges if and only if it is Cauchy.*

Proof Let $\{a_n\}$ be a sequence in $\mathbb{R}^m$ converging to $L \in \mathbb{R}^m$. We shall prove that the sequence is Cauchy. Let $\varepsilon > 0$. Since $a_n \to L$, there is an N such that for all $n > N$, we have $|a_n - L| < \frac{\varepsilon}{2}$. Hence, by the triangle inequality,

$$k, n > N \implies |a_k - a_n| \le |a_k - L| + |L - a_n| < \frac{\varepsilon}{2} + \frac{\varepsilon}{2} = \varepsilon.$$

This proves that a convergent sequence is also Cauchy.

Now let $\{a_n\}$ be Cauchy. We shall prove that this sequence also converges. We first prove that the sequence is bounded. To see this, let us put $\varepsilon = 1$ in the definition of being a Cauchy sequence; then there is an N such that for all $k, n > N$, we have $|a_k - a_n| < 1$. Fix $k > N$. Then by the triangle inequality, for every $n > k$,

$$|a_n| = |a_n - a_k + a_k| \le |a_n - a_k| + |a_k| < 1 + |a_k|.$$

It follows that for every natural number n, we have

$$|a_n| \le \max\{|a_1|,\ |a_2|,\ |a_3|,\ \ldots,\ |a_{k-1}|,\ 1 + |a_k|\}.$$

This shows that the sequence $\{a_n\}$ is bounded. The Bolzano–Weierstrass theorem (Theorem 3.14 on p. 179) now implies that $\{a_n\}$ has a convergent subsequence. Lemma 3.15 then guarantees that the whole sequence $\{a_n\}$ converges.   ■

Because every Cauchy sequence in $\mathbb{R}^m$ converges in $\mathbb{R}^m$, we say that $\mathbb{R}^m$ is **complete**. This property of $\mathbb{R}^m$ is essential to many objects in analysis, e.g., series, differentiation, integration, all of which use limit processes. The rationals, $\mathbb{Q}$, is an example of an incomplete space.

Example 3.29 The sequence

$$1,\ 1.4,\ 1.41,\ 1.414,\ 1.4142,\ 1.41421, \ldots$$

is a Cauchy sequence of rational numbers, but its limit (which is supposed to be $\sqrt{2}$) does not exist as a rational number! (By the way, we'll study decimal expansions of real numbers in Section 3.8, starting on p. 226.)

From this example you can imagine the difficulties the incompleteness of $\mathbb{Q}$ can cause when you are trying to do analysis with strictly rational numbers.

3.4.3 Contractive Sequences

Cauchy's criterion is important because it allows us to determine whether a sequence converges or diverges by proving instead that the sequence is or is not Cauchy. Unfortunately, to determine convergence by appealing directly to the definition of a Cauchy sequence is not always easy. The goal of this subsection is to present a simple condition that we could check that guarantees that a given sequence is Cauchy and hence converges.

A sequence $\{a_n\}$ in $\mathbb{R}^m$ is a said to be **contractive** if there is a real number r with $0 < r < 1$ such that for all n,

$$|a_{n+1} - a_n| \le r\,|a_n - a_{n-1}|. \tag{3.30}$$

This inequality says that the distance between adjacent members of the sequence $\{a_n\}$ shrinks (or contracts) by at least a factor of r at each step.

Contractive sequence theorem

> **Theorem 3.17** *If a sequence is contractive, then it converges.*

Proof Let $\{a_n\}$ be a contractive sequence. Then with $n = 2$ in (3.30), we see that

$$|a_3 - a_2| \le r\,|a_2 - a_1| = Cr^2, \quad \text{where} \ \ C = r^{-1}|a_2 - a_1|.$$

With $n = 3$, we get

$$|a_4 - a_3| \le r\,|a_3 - a_2| \le r \cdot Cr^2 = Cr^3.$$

By induction, for $n \ge 2$ we get

$$|a_{n+1} - a_n| \le C\,r^n. \tag{3.31}$$

To prove that $\{a_n\}$ converges, all we have to do is prove that the sequence is Cauchy. Let $k, n \ge 2$ with $k \le n$, say $n = k + j$, where $j \ge 0$. Then according to (3.31), the triangle inequality, and the geometric sum formula from Eq. (2.3) on p. 41, we can write

$$|a_n - a_k| = \left| (a_n - a_{n-1}) + (a_{n-1} - a_{n-2}) + \cdots + (a_{k+1} - a_k) \right|$$
$$\leq |a_n - a_{n-1}| + |a_{n-1} - a_{n-2}| + |a_{n-2} - a_{n-3}| + \cdots + |a_{k+1} - a_k|$$
$$= C\,r^{n-1} + C\,r^{n-2} + C\,r^{n-3} + \cdots + C\,r^k$$
$$= C\,r^k \left[1 + r + r^2 + \cdots + r^{j-1} \right] \quad (\text{since } n = k + j)$$
$$= C\,r^k \frac{1 - r^j}{1 - r} \leq M\,r^k, \quad \text{where } M = \frac{C}{1 - r}.$$

We are now ready to prove that the sequence $\{a_n\}$ is Cauchy. Let $\varepsilon > 0$. Since $r < 1$, we know that $M\,r^k \to 0$ as $n \to \infty$. Therefore, there is an $N > 1$ such that for all $n > N$, $M\,r^k < \varepsilon$. Let $k, n > N$. Then $k, n \geq 2$, and by symmetry, we may assume that $k \leq n$ (otherwise, just switch k and n in what follows). Hence, by the above calculation, we find that
$$|a_n - a_k| \leq M\,r^k < \varepsilon.$$

This proves that the sequence $\{a_n\}$ is Cauchy.   ∎

We remark that by the tails theorem (Theorem 3.3 on p. 162), a sequence $\{a_n\}$ will converge as long as (3.30) holds for sufficiently large n. We also remark that the converse of the contractive sequence theorem is false (see Problem 7). We now consider an example.

Example 3.30 Define

$$a_1 = 1 \quad \text{and} \quad a_{n+1} = \sqrt{9 - 2a_n}, \quad n \geq 1. \tag{3.32}$$

Here is a plot of the first few points in this sequence:

Fig. 3.11 In addition to a_1, using a calculator we obtain $a_2 = \sqrt{7} \approx 2.65$, $a_3 \approx 1.93$, $a_4 \approx 2.27$, $a_5 = 2.11$, and $a_6 \approx 2.19$

It is not a priori obvious that this sequence is well defined; how do we know that $9 - 2a_n \geq 0$ for all n, so that we can take the square root $\sqrt{9 - 2a_n}$ to define a_{n+1}? Thus, we need to show that a_n cannot get so big that $9 - 2a_n$ becomes negative. This is accomplished through the following claim:

For all $n \in \mathbb{N}$, a_n is defined and $1 \leq a_n \leq 3$. $\tag{3.33}$

Of course, from Fig. 3.11 we could (rightly) guess that $1 \leq a_n \leq \sqrt{7}$ for all n, but it turns out that the slightly larger number 3 is enough to prove what we want. Now to prove (3.33), observe that it holds for $n = 1$. If (3.33) holds for a_n, then by algebra, we get $3 \leq 9 - 2a_n \leq 7$. In particular, the square root $a_{n+1} = \sqrt{9 - 2a_n}$ is well defined, and

$$\sqrt{3} \leq a_{n+1} = \sqrt{9 - 2a_n} \leq \sqrt{7}.$$

Since $1 = \sqrt{1} \leq \sqrt{3}$ and $\sqrt{7} \leq \sqrt{9} = 3$, (3.33) holds for a_{n+1}. Therefore, (3.33) holds for every n. Now multiplying by conjugates, for $n \geq 2$ we obtain

$$
\begin{aligned}
a_{n+1} - a_n &= \left(\sqrt{9 - 2a_n} - \sqrt{9 - 2a_{n-1}} \right) \frac{\sqrt{9 - 2a_n} + \sqrt{9 - 2a_{n-1}}}{\sqrt{9 - 2a_n} + \sqrt{9 - 2a_{n-1}}} \\
&= \frac{-2a_n + 2a_{n-1}}{\sqrt{9 - 2a_n} + \sqrt{9 - 2a_{n-1}}} \\
&= \frac{2}{\sqrt{9 - 2a_n} + \sqrt{9 - 2a_{n-1}}} \cdot (-a_n + a_{n-1}).
\end{aligned}
$$

The smallest the denominator can possibly be occurs when a_{n-1} and a_n are the largest they can be, which according to (3.33), is no more than 3. It follows that for every $n \geq 2$,

$$\frac{2}{\sqrt{9 - 2a_n} + \sqrt{9 - 2a_{n-1}}} \leq \frac{2}{\sqrt{9 - 2 \cdot 3} + \sqrt{9 - 2 \cdot 3}} = \frac{2}{\sqrt{3} + \sqrt{3}} = r,$$

where $r = \frac{1}{\sqrt{3}} < 1$. Thus, for every $n \geq 2$,

$$|a_{n+1} - a_n| = \frac{2}{\sqrt{9 - 2a_n} + \sqrt{9 - 2a_{n-1}}} |a_n - a_{n-1}| \leq r |a_n - a_{n-1}|.$$

This proves that the sequence $\{a_n\}$ is contractive, and therefore $a_n \to L$ for some real number L. Because $a_n \geq 0$ for all n and limits preserve inequalities we must have $L \geq 0$ too. Moreover, by (3.32), we have

$$L^2 = \lim a_{n+1}^2 = \lim(9 - 2a_n) = 9 - 2L,$$

which implies that $L^2 + 2L - 9 = 0$. Solving this quadratic equation and taking the positive root, we obtain $L = \sqrt{10} - 1$.

▶ **Exercises 3.4**

1. Prove directly, via the definition, that the following sequences are Cauchy.

$$(a) \ \left\{ 10 + \frac{(-1)^n}{\sqrt{n}} \right\}, \quad (b) \ \left\{ \left(7 + \frac{3}{n} \right)^2 \right\}, \quad (c) \ \left\{ \frac{n^2}{n^2 - 5} \right\}.$$

2. Negate the statement that a sequence $\{a_n\}$ is Cauchy. With your negation, prove that the following sequences are not Cauchy (and hence cannot converge).

$$(a) \ \{(-1)^n\}, \qquad (b) \ \left\{ a_n = \sum_{k=0}^{n} (-1)^n \right\}, \qquad (c) \ \{i^n + 1/n\}.$$

3. Prove that the following sequences are contractive; then find their limits.

   (a) Let $a_1 = 0$ and $a_{n+1} = (2a_n - 3)/4$.
   (b) Let $a_1 = 1$ and $a_{n+1} = \frac{1}{5}a_n^2 - 1$.
   (c) Let $a_1 = 0$ and $a_{n+1} = \frac{1}{8}a_n^3 + \frac{1}{4}a_n + \frac{1}{2}$.
   (d) Let $a_1 = 1$ and $a_{n+1} = \frac{1}{1+3a_n}$. Suggestion: Prove that $\frac{1}{4} \le a_n \le 1$ for all n.
   (e) (Cf. Example 3.28.) Let $a_1 = 0$, $a_2 = 1$, and $a_n = \frac{2}{3}a_{n-2} + \frac{1}{3}a_{n-1}$ for $n > 2$.
   (f) (Cf. Example 3.30.) Let $a_1 = 1$ and $a_{n+1} = \sqrt{5 - 2a_n}$.
   (g) Let $a_1 = 1$ and $a_{n+1} = \frac{a_n}{2} + \frac{1}{a_n}$.

4. Let $f : \mathbb{R}^m \to \mathbb{R}^m$ be **contractive**, which means that there is $0 < r < 1$ such that
$$|f(x) - f(y)| \le r\,|x - y| \quad \text{for all } x, y \in \mathbb{R}^m.$$

   Let $a \in \mathbb{R}^m$ and define the sequence $\{a_n\}$ by $a_1 = a$ and $a_{n+1} = f(a_n)$ for $n = 1, 2, 3, \ldots$.

   (i) Prove that $\{a_n\}$ is contractive.
   (ii) Prove that the limit $L = \lim a_n$ satisfies $f(L) = L$.

5. **(Roots of polynomials)** We can use Cauchy sequences to obtain roots of polynomials. Using a graphing calculator, we see that $x^3 - 4x + 2$ has exactly one root, call it a, in the interval $[0, 1]$. This root is irrational by the rational zeros theorem on p. 84. We can express this root as a sequence of rational numbers as follows. Define the sequence $\{a_n\}$ recursively by $a_{n+1} = \frac{1}{4}(a_n^3 + 2)$ with $a_1 = 0$. Prove that $\{a_n\}$ is contractive and converges to a.

6. Here are some Cauchy limit theorems. Let $\{a_n\}$ be a sequence in $\mathbb{R}^m$.

   (a) Prove that $\{a_n\}$ is Cauchy if and only if for every $\varepsilon > 0$, there is a number N such that for all $n > N$ and $k \ge 1$, $|a_{n+k} - a_n| < \varepsilon$.
   (b) Given a sequence $\{b_n\}$ of natural numbers, we call the sequence $\{d_n\}$, where $d_n = a_{n+b_n} - a_n$, a **difference sequence**. Prove that $\{a_n\}$ is Cauchy if and only if *every* difference sequence converges to zero (that is, is a null sequence). Suggestion: To prove the "if" part, instead prove the contrapositive: If $\{a_n\}$ is not Cauchy, then there is a difference sequence that does not converge to zero.

7. Give an example of a convergent sequence that is not contractive.

8. **(Continued fractions**—see Chapter 8 for more on this amazing topic!) In this problem we investigate the **continued fraction**

$$\sqrt{2} = 1 + \cfrac{1}{2 + \cfrac{1}{2 + \cdots}}.$$

We interpret the infinite fraction on the right as the limit of the fractions

$$a_1 = 1, \quad a_2 = 1 + \frac{1}{2}, \quad a_3 = 1 + \frac{1}{2 + \frac{1}{2}}, \quad a_4 = 1 + \frac{1}{2 + \frac{1}{2 + \frac{1}{2}}}, \ldots,$$

where this sequence is defined recursively by $a_1 = 1$ and $a_{n+1} = 1 + 1/(1 + a_n)$ for $n \geq 1$. Prove that $\{a_n\}$ converges and its limit is $\sqrt{2}$. Here's a related example: Prove that

$$\Phi = 1 + \cfrac{1}{1 + \cfrac{1}{1 + \cfrac{1}{1 + \cdots}}} \qquad (3.34)$$

in the sense that the right-hand continued fraction converges with value Φ, where Φ is the golden ratio defined in (3.25). In other words, prove that $\Phi = \lim \phi_n$, where $\{\phi_n\}$ is the sequence defined by $\phi_1 := 1$ and $\phi_{n+1} = 1 + 1/\phi_n$ for $n \in \mathbb{N}$.

9. The Fibonacci sequence was defined in Problem 9 on p. 193. Prove that the sequence of ratios $\{F_{n+1}/F_n\}$ converges with limit the golden ratio:

$$\Phi = \lim_{n \to \infty} \frac{F_{n+1}}{F_n}.$$

3.5 Baby Infinite Series

Imagine taking a stick of unit length and cutting it in half, getting two sticks of length $1/2$. We then take one of the halves and cut that piece in half, getting two sticks of length $1/4 = 1/2^2$. We now take one of these fourths and cut it in half, getting two sticks of length $1/8 = 1/2^3$. We continue this process indefinitely as seen here:

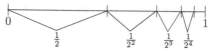

Then the sum of all the lengths of all the sticks formed is 1:

$$1 = \frac{1}{2} + \frac{1}{2^2} + \frac{1}{2^3} + \frac{1}{2^4} + \frac{1}{2^5} + \cdots.$$

Another geometric proof of the sum of this series is to take a square of side length 1 and divide it into halves infinitely many times:

The aim of this section is to rigorously define and study infinite series.[8]

3.5.1 Basic Results on Infinite Series

Given a sequence $\{a_n\}_{n=1}^{\infty}$ of complex numbers, we want to attach a meaning to $\sum_{n=1}^{\infty} a_n$, which we often write as $\sum a_n$ for simplicity. To this end, we define the *n*th **partial sum**, s_n, of the series to be

$$s_n = \sum_{k=1}^{n} a_k = a_1 + a_2 + \cdots + a_n.$$

Of course, here there are only finitely many numbers being summed, so the right-hand side has a clear definition. If the sequence $\{s_n\}$ of partial sums converges, then we say that the infinite series $\sum a_n$ **converges**, and we define

$$\sum a_n = \sum_{n=1}^{\infty} a_n = a_1 + a_2 + a_3 + \cdots$$

as the limit

$$\lim s_n = \lim_{n \to \infty} \sum_{k=1}^{n} a_k.$$

If the sequence of partial sums does not converge, then we say that the series **diverges**. Since $\mathbb{R} \subseteq \mathbb{C}$, restricting to real sequences $\{a_n\}$, we already have built in to the above definition the convergence of a series of real numbers. We remark that a series of nonnegative real numbers can be interpreted as the area of infinitely many rectangles, as seen here (Fig. 3.12):

---

[8]"If you disregard the very simplest cases, there is in all of mathematics not a single infinite series whose sum has been rigorously determined. In other words, the most important parts of mathematics stand without a foundation." Niels Abel (1802–1829) [225]. (Of course, nowadays series are "rigorously determined"—this is the point of this section!)

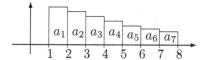

Fig. 3.12 At each natural number n, draw a rectangle between n and $n + 1$ with height a_n. The base of each rectangle has length 1, and the height is a_n, so the area of each rectangle is a_n. Thus, $\sum a_n$ is just the sum of the areas of the rectangles

There is a similar interpretation when some terms in the series are negative; if a_n is negative, we draw a rectangle, but now below the horizontal axis. The sum $\sum a_n$ is then interpreted as the sum of the areas of the rectangles above the horizontal axis minus the sum of the areas of the rectangles below the horizontal axis.

We remark that just as a sequence can be indexed so that its starting value is a_0 or a_{-7} or a_{1234}, we can do a similar thing with series, such as

$$\sum_{n=0}^{\infty} a_n, \qquad \sum_{n=-7}^{\infty} a_n, \qquad \sum_{n=1234}^{\infty} a_n.$$

For convenience, in our proofs we most of the time work with series starting at $n = 1$, although all the results we discuss hold for series starting with any index.

Example 3.31 Consider the series

$$\sum_{n=0}^{\infty} (-1)^n = 1 - 1 + 1 - 1 + - \cdots .$$

Observe that $s_1 = 1$, $s_2 = 1 - 1 = 0$, $s_3 = 1 - 1 + 1 = 1$, and in general, $s_n = 1$ if n is odd, and $s_n = 0$ if n is even. Since $\{s_n\}$ diverges, the series $\sum_{n=0}^{\infty} (-1)^n$ diverges.

Example 3.32 Consider the following series:

$$\sum_{n=1}^{\infty} \frac{1}{n(n+1)} = \frac{1}{1 \cdot 2} + \frac{1}{2 \cdot 3} + \cdots + \frac{1}{n(n+1)} + \cdots .$$

To analyze this series, we use the *method of partial fractions*[9] and note that

---

[9]If $p(x)$ and $q(x) = (x - r_1)(x - r_2) \cdots (x - r_n)$ are polynomials in which the r_i are distinct constants and the degree of p is less than n, the *method of partial fractions* supposes that

$$\frac{p(x)}{q(x)} = \frac{c_1}{x - r_1} + \frac{c_2}{x - r_2} + \cdots + \frac{c_n}{x - r_n}$$

and then solves for the constants $c_1, \ldots, c_n$. If there is a repeated factor, say $(x - r_1)^2$ in $q(x)$, then the term $c_1'/(x - r_1)^2$ is added. It would be advantageous to review the *method of partial fractions* from a calculus book. In the above example, $p(x) = 1$ and $q(x) = x(x + 1)$.

$$\frac{1}{k(k+1)} = \frac{1}{k} - \frac{1}{k+1}.$$

Using this formula, we see that the adjacent terms in s_n cancel (except for the first and the last):

$$s_n = \frac{1}{1 \cdot 2} + \frac{1}{2 \cdot 3} + \cdots + \frac{1}{n(n+1)} \tag{3.35}$$

$$= \left(\frac{1}{1} - \frac{1}{2}\right) + \left(\frac{1}{2} - \frac{1}{3}\right) + \cdots + \left(\frac{1}{n} - \frac{1}{n+1}\right) = 1 - \frac{1}{n+1}.$$

It follows that $s_n = 1 - 1/(n+1) \to 1$, and therefore $\displaystyle\sum_{n=1}^{\infty} \frac{1}{n(n+1)} = 1$.

There are two very simple tests that will help determine the convergence or divergence of a series. The first test might also be called the *fundamental test*, because it is the test that one should always try first when given a series.

nth term test

Theorem 3.18 *If $\sum a_n$ converges, then $a_n \to 0$. Stated another way, if $a_n \not\to 0$, then $\sum a_n$ diverges.*

Proof Assume that $\sum a_n$ converges, let $s = \sum a_n$, and let s_n denote the nth partial sum of the series. Observe that

$$s_n - s_{n-1} = (a_1 + a_2 + \cdots + a_{n-1} + a_n) - (a_1 + a_2 + \cdots + a_{n-1}) = a_n.$$

By definition of convergence of $\sum a_n$, we have $s_n \to s$. Therefore, $s_{n-1} \to s$ as well, whence $a_n = s_n - s_{n-1} \to s - s = 0$. ∎

This theorem is somewhat "obvious"; for example, supposing that $a_n \to 1$, which says that $a_n \approx 1$ for all n large, then the sum $a_1 + a_2 + a_3 + \cdots$ would be adding infinitely many numbers close to 1, which would make the sum diverge.

Example 3.33 The series $\sum_{n=0}^{\infty}(-1)^n = 1 - 1 + 1 - 1 + - \cdots$ and $\sum_{n=1}^{\infty} n = 1 + 2 + 3 + \cdots$ cannot converge, since their nth terms do not tend to zero.

The converse of the nth term test is false; that is, even though $\lim a_n = 0$, it may not follow that $\sum a_n$ exists.[10] Consider the following example.

---

[10]"The sum of an infinite series whose final term vanishes perhaps is infinite, perhaps finite." Jacob Bernoulli (1654–1705) Ars conjectandi.

Example 3.34 (The harmonic series diverges, Proof I) Consider

$$\sum_{n=1}^{\infty} \frac{1}{n} = 1 + \frac{1}{2} + \frac{1}{3} + \frac{1}{4} + \cdots.$$

This series is called the **harmonic series**; see [134] for "what's harmonic about the harmonic series." To see that the harmonic series does not converge, observe that

$$s_{2n} = 1 + \frac{1}{2} + \left(\frac{1}{3} + \frac{1}{4}\right) + \left(\frac{1}{5} + \frac{1}{6}\right) + \cdots + \left(\frac{1}{2n-1} + \frac{1}{2n}\right)$$

$$\geq 1 + \frac{1}{2} + \left(\frac{1}{4} + \frac{1}{4}\right) + \left(\frac{1}{6} + \frac{1}{6}\right) + \cdots + \left(\frac{1}{2n} + \frac{1}{2n}\right)$$

$$= 1 + \frac{1}{2} + \left(\frac{1}{2}\right) + \left(\frac{1}{3}\right) + \cdots + \left(\frac{1}{n}\right)$$

$$= \frac{1}{2} + s_n.$$

Thus, $s_{2n} \geq 1/2 + s_n$. Now if the harmonic series did converge, say to some real number s, that is, $s_n \to s$, then we would also have $s_{2n} \to s$. However, the inequality $s_{2n} \geq 1/2 + s_n$ would imply that $s \geq 1/2 + s$, which is an impossibility. Therefore, the harmonic series does not converge. See Problem 5 for more proofs.

Using the inequality $s_{2n} \geq 1/2 + s_n$, one can show (and we encourage you to do it!) that the partial sums of the harmonic series are unbounded. Then one can deduce that the harmonic series must diverge by the following very useful test.

Nonnegative series test

> **Theorem 3.19** *A series $\sum a_n$ of nonnegative real numbers converges if and only if the sequence $\{s_n\}$ of partial sums is bounded, in which case $s_n \leq s$ for all n, where $s = \sum a_n := \lim s_n$.*

Proof Since $a_n \geq 0$ for all n, we have

$$s_n = a_1 + a_2 + \cdots + a_n \leq a_1 + a_2 + \cdots + a_n + a_{n+1} = s_{n+1},$$

so the sequence of partial sums $\{s_n\}$ is nondecreasing: $s_1 \leq s_2 \leq \cdots \leq s_n \leq \cdots$. By the monotone criterion for sequences, the sequence of partial sums converges if and only if it is bounded. To see that $s_n \leq s := \sum_{m=1}^{\infty} a_m$ for all n, fix $n \in \mathbb{N}$ and note that $s_n \leq s_k$ for all $k \geq n$, because the partial sums are nondecreasing. Taking $k \to \infty$ and using that limits preserve inequalities gives $s_n \leq s$. ∎

Example 3.35 If the sum of the reciprocals of the natural numbers diverges, what about the sum of the reciprocals of the squares (called the 2-series):

$$\sum_{n=1}^{\infty} \frac{1}{n^2} = 1 + \frac{1}{2^2} + \frac{1}{3^2} + \frac{1}{4^2} + \cdots .$$

To investigate the convergence of the 2-series, using (3.35) in Example 3.32, we note that

$$s_n = 1 + \frac{1}{2 \cdot 2} + \frac{1}{3 \cdot 3} + \frac{1}{4 \cdot 4} + \cdots + \frac{1}{n \cdot n}$$
$$\leq 1 + \frac{1}{1 \cdot 2} + \frac{1}{2 \cdot 3} + \frac{1}{3 \cdot 4} + \cdots + \frac{1}{(n-1) \cdot n} \leq 1 + 1 = 2.$$

Since the partial sums of the 2-series are bounded, the 2-series converges. Now, what is the value of the 2-series? This question was the famous *Basel problem*, answered by Leonhard Euler (1707–1783) in 1734. Starting in Section 5.2 on p. 393, we shall rigorously prove, in 11 ways in this book, that the value of the 2-series is $\pi^2/6$! (Now, what does π have to do with reciprocals of squares of natural numbers???)

3.5.2 Some Properties of Series

It is important to understand that the convergence or divergence of a series depends only on the *tails* of the series.

Tails theorem for series

Theorem 3.20 *A series $\sum a_n$ converges if and only if there is an index m such that $\sum_{n=m}^{\infty} a_n$ converges.*

Proof Let s_n denote the nth partial sum of $\sum a_n$, and t_n that of an arbitrary "m-tail" $\sum_{n=m}^{\infty} a_n$. Then with $a = \sum_{k=1}^{m-1} a_k$, we see that $s_n = a + t_n$. It follows that $\{s_n\}$ converges if and only if $\{t_n\}$ converges, and our theorem is proved. ∎

Here's a theorem on linear combinations of series.

Arithmetic properties of series

> **Theorem 3.21** *If $\sum a_n$ and $\sum b_n$ converge, then given complex numbers, c, d, the series $\sum (c\, a_n + d\, b_n)$ converges, and*
>
> $$\sum (c\, a_n + d\, b_n) = c \sum a_n + d \sum b_n.$$
>
> *Moreover, we can group the terms in the series*
>
> $$a_1 + a_2 + a_3 + a_4 + a_5 + \cdots$$
>
> *inside parentheses in any way we wish as long as we do not change the ordering of the terms and the resulting series still converges with sum $\sum a_n$; in other words, the associative law holds for convergent infinite series.*

Proof The nth partial sum of $\sum (c\, a_n + d\, b_n)$ is

$$\sum_{k=1}^{n} (c\, a_k + d\, b_k) = c \sum_{k=1}^{n} a_k + d \sum_{k=1}^{n} b_k = c\, s_n + d\, t_n,$$

where s_n and t_n are the nth partial sums of $\sum a_n$ and $\sum b_n$, respectively. Since $s_n \to \sum a_n$ and $t_n \to \sum b_n$, the first statement of our theorem follows.

Let $1 = \nu_1 < \nu_2 < \nu_3 < \cdots$ be any strictly increasing sequence of integers. We must show that the infinite series

$$(a_1 + a_2 + \cdots + a_{\nu_2 - 1}) + (a_{\nu_2} + a_{\nu_2 + 1} + \cdots + a_{\nu_3 - 1})$$
$$+ (a_{\nu_3} + a_{\nu_3 + 1} + \cdots + a_{\nu_4 - 1}) + (a_{\nu_4} + a_{\nu_4 + 1} + \cdots + a_{\nu_5 - 1}) + \cdots$$

converges with sum $\sum_{n=1}^{\infty} a_n$. In other words, if $\{S_n\}$ denotes the partial sums of this series with parentheses inserted, and if $\{s_n\}$ denotes the partial sums of $\sum_{n=1}^{\infty} a_n$, then we need to show that $\lim S_n = \lim s_n$. However, observe that

$$S_n = (a_1 + a_2 + \cdots + a_{\nu_2 - 1}) + (a_{\nu_2} + a_{\nu_2 + 1} + \cdots + a_{\nu_3 - 1}) + \cdots$$
$$\cdots + (a_{\nu_n} + a_{\nu_n + 1} + \cdots + a_{\nu_{n+1} - 1}) = s_{\nu_{n+1} - 1},$$

since the associative law holds for finite sums (so we can drop the parentheses). Therefore, $\{S_n\}$ is just a subsequence of $\{s_n\}$ and hence has the same limit. ∎

In Section 6.6, we'll see that the commutative law may *not* hold; see Riemann's rearrangement theorem on p. 484! It is worth remembering that the associative law does not work in reverse.

Example 3.36 For instance, the series

$$0 = 0 + 0 + 0 + 0 + \cdots = (1 - 1) + (1 - 1) + (1 - 1) + \cdots$$

certainly converges, but we cannot omit the parentheses and conclude that $1 - 1 + 1 - 1 + 1 - 1 + \cdots$ converges, which we have already shown does not.

3.5.3 Telescoping Series

As seen in Example 3.32, the value of the series $\sum 1/n(n+1)$ was very easy to find because in writing out its partial sums, we saw that the sum "telescoped" to give a simple expression (Fig. 3.13)

Fig. 3.13 Ever see a pirate telescope collapse?

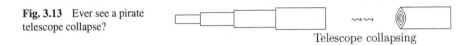

Telescope collapsing

In general, it is very difficult to find the value of a convergent series, but for telescoping series, the sums are quite straightforward to find.

Telescoping series theorem

> **Theorem 3.22** *Let $\{x_n\}$ be a sequence of complex numbers and let $\sum_{n=0}^{\infty} a_n$ be the series with nth term $a_n = x_n - x_{n+1}$. Then $\lim x_n$ exists if and only if $\sum a_n$ converges, in which case*
>
> $$\sum_{n=0}^{\infty} a_n = x_0 - \lim x_n.$$

Proof Just like the pirate telescope collapsing, observe that adjacent terms of the following partial sum cancel, collapsing to just two terms:

$$s_n = (x_0 - x_1) + (x_1 - x_2) + \cdots + (x_{n-1} - x_n) + (x_n - x_{n+1}) = x_0 - x_{n+1}.$$

If $x := \lim x_n$ exists, we have $x = \lim x_{n+1}$ as well, and therefore $\sum a_n := \lim s_n$ exists with sum $x_0 - x$. Conversely, if $s = \lim s_n$ exists, then $s = \lim s_{n-1}$ as well, and since $x_n = x_0 - s_{n-1}$, it follows that $\lim x_n$ exists. ∎

Example 3.37 Let a be a nonzero complex number that is not a negative integer. We claim that

$$\sum_{n=0}^{\infty} \frac{1}{(n+a)(n+a+1)} = \frac{1}{a}.$$

Indeed, we can use the *method of partial fractions* to write

$$a_n = \frac{1}{(n+a)(n+a+1)} = \frac{1}{(n+a)} - \frac{1}{n+a+1} = x_n - x_{n+1}, \quad \text{where } x_n = \frac{1}{(n+a)}.$$

Since $\lim x_n = 0$ and $x_0 = 1/(0 + a) = 1/a$, we get our claim from the telescoping series theorem.

Example 3.38 More generally, given a natural number k, we claim that

$$\sum_{n=0}^{\infty} \frac{1}{(n+a)(n+a+1)\cdots(n+a+k)} = \frac{1}{k} \frac{1}{a(a+1)\cdots(a+k-1)}. \quad (3.36)$$

Indeed, observe that we can write the fraction as

$$\frac{1}{(n+a)(n+a+1)\cdots(n+a+k)}$$
$$= \underbrace{\frac{1}{k(n+a)\cdots(n+a+k-1)}}_{x_n} - \underbrace{\frac{1}{k(n+a+1)\cdots(n+a+k)}}_{x_{n+1}}.$$

Since $\lim x_n = 0$ (why?) and $x_0 = \dfrac{1}{k}\dfrac{1}{a(a+1)\cdots(a+k-1)}$, our claim follows from the telescoping series theorem. For a specific example, if we put $a = 1/2$ and $k = 2$ in (3.36), we obtain (after a little algebra)

$$\frac{1}{1\cdot 3\cdot 5} + \frac{1}{3\cdot 5\cdot 7} + \frac{1}{5\cdot 7\cdot 9} + \cdots = \frac{1}{12}.$$

With $a = 1/3$ and $k = 2$ in (3.36), we obtain another beautiful sum:

$$\frac{1}{1\cdot 4\cdot 7} + \frac{1}{4\cdot 7\cdot 10} + \frac{1}{7\cdot 10\cdot 13} + \cdots = \frac{1}{24}.$$

More examples of telescoping series can be found in the article [197]. Not only can the telescoping series theorem quickly find sums of certain series, it can also construct series with any specified sum, as the following corollary shows.

Corollary 3.23 *Let s be a complex number and let $\{x_n\}_{n=0}^{\infty}$ be a null sequence (that is, $\lim x_n = 0$) such that $x_0 = s$. Then if we set $a_n = x_n - x_{n+1}$, the series $\sum_{n=0}^{\infty} a_n$ converges to s.*

Example 3.39 For example, let $s = 1$. Then $x_n = 1/2^n$ defines a null sequence with $1/2^0 = 1$. Thus, by Corollary 3.23, $\sum_{n=0}^{\infty} a_n = 1$ with

$$a_n = x_n - x_{n+1} = \frac{1}{2^n} - \frac{1}{2^{n+1}} = \frac{1}{2^{n+1}}.$$

Hence, $\displaystyle\sum_{n=0}^{\infty} \frac{1}{2^{n+1}} = 1$. Shifting the summation index, we obtain

$$\sum_{n=1}^{\infty} \frac{1}{2^n} = 1,$$

just as hypothesized in the introduction to this section!

Example 3.40 The sequence with $x_n = 1/(n+1)$ defines a null sequence with $x_0 = 1$. In this case,

$$a_n = x_n - x_{n+1} = \frac{1}{n+1} - \frac{1}{n+2} = \frac{1}{(n+1)(n+2)},$$

so $\displaystyle\sum_{n=0}^{\infty} \frac{1}{(n+1)(n+2)} = 1$. Shifting the summation index, we obtain the result from Example 3.32:

$$\sum_{n=1}^{\infty} \frac{1}{n(n+1)} = 1.$$

What fancy formulas for 1 do you get when you apply Corollary 3.23 to the sequences $x_n = 1/(n+1)^2$ and $x_n = 1/\sqrt{n+1}$?

We end this section with the all-important geometric series. A **geometric series** is a series of the form $\sum a^n$, where a is a complex number. The following theorem is usually proved, say in a calculus course, using the formula for a geometric sum (see Eq. (2.3) on p. 41). However, we shall prove it using the telescoping series theorem.

Geometric series test

> **Theorem 3.24** *For every nonzero complex number a and $k \in \mathbb{Z}$, the geometric series $\sum_{n=k}^{\infty} a^n$ converges if and only if $|a| < 1$, in which case*
>
> $$\sum_{n=k}^{\infty} a^n = a^k + a^{k+1} + a^{k+2} + a^{k+3} + \cdots = \frac{a^k}{1-a}.$$

Proof If $|a| \geq 1$ and $n \geq 0$, then $|a|^n \geq 1$, so the terms of the geometric series do not tend to zero, and therefore the geometric series cannot converge, by the nth term test. Thus, we may henceforth assume that $|a| < 1$. Observe that

$$a^n = \frac{1}{1-a}(a^n - a^{n+1}),$$

by factoring out an a^n from $a^n - a^{n+1}$ on the right. Thus,

$$a^n = x_n - x_{n+1}, \quad \text{where} \quad x_n = \frac{a^n}{1-a}.$$

Since $|a| < 1$, we have $\lim x_n = 0$, so by the telescoping series theorem starting from the index k instead of the index 0, it follows that $\sum_{n=k}^{\infty} a^n$ converges and equals $x_k = a^k/(1-a)$. ∎

We remark that if $a = 0$, then the geometric series $a^k + a^{k+1} + a^{k+2} + \cdots$ is not defined if $k = 0, -1, -2, \ldots$, and equals zero if $k = 1, 2, 3, \ldots$.

Example 3.41 If we put $a = 1/2 < 1$ in the geometric series theorem, then we have

$$\sum_{n=1}^{\infty} \frac{1}{2^n} = \frac{1}{2} + \frac{1}{2^2} + \frac{1}{2^3} + \cdots = \frac{1/2}{1-1/2} = 1,$$

a fact that we already knew from Example 3.39.

▶ **Exercises 3.5**

1. Determine the convergence of each of the following series. If the series converges, find the sum.

(a) $\sum_{n=1}^{\infty} \left(1 + \frac{1}{n}\right)^n$, (b) $\sum_{n=1}^{\infty} \left(\frac{i}{2}\right)^n$ (where i is the imaginary unit), (c) $\sum_{n=1}^{\infty} \frac{1}{n^{1/n}}$.

2. Let $\{a_n\}$ be a sequence of complex numbers.

(a) Assume that $\sum a_n$ converges. Prove that the sum of the even terms $\sum_{n=1}^{\infty} a_{2n}$ converges if and only if the sum of the odd terms $\sum_{n=1}^{\infty} a_{2n-1}$ converges, in which case $\sum a_n = \sum a_{2n} + \sum a_{2n-1}$.

(b) Let $\sum c_n$ be a series obtained from $\sum a_n$ by modifying at most finitely many terms. Show that $\sum a_n$ converges if and only if $\sum c_n$ converges.

(c) Assume that $\lim a_n = 0$. Fix $\alpha, \beta \in \mathbb{C}$ with $\alpha + \beta \neq 0$. Prove that $\sum_{n=1}^{\infty} a_n$ converges if and only if $\sum_{n=1}^{\infty} (\alpha a_n + \beta a_{n+1})$ converges.

3. Using Problem 3d on p. 44 to simplify the partial sums, prove that

$$\sum_{n=1}^{\infty} \frac{n}{2^n} = \frac{1}{2} + \frac{2}{2^2} + \frac{3}{2^3} + \cdots = 2.$$

4. Let a be a complex number. Using the telescoping series theorem, show that

$$\frac{a}{1-a^2} + \frac{a^2}{1-a^4} + \frac{a^4}{1-a^8} + \frac{a^8}{1-a^{16}} + \cdots = \begin{cases} \frac{a}{1-a} & |a| < 1, \\ \frac{1}{1-a} & |a| > 1. \end{cases}$$

Suggestion: Using the identity $\frac{x}{1-x^2} = \frac{1}{1-x} - \frac{1}{1-x^2}$ for $x \neq \pm 1$ (you should prove this identity!), write $\frac{a^{2^n}}{1-a^{2^{n+1}}}$ as the difference $x_n - x_{n+1}$, where $x_n = 1/(1 - a^{2^n})$.

5. ($\sum_{n=1}^{\infty} \frac{1}{n}$ **diverges, Proofs II–IV**) For more proofs, see [124]. Let $s_n = \sum_{k=1}^{n} \frac{1}{k}$.

   (a) Using that $1 + \frac{1}{n} < e^{1/n}$ for all $n \in \mathbb{N}$, which is from (3.26), show that

$$\left(1 + \frac{1}{1}\right)\left(1 + \frac{1}{2}\right)\left(1 + \frac{1}{3}\right) \cdots \left(1 + \frac{1}{n}\right) \leq e^{s_n}, \quad \text{for all } n \in \mathbb{N}.$$

   Show that the left-hand side equals $n + 1$ and conclude that $\{s_n\}$ cannot converge.

   (b) Show that for every $k \in \mathbb{N}$ with $k \geq 3$, we have

$$\frac{1}{k-1} + \frac{1}{k} + \frac{1}{k+1} \geq \frac{3}{k}.$$

   Using this inequality, prove that for every $n \in \mathbb{N}$, $s_{3n+1} \geq 1 + s_n$ by grouping the terms of s_{3n+1} into threes (starting from $\frac{1}{2}$). Now show that $\{s_n\}$ cannot converge.

   (c) Prove the inequality

$$\frac{1}{(k-1)! + 1} + \frac{1}{(k-1)! + 2} + \cdots + \frac{1}{k!} \geq 1 - \frac{1}{k}$$

   for every $k \in \mathbb{N}$ with $k \geq 2$. Writing $s_{n!}$ in groups of the form given on the left-hand side of this inequality, prove that $s_{n!} \geq 1 + n - s_n$. Conclude that $\{s_n\}$ cannot converge.

6. We shall prove that $\sum_{n=1}^{\infty} nz^{n-1}$ converges if and only if $|z| < 1$, in which case

$$\frac{1}{(1-z)^2} = \sum_{n=1}^{\infty} nz^{n-1}. \tag{3.37}$$

   (i) If $\sum_{n=1}^{\infty} nz^{n-1}$ converges, prove that $|z| < 1$.

   (ii) Prove that $(1-z) \sum_{k=1}^{n} kz^{k-1}$ can be written as $(1-z) \sum_{k=1}^{n} kz^{k-1} = \frac{1-z^n}{1-z} - nz^n$.

   (iii) Using Problem 4 on p. 160, prove that $1/(1-z)^2 = \sum_{n=1}^{\infty} nz^{n-1}$ for all $|z| < 1$.

   (iv) Solve Problem 3 using (3.37).

   (v) Can you prove that $\frac{2}{(1-z)^3} = \sum_{n=2}^{\infty} n(n-1)z^{n-2}$ for $|z| < 1$ using a similar technique? (Do this problem if you are feeling extra confident!)

7. The Fibonacci sequence was defined in Problem 9 on p. 47. Prove the interesting series formulas

$$(a) \sum_{n=2}^{\infty} \frac{1}{F_{n-1} F_{n+1}} = 1, \quad (b) \sum_{n=2}^{\infty} \frac{F_n}{F_{n-1} F_{n+1}} = 2, \quad (c) \sum_{n=1}^{\infty} \frac{F_n}{3^n} = \frac{3}{5}.$$

Suggestion: Think telescoping series for (a) and (b), and for (c) find a formula for s_n.

8. Here is a generalization of the telescoping series theorem.

(i) Let $x_n \to x$ and let $k \in \mathbb{N}$. Prove that $\sum_{n=0}^{\infty} a_n$, with $a_n = x_n - x_{n+k}$, converges, and

$$\sum_{n=0}^{\infty} a_n = x_0 + x_1 + \cdots + x_{k-1} - k x.$$

Using this formula, find $\sum_{n=0}^{\infty} \frac{1}{(2n+1)(2n+7)}$ and $\sum_{n=0}^{\infty} \frac{1}{(3n+1)(3n+7)}$.

(ii) Let a be a nonzero complex number not equal to a negative integer, and let $k \in \mathbb{N}$. Using (i), prove that

$$\boxed{\sum_{n=0}^{\infty} \frac{1}{(n+a)(n+a+k)} = \frac{1}{k} \left[\frac{1}{a} + \frac{1}{a+1} + \cdots + \frac{1}{a+k-1} \right].}$$

With $a = 1$ and $k = 2$, derive a beautiful expression for $3/4$.

(iii) Here's a fascinating result: Given another natural number m, prove that

$$\boxed{\sum_{n=0}^{\infty} \frac{1}{(n+a)(n+a+m) \cdots (n+a+km)} = \frac{1}{km} \sum_{n=0}^{m-1} \frac{1}{(n+a)(n+a+m) \cdots (n+a+(k-1)m)}.}$$

Find a beautiful series when $a = 1$ and $k = m = 2$.

9. Let $x_n \to x$, and let $c_1, \ldots, c_k$ be $k \geq 2$ numbers such that $c_1 + \cdots + c_k = 0$.

(i) Prove that $\sum_{n=0}^{\infty} a_n$ with $a_n = c_1 x_{n+1} + c_2 x_{n+2} + \cdots + c_k x_{n+k}$ converges, and

$$\sum_{n=0}^{\infty} a_n = c_1 x_1 + (c_1 + c_2) x_2 + \cdots + (c_1 + c_2 + \cdots + c_{k-1}) x_{k-1}$$

$$+ (c_2 + 2c_3 + 3c_4 + \cdots + (k-1) c_k) x.$$

(ii) Using (i), find $\dfrac{5}{5 \cdot 7 \cdot 9} + \dfrac{11}{7 \cdot 9 \cdot 11} + \dfrac{17}{9 \cdot 11 \cdot 13} + \cdots + \dfrac{6n+5}{(2n+5)(2n+7)(2n+9)} + \cdots$.

10. Let a be a complex number not equal to $0, -1, -1/2, -1/3, \ldots$. Prove that

$$\sum_{n=1}^{\infty} \frac{n}{(a+1)(2a+1)\cdots(na+1)} = \frac{1}{a}.$$

3.6 Absolute Convergence and a Potpourri of Convergence Tests

We now give some important tests that guarantee when certain series converge.

3.6.1 Various Tests for Convergence

The first test is the series version of Cauchy's criterion for sequences.

Cauchy's criterion for series

> **Theorem 3.25** *The series $\sum a_k$ converges if and only if for every $\varepsilon > 0$, there is an N such that for all $n > m > N$, we have*
>
> $$\left| \sum_{k=m+1}^{n} a_k \right| = |a_{m+1} + a_{m+2} + \cdots + a_n| < \varepsilon,$$
>
> *in which case, for every $m > N$, $\left| \sum_{k=m+1}^{\infty} a_k \right| \le \varepsilon$. In particular, for a convergent series $\sum a_k$, we have*
>
> $$\lim_{m \to \infty} \sum_{k=m+1}^{\infty} a_k = 0.$$

Proof Let s_n denote the nth partial sum of $\sum a_k$. Then to say that the series $\sum a_k$ converges means that the sequence $\{s_n\}$ converges. Cauchy's criterion for sequences states that $\{s_n\}$ converges if and only if for every $\varepsilon > 0$, there is an N such that for all $n, m > N$, we have $|s_n - s_m| < \varepsilon$. Since $|s_n - s_m| = |s_m - s_n|$, in the Cauchy criterion we can assume that $n > m$. Now observe that for $n > m$,

$$s_n - s_m = \sum_{k=1}^{n} a_k - \sum_{k=1}^{m} a_k = \sum_{k=m+1}^{n} a_k.$$

Hence, the Cauchy criterion is equivalent to $|\sum_{k=m+1}^{n} a_k| < \varepsilon$ for all $n > m > N$. Taking $n \to \infty$ shows that $|\sum_{k=m+1}^{\infty} a_k| \le \varepsilon$. ∎

The following "comparison test" is obvious from looking at the picture

Assuming $0 \leq a_n \leq b_n$ for all n, the comparison test says that if the sum of the areas of the rectangles with heights b_n is finite, then the same holds for the smaller rectangles with heights a_n; if the sum of the areas of the rectangles with heights a_n is infinite, then the same holds for the larger rectangles with heights b_n.

Comparison test

Theorem 3.26 *Let $\{a_n\}$ and $\{b_n\}$ be real sequences and suppose that for n sufficiently large, say for all $n \geq m$ for some $m \in \mathbb{N}$, we have*

$$0 \leq a_n \leq b_n.$$

If $\sum b_n$ converges, then $\sum a_n$ converges. Equivalently, if $\sum a_n$ diverges, then $\sum b_n$ diverges. In the case of convergence, $\sum_{n=m}^{\infty} a_n \leq \sum_{n=m}^{\infty} b_n$.

Proof By the tails theorem for series (Theorem 3.20 on p. 198), $\sum a_n$ and $\sum b_n$ converge if and only if $\sum_{n=m}^{\infty} a_n$ and $\sum_{n=m}^{\infty} b_n$ converge. By working with these series instead of the original ones, we may assume that $0 \leq a_k \leq b_k$ holds for every k. Summing from $k = 1$ to $k = n$, we conclude that

$$s_n \leq t_n \quad \text{for all } n,$$

where s_n, respectively t_n, denotes the nth partial sum for $\sum a_n$, respectively $\sum b_n$. Assume that $\sum b_n$ converges and let $t = \sum b_n$. By the nonnegative series test (Theorem 3.19 on p. 197), $t_n \leq t$ for all n. Hence, $s_n \leq t$ for all n. Again by the nonnegative series test, it follows that $\sum a_n$ converges, and taking $n \to \infty$ in $s_n \leq t$ shows that $\sum a_n \leq t$, which is to say, that $\sum a_n \leq \sum b_n$. ∎

Example 3.42 (*The p-series*) We claim that the *p*-**series**

$$\sum_{n=1}^{\infty} \frac{1}{n^p} = 1 + \frac{1}{2^p} + \frac{1}{3^p} + \cdots,$$

where p is a rational number, converges for $p \geq 2$ and diverges for $p \leq 1$. To see this, assume first that $p \leq 1$. Then

$$\frac{1}{n} \leq \frac{1}{n^p},$$

because this inequality is equivalent to $1 \leq n^{1-p}$, which holds by raising both sides of the inequality $1 \leq n$ to the nonnegative power $1 - p$ (see the power rules theorem,

Theorem 2.34 on p. 95). Since the harmonic series diverges, so does the p-series for $p \leq 1$ by the comparison test. If $p \geq 2$, then by a similar argument, we have

$$\frac{1}{n^p} \leq \frac{1}{n^2}.$$

In the last section, we showed that the 2-series $\sum 1/n^2$ converges, so by the comparison test, the p-series for $p \geq 2$ converges. Now what about for $1 < p < 2$? To answer this question we shall appeal to Cauchy's condensation test below.

3.6.2 Cauchy Condensation Test

The following test is usually not found in elementary calculus textbooks, but it's very useful.

Cauchy condensation test

> **Theorem 3.27** *If $\{a_n\}$ is a nonincreasing sequence of nonnegative real numbers, then the infinite series $\sum a_n$ converges if and only if*
>
> $$\sum_{n=0}^{\infty} 2^n a_{2^n} = a_1 + 2a_2 + 4a_4 + 8a_8 + \cdots$$
>
> *converges.*

Proof Let the partial sums of $\sum a_n$ be denoted by s_n and those of $\sum_{n=0}^{\infty} 2^n a_{2^n}$ by t_n. Then by the nonnegative series test (Theorem 3.19 on p. 197), $\sum a_n$ converges if and only if $\{s_n\}$ is bounded, and $\sum 2^n a_{2^n}$ converges if and only if $\{t_n\}$ is bounded. Therefore, we just have to prove that $\{s_n\}$ is bounded if and only if $\{t_n\}$ is bounded.

Consider the "if" part: Assume that $\{t_n\}$ is bounded; we shall prove that $\{s_n\}$ is bounded. To prove this, we note that $s_n \leq s_{2^n-1}$, and we can write

$$s_n \leq s_{2^n-1} = a_1 + (a_2 + a_3) + (a_4 + a_5 + a_6 + a_7) + \cdots + (a_{2^{n-1}} + \cdots + a_{2^n-1}),$$

where in the kth parentheses, we group the terms of the series with index running from 2^k to $2^{k+1} - 1$. Since the a_n are nonincreasing (that is, $a_n \geq a_{n+1}$ for all n), replacing each number in parentheses by the first term in the parentheses cannot decrease the value of the sum, so

$$s_n \leq a_1 + (a_2 + a_2) + (a_4 + a_4 + a_4 + a_4) + \cdots + (a_{2^{n-1}} + \cdots + a_{2^{n-1}})$$
$$= a_1 + 2a_2 + 4a_4 + \cdots + 2^{n-1}a_{2^{n-1}} = t_{n-1}.$$

Since $\{t_n\}$ is bounded, it follows that $\{s_n\}$ is bounded as well. Here's a picture of this bound showing $s_{2^n-1} \le t_{n-1}$ in the case $n = 3$:

$$\le$$

Now the "only if" part: Assume that $\{s_n\}$ is bounded; we shall prove that $\{t_n\}$ is bounded. Here's a picture to consider:

$$\le$$

From this picture, we see (and it's easy to prove algebraically) that

$$a_2 + 2a_4 + 4a_8 + \cdots + 2^{n-1}a_{2^n}$$

$$\le a_2 + (a_3 + a_4) + (a_5 + a_6 + a_7 + a_8) + \cdots + (a_{2^{n-1}+1} + \cdots + a_{2^n}).$$

Another way to write this is

$$\frac{1}{2}(t_n - a_1) \le s_{2^n} - a_1,$$

or stated another way, $\frac{1}{2}a_1 + \frac{1}{2}t_n \le s_{2^n}$. Since $\{s_n\}$ is bounded, it follows that $\{t_n\}$ is bounded as well. This completes our proof.   ∎

See Problems 4 and 11 for other condensation tests.

Example 3.43 (*The p-series revisited*) Consider the p-series ($p \ge 0$ is rational)

$$\sum_{n=1}^{\infty} \frac{1}{n^p} = 1 + \frac{1}{2^p} + \frac{1}{3^p} + \cdots.$$

With $a_n = \frac{1}{n^p}$, this series converges, by Cauchy's condensation test, if and only if

$$\sum_{n=0}^{\infty} 2^n a_{2^n} = \sum_{n=1}^{\infty} \frac{2^n}{(2^n)^p} = \sum_{n=1}^{\infty} \left(\frac{1}{2^{p-1}}\right)^n$$

converges. This series is a geometric series, so the series converges if and only if $\frac{1}{2^{p-1}} < 1$, which holds if and only if $p > 1$. Summarizing, we get

$$\boxed{\;\; p\text{-}\textbf{test}:\quad \sum_{n=1}^{\infty} \frac{1}{n^p} \quad \begin{cases} \text{converges for } p > 1, \\ \text{diverges for } p \le 1. \end{cases} \;\;}$$

Once we develop the theory of real exponents, we shall see that the same p-test holds for p real. By the way, $\sum_{n=1}^{\infty} 1/n^p$ is also denoted by $\zeta(p)$, the **zeta function** at p:

$$\zeta(p) := \sum_{n=1}^{\infty} \frac{1}{n^p}.$$

We'll come across this function again in Section 4.7 on p. 300.

Cauchy's condensation test is especially useful in dealing with series involving logarithms; see the problems. Although we technically haven't introduced the logarithm function, we'll *thoroughly* develop this function in Section 4.7, starting on p. 300, so for now, we'll assume that you know properties of $\log x$ for $x > 0$. Actually, for the particular example below, we just need to know that $\log x^k = k \log x$ for all $k \in \mathbb{Z}$, $\log x > 0$ for $x > 1$, and that $\log x$ is increasing with x.

Example 3.44 Consider the series

$$\sum_{n=2}^{\infty} \frac{1}{n \log n}.$$

At first glance, it may seem difficult to determine the convergence of this series, but Cauchy's condensation test gives the answer quickly:

$$\sum_{n=1}^{\infty} 2^n \cdot \frac{1}{2^n \log 2^n} = \frac{1}{\log 2} \sum_{n=1}^{\infty} \frac{1}{n},$$

which diverges. (You should check that $1/(n \log n)$ is nonincreasing.) Therefore, by Cauchy's condensation test, $\sum_{n=2}^{\infty} \frac{1}{n \log n}$ also diverges.[11]

3.6.3 Absolute Convergence

A series $\sum a_n$ is said to be **absolutely convergent** if the series of absolute values $\sum |a_n|$ converges. Part *(1)* of the following theorem says that if the series of absolute values converges, then we can omit taking the absolute values and still have a convergent series.

---

[11] This series is usually handled in elementary calculus courses using the technologically advanced (mathematically speaking) *integral test*, but Cauchy's condensation test gives one way to handle such series without knowing any calculus!

Absolute convergence

> **Theorem 3.28** *Let $\sum a_n$ be an infinite series.*
>
> *(1) If $\sum |a_n|$ converges, then $\sum a_n$ also converges, and*
>
> $$\left| \sum a_n \right| \leq \sum |a_n| \qquad \text{(\textbf{\textit{triangle inequality for series}})}. \qquad (3.38)$$
>
> *(2) Every linear combination of absolutely convergent series is absolutely convergent.*

Proof Suppose that $\sum |a_n|$ converges. We shall prove that $\sum a_n$ converges and (3.38) holds. To prove convergence, we use Cauchy's criterion, so let $\varepsilon > 0$. Since $\sum |a_n|$ converges, there is an N such that for all $n > m > N$, we have

$$\sum_{k=m+1}^{n} |a_k| < \varepsilon.$$

By the usual triangle inequality, for $n > m > N$, we have

$$\left| \sum_{k=m+1}^{n} a_k \right| \leq \sum_{k=m+1}^{n} |a_k| < \varepsilon.$$

Thus by Cauchy's criterion for series, $\sum a_n$ converges. To prove (3.38), let s_n denote the nth partial sum of $\sum a_n$. Then,

$$|s_n| = \left| \sum_{k=1}^{n} a_k \right| \leq \sum_{k=1}^{n} |a_k| \leq \sum_{k=1}^{\infty} |a_k|.$$

Since $|s_n| \to |\sum a_k|$ and limits preserve inequalities, it follows that $|\sum a_k| \leq \sum |a_k|$.

If $\sum |a_n|$ and $\sum |b_n|$ converge, then given complex numbers c, d, since $|ca_n + db_n| \leq |c||a_n| + |d||b_n|$, it follows by the comparison theorem that $\sum |ca_n + db_n|$ also converges. ∎

Example 3.45 Since the 2-series $\sum 1/n^2$ converges, each of the following series is absolutely convergent:

$$\sum_{n=1}^{\infty} \frac{(-1)^n}{n^2}, \quad \sum_{n=1}^{\infty} \frac{i^n}{n^2}, \quad \sum_{n=1}^{\infty} \frac{(-i)^n}{n^2}.$$

It is possible to have a convergent series that is not absolutely convergent.

Example 3.46 Although the harmonic series $\sum 1/n$ diverges, the **alternating harmonic series**

$$\sum_{n=1}^{\infty} \frac{(-1)^{n-1}}{n} = 1 - \frac{1}{2} + \frac{1}{3} - \frac{1}{4} + \frac{1}{5} - \frac{1}{6} + - \cdots$$

converges. To see this, observe that if n is even, then

$$
\begin{aligned}
s_n &= \left(1 - \frac{1}{2}\right) + \left(\frac{1}{3} - \frac{1}{4}\right) + \left(\frac{1}{5} - \frac{1}{6}\right) + \cdots + \left(\frac{1}{n-1} - \frac{1}{n}\right) \\
&= \frac{1}{1 \cdot 2} + \frac{1}{3 \cdot 4} + \frac{1}{5 \cdot 6} + \cdots + \frac{1}{(n-1)n}.
\end{aligned}
$$

Similarly, if n is odd, then

$$s_n = \frac{1}{1 \cdot 2} + \frac{1}{3 \cdot 4} + \frac{1}{5 \cdot 6} + \cdots + \frac{1}{(n-2)(n-1)} + \frac{1}{n}.$$

These expressions for s_n suggest that we consider the infinite series

$$\frac{1}{1 \cdot 2} + \frac{1}{3 \cdot 4} + \frac{1}{5 \cdot 6} + \cdots, \tag{3.39}$$

which converges by comparison with the 2-series $1/1^2 + 1/2^2 + 1/3^2 + \cdots$. Let $b_2, b_3, b_4, \ldots$ denote the partial sums of (3.39) repeated twice in a row:

$$b_2 = \frac{1}{1 \cdot 2}, \quad b_3 = \frac{1}{1 \cdot 2}, \quad b_4 = \frac{1}{1 \cdot 2} + \frac{1}{3 \cdot 4}, \quad b_5 = \frac{1}{1 \cdot 2} + \frac{1}{3 \cdot 4}, \cdots.$$

After some thought, we observe that for $n \geq 2$,

$$s_n = b_n + c_n,$$

where

$$c_n = 0 \ \text{if } n \text{ is even} \quad \text{and} \quad c_n = \frac{1}{n} \ \text{if } n \text{ is odd}.$$

Since $c_n \to 0$ and $\{b_n\}$ converges, it follows that $\{s_n\}$ also converges, with limit equal to $\lim b_n$. In other words,

$$\sum_{n=1}^{\infty} \frac{(-1)^{n-1}}{n} = \frac{1}{1 \cdot 2} + \frac{1}{3 \cdot 4} + \frac{1}{5 \cdot 6} + \cdots.$$

Another way to prove convergence is to use the *alternating series test*, a subject that we will study thoroughly in Section 6.1 on p. 428. Later on, in Section 4.7 on p. 311, we'll prove that $\sum_{n=1}^{\infty} \frac{(-1)^{n-1}}{n}$ equals log 2.

▶ **Exercises 3.6**

1. For this problem, assume all the "well-known" high school properties of $\log x$ (e.g., $\log x^k = k \log x, \log(xy) = \log x + \log y$). Using the Cauchy condensation test, determine the convergence of the following series:

$$(a) \sum_{n=2}^{\infty} \frac{1}{n(\log n)^2} \quad , \quad (b) \sum_{n=2}^{\infty} \frac{1}{n(\log n)^p} \quad , \quad (c) \sum_{n=2}^{\infty} \frac{1}{n(\log n)(\log(\log n))}.$$

   For *(b)*, state which p give convergent/divergent series.

2. Prove that

$$(a) \sum_{n=1}^{\infty} \frac{(-1)^{n-1}}{n} = 1 - \frac{1}{2 \cdot 3} - \frac{1}{4 \cdot 5} - \frac{1}{6 \cdot 7} - \cdots ,$$

$$(b) \sum_{n=1}^{\infty} \frac{(-1)^{n-1}}{n^2} = \frac{1}{1^2 \cdot 2^2}(1+2) + \frac{1}{3^2 \cdot 4^2}(3+4) + \frac{1}{5^2 \cdot 6^2}(5+6) + \cdots .$$

3. For what $x \in \mathbb{R}$ do the following series converge?

$$(a) \sum_{n=1}^{\infty} \frac{x^{2n}}{1+x^{2n}} \quad , \quad (b) \sum_{n=1}^{\infty} \left(\frac{nx^2}{1+n}\right)^n \quad , \quad (c) \sum_{n=1}^{\infty} \frac{1}{(1+x^{2n^2})^{1/n}}.$$

4. We consider various (unrelated) properties of real series $\sum a_n$ with $a_n \geq 0$ for all n.

   (a) **General Cauchy condensation test**: If the a_n are nonincreasing, then given a natural number $b > 1$, prove that $\sum a_n$ converges or diverges with the series

$$\sum_{n=0}^{\infty} b^n a_{b^n} = a_1 + b\,a_b + b^2\,a_{b^2} + b^3\,a_{b^3} + \cdots .$$

   Thus, the Cauchy condensation test is just this test with $b = 2$.

   (b) If $\sum a_n$ converges, prove that for every $k \in \mathbb{N}$, the series $\sum a_n^k$ also converges.
   (c) If $\sum a_n$ converges, give an example showing that $\sum \sqrt{a_n}$ may not converge. However, prove that the series $\sum \sqrt{a_n}/n$ does converge. Suggestion: Use the inequality $ab \leq (a^2 + b^2)/2$ for all $a, b \in \mathbb{R}$ from Eq. (2.33) on p. 125.

(d) If $a_n > 0$ for all n, prove that $\sum a_n$ converges if and only if for *every* sequence $\{b_n\}$ of nonnegative real numbers, $\sum (a_n^{-1} + b_n)^{-1}$ converges.

(e) If $\sum b_n$ is another series of nonnegative real numbers, prove that $\sum a_n$ and $\sum b_n$ converge if and only if $\sum \sqrt{a_n^2 + b_n^2}$ converges.

(f) Prove that $\sum a_n$ converges if and only if $\sum \frac{a_n}{1+a_n}$ converges.

(g) For each $n \in \mathbb{N}$, define $b_n = \sqrt{\sum_{k=n}^{\infty} a_k} = \sqrt{a_n + a_{n+1} + a_{n+2} + \cdots}$. Prove that if $a_n > 0$ for all n and $\sum a_n$ converges, then $\sum \frac{a_n}{b_n}$ converges. Suggestion: Show that $a_n = b_n^2 - b_{n+1}^2$, and using this fact, show that $\frac{a_n}{b_n} \le 2(b_n - b_{n+1})$ for all n.

5. We already know that if $\sum a_n$ (of complex numbers) converges, then $\lim a_n = 0$. When the a_n form a nonincreasing sequence of nonnegative real numbers, then prove the following astonishing fact (called **Pringsheim's theorem**): If $\sum a_n$ converges, then $n\, a_n \to 0$. Suggestion: Let $\varepsilon > 0$ and choose N such that $n > m > N$ implies

$$a_{m+1} + a_{m+2} + \cdots + a_n < \frac{\varepsilon}{2}.$$

Take $n = 2m$ and then $n = 2m+1$. A slicker proof uses the Cauchy condensation test.

6. (**Limit comparison test**) Let $\{a_n\}$ and $\{b_n\}$ be nonzero complex sequences and suppose that the following limit exists: $L := \lim \left| \frac{a_n}{b_n} \right|$. Prove that:

   (i) If $L \ne 0$, then $\sum a_n$ is absolutely convergent if and only if $\sum b_n$ is absolutely convergent.

   (ii) If $L = 0$ and $\sum b_n$ is absolutely convergent, then $\sum a_n$ is absolutely convergent.

7. Here's an alternative method to prove that the alternating harmonic series converges.

   (i) Let $\{b_n\}$ be a sequence in $\mathbb{R}^m$ and suppose that the even and odd subsequences $\{b_{2n}\}$ and $\{b_{2n-1}\}$ both converge and have the same limit L. Prove that the original sequence $\{b_n\}$ converges and has limit L.

   (ii) Show that the subsequences of even and odd partial sums of the alternating harmonic series both converge and have the same limit.

8. (**Telescoping comparison test**) Let $\{a_n\}$ be a sequence of positive numbers. Prove that $\sum a_n$ converges if and only if there exist a constant $c > 0$ and a sequence $\{x_n\}$ of positive numbers such that

$$a_n \le c\,(x_n - x_{n+1}) \quad \text{for all } n \text{ sufficiently large.}$$

Suggestion: For the "only if" part, consider $x_n = s - (a_1 + a_2 + \cdots + a_{n-1})$, where $s = \sum a_n$, which is assumed to converge.

9. **(Ratio comparison test)** Let $\{a_n\}$ and $\{b_n\}$ be sequences of positive numbers and suppose that $\frac{a_{n+1}}{a_n} \le \frac{b_{n+1}}{b_n}$ for all n. If $\sum b_n$ converges, prove that $\sum a_n$ also converges. (Equivalently, if $\sum a_n$ diverges, then $\sum b_n$ also diverges.)

10. (Cf. [113, 122, 253]) We already know that the harmonic series $\sum 1/n$ diverges. It turns out that omitting certain numbers from this sum makes the sum converge. Fix a natural number $b \ge 2$. Recall (see Section 2.5) that we can write every natural number n uniquely as $n = a_k a_{k-1} \ldots a_0$, where $0 \le a_j \le b - 1$, $j = 0, \ldots, k$, are called digits, and where the notation $a_k \ldots a_0$ means that

$$a = a_k b^k + a_{k-1} b^{k-1} + \cdots + a_1 b + a_0.$$

Prove that the following sum converges:

$$\sum_{n \text{ has no 0 digit}} \frac{1}{n}.$$

Suggestion: For each $k = 0, 1, 2, \ldots$, let c_k be the sum over all numbers of the form $\frac{1}{n}$ where $n = a_k a_{k-1} \ldots a_0$ with none of the a_j equal to zero. Show that there are at most $(b - 1)^{k+1}$ such n and that $n \ge b^k$ and use these facts to show that $c_k \le \frac{(b-1)^{k+1}}{b^k}$. Prove that $\sum_{k=0}^{\infty} c_k$ converges and use this to prove that the desired sum converges.

11. **(Hui Lin's condensation test)**[12] Let $b > 1$ be a natural number and define, for $n = 0, 1, 2, \ldots, \beta_n = 1 + \frac{b^n - 1}{b - 1}$. Note that $\beta_0 = 1$, $\beta_1 = 2$, and for $n > 1$,

$$\beta_n = 2 + b + b^2 + b^3 + \cdots + b^{n-1},$$

where we used the sum of a geometric progression (2.3) on p. 41. If $\{a_n\}$ is a nonincreasing sequence of nonnegative real numbers, prove that $\sum a_n$ converges or diverges with the series

$$\sum_{n=0}^{\infty} b^n a_{\beta_n} = a_1 + b\, a_2 + b^2\, a_{2+b} + b^3\, a_{2+b+b^2} + b^4\, a_{2+b+b^2+b^3} + \cdots.$$

---

[12]Hui Lin was a student in my fall 2014 real analysis course, and he discovered this very interesting test. I thank him for allowing me to present his work.

3.7   Tannery's Theorem and Defining the Exponential Function exp(z)

Tannery's theorem (named after Jules Tannery (1848–1910)) is a little known but fantastic theorem that I learned from [29, 30, 40, 79]. Tannery's theorem is really a special case of the *Weierstrass M-test* [40, p. 124] or the *Lebesgue dominated convergence theorem* [148], which is why it probably doesn't get much attention. We shall use Tannery's theorem quite a bit in the sequel. In particular, we shall use it to derive certain properties of the complex exponential function, which is undoubtedly the most important function in analysis and arguably all of mathematics. In this section we derive some properties of the exponential function, including its relationship to the number e defined on p. 310 in Section 3.3.

3.7.1   Tannery's Theorem for Series

Tannery has two theorems, one for series and the other for products; we'll cover his theorem for products on p. 547 in Section 7.3. Here is the one for series.

Tannery's theorem for series

> **Theorem 3.29** *For each natural number n, let $\sum_{k=1}^{m_n} a_k(n)$ be a finite sum such that $m_n \to \infty$ as $n \to \infty$. If for each k, $\lim_{n\to\infty} a_k(n)$ exists, and there is a convergent series $\sum_{k=1}^{\infty} M_k$ of nonnegative real numbers such that $|a_k(n)| \le M_k$ for all $n \in \mathbb{N}$ and $1 \le k \le m_n$, then*
>
> $$\lim_{n\to\infty} \sum_{k=1}^{m_n} a_k(n) = \sum_{k=1}^{\infty} \lim_{n\to\infty} a_k(n);$$
>
> *that is, both sides are well defined (the limits and sums converge) and are equal.*

Proof First of all, we remark that the series on the right converges. Indeed, if we put $a_k := \lim_{n\to\infty} a_k(n)$ (the limit exists by assumption), then taking $n \to \infty$ in the inequality $|a_k(n)| \le M_k$, we have $|a_k| \le M_k$ as well. Therefore, by the comparison test, $\sum_{k=1}^{\infty} |a_k|$ converges, and hence $\sum_{k=1}^{\infty} a_k$ converges as well.

Now to prove our theorem, let $\varepsilon > 0$ be given. It follows from Cauchy's criterion for series that there is an ℓ such that

$$M_{\ell+1} + M_{\ell+2} + \cdots < \frac{\varepsilon}{3}.$$

Since $m_n \to \infty$ as $n \to \infty$, we can choose N_1 such that for all $n > N_1$, we have $m_n > \ell$. Then using that $|a_k(n)| \le M_k$ and $|a_k| \le M_k$, observe that for every $n > N_1$,

$$\left| \sum_{k=1}^{m_n} a_k(n) - \sum_{k=1}^{\infty} a_k \right| = \left| \sum_{k=1}^{\ell} (a_k(n) - a_k) + \sum_{k=\ell+1}^{m_n} a_k(n) - \sum_{k=\ell+1}^{\infty} a_k \right|$$

$$\leq \sum_{k=1}^{\ell} |a_k(n) - a_k| + \sum_{k=\ell+1}^{m_n} M_k + \sum_{k=\ell+1}^{\infty} M_k$$

$$< \sum_{k=1}^{\ell} |a_k(n) - a_k| + \frac{\varepsilon}{3} + \frac{\varepsilon}{3} = \sum_{k=1}^{\ell} |a_k(n) - a_k| + \frac{2\varepsilon}{3}.$$

Since for each k, $\lim_{n\to\infty} a_k(n) = a_k$, there is an N_2 such that for each $k = 1, 2, \ldots, \ell$ and for $n > N_2$, we have $|a_k(n) - a_k| < \varepsilon/(3\ell)$. Thus, if $n > \max\{N_1, N_2\}$, then

$$\left| \sum_{k=1}^{m_n} a_k(n) - \sum_{k=1}^{\infty} a_k \right| < \sum_{k=1}^{\ell} \frac{\varepsilon}{3\ell} + \frac{2\varepsilon}{3} = \frac{\varepsilon}{3} + \frac{2\varepsilon}{3} = \varepsilon.$$

This completes the proof.   ∎

Tannery's theorem, $\lim_{n\to\infty} \sum_{k=1}^{m_n} a_k(n) = \sum_{k=1}^{\infty} \lim_{n\to\infty} a_k(n)$, states that under certain conditions, we can switch the position of limits and of summations that become infinite series. (Of course, by the algebra of limits, we can always switch limits and summations with a fixed *finite* number of terms, but summations that become infinite series is a whole other matter.) In Problem 8, using an almost identical argument as that used in the proof above, you will prove that Tannery's theorem holds even when all the m_n are infinite:

$$\lim_{n\to\infty} \sum_{k=1}^{\infty} a_k(n) = \sum_{k=1}^{\infty} \lim_{n\to\infty} a_k(n).$$

See Problem 9 for an application to double series.

Example 3.47 We shall derive the formula

$$\frac{1}{2} = \lim_{n\to\infty} \left\{ \frac{1+2^n}{2^n 3 + 4} + \frac{1+2^n}{2^n 3^2 + 4^2} + \frac{1+2^n}{2^n 3^3 + 4^3} + \cdots + \frac{1+2^n}{2^n 3^n + 4^n} \right\}.$$

To prove this, we write the right-hand side as

$$\lim_{n\to\infty} \left\{ \frac{1+2^n}{2^n 3 + 4} + \frac{1+2^n}{2^n 3^2 + 4^2} + \frac{1+2^n}{2^n 3^3 + 4^3} + \cdots + \frac{1+2^n}{2^n 3^n + 4^n} \right\} = \lim_{n\to\infty} \sum_{k=1}^{m_n} a_k(n),$$

where $m_n = n$ and

$$a_k(n) = \frac{1+2^n}{2^n 3^k + 4^k}.$$

Observe that for each $k \in \mathbb{N}$,

$$\lim_{n \to \infty} a_k(n) = \lim_{n \to \infty} \frac{1 + 2^n}{2^n 3^k + 4^k} = \lim_{n \to \infty} \frac{\frac{1}{2^n} + 1}{3^k + \frac{4^k}{2^n}} = \frac{1}{3^k}.$$

Also,

$$|a_k(n)| = \frac{1 + 2^n}{2^n 3^k + 4^k} \le \frac{2^n + 2^n}{2^n 3^k} = \frac{2 \cdot 2^n}{2^n 3^k} = \frac{2}{3^k} =: M_k.$$

By the geometric series test, we know that $\sum_{k=1}^{\infty} M_k$ converges. Hence by Tannery's theorem, we have

$$\lim_{n \to \infty} \left\{ \frac{1 + 2^n}{2^n 3 + 4} + \frac{1 + 2^n}{2^n 3^2 + 4^2} + \frac{1 + 2^n}{2^n 3^3 + 4^3} + \cdots + \frac{1 + 2^n}{2^n 3^n + 4^n} \right\}$$

$$= \lim_{n \to \infty} \sum_{k=1}^{m_n} a_k(n) = \sum_{k=1}^{\infty} \lim_{n \to \infty} a_k(n) = \sum_{k=1}^{\infty} \frac{1}{3^k} = \frac{1/3}{1 - 1/3} = \frac{1}{2}.$$

If the hypotheses of Tannery's theorem are not met, then the conclusion of Tannery's theorem may not hold, as the following example illustrates.

Example 3.48 Here's a nonexample of Tannery's theorem.[13] For each $k, n \in \mathbb{N}$, let $a_k(n) := 1/n$ and let $m_n = n$. Then

$$\lim_{n \to \infty} a_k(n) = \lim_{n \to \infty} \frac{1}{n} = 0 \implies \sum_{k=1}^{\infty} \lim_{n \to \infty} a_k(n) = \sum_{k=1}^{\infty} 0 = 0.$$

On the other hand,

$$\sum_{k=1}^{m_n} a_k(n) = \sum_{k=1}^{n} \frac{1}{n} = \frac{1}{n} \cdot \sum_{k=1}^{n} 1 = 1 \implies \lim_{n \to \infty} \sum_{k=1}^{m_n} a_k(n) = \lim_{n \to \infty} 1 = 1.$$

Thus, for this example,

$$\lim_{n \to \infty} \sum_{k=1}^{m_n} a_k(n) \ne \sum_{k=1}^{\infty} \lim_{n \to \infty} a_k(n).$$

It turns out there is no constant M_k such that $|a_k(n)| \le M_k$ where the series $\sum_{k=1}^{\infty} M_k$ converges. Indeed, the inequality $|a_k(n)| = 1/n \le M_k$ for $1 \le k \le n$ implies (set $k = n$) that $1/k \le M_k$ for all k. Since $\sum_{k=1}^{\infty} 1/k$ diverges, the series $\sum_{k=1}^{\infty} M_k$ must also diverge.

---

[13]There are many other nonexamples, such as $a_k(n) = 1/(n + k)$.

3.7.2 The Exponential Function

The **exponential function** $\exp : \mathbb{C} \to \mathbb{C}$ is the function defined by

$$\exp(z) := \sum_{n=0}^{\infty} \frac{z^n}{n!}, \qquad \text{for } z \in \mathbb{C}.$$

Of course, we need to show that the right-hand side converges for each $z \in \mathbb{C}$. In fact, we claim that the series defining $\exp(z)$ is absolutely convergent. To prove this, fix $z \in \mathbb{C}$ and then choose $N \in \mathbb{N}$ such that $|z| \le \frac{1}{2}N$. (Just as a reminder, recall that such a k exists, since $\mathbb{N}$ is not bounded above.) Then for every $n \ge N$, we have

$$\frac{|z|^n}{n!} = \left[\left(\frac{|z|}{1} \right) \cdot \left(\frac{|z|}{2} \right) \cdots \left(\frac{|z|}{N} \right) \right] \cdot \left[\left(\frac{|z|}{(N+1)} \right) \cdots \left(\frac{|z|}{n} \right) \right]$$

$$\le |z|^N \left(\frac{1}{2} \cdot \frac{N}{N+1} \right) \cdot \left(\frac{1}{2} \cdot \frac{N}{N+2} \right) \cdots \left(\frac{1}{2} \cdot \frac{N}{n} \right)$$

$$\le \left(\frac{1}{2} N \right)^N \left(\frac{1}{2} \right) \cdot \left(\frac{1}{2} \right) \cdots \left(\frac{1}{2} \right) = N^N \frac{1}{2^n}.$$

Thus, for $n \ge N$, $\frac{|z|^n}{n!} \le \frac{C}{2^n}$, where C is the constant N^N. Since the geometric series $\sum 1/2^n$ converges, by the comparison test, the series defining $\exp(z)$ is absolutely convergent for every $z \in \mathbb{C}$. In the following theorem, we relate the exponential function to Euler's number e defined on p. 181 in Section 3.3. The proof of Property *(1)* in this theorem is a beautiful application of Tannery's theorem.

Properties of the complex exponential

Theorem 3.30 *The exponential function has the following properties:*

(1) For every $z \in \mathbb{C}$ and sequence $z_n \to z$, we have

$$\exp(z) = \lim_{n \to \infty} \left(1 + \frac{z_n}{n} \right)^n.$$

In particular, setting $z_n = z$ for all n yields

$$\exp(z) = \lim_{n \to \infty} \left(1 + \frac{z}{n} \right)^n,$$

and setting $z = 1$, we get

$$\exp(1) = \lim_{n \to \infty} \left(1 + \frac{1}{n} \right)^n = e.$$

(2) For all complex numbers z and w,

$$\exp(z) \cdot \exp(w) = \exp(z + w).$$

(3) $\exp(z)$ is never zero for any complex number z, and

$$\frac{1}{\exp(z)} = \exp(-z).$$

Proof To prove (1), let $z \in \mathbb{C}$ and let $\{z_n\}$ be a complex sequence and suppose that $z_n \to z$; we need to show that $\lim_{n \to \infty}(1 + z_n/n)^n = \exp(z)$. To begin, we expand $(1 + z_n/n)^n$ using the binomial theorem:

$$\left(1 + \frac{z_n}{n}\right)^n = \sum_{k=0}^{n} \binom{n}{k} \frac{z_n^k}{n^k} \quad \Longrightarrow \quad \left(1 + \frac{z_n}{n}\right)^n = \sum_{k=0}^{n} a_k(n),$$

where $a_k(n) = \binom{n}{k} \frac{z_n^k}{n^k}$. Hence, we are aiming to prove that

$$\lim_{n \to \infty} \sum_{k=0}^{n} a_k(n) = \exp(z).$$

Of course, written in this way, we have the perfect setup for Tannery's theorem! However, before going to Tannery's theorem, we note that by the definition of $a_k(n)$, we have $a_0(n) = 1$ and $a_1(n) = z_n$. Therefore, since $z_n \to z$,

$$\lim_{n \to \infty} \sum_{k=0}^{n} a_k(n) = \lim_{n \to \infty} \left(1 + z_n + \sum_{k=2}^{n} a_k(n)\right) = 1 + z + \lim_{n \to \infty} \sum_{k=2}^{n} a_k(n).$$

Thus, we just have to apply Tannery's theorem to the sum starting from $k = 2$; for this reason, we henceforth assume that $k, n \geq 2$. Now observe that for $2 \leq k \leq n$, we have

$$\binom{n}{k} \frac{1}{n^k} = \frac{n!}{k!(n-k)!} \frac{1}{n^k} = \frac{1}{k!} n(n-1)(n-2)\cdots(n-k+1) \frac{1}{n^k}$$

$$= \frac{1}{k!} \left(1 - \frac{1}{n}\right)\left(1 - \frac{2}{n}\right)\cdots\left(1 - \frac{k-1}{n}\right).$$

Thus, for $2 \leq k \leq n$,

$$a_k(n) = \frac{1}{k!}\left[\left(1 - \frac{1}{n}\right)\left(1 - \frac{2}{n}\right)\cdots\left(1 - \frac{k-1}{n}\right)\right] z_n^k.$$

Using this expression for $a_k(n)$, we can easily verify the hypotheses of Tannery's theorem. First, since $z_n \to z$,

$$\lim_{n\to\infty} a_k(n) = \lim_{n\to\infty} \frac{1}{k!}\left[\left(1-\frac{1}{n}\right)\left(1-\frac{2}{n}\right)\cdots\left(1-\frac{k-1}{n}\right)\right]z_n^k = \frac{z^k}{k!}.$$

Second, since $\{z_n\}$ is a convergent sequence, it must be bounded, say by a constant C, so that $|z_n| \le C$ for all n. Then for $2 \le k \le n$,

$$
\begin{aligned}
|a_k(n)| &= \left|\frac{1}{k!}\left[\left(1-\frac{1}{n}\right)\left(1-\frac{2}{n}\right)\cdots\left(1-\frac{k-1}{n}\right)\right]z_n^k\right| \\
&\le \frac{1}{k!}\left[\left(1-\frac{1}{n}\right)\left(1-\frac{2}{n}\right)\cdots\left(1-\frac{k-1}{n}\right)\right]C^k \le \frac{C^k}{k!} =: M_k,
\end{aligned}
$$

where we used that the term in brackets is a product of positive numbers ≤ 1, so the product is also ≤ 1. Note that $\sum_{k=2}^{\infty} M_k = \sum_{k=2}^{\infty} C^k/k!$ converges (its sum equals $\exp(C) - 1 - C$, but this isn't important). Hence by Tannery's theorem,

$$
\begin{aligned}
\lim_{n\to\infty}\left(1+\frac{z_n}{n}\right)^n &= 1 + z + \lim_{n\to\infty}\sum_{k=2}^{n} a_k(n) = 1 + z + \sum_{k=2}^{\infty}\lim_{n\to\infty} a_k(n) \\
&= 1 + z + \sum_{k=2}^{\infty}\frac{z^k}{k!} = \exp(z).
\end{aligned}
$$

To prove *(2)*, observe that

$$\exp(z)\cdot\exp(w) = \lim_{n\to\infty}\left(1+\frac{z}{n}\right)^n\left(1+\frac{w}{n}\right)^n = \lim_{n\to\infty}\left(1+\frac{z_n}{n}\right)^n,$$

where $z_n = z+w+(z\,w)/n$. Since $z_n \to z+w$, we get $\exp(z)\cdot\exp(w) = \exp(z+w)$. In particular,

$$\exp(z)\cdot\exp(-z) = \exp(z-z) = \exp(0) = 1,$$

which implies *(3)*.   ∎

An easy induction argument using Property *(2)* in Theorem 3.30 shows that for all complex numbers $z_1, \ldots, z_n$, we have

$$\exp(z_1 + \cdots + z_n) = \exp(z_1)\cdots\exp(z_n).$$

We remark that Tannery's theorem can also be used to establish formulas for sine and cosine; see Problem 2. Also, in Theorem 4.33 on p. 305 we'll see that $\exp(z) = e^z$; however, at this point, we don't even know what e^z ("e to the power z") means, since we haven't defined what it means to raise a real number to a complex power.

3.7.3 Approximation and Irrationality of *e*

We now turn to the question of approximating e. Because $n!$ grows very large as $n \to \infty$, we can use the series for the exponential function to calculate e quite easily. If s_n denotes the nth partial sum for $e = \exp(1)$, then

$$
\begin{aligned}
e &= s_n + \frac{1}{(n+1)!} + \frac{1}{(n+2)!} + \frac{1}{(n+3)!} + \cdots \\
&= s_n + \frac{1}{(n+1)!} + \frac{1}{(n+1)!(n+2)} + \frac{1}{(n+1)!(n+2)(n+3)} + \cdots \\
&< s_n + \frac{1}{(n+1)!} \left\{ 1 + \frac{1}{(n+1)} + \frac{1}{(n+1)^2} + \cdots \right\} \\
&= s_n + \frac{1}{(n+1)!} \cdot \frac{1}{1 - \dfrac{1}{n+1}} \\
&= s_n + \frac{1}{n!\, n}.
\end{aligned}
$$

Thus, we get the following useful estimate for e:

$$
\boxed{\; s_n < e < s_n + \frac{1}{n!\, n}. \;} \tag{3.40}
$$

Example 3.49 In particular, with $n = 1$ we have $s_1 = 2$ and $1/(1!\,1) = 1$, and therefore $2 < e < 3$. Of course, we can get a much more precise estimate with higher values of n: with $n = 10$, we obtain (in common decimal notation; see Section 3.8)

$$
2.718281801 < e < 2.718281829.
$$

This is quite an accurate approximation!

The estimate (3.40) also gives an easy proof that the number e is irrational, a fact first proved by Euler in 1737 [35, p. 463].

Irrationality of *e*

Theorem 3.31 *The number e is irrational.*

Proof Indeed, by way of contradiction, suppose that $e = p/q$, where p and q are positive integers with $q > 1$. Then (3.40) with $n = q$ implies that

$$
s_q < \frac{p}{q} < s_q + \frac{1}{q!\, q}. \tag{3.41}
$$

Since $s_q = 2 + \frac{1}{2!} + \cdots + \frac{1}{q!}$, the number $q! s_q$ is an integer (this is because $q! = 1 \cdot 2 \cdots k \cdot (k+1) \cdots q$ contains a factor of $k!$ for each $1 \leq k \leq q$). Then multiplying the inequalities in (3.41) by $q!$ and putting $m = q! s_q$ (which, as we mentioned, is an integer), we obtain

$$m < p\,(q-1)! < m + \frac{1}{q}.$$

Since $q > 1$, we have $m + 1/q < m + 1$, so $m < p\,(q-1)! < m + 1$. Hence, the integer $p\,(q-1)!$ lies strictly between the two consecutive integers m and $m+1$, which of course is absurd. ∎

We end with the following neat *infinite nested product* formula for e:

$$\boxed{e = 1 + \frac{1}{1} + \frac{1}{2}\Big(1 + \frac{1}{3}\Big(1 + \frac{1}{4}\Big(1 + \frac{1}{5}\big(\cdots\big)\Big)\Big)\Big);} \qquad (3.42)$$

see Problem 6.

▶ **Exercises 3.7**

1. Find the following limits, where for (c) and (d), prove that the limits are $\sum_{k=1}^{\infty} \frac{1}{k^2}$:

(a) $\displaystyle \lim_{n \to \infty} \left\{ \frac{1+n}{(1+2n)} + \frac{2^2 + n^2}{(1+2n)^2} + \cdots + \frac{n^n + n^n}{(1+2n)^n} \right\}$,

(b) $\displaystyle \lim_{n \to \infty} \left\{ \frac{n}{\sqrt{1 + (1 \cdot 2 \cdot n)^2}} + \frac{n}{\sqrt{1 + (2 \cdot 3 \cdot n)^2}} + \cdots + \frac{n}{\sqrt{1 + (n \cdot (n+1) \cdot n)^2}} \right\}$,

(c) $\displaystyle \lim_{n \to \infty} \left\{ \frac{1 + n^2}{1 + (1 \cdot n)^2} + \frac{2^2 + n^2}{1 + (2 \cdot n)^2} + \cdots + \frac{n^2 + n^2}{1 + (n \cdot n)^2} \right\}$,

(d) $\displaystyle \lim_{n \to \infty} \left\{ \Big(\frac{n}{1 + 1^{2n}}\Big)^{\frac{1}{n}} + \Big(\frac{n}{1 + 2^{2n}}\Big)^{\frac{1}{n}} + \cdots + \Big(\frac{n}{1 + n^{2n}}\Big)^{\frac{1}{n}} \right\}$,

(e) $\displaystyle \lim_{n \to \infty} \left\{ \Big(\frac{n}{n+1}\Big)^{n} + \Big(\frac{n}{n+1}\Big)^{2n} + \cdots + \Big(\frac{n}{n+1}\Big)^{n \cdot n} \right\}$,

(f) $\displaystyle \lim_{n \to \infty} \left\{ \frac{n + 1^n}{n + 2^1 \cdot 1^n} + \frac{n + 2^n}{n + 2^2 \cdot 2^n} + \cdots + \frac{n + n^n}{n + 2^n \cdot n^n} \right\}$.

2. For each $z \in \mathbb{C}$, define the **cosine** of z by

$$\cos z = \lim_{n \to \infty} \frac{1}{2}\left[\Big(1 + \frac{iz}{n}\Big)^{n} + \Big(1 - \frac{iz}{n}\Big)^{n} \right]$$

and the **sine** of z by

$$\sin z = \lim_{n \to \infty} \frac{1}{2i} \left\{ \left(1 + \frac{iz}{n} \right)^n - \left(1 - \frac{iz}{n} \right)^n \right\}.$$

(a) Use Tannery's theorem in a similar way as we did in the proof of Property
(1) in Theorem 3.30 to prove that the limits defining $\cos z$ and $\sin z$ exist and
moreover,

$$\cos z = \sum_{k=0}^{\infty} (-1)^k \frac{z^{2k}}{(2k)!} \quad \text{and} \quad \sin z = \sum_{k=0}^{\infty} (-1)^k \frac{z^{2k+1}}{(2k+1)!}.$$

(b) Following the proof that e is irrational, prove that $\cos 1$ (or $\sin 1$ if you prefer)
is irrational.

3. Following [152], we prove that for every $m \geq 3$,

$$\sum_{n=0}^{m} \frac{1}{n!} - \frac{3}{2m} < \left(1 + \frac{1}{m} \right)^m < \sum_{n=0}^{m} \frac{1}{n!}.$$

Taking $m \to \infty$ gives an alternative proof that $\exp(1) = e$. Fix $m \geq 3$.

(i) Prove that for all $2 \leq k \leq m$, we have

$$1 - \frac{k(k-1)}{2m} = 1 - \frac{(1 + 2 + \cdots + k - 1)}{m} \leq \left(1 - \frac{1}{m} \right) \cdots \left(1 - \frac{k-1}{m} \right) < 1.$$

(ii) Using (i), prove that

$$\sum_{n=0}^{m} \frac{1}{n!} - \frac{1}{2m} \sum_{n=2}^{m} \frac{1}{(n-2)!} \leq \left(1 + \frac{1}{m} \right)^m < \sum_{n=0}^{m} \frac{1}{n!}.$$

Now prove the formula. Suggestion: Use the binomial theorem on $(1 + \frac{1}{m})^m$.

4. Let $\{a_n\}$ be a sequence of rational numbers tending to $+\infty$, that is, a sequence
such that given $M > 0$, there is an N such that for all $n > N$, we have $a_n > M$.
In this problem we show that

$$e = \lim \left(1 + \frac{1}{a_n} \right)^{a_n}. \tag{3.43}$$

This formula also holds when the a_n are real numbers, but as of now, we've only
defined rational powers (we'll consider real powers in Section 4.7 on p. 300).

(i) If the rational numbers a_n are all integers tending to $+\infty$, prove (3.43).

(ii) Back to the case of a rational sequence $\{a_n\}$, for each n, let m_n be the unique integer such that $m_n - 1 \le a_n < m_n$ (thus, $m_n = \lfloor a_n \rfloor - 1$, where $\lfloor a_n \rfloor$ is the greatest integer function). Prove that if $m_n \ge 1$, then

$$\left(1 + \frac{1}{m_n}\right)^{m_n - 1} \le \left(1 + \frac{1}{a_n}\right)^{a_n} \le \left(1 + \frac{1}{m_n - 1}\right)^{m_n}.$$

Now prove (3.43).

5. Let $\{b_n\}$ be any null sequence of positive rational numbers. Prove that

$$e = \lim\, (1 + b_n)^{\frac{1}{b_n}}.$$

6. Prove that for every $n \in \mathbb{N}$,

$$1 + \frac{1}{1!} + \frac{1}{2!} + \cdots + \frac{1}{n!} = 1 + \frac{1}{1} + \frac{1}{2}\left(1 + \frac{1}{3}\left(1 + \frac{1}{4}\left(\cdots \left(1 + \frac{1}{n-1}\left(1 + \frac{1}{n}\right)\right)\cdots\right)\right)\right).$$

The infinite nested sum in (3.42) denotes the limit as $n \to \infty$ of this expression.

7. Trying to imitate the proof that e is irrational, prove that fore every $m \in \mathbb{N}$, $\exp(1/m)$ is irrational. After doing this, show that $\cos(1/m)$ (or $\sin(1/m)$ if you prefer) is irrational, where cosine and sine are as defined in Problem 2. (If you're interested in more irrationality proofs, see the article [189].)

8. (**Tannery's theorem II**) For each natural number n, let $\sum_{k=1}^{\infty} a_k(n)$ be a convergent series. Suppose that for each k, $\lim_{n \to \infty} a_k(n)$ exists and there is a convergent series $\sum_{k=1}^{\infty} M_k$ of nonnegative real numbers such that $|a_k(n)| \le M_k$ for all k, n. Imitating the proof of the Tannery's theorem, prove that

$$\boxed{\lim_{n \to \infty} \sum_{k=1}^{\infty} a_k(n) = \sum_{k=1}^{\infty} \lim_{n \to \infty} a_k(n).}$$

9. (**Tannery's theorem and Cauchy's double series theorem**; see Section 6.5 for more on double series!) In this problem we relate Tannery's theorem to double series. A **double sequence** is just a map $a : \mathbb{N} \times \mathbb{N} \to \mathbb{C}$; for $m, n \in \mathbb{N}$ we denote $a(m, n)$ by a_{mn}. We say that the **iterated series** $\sum_{m=1}^{\infty} \sum_{n=1}^{\infty} a_{mn}$ converges if for each $m \in \mathbb{N}$, the series $\sum_{n=1}^{\infty} a_{mn}$ converges (call the sum α_m) and the series $\sum_{m=1}^{\infty} \alpha_m$ converges. Similarly, we say that the iterated series $\sum_{n=1}^{\infty} \sum_{m=1}^{\infty} a_{mn}$ converges if for each $n \in \mathbb{N}$, the series $\sum_{m=1}^{\infty} a_{mn}$ converges (call the sum β_n) and the series $\sum_{n=1}^{\infty} \beta_n$ converges.

The object of this problem is to prove that given a double sequence $\{a_{mn}\}$ of complex numbers such that either

$$\sum_{m=1}^{\infty}\sum_{n=1}^{\infty}|a_{mn}| \text{ converges} \quad \text{or} \quad \sum_{n=1}^{\infty}\sum_{m=1}^{\infty}|a_{mn}| \text{ converges,} \qquad (3.44)$$

then

$$\sum_{m=1}^{\infty}\sum_{n=1}^{\infty} a_{mn} = \sum_{n=1}^{\infty}\sum_{m=1}^{\infty} a_{mn} \qquad (3.45)$$

in the sense that both iterated sums converge and are equal. The implication (3.44) $\implies$ (3.45) is called **Cauchy's double series theorem**; see Theorem 6.26 on p. 471 in Section 6.5 for the full story. To prove this, you may proceed as follows.

(i) Assume that $\sum_{m=1}^{\infty}\sum_{n=1}^{\infty}|a_{mn}|$ converges; we must prove the equality (3.45). To do so, for each $k \in \mathbb{N}$, define $M_k = \sum_{j=1}^{\infty}|a_{kj}|$, which converges by assumption. Then $\sum_{k=1}^{\infty} M_k$ also converges by assumption. Define $a_k(n) = \sum_{j=1}^{n} a_{kj}$. Prove that Tannery's theorem II, from the previous problem, can be applied to these $a_k(n)$, and in doing so, establish the equality (3.45).

(ii) Assuming that $\sum_{n=1}^{\infty}\sum_{m=1}^{\infty}|a_{mn}|$ converges, prove the equality (3.45).

(iii) Cauchy's double series theorem can be used to prove neat and nonobvious identities. For example, prove that for every $k \in \mathbb{N}$ and $z \in \mathbb{C}$ with $|z| < 1$, we have

$$\sum_{n=1}^{\infty} \frac{z^{n(k+1)}}{1-z^n} = \sum_{m=1}^{\infty} \frac{z^{m+k}}{1-z^{m+k}};$$

that is,

$$\frac{z^{k+1}}{1-z} + \frac{z^{2(k+1)}}{1-z^2} + \frac{z^{3(k+1)}}{1-z^3} + \cdots = \frac{z^{1+k}}{1-z^{1+k}} + \frac{z^{2+k}}{1-z^{2+k}} + \frac{z^{3+k}}{1-z^{3+k}} + \cdots.$$

Suggestion: Apply Cauchy's double series to $\{a_{mn}\}$, where $a_{mn} = z^{n(m+k)}$.

3.8 Decimals and "Most" Real Numbers Are Irrational

Since grade school, we have represented real numbers in *base* 10.[14] In this section we continue the discussion initiated in Section 2.5 on p. 69 on the use of arbitrary bases. We also look at the celebrated *Cantor's diagonal argument*.

---

[14]"To what heights would science now be raised if Archimedes had made that discovery! [= the decimal system of numeration or its equivalent (with some base other than 10)]." Carl Gauss (1777–1855).

3.8.1   Decimal and b-Adic Representations of Real Numbers

We are all familiar with the common decimal or base 10 notation, which we used without mention in the last section concerning the estimate $2.718281801 <$ $e < 2.718281829$. Here, we know that the decimal (also called base 10) *notation* 2.718281801 represents the *number*

$$2 + \frac{7}{10} + \frac{1}{10^2} + \frac{8}{10^3} + \frac{2}{10^4} + \frac{8}{10^5} + \frac{1}{10^6} + \frac{8}{10^7} + \frac{0}{10^8} + \frac{1}{10^9},$$

that is, this real number gives meaning to the symbol 2.718281801. More generally, the *symbol* $\alpha_k \alpha_{k-1} \ldots \alpha_0.a_1 a_2 a_3 a_4 a_5 \ldots$, where the α_n and a_n are integers in $0, 1, \ldots, 9$, represents the *number*

$$\alpha_k \cdot 10^k + \cdots + \alpha_1 \cdot 10 + \alpha_0 + \frac{a_1}{10} + \frac{a_2}{10^2} + \frac{a_3}{10^3} + \cdots = \sum_{n=0}^{k} \alpha_n \cdot 10^n + \sum_{n=1}^{\infty} \frac{a_n}{10^n}.$$

Notice that the infinite series $\sum_{n=1}^{\infty} \frac{a_n}{10^n}$ converges, because $0 \le a_n \le 9$ for all n, so we can compare this series with $\sum_{n=1}^{\infty} \frac{9}{10^n} = 1 < \infty$. In particular, the number $\sum_{n=1}^{\infty} \frac{a_n}{10^n}$ lies in $[0, 1]$.

More generally, instead of restricting to base 10, we can use other bases. Let $b > 1$ be an integer (the **base**). Then the *symbol* $\alpha_k \alpha_{k-1} \ldots \alpha_0._b a_1 a_2 a_3 a_4 a_5 \ldots$, where the α_n and a_n are integers in $0, 1, \ldots, b - 1$, represents the real *number*

$$a = \alpha_k \cdot b^k + \cdots + \alpha_1 \cdot b + \alpha_0 + \frac{a_1}{b} + \frac{a_2}{b^2} + \frac{a_3}{b^3} + \cdots = \sum_{n=0}^{k} \alpha_n \cdot b^n + \sum_{n=1}^{\infty} \frac{a_n}{b^n}.$$

The infinite series $\sum_{n=1}^{\infty} \frac{a_n}{b^n}$ converges, because $0 \le a_n \le b - 1$ for all n, so we can compare this series with

$$\sum_{n=1}^{\infty} \frac{b-1}{b^n} = (b-1) \cdot \sum_{n=1}^{\infty} \left(\frac{1}{b}\right)^n = (b-1) \frac{\frac{1}{b}}{1 - \frac{1}{b}} = 1 < \infty.$$

The symbol $\alpha_k \alpha_{k-1} \ldots \alpha_0._b a_1 a_2 a_3 a_4 a_5 \ldots$ is called the **b-adic representation** or **b-adic expansion** of a. The α_n and a_n are called **digits**. A natural question is this: Does every nonnegative real number have such a representation? The answer is yes. If a is negative, its b-adic representation is, by definition, negative the b-adic representation of the positive number $-a$.

Now given $x \ge 0$, to prove that it has a b-adic expansion, write $x = m + a$, where $m = \lfloor x \rfloor$ is the integer part of x and $0 \le a < 1$. We already know that m has a b-adic expansion, so we can focus on writing a in a b-adic expansion.

Theorem 3.32 *Let $b \in \mathbb{N}$ with $b > 1$. Then for every $a \in [0, 1]$, there exists a sequence of integers $\{a_n\}_{n=1}^{\infty}$ with $0 \leq a_n \leq b - 1$ for all n such that*

$$a = \sum_{n=1}^{\infty} \frac{a_n}{b^n},$$

where infinitely many of the a_n are nonzero if $a \neq 0$.

Proof If $a = 0$, we take all the a_n to be zero, so we henceforth assume that $0 < a \leq 1$; we find the a_n as follows. First, we divide $(0, 1]$ into b nonoverlapping intervals:

$$\left(0, \frac{1}{b}\right], \ \left(\frac{1}{b}, \frac{2}{b}\right], \ \left(\frac{2}{b}, \frac{3}{b}\right], \ \dots, \ \left(\frac{b-1}{b}, 1\right],$$

as seen here (Fig. 3.14):

Fig. 3.14 We divide $(0, 1]$ into b subintervals and find in which interval a lies. This uniquely determines a_1; in this example, $a_1 = 1$

Since $a \in (0, 1]$, a must lie in one of these intervals, so there is an integer a_1 with $0 \leq a_1 \leq b - 1$ such that

$$a \in \left(\frac{a_1}{b}, \frac{a_1 + 1}{b}\right] \quad \Longleftrightarrow \quad \frac{a_1}{b} < a \leq \frac{a_1 + 1}{b}.$$

Second, we divide $\left(\frac{a_1}{b}, \frac{a_1+1}{b}\right]$ into b nonoverlapping subintervals. Since the length of $\left(\frac{a_1}{b}, \frac{a_1+1}{b}\right]$ is $\frac{a_1+1}{b} - \frac{a_1}{b} = \frac{1}{b}$, we are dividing the interval $\left(\frac{a_1}{b}, \frac{a_1+1}{b}\right]$ into b subintervals of length $(1/b)/b = 1/b^2$, namely into the intervals

$$\left(\frac{a_1}{b}, \frac{a_1}{b} + \frac{1}{b^2}\right], \ \left(\frac{a_1}{b} + \frac{1}{b^2}, \frac{a_1}{b} + \frac{2}{b^2}\right], \ \dots, \ \left(\frac{a_1}{b} + \frac{b-1}{b^2}, \frac{a_1}{b} + \frac{1}{b}\right].$$

Here's a picture (Fig. 3.15):

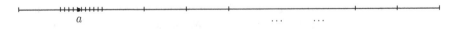

Fig. 3.15 We divide $(a_1/b, (a_1 + 1)/b]$ into b subintervals and find in which interval a lies. This will then uniquely determine a_2

Now $a \in \left(\frac{a_1}{b}, \frac{a_1+1}{b}\right]$, so a must lie in one of these intervals. Thus, there is an integer a_2 with $0 \leq a_2 \leq b - 1$ such that

$$a \in \left(\frac{a_1}{b} + \frac{a_2}{b^2}, \frac{a_1}{b} + \frac{a_2+1}{b^2}\right] \iff \frac{a_1}{b} + \frac{a_2}{b^2} < a \leq \frac{a_1}{b} + \frac{a_2+1}{b^2}.$$

Continuing this process (slang for "by induction"), we can find a sequence of integers $\{a_n\}$ such that $0 \leq a_n \leq b - 1$ for all n and

$$\frac{a_1}{b} + \frac{a_2}{b^2} + \cdots + \frac{a_{n-1}}{b^{n-1}} + \frac{a_n}{b^n} < a \leq \frac{a_1}{b} + \cdots + \frac{a_{n-1}}{b^{n-1}} + \frac{a_n+1}{b^n}. \tag{3.46}$$

Let $y := \sum_{n=1}^{\infty} \frac{a_n}{b^n}$; this series converges, because its partial sums are bounded by a according to the left-hand inequality in (3.46). Since $1/b^n \to 0$ as $n \to \infty$, by taking $n \to \infty$ in (3.46) and using the squeeze theorem, we see that $y \leq a \leq y$. This shows that $a = y = \sum_{n=1}^{\infty} \frac{a_n}{b^n}$. We claim that infinitely many of the a_n are nonzero. Indeed, if there were at most finitely many nonzero a_n, say for some m we had $a_n = 0$ for all $n > m$, then we would have $a = \sum_{n=1}^{m} \frac{a_n}{b^n}$. Now setting $n = m$ in (3.46) and looking at the left-hand inequality shows that $a < a$. This is impossible, so there must be infinitely many nonzero a_n. ∎

Here's another question: If a b-adic representation exists, is it unique? The answer to this question is no.

Example 3.50 Consider, for example, the number $1/2$, which has two decimal (base 10) expansions:

$$\frac{1}{2} = 0.50000000\ldots \quad \text{and} \quad \frac{1}{2} = 0.49999999\ldots.$$

Notice that the first decimal expansion terminates.

You might remember from high school that the only numbers with two different decimal expansions are the ones that have a terminating expansion. In general, a b-adic expansion $0._b a_1 a_2 a_3 a_4 a_5 \ldots$ is said to **terminate** if all the a_n equal zero for n sufficiently large.

Theorem 3.33 *Let $b \in \mathbb{N}$ with $b > 1$. Then every real number in $(0, 1]$ has a unique b-adic expansion except those numbers that can be represented by a terminating expansion. The numbers with terminating expansions also have a b-adic expansion where $a_n = b - 1$ for all n sufficiently large.*

Proof For $a \in (0, 1]$, let $a = \sum_{n=1}^{\infty} \frac{a_n}{b^n}$ be its b-adic expansion found in Theorem 3.32, so there are infinitely many nonzero a_n. Suppose that $\{c_n\}$ is another sequence of integers, not equal to the sequence $\{a_n\}$, such that $0 \leq c_n \leq b - 1$ for all

n and such that $a = \sum_{n=1}^{\infty} \frac{c_n}{b^n}$. Since $\{a_n\}$ and $\{c_n\}$ are not the same sequence, there is at least one n such that $a_n \neq c_n$. Let m be the smallest natural number such that $a_m \neq c_m$. Then $a_n = c_n$ for $n = 1, 2, \ldots, m-1$, so

$$\sum_{n=1}^{\infty} \frac{a_n}{b^n} = \sum_{n=1}^{\infty} \frac{c_n}{b^n} \implies \sum_{n=m}^{\infty} \frac{a_n}{b^n} = \sum_{n=m}^{\infty} \frac{c_n}{b^n}.$$

Since there are infinitely many nonzero a_n, we have

$$\frac{a_m}{b^m} < \sum_{n=m}^{\infty} \frac{a_n}{b^n} = \sum_{n=m}^{\infty} \frac{c_n}{b^n} = \frac{c_m}{b^m} + \sum_{n=m+1}^{\infty} \frac{c_n}{b^n}$$

$$\leq \frac{c_m}{b^m} + \sum_{n=m+1}^{\infty} \frac{b-1}{b^n} = \frac{c_m}{b^m} + \frac{1}{b^m}.$$

Multiplying the extremities of these inequalities by b^m, we obtain $a_m < c_m + 1$, so $a_m \leq c_m$. Since $a_m \neq c_m$, we must actually have $a_m < c_m$. Now

$$\frac{c_m}{b^m} \leq \sum_{n=m}^{\infty} \frac{c_n}{b^n} = \sum_{n=m}^{\infty} \frac{a_n}{b^n} = \frac{a_m}{b^m} + \sum_{n=m+1}^{\infty} \frac{a_n}{b^n}$$

$$\leq \frac{a_m}{b^m} + \sum_{n=m+1}^{\infty} \frac{b-1}{b^n} = \frac{a_m}{b^m} + \frac{1}{b^m} \leq \frac{c_m}{b^m}.$$

Since the ends are equal, all the inequalities in between must be equalities. In particular, making the first inequality into an equality shows that $c_n = 0$ for all $n = m+1, m+2, m+3, \ldots$, and making the middle inequality into an equality shows that $a_n = b-1$ for all $n = m+1, m+2, m+3, \ldots$. It follows that a has exactly one b-adic expansion, except when we can write a as

$$a = 0._b c_1 c_2 \ldots c_m = 0._b a_1 \ldots a_m (b-1)(b-1)(b-1)\ldots,$$

a terminating expansion and an expansion that has repeating $b-1$.   ∎

3.8.2 Rational Numbers

We now consider periodic decimals, such as

$$\frac{1}{3} = 0.3333333\ldots, \qquad \frac{3526}{495} = 7.1232323\ldots, \qquad \frac{611}{495} = 1.2343434\ldots.$$

As you well know, we usually write these decimals with bars to indicate a repeating pattern:

$$\frac{1}{3} = 0.\overline{3}\ldots, \quad \frac{3526}{495} = 7.1\overline{23}\ldots, \quad \frac{611}{495} = 1.2\overline{34}\ldots.$$

For general b-adic expansions, we say that $\alpha_k \ldots \alpha_0._b a_1 a_2 a_3 \ldots$ is **periodic** if there exists an $\ell \in \mathbb{N}$ (called a **period**) such that $a_n = a_{n+\ell}$ for all n sufficiently large.

Example 3.51 In the base 10 expansion of $\frac{3526}{495} = 7.1232323\ldots = 7.a_1 a_2 a_3 \cdots$, we have $a_n = a_{n+2}$ for all $n \geq 2$.

Example 3.52 We can actually see how the periodic pattern appears by reviewing *high school long division*! Indeed, let's compute $41/333$ by long dividing 333 into 41. We first multiply 41 by 10 to start the process:

```
        .123
  333 )410
       333
       ‾‾‾
       770
       666
      ‾‾‾‾
      1040
       999
      ‾‾‾‾
        41
```

At this point, we get the remainder 41, and after multiplication by 10 we are back at the beginning. Thus, by continuing this process of long division, we are going to repeat the pattern $1, 2, 3$, obtaining $41/333 = 0.\overline{123}$. Now let's review *how* long division is done. Let $p = 41, q = 333$, and $b = 10$. We first multiply p by the base b to get 410 and then divide by q, getting the quotient 1 (which we write on the top line) and remainder 77. Second, we multiply 77 by the base b to get 770 and then divide by q, getting the quotient 2 (which we write on the top line) and remainder 104. We then multiply 104 by the base b to get 1040, and so on. The pattern is to repeat the process of multiplying by the base b, then dividing the result by q to get a quotient, which becomes a decimal digit, and a remainder with which we continue the process. This procedure is exactly what we do in **Step 1** of the proof of Theorem 3.34 below.

Example 3.53 Now suppose we are given a periodic decimal, such as $x := 0.\overline{123}$; how do we show that it represents a rational number? The trick, often learned in high school, is to multiply x by 10^3 and note that

$$10^3 x = 123.\overline{123}.$$

It follows that $1000x = 123 + x$. Solving for x, we get $x = 123/999 = 41/333$. We use this technique in **Step 2** of the proof of Theorem 3.34 below.

Theorem 3.34 *Let $b \in \mathbb{N}$ with $b > 1$. A real number is rational if and only if its b-adic expansion is periodic.*

Proof We first prove the "only if" statement, then the "if" statement.

Step 1: We prove "only if": Given integers p, q with $q > 0$, we show that p/q has a periodic b-adic expansion. By the division algorithm (Theorem 2.16 on p. 60), we have $p/q = q' + r/q$, where $q' \in \mathbb{Z}$ and $0 \leq r < q$. Since integers have b-adic expansions, we just have to prove that r/q has a periodic b-adic expansion. By relabeling r with the letter p, we shall assume that $0 \leq p < q$, so that $p/q < 1$. Proceeding via *high school long division* reviewed in the examples above, we construct the decimal expansion of p/q.

First, using the division algorithm, we divide bp by q, obtaining a unique integer a_1 such that $bp = a_1 q + r_1$, where $0 \leq r_1 < q$. Observe that

$$\frac{p}{q} - \frac{a_1}{b} = \frac{bp - a_1 q}{bq} = \frac{r_1}{bq} \geq 0.$$

Thus,

$$\frac{a_1}{b} \leq \frac{p}{q} < 1,$$

which implies that $0 \leq a_1 < b$.

Next, using the division algorithm, we divide $b\,r_1$ by q, obtaining a unique integer a_2 such that $br_1 = a_2 q + r_2$, where $0 \leq r_2 < q$. Observe that

$$\frac{p}{q} - \frac{a_1}{b} - \frac{a_2}{b^2} = \frac{r_1}{bq} - \frac{a_2}{b^2} = \frac{br_1 - a_2 q}{b^2 q} = \frac{r_2}{b^2 q} \geq 0.$$

Thus,

$$\frac{a_2}{b^2} \leq \frac{r_1}{bq} < \frac{q}{bq} = \frac{1}{b},$$

which implies that $0 \leq a_2 < b$.

Once more using the division algorithm, we divide $b\,r_2$ by q, obtaining a unique integer a_3 such that $br_2 = a_3 q + r_3$, where $0 \leq r_3 < q$. Observe that

$$\frac{p}{q} - \frac{a_1}{b} - \frac{a_2}{b^2} - \frac{a_3}{b^3} = \frac{r_2}{b^2 q} - \frac{a_3}{b^3} = \frac{br_2 - a_3 q}{b^3 q} = \frac{r_3}{b^3 q} \geq 0.$$

Thus,

$$\frac{a_3}{b^3} \leq \frac{r_2}{b^2 q} < \frac{q}{b^2 q} = \frac{1}{b^2},$$

which implies that $0 \leq a_3 < b$. Continuing by induction, for each n we construct integers a_n, r_n such that $0 \leq a_n < b, 0 \leq r_n < q, br_n = a_{n+1} q + r_{n+1}$, and

$$\frac{p}{q} - \frac{a_1}{b} - \frac{a_2}{b^2} - \frac{a_3}{b^3} - \cdots - \frac{a_n}{b^n} = \frac{r_n}{b^n q}.$$

Since $0 \le r_n < q$ for each n, it follows that $\frac{r_n}{b^n q} \to 0$ as $n \to \infty$, so we can write

$$\frac{p}{q} = \sum_{n=1}^{\infty} \frac{a_n}{b^n}, \quad \text{which is say,} \quad \frac{p}{q} = 0._b a_1 a_2 a_3 a_4 a_5 \ldots. \tag{3.47}$$

Now one of two things holds: Either some remainder r_n is equal to zero or none of the r_n are zero. Suppose that we are in the first case, that some r_n equals zero. By construction, we divide br_n by q using the division algorithm to get $br_n = a_{n+1}q + r_{n+1}$. Since $r_n = 0$ and quotients and remainders are unique, we must have $a_{n+1} = 0$ and $r_{n+1} = 0$. By construction, we divide br_{n+1} by q using the division algorithm to get $br_{n+1} = a_{n+2}q + r_{n+2}$. Since $r_{n+1} = 0$ and quotients and remainders are unique, we must have $a_{n+2} = 0$ and $r_{n+2} = 0$. Continuing this procedure, we see that all a_k with $k > n$ are zero. This, in view of (3.47), shows that the b-adic expansion of p/q has repeating zeros, so in particular, it is periodic.

Suppose that we are in the second case, that no r_n is zero. Consider the $q + 1$ remainders $r_1, r_2, \ldots, r_{q+1}$. Since $0 \le r_n < q$, each r_n can take on at most q different values (namely $0, 1, 2, \ldots, q - 1$, "q pigeonholes"), so by the pigeonhole principle (see p. 136), two of these remainders must have the same value ("be in the same pigeonhole"). Thus, $r_k = r_{k+\ell}$ for some k and ℓ. We now show that $a_{k+1} = a_{k+\ell+1}$. Indeed, a_{k+1} was defined by dividing br_k by q, so that $br_k = a_{k+1}q + r_{k+1}$. On the other hand, $a_{k+\ell+1}$ was defined by dividing $br_{k+\ell}$ by q, so that $br_{k+\ell} = a_{k+\ell+1}q + r_{k+\ell+1}$. Now the division algorithm states that the quotients and remainders are unique. Since $br_k = br_{k+\ell}$, it follows that $a_{k+1} = a_{k+\ell+1}$ and $r_{k+1} = r_{k+\ell+1}$. Repeating this same argument shows that $a_{k+n} = a_{k+\ell+n}$ for all $n \ge 0$; that is, $a_n = a_{n+\ell}$ for all $n \ge k$. Thus, p/q has a periodic b-adic expansion.

Step 2: We now prove the "if" portion: A number with a periodic b-adic expansion is rational. Let a be a real number and suppose that its b-adic decimal expansion is periodic. Since a is rational if and only if its noninteger part is rational, we may assume that the integer part of a is zero. Let

$$a = 0._b a_1 a_2 \ldots a_k \overline{b_1 \ldots b_\ell}$$

have a periodic b-adic expansion, where the bar means that the block $b_1 \cdots b_\ell$ repeats. Observe that in an expansion $\alpha_m \alpha_{m-1} \cdots \alpha_0._b \beta_1 \beta_2 \beta_3 \ldots$, multiplication by b^n for $n \in \mathbb{N}$ moves the decimal point n places to the right. (Try to prove this; think about the familiar base 10 case first.) In particular,

$$b^{k+\ell} a = a_1 a_2 \ldots a_k b_1 \ldots b_\ell._b \overline{b_1 \ldots b_\ell} = a_1 a_2 \ldots a_k b_1 \ldots b_\ell + 0._b \overline{b_1 \ldots b_\ell}$$

and

$$b^k a = a_1 a_2 \ldots a_k._b \overline{b_1 \ldots b_\ell} = a_1 a_2 \ldots a_k + 0._b \overline{b_1 \ldots b_\ell}.$$

Subtracting, we see that the numbers given by $0._b \overline{b_1 \ldots b_\ell}$ cancel, so $b^{k+\ell}a - b^k a = p$, where p is an integer. Hence, $a = p/q$, where $q = b^{k+\ell} - b^k$. Thus a is rational. ∎

3.8.3 Cantor's Diagonal Argument

Now that we know about decimal expansions, we can present Cantor's second proof that the irrationals and nontrivial intervals are uncountable. His first proof appeared in Section 2.11 on p. 141.

Cantor's second proof

> **Theorem 3.35** $\mathfrak{c} \neq \aleph_0$.

Proof Since $(0, 1)$ has cardinality $\mathfrak{c}$, we just have to prove that $(0, 1)$ is not countable. Assume, to get a contradiction, that there is a bijection $f : \mathbb{N} \longrightarrow (0, 1)$. Let us write the images of f as decimals (base 10):

$$1 \longleftrightarrow f(1) = .a_{11}\, a_{12}\, a_{13}\, a_{14} \cdots$$
$$2 \longleftrightarrow f(2) = .a_{21}\, a_{22}\, a_{23}\, a_{24} \cdots$$
$$3 \longleftrightarrow f(3) = .a_{31}\, a_{32}\, a_{33}\, a_{34} \cdots$$
$$4 \longleftrightarrow f(4) = .a_{41}\, a_{42}\, a_{43}\, a_{44} \cdots$$
$$\vdots \qquad \vdots \qquad ,$$

where we assume that in each of these expansions there is never an infinite run of 9's. Recall from Theorem 3.33 that every real number in $(0, 1)$ has a *unique* such representation. Now let us define a real number $a = .a_1\, a_2\, a_3 \ldots$, where

$$a_n := \begin{cases} 3 & \text{if } a_{nn} \neq 3 \\ 7 & \text{if } a_{nn} = 3. \end{cases}$$

(The choice of 3 and 7 is arbitrary—you can choose another pair of unequal integers in $0, \ldots, 9$ if you like!) Notice that $a_n \neq a_{nn}$ for all n. In particular, $a \neq f(1)$, because a and $f(1)$ differ in the first digit. On the other hand, $a \neq f(2)$, because a and $f(2)$ differ in the second digit. Similarly, $a \neq f(n)$ for every n, since a and $f(n)$ differ in the nth digit. This contradicts that $f : \mathbb{N} \to (0, 1)$ is onto. ∎

Cantor's diagonal argument is not only elegant, it is useful, for it can be used to *generate* transcendental numbers; see [94].

▶ **Exercises 3.8**

1. Find the numbers with the b-adic expansions (here $b = 10, 2, 3$, respectively):

$(a)\ 0.010101\ldots, \quad (b)\ 0._2010101\ldots, \quad (c)\ 0._3010101\ldots.$

2. Prove that a real number $a \in (0, 1)$ has a terminating decimal expansion if and only if $2^m 5^n a \in \mathbb{Z}$ for some nonnegative integers m, n.

3. (s-**adic expansions**) Let $s = \{b_n\}$ be a sequence of integers with $b_n > 1$ for all n and let $0 < a \le 1$. Prove that there is a sequence of integers $\{a_n\}_{n=1}^\infty$ with $0 \le a_n \le b_n - 1$ for all n and with infinitely many nonzero a_n such that

$$a = \sum_{n=1}^\infty \frac{a_n}{b_1 \cdot b_2 \cdot b_3 \cdots b_n}.$$

Suggestion: Can you imitate the proof of Theorem 3.32?

4. (**Cantor's original diagonal argument**) Let g and c be two distinct objects and let G be the set consisting of all functions $f : \mathbb{N} \longrightarrow \{g, c\}$. Let $f_1, f_2, f_3, \ldots$ be an infinite sequence of elements of G. Prove that there is an element f in G that is not in this list. From this prove that G is uncountable. Conclude that the set of all sequences of 0's and 1's is uncountable; in the next problem we find its cardinality.

5. (**Real numbers and sequences of 0's and 1's**) Let $2^{\mathbb{N}}$ denote[15] the set of all sequences of 0's and 1's, or what's the same thing, the set of functions from $\mathbb{N}$ into $\{0, 1\}$. We denote the cardinality of $2^{\mathbb{N}}$ by, what else, $2^{\aleph_0}$. We shall prove that $2^{\aleph_0} = \mathfrak{c}$. (In particular, $2^{\mathbb{N}}$ and $\mathbb{R}$ are in a one-to-one correspondence.)

   (i) Prove that $B := \{(b_1, b_2, \ldots) \in 2^{\mathbb{N}} ; b_i = 0 \text{ for all } i \text{ sufficiently large}\}$ is countable.

   (ii) Let $A = 2^{\mathbb{N}} \setminus B$, so that $2^{\mathbb{N}} = A \cup B$. Define $f : A \longrightarrow (0, 1]$ by associating a sequence $(a_1, a_2, a_3, \ldots) \in A$ with the real number having binary expansion $0._2a_1a_2a_3 \ldots \in (0, 1]$. Prove that f is a bijection.

   (iii) Conclude that $2^{\aleph_0} = \mathfrak{c}$.

6. (**Cantor's unbelievable theorem**) Let $X = 2^{\mathbb{N}}$.

   (i) Prove that X^2 has the same cardinality as X.

   (ii) Generalizing, prove that for every $m \in \mathbb{N}$, the m-fold product X^m has the same cardinality as X. Consequently, $\mathbb{R}^m$ has the same cardinality as $\mathbb{R}$, to which Cantor said "I see it, but I don't believe it!"

   (iii) We now generalize even further. Let $X^{\mathbb{N}}$ denote the collection of all functions from $\mathbb{N}$ into X. In other words, $X^{\mathbb{N}}$ denotes the collection of all sequences $(x_1, x_2, x_3, \ldots)$, where $x_k \in X$ for each k (thus, x_k is itself a sequence, of 0's and 1's). We remark that it is common to denote $X^{\mathbb{N}}$ by X^∞, because $X^{\mathbb{N}}$ can be viewed as the infinite Cartesian product $X \times X \times X \times \cdots$. Prove:

---

[15]In the notation from Problem 5 on p. 145, $2^{\mathbb{N}}$ would be denoted by $\{0, 1\}^{\mathbb{N}}$, so "2" is shorthand for the two-element set $\{0, 1\}$.

Cantor's unbelievable theorem: $X^{\mathbb{N}}$ has the same cardinality as X.

In particular, $\mathbb{R}^\infty$, the collection of all infinite sequences of real numbers, has the same cardinality as $\mathbb{R}$. Suggestion: An element of X is a function from $\mathbb{N}$ into $\{0, 1\}$, while an element of $X^{\mathbb{N}}$ can be considered a function from $\mathbb{N} \times \mathbb{N}$ into $\{0, 1\}$ (try to see why). Use a bijection between $\mathbb{N}$ and $\mathbb{N} \times \mathbb{N}$ to get a bijection between X and $X^{\mathbb{N}}$.

Chapter 4
Limits, Continuity, and Elementary Functions

One merit of mathematics few will deny: it says more in fewer words than any other science. The formula, $e^{i\pi} = -1$ expressed a world of thought, of truth, of poetry, and of the religious spirit "God eternally geometrizes."
David Eugene Smith (1860–1944) [201].

In this chapter we study what are without doubt the most important functions in all of analysis and topology, the continuous functions. In particular, we study the continuity properties of "the most important function in mathematics" [205, p. 1]: $\exp(z) = \sum_{n=0}^{\infty} \frac{z^n}{n!}, z \in \mathbb{C}$. From this single function arise just about every important function and number you can think of: the logarithm function, powers, roots, the trigonometric functions, the hyperbolic functions, the number e, the number π, ..., and the famous formula displayed in the above quotation!

What do the Holy Bible, squaring the circle, House bill No. 246 of the Indiana state legislature in 1897, square-free natural numbers, coprime natural numbers, the sentence

$$\textit{May I have a large container of coffee? Thank you,} \tag{4.1}$$

the mathematicians Archimedes, William Jones, Leonhard Euler, Johann Lambert, Ferdinand von Lindemann, John Machin, and Yasumasa Kanada have to do with one another? The answer (drum roll please): They all have been involved in the life of the remarkable number π! This fascinating number is defined and some of its amazing and death-defying properties and formulas are studied in this chapter! By the way, the sentence (4.1) is a mnemonic device to remember the digits of π. The number of letters in each word represents a digit of π; e.g., "May" represents 3, "I" 1, etc. The sentence (4.1) gives ten digits of π: 3.141592653.[1]

---

[1] Using mnemonics to memorize digits of π isn't a good idea if you want to beat Chao Lu's result of reciting 67,890 digits from memory! (see http://www.pi-world-ranking-list.com).

© Paul Loya 2017
P. Loya, *Amazing and Aesthetic Aspects of Analysis*,
https://doi.org/10.1007/978-1-4939-6795-7_4

In Section 4.1, we begin our study of continuity by learning about limits of functions; in Section 4.2, we study some useful limit properties; and then in Section 4.3, we discuss continuous functions in terms of limits of functions. In Section 4.4, we study some fundamental properties of continuous functions, and in Section 4.5, we give many neat applications of continuity. A special class of functions, called monotone functions, have many special properties, which are investigated in Section 4.6. In Section 4.7, we study "the most important function in mathematics," and we also study its inverse, the logarithm function, and then we use the logarithm function to define powers. We also define the Riemann zeta function and the Euler–Mascheroni constant γ,

$$\gamma := \lim_{n \to \infty} \left(1 + \frac{1}{2} + \cdots + \frac{1}{n} - \log n \right),$$

a constant that will come up again and again (see the book [104], which is devoted to this number); and we'll prove that the alternating harmonic series has sum $\log 2$:

$$\log 2 = 1 - \frac{1}{2} + \frac{1}{3} - \frac{1}{4} + \frac{1}{5} - \frac{1}{6} + - \cdots,$$

another fact that will come up often. Here's an interesting question: Does the series

$$\sum_{p \text{ is prime}} \frac{1}{p} = \frac{1}{2} + \frac{1}{3} + \frac{1}{5} + \frac{1}{7} + \frac{1}{11} + \frac{1}{13} + \frac{1}{17} + \frac{1}{19} + \frac{1}{23} + \frac{1}{29} + \cdots$$

converge or diverge? For the answer, see Section 4.8. In Section 4.9, we use the exponential function to define the trigonometric functions, and we define π, the fundamental constant of geometry. In Section 4.10 we study roots of complex numbers, and we give fairly elementary proofs of the fundamental theorem of algebra. In Section 4.11 we study the inverse trigonometric functions. The calculation of the incredible number π and (we hope) the imparting of the sense of great fascination that this number arouses are the highlights of Section 4.12.

CHAPTER 4 OBJECTIVES: THE STUDENT WILL BE ABLE TO . . .

- Apply the rigorous ε-δ definition of limits for functions and continuity.
- Apply, and know how to prove, the fundamental theorems of continuous functions.
- Define the elementary functions (exponential, trigonometric, and their inverses) and the number π.
- Explain three related proofs of the fundamental theorem of algebra.

4.1   Continuity and ε-δ Arguments for Limits of Functions

Simply put, "continuity" basically says that "this world works as it should." More precisely, in nature we observe the following

Law of continuity: *The precision of a continuous function's outputs can be controlled by restricting its inputs to within some safety margin.*

For example, if the temperature (output) is 90°F at 1 PM (time = input), then the "law of continuity" says that the temperature will be near 90°F for some time around 1 PM. Consider the following examples:

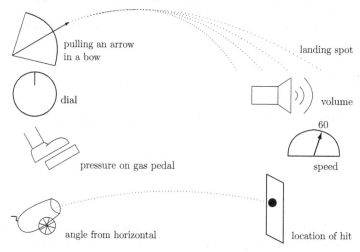

In all these examples, the precision of the output (landing spot of an arrow, volume of a radio, speed, and the location of a cannonball hit) can be controlled by small adjustments, or safety margins, in your input (depth of draw, turn of dial, pressure on pedal, and the angle of the cannon). The goal of this section is to make this "precision–safety margin" idea *precise* using ε's and δ's.

4.1.1   Limit Points

Before reading on, it might benefit the reader to reread the material on open balls in Section 2.9 on p. 126. If $A \subseteq \mathbb{R}^m$, then a point $c \in \mathbb{R}^m$ is said to be a **limit point** of A if every open ball centered at c contains a point of A different from c. In other words, given $r > 0$, there is a point $x \in A$ such that $x \in B_r(c)$ and $x \neq c$. Let's box this definition:

---

c is a limit point of A $\iff$ for each $r > 0$, there's an $x \in A$ such that $0 < |x - c| < r$.

The inequality $0 < |x - c|$ just means that $x \neq c$, while the inequality $|x - c| < r$ just means that $x \in B_r(c)$. We remark that the point c may or may not belong to A. Here's a picture to keep in mind (Fig. 4.1):

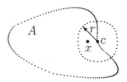

Fig. 4.1 Consider the region A inside a closed curve (not including the curve itself, which is why we draw a *dotted curve*). Every point on the curve or inside the curve is a limit point of A. This figure shows a point c on the curve. No matter how small $r > 0$ is, there is always a point x of A within distance r of c. Thus, c is a limit point of A

Note that if $m = 1$, then c is a limit point of $A \subseteq \mathbb{R}$ if

for every $r > 0$, there is an $x \in A$ such that $x \in (c - r, c + r)$ and $x \neq c$.

Example 4.1 Consider the interval $A = [0, 1)$:

Observe that 0 and 1 are limit points of A; here, 0 belongs to A, while 1 does not. Moreover, as the reader can verify, every point in the closed interval $[0, 1]$ is a limit point of A. Here's a picture showing why 1 is a limit point of A:

For every $r > 0$, there is always a point x of A whose distance from 1 is less than r.

Example 4.2 Let $A = \{1/n \, ; \, n \in \mathbb{N}\}$, a sequence of points converging to zero:

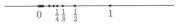

The diligent reader will verify that 0 is the only limit point of A; note that for every $r > 0$, there is always an n such that $1/n \in (-r, r)$ by the $1/n$-principle:

The name "limit point" fits because the following lemma states that limit points are exactly that, limits of sequences of points in A.

Limit points and sequences

Lemma 4.1 *A point $c \in \mathbb{R}^m$ is a limit point of a set $A \subseteq \mathbb{R}^m$ if and only if $c = \lim a_n$ for some sequence $\{a_n\}$ contained in A with $a_n \neq c$ for each n.*

Proof Assume that c is a limit point of A. For each n, by definition of limit point (put $r = 1/n$ in the definition), there is a point $a_n \in A$ such that $0 < |a_n - c| < 1/n$. By Theorem 3.1 on p. 253, on limit recipes, we have $a_n \to c$.

Conversely, suppose that $c = \lim a_n$ for some sequence $\{a_n\}$ contained in A with $a_n \neq c$ for each n. We shall prove that c is a limit point for A. Let $r > 0$. Then by definition of convergence for $a_n \to c$, there is an n sufficiently large such that $|a_n - c| < r$. Since $a_n \neq c$ by assumption, we have $0 < |x - c| < r$ with $x = a_n \in A$, so c is a limit point of A. ∎

4.1.2 Continuity, ε's, and δ's

Consider our airplane saga that we've seen with sequences. Suppose you're in an airplane above Chicago at 5 PM; then obviously, the plane is still near Chicago for some time interval around 5 PM. To make this "mathy," let $f(t)$ equal the position of the plane at time t, and consider the following picture:

✈ At 5PM   ✈ t around 5PM
$f(5)$ $f(t)$

● Chicago   ● Chicago

Our world follows the "law of continuity": if $f(5)$ is above Chicago (left), then $f(t)$ is still near Chicago for some time interval around 5 PM (right). We now make this "near" and "time interval around" precise. Following our work on sequences, it's not difficult to do this! Take any radius of $f(5)$, say a radius $\varepsilon > 0$, as shown here:

✈ $f(5)$ ε $f(5)$ = position of plane at 5PM

● Chicago

Then "nearness" is quantified by saying that $f(t)$ is inside the ball of the given radius for some time interval around 5; that is,

$$|f(t) - f(5)| < \varepsilon \text{ for some time interval centered on 5 .}$$

That is, there is some distance, historically it has been denoted by $\delta > 0$, such that $|f(t) - f(5)| < \varepsilon$ holds for $|t - 5| < \delta$. To conclude this rather long motivational speech, here's our "mathy" way to describe our flight:

For every $\varepsilon > 0$, for some $\delta > 0$, we have $|f(t) - f(5)| < \varepsilon$ if $|t - 5| < \delta$.

Here is an equivalent rewording:

For every $\varepsilon > 0$ there is a $\delta > 0$ such that if $|t - 5| < \delta$, then $|f(t) - f(5)| < \varepsilon$.

Every law of continuity example can be described via ε's and δ's; e.g., consider our cannon:

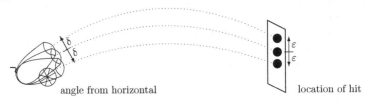

angle from horizontal   location of hit

That is, if you hit the wall at a certain location at a particular angle, then given any precision ε, you can hit the wall within that precision of the original hit for angles within some safety margin δ of the original angle.

We now generalize to arbitrary dimensions. Let $m, p \in \mathbb{N}$ and let $f : D \longrightarrow \mathbb{R}^m$, where $D \subseteq \mathbb{R}^p$. The function $f : D \longrightarrow \mathbb{R}^m$ is **continuous at a point** $c \in D$ if for each $\varepsilon > 0$, there is a $\delta > 0$ such that

$$\boxed{x \in D \quad \text{and} \quad |x - c| < \delta \quad \Longrightarrow \quad |f(x) - f(c)| < \varepsilon.} \qquad (4.2)$$

In the case $p = m = 1$ (the easiest to draw!), here's a picture of continuity (Fig. 4.2):

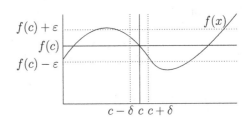

Fig. 4.2 Given $\varepsilon > 0$, there is a $\delta > 0$ such that if $x \in D$ is within the distance δ of c, then $f(x)$ is within the radius ε of $f(c)$

Just like our airplane and cannon examples, a function is continuous at a point c if *the precision of its outputs ($f(x)$ is near $f(c)$) can be controlled by restricting its inputs to within some safety margin around c* (explicitly, $|f(x) - f(c)| < \varepsilon$ if $|x - c| < \delta$). Continuity is intimately related to the notion of limits, our next topic, which is the main focus of this and the next section; we shall return to continuity in Section 4.3.

4.1.3 The ε-δ Definition of Limit

The definition of limit is very similar to that of continuity: It says that the precision of a function's outputs "near a value L" can be controlled by restricting its inputs to within some safety margin "around c," *but not equal to c* (explicitly, $|f(x) - L| < \varepsilon$ if $|x - c| < \delta$ and $x \neq c$). It's important in many applications[2] to add the extra condition "not equal to c." Note that if $x \neq c$, we have $|x - c| > 0$, so we are in effect saying that $0 < |x - c| < \delta$. Finally, since x is not allowed to equal c, we do not need c to be in the domain of f; we'll see in a moment that it's enough to take c to be a limit point of D.

With this discussion as a backdrop, here's a precise definition. A function $f : D \longrightarrow \mathbb{R}^m$ is said to have a **limit** L at a limit point c of D if for each $\varepsilon > 0$ there is a $\delta > 0$ such that

$$\boxed{x \in D \quad \text{and} \quad 0 < |x - c| < \delta \quad \Longrightarrow \quad |f(x) - L| < \varepsilon.} \tag{4.3}$$

Note: Since c is a limit point of D, there always exist points $x \in D$ with $0 < |x - c| < \delta$, so this implication is not empty! If (4.3) holds, we write

$$L = \lim_{x \to c} f \quad \text{or} \quad L = \lim_{x \to c} f(x),$$

or we sometimes write $f \to L$ or $f(x) \to L$ as $x \to c$. For $p = m = 1$, the condition (4.3) simplifies as follows: $f : D \longrightarrow \mathbb{R}$ has limit L at a limit point c of $D \subseteq \mathbb{R}$ if for each $\varepsilon > 0$, there is a $\delta > 0$ such that

$$\boxed{x \in D \text{ with } c - \delta < x < c + \delta \text{ and } x \neq c \quad \Longrightarrow \quad L - \varepsilon < f(x) < L + \varepsilon.}$$

Here's an illustration of the limit concept in the case $p = m = 1$, where there are three similar functions, f_1, f_2, f_3, a point c, and an L (Fig. 4.3):

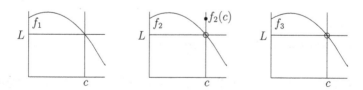

Fig. 4.3 In the first picture, $L = f_1(c)$, in the second, $f_2(c) \neq L$, and in the third, $f_3(c)$ is not defined (leaving a hole in the graph)

In all three cases, we have $\lim_{x \to c} f = L$, where $f = f_1, f_2, f_3$ (Fig. 4.4):

---

[2]For example, the definition of derivative in calculus; a more everyday example deals with paying back loans, as we'll see in Example 4.4.

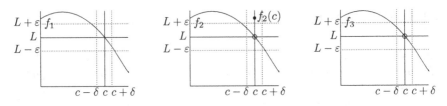

Fig. 4.4 Given $\varepsilon > 0$, there is a $\delta > 0$ such that if we take any $x \neq c$ with $c - \delta < x < c + \delta$, we have $L - \varepsilon < f(x) < L + \varepsilon$

For general p and m, here's an abstract picture of a function $f : D \to \mathbb{R}^m$ with limit L at a point c (Fig. 4.5):

Fig. 4.5 Given $\varepsilon > 0$, there is a $\delta > 0$ such that if $x \in D$ is within distance δ of c (and $x \neq c$), then $f(x)$ is within distance ε of L

We end by remarking that an alternative definition of limit involves open balls. Indeed, observe that a function $f : D \longrightarrow \mathbb{R}^m$ has limit L at a limit point c of $D \subseteq \mathbb{R}^p$ if for each $\varepsilon > 0$, there is a $\delta > 0$ such that

$$x \in D \cap B_\delta(c) \quad \text{and} \quad x \neq c \implies f(x) \in B_\varepsilon(L).$$

This open ball viewpoint is useful in the subject of topology.

In this section we focus on limits, and in Section 4.3 we focus on continuity.

4.1.4   Working with the ε-δ Definition

Techniques we learned back in Section 3.1.2 (starting on p. 152) to prove that sequences converge, such as the "limit recipe" and the "fraction fact," are useful for ε-δ arguments. Here is a recipe on how to attack many limit problems:

Precision-safety margin recipe:
(1) Always start your proof with "Let $\varepsilon > 0$."
(2) "Massage" output precision $|f(x) - L|$ in terms of input margin $|x - c|$:
to get an inequality that looks like
$$|f(x) - L| \leq \text{constant} \, |x - c|.$$
   This may require an initial choice of safety margin (see examples).
(3) Find the safety margin δ.

We take the convention that if not explicitly mentioned, the domain D of a function is always taken to be the set of all points for which the function makes sense.

Example 4.3 Let us prove that

$$\lim_{x \to 2} \left(3x^2 - 10\right) = 2.$$

(Here, the domain D of $3x^2 - 10$ is assumed to be all of $\mathbb{R}$.) Let $\varepsilon > 0$ be given. We need to prove that there is a real number $\delta > 0$ such that

$$0 < |x - 2| < \delta \implies \left|3x^2 - 10 - 2\right| = \left|3x^2 - 12\right| < \varepsilon.$$

We now massage our output precision in terms of the input margin as follows:

$$\left|3x^2 - 12\right| = 3\left|x^2 - 4\right| = 3\left|x + 2\right| \cdot |x - 2|.$$

We now bound $3\,|x + 2|$. To do so, let us restrict our inputs x so that $|x - 2| < 1$; that is, $-1 < x - 2 < 1$, or $1 < x < 3$. In this case,

$$|x + 2| \leq 3 + 2 = 5.$$

Thus, if $|x - 2| < 1$, then $3|x + 2| \leq 15$. Hence,

$$|x - 2| < 1 \implies \left|3x^2 - 12\right| \leq 15\,|x - 2|. \tag{4.4}$$

Now

$$15\,|x - 2| < \varepsilon \iff |x - 2| < \frac{\varepsilon}{15}. \tag{4.5}$$

For this reason, let us pick δ to be the minimum of 1 and $\varepsilon/15$. Then $|x - 2| < \delta$ implies $|x - 2| < 1$ and $|x - 2| < \varepsilon/15$, and therefore, according to (4.4) and (4.5), we have

$$0 < |x - 2| < \delta \implies \left|3x^2 - 12\right| \overset{\text{by (4.4)}}{\leq} 15\,|x - 2| \overset{\text{by (4.5)}}{<} \varepsilon.$$

Thus, by definition of limit, $\lim_{x \to 2}(3x^2 - 10) = 2$.

Example 4.4 (*Loan amortization*) Suppose you borrow $B. If each month your
balance is multiplied by a factor[3] $x > 0$, how much is your monthly payment p if
you want to pay the loan back in M months, where $M \geq 2$? Assuming $x \neq 1$, the
answer is (can you prove it?) $p = x^M B / f(x)$, where

$$f(x) = \frac{x^M - 1}{x - 1}.$$

The function f is not defined at $x = 1$, however, we claim that its limit at 1 exists:

$$\lim_{x \to 1} f(x) = M.$$

Although f has a "hole" at $x = 1$, it can be "filled":

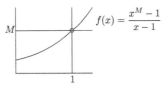

Let $\varepsilon > 0$ be given. We need to prove that there is a real number $\delta > 0$ such that

$$0 < |x - 1| < \delta \quad \Longrightarrow \quad |f(x) - M| < \varepsilon.$$

To do so, note that $x^M - 1 = (x - 1)(1 + x + x^2 + \cdots + x^{M-1})$, by the sum for a
geometric progression (or by just multiplying out). Using this fact, and some algebra
(which we leave to the interested reader), one can show that

$$f(x) - M = g(x)(x - 1),$$

where $g(x)$ is the polynomial

$$g(x) = M - 1 + (M - 2)x + (M - 3)x^2 + \cdots + 2x^{M-3} + x^{M-2}.$$

Thus,

$$|f(x) - M| = |g(x)| \cdot |x - 1|.$$

Let us tentatively restrict x so that $|x - 1| < 1$. This implies $0 < x < 2$, so by looking
at the formula for $g(x)$, we see that

$$|x - 1| < 1 \quad \Longrightarrow \quad |g(x)| \leq C,$$

---

[3]If $x > 1$, the lender is charging you interest, while if $0 < x < 1$, the lender is actually paying you
interest. Note that the limit $\lim_{x \to 1} f(x) = M$ is "obvious," for $x = 1$ corresponds to not charging
any interest, so if our loan is $B, to pay it back in M months, we just need to pay $p = \$B/M$ each
month.

where $C = M - 1 + (M - 2) \cdot 2 + (M - 3) \cdot 2^2 + \cdots + 2 \cdot 2^{M-3} + 2^{M-2}$. Hence,

$$|x - 1| < 1 \quad \Longrightarrow \quad |f(x) - M| \leq C |x - 1|. \tag{4.6}$$

Now

$$C |x - 1| < \varepsilon \quad \Longleftrightarrow \quad |x - 1| < \frac{\varepsilon}{C}. \tag{4.7}$$

For this reason, let us pick δ to be the minimum of 1 and ε/C. Then $|x - 1| < \delta$ implies $|x - 1| < 1$ and $|x - 1| < \varepsilon/C$, and therefore, according to (4.6) and (4.7), we have

$$0 < |x - 1| < \delta \quad \Longrightarrow \quad |f(x) - M| \overset{\text{by (4.6)}}{\leq} c |x - 1| \overset{\text{by (4.7)}}{<} \varepsilon.$$

Thus, by definition of limit, $\lim_{x \to 1} f(x) = M$.

Example 4.5 Now let $a > 0$ be a positive real number and let us show that

$$\lim_{x \to 0} \frac{\sqrt{a + x} - \sqrt{a}}{x} = \frac{1}{2\sqrt{a}}.$$

Here, the domain is $D = [-a, 0) \cup (0, \infty)$. You've seen this limit before in calculus: It's just the derivative of the square root function at a; see Fig. 4.6. Let $\varepsilon > 0$. We need to prove that there is a real number $\delta > 0$ such that

$$0 < |x| < \delta \quad \Longrightarrow \quad \left| \frac{\sqrt{a + x} - \sqrt{a}}{x} - \frac{1}{2\sqrt{a}} \right| < \varepsilon.$$

To establish this result, we "massage" the absolute value with the "multiply by conjugate trick":

$$\sqrt{a + x} - \sqrt{a} = \frac{\sqrt{a + x} - \sqrt{a}}{1} \cdot \frac{\sqrt{a + x} + \sqrt{a}}{\sqrt{a + x} + \sqrt{a}} = \frac{x}{\sqrt{a + x} + \sqrt{a}}, \tag{4.8}$$

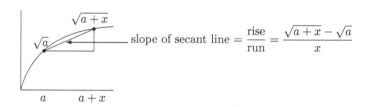

Fig. 4.6 Graph of the square root function. The height of the *triangle* is $\sqrt{a + x} - \sqrt{a}$ and the width is x

where we noticed that the top of the middle expression is a difference of squares. Therefore,

$$\left| \frac{\sqrt{a+x} - \sqrt{a}}{x} - \frac{1}{2\sqrt{a}} \right| = \left| \frac{1}{\sqrt{a+x} + \sqrt{a}} - \frac{1}{2\sqrt{a}} \right| = \left| \frac{\sqrt{a+x} - \sqrt{a}}{2\sqrt{a}\,(\sqrt{a+x} + \sqrt{a})} \right|.$$

Applying (4.8) to the far right numerator, we get

$$\left| \frac{\sqrt{a+x} - \sqrt{a}}{x} - \frac{1}{2\sqrt{a}} \right| = \frac{|x|}{2\sqrt{a}\,(\sqrt{a+x} + \sqrt{a})^2}.$$

We can make the fraction on the right bigger by making the denominator smaller. To this end, observe that $(\sqrt{a+x} + \sqrt{a})^2 \geq (\sqrt{a})^2 = a$, so

$$\frac{1}{2\sqrt{a}\,(\sqrt{a+x} + \sqrt{a})^2} \leq \frac{1}{2\sqrt{a}\cdot a} = \frac{1}{2a^{3/2}} \implies \left| \frac{\sqrt{a+x} - \sqrt{a}}{x} - \frac{1}{2\sqrt{a}} \right| \leq \frac{|x|}{2a^{3/2}},$$

such a simple expression! Now

$$\frac{|x|}{2a^{3/2}} < \varepsilon \iff |x| < 2a^{3/2}\,\varepsilon.$$

With this in mind, we choose $\delta = 2a^{3/2}\,\varepsilon$, and with this choice of δ, we obtain our desired inequality:

$$x \in D \quad \text{and} \quad 0 < |x| < \delta \implies \left| \frac{\sqrt{a+x} - \sqrt{a}}{x} - \frac{1}{2\sqrt{a}} \right| < \varepsilon.$$

Example 4.6 We now give an example involving complex numbers. Let c be a nonzero complex number, and let us show that

$$\lim_{z \to c} \frac{1}{z} = \frac{1}{c}.$$

Here, $f : D \longrightarrow \mathbb{C}$ is the function $f(z) = 1/z$ with $D \subseteq \mathbb{C}$ consisting of all nonzero complex numbers. (Recall that $\mathbb{C} = \mathbb{R}^2$, so D is a subset of $\mathbb{R}^2$, and in terms of our original definition (4.3), $D \subseteq \mathbb{R}^p$ and $f : D \longrightarrow \mathbb{R}^m$ with $p = m = 2$.) Let $\varepsilon > 0$. We need to prove that there is a real number $\delta > 0$ such that

$$0 < |z - c| < \delta \implies \left| \frac{1}{z} - \frac{1}{c} \right| < \varepsilon.$$

Observe that

$$\left| \frac{1}{z} - \frac{1}{c} \right| = \left| \frac{c - z}{zc} \right| = \frac{1}{|zc|}\,|z - c|.$$

We now make the denominator smaller, although not too small, for otherwise, if z is too close to zero, then $1/|zc|$ can be very large. To this end, we tentatively restrict z so that $|z - c| < \frac{|c|}{2}$. In this case, as seen in Fig. 4.7, we also have $|z| > \frac{|c|}{2}$. Here is a proof if you like:

$$|c| = |c - z + z| \le |c - z| + |z| < \frac{|c|}{2} + |z| \implies \frac{|c|}{2} < |z|.$$

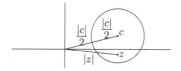

Fig. 4.7 If $|z - c| < \frac{|c|}{2}$, then $|z| > \frac{|c|}{2}$

Therefore, if $|z - c| < \frac{|c|}{2}$, then $|zc| > \frac{|c|}{2} \cdot |c| = \frac{1}{2}|c|^2 = b$, where $b = |c|^2/2$ is a positive number. Thus,

$$|z - c| < \frac{|c|}{2} \implies \left| \frac{1}{z} - \frac{1}{c} \right| \le \frac{1}{b}|z - c|. \tag{4.9}$$

Now

$$\frac{|z - c|}{b} < \varepsilon \iff |z - c| < b\varepsilon. \tag{4.10}$$

For this reason, let us pick δ to be the minimum of $|c|/2$ and $b\varepsilon$. Then $|z - c| < \delta$ implies $|z - c| < |c|/2$ and $|z - c| < b\varepsilon$. Therefore, according to (4.9) and (4.10), we have

$$0 < |z - c| < \delta \implies \left| \frac{1}{z} - \frac{1}{c} \right| \overset{\text{by (4.9)}}{\le} \frac{1}{b}|z - c| \overset{\text{by (4.10)}}{<} \varepsilon.$$

Thus, by definition of limit, $\lim_{z \to c} 1/z = 1/c$.

Example 4.7 Here is one last example. Define $f : \mathbb{R}^2 \backslash \{0\} \longrightarrow \mathbb{R}$ by

$$f(x) = \frac{x_1^2 x_2}{x_1^2 + x_2^2}, \quad x = (x_1, x_2).$$

We shall prove that $\lim_{x \to 0} f = 0$. (In the subscript "$x \to 0$," 0 denotes the zero vector $(0, 0)$ in $\mathbb{R}^2$, while on the right of $\lim_{x \to 0} f = 0$, 0 denotes the real number zero; it should always be clear from context what "0" means.) Before our actual proof, we first recall Eq. (2.33) on p. 125:

$$\boxed{a\,b \le \frac{1}{2}(a^2 + b^2).}$$

This inequality is worth committing to memory (just expand $(a-b)^2 \geq 0$). It follows that

$$
|f(x)| = \frac{|x_1|}{x_1^2 + x_2^2} \cdot |x_1 x_2|
$$
$$
\leq \frac{|x_1|}{x_1^2 + x_2^2} \cdot \frac{1}{2}(x_1^2 + x_2^2) = \frac{|x_1|}{2} \leq \frac{1}{2}|x|.
$$

Given $\varepsilon > 0$, choose $\delta = 2\varepsilon$. Then

$$
0 < |x| < \delta \quad \Longrightarrow \quad |f(x)| \leq \frac{1}{2}|x| < \frac{1}{2}(2\varepsilon) = \varepsilon.
$$

This proves that $\lim_{x \to 0} f = 0$.

4.1.5 The Sequence Definition of Limit

It turns out that we can relate limits of functions to limits of sequences, which was studied in Chapter 3, so we can use much of the theory developed in that chapter to analyze limits of functions. In particular, take note of the following important theorem!

Sequence criterion for limits

Theorem 4.2 *Let $f : D \longrightarrow \mathbb{R}^m$, let c be a limit point of D, and let $L \in \mathbb{R}^m$. Then the following two statements are equivalent:*

(1) $L = \lim_{x \to c} f$.
(2) For every sequence $\{a_n\}$ of points in $D \backslash \{c\}$ with $c = \lim a_n$, the sequence $\{f(a_n)\}$ converges to L.

Proof Assume *(1)* and let $\{a_n\}$ be a sequence of points in $D \backslash \{c\}$ converging to c; we will show that $\{f(a_n)\}$ converges to L, as visualized in Fig. 4.8 by the dotted arrows:

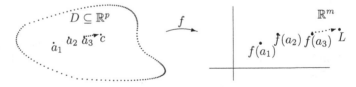

Fig. 4.8 We need to show that f takes a sequence converging to c to a sequence converging to L

Let $\varepsilon > 0$. Since f has limit L at c, there is a $\delta > 0$ such that

$$x \in D \quad \text{and} \quad 0 < |x - c| < \delta \quad \Longrightarrow \quad |f(x) - L| < \varepsilon.$$

Since $a_n \to c$ and $a_n \neq c$ for every n, it follows that there is an N such that

$$n > N \quad \Longrightarrow \quad 0 < |a_n - c| < \delta.$$

The definition of f having limit L at c now implies that

$$n > N \quad \Longrightarrow \quad |f(a_n) - L| < \varepsilon.$$

Thus, $L = \lim f(a_n)$.

To prove *(2)* $\Longrightarrow$ *(1)*, we shall instead prove the logically equivalent contrapositive; that is, assuming $L \neq \lim_{x \to c} f$, we prove that there is a sequence $\{a_n\}$ of points in $D \backslash \{c\}$ converging to c such that $\{f(a_n)\}$ does not converge to L. Now $L \neq \lim_{x \to c} f$ means (negating the definition $L = \lim_{x \to c} f$) that there is an $\varepsilon > 0$ such that for all $\delta > 0$, there is an $x \in D$ with $0 < |x - c| < \delta$ and $|f(x) - L| \geq \varepsilon$. Since this statement is true for all $\delta > 0$, it is in particular true for $\delta = 1/n$ for each $n \in \mathbb{N}$. Thus, for each $n \in \mathbb{N}$, there is a point $a_n \in D$ with $0 < |a_n - c| < 1/n$ and $|f(a_n) - L| \geq \varepsilon$. It follows that $\{a_n\}$ is a sequence of points in $D \backslash \{c\}$ converging to c, and $\{f(a_n)\}$ does not converge to L. This completes the proof of the contrapositive. ∎

This theorem can be used to give easy proofs that limits do not exist (Fig. 4.8).

Example 4.8 Recall from Example 1.27 on p. 23 the **Dirichlet function**, named after Lejeune Dirichlet (1805–1859):

$$\mathcal{D} : \mathbb{R} \longrightarrow \mathbb{R} \quad \text{is defined by} \quad \mathcal{D}(x) = \begin{cases} 1 & \text{if } x \text{ is rational,} \\ 0 & \text{if } x \text{ is irrational,} \end{cases}$$

an approximate graph of which is shown here:

Let $c \in \mathbb{R}$. Then as we saw in Example 3.13 on p. 166, there is a sequence $\{a_n\}$ of rational numbers converging to c with $a_n \neq c$ for all n. Since a_n is rational, we have $\mathcal{D}(a_n) = 1$ for all $n \in \mathbb{N}$, so

$$\lim \mathcal{D}(a_n) = \lim 1 = 1.$$

Similarly, there is a sequence $\{b_n\}$ of irrational numbers converging to c with $b_n \neq c$ for all n, in which case

$$\lim \mathcal{D}(b_n) = \lim 0 = 0.$$

Therefore, according to our sequence criterion, $\lim_{x \to c} \mathcal{D}$ cannot exist.

Example 4.9 Consider the function $f : \mathbb{R}^2 \setminus \{0\} \longrightarrow \mathbb{R}$ defined by

$$f(x) = \frac{x_1 x_2}{x_1^2 + x_2^2}, \qquad x = (x_1, x_2) \neq 0.$$

We claim that $\lim_{x \to 0} f$ does not exist. To see this, observe that $f(x_1, 0) = 0$ for every $x_1 \neq 0$, so $f(a_n) \to 0$ for every sequence a_n along the x_1-axis that is approaching, but never equaling, 0. On the other hand, since

$$f(x_1, x_1) = \frac{x_1^2}{x_1^2 + x_1^2} = \frac{1}{2},$$

it follows that $f(a_n) \to 1/2$ for every sequence a_n along the diagonal $x_1 = x_2$ that is approaching, but never equaling, 0. Therefore $\lim_{x \to 0} f$ does not exist.

▶ **Exercises 4.1**

1. Using the ε-δ definition of limit, prove that (where z is a complex variable)

   (a) $\lim_{z \to 1} (z^2 + 2z) = 3$,   (b) $\lim_{z \to 2} z^3 = 8$,   (c) $\lim_{z \to 2} \dfrac{1}{z^2} = \dfrac{1}{4}$,   (d) $\lim_{z \to 2} \dfrac{3z}{z+1} = 2$.

   Suggestion: For (b), can you factor $z^3 - 8$?

2. Using the ε-δ definition of limit, prove that (where x, a are real variables and where in (b) and (c), $a > 0$)

   (a) $\lim_{x \to a} \dfrac{x^2 - x - a^2}{x + a} = -\dfrac{1}{2}$,   (b) $\lim_{x \to a} \dfrac{1}{\sqrt{x}} = \dfrac{1}{\sqrt{a}}$,   (c) $\lim_{x \to 0} \dfrac{\sqrt{a^2 + 6x^2} - a}{x^2} = \dfrac{3}{a}$.

3. Prove that the limits (a) and (b) do not exist, while (c) does exist:

   (a) $\lim_{x \to 0} \dfrac{x_1^2 + x_2}{\sqrt{x_1^2 + x_2^2}}$,   (b) $\lim_{x \to 0} \dfrac{x_1^2 + x_2}{x_1^2 + x_2^2}$,   (c) $\lim_{x \to 0} \dfrac{x_1^2 x_2^2}{x_1^2 + x_2^2}$.

4. Here are problems involving functions similar to Dirichlet's function. Define

   $$f(x) = \begin{cases} x & \text{if } x \text{ is rational,} \\ 0 & \text{if } x \text{ is irrational,} \end{cases} \qquad g(x) = \begin{cases} x & \text{if } x \text{ is rational,} \\ 1 - x & \text{if } x \text{ is irrational.} \end{cases}$$

   (a) Prove that $\lim_{x \to 0} f = 0$, but $\lim_{x \to c} f$ does not exist for $c \neq 0$.
   (b) Prove that $\lim_{x \to 1/2} g = 1/2$, but $\lim_{x \to c} g$ does not exist for $c \neq 1/2$.

5. Let $f : D \longrightarrow \mathbb{R}$ with $D \subseteq \mathbb{R}^p$ and let $L = \lim_{x \to c} f$. Assume that $f(x) \geq 0$ for all $x \neq c$ sufficiently close to c. Prove that $L \geq 0$ and $\sqrt{L} = \lim_{x \to c} \sqrt{f(x)}$.

4.2 A Potpourri of Limit Properties for Functions

Now that we have a working knowledge of the ε-δ definition of limit, we turn to studying the properties of limits that will be used throughout the rest of the book.

4.2.1 Limit Theorems

As we already mentioned in Section 4.1.5 on p. 250, combining the sequence criterion for limits (Theorem 4.2 on p. 250) with the limit theorems in Chapter 3, we can easily prove results concerning limits. Here are some examples, beginning with the following companion to the uniqueness theorem (Theorem 3.2 on p. 162) for sequences.

Uniqueness of limits

Theorem 4.3 *A function can have at most one limit at a given limit point of its domain.*

Proof If $\lim_{x \to c} f$ equals both L and L', then according to the sequence criterion, for all sequences $\{a_n\}$ of points in $D \setminus \{c\}$ converging to c, we have $\lim f(a_n) = L$ and $\lim f(a_n) = L'$. Since we know that limits of sequences are unique (Theorem 3.2 on p. 162), we conclude that $L = L'$. ∎

If $f : D \longrightarrow \mathbb{R}^m$, then we can write f in terms of its components as

$$f = (f_1, \ldots, f_m),$$

where for $k = 1, 2, \ldots, m$, $f_k : D \longrightarrow \mathbb{R}$, are the **component functions** of f. In particular, if $f : D \longrightarrow \mathbb{C}$, then we can always break up f as

$$f = (f_1, f_2) \quad \Longleftrightarrow \quad f = f_1 + i f_2,$$

where $f_1, f_2 : D \longrightarrow \mathbb{R}$.

Example 4.10 For instance, if $f : \mathbb{C} \longrightarrow \mathbb{C}$ is defined by $f(z) = z^2$, then we can write this function as $f(x+iy) = (x+iy)^2 = x^2 - y^2 + i2xy$. Therefore, $f = f_1 + i f_2$, where if $z = x + iy$, then

$$f_1(z) = x^2 - y^2, \quad f_2(z) = 2xy.$$

The following theorem is a companion to the component theorem (Theorem 3.4 on p. 163) for sequences.

Component theorem

Theorem 4.4 *A function converges to $L \in \mathbb{R}^m$ (at a given limit point of the domain) if and only if each component function converges in $\mathbb{R}$ to the corresponding component of L.*

Proof Let $f : D \longrightarrow \mathbb{R}^m$. Then $\lim_{x \to c} f = L$ if and only if for every sequence $\{a_n\}$ of points in $D \backslash \{c\}$ converging to c, we have $\lim f(a_n) = L$. According to the component theorem for sequences, $\lim f(a_n) = L$ if and only if for each $k = 1, 2, \ldots, m$, we have $\lim_{n \to \infty} f_k(a_n) = L_k$. This shows that $\lim_{x \to c} f = L$ if and only if for each $k = 1, 2, \ldots, m$, $\lim_{x \to c} f_k = L_k$. This completes our proof. ∎

The following theorem is a function analogue of the "algebra of limits" studied in Section 3.2, and it follows from the corresponding theorems for sequences in Theorems 3.9 and 3.10 on pp. 169 and 170, so we won't bother with the proof.

Algebra of limits

Theorem 4.5 *If f and g both have limits as $x \to c$, then*

(1) $\lim_{x \to c} |f| = |\lim_{x \to c} f|$.
(2) $\lim_{x \to c} (af + bg) = a \lim_{x \to c} f + b \lim_{x \to c} g$, *for every real a, b.*
    If f and g take values in $\mathbb{C}$, then
(3) $\lim_{x \to c} fg = (\lim_{x \to c} f)(\lim_{x \to c} g)$.
(4) $\lim_{x \to c} f/g = \lim_{x \to c} f / \lim_{x \to c} g$, *provided that $\lim_{x \to c} g$ is nonzero.*

Note that if $\lim_{x \to c} g$ is nonzero, then one can show that $g(x) \neq 0$ for x sufficiently close to c; thus, the quotient $f(x)/g(x)$ is defined for x sufficiently close to c. Also note that by induction, we can use the algebra of limits on finite sums and finite products of functions.

Example 4.11 It is easy to show that at every point $c \in \mathbb{C}$, $\lim_{z \to c} z = c$. Therefore, by our algebra of limits, for every complex number a and natural number n, we have

$$\lim_{z \to c} az^n = a \lim_{z \to c} \underbrace{z \cdot z \cdots z}_{n \text{ z's}} = a \left(\lim_{z \to c} z \right) \cdot \left(\lim_{z \to c} z \right) \cdots \left(\lim_{z \to c} z \right) = a c \cdot c \cdots c = a c^n.$$

Therefore, given a polynomial

$$p(z) = a_n z^n + a_{n-1} z^{n-1} + \cdots + a_1 z + a_0,$$

then for every $c \in \mathbb{C}$, the algebra of limits implies that

$$\lim_{z \to c} p(z) = \lim_{z \to c} a_n z^n + \lim_{z \to c} a_{n-1} z^{n-1} + \cdots + \lim_{z \to c} a_1 z + \lim_{z \to c} a_0$$
$$= a_n c^n + a_{n-1} c^{n-1} + \cdots + a_1 c + a_0;$$

that is,

$$\lim_{z \to c} p(z) = p(c).$$

Example 4.12 Now let $q(z)$ be another polynomial and suppose that $q(c) \neq 0$. Then by our algebra of limits, we have

$$\lim_{z \to c} \frac{p(z)}{q(z)} = \frac{\lim_{z \to c} p(z)}{\lim_{z \to c} q(z)} = \frac{p(c)}{q(c)}.$$

The following theorem is useful in dealing with compositions of functions.

Composition of limits

Theorem 4.6 *Let $f : D \longrightarrow \mathbb{R}^m$ and $g : C \longrightarrow \mathbb{R}^p$, where $D \subseteq \mathbb{R}^p$ and $C \subseteq \mathbb{R}^q$ and suppose that $g(C) \subseteq D$, so that $f \circ g : C \longrightarrow \mathbb{R}^m$ is defined. Let d be a limit point of D, and c a limit point of C, and assume that*

(1) $d = \lim_{x \to c} g(x)$.
(2) $L = \lim_{y \to d} f(y)$.
(3) Either $f(d) = L$ or $d \neq g(x)$ for all $x \neq c$ sufficiently near c.

Then

$$L = \lim_{x \to c} f \circ g.$$

Proof Although the statement of this theorem is complicated at first glance, the proof is very simple. Let $\{a_n\}$ be a sequence in $C \backslash \{c\}$ with $a_n \to c$; we must show that $f(g(a_n)) \to L$. Figure 4.9 shows an abstract picture of what's going on.

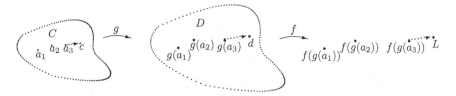

Fig. 4.9 Since $d = \lim_{x \to c} g(x)$, the function g takes the sequence $\{a_n\}$ to a sequence $\{g(a_n)\}$ with $g(a_n) \to d$. Using that $L = \lim_{y \to d} f(y)$, we must show that f takes the sequence $\{g(a_n)\}$ to a sequence $\{f(g(a_n))\}$ with $f(g(a_n)) \to L$

The first statement in Fig. 4.9 is just the sequence criterion for limits (Theorem 4.2 on p. 250) applied to g. It's the last statement in Fig. 4.9 that requires some thought, and it is there where assumption *(3)* comes in. First, if $g(x) \neq d$ for all $x \neq c$ sufficiently near c, then a tail of the sequence $\{g(a_n)\}$ is a sequence in $D \backslash \{d\}$ converging to d. Thus, by the sequence criterion applied to f, we indeed have $f(g(a_n)) \to L$. Second, suppose that $f(d) = L$. Then we leave you to think about why for *every* sequence

$\{d_n\}$ in D converging to d (whether or not $d_n \neq d$ for all n), we have $f(d_n) \to L$. Therefore, in this case we also have $f(g(a_n)) \to L$. This completes our proof. ∎

We now finish our limit theorems by considering limits and inequalities. In the following two theorems, all functions map a subset $D \subseteq \mathbb{R}^p$ to $\mathbb{R}$.

Squeeze theorem

> **Theorem 4.7** *Let f, g, and h be such that $f(x) \leq g(x) \leq h(x)$ for all x sufficiently close to a limit point c in D and such that both limits $\lim_{x \to c} f$ and $\lim_{x \to c} h$ exist and are equal. Then the limit $\lim_{x \to c} g$ also exists, and*
>
> $$\lim_{x \to c} f = \lim_{x \to c} g = \lim_{x \to c} h.$$

The squeeze theorem for functions is a direct consequence of the sequence criterion and the corresponding squeeze theorem for sequences (Theorem 3.6 on p. 165), and therefore we shall omit the proof. The next theorem follows from (as you might have guessed) the sequence criterion and the corresponding preservation of inequalities theorem for sequences (Theorem 3.7 on p. 166).

Preservation of inequalities

> **Theorem 4.8** *Suppose that $\lim_{x \to c} f$ and $\lim_{x \to c} g$ exist.*
>
> *(1) If $f(x) \leq g(x)$ for $x \neq c$ sufficiently close to c, then $\lim_{x \to c} f \leq \lim_{x \to c} g$.*
> *(2) If for some real numbers a and b, we have $a \leq f(x) \leq b$ for $x \neq c$ sufficiently close to c, then $a \leq \lim_{x \to c} f \leq b$.*

4.2.2 Limits, Limits, Limits, and More Limits

When the domain is a subset of $\mathbb{R}$, there are various extensions of the limit idea. We begin with left- and right-hand limits. For the rest of this section we consider functions $f : D \longrightarrow \mathbb{R}^m$, where $D \subseteq \mathbb{R}$ (later we'll further restrict to $m = 1$).

Suppose that c is a limit point of the set $D \cap (-\infty, c)$. Then $f : D \longrightarrow \mathbb{R}^m$ is said to have a **left-hand limit** L at c if for each $\varepsilon > 0$, there is a $\delta > 0$ such that

$$\boxed{x \in D \quad \text{and} \quad c - \delta < x < c \quad \Longrightarrow \quad |f(x) - L| < \varepsilon.} \qquad (4.11)$$

In a similar way we define a right-hand limit: Suppose that c is a limit point of the set $D \cap (c, \infty)$. Then f is said to have a **right-hand limit** L at c if for each $\varepsilon > 0$, there is $\delta > 0$ such that

$$\boxed{x \in D \quad \text{and} \quad c < x < c + \delta \quad \Longrightarrow \quad |f(x) - L| < \varepsilon.} \qquad (4.12)$$

We express left-hand limits in one of several ways:

$$L = \lim_{x \to c-} f \, , \quad L = \lim_{x \to c-} f(x) \, , \quad L = f(c-) \, , \quad \text{or } f(x) \to L \text{ as } x \to c- \, ,$$

with similar expressions with $c+$ replacing $c-$ for right-hand limits. For example, here's a picture with $f(0-) = 4$, $f(0+) = -2$, and $f(0) = 2$:

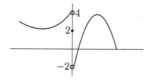

The following result relates one-sided limits and regular limits. Its proof is straightforward, and we leave it to the reader as a good exercise.

Theorem 4.9 *Let* $f : D \longrightarrow \mathbb{R}^m$ *with* $D \subseteq \mathbb{R}$ *and suppose that* c *is a limit point of the sets* $D \cap (-\infty, c)$ *and* $D \cap (c, \infty)$. *Then*

$$L = \lim_{x \to c} f \quad \Longleftrightarrow \quad L = f(c-) \text{ and } L = f(c+).$$

If only one of $f(c-)$ and $f(c+)$ makes sense, then $L = \lim_{x \to c} f$ if and only if $L = f(c-)$ (when c is a limit point only of $D \cap (-\infty, c)$) or $L = f(c+)$ (when c is a limit point only of $D \cap (c, \infty)$), whichever makes sense.

We now describe limits at infinity. Suppose that for every real number N there is a point $x \in D$ such that $x > N$. A function $f : D \longrightarrow \mathbb{R}^m$ is said to have a **limit** L as $x \to \infty$ if for each $\varepsilon > 0$, there is an $N \in \mathbb{R}$ such that

$$\boxed{x \in D \quad \text{and} \quad x > N \quad \Longrightarrow \quad |f(x) - L| < \varepsilon.} \tag{4.13}$$

This definition says that given a precision $\varepsilon > 0$, we can make $f(x)$ within ε of L by taking x sufficiently large ($x > N$ for some N); here's a picture when $m = 1$:

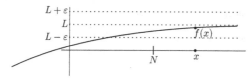

Now suppose that for every real number N there is a point $x \in D$ such that $x < N$. A function $f : D \longrightarrow \mathbb{R}^m$ is said to have a **limit** L as $x \to -\infty$ if for each $\varepsilon > 0$, there is an $N \in \mathbb{R}$ such that

$$\boxed{x \in D \quad \text{and} \quad x < N \quad \Longrightarrow \quad |f(x) - L| < \varepsilon.} \tag{4.14}$$

To express limits as $x \to \pm\infty$, we often write (sometimes with ∞ replaced by $+\infty$)

$$L = \lim_{x\to\infty} f \; , \quad L = \lim_{x\to\infty} f(x) \; , \quad f \to L \text{ as } x \to \infty \; , \quad \text{or } f(x) \to L \text{ as } x \to \infty,$$

with similar expressions when $x \to -\infty$.

Finally, we discuss infinite limits, which are also called properly divergent limits of functions.[4] We now let $m = 1$ and consider functions $f : D \longrightarrow \mathbb{R}$ with $D \subseteq \mathbb{R}$. Suppose that for every real number N there is a point $x \in D$ such that $x > N$. Then f is said to **diverge** to ∞ as $x \to \infty$ if for every real number $M > 0$, there is an $N \in \mathbb{R}$ such that

$$\boxed{x \in D \text{ and } x > N \;\; \Longrightarrow \;\; M < f(x).} \tag{4.15}$$

Here's a picture to keep in mind:

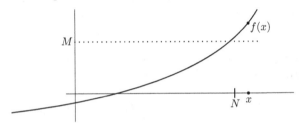

This picture says that given $M > 0$, however large, we can make $f(x) > M$ by taking x sufficiently large ($x > N$ for some N).

Also, f is said to **diverge to** $-\infty$ as $x \to \infty$ if for every real number $M < 0$, there is an $N \in \mathbb{R}$ such that

$$\boxed{x \in D \text{ and } x > N \;\; \Longrightarrow \;\; f(x) < M.} \tag{4.16}$$

If f diverges to ∞ or $-\infty$, we say that f is **properly divergent** as $x \to \infty$. When f is properly divergent to ∞, we write

$$\infty = \lim_{x\to\infty} f \; , \quad \infty = \lim_{x\to\infty} f(x) \; , \quad f \to \infty \text{ as } x \to \infty \; , \quad \text{or } f(x) \to \infty \text{ as } x \to \infty;$$

with similar expressions when f properly diverges to $-\infty$.

In a very similar manner we can define properly divergent limits of functions as $x \to -\infty$, as $x \to c$, as $x \to c-$, and $x \to c+$; we leave these other definitions for the reader to formulate.

Let us now consider an example.

---

[4]"I protest against the use of infinite magnitude as something completed, which in mathematics is never permissible. Infinity is merely a façon de parler, the real meaning being a limit which certain ratios approach indefinitely near, while others are permitted to increase without restriction." Carl Friedrich Gauss (1777–1855).

Example 4.13 Let $a > 1$ and let $f : \mathbb{Q} \longrightarrow \mathbb{R}$ be defined by $f(x) = a^x$ (therefore in this case, $D = \mathbb{Q}$). Here, we recall that a^x is defined for every rational number x (see Section 2.7 on p. 89). We shall prove that

$$\lim_{x \to \infty} f = \infty \quad \text{and} \quad \lim_{x \to -\infty} f = 0.$$

In Section 4.7 we shall define a^x for every $x \in \mathbb{R}$ (in fact, for every complex power), and we shall establish these same limits with $D = \mathbb{R}$. Before proving these results, we claim that

$$\text{for every rational } p < q, \text{ we have } a^p < a^q. \tag{4.17}$$

Indeed, $1 < a$ and $q - p > 0$, so by our power rules on p. 95,

$$1 = 1^{q-p} < a^{q-p},$$

which, after multiplication by a^p, gives our claim. We now prove that $f \to \infty$ as $x \to \infty$. To prove this, we note that since $a > 1$, we can write $a = 1 + b$ for some $b > 0$, so by Bernoulli's inequality on p. 42, for every $n \in \mathbb{N}$,

$$a^n = (1 + b)^n \geq 1 + nb > nb. \tag{4.18}$$

Now fix $M > 0$. By the Archimedean principle on p. 95 (or from the fact that $\mathbb{N}$ is not bounded above by $\mathbb{R}$ on p. 95), we can choose $N \in \mathbb{N}$ such that $Nb > M$, and therefore by (4.17) and (4.18),

$$x \in \mathbb{Q} \quad \text{and} \quad x > N \quad \Longrightarrow \quad M < Nb < a^N < a^x.$$

This proves that $f \to \infty$ as $x \to \infty$. We now show that $f \to 0$ as $x \to -\infty$. Let $\varepsilon > 0$. Also by the Archimedean principle, there is an $N \in \mathbb{N}$ such that $1/(b\varepsilon) < N$. Since

$$\frac{1}{b\varepsilon} < N \quad \Longrightarrow \quad \frac{1}{Nb} < \varepsilon,$$

by (4.17) and (4.18) it follows that

$$x \in \mathbb{Q} \quad \text{and} \quad x < -N \quad \Longrightarrow \quad 0 < a^x < a^{-N} = \frac{1}{a^N} < \frac{1}{Nb} < \varepsilon.$$

This proves that $f \to 0$ as $x \to -\infty$.

Many of the limit theorems in Section 4.2.1 that we have worked out for "regular limits" also hold for left- and right-hand limits, limits at infinity, and infinite limits. To avoid repeating these limit theorems in each of our new contexts (which will take up a few pages at least!), we shall make the following general comment:

> *The sequence criterion, uniqueness of limits, component theorem, algebra and composition of limits, squeeze theorem, and preservation of inequalities for standard limits, have analogous statements for left/right-hand limits, limits at infinity, and infinite (properly divergent) limits.*

We encourage the reader to think about these analogous statements, and we shall make use of these extended versions without much comment in the sequel. Of course, some statements don't hold when we consider infinite limits; for example, we cannot subtract infinities or divide them, nor can we multiply zero and infinity. In the following examples we focus on the extended composition of limits theorem.

Example 4.14 For an example of this general comment, suppose that a function f has a limit $L = \lim_{y \to \infty} f(y)$. We leave the reader to verify that $\lim_{x \to 0^+} 1/x = \infty$. Therefore, according to the extended composition of limits theorem, we have

$$L = \lim_{x \to 0^+} f\left(\frac{1}{x}\right).$$

Similarly, since it's easy to check that $\lim_{x \to -\infty} -x = \infty$, by the extended composition of limits theorem, we have

$$L = \lim_{x \to -\infty} f(-x).$$

More generally, if g is a function with $\lim_{x \to c} g(x) = \infty$, where c is either a real number, ∞, or $-\infty$, then by our composition of limits theorem,

$$L = \lim_{x \to c} f \circ g.$$

Example 4.15 Suppose as above that $\lim_{x \to c} g(x) = \infty$, where c is either a real number, ∞, or $-\infty$. Since $\lim_{y \to \infty} 1/y = 0$, by the extended composition of limits theorem, we have

$$\lim_{x \to c} \frac{1}{g(x)} = 0.$$

▶ **Exercises 4.2**

1. Prove (use the ε-δ definition of left/right-hand limit for (a) and (b))

   (a) $\lim_{x \to 0^-} \dfrac{x}{|x|} = -1,$ (b) $\lim_{x \to 0^+} \dfrac{x}{|x|} = 1,$ (c) $\lim_{x \to 0} \dfrac{x}{|x|}$ does not exist.

2. Using the ε-N definition of limits at infinity, prove

   (a) $\lim_{x \to \infty} \dfrac{x^2 + x + 1}{2x^2 - 1} = \dfrac{1}{2},$ (b) $\lim_{x \to \infty} \sqrt{x^2 + 1} - x = 0,$ (c) $\lim_{x \to \infty} \sqrt{x^2 + x} - x = \dfrac{1}{2}.$

3. Let $p(x) = a_n x^n + \cdots + a_1 x + a_0$ and $q(x) = b_m x^m + \cdots + b_1 x + b_0$ be polynomials with real coefficients and with $a_n \neq 0$ and $b_m \neq 0$.

    (a) Prove that for every natural number k, $\lim_{x \to \infty} 1/x^k = 0$.

    (b) If $n < m$, prove that $\lim_{x \to \infty} p(x)/q(x) = 0$.

    (c) If $n = m$, prove that $\lim_{x \to \infty} p(x)/q(x) = a_n/b_n$.

    (d) If $n > m$, prove that if $a_n/b_m > 0$, then $\lim_{x \to \infty} p(x)/q(x) = \infty$, and on the other hand, if $a_n/b_m < 0$, then $\lim_{x \to \infty} p(x)/q(x) = -\infty$.

4. Let $f, g : D \longrightarrow \mathbb{R}$ with $D \subseteq \mathbb{R}$, $\lim_{x \to \infty} f = \infty$, and $g(x) \neq 0$ for all $x \in D$. Suppose that for some real number L, we have

$$\lim_{x \to \infty} \frac{f}{g} = L.$$

    (a) If $L > 0$, prove that $\lim_{x \to \infty} g = +\infty$.

    (b) If $L < 0$, prove that $\lim_{x \to \infty} g = -\infty$.

    (c) If $L = 0$, can you draw any conclusions about $\lim_{x \to \infty} g$?

4.3 Continuity, Thomae's Function, and Volterra's Theorem

In this section we continue our study of the most important functions in all of analysis and topology, continuous functions. Perhaps one of the most fascinating functions you'll ever run across is the modified Dirichlet function, or Thomae's function, which has the perplexing and pathological property that it is continuous on the irrational numbers and discontinuous on the rational numbers! We'll see that there is no function opposite to Thomae's, that is, continuous on the rationals and discontinuous on the irrationals; this was proved by Vito Volterra (1860–1940) in 1881. For an interesting account of Thomae's function and its relation to Volterra's theorem, see [62].

4.3.1 Continuous Functions

Let $D \subseteq \mathbb{R}^p$. Recall that $f : D \longrightarrow \mathbb{R}^m$ is **continuous at a point** $c \in D$ if for each $\varepsilon > 0$, there is a $\delta > 0$ such that

$$\boxed{x \in D \quad \text{and} \quad |x - c| < \delta \quad \Longrightarrow \quad |f(x) - f(c)| < \varepsilon.} \tag{4.19}$$

We can relate this definition to the definition of limit. Suppose that $c \in D$ is a limit point of D. Then comparing (4.19) with the definition of limit, we see that for a limit point c of D such that $c \in D$, we have

$$\boxed{f \text{ is continuous at } c \iff f(c) = \lim_{x \to c} f.}$$

Technically speaking, when we compare (4.19) to the definition of limit, for a limit we actually require that $0 < |x - c| < \delta$, but in the case that $|x - c| = 0$, that is, $x = c$, we have $|f(x) - f(c)| = |f(c) - f(c)| = 0$, which is automatically less than ε, so the condition that $0 < |x - c|$ can be dropped. What if $c \in D$ is not a limit point of D? In this case, c is called an **isolated point** in D and by definition of (not being a) limit point, there is an open ball $B_\delta(c)$ such that $B_\delta(c) \cap D = \{c\}$; that is, the only point of D inside $B_\delta(c)$ is c itself. Hence, with this δ, for every $\varepsilon > 0$, the condition (4.19) is automatically satisfied:

$$x \in D \quad \text{and} \quad |x - c| < \delta \quad \Longrightarrow \quad x = c \quad \Longrightarrow \quad |f(x) - f(c)| = 0 < \varepsilon.$$

Therefore, at isolated points of D, the function f is automatically continuous by default. For this reason, if we want to prove theorems concerning the continuity of $f : D \longrightarrow \mathbb{R}^m$ at a point $c \in D$, we can always assume that c is a limit point of D; in this case, we have all the limit theorems from the last section at our disposal. This is exactly why we spent so much time on learning limits during the last two sections!

If f is continuous at every point in a subset $A \subseteq D$, we say that f is **continuous on** A. If f is continuous at every point of D, we usually just say that "f is **continuous**" instead of "f is **continuous on** D." Thus,

$$\boxed{f \text{ is continuous} \iff \text{for all } c \in D, f \text{ is continuous at } c.}$$

Example 4.16 (*Loan amortization*) In view of our loan amortization Example 4.4 on p. 246, let us define $f : \mathbb{R} \longrightarrow \mathbb{R}$ by

$$f(x) = \begin{cases} \dfrac{x^M - 1}{x - 1} & x \neq 1, \\ M & x = 1. \end{cases}$$

From Example 4.4 we know that $\lim_{x \to 1} f = 1$, so f is continuous at 1. In fact, from Example 4.12 on p. 255 it follows that f is continuous on all of $\mathbb{R}$.

Example 4.17 Dirichlet's function is discontinuous at every point in $\mathbb{R}$, since in Example 4.8 on p. 251 we already proved that $\lim_{x \to c} D(x)$ does not exist at any $c \in \mathbb{R}$.

Example 4.18 Define $f : \mathbb{R}^2 \longrightarrow \mathbb{R}$ by

$$f(x_1, x_2) = \begin{cases} \dfrac{x_1^2 x_2}{x_1^2 + x_2^2} & (x_1, x_2) \neq 0, \\ 0 & (x_1, x_2) = 0. \end{cases}$$

From Example 4.7 on p. 249, we know that $\lim_{x \to 0} f = 0$, so f is continuous at 0.

Example 4.19 If we define $f : \mathbb{R}^2 \longrightarrow \mathbb{R}$ by

$$f(x_1, x_2) = \begin{cases} \dfrac{x_1 x_2}{x_1^2 + x_2^2} & (x_1, x_2) \neq 0, \\ 0 & (x_1, x_2) = 0, \end{cases}$$

then we already proved that $\lim_{x \to 0} f$ does not exist in Example 4.9 on p. 252. In particular, f is not continuous at 0.

Example 4.20 From Example 4.11 on p. 254, every polynomial function $p : \mathbb{C} \to \mathbb{C}$ is continuous (that is, continuous at every point $c \in \mathbb{C}$). From Example 4.12 on p. 255, every rational function $p(z)/q(z)$, a quotient of polynomials, is continuous at every point $c \in \mathbb{C}$ such that $q(c) \neq 0$.

4.3.2 Continuity Theorems

We now state some theorems on continuity. These theorems follow almost immediately from our limit theorems in Section 4.2, so we shall omit all the proofs. First we note that the sequence criterion for limits of functions (Theorem 4.2 on p. 250) implies the following important theorem.

Sequence criterion for continuity

Theorem 4.10 *A function $f : D \longrightarrow \mathbb{R}^m$ is continuous at $c \in D$ if and only if for every sequence $\{a_n\}$ in D with $c = \lim a_n$, we have $f(c) = \lim f(a_n)$.*

We can write the last equality as $f(\lim a_n) = \lim f(a_n)$, since $c = \lim a_n$. Thus,

$$f \text{ is continuous at } c \iff f\left(\lim_{n \to \infty} a_n\right) = \lim_{n \to \infty} f(a_n),$$

for all sequences $\{a_n\}$ in D with $c = \lim a_n$. Thus, for continuous functions, limits can be "pulled out" so to speak. Here's another way to state continuity:

$$f \text{ is continuous at } c \iff f(a_n) \to f(c) \text{ for all sequences } \{a_n\} \text{ in } D \text{ with } a_n \to c.$$

Example 4.21 **Question**: Suppose that $f, g : \mathbb{R} \to \mathbb{R}^m$ are continuous and $f(r) = g(r)$ for all rational numbers r; must $f(x) = g(x)$ for all irrational numbers x? The answer is yes, for let c be an irrational number. Then (see Example 3.13 on p. 166) there is a sequence of rational numbers $\{r_n\}$ converging to c. Since f and g are both continuous and $f(r_n) = g(r_n)$ for all n, we have

$$f(c) = \lim f(r_n) = \lim g(r_n) = g(c).$$

Note that the answer would be false if either f or g were not continuous. For example, with D denoting Dirichlet's function, $D(r) = 1$ for all rational numbers, but $D(x) \neq 1$ for all irrational numbers x. See Problem 2 for a related problem.

By the component theorem (Theorem 4.4 on p. 254), it follows that a function $f = (f_1, \ldots, f_m)$ is continuous at a point c if and only if every component f_k is continuous at c. Thus, we have the following theorem.

Component criterion for continuity

> **Theorem 4.11** *A function is continuous at c if and only if all of its component functions are continuous at c.*

Next, the composition of limits theorem (Theorem 4.6 on p. 255) implies the following result.

> **Theorem 4.12** *Let $f : D \longrightarrow \mathbb{R}^m$ and $g : C \longrightarrow \mathbb{R}^p$, where $D \subseteq \mathbb{R}^p$ and $C \subseteq \mathbb{R}^q$, and suppose that $g(C) \subseteq D$, so that $f \circ g : C \longrightarrow \mathbb{R}^m$ is defined. If g is continuous at c and f is continuous at $g(c)$, then the composite function $f \circ g$ is continuous at c.*

More simply: *The composition of continuous functions is continuous.* Finally, our algebra of limits theorem (Theorem 4.5 on p. 254) implies the following.

> **Theorem 4.13** *If $f, g : D \longrightarrow \mathbb{R}^m$ are both continuous at c, then*
>
> *(1) $|f|$ and $af + bg$ are continuous at c, for any constants a, b.*
>   *If f and g take values in $\mathbb{C}$, then*
> *(2) fg and (provided $g(c) \neq 0$) f/g are continuous at c.*

In simple language: Linear combinations of $\mathbb{R}^m$-valued continuous functions are continuous. Products, norms, and quotients of real- or complex-valued continuous functions are continuous (provided that the denominator functions are not zero). Finally, the left- and right-hand limit theorem (Theorem 4.9 on p. 257) implies the following.

> **Theorem 4.14** *Let $f : D \longrightarrow \mathbb{R}^m$ with $D \subseteq \mathbb{R}$ and let $c \in D$ be a limit point of the sets $D \cap (-\infty, c)$ and $D \cap (c, \infty)$. Then f is continuous at c if and only if $f(c) = f(c+) = f(c-)$.*

If only one of $f(c-)$ and $f(c+)$ makes sense, then f is continuous at c if and only if $f(c) = f(c-)$ or $f(c) = f(c+)$, whichever makes sense.

4.3.3 Thomae's Function and Volterra's Theorem

We now define a fascinating function that I found on page 14 of Carl Thomae's (1840–1921) 1875 book [240]. This function, called Thomae's function [16, p. 123] or the (modified) Dirichlet function [256], has the perplexing property that it is continuous at every irrational number and discontinuous at every rational number. (See Problem 7 on p. 299 for a generalization.)

We define **Thomae's function**, aka (also known as) the **modified Dirichlet function**, $T : \mathbb{R} \longrightarrow \mathbb{R}$ by

$$T(x) = \begin{cases} 1/q & \text{if } x \in \mathbb{Q} \text{ and } x = p/q \text{ in lowest terms and } q > 0, \\ 0 & \text{if } x \text{ is irrational}. \end{cases}$$

Here, we interpret 0 as $0/1$ in lowest terms, so $T(0) = 1/1 = 1$. Here's a graph of this "pathological function" (Fig. 4.10):

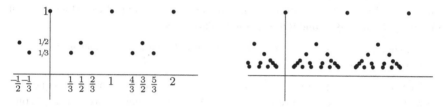

Fig. 4.10 The left-hand side shows plots of $T(p/q)$ for q at most 3, and the right-hand side shows plots of $T(p/q)$ for q at most 7

To see that T is discontinuous on rational numbers, let $c \in \mathbb{Q}$ and let $\{a_n\}$ be a sequence of irrational numbers converging to c, e.g., $a_n = c + \sqrt{2}/n$ works (or see Example 3.13 on p. 166). Then $\lim T(a_n) = \lim 0 = 0$, while $T(c) > 0$; hence T is discontinuous at c.

To see that T is continuous at each irrational number, let c be irrational and let $\varepsilon > 0$. Consider the case $c > 0$ (the case $c < 0$ is analogous) and choose $m \in \mathbb{N}$ with $c < m$. Let $0 < x \le m$ and let's consider the veracity of the inequality

$$|T(x) - T(c)| < \varepsilon, \quad \text{that is,} \quad T(x) < \varepsilon, \tag{4.20}$$

where we used that $T(c) = 0$ and $T(x) \ge 0$. If x is irrational, then $T(x) = 0 < \varepsilon$ holds. If $x = p/q$ is rational in lowest terms and $q \ge 1$, then $T(x) = 1/q < \varepsilon$ holds if and only if $q > 1/\varepsilon$. Stated another way, the inequality (4.20) holds for all $0 < x \le m$ *except* for rational numbers $x = p/q$ with $q \le 1/\varepsilon$. Thus, if we let q_1 equal the greatest integer $\le 1/\varepsilon$, then the only rational numbers p/q with $0 < p/q \le m$ for which (4.20) fails are those whose denominators q equal $1, 2, 3, \ldots,$ or q_1. Note that the numerator p can be at most $q_1 m$ (since $p/q \le m$ implies $p \le qm$, which implies $p \leq q_1 m$). There are at most finitely many rationals with this property (there are at most $q_1 m$ numerators and q_1 denominators, for no more than $q_1^2 m$ rationals). In particular, we can choose $\delta > 0$ such that the interval $(c - \delta, c + \delta)$ is contained in $(0, m)$ and contains none of the above rationals, as illustrated in Fig. 4.11. Since all rationals p/q (in lowest terms) in $(c - \delta, c + \delta)$ satisfy $q \geq 1/\varepsilon$, we have

$$x \in (c - \delta, c + \delta) \quad \Longrightarrow \quad |T(x) - T(c)| < \varepsilon,$$

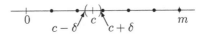

Fig. 4.11 The dots represent the finitely many rational numbers $0 < p/q \leq m$ with $q = 1, 2, 3, \ldots, q_1$

which proves that T is continuous at c. Thus, $T(x)$ is discontinuous at every *rational* number and continuous at every *irrational* number. The inquisitive student might ask whether there is a function opposite to Thomae's function.

Is there a function that is continuous at every *rational* point and discontinuous at every *irrational* point?

The answer is no. There are many ways to prove this; one can answer this question using the Baire category theorem (cf. [1, p. 128]), but we shall answer it using "compactness" arguments originating with Vito Volterra's (1860–1940) first publication in 1881 (before he was twenty!) [3]. To state his theorem, we need some terminology.

Let $D \subseteq \mathbb{R}^p$. A subset $A \subseteq D$ is said to be **dense** in D if for each point $c \in D$ and each $r > 0$, there is a point $x \in A$ such that $|x - c| < r$. For $p = 1$, $A \subseteq D$ is dense if for all $c \in D$ and $r > 0$,

$$(c - r, c + r) \cap A \neq \varnothing.$$

An equivalent definition of **dense** is the statement that every point in D is either in A or a limit point of A, which is equivalent to the condition that given any $c \in D$, there is a sequence $\{a_n\}$ in A such that $a_n \to c$.

Example 4.22 The set of rational numbers $\mathbb{Q}$ is dense in $\mathbb{R}$, because (by the density of the (ir)rationals in $\mathbb{R}$, Theorem 2.38 on p. 98) for every $c \in \mathbb{R}$ and $r > 0$, $(c - r, c + r) \cap \mathbb{Q}$ is never empty. Here's an attempted picture of $\mathbb{Q}$ in $\mathbb{R}$ as a densely packed bunch of points:

Similarly, the set $\mathbb{Q}^c$, the irrational numbers, is also dense in $\mathbb{R}$.

Given a function $f : D \longrightarrow \mathbb{R}^m$, we denote by $C_f \subseteq D$ the set of points in D at which f is continuous. Explicitly,

$$C_f := \{c \in D ; \ f \text{ is continuous at } c\}.$$

The function f is said to be **pointwise discontinuous** if C_f is dense in D.

Volterra's theorem

Theorem 4.15 *Two pointwise discontinuous functions defined on a nonempty open interval always have a point of continuity in common.*

Proof Let f and g be pointwise discontinuous functions on an open interval I. We prove our theorem in three steps.

Step 1: A closed interval $[\alpha, \beta]$ is said to be **nontrivial** if $\alpha < \beta$. Let $\varepsilon > 0$ and let $(a, b) \subseteq I$ be a nonempty open interval. We shall prove that there is a nontrivial closed interval $J \subseteq (a, b)$ such that for all $x, y \in J$,

$$|f(x) - f(y)| < \varepsilon \quad \text{and} \quad |g(x) - g(y)| < \varepsilon.$$

Indeed, since the continuity points of f are dense in I, there is a point $c \in (a, b)$ at which f is continuous. It follows that for some $\delta > 0$, $x \in I \cap [c - \delta, c + \delta]$ implies that $|f(x) - f(c)| < \varepsilon/2$. We can choose $\delta > 0$ smaller if necessary so that $J' = [c - \delta, c + \delta]$ is contained in (a, b). Then for all $x, y \in J'$, we have

$$|f(x) - f(y)| = |(f(x) - f(c)) + (f(c) - f(y))| \le |f(x) - f(c)| + |f(c) - f(y)| < \varepsilon.$$

Using the same argument for g but with $(c - \delta, c + \delta)$ in place of (a, b) shows that there is a nontrivial closed interval $J \subseteq J'$ such that $x, y \in J$ implies that $|g(x) - g(y)| < \varepsilon$. Since $J \subseteq J'$, the function f automatically satisfies $|f(x) - f(y)| < \varepsilon$ for $x, y \in J$. This completes the proof of **Step 1**.

Step 2: With $\varepsilon = 1$ and $(a, b) = I$ in **Step 1**, there is a nontrivial closed interval $[a_1, b_1] \subseteq I$ such that $x, y \in [a_1, b_1]$ implies that

$$|f(x) - f(y)| < 1 \quad \text{and} \quad |g(x) - g(y)| < 1.$$

Now with $\varepsilon = 1/2$ and $(a, b) = (a_1, b_1)$ in **Step 1**, there is a nontrivial closed interval $[a_2, b_2] \subseteq (a_1, b_1)$ such that $x, y \in [a_2, b_2]$ implies that

$$|f(x) - f(y)| < \frac{1}{2} \quad \text{and} \quad |g(x) - g(y)| < \frac{1}{2}.$$

Continuing by induction, we construct a sequence of nontrivial closed intervals $\{[a_n, b_n]\}$ such that $[a_{n+1}, b_{n+1}] \subseteq (a_n, b_n)$ for each n and $x, y \in [a_n, b_n]$ implies that

$$|f(x) - f(y)| < \frac{1}{n} \quad \text{and} \quad |g(x) - g(y)| < \frac{1}{n}. \tag{4.21}$$

By the nested intervals theorem on p. 100, there is a point c contained in every interval $[a_n, b_n]$.

Step 3: We now complete the proof. We claim that both f and g are continuous at c. To prove continuity, let $\varepsilon > 0$. Choose $n \in \mathbb{N}$ with $1/n < \varepsilon$. Since $[a_{n+1}, b_{n+1}] \subseteq (a_n, b_n)$, we have $c \in (a_n, b_n)$. Now choose $\delta > 0$ such that $(c - \delta, c + \delta) \subseteq (a_n, b_n)$. With this choice of $\delta > 0$, in view of (4.21) and the fact that $1/n < \varepsilon$, we obtain

$$|x - c| < \delta \implies |f(x) - f(c)| < \varepsilon \quad \text{and} \quad |g(x) - g(c)| < \varepsilon.$$

Thus, f and g are continuous at c, and our proof is complete.   ∎

Thus, there cannot be a function $f : \mathbb{R} \longrightarrow \mathbb{R}$ that is continuous at every rational point and discontinuous at every irrational point. Indeed, if it were so, then f would be pointwise discontinuous (because $\mathbb{Q}$ is dense in $\mathbb{R}$) and the function f and Thomae's function wouldn't have any continuity points in common, contradicting Volterra's theorem.

▶ **Exercises 4.3**

1. Recall that $\lfloor x \rfloor$ denotes the greatest integer less than or equal to x. Find the set of continuity points for the following functions:

$$(a)\ f(x) = \lfloor x \rfloor, \quad (b)\ g(x) = x\lfloor x \rfloor, \quad (c)\ h(x) = \lfloor 1/x \rfloor,$$

   where the domains are $\mathbb{R}$, $\mathbb{R}$, and $(0, \infty)$, respectively. Are the functions continuous on the domains $(-1, 1)$, $(-1, 1)$, and $(1, \infty)$, respectively?

2. In this problem we deal with zero sets of functions. Let $f : D \longrightarrow \mathbb{R}^m$ with $D \subseteq \mathbb{R}^p$. The **zero set** of f is the set $Z(f) := \{x \in D ; \ f(x) = 0\}$.

   (a) If f is continuous and $c \in D$ is a limit point of $Z(f)$, prove that $f(c) = 0$.
   (b) If f is continuous and $Z(f)$ is dense in D, prove that f is the zero function, that is, $f(x) = 0$ for all $x \in D$.
   (c) Using (b), prove that if $f, g : D \longrightarrow \mathbb{R}^m$ are continuous and $f(x) = g(x)$ on a dense subset of D, then $f = g$, that is, $f(x) = g(x)$ for all $x \in D$.

3. In this problem we look at additive functions. Let $f : \mathbb{R}^p \longrightarrow \mathbb{R}^m$ be **additive** in the sense that $f(x + y) = f(x) + f(y)$ for all $x, y \in \mathbb{R}^p$.

   (a) Prove that $f(0) = 0$ and $f(x - y) = f(x) - f(y)$ for all x, y.
   (b) Prove that if f is continuous at some point x_0, then f is continuous at all points.
   (c) Assume now that $p = 1$, so that $f : \mathbb{R} \longrightarrow \mathbb{R}^m$ is additive (no continuity assumptions at this point). Prove that $f(r) = f(1) r$ for all $r \in \mathbb{Q}$.
   (d) (Cf. [269]) Prove that if $f : \mathbb{R} \longrightarrow \mathbb{R}^m$ is continuous, then $f(x) = f(1) x$ for all $x \in \mathbb{R}$.

4. Let $f : \mathbb{R}^p \longrightarrow \mathbb{C}$ be **multiplicative** in the sense that $f(x + y) = f(x) f(y)$ for all $x, y \in \mathbb{R}^p$. Assume that f is not the zero function.

(a) Prove that $f(x) \neq 0$ for all x.

(b) Prove that $f(0) = 1$ and $f(-x) = 1/f(x)$.

(c) Prove that if f is continuous at some point x_0, then f is continuous at all points.

We show in Problem 12 on p. 313 that when $p = 1$ and f is real-valued and continuous, then f is given by an "exponential function."

5. Let $f : I \longrightarrow \mathbb{R}$ be a continuous function on a closed and bounded interval I. Suppose there is $0 < r < 1$ having the property that for each $x \in I$, there is a point $y \in I$ with $|f(y)| \leq r|f(x)|$. Prove that f must have a root, that is, there is a point $c \in I$ such that $f(c) = 0$.

6. Consider the following function related to Thomae's function:

$$t(x) := \begin{cases} q & \text{if } x = p/q \text{ in lowest terms and } q > 0, \\ 0 & \text{if } x \text{ is irrational.} \end{cases}$$

Prove that $t : \mathbb{R} \longrightarrow \mathbb{R}$ is discontinuous at every point in $\mathbb{R}$.

7. Here are some fascinating questions related to Volterra's theorem.

(a) Are there functions $f, g : \mathbb{R} \longrightarrow \mathbb{R}$ that don't have any continuity points in common such that f is pointwise discontinuous and g is not pointwise continuous but has at least one continuity point? Give an example or prove there are no such functions.

(b) Is there a continuous function $f : \mathbb{R} \longrightarrow \mathbb{R}$ that maps rationals to irrationals? Give an example or prove that there is no such function.

(c) Is there a continuous function $f : \mathbb{R} \longrightarrow \mathbb{R}$ that maps rationals to irrationals *and* irrationals to rationals? Suggestion: Suppose there is such a function and consider the function $T \circ f$, where T is Thomae's function.

4.4 Compactness, Connectedness, and Continuous Functions

Consider the continuous function $f(x) = 1/x$ with domain $D = \mathbb{R} \backslash \{0\}$, the real line with a "hole":

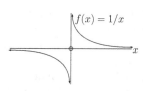

Notice that f has the following properties: f is not bounded on D, in particular f does not attain a maximum or minimum value on D, and that although the range of f contains both positive and negative values, f never takes on the intermediate value of

0. In this section we prove that the unboundedness property is absent when the domain is a closed and bounded interval and the property of not having an intermediate value is absent when the domain is an interval. We shall prove these results using two rather distinct viewpoints:

(I) A somewhat concrete *analytical* approach that uses only concepts we've covered in previous sections.

(II) A somewhat abstract *topological* approach based on the topological lemmas presented in Section 4.4.1.

If you're interested only in the easier analytical approach, skip Section 4.4.1 and also skip the **Proof II**'s in Theorems 4.19, 4.20, and 4.23. For an interesting and different approach using the concept of "tagged partitions," see [91].

4.4.1 Some Fundamental Topological Lemmas

Let $A \subseteq \mathbb{R}$. A collection $\mathcal{U}$ of subsets of $\mathbb{R}$ is called a **cover** of A if the union of all the sets in $\mathcal{U}$ contains A, in which case we also say that $\mathcal{U}$ **covers** A. Explicitly, $\mathcal{U} = \{\mathcal{U}_\alpha\}$ covers A if $A \subseteq \bigcup_\alpha \mathcal{U}_\alpha$. We are mostly interested in coverings by open intervals, that is, coverings in which each $\mathcal{U}_\alpha$ is an open interval.

Example 4.23 The interval $(0, 1)$ is covered by $\mathcal{U} = \{U_n = (1/n, 1)\}$, because $(0, 1) \subseteq \bigcup_{n=1}^{\infty}(1/n, 1)$. Here's a picture of the cover:

As n gets larger and larger, the intervals $(1/n, 1)$ "fill up" the interval $(0, 1)$.

Example 4.24 $[0, 1]$ is covered by $\mathcal{V} = \{V_n = (-1/n, 1 + 1/n)\} = \{(-1, 2),$ $(-1/2, 3/2), (-1/3, 4/3), \ldots\}$, because $[0, 1] \subseteq \bigcup_{n=1}^{\infty}(-1/n, 1 + 1/n)$. Here's a picture of the cover:

It's interesting to notice that the cover $\mathcal{U}$ in Example 4.23 does not have a **finite subcover**, that is, there are not finitely many elements of $\mathcal{U}$ that will still cover $(0, 1)$. To see this, let $\{U_{n_1}, \ldots, U_{n_k}\}$ be a finite subcollection of elements of $\mathcal{U}$. By relabeling, we may assume that $n_1 < n_2 < \cdots < n_k$. Since n_k is the largest of these k numbers, we have $\bigcup_{j=1}^{k} U_{n_j} = (\frac{1}{n_k}, 1)$, which does not cover $(0, 1)$, because there is a "gap" between 0 and $1/n_k$ as seen here (Fig. 4.12):

Fig. 4.12 The union of the finite subcover $\{U_{n_1}, \ldots, U_{n_k}\}$ equals $(1/n_k, 1)$, and in particular, does not cover all of $(0, 1)$

On the other hand, the cover $\mathcal{V}$ in Example 4.24 does have a finite subcover, that is, there are finitely many elements of $\mathcal{V}$ that will cover $[0, 1]$. Indeed, $[0, 1]$ is covered by the single element V_1 of $\mathcal{V}$, because $[0, 1] \subseteq (-1, 2)$. It turns out that *every* cover of $[0, 1]$ by open intervals will *always* have a finite subcover. In fact, this is a general phenomenon for closed and bounded intervals.

Compactness lemma

Lemma 4.16 *Every cover of a closed and bounded interval by open intervals has a finite subcover.*

Proof Let $\mathcal{U}$ be a cover of a closed and bounded interval $[a, b]$ by open intervals. We must show that there are finitely many elements of $\mathcal{U}$ that still cover $[a, b]$. Let A be the set of all numbers x in $[a, b]$ such that the interval $[a, x]$ is contained in the union of finitely many sets in $\mathcal{U}$. Since $[a, a]$ is the set containing the single point a and $\mathcal{U}$ covers $[a, b]$, the interval $[a, a]$ is contained in at least one set in $\mathcal{U}$, so A is not empty. Being a nonempty subset of $\mathbb{R}$ bounded above by b, A has a supremum, say $\xi \leq b$. Since ξ belongs to the interval $[a, b]$ and $\mathcal{U}$ covers $[a, b]$, ξ belongs to some open interval (c, d) in the collection $\mathcal{U}$. Since $c < \xi < d$, we can choose a real number η with $c < \eta < \xi$. Then η is less than ξ, the supremum of A, so by definition of the supremum of A, it follows that the interval $[a, \eta]$ must be covered by finitely many sets in $\mathcal{U}$, say $[a, \eta] \subseteq U_1 \cup \cdots \cup U_k$. Adding $U_{k+1} := (c, d)$ to this collection, it follows that the interval $[a, d)$ is covered by the finitely many sets $U_1, \ldots, U_{k+1}$ in $\mathcal{U}$. In particular, unless $\xi = b$, the set A would contain a number greater than ξ, as shown here (Fig. 4.13):

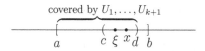

Fig. 4.13 If $\xi < b$, then every $x \in [a, b]$ with $\xi < x < d$, would belong to A, contradicting that ξ is an upper bound for A

Hence, $\xi = b$, and $[a, b]$ can be covered by finitely many sets in $\mathcal{U}$. ∎

Because closed and bounded intervals have this *finite* subcover property, and therefore behave somewhat like finite sets (which are "compact"—take up little space), we call closed and bounded intervals **compact**. We now move to the subject of open sets. A set $A \subseteq \mathbb{R}^m$ is **open** in $\mathbb{R}^m$ if each point in A is contained in some open ball that's entirely contained in A. That is, A is open if for each point $c \in A$, there is an $r > 0$ such that $B_r(c) \subseteq A$. We also consider the empty set $\varnothing$ to be open. Here's a picture to understand this concept (Fig. 4.14):

Fig. 4.14 *Left* the region inside a closed curve, not including the curve itself, is open. Every point inside the curve is contained in an open ball that's contained in the region. *Right* if we include the curve, the new set is *not* open, because every open ball centered at a point on the curve will contain points outside the set

For subsets of the real line, it turns out that open sets are just unions of open intervals; see Problem 9. Explicitly, $A \subseteq \mathbb{R}$ is open if and only if $A = \bigcup_\alpha U_\alpha$ for some open intervals U_α. We shall focus on subsets of $\mathbb{R}$.

Example 4.25 $\mathbb{R} = (-\infty, \infty)$ is an open interval, so $\mathbb{R}$ is open.

Example 4.26 Every open interval (a, b) is open, because it's a union consisting of just itself.

Example 4.27 $\mathbb{R}\backslash\mathbb{Z}$ is also open, because $\mathbb{R}\backslash\mathbb{Z} = \bigcup_{n\in\mathbb{Z}}(n, n + 1)$:

$$\overset{\circ}{\underset{-3}{}}\ \overset{\circ}{\underset{-2}{}}\ \overset{\circ}{\underset{-1}{}}\ \overset{\circ}{\underset{0}{}}\ \overset{\circ}{\underset{1}{}}\ \overset{\circ}{\underset{2}{}}\ \overset{\circ}{\underset{3}{}}$$

On the other hand, $\mathbb{Z}$ is not open, because every open ball (interval in this case) centered at an integer will contain noninteger numbers.

A set $A \subseteq \mathbb{R}$ is **disconnected** if there are open sets $\mathcal{U}$ and $\mathcal{V}$ such that $A \cap \mathcal{U}$ and $A \cap \mathcal{V}$ are nonempty, disjoint, and have union A. To have union A, we mean $A = (A \cap \mathcal{U}) \cup (A \cap \mathcal{V})$, which is actually equivalent to saying that

$$A \subseteq \mathcal{U} \cup \mathcal{V}.$$

Here's a picture of a disconnected set (Fig. 4.15):

Fig. 4.15 The set A consists of the *dark colored points* and *lines*

A set $A \subseteq \mathbb{R}$ is **connected** if it's not disconnected.

Example 4.28 $A = [-1, 0) \cup (0, 1)$ is disconnected:

$$\begin{array}{ccc} \bullet & \circ & \circ \\ -1 & 0 & 1 \end{array}$$

Indeed, $\mathcal{U} = (-2, 0)$ and $\mathcal{V} = (0, 1)$ are open, and $A \cap \mathcal{U} = [-1, 0)$ and $A \cap \mathcal{V} = (0, 1)$ are nonempty and disjoint, and their union is A. The set of integers:

$$\begin{array}{ccccccc} \bullet & \bullet & \bullet & \bullet & \bullet & \bullet & \bullet \\ -3 & -2 & -1 & 0 & 1 & 2 & 3 \end{array}$$

is also disconnected, since if $\mathcal{U} = (-\infty, 1/2)$ and $\mathcal{V} = (1/2, \infty)$, then $\mathbb{Z} \cap \mathcal{U}$ and $\mathbb{Z} \cap \mathcal{V}$ are nonempty and disjoint and have union $\mathbb{Z}$.

Intuitively, intervals should always be connected. This is in fact the case.

Connectedness lemma

Lemma 4.17 *Intervals (open, closed, bounded, unbounded, etc.) are connected.*

Proof Let I be an interval and suppose, for the sake of contradiction, that it is disconnected. Then there are open sets $\mathcal{U}$ and $\mathcal{V}$ such that $I \cap \mathcal{U}$ and $I \cap \mathcal{V}$ are disjoint, nonempty, and have union I. Let $a, b \in I$ with $a \in \mathcal{U}$ and $b \in \mathcal{V}$. By relabeling if necessary, we may assume that $a < b$. Since I is an interval, we have $[a, b] \subseteq I$. Thus, the sets $[a, b] \cap \mathcal{U}$ and $[a, b] \cap \mathcal{V}$ are disjoint (since $I \cap \mathcal{U}$ and $I \cap \mathcal{V}$ are), nonempty (since $a \in \mathcal{U}$ and $b \in \mathcal{V}$), and have union $[a, b]$ (since $I \cap \mathcal{U}$ and $I \cap \mathcal{V}$ have union I). We shall henceforth work with $[a, b]$ to get our contradiction. Define

$$c = \sup\big([a, b] \cap \mathcal{U}\big).$$

This number exists because $[a, b] \cap \mathcal{U}$ contains a and is bounded above by b. In particular, $c \in [a, b]$. Since $[a, b] \subseteq \mathcal{U} \cup \mathcal{V}$, the point c must belong to either $\mathcal{U}$ or $\mathcal{V}$. Thus, we just have to derive a contradiction in both situations. We shall assume that $c \in \mathcal{U}$ and leave for your enjoyment the case $c \in \mathcal{V}$. Note that $[a, b] \cap \mathcal{U}$ and $[a, b] \cap \mathcal{V}$ are disjoint. Thus, recalling that $c \in \mathcal{U}$ and $b \in \mathcal{V}$, it follows that $c \neq b$, so $c < b$. The set $\mathcal{U}$ is open, so it's a union of open intervals, and therefore, since $c \in \mathcal{U}$, we have $c \in (\alpha, \beta)$ for some open interval $(\alpha, \beta) \subseteq \mathcal{U}$. Here's what the situation could look like:

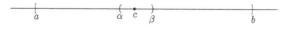

Note that the sets (c, b) and (c, β) are nonempty (because $c < b$ and $c < \beta$), and $(c, b) \subseteq [a, b]$, and $(c, \beta) \subseteq \mathcal{U}$. Therefore, if $b' = \min\{b, \beta\}$, then (c, b') is nonempty and $(c, b') \subseteq [a, b] \cap \mathcal{U}$. Now, every number in the nonempty interval (c, b') is larger than c and belongs to $[a, b] \cap \mathcal{U}$. This, however, contradicts the fact that c is an upper bound for $[a, b] \cap \mathcal{U}$. ∎

4.4.2　The Boundedness Theorem

The geometric content of the boundedness theorem is that the graph of a continuous function f on a closed and bounded interval lies between two horizontal lines, that is, there is a constant M such that $|f(x)| \le M$ for all x in the interval. Therefore, the graph does not extend infinitely up or down. Figure 4.16 shows a graph illustrating these ideas (the dots and the point c in the figure have to do with the max/min value and intermediate value theorems we'll discuss later).

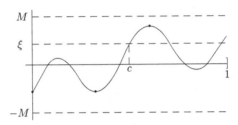

Fig. 4.16 Illustrations of the boundedness, max/min value, and intermediate value theorems for a function f on $[0, 1]$

The function $f(x) = 1/x$ on $(0, 1]$ or $f(x) = x$ on an unbounded interval shows that the boundedness theorem does not hold when the interval is not closed and bounded. Before proving the boundedness theorem, we need the following lemma.

Inequality lemma

> **Lemma 4.18** *Let* $f : I \longrightarrow \mathbb{R}$ *be a continuous map on an interval* I, *let* $c \in I$, *and suppose that* $|f(c)| < d$ *for some* $d \in \mathbb{R}$. *Then there is an open interval* I_c *containing* c *such that for all* $x \in I$ *with* $x \in I_c$, *we have* $|f(x)| < d$.

Proof Let $\varepsilon = d - |f(c)|$, and using the definition of continuity, choose $\delta > 0$ such that

$$x \in I \text{ and } |x - c| < \delta \implies |f(x) - f(c)| < \varepsilon.$$

Let $I_c = (c - \delta, c + \delta)$. Then given $x \in I$ with $x \in I_c$, we have $|x - c| < \delta$, so

$$|f(x)| = |(f(x) - f(c)) + f(c)| \le |f(x) - f(c)| + |f(c)|$$
$$< \varepsilon + |f(c)|$$
$$= (d - |f(c)|) + |f(c)| = d.$$

This proves our lemma. ∎

An analogous proof shows that if $\alpha < f(c) < \beta$, then there is an open interval I_c containing c such that $x \in I$ and $x \in I_c$ implies $\alpha < f(x) < \beta$. Another analogous proof shows that if $f : D \longrightarrow \mathbb{R}^m$ is continuous with $D \subseteq \mathbb{R}^p$ and $|f(c)| < d$, then there is an open ball B containing c such that $x \in D$ and $x \in B$ implies $|f(x)| < d$. We'll leave these generalizations to the interested reader. The following theorem is the first of our **fundamental theorems on continuous functions**.

Boundedness theorem

> **Theorem 4.19** *A continuous real-valued function on a closed and bounded interval is bounded.*

Proof Let f be a continuous function on a closed and bounded interval I.

Proof I: Assume that f is unbounded; we shall prove that f is not continuous. Since f is unbounded, for each natural number n there is a point x_n in I such that $|f(x_n)| \geq n$. By the Bolzano–Weierstrass theorem on p. 313, the sequence $\{x_n\}$ has a convergent subsequence, say $\{x_n'\}$, that converges to some c in I. By the way the numbers x_n were chosen, it follows that $|f(x_n')| \to \infty$, which shows that $f(x_n') \not\to f(c)$, for if $f(x_n') \to f(c)$, then we would have $|f(c)| = \lim |f(x_n')| = \infty$, an impossibility. Thus, by the sequence criterion on p. 263, f is not continuous at c.

Proof II: Given any arbitrary point c in I, we have $|f(c)| < |f(c)| + 1$, so by our inequality lemma there is an open interval I_c containing c such that for each $x \in I_c$, we have $|f(x)| < |f(c)| + 1$. The collection of all such open intervals $\mathcal{U} = \{I_c ; c \in I\}$ covers I, so by the compactness lemma, there are finitely many open intervals in $\mathcal{U}$ that cover I, say $I_{c_1}, \ldots, I_{c_n}$. Let M be the largest of the values $|f(c_1)| + 1, \ldots, |f(c_n)| + 1$. We claim that f is bounded by M on all of I. Indeed, given $x \in I$, since $I_{c_1}, \ldots, I_{c_n}$ cover I, there is an interval I_{c_k} containing x. Then by definition of I_{c_k}, we have $|f(x)| < |f(c_k)| + 1$. By definition of M, $|f(c_k)| + 1 \leq M$, so f is bounded by M as claimed. ∎

4.4.3 The Max/Min Value Theorem

The geometric content of our second **fundamental theorem on continuous functions** is that the graph of a continuous function on a closed and bounded interval must have highest (maximum) and lowest (minimum) points. The dots in Fig. 4.16 on p. 274 show such extreme points; note that there are two lowest points in the figure, so maxima/minima need not be unique. The simple example $f(x) = x$ on $(0, 1)$ shows that the max/min theorem does not hold when the interval is not closed and bounded.

Max/min value theorem

> **Theorem 4.20** *A continuous real-valued function on a closed and bounded interval achieves its maximum and minimum values. That is, if $f : I \longrightarrow \mathbb{R}$ is a continuous function on a closed and bounded interval I, then for some values c and d in the interval I, we have*
>
> $$f(c) \leq f(x) \leq f(d) \quad \text{for all } x \text{ in } I.$$

Proof Define
$$M = \sup\{f(x) \,;\, x \in I\}.$$

This number is finite by the boundedness theorem. We shall prove that there is a number d in $[a, b]$ such that $f(d) = M$. This proves that f achieves its maximum; a related proof shows that f achieves its minimum.

Proof I: By the definition of supremum, for each natural number n, there exists an x_n in I such that

$$M - \frac{1}{n} < f(x_n) \leq M, \tag{4.22}$$

for otherwise, the value $M - 1/n$ would be a smaller upper bound for $\{f(x) \,;\, x \in I\}$. By the Bolzano–Weierstrass theorem on p. 179, the sequence $\{x_n\}$ has a convergent subsequence $\{x_n'\}$; let's say that $x_n' \to d$, where d is in $[a, b]$. By the sequence criterion on p. 263, we have $f(x_n') \to f(d)$. On the other hand, by (4.22) and the squeeze theorem (see p. 164), we have $f(x_n) \to M$, so $f(x_n') \to M$ as well. By uniqueness of limits, $f(d) = M$.

Proof II: Assume, for the sake of contradiction, that $f(x) < M$ for all x in I. Let c be a point in I. Since $f(c) < M$ by assumption, we can choose $\varepsilon_c > 0$ such that $f(c) + \varepsilon_c < M$, so by our inequality lemma, there is an open interval I_c containing c such that for all $x \in I_c$, $|f(x)| < M - \varepsilon_c$. The collection $\mathscr{U} = \{I_c \,;\, c \in I\}$ covers I, so by the compactness lemma, there are finitely many intervals, say $I_{c_1}, \dots, I_{c_n}$, that cover I. Let m be the largest of the finitely many values $M - \varepsilon_{c_k}, k = 1, \dots, n$. Then $m < M$, and given $x \in I$, since $I_{c_1}, \dots, I_{c_n}$ cover I, there is an interval I_{c_k} containing x, which shows that

$$|f(x)| < M - \varepsilon_{c_k} \leq m < M.$$

This implies that M cannot be the supremum of f over I, since m is a smaller upper bound for f. This gives a contradiction to the definition of M. ∎

An almost identical argument as in **Proof I** gives the following result:

Corollary 4.21 *A continuous real-valued function on a closed ball in $\mathbb{R}^m$ achieves its maximum and minimum values. That is, if $f : B \longrightarrow \mathbb{R}$ is a continuous function on a closed ball $B \subseteq \mathbb{R}^m$, then for some $c, d \in B$, we have*

$$f(c) \leq f(x) \leq f(d) \quad \text{for all } x \text{ in } B.$$

4.4.4 The Intermediate Value Theorem

A real-valued function f on an interval I is said to have the **intermediate value property** if it attains all its intermediate values in the sense that if $a \leq b$ both belong to I, then for each real number ξ between $f(a)$ and $f(b)$, there is a c in $[a, b]$ such that $f(c) = \xi$. By "between" we mean that either $f(a) \leq \xi \leq f(b)$ or $f(b) \leq \xi \leq f(a)$. Geometrically, this means that the graph of f can be drawn without "jumps," that is, without ever lifting up the pencil. The intermediate value theorem (Theorem 4.23) states that every continuous function on an interval has the intermediate value property. See the previous Fig. 4.16 on p. 274 for an example in which we take, for instance, $a = 0$ and $b = 1$; note for this example that the point c need not be unique (there is another c' such that $f(c') = \xi$). The function $f(x) = 1/x$ with domain $D = \mathbb{R}\backslash\{0\}$ shows that the intermediate value theorem fails when the domain is not an interval.

Before proving the intermediate value theorem, we first think a little about intervals. Note that every interval I (bounded or unbounded, open, closed, etc.) has the following property: given points $a, b \in I$ with $a < b$, every point c between a and b is also in I. This property in fact characterizes intervals as shown in Lemma 4.22. We shall leave its proof to the interested reader.

Lemma 4.22 *A set A in $\mathbb{R}$ is an interval if and only if given points $a < b$ in A, we have $[a, b] \subseteq A$. That is, A is an interval if and only if given points a, b in A with $a < b$, all points between a and b also lie in A.*

(In fact, some mathematicians might even take this lemma as the *definition* of interval.) We are now ready to prove our third important **fundamental theorem on continuous functions**.

Intermediate value theorem (IVT)

> **Theorem 4.23** *Every real-valued continuous function on an interval (of any type, bounded, unbounded, open, closed, . . .) has the intermediate value property. Moreover, the range of such a function is also an interval.*

Proof Let f be a real-valued continuous function on an interval I and let ξ be between $f(a)$ and $f(b)$, where $a < b$ and $a, b \in I$. We shall prove that there is a c in $[a, b]$ such that $f(c) = \xi$. Assume that $f(a) \leq \xi \leq f(b)$; the reverse inequalities have a related proof. Note that if $\xi = f(a)$, then $c = a$ works, and if $\xi = f(b)$, then $c = b$ works. Therefore, we henceforth assume that $f(a) < \xi < f(b)$. We now prove that f has the intermediate value property, or IVP for short.

Proof I: To prove that f has the IVP, we don't care about f outside of $[a, b]$, so let's (re)define f outside of the interval $[a, b]$ such that f is equal to the constant value $f(a)$ on $(-\infty, a)$ and $f(b)$ on (b, ∞). This gives us a continuous function, which we again denote by f, that has domain $\mathbb{R}$, as shown here:

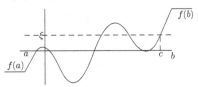

Define
$$A = \{x \in \mathbb{R} \, ; \, f(x) \leq \xi\}.$$

Since $f(a) < \xi$, we see that $a \in A$, so A is not empty, and since $\xi < f(b)$, we see that A is bounded above by b. In particular, $c := \sup A$ exists and $a \leq c \leq b$. We shall prove that $f(c) = \xi$, which is "obvious" from the above figure. To prove this rigorously, observe that by the definition of *least* upper bound, for every $n \in \mathbb{N}$, there is a point $x_n \in A$ such that $c - \frac{1}{n} < x_n \leq c$. As $n \to \infty$, we have $x_n \to c$, so by the sequence criterion on p. 263, $f(x_n) \to f(c)$. Since $f(x_n) \leq \xi$, because $x_n \in A$ and limits preserve inequalities, we have $f(c) \leq \xi$. On the other hand, by the definition of upper bound, for every $n \in \mathbb{N}$, we must have $f(c + \frac{1}{n}) > \xi$. Taking $n \to \infty$ and using the sequence criterion and that limits preserve inequalities, we see that $f(c) \geq \xi$. It follows that $f(c) = \xi$.

Proof II: To prove that f has the IVP using topology, suppose that $f(x) \neq \xi$ for every x in I. Let c be a point in I. If $f(c) < \xi$, then by the discussion after our inequality lemma, there is an open interval I_c containing c such that if $x \in I$ with $x \in I_c$, we have $f(x) < \xi$. Similarly, if $\xi < f(c)$, there is an open interval I_c containing c such that if $x \in I$ with $x \in I_c$, we have $\xi < f(x)$. In summary, we have assigned to each point $c \in I$ an open interval I_c that contains c such that either $f(x) < \xi$ or $\xi < f(x)$ for all $x \in I$ with $x \in I_c$. Let $\mathcal{U}$ be the union of all the I_c such that $f(c) < \xi$, and let $\mathcal{V}$ be the union of all the I_c such that $\xi < f(c)$. Then $\mathcal{U}$ and $\mathcal{V}$ are unions of open intervals and so are open sets by definition, and $a \in \mathcal{U}$ since

$f(a) < \xi$, and $b \in \mathcal{V}$ since $\xi < f(b)$. Notice that $\mathcal{U}$ and $\mathcal{V}$ are disjoint, because $\mathcal{U}$ has the property that if $x \in \mathcal{U}$, then $f(x) < \xi$, and $\mathcal{V}$ has the property that if $x \in \mathcal{V}$, then $\xi < f(x)$. Thus, $\mathcal{U}$ and $\mathcal{V}$ are disjoint, nonempty, and $I \subseteq \mathcal{U} \cup \mathcal{V}$. This contradicts the fact that intervals are connected.

We now prove that $f(I)$ is an interval. By our lemma, $f(I)$ is an interval if and only if given points α, β in $f(I)$ with $\alpha < \beta$, all points between α and β also lie in $f(I)$. Since $\alpha, \beta \in f(I)$, we can write $\alpha = f(x)$ and $\beta = f(y)$. Now let $f(x) < \xi < f(y)$. We need to show that $\xi \in f(I)$. However, according to the intermediate value property, there is a c in I such that $f(c) = \xi$. Thus, ξ is in $f(I)$, and our proof is complete.   ∎

A **root**, or **zero**, of a function f on a domain D is point $c \in D$ such that $f(c) = 0$.

Corollary 4.24 *Let f be a real-valued continuous function on an interval and let $a < b$ be points in the interval such that $f(a)$ and $f(b)$ have opposite signs (that is, $f(a) > 0$ and $f(b) < 0$, or $f(a) < 0$ and $f(b) > 0$). Then f has a root in the open interval (a, b).*

Proof Since 0 is between $f(a)$ and $f(b)$, by the intermediate value theorem there is a point c in $[a, b]$ such that $f(c) = 0$; since $f(a)$ and $f(b)$ are nonzero, c must lie strictly between a and b.   ∎

4.4.5 The Fundamental Theorems of Continuous Functions in Action

Example 4.29 The intermediate value theorem can be used to prove that every non-negative real number has a square root. To see this, let $a \geq 0$ and consider the function $f(x) = x^2$. Then f is continuous on $\mathbb{R}$, $f(0) = 0$, and

$$f(a+1) = (a+1)^2 = a^2 + 2a + 1 \geq 2a \geq a.$$

Therefore, $f(0) \leq a \leq f(a+1)$. Hence, by the intermediate value theorem, there is a point c with $0 \leq c \leq a+1$ such that $f(c) = a$, which is to say that $c^2 = a$. This proves that a has a square root. (The uniqueness of c follows from the last power rule in Theorem 2.23 on p. 80.) Of course, considering the function $f(x) = x^n$, we can prove that every nonnegative real number has a unique nth root.

Example 4.30 **Question:** Is there a continuous function $f : [0, 1] \longrightarrow \mathbb{R}$ that assumes each value in its range exactly twice? In other words, for each $y \in f([0, 1])$, are there exactly two points $x_1, x_2 \in [0, 1]$ such that $y = f(x_1) = f(x_2)$? Such a function is said to be "two-to-one." The answer is no. (See Problem 7 for generalizations of this example.) To see this, assume, by way of contradiction, that there is such a two-to-one function. Let y_0 be the maximum value of f, which exists by the max/min value theorem. Then there are exactly two points $a, b \in [0, 1]$, say $0 \le a < b \le 1$, such that $y_0 = f(a) = f(b)$. Note that all other points $x \in [0, 1]$ besides a, b must satisfy $f(x) < y_0$. This is because if $x \ne a, b$ yet $f(x) = y_0 = f(a) = f(b)$, then there would be three points $x, a, b \in [0, 1]$ assuming the same value, contradicting the two-to-one property. We claim that $a = 0$. Indeed, suppose that $0 < a$ and choose $c \in (a, b)$; then $0 < a < c < b$. Since $f(0) < y_0$ and $f(c) < y_0$, we can choose $\xi \in \mathbb{R}$ such that $f(0) < \xi < y_0$ and $f(c) < \xi < y_0$. Therefore,

$$f(0) < \xi < f(a), \quad f(c) < \xi < f(a), \quad f(c) < \xi < f(b).$$

By the intermediate value theorem, there are points

$$0 < c_1 < a, \quad a < c_2 < c, \quad c < c_3 < b$$

such that $\xi = f(c_1) = f(c_2) = f(c_3)$. Note that c_1, c_2, c_3 are all distinct and ξ is assumed at least three times by f. This contradicts the two-to-one property, so $a = 0$. Thus, f achieves its maximum at 0. Since $-f$ is also two-to-one, it follows that $-f$ also achieves its maximum at 0, which is the same as saying that f achieves its minimum at 0. However, if $y_0 = f(0)$ is both the maximum and minimum of f, then f must be the constant function $f(x) = y_0$ for all $x \in [0, 1]$, contradicting the two-to-one property of f.

▶ **Exercises 4.4**

1. Are the following subsets of $\mathbb{R}$ open, compact, or connected? If a set does not have a certain property, explain why.

   (a) $\mathbb{Q}$; (b) $[0, 5] \cup [6, 7]$; (c) $(0, 5) \cup (6, 7)$; (d) $\left\{ 1 - \dfrac{1}{n} ; n \in \mathbb{N} \right\}$;

   (e) $\{1\} \cup \left\{ 1 - \dfrac{1}{n} ; n \in \mathbb{N} \right\}$.

2. Is there a *nonconstant* continuous function $f : \mathbb{R} \longrightarrow \mathbb{R}$ that takes on only rational values (that is, whose range is contained in $\mathbb{Q}$)? What about only irrational values?

3. In this problem we investigate real roots of real-valued *odd*-degree polynomials.

   (a) Let $p(x) = x^n + a_{n-1} x^{n-1} + \cdots + a_1 x + a_0$ be a polynomial with each a_k real and $n \ge 1$ (not necessarily odd). Prove that there is a real number $a > 0$ such that
   $$\frac{1}{2} \le 1 + \frac{a_{n-1}}{x} + \cdots + \frac{a_0}{x^n}, \qquad \text{for all } |x| \ge a. \tag{4.23}$$

   (b) Using (4.23), prove that if n is odd, there is $c \in [-a, a]$ with $p(c) = 0$.

   (c) **Puzzle**: Does there exist a real number that is one more than its cube?

4. In this problem we investigate real roots of real-valued *even*-degree polynomials. Let $p(x) = x^n + a_{n-1} x^{n-1} + \cdots + a_1 x + a_0$ with each a_k real and $n \geq 2$ even.

   (a) Let $b > 0$ with $b^n \geq \max\{2a_0, a^n\}$, where a is given in (4.23). Prove that if $|x| \geq b$, then $p(x) \geq a_0 = p(0)$.

   (b) Prove that there is $c \in \mathbb{R}$ such that for all $x \in \mathbb{R}$, $p(c) \leq p(x)$. That is, $p : \mathbb{R} \longrightarrow \mathbb{R}$ achieves a minimum value. Is this statement true for odd-degree polynomials?

   (c) Show that there exists $d \in \mathbb{R}$ such that the equation $p(x) = \xi$ has a solution $x \in \mathbb{R}$ if and only if $\xi \geq d$. In particular, p has a real root if and only if $d \leq 0$.

5. Here is a variety of continuity problems. Let $f, g : [0, 1] \longrightarrow \mathbb{R}$ be continuous.

   (a) If f is one-to-one, prove that f achieves its maximum and minimum values at 0 or 1; that is, the maximum and minimum values of f cannot occur at points in $(0, 1)$.

   (b) If f is one-to-one and $f(0) < f(1)$, prove that f is strictly increasing; that is, for all $a, b \in [0, 1]$ with $a < b$, we have $f(a) < f(b)$.

   (c) (In (c) and (d) we do not assume that f is one-to-one.) If $f(0) = f(1)$, prove that there are points $a, b \in (0, 1)$ with $a \neq b$ such that $f(a) = f(b)$.

   (d) If $f(0) < g(0)$ and $g(1) < f(1)$, prove that there is a point $c \in (0, 1)$ such that $f(c) = g(c)$.

   (e) If f and g have the same maximum value on $[0, 1]$, prove that there is $c \in [0, 1]$ such that $f(c) = g(c)$.

6. Let $f : I \longrightarrow \mathbb{R}$ and $g : I \longrightarrow \mathbb{R}$ be continuous functions on a closed and bounded interval I and suppose that $f(x) < g(x)$ for all x in I.

   (a) Prove that there is a constant $\alpha > 0$ such that $f(x) + \alpha < g(x)$ for all $x \in I$.

   (b) Prove that there is a constant $\beta > 1$ such that $\beta f(x) < g(x)$ for all $x \in I$.

   (c) Do properties (a) and (b) hold if I is bounded but not closed (e.g., $I = (0, 1)$ or $I = (0, 1]$) or unbounded (e.g., $I = \mathbb{R}$ or $I = [1, \infty)$)? In each of these two cases, either prove (a) and (b) or give counterexamples.

7. (**n-to-one functions**) This problem is a continuation of Example 4.30.

   (a) Define a (necessarily discontinuous) function $f : [0, 1] \longrightarrow \mathbb{R}$ that takes on each value in its range exactly two times.

   (b) Prove that there does *not* exist a continuous function $f : [0, 1] \longrightarrow \mathbb{R}$ that takes on each value in its range exactly n times, where $n \in \mathbb{N}$ with $n \geq 2$.

   (c) Now what about a function with domain $\mathbb{R}$ instead of $[0, 1]$? Prove that there does *not* exist a continuous function $f : \mathbb{R} \longrightarrow \mathbb{R}$ that takes on each value in its range exactly two times.

   (d) If $n \in \mathbb{N}$ is even, prove there does *not* exist a continuous function $f : \mathbb{R} \longrightarrow \mathbb{R}$ that takes on each value in its range exactly n times. If n is odd, there does exist such a function! Draw such a function when $n = 3$ (try to draw a "zigzag"-type function). If you're interested in a formula for a continuous n-to-one function for arbitrary odd n, try to come up with one or see [260].

8. Show that a function $f : \mathbb{R} \longrightarrow \mathbb{R}$ can have at most a countable number of strict maxima. Here, a **strict maximum** is a point c such that $f(x) < f(c)$ for all x sufficiently close to c. Suggestion: At each point c where f has a strict maximum, consider a small interval (p, q) containing c where $p, q \in \mathbb{Q}$.

9. Prove that a subset of $\mathbb{R}$ is open if and only if the set is a union of open intervals.

The remaining exercises give alternative proofs of the boundedness, max/min, and intermediate value theorems.

10. (**Boundedness, Proof III**) Let $f : \mathbb{R} \longrightarrow \mathbb{R}$ be continuous such that f is constant outside some bounded interval (a, b); we shall prove that f is bounded. Since every continuous function on a closed and bounded interval can be extended to such a function with domain $\mathbb{R}$ (see the proof of the intermediate value theorem), this proves the boundedness theorem. Define

$$A = \{c \in (-\infty, b] \,;\, f \text{ is a bounded on } (-\infty, c]\}.$$

(i) Show that $d := \sup A$ exists.
(ii) Show there is a $\delta > 0$ such that f is bounded on $(d - \delta, d + \delta)$.
(iii) Show that f is bounded on $(-\infty, d+\delta)$, $d = b$, and prove that f is bounded.

11. (**Max/min, Proof III**) We give another proof of the max/min value theorem as follows. Let M be the supremum of a real-valued continuous function f on a closed and bounded interval I. Assume that $f(x) < M$ for all x in I and define

$$g(x) = \frac{1}{M - f(x)}.$$

Show that g is a continuous unbounded function on I. Now use the boundedness theorem to arrive at a contradiction.

12. (**Max/min, Proof IV**) We prove the max/min value theorem for a continuous function $f : \mathbb{R} \longrightarrow \mathbb{R}$ such that f is constant outside some bounded interval (a, b). This proves the max/min theorem, since every continuous function on a closed and bounded interval can be extended to such a function with domain $\mathbb{R}$ (see the proof of the intermediate value theorem). For each $c \in \mathbb{R}$, let

$$M_c = \sup\{f(x) \,;\, x \in (-\infty, c]\}.$$

This number is finite by Problem 10. Let M be supremum of f over all of $\mathbb{R}$. We shall prove that there is a d such that $f(d) = M$. This proves that f achieves its maximum; a related proof shows that f achieves its minimum. Let

$$A = \{c \in \mathbb{R}, \,;\, M_c < M\}.$$

(i) If $A = \varnothing$, prove that f achieves it maximum.

(ii) If A is not empty, prove that $d := \sup A$ exists.

(iii) We claim that $f(d) = M$. By way of contradiction, suppose that $f(d) < M$ and derive a contradiction.

13. **(IVP, Proof III)** Here's another proof of the intermediate value theorem. Let f be a real-valued continuous function on an interval $[a, b]$ and suppose that $f(a) < \xi < f(b)$.

(i) Define
$$A = \{x \in [a, b]\,;\ f(x) < \xi\}.$$

Show that $c := \sup A$ exists and $c < b$. We shall prove that $f(c) = \xi$ by showing that $f(c) < \xi$ or $f(c) > \xi$ gives rise to contradictions.

(ii) If $f(c) < \xi$, show that c is not an upper bound for A.

(iii) If $f(c) > \xi$, show that c is not the least upper bound for A.

14. **(IVP, Proof IV)** In this problem we prove the intermediate value theorem using the compactness lemma. Let f be a real-valued continuous function on $[a, b]$ and let $f(a) < \xi < f(b)$. Suppose that $f(x) \neq \xi$ for all x in $[a, b]$.

(i) Show that there are finitely many intervals, say $(a_1, b_1), \ldots, (a_n, b_n)$, that cover $I = [a, b]$, where each interval (a_k, b_k) has nonempty intersection with I, and either $f(x) < \xi$ or $f(x) > \xi$, for all $x \in (a_k, b_k) \cap I$. Suggestion: Define an open cover of $[a, b]$ consisting of all the open intervals I_c found in "Proof II" of Theorem 4.23.

(ii) We may assume that $a_1 \leq a_2 \leq a_3 \leq \ldots \leq a_n$ by reordering the a_k if necessary. Using induction, prove that $f(x) < \xi$ for all x in $(a_k, b_k) \cap I$, $k = 1, \ldots, n$. Derive a contradiction, thereby proving the intermediate value theorem.

15. **(IVP, Proof V)** We give one last proof of the intermediate value theorem called the *bisection method*. Let f be a continuous function on an interval and suppose that $f(a) < \xi < f(b)$, where $a < b$.

(i) Let $a_1 = a$ and $b_1 = b$ and let c_1 be the midpoint of $[a_1, b_1]$ and define the numbers a_2 and b_2 by $a_2 = a_1$ and $b_2 = c_1$ if $\xi \leq f(c_1)$, and $a_2 = c_1$ and $b_2 = b_1$ if $f(c_1) < \xi$. Prove that in either case, we have $[a_2, b_2] \subseteq [a_1, b_1]$ and $f(a_2) < \xi \leq f(b_2)$.

(ii) Construct a nested sequence of closed and bounded intervals $[a_n, b_n]$ such that $f(a_n) < \xi \leq f(b_n)$ for each n.

(iii) Using the nested intervals theorem on p. 100, show that the intersection of all $[a_n, b_n]$ is a single point, call it c, and show that $f(c) = \xi$.

16. We prove the connectedness lemma using the notion of "chains." Let $\mathcal{U}$ and $\mathcal{V}$ be open sets and suppose, by way of contradiction, that $[a, b] \cap \mathcal{U}$ and $[a, b] \cap \mathcal{V}$ are disjoint, nonempty, and have union $[a, b]$, where $a \in \mathcal{U}$ and $b \in \mathcal{V}$. We say that a point $c \in [a, b]$ is joined to a by a **chain** in $\mathcal{U}$ if there are finitely many open

intervals in $\mathcal{U}$, say $I_1, \ldots, I_n \subseteq \mathcal{U}$ for some n, such that $a \in I_1$, $I_k \cap I_{k+1} \neq \varnothing$ for $k = 1, \ldots, n - 1$, and $c \in I_n$. Let

$$A = \{ c \in [a, b] \, ; \, c \text{ is joined to } a \text{ by a chain in } \mathcal{U}. \}$$

(i)   Show that $d = \sup A$ exists, where $a < d < b$. Then $d \in \mathcal{U}$ or $d \in \mathcal{V}$.
(ii) However, show that $d \notin \mathcal{U}$ by assuming $d \in \mathcal{U}$ and deriving a contradiction.
(iii) However, show that $d \notin \mathcal{V}$ by assuming $d \in \mathcal{V}$ and deriving a contradiction.

4.5 ★ Amazing Consequences of Continuity

This section is devoted to answering interesting questions using the intermediate value theorem, or IVT. For example, is it always possible to stabilize a wobbly square table? The answer is yes. Using the IVT, we can prove that a square table can be stabilized by a rotation! **Remark:** Some of the constructions in this section are of an intuitive/geometric flavor; for example, although we have not defined *angles* or *area* rigorously, we use them and assume properties of them that agree with intuition. Thus, the "proofs" in this section are really convincing arguments. This is the only section of the book where convincing arguments are acceptable!

4.5.1 *Mountain Pass Theorem*

The intermediate value theorem helps us to solve the following puzzle [234, p. 239]. At 1 o'clock in the afternoon, a man starts walking up a mountain, arriving at 10 o'clock in the evening at his cabin. At 1 o'clock the next afternoon he walks back down by the exact same route, and happens to arrive at the bottom of the mountain at 10 o'clock; see Fig. 4.17.

Fig. 4.17 *Left* walking up on day 1. *Right* walking down on day 2

Is there a time when he is at the same place on the mountain on both days? To solve this puzzle, let $d_1(x)$ and $d_2(x)$ be the distance the man is from his cabin, measured along his route, at time x on days one and two, respectively. Then the functions $d_1 : [1, 10] \longrightarrow \mathbb{R}$ and $d_2 : [1, 10] \longrightarrow \mathbb{R}$ are continuous. We need to show that

$d_1(x) = d_2(x)$ at some time x. To see this, we use the **difference trick**, by which we mean to consider the continuous function of the difference,

$$f(x) = d_1(x) - d_2(x).$$

Observe that $f(1) = d_1(1) > 0$ and $f(10) = -d_2(10) < 0$. The IVT implies there is some point t where $f(t) = 0$. This t is a time that solves our puzzle. A physical way to solve this problem is to have another person walk down the mountain while the man is walking up. At some moment, the two will meet.

4.5.2 Pancake Theorem

A **region** is an open, connected, nonempty subset of Euclidean space. Fix an angle θ and consider a bounded region in the plane (think of the region as a "pancake"). Can we slice the region with a line making the angle θ with the positive horizontal axis such that the areas of the pieces on each side of the slice are equal? See Fig. 4.18.

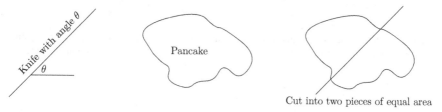

Cut into two pieces of equal area

Fig. 4.18 Holding a knife at an angle θ, can we slice a pancake into parts of equal area? The IVT, and everyday experience, says yes!

To answer this question, we first discuss directed lines. For $v, b \in \mathbb{R}^2$ with $v \neq 0$, the **directed line** through b with direction v is the line

$$\{tv + b \, ; \, t \in \mathbb{R}\}.$$

See Fig. 4.19.

Fig. 4.19 The directed line through b with direction v

The line is *directed*, because the vector v gives the line a direction. In particular, referring to the vector v, a directed line divides the plane into a left half and a right half. Returning to the pancake problem, let u be the unit vector that makes the angle $\theta+90°$ with the horizontal, and for each $t \in \mathbb{R}$, consider the directed line $\ell(t)$ through the point tu with a direction vector that makes the angle θ with the positive horizontal. For each $t \in \mathbb{R}$, let $f(t)$ be the area of the region to the *right* of the directed line $\ell(t)$; see Fig. 4.20. The function $f : \mathbb{R} \longrightarrow \mathbb{R}$ is a continuous function. Moreover, since the region is bounded, there is an α sufficiently negative such that $f(\alpha) = 0$ (we can choose α such that no part of the region lies to the right of $\ell(\alpha)$). Also, there is a β sufficiently positive such that $f(\beta) = A$, which equals the area of the region (in other words, the whole of the region lies to the right of $\ell(\beta)$). Since

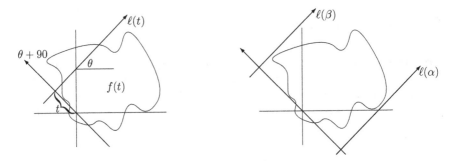

Fig. 4.20 *Left* the function $f(t)$ gives the area of the region to the *right* of $\ell(t)$. *Right* for α negative enough, we have $f(\alpha) = 0$, and for β positive enough, $f(\beta)$ is equal to the area of the region

$$f(\alpha) = 0 \le \frac{A}{2} \le A = f(\beta),$$

the IVT says that there is a t such that $f(t) = A/2$. This proves the pancake theorem. In Problem 6 you will prove that the line is unique; that is, there is exactly one line with angle θ that cuts the pancake into two parts of equal area.

4.5.3 Over Easy Egg Theorem

Consider two bounded regions in the plane, which could overlap. For example, we could consider an *over easy egg* as seen in Fig. 4.21. With a single straight slice, it turns out that we can simultaneously cut each region into parts of equal area! To prove this, let's call the regions *region 1* and *region 2*. By the pancake theorem, we know that for each angle θ, there is a unique directed line $\ell(\theta)$ whose direction vector makes the angle θ with the positive horizontal and that divides region 1 into two parts of equal area. See Fig. 4.22. Our goal is to find an angle such that the line $\ell(\theta)$ also cuts region 2 into two parts of equal area. To do so, we again use the difference trick and define the continuous function (see Fig. 4.22)

$$f(\theta) = L(\theta) - R(\theta),$$

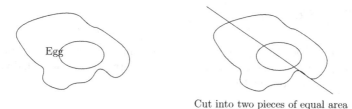

Cut into two pieces of equal area

Fig. 4.21 Is there a line that simultaneously divides the egg yolk and the egg white into parts of equal area? The IVT says yes!

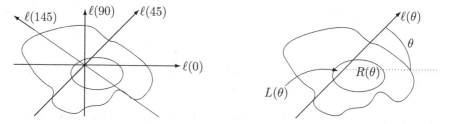

Fig. 4.22 *Left* by the pancake theorem, for each angle θ there is a directed line $\ell(\theta)$ whose direction vector has angle θ and that divides *region 1* (the egg *white*) into two parts of equal area. Shown are the lines with direction vectors having angles $0°, 45°, 90°$, and $145°$. *Right* let $L(\theta)$ equal the area of *region 2* (the egg *yolk*) to the *left* of $\ell(\theta)$, and let $R(\theta)$ equal the area of *region 2* to the *right* of $\ell(\theta)$

where $L(\theta)$ (respectively $R(\theta)$) is the area of region 2 to the left (respectively right) of the line $\ell(\theta)$. We need to show that there is an angle θ such that $f(\theta) = 0$. However, as seen in Fig. 4.23, we have $L(0) = R(180)$ and $L(180) = R(0)$. Hence,

$$f(0) = L(0) - R(0) = R(180) - L(180) = -f(180).$$

Thus, $f(0)$ and $f(180)$ have opposite signs! By the IVT (see Corollary 4.24 on p. 279), there is an angle θ such that $f(\theta) = 0$. This completes our proof.

Fig. 4.23 Observe that $L(0) = R(180)$ and $R(0) = L(180)$

4.5.4 Pizza Theorem

Consider again a bounded region in the plane (this time let's think of the region as a "pizza"). Using exactly two cuts at right angles, it turns out that we can divide the region into four parts of equal area!

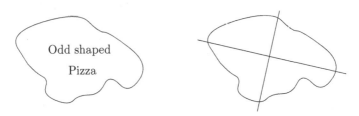

Fig. 4.24 Using *right-angle* cuts, can we cut a pizza, no matter what its shape is, into four parts of equal area? The IVT says yes!

See Fig. 4.24. To prove this, for each angle θ, let $\ell(\theta)$ be the unique directed line whose direction vector makes the angle θ with the positive horizontal and that divides the region into two parts of equal area. Turning the knife $90°$, the line $\ell(\theta + 90)$ also cuts the region into two parts of equal area. Consider the regions $A_1(\theta)$, $A_2(\theta)$, $A_3(\theta)$, and $A_4(\theta)$ shown here (Fig. 4.25):

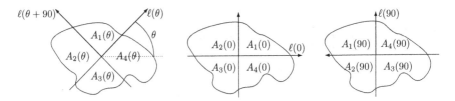

Fig. 4.25 $A_1(\theta)$ is the part of the region to the *left* of $\ell(\theta)$ and the *right* of $\ell(\theta+90)$. $A_2(\theta)$, $A_3(\theta)$, and $A_4(\theta)$ have similar definitions

We need to find a θ such that

$$A_1(\theta) = A_2(\theta) = A_3(\theta) = A_4(\theta).$$

We prove this in two steps.

   Step 1: We claim that the opposite diagonal regions are always equal; that is, we claim that for all θ,

$$A_1(\theta) = A_3(\theta) \quad \text{and} \quad A_2(\theta) = A_4(\theta).$$

Indeed, for every angle θ, we have two equations,

   (1) $A_1(\theta) + A_2(\theta) = A_3(\theta) + A_4(\theta)$ by definition of $\ell(\theta)$, and

$\quad$ **(2)** $A_1(\theta) + A_4(\theta) = A_2(\theta) + A_3(\theta)$ $\quad$ by definition of $\ell(\theta + 90)$.

Subtracting the second equation from the first, we obtain

$$A_2(\theta) - A_4(\theta) = A_4(\theta) - A_2(\theta).$$

This implies $A_2(\theta) = A_4(\theta)$. Canceling $A_2(\theta)$ and $A_4(\theta)$ from either the first or second equation then gives $A_1(\theta) = A_3(\theta)$.

$\quad$ **Step 2**: To prove that there is some angle such that $A_1(\theta) = A_2(\theta) = A_3(\theta) = A_4(\theta)$, by **Step 1** we just need to show that there is an angle such that $A_1(\theta) = A_2(\theta)$. We again use the *difference trick* and define the continuous function

$$f(\theta) = A_1(\theta) - A_2(\theta).$$

Observe that

$$\begin{aligned}
f(0) &= A_1(0) - A_2(0) \\
&= A_4(90) - A_1(90) \quad \text{(see Fig. 4.25)} \\
&= A_2(90) - A_1(90) \quad \text{(since } A_2(90) = A_4(90) \text{ by \textbf{Step1})} \\
&= -f(90).
\end{aligned}$$

Since $f(0)$ and $f(90)$ have opposite signs, there is, by the IVT, an angle θ such that $f(\theta) = 0$. This completes the proof of the pizza division theorem.

4.5.5 Wobbly Table Theorem

I've saved the best for last [77]. Consider a square table with legs of equal length placed on an uneven smooth floor. The table may or may not be wobbling. If it's wobbling, we claim that by rotating the table about its center no more than 90°, we can make it perfectly stable (all four legs will rest on the ground)! To prove this,

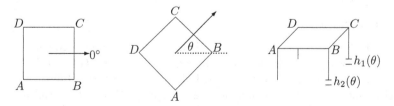

Fig. 4.26 *Left* the wobbly table not yet rotated. *Middle* we rotate the table about its center by the angle θ. *Right* the functions $h_1(\theta)$ and $h_2(\theta)$ are the heights of the wobble along the diagonals

label the corners of the square table A, B, C, D as shown in Fig. 4.26. We now start rotating the table about its center.

Rotating the table by an angle θ, we let

$$h_1(\theta) = \text{height of the wobble along the diagonal } AC$$

and

$$h_2(\theta) = \text{height of the wobble along the diagonal } BD.$$

The important point to observe is that when a table wobbles, it wobbles along exactly one diagonal; two legs along a diagonal are firmly on the ground while the other two legs can wobble like a seesaw or teeter-totter. Thus, for every angle θ, exactly one of three cases holds:

$$h_1(\theta) > 0 \text{ and } h_2(\theta) = 0 \quad \text{(table wobbles along } AC\text{)};$$
$$h_1(\theta) = 0 \text{ and } h_2(\theta) > 0 \quad \text{(table wobbles along } BD\text{)};$$
$$h_1(\theta) = 0 \text{ and } h_2(\theta) = 0 \quad \text{(table is stable)}.$$

In particular, $h_1(\theta) = h_2(\theta)$ if and only if the table is stable. For this reason, we yet again use the *difference trick* and define the continuous function

$$f(\theta) = h_1(\theta) - h_2(\theta).$$

As seen here,

we have

$$f(0) = h_1(0) - h_2(0) = h_2(90) - h_1(90) = -f(90).$$

Since $f(0)$ and $f(90)$ have opposite signs, there is, by the IVT, an angle θ such that $f(\theta) = 0$. This completes the proof of the wobbly table theorem!

▶ **Exercises 4.5**

1. **(Brouwer's fixed point theorem)**

   (i) If $f : [a, b] \longrightarrow [a, b]$ is a continuous function on a closed and bounded interval, prove that there is a point $c \in [a, b]$ such that $f(c) = c$. This result is a special case of a theorem by Luitzen Brouwer (1881–1966).

   (ii) **Puzzle**: You have a straight wire lying perpendicular to a wall that it is touching. You pick it up and bend it into any shape. Then you put it down so it touches the wall. Is there a point on the bent wire whose distance to the wall is exactly the same as it was originally (see Fig. 4.27)?

Fig. 4.27 The point shown is the same distance from the wall before and after the wire was bent

2. (**Bobble dog theorem**) A **bobble dog** is attached to a dashboard in a car, and when the car moves, the head amusingly moves back and forth. Assume that we have a boggle dog whose head can move only along a one-dimensional angle back and forth between two angles $\pm\theta_0$, as seen in the left picture in Fig. 4.28.

Fig. 4.28 *Left* the head of a bobble dog can move between the angles $-\theta_0$ and θ_0. *Middle* the position of the head at point A. *Right* the position of the head at point B. We want to prove that there is a θ such that $f(\theta) = \theta$

Suppose you take a drive through a city from a point A to a point B. Prove that there is an angle such that if the dog's head starts at that angle at the instant the car starts at point A, then the instant the car makes it to point B, the head is at the same angle it started at.

3. (**Cocktail theorem**) Take a glass with a straw fixed at a certain position along the rim, as shown in the left picture in Fig. 4.29.

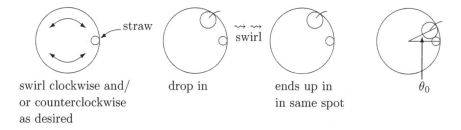

swirl clockwise and/   drop in   ends up in θ_0
or counterclockwise   in same spot
as desired

Fig. 4.29 The cocktail theorem

Swirl the drink clockwise and/or counterclockwise as desired for a minute (not moving the location of the straw). Prove that there is a location along the rim such that if you drop a cherry (assumed to be a perfect sphere) at that location, after one minute of swirling, the cherry ends up exactly where it started. Assume that the cherry can move only along the rim and if it hits the straw, it cannot pass it. Thus, if θ_0 is the angle shown in the far-right picture in Fig. 4.29, then the cherry is confined to the rim in such a way that the angle from the center of the glass to the center of the cherry is between θ_0 and $360° - \theta_0$.

4. (**Antipodal point puzzle**) In this problem we prove that there are, at any given moment, antipodal (opposite) points on the earth's equator that have the same temperature. (Instead of temperature, we could use pressure, elevation, or any other quantity that varies continuously with position.)

    (i) Let $a > 0$ and let $f : [0, a] \longrightarrow \mathbb{R}$ be a continuous function with $f(0) = f(a)$. Show there exists a point $c \in [0, a]$ such that $f(c) = f(c + a/2)$.

   (ii) Using (i), solve the antipodal point puzzle.

  (iii) Prove that there exist two days, exactly six months apart, with exactly the same number of hours of daylight. (Note that we don't have to be an astronomer to solve this problem!)

5. (**Birthday cake theorem**) Suppose you have a circular birthday cake with a lot of sweet topping on it. The baker, however, was not careful and the topping was put on very unevenly. Prove that you can cut the cake through its exact center so that the two halves have exactly the same amount sweet topping.

6. For the pancake theorem, prove that the line is unique. Suggestion: Argue that if $t_1 < t_2$ and $0 < f(t_1) < A$, then $f(t_1) < f(t_2)$.

7. Consider a bounded region in the plane with a finite boundary perimeter. Prove that there exists a line that divides the figure into two parts, each part having the same *area* and having the same *boundary perimeter.*

8. (**Photo cropping theorem**) Fixing a bounded region in the plane, in this problem we show that we can always crop it using a (possibly rotated) *square* (Fig. 4.30). Here, *crop* means to enclose the region in a rectangle such that each edge of the rectangle touches an edge of the region. Proceed as follows.

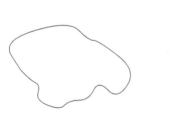
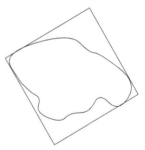

Fig. 4.30 Given a region, can we crop it with a (possibly rotated) *square*? The IVT says yes!

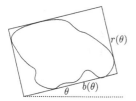 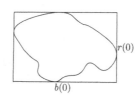 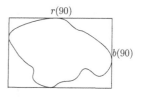

Fig. 4.31 *Left* for every θ, there exists a unique *rectangle* with angle θ cropping the region. Here, $b(\theta)$ and $r(\theta)$ are the *bottom* and *right*, respectively, of the cropping *rectangle*. *Middle/right* Note that $b(0) = r(90)$ and $r(0) = b(90)$

(i) For every angle θ, prove that there exists a unique directed line $L(\theta)$ such that the region lies to the left of the directed line and an edge of the region touches the line. Observe that a cropping rectangle is formed by the lines $L(\theta), L(\theta + 90), L(\theta + 180)$, and $L(\theta + 270)$ (Fig. 4.31).

(ii) Let $b(\theta) = L(\theta)$ and $r(\theta) = L(\theta+90)$ be the bottom and right, respectively, of the cropping rectangle. Show that there is a θ such that $b(\theta) = r(\theta)$. This proves the photo cropping theorem.

4.6 Monotone Functions and Their Inverses

In this section we study monotone functions on intervals and their continuity properties. In particular, we prove the following fascinating fact: Every monotone function on an interval (no other assumptions besides monotonicity) is continuous everywhere on the interval except perhaps at countably many points. With the monotonicity assumption dropped, anything can happen, for instance, recall that Dirichlet's function is nowhere continuous.

4.6.1 Continuous and Discontinuous Monotone Functions

Let $I \subseteq \mathbb{R}$ be an interval. A function $f : I \longrightarrow \mathbb{R}$ is said to be **nondecreasing** if $a \leq b$ (where $a, b \in I$) implies $f(a) \leq f(b)$, **(strictly) increasing** if $a < b$ implies $f(a) < f(b)$, **nonincreasing** if $a \leq b$ implies $f(a) \geq f(b)$, and **(strictly) decreasing** if $a < b$ implies $f(a) > f(b)$. The function is **monotone** if it is one of these four types. (Actually only two types, because increasing and decreasing functions are special cases of nondecreasing and nonincreasing functions, respectively.)

Example 4.31 A neat example of a monotone (nondecreasing) function is **Zeno's function** $Z : [0, 1] \longrightarrow \mathbb{R}$, named after Zeno of Elea (490 B.C.–425 B.C.):

$$Z(x) = \begin{cases} 0 & x = 0, \\ 1/2 & 0 < x \le 1/2, \\ 1/2 + 1/2^2 = 3/4 & 1/2 < x \le 3/4, \\ 1/2 + 1/2^2 + 1/2^3 = 7/8 & 3/4 < x \le 7/8, \\ \cdots \text{ etc. } \cdots & \cdots \\ 1 & x = 1. \end{cases}$$

Here's an attempted picture of Zeno's function:

This function is called Zeno's function because as described by Aristotle (384–322 B.C.), Zeno argued that "there is no motion because that which is moved must arrive at the middle of its course before it arrives at the end" (you can read about this in [108]). Zeno's function moves from 0 to 1 via half-way stops. Observe that the left-hand limits of Zeno's function exist at each point of [0, 1] except at $x = 0$, where the left-hand limit is not defined, and the right-hand limits exist at each point of [0, 1] except at $x = 1$, where the right-hand limit is not defined. Also observe that Zeno's function has discontinuity points exactly at the (countably many) points $x = (2^k - 1)/2^k$ for $k = 0, 1, 2, 3, 4, \ldots$.

It's an amazing fact that Zeno's function is typical: *Every* monotone function on an interval has left- and right-hand limits at every point of the interval except at the endpoints, where a left- or right-hand limit may not be defined, and has at most countably many discontinuities. Now for simplicity . . .

To avoid worrying about endpoints, in this section we consider only monotone functions with domain $\mathbb{R}$. *However, every result we prove has an analogous statement for domains that are intervals.*

We repeat, every statement that we mention in this section holds for monotone functions on intervals (open, closed, half-open, etc.) as long as we make suitable modifications of these statements at endpoints.

Lemma 4.25 *Let* $f : \mathbb{R} \longrightarrow \mathbb{R}$ *be nondecreasing. Then for all* $c \in \mathbb{R}$, *the left- and right-hand limits* $f(c-) = \lim_{x \to c-} f(x)$ *and* $f(c+) = \lim_{x \to c+} f(x)$ *exist. Moreover,*

$$f(c-) \le f(c) \le f(c+),$$

and if $c < d$, *then*

$$f(c+) \le f(d-). \tag{4.24}$$

Proof Fix $c \in \mathbb{R}$. We first show that $f(c-)$ exists. Since f is nondecreasing, for all $x \leq c$, $f(x) \leq f(c)$, so the set $\{f(x); x < c\}$ is bounded above by $f(c)$. Hence, $b := \sup\{f(x); x < c\}$ exists and $b \leq f(c)$. Given $\varepsilon > 0$, by definition of the supremum there is a $y < c$ such that $b - \varepsilon < f(y)$. Let $\delta = c - y$. Then $c - \delta < x < c$ implies that $y < x < c$, which implies that

$$|b - f(x)| = b - f(x) \quad \text{(since } f(x) \leq b \text{ by definition of supremum)}$$
$$\leq b - f(y) \quad \text{(since } f(y) \leq f(x))$$
$$< \varepsilon.$$

This shows that $f(c-) := \lim_{x \to c-} f(x)$ exists and equals b, that is,

$$f(c-) = \sup\{f(x); x < c\}.$$

As already noted, $b \leq f(c)$, so $f(c-) \leq f(c)$. By considering the set of values $\{f(x); c < x\}$, one can similarly prove that

$$f(c+) = \inf\{f(x); c < x\},$$

and $f(c) \leq f(c+)$. Let $c < d$. Then $\{f(x); x < c\} \subseteq \{f(x); x < d\}$, so by the definition of infimum and supremum, we have

$$\inf\{f(x); x < c\} \leq \sup\{f(x); x < d\}.$$

In other words, $f(c+) \leq f(d-)$, and our proof is now complete. ∎

Recall from Theorem 4.14 on p. 264 that a function f is continuous at a point c if and only if $f(c-) = f(c) = f(c+)$. Lemma 4.25 shows that for a nondecreasing function f, at discontinuity points one of the three possibilities in Fig. 4.32 occurs.

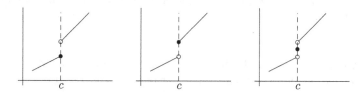

Fig. 4.32 *Left* $f(c-) = f(c) < f(c+)$. *Middle* $f(c-) < f(c) = f(c+)$. *Right* $f(c-) < f(c) < f(c+)$

Of course, there is a corresponding picture for nonincreasing functions in which the inequalities are reversed. In particular, notice that at a point of discontinuity, the function "jumps" by the value $f(c+) - f(c-)$. More generally, every function $f : D \longrightarrow \mathbb{R}$ with $D \subseteq \mathbb{R}$ is said to have a **jump discontinuity** at a point $c \in D$, where c is a limit point of $D \cap (c, \infty)$ and $D \cap (-\infty, c)$, if f is discontinuous at c

but both the left- and right-hand limits $f(c\pm)$ exist; the number $f(c+) - f(c-)$ is then called the **jump** of f at c. If c is a limit point of only one of the sets $D \cap (c, \infty)$ and $D \cap (-\infty, c)$, then we require only the corresponding right- or left-hand limit to exist. We now prove that every monotone function has at most countably many discontinuities (each of which is a jump discontinuity); see Problem 2 for another proof.

Theorem 4.26 *A monotone function on $\mathbb{R}$ has uncountably many points of continuity and at most countably many discontinuities, each discontinuity being a jump discontinuity.*

Proof Assume that f is nondecreasing, since the case for a nonincreasing function is proved in an analogous manner. We know that f is discontinuous at a point x if and only if $f(x+) - f(x-) > 0$. Given such a discontinuity point, choose a rational number r_x in the interval $(f(x-), f(x+))$. Since f is nondecreasing, given any two such discontinuity points $x < y$, we have (see (4.24)) $f(x+) \le f(y-)$, so the intervals $(f(x-), f(x+))$, and $(f(y-), f(y+))$ are disjoint. Thus, $r_x \ne r_y$, so different discontinuity points are associated with different rational numbers. It follows that the set of all discontinuity points of f is in one-to-one correspondence with a subset of the rationals. Therefore, since a subset of a countable set is countable, the set of all discontinuity points of f is countable. Since $\mathbb{R}$, which is uncountable, is the union of the continuity points of f and the discontinuity points of f, the continuity points of f must be uncountable. ∎

The following is a very simple and useful characterization of continuous monotone functions on intervals.

Theorem 4.27 *A monotone function on $\mathbb{R}$ is continuous on $\mathbb{R}$ if and only if its range is an interval.*

Proof By the intermediate value theorem, we already know that the range of every (in particular, a monotone) continuous function on $\mathbb{R}$ is an interval. Let $f : \mathbb{R} \longrightarrow \mathbb{R}$ be monotone and suppose, for concreteness, that f is nondecreasing, since the case for a nonincreasing function is similar. It remains to prove that if the range of f is an interval, then f is continuous. We shall prove the contrapositive, so assume that f is not continuous at some point c. Then one of the equalities in

$$f(c-) = f(c) = f(c+)$$

must fail. Since f is nondecreasing, we have $f(c-) \le f(c) \le f(c+)$, and therefore, one of the intervals $(f(c-), f(c))$, $(f(c), f(c+))$ is nonempty. Whichever interval is nonempty is not contained in the range of f. By Lemma 4.22 on p. 277, the range of f cannot be an interval.   ■

4.6.2   Monotone Inverse Theorem

Recall from Section 1.3 that a function has an inverse if and only if the function is injective, that is, one-to-one. Notice that a strictly monotone function $f : \mathbb{R} \longrightarrow \mathbb{R}$ is one-to-one, since, for instance, if f is strictly increasing, then $x \neq y$, say $x < y$, implies that $f(x) < f(y)$, which in particular says that $f(x) \neq f(y)$. Thus, a strictly monotone function is one-to-one. The last result in this section states that a one-to-one continuous function is automatically strictly monotone. This result makes intuitive sense, for if the graph of the function had a dip in it, the function would not pass the so-called *horizontal line test* learned in high school.

Monotone inverse theorem

> **Theorem 4.28** *A one-to-one continuous function* $f : \mathbb{R} \longrightarrow \mathbb{R}$ *is strictly monotone, its range is an interval, and it has a continuous strictly monotone inverse (with the same monotonicity as* f *).*

Proof Let $f : \mathbb{R} \longrightarrow \mathbb{R}$ be a one-to-one continuous function. We shall prove that f is strictly monotone. Fix points $x_0 < y_0$. Then $f(x_0) \neq f(y_0)$, so either $f(x_0) < f(y_0)$ or $f(x_0) > f(y_0)$. For concreteness, assume that $f(x_0) < f(y_0)$; the other case $f(x_0) > f(y_0)$ can be dealt with analogously. We claim that f is strictly increasing. Indeed, if this were not the case, then there would exist points $x_1 < y_1$ such that $f(y_1) < f(x_1)$. Now consider the function $g : [0, 1] \to \mathbb{R}$ defined by

$$g(t) = f(ty_0 + (1 - t)y_1) - f(tx_0 + (1 - t)x_1).$$

Since f is continuous, g is continuous, and

$$g(0) = f(y_1) - f(x_1) < 0 \quad \text{and} \quad g(1) = f(y_0) - f(x_0) > 0.$$

Hence by the IVT, there is a $c \in [0, 1]$ such that $g(c) = 0$. This implies that $f(a) = f(b)$, where $a = cx_0 + (1 - c)x_1$ and $b = cy_0 + (1 - c)y_1$. Since f is one-to-one, we must have $a = b$; however, this is impossible, since $x_0 < y_0$ and $x_1 < y_1$ implies $a < b$. This contradiction shows that f must be strictly monotone.

Now let $f : \mathbb{R} \longrightarrow \mathbb{R}$ be a continuous strictly monotone function and let $I = f(\mathbb{R})$. By Theorem 4.27, we know that I is an interval too. We shall prove that $f^{-1} : I \longrightarrow \mathbb{R}$ is also a strictly monotone function; then Theorem 4.27 implies that f^{-1} is continuous. Now suppose, for instance, that f is strictly increasing; we shall prove that f^{-1} is also strictly increasing. If $x < y$ both belong I, then $x = f(\xi)$ and $y = f(\eta)$ for some ξ and η. Since f is strictly increasing it must be that $\xi < \eta$, and hence, since $\xi = f^{-1}(x)$ and $\eta = f^{-1}(y)$, we have $f^{-1}(x) < f^{-1}(y)$. Thus, f^{-1} is strictly increasing, and our proof is complete.   ∎

Here is a nice application of the monotone inverse theorem.

Example 4.32 Note that for every $n \in \mathbb{N}$, the function $f(x) = x^n$ is strictly increasing on $[0, \infty)$. Therefore, $f^{-1}(x) = x^{1/n}$ is continuous. In particular, for every $m \in \mathbb{N}$, $g(x) = x^{m/n} = (x^{1/n})^m$ is continuous on $[0, \infty)$, being a composition of the continuous functions f^{-1} and the mth power. Similarly, the function $x \mapsto x^{m/n}$ for $m \in \mathbb{Z}$ with $m < 0$ is continuous on $(0, \infty)$. Therefore, for every $r \in \mathbb{Q}$, $x \mapsto x^r$ is continuous on $[0, \infty)$ if $r > 0$ and on $(0, \infty)$ if $r < 0$.

▶ **Exercises 4.6**

1. Prove the following algebraic properties of nondecreasing functions:

    (a) If f and g are nondecreasing, then $f + g$ is nondecreasing.
    (b) If f and g are nondecreasing and nonnegative, then $f g$ is nondecreasing.
    (c) Does (b) hold for any (not necessarily nonnegative) nondecreasing functions? Either prove this or give a counterexample.

2. Here is a different way to prove that a monotone function has at most countably many discontinuities. Let $f : [a, b] \longrightarrow \mathbb{R}$ be nondecreasing.

    (i) Given a finite number $x_1, \dots, x_k$ of points in (a, b), prove that

    $$j(x_1) + \cdots + j(x_k) \le f(b) - f(a), \qquad \text{where } j(x) = f(x+) - f(x-).$$

    (ii) Given $n \in \mathbb{N}$, prove that there is at most a finite number of points $c \in (a, b)$ such that $f(c+) - f(c-) > 1/n$.
    (iii) Now prove that f can have at most countably many discontinuities.

3. Let $f : \mathbb{R} \longrightarrow \mathbb{R}$ be a monotone function. If f happens also to be additive (see Problem 3 on p. 268), prove that f is continuous. Thus, every additive monotone function is continuous.

4. In this problem we investigate jump functions. Let $x_1, x_2, \dots$ be countably many points on the real line and let $c_1, c_2, \dots$ be nonzero complex numbers such that $\sum c_n$ is absolutely convergent. For $x \in \mathbb{R}$, the functions

$$\varphi_\ell(x) = \sum_{x_n < x} c_n \quad \text{and} \quad \varphi_r(x) = \sum_{x_n \le x} c_n \qquad (4.25)$$

are called a **(left-continuous) jump function** and **(right-continuous) jump function**, respectively. By the sums defining φ_ℓ and φ_r we define $\varphi_\ell(x) = \lim s_n(x)$ and $\varphi_r(x) = \lim t_n(x)$, where

$$s_n(x) = \sum_{k \le n,\, x_k < x} c_k \quad \text{and} \quad t_n(x) = \sum_{k \le n,\, x_k \le x} c_k ;$$

here, we sum only over those $k \leq n$ such that $x_k < x$ for $s_n(x)$, and $x_k \leq x$ for $t_n(x)$.

(a) Prove that $\varphi_\ell, \varphi_r : \mathbb{R} \longrightarrow \mathbb{C}$ are well defined for all $x \in \mathbb{R}$ (that is, the two infinite series (4.25) converge for all $x \in \mathbb{R}$).
(b) If all the c_n are nonnegative real numbers, prove that φ_ℓ and φ_r are nondecreasing functions on $\mathbb{R}$.
(c) If all the c_n are nonpositive real numbers, prove that φ_ℓ and φ_r are nonincreasing functions on $\mathbb{R}$.

5. In this problem we prove that φ_r in (4.25) is right-continuous having jump discontinuities only at $x_1, x_2, \ldots$ with the jump at x_n equal to c_n. To this end, let $\varepsilon > 0$. Since $\sum |c_n|$ converges, by Cauchy's criterion for series, we can choose N such that

$$\sum_{n \geq N+1} |c_n| < \varepsilon. \tag{4.26}$$

(i) Prove that for every $\delta > 0$,

$$\varphi_r(x + \delta) - \varphi_r(x) = \sum_{x < x_n \leq x+\delta} c_n.$$

Using (4.26), prove that for $\delta > 0$ sufficiently small, $|\varphi_r(x + \delta) - \varphi_r(x)| < \varepsilon$.
(ii) Prove that for every $\delta > 0$,

$$\varphi_r(x) - \varphi_r(x - \delta) = \sum_{x-\delta < x_n \leq x} c_n.$$

If x is not one of the points $x_1, \ldots, x_N$, then using (4.26), prove that for $\delta > 0$ sufficiently small, $|\varphi_r(x) - \varphi_r(x - \delta)| < \varepsilon$.
(iii) If $x = x_k$ for some $1 \leq k \leq N$, prove that $|\varphi_r(x) - \varphi_r(x - \delta) - c_k| < \varepsilon$.
(iv) Finally, prove that φ_r is right-continuous, having jump discontinuities only at $x_1, x_2, \ldots$ with the jump at x_n equal to c_n.

6. Prove that φ_ℓ is left-continuous having jump discontinuities only at $x_1, x_2, \ldots$, where the jump at x_n is equal to c_n, with the notation given in (4.25).
7. (**Generalized Thomae functions**) In this problem we generalize Thomae's function to arbitrary countable sets. Let $A \subseteq \mathbb{R}$ be a countable set.

(a) Define a nondecreasing function on $\mathbb{R}$ that is discontinuous exactly on A.
(b) Suppose that A is dense. (Dense is defined on p. 266.) Prove there does not exist a continuous function on $\mathbb{R}$ that is discontinuous exactly on A^c.

4.7 Exponentials, Logs, Euler and Mascheroni, and the ζ-Function

We now come to a very fun part of real analysis: We apply our work done in the preceding chapters and sections to study the so-called *elementary transcendental functions*, the exponential, logarithmic, and in Section 4.9, trigonometric functions. In particular, we develop the properties of undoubtedly the most important function in all of analysis, the exponential function. We also study logarithms and (complex) powers and derive some of their main properties. For another approach to defining logarithms, see the interesting article [11], and for a brief history, [194].

4.7.1 The Exponential Function

Recall that (see p. 216 in Section 3.7) the exponential function is defined by

$$\exp(z) = \sum_{n=0}^{\infty} \frac{z^n}{n!}, \qquad z \in \mathbb{C}.$$

Some properties of the exponential function are found in Theorem 3.30 on p. 219. Here's another important property, which follows easily from Tannery's theorem.

Theorem 4.29 *The exponential function* $\exp : \mathbb{C} \longrightarrow \mathbb{C}$ *is continuous.*

Proof Let $c \in \mathbb{C}$ and let $z_n \to c$; we must show that $\exp(z_n) \to \exp(c)$. To see this, write

$$\exp(z_n) = \sum_{k=0}^{\infty} a_k(n), \quad \text{where } a_k(n) = \frac{z_n^k}{k!}.$$

For each $k \in \mathbb{N}$ we have $\lim_{n\to\infty} a_k(n) = c^k/k!$. Also, since convergent sequences are bounded, there is a constant $C > 0$ such that $|z_n| \le C$ for all $n \in \mathbb{N}$. Thus,

$$|a_k(n)| \le \frac{C^k}{k!}.$$

Since $\sum_{k=0}^{\infty} C^k/k!$ converges, the extended Tannery's theorem (see the remark after the proof of Tannery's theorem (Theorem 3.29 on p. 216) and see Problem 8 on p. 225) implies that

$$\lim_{n\to\infty} \exp(z_n) = \lim_{n\to\infty} \sum_{k=0}^{\infty} a_k(n) = \sum_{k=0}^{\infty} \frac{c^k}{k!} = \exp(c).$$

This completes the proof of the theorem. ∎

We now restrict the exponential function to real variables $z = x \in \mathbb{R}$:

$$\exp(x) = \sum_{n=0}^{\infty} \frac{x^n}{n!}, \qquad x \in \mathbb{R}.$$

In particular, the right-hand side, being a sum of real numbers, is a real number, so $\exp : \mathbb{R} \longrightarrow \mathbb{R}$. Of course, this *real* exponential function shares all of the properties of the complex exponential explained in Theorem 3.30 on p. 219. In the following theorem we show that this real-valued exponential function has the increasing/decreasing properties you learned about in elementary calculus; see Fig. 4.33.

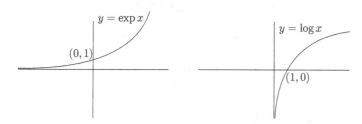

Fig. 4.33 The graph of $\exp : \mathbb{R} \longrightarrow (0, \infty)$ looks like the graph you learned about in high school! Being strictly increasing, it has an inverse function $\exp^{-1}$, which we call the logarithm, $\log : (0, \infty) \longrightarrow \mathbb{R}$

Properties of the real exponential

Theorem 4.30 *The real exponential function has the following properties:*

(1) $\exp : \mathbb{R} \longrightarrow (0, \infty)$ *and is a strictly increasing continuous bijection. Moreover,*

$$\lim_{x\to\infty} \exp(x) = \infty \quad and \quad \lim_{x\to-\infty} \exp(x) = 0.$$

(2) For every $x \in \mathbb{R}$, we have

$$1 + x \leq \exp(x),$$

with strict inequality for $x \neq 0$, that is, $1 + x < \exp(x)$ for $x \neq 0$.

Proof Observe that for $x \geq 0$,

$$\exp(x) = 1 + x + \frac{x^2}{2!} + \frac{x^3}{3!} + \cdots \geq 1 + x,$$

with strict inequalities for $x > 0$. In particular, $\exp(x) > 0$ for $x \geq 0$, and the inequality $\exp(x) \geq 1 + x$ shows that $\lim_{x \to \infty} \exp(x) = \infty$. If $x < 0$, then $-x > 0$, so $\exp(-x) > 0$, and therefore by Property *(3)*, of Theorem 3.30 on p. 219,

$$\exp(x) = \frac{1}{\exp(-x)} > 0.$$

Thus, $\exp(x)$ is positive for all $x \in \mathbb{R}$, ad recalling Example 4.15 on p. 260, we see that

$$\lim_{x \to -\infty} \exp(x) = \lim_{x \to -\infty} \frac{1}{\exp(-x)} = \lim_{x \to \infty} \frac{1}{\exp(x)} = 0.$$

As a side remark, we can also get $\exp(x) \geq 0$ for all $x \in \mathbb{R}$ by noting that

$$\exp(x) = \exp(x/2) \cdot \exp(x/2) = (\exp(x/2))^2,$$

which is a square, and so $\exp(x) \geq 0$. Back to our proof, note that if $x < y$, then $y - x > 0$, so $\exp(y - x) \geq 1 + (y - x) > 1$, and thus

$$\exp(x) = 1 \cdot \exp(x) < \exp(y - x) \cdot \exp(x)$$
$$= \exp(y - x + x) = \exp(y).$$

Thus, exp is strictly increasing on $\mathbb{R}$. The continuity property of exp implies that $\exp(\mathbb{R})$ is an interval, and then the limit properties of exp imply that the interval must be $(0, \infty)$; that is, the range of exp is $(0, \infty)$. Since exp is strictly increasing, it's injective, and therefore $\exp : \mathbb{R} \longrightarrow (0, \infty)$ is a continuous bijection.

We now verify *(2)*. We already know that $\exp(x) \geq 1 + x$ for $x \geq 0$. If $x \leq -1$, then $1 + x \leq 0$, so our inequality is automatically satisfied, since $\exp(x) > 0$. If $-1 < x < 0$, then by the series expansion for exp, we have

$$\exp(x) - (1 + x) = \left(\frac{x^2}{2!} + \frac{x^3}{3!} \right) + \left(\frac{x^4}{4!} + \frac{x^5}{5!} \right) + \cdots,$$

where we group the terms in pairs. A typical term in parentheses is of the form

$$\left(\frac{x^{2k}}{(2k)!} + \frac{x^{2k+1}}{(2k+1)!} \right) = \frac{x^{2k}}{(2k)!} \left(1 + \frac{x}{(2k+1)} \right), \quad k = 1, 2, 3, \ldots.$$

For $-1 < x < 0$, $1 + \frac{x}{(2k+1)}$ is positive, and so is x^{2k} (being the square of x^k). Hence, being a sum of positive numbers, $\exp(x) - (1 + x)$ is positive for $-1 < x < 0$. ■

The inequality $1 + x \leq \exp(x)$ is quite useful, and we will find many opportunities to use it in the sequel; see Problem 4 for a nice application to the arithmetic–geometric mean inequality (AGMI).

4.7.2 Existence and Properties of Logarithms

Since $\exp : \mathbb{R} \longrightarrow (0, \infty)$ is a strictly increasing continuous bijection (so in particular is one-to-one), by the monotone inverse theorem (Theorem 4.28 on p. 297), this function has a strictly increasing continuous bijective inverse, $\exp^{-1} : (0, \infty) \longrightarrow \mathbb{R}$. This function is called the **logarithm** function[5] and is denoted by $\log$,

$$\boxed{\log = \exp^{-1} : (0, \infty) \longrightarrow \mathbb{R}.}$$

By definition of the inverse function, $\log$ satisfies

$$\exp(\log x) = x, \quad x \in (0, \infty) \quad \text{and} \quad \log(\exp x) = x, \quad x \in \mathbb{R}. \tag{4.27}$$

In high school, the logarithm is usually introduced as follows. If $a > 0$, then the unique real number ξ having the property that

$$\exp(\xi) = a$$

is called the **logarithm** of a, where ξ is unique, because $\exp : \mathbb{R} \longrightarrow (0, \infty)$ is bijective. This high school definition is equivalent to ours, because we can see that $\xi = \log a$ by putting ξ for x in the second equation in (4.27):

$$\xi = \log(\exp(\xi)) = \log a.$$

The limit properties of $\log$ in the next theorem follow directly from the limit properties of the exponential function in Part *(1)* of Theorem 4.30.

Theorem 4.31 *The logarithm* $\log : (0, \infty) \longrightarrow \mathbb{R}$ *is a strictly increasing continuous bijection. Moreover,*

$$\lim_{x \to \infty} \log x = \infty \quad \text{and} \quad \lim_{x \to 0^+} \log x = -\infty.$$

---

[5] In elementary calculus classes, our logarithm function is denoted by ln and is called the **natural logarithm** function, the notation log usually referring to the "base 10" logarithm. However, in more advanced mathematics, log always refers to the natural logarithm function: "Mathematics is the art of giving the same name to different things." Henri Poincaré (1854–1912). [As opposed to the quotation: "Poetry is the art of giving different names to the same thing."]

The following theorem lists some of the well-known properties of log.

Properties of the logarithm

Theorem 4.32 *The logarithm has the following properties:*

(1) $\exp(\log x) = x$ *and* $\log(\exp x) = x$.
(2) $\log(xy) = \log x + \log y$.
(3) $\log 1 = 0$ *and* $\log e = 1$.
(4) $\log(x/y) = \log x - \log y$.
(5) $\log x < \log y$ *if and only if* $x < y$.
(6) $\log x > 0$ *if* $x > 1$ *and* $\log x < 0$ *if* $x < 1$.

Proof We shall leave most of these proofs to the reader. The property *(1)* is just a restatement of the equations in (4.27). Consider now the proof of *(2)*. We have

$$\exp(\log(xy)) = xy.$$

On the other hand,

$$\exp(\log x + \log y) = \exp(\log x)\exp(\log y) = xy,$$

so

$$\exp(\log(xy)) = \exp(\log x + \log y).$$

Since exp is one-to-one, we must have $\log(xy) = \log x + \log y$. To prove *(3)*, observe that

$$\exp(0) = 1 = \exp(\log 1),$$

so because exp is one-to-one, $0 = \log 1$. Also, since

$$\exp(1) = e = \exp(\log e),$$

we get $1 = \log e$. We leave the rest of the properties to the reader. ∎

4.7.3 Powers and Roots of Real Numbers

Recall that in Section 2.7 on p. 89, we defined the meaning of a^r for $a > 0$ and $r \in \mathbb{Q}$; namely, if $r = m/n$ with $m \in \mathbb{Z}$ and $n \in \mathbb{N}$, then $a^r = \left(\sqrt[n]{a}\right)^m$. We also proved that these rational powers satisfy all the "power rules" that we learned in high school (see Theorem 2.34 on p. 89). We now ask: Can we define a^x for x an arbitrary irrational number. In fact, we shall now define a^z for z an arbitrary *complex* number!

Given a positive real number a and complex number z, we define

$$a^z = \exp(z \log a).$$

The number a is called the **base** and z is called the **exponent**. The astute student might ask: What if $z = k$ is an integer? Does this definition of a^k agree with our usual definition of k products of a? What about if $z = p/q \in \mathbb{Q}$? Then is the definition of $a^{p/q}$ as $\exp((p/q) \log a)$ in agreement with our previous definition as $\sqrt[q]{a^p}$? We answer these questions and more in the following theorem.

Generalized power rules

Theorem 4.33 *The following laws of exponents hold:*

(1) For $a > 0$, $a^k = \underbrace{a \cdot a \cdots a}_{k \text{ times}}$ for every integer k.

(2) $e^z = \exp z$ for all $z \in \mathbb{C}$.

(3) $\log x^y = y \log x$ for all $x, y > 0$.

(4) For every $a > 0$, $x \in \mathbb{R}$, and $z, w \in \mathbb{C}$,

$$a^z \cdot a^w = a^{z+w}; \quad a^z \cdot b^z = (ab)^z; \quad (a^x)^z = a^{xz}.$$

(5) If $p/q \in \mathbb{Q}$, then

$$a^{p/q} = \sqrt[q]{a^p}.$$

(6) If $a > 1$, then $x \mapsto a^x$ is a strictly increasing continuous bijection of $\mathbb{R}$ onto $(0, \infty)$, and $\lim_{x \to \infty} a^x = \infty$ and $\lim_{x \to -\infty} a^x = 0$. On the other hand, if $0 < a < 1$, then $x \mapsto a^x$ is a strictly decreasing continuous bijection of $\mathbb{R}$ onto $(0, \infty)$, and $\lim_{x \to \infty} a^x = 0$ and $\lim_{x \to -\infty} a^x = \infty$.

(7) If $a, b > 0$ and $x > 0$, then $a < b$ if and only if $a^x < b^x$.

Proof By definition of a^k and the additive property of the exponential,

$$a^k = \exp(k \log a) = \exp(\underbrace{\log a + \cdots + \log a}_{k \text{ times}}) = \underbrace{\exp(\log a) \cdots \exp(\log a)}_{k \text{ times}} = \underbrace{a \cdots a}_{k \text{ times}},$$

which proves *(1)*. Property *(2)* follows from the fact that $\log e = 1$:

$$e^z := \exp(z \log e) = \exp(z).$$

To prove *(3)*, observe that

$$\exp(\log(x^y)) = x^y = \exp(y \log x).$$

Since the exponential is one-to-one, we have $\log(x^y) = y \log x$.

If $x \in \mathbb{R}$ and $z, w \in \mathbb{C}$, then the following computations prove *(4)*:

$$a^z \cdot a^w = \exp(z \log a) \exp(w \log a) = \exp(z \log a + w \log a)$$
$$= \exp\left((z + w) \log a\right) = a^{z+w},$$

$$a^z \cdot b^z = \exp(z \log a) \exp(z \log b) = \exp(z \log a + z \log b)$$
$$= \exp\left(z \log(ab)\right) = (ab)^z,$$

and

$$(a^x)^z = \exp(z \log a^x) = \exp(xz \log a) = a^{xz}.$$

To prove *(5)*, observe that by the last formula in *(4)*,

$$\left(a^{p/q}\right)^p = a^{(p/q)q} = a^p.$$

Therefore, since $a^{p/q} > 0$, by the uniqueness of roots (Theorem 2.32 on p. 93), $a^{p/q} = \sqrt[q]{a^p}$.

We leave the reader to verify that since $\exp : \mathbb{R} \longrightarrow (0, \infty)$ is a strictly increasing bijection with the limits $\lim_{x \to \infty} \exp(x) = \infty$ and $\lim_{x \to -\infty} \exp(x) = 0$, then for every $b > 0$, $\exp(bx)$ is also a strictly increasing continuous bijection of $\mathbb{R}$ onto $(0, \infty)$, and $\lim_{x \to \infty} \exp(bx) = \infty$ and $\lim_{x \to -\infty} \exp(bx) = 0$. On the other hand, if $b < 0$, say $b = -c$, where $c > 0$, then these properties are reversed: $\exp(-cx)$ is a strictly decreasing continuous bijection of $\mathbb{R}$ onto $(0, \infty)$, and $\lim_{x \to \infty} \exp(-cx) = 0$ and $\lim_{x \to -\infty} \exp(-cx) = \infty$. Keeping this discussion fresh in our memory, we can prove *(6)*. First, note that if $a > 1$, then $\log a > 0$ (Property *(6)* of Theorem 4.32), so $a^x = \exp(x \log a) = \exp(bx)$, where $b = \log a$, has the required properties in *(6)*. On the other hand, if $0 < a < 1$, then $\log a < 0$, so $a^x = \exp(x \log a) = \exp(-cx)$, where $c = -\log a > 0$, has the required properties in *(6)*.

Finally, to verify *(7)*, observe that for $a, b > 0$ and $x > 0$, using the fact that log and exp are strictly increasing, we obtain

$$a < b \iff \log a < \log b \iff x \log a < x \log b$$
$$\iff a^x = \exp(x \log a) < \exp(x \log b) = b^x.$$

∎

Example 4.33 Using Tannery's theorem, we shall prove the pretty formula

$$\frac{e}{e-1} = \lim_{n \to \infty} \left\{ \left(\frac{n}{n}\right)^n + \left(\frac{n-1}{n}\right)^n + \left(\frac{n-2}{n}\right)^n + \cdots + \left(\frac{1}{n}\right)^n \right\}.$$

To prove this, we write the right-hand side as $\lim\limits_{n\to\infty} \sum\limits_{k=0}^{n-1} a_k(n)$, where for $0 \leq k \leq n-1$,

$$a_k(n) := \left(\frac{n-k}{n}\right)^n = \left(1 - \frac{k}{n}\right)^n.$$

Observe that

$$\lim_{n\to\infty} a_k(n) = \lim_{n\to\infty} \left(1 - \frac{k}{n}\right)^n = e^{-k},$$

and for $0 \leq k \leq n-1$,

$$|a_k(n)| = \left(1 - \frac{k}{n}\right)^n \leq \left(e^{-k/n}\right)^n = e^{-k},$$

where we used that $1 + x \leq e^x$ for all $x \in \mathbb{R}$ from Theorem 4.30. Thus, $|a_k(n)| \leq M_k$, where $M_k = e^{-k}$. Since $e^{-1} < 1$, by the geometric series test, $\sum_{k=0}^{\infty} M_k = \sum_{k=0}^{\infty} (e^{-1})^k < \infty$. Hence by Tannery's theorem, we have

$$\lim_{n\to\infty} \left\{ \left(\frac{n}{n}\right)^n + \left(\frac{n-1}{n}\right)^n + \cdots + \left(\frac{1}{n}\right)^n \right\}$$
$$= \lim_{n\to\infty} \sum_{k=0}^{n-1} a_k(n) = \sum_{k=0}^{\infty} \lim_{n\to\infty} a_k(n) = \sum_{k=0}^{\infty} e^{-k} = \frac{1}{1 - 1/e} = \frac{e}{e-1}.$$

Example 4.34 Here's a **Puzzle**: Do there exist *rational* numbers α and β such that α^β is *irrational*? You should be able to answer this in the affirmative! Here's a harder question [117]: Do there exist *irrational* numbers α and β such that α^β is *rational*? Here's a very cool argument to the affirmative. Consider $\alpha = \sqrt{2}$ and $\beta = \sqrt{2}$, both of which are irrational. Then there are two cases: either α^β rational or irrational. If α^β is rational, then we have answered our question in the affirmative. However, in the case that $\alpha' := \alpha^\beta$ is irrational, then by our rule *(4)* of exponents,

$$(\alpha')^\beta = \left(\alpha^\beta\right)^\beta = \alpha^{\beta^2} = \sqrt{2}^2 = 2$$

is rational, so our answer is affirmative in this case as well.[6] Do there exist *irrational* numbers α and β such that α^β is *irrational*? For the answer, see Problem 6.

---

[6]If you're wondering, $\sqrt{2}^{\sqrt{2}}$ is in fact irrational, by the Gelfond–Schneider theorem.

4.7.4   The Riemann Zeta Function

The last two subsections are applications of what we've learned about exponentials, logs, and powers. We begin with the Riemann zeta function, which is involved in one of the most renowned unsolved problems in all of mathematics: *The Riemann hypothesis*, which we'll explain in a moment. Here, the Riemann zeta function is simply a "generalized *p*-series," where instead of using *p*, a rational number, we use a complex number:

$$\zeta(z) := \sum_{n=1}^{\infty} \frac{1}{n^z} = 1 + \frac{1}{2^z} + \frac{1}{3^z} + \frac{1}{4^z} + \cdots .$$

The Riemann zeta function

Theorem 4.34 *The Riemann zeta function converges absolutely for all $z \in \mathbb{C}$ with* $\operatorname{Re} z > 1$.

Proof Let *p* be an arbitrary rational number with $p > 1$; then we just have to prove that $\zeta(z)$ converges absolutely for all $z \in \mathbb{C}$ with $\operatorname{Re} z \geq p$. To see this, let $z = x + iy$ with $x \geq p$ and observe that $n^z = e^{z \log n} = e^{x \log n} \cdot e^{iy \log n}$. In Problem 4 you'll prove that $|e^{i\theta}| = 1$ for every real θ, so $|e^{iy \log n}| = 1$, and hence

$$|n^z| = |e^{x \log n} \cdot e^{iy \log n}| = e^{x \log n} \geq e^{p \log n} = n^p.$$

Therefore, $|1/n^z| \leq 1/n^p$, so by comparison with the *p*-series $\sum 1/n^p$, it follows that $\sum |1/n^z|$ converges. This completes our proof. ∎

We now state the Riemann hypothesis. It turns out using techniques from *complex analysis* that it's possible to define $\zeta(z)$ not just for $\operatorname{Re} z > 1$ as we have shown, but also for all complex numbers $z \in \mathbb{C}$ except for $z = 1$. Consider the so-called *critical strip*, which is the set of $z \in \mathbb{C}$ such that $0 < \operatorname{Re} z < 1$, and consider the *critical line*, the set of $z \in \mathbb{C}$ such that $\operatorname{Re} z = 1/2$, as shown here (Fig. 4.34):

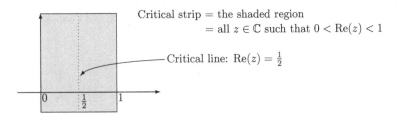

Critical strip = the shaded region
= all $z \in \mathbb{C}$ such that $0 < \operatorname{Re}(z) < 1$

Critical line: $\operatorname{Re}(z) = \frac{1}{2}$

Fig. 4.34 The shaded region continues *vertically up* and *down*

Many mathematicians, starting with Riemann, have found zeros of the ζ-function (that is, points z such that $\zeta(z) = 0$) on the critical line. (In fact, the ζ-function has infinitely many zeros on the critical line.) On the other hand, no one has found a single zero of the ζ-function in the critical strip not on the critical line. The following conjecture seems natural to make:

Riemann hypothesis: If z is in the critical strip and $\zeta(z) = 0$, then z is on the critical line.

Although simple to state, this conjecture is so difficult to solve (either in favor or against) that there is a $1 million reward for its solution! For more on the Riemann hypothesis, see http://www.claymath.org/millennium/, and see [53] for ideas on how to solve this conjecture. By the way, the Riemann hypothesis has profound implications to prime numbers; to understand why, see Section 7.6, where we derive a formula for the ζ-function in terms of primes.

4.7.5 The Euler–Mascheroni Constant

The constant

$$\gamma := \lim_{n\to\infty}\left(1 + \frac{1}{2} + \cdots + \frac{1}{n} - \log n\right)$$

is called the **Euler–Mascheroni constant**. This constant was calculated to 16 digits in 1781 by Euler, who used the notation C for γ. The symbol γ was first used by Lorenzo Mascheroni (1750–1800) in 1790, when he computed γ to 32 decimal places, although only the first 19 places were correct (cf. [104, pp. 90–91]). To prove that the limit on the right of γ exists, consider the sequence

$$\gamma_n = 1 + \frac{1}{2} + \cdots + \frac{1}{n} - \log n, \qquad n = 2, 3, 4, \ldots.$$

We shall prove that γ_n is nonincreasing and bounded below, and hence the Euler–Mascheroni constant is defined. In our proof, we shall see that γ is between 0 and 1; its value up to ten digits in base 10 is $\gamma = 0.5772156649\ldots$. Here's a mnemonic to remember the digits of γ [254]:

These numbers proceed to a limit Euler's subtle mind discerned. (4.28)

The number of letters in each word represents a digit of γ; e.g., "These" represents 5, "numbers" represents 7, etc. The sentence (4.28) gives ten digits of γ: 0.5772156649. By the way, it is not known[7] whether γ is rational or irrational, let alone transcendental!

---

[7]"Unfortunately, what is little recognized is that the most worthwhile scientific books are those in which the author clearly indicates what he does not know; for an author most hurts his readers by concealing difficulties." Evariste Galois (1811–1832) [201].

To prove that $\{\gamma_n\}$ is a bounded monotone sequence, we shall need the following inequality proved in Section 3.3 (see Eq. (3.26) on p. 181):

$$\left(\frac{n+1}{n}\right)^n < e < \left(\frac{n+1}{n}\right)^{n+1} \qquad \text{for all } n \in \mathbb{N}.$$

A little algebra reveals another way to write these inequalities as

$$e^{1/(n+1)} < \frac{n+1}{n} < e^{1/n}.$$

Taking logarithms, we obtain

$$\frac{1}{n+1} < \log(n+1) - \log n < \frac{1}{n}. \tag{4.29}$$

Using the definition of γ_n and the first inequality in (4.29), we see that

$$\gamma_n = 1 + \frac{1}{2} + \cdots + \frac{1}{n} - \log n = \gamma_{n+1} - \frac{1}{n+1} + \log(n+1) - \log n > \gamma_{n+1},$$

so the sequence $\{\gamma_n\}$ is strictly decreasing. In particular, $\gamma_n < \gamma_1 = 1$ for all n. We now show that γ_n is bounded below by zero. We already know that $\gamma_1 = 1 > 0$. Using the second inequality in (4.29) with $n = 2, n = 3, \ldots, n = n$, we obtain

$$\gamma_n = 1 + \frac{1}{2} + \frac{1}{3} + \cdots + \frac{1}{n} - \log n > 1 + \left(\log 3 - \log 2\right) + \left(\log 4 - \log 3\right)$$

$$+ \left(\log 5 - \log 4\right) + \cdots + \left(\log n - \log(n-1)\right) + \left(\log(n+1) - \log n\right) - \log n$$

$$= 1 - \log 2 + \log(n+1) - \log n > 1 - \log 2 > 0.$$

Here, we used that $\log 2 < 1$ because $2 < e$. Thus, $\{\gamma_n\}$ is strictly decreasing and bounded below by $1 - \log 2 > 0$, so γ is well defined and $0 < \gamma < 1$.

We can now show that the value of the alternating harmonic series

$$\sum_{n=1}^{\infty} \frac{(-1)^{n-1}}{n} = 1 - \frac{1}{2} + \frac{1}{3} - \frac{1}{4} + \frac{1}{5} - \frac{1}{6} + - \cdots$$

is $\log 2$. Indeed, by the definition of γ, we can write

$$\gamma = \lim_{n \to \infty} \left(1 + \frac{1}{2} + \frac{1}{3} + \frac{1}{4} + \cdots + \frac{1}{2n} - \log 2n\right)$$

and also

$$\gamma = \lim_{n \to \infty} 2\left(\frac{1}{2} + \frac{1}{4} + \cdots + \frac{1}{2n}\right) - \log n.$$

Subtracting, we obtain

$$0 = \lim_{n \to \infty} \left(1 - \frac{1}{2} + \frac{1}{3} - \frac{1}{4} + - \cdots - \frac{1}{2n} \right) - \log 2,$$

which proves that

$$\log 2 = \sum_{n=1}^{\infty} (-1)^{n-1} \frac{1}{n} = 1 - \frac{1}{2} + \frac{1}{3} - \frac{1}{4} + \frac{1}{5} - \frac{1}{6} + - \cdots .$$

Using a similar technique, one can find series representations for log 3; see Problem 7. Using the above formula for log 2, in Problem 7 you are asked to derive the following striking expression:

$$2 = \frac{e^1}{e^{1/2}} \cdot \frac{e^{1/3}}{e^{1/4}} \cdot \frac{e^{1/5}}{e^{1/6}} \cdot \frac{e^{1/7}}{e^{1/8}} \cdot \frac{e^{1/9}}{e^{1/10}} \cdots . \tag{4.30}$$

▶ **Exercises 4.7**

1. Establish the following properties of exponential functions.

   (a) If $z_n \to z$ and $a_n \to a$ (with z_n, z complex and $a_n, a > 0$), then $a_n^{z_n} \to a^z$.
   (b) If $a, b > 0$, then for every $x < 0$, we have $a < b$ if and only if $a^x > b^x$.
   (c) If $a, b > 0$, then for every complex number z, we have $a^{-z} = 1/a^z$ and $(a/b)^z = a^z/b^z$.
   (d) For every $z \in \mathbb{C}$, we have $\overline{\exp(z)} = \exp(\bar{z})$; that is, the complex conjugate of $\exp(z)$ is $\exp$ of the complex conjugate of z.
   (e) Prove that for every $\theta \in \mathbb{R}$, $|e^{i\theta}| = 1$.

2. Let $a \in \mathbb{R}$ with $a \neq 0$ and define $f(x) = x^a$.

   (a) If $a > 0$, prove that $f : [0, \infty) \longrightarrow \mathbb{R}$ is continuous and strictly increasing, $\lim_{x \to 0+} f = 0$, and $\lim_{x \to \infty} f = \infty$.
   (b) If $a < 0$, prove that $f : (0, \infty) \longrightarrow \mathbb{R}$ is continuous and strictly decreasing, $\lim_{x \to 0+} f = \infty$, and $\lim_{x \to \infty} f = 0$.

3. Establish the following limit properties of the exponential function.

   (a) Show that for every $n \in \mathbb{N}$ and $x \in \mathbb{R}$ with $x > 0$ we have $e^x > \dfrac{x^{n+1}}{(n+1)!}$. Use this inequality to prove that for every natural number n, we have $\lim_{x \to \infty} \dfrac{x^n}{e^x} = 0$.
   (b) Using (a), prove that for every $a \in \mathbb{R}$ with $a > 0$, we have $\lim_{x \to \infty} \dfrac{x^a}{e^x} = 0$. It follows that e^x grows faster than every power (no matter how large) of x. This limit is usually derived in elementary calculus using L'Hospital's rule.

4. Consider the real numbers $a_1, \ldots, a_n \geq 0$. Recall from Problem 7 on p. 46 that the **arithmetic–geometric mean inequality** (AGMI) is the inequality

$$(a_1 \cdot a_2 \cdots a_n)^{1/n} \leq \frac{a_1 + \cdots + a_n}{n}.$$

Prove the AGMI by considering the inequality $1 + x \leq e^x$ for $x = x_k$, $k = 1, \ldots, n$, where $x_k = -1 + a_k/a$ (so that $a_k/a = 1 + x_k$) with $a = (a_1 + \cdots + a_n)/n$.

5. Derive the following remarkable formula for all $x > 0$:

$$\log x = \lim_{n \to \infty} n\left(\sqrt[n]{x} - 1\right) \quad \textbf{(Halley's formula)},$$

named after the famous Edmond Halley (1656–1742) of Halley's comet. Suggestion: Write $\sqrt[n]{x} = e^{\log x/n}$ and write $e^{\log x/n}$ as a series in $\log x/n$.

6. (Cf. [117]) **Puzzle**: Do there exist *irrational* numbers α and β such that α^β is *irrational*? Suggestion: Consider α^β and $\alpha^{\beta'}$, where $\alpha = \beta = \sqrt{2}$ and $\beta' = \sqrt{2} + 1$.

7. In this fun problem, we derive some interesting formulas.

   (a) Prove that each of the following sums equals γ, the Euler–Mascheroni constant:

   $$\sum_{n=1}^{\infty} \left[\frac{1}{n} - \log\left(1 + \frac{1}{n}\right)\right], \; 1 + \sum_{n=2}^{\infty} \left[\frac{1}{n} + \log\left(1 - \frac{1}{n}\right)\right],$$

   $$1 + \sum_{n=1}^{\infty} \left[\frac{1}{n+1} + \log\left(1 + \frac{1}{n}\right)\right].$$

   Suggestion: Think telescoping series.

   (b) Using a technique similar to how we derived our formula for $\log 2$, prove that

   $$\log 3 = 1 + \frac{1}{2} - \frac{2}{3} + \frac{1}{4} + \frac{1}{5} - \frac{2}{6} + \frac{1}{7} + \frac{1}{8} - \frac{2}{9} + + - \cdots.$$

   Can you find a series representation for $\log 4$?

   (c) Define $a_n = (e^1/e^{1/2}) \cdots (e^{1/(2n-1)}/e^{1/(2n)})$. Prove that $2 = \lim a_n$.

8. Show that the sequence $\{a_n\}$, with $a_n = \sum_{k=1}^{n} \frac{1}{n+k} = \frac{1}{n+1} + \frac{1}{n+2} + \cdots + \frac{1}{n+n}$, converges to $\log 2$. Suggestion: Can you relate a_n to the $2n$th partial sum of the alternating harmonic series?

9. (Cf. [41, 93]) In this problem we establish a "well-known" limit from calculus, but without using calculus!

   (i) Show that $\log x < x$ for all $x > 0$.
   (ii) Show that $(\log x)/x < 2/x^{1/2}$ for $x > 0$. Suggestion: $\log x = 2 \log x^{1/2}$.

(iii) Show that $\lim_{x \to \infty} \dfrac{\log x}{x} = 0$. This limit is usually derived in elementary calculus using L'Hospital's rule.

(iv) Now let $a \in \mathbb{R}$ with $a > 0$. Prove that $\lim_{x \to \infty} \dfrac{\log x}{x^a} = 0$. Thus, $\log x$ grows slower than every power (no matter how small) of x.

10. In this problem we get an inequality for $\log(1 + x)$ and use it to obtain a nice formula.

   (i) Prove that for all $x \in [0, 1]$, we have $e^{\frac{1}{2}x} \leq 1 + x$. Conclude that for all $x \in [0, 1]$, we have $\log(1 + x) \geq x/2$.

   (ii) Using Tannery's theorem, prove that

$$\zeta(2) = \lim_{n \to \infty} \left\{ \frac{1}{n^2 \log\left(1 + \frac{1^2}{n^2}\right)} + \frac{1}{n^2 \log\left(1 + \frac{2^2}{n^2}\right)} + \cdots + \frac{1}{n^2 \log\left(1 + \frac{n^2}{n^2}\right)} \right\}.$$

11. In high school you probably learned logarithms with other "bases" besides e. Let $a \in \mathbb{R}$ with $a > 0$ and $a \neq 1$. For $x > 0$, we define

$$\log_a x := \frac{\log x}{\log a},$$

called the **logarithm of** x **to the base** a. Note that if $a = e$, then $\log_e = \log$, our usual logarithm. Here are some of the well-known properties of $\log_a$.

   (a) Prove that $x \mapsto \log_a x$ is the inverse function of $x \mapsto a^x$.
   (b) Prove that for every $x, y > 0$, $\log_a xy = \log_a x + \log_a y$.
   (c) Prove that if $b > 0$ with $b \neq 1$ is another base, then for all $x > 0$,

$$\log_a x = \left(\frac{\log b}{\log a} \right) \log_b x \qquad \textbf{(Change of base formula)}.$$

12. Part (a) of this problem states that a "function that looks like an exponential function is an exponential function," while (b) says the same for the logarithm function.

   (a) Let $f : \mathbb{R} \longrightarrow \mathbb{R}$ satisfy $f(x+y) = f(x) f(y)$ for all $x, y \in \mathbb{R}$ (cf. Problem 4 on p. 268). Assume that f is not the zero function. If f is continuous, prove that

$$f(x) = a^x \quad \text{for all } x \in \mathbb{R}, \text{ where } a = f(1).$$

   Suggestion: Show that $f(x) > 0$ for all x. Now there are several ways to proceed. One way is to first prove that $f(r) = (f(1))^r$ for all rational r (to prove this, you do not require the continuity assumption). The second way is to define $h(x) = \log f(x)$. Prove that h is linear, then apply Problem 3 on p. 268.

(b) Let $g : (0, \infty) \longrightarrow \mathbb{R}$ satisfy $g(x \cdot y) = g(x) + g(y)$ for all $x, y > 0$. Prove that if g is continuous, then there exists a unique real number c such that

$$g(x) = c \, \log x \quad \text{for all } x \in (0, \infty).$$

13. (**Exponentials the "old-fashioned way"**) Fix $a > 0$ and $x \in \mathbb{R}$. In this section we defined $a^x := \exp(x \log a)$. However, in this problem we shall define a^x the "old-fashioned way" via rational sequences. **In this problem we assume knowledge of rational powers as defined in Section 2.7 on p. 89, and we proceed to define them for real powers.**

(i) From Example 3.6 on p. 157, we know that $a^{1/m} \to 1$ and $a^{-1/m} = (a^{-1})^{1/m} \to 1$ as $m \to \infty$. Thus, given $\varepsilon > 0$ we can choose $m \in \mathbb{N}$ such that $1 - \varepsilon < a^{\pm 1/m} < 1 + \varepsilon$. If r is a rational number with $|r| < 1/m$, prove that $|a^r - 1| < \varepsilon$.

(ii) Let $\{r_n\}$ be a sequence of rational numbers converging to x. Prove that $\{a^{r_n}\}$ is a Cauchy sequence; hence it converges to a real number, say ξ. We define

$$a^x := \xi.$$

Prove that this definition makes sense, that is, if $\{r_n'\}$ is any other sequence of rational numbers converging to x, then $\{a^{r_n'}\}$ also converges to ξ.
(Of course, this alternative definition of powers agrees with the definition of a^x found in this section!)

14. (**Logarithms the "old-fashioned way"**) In this problem we show how to define the logarithm using rational sequences and not as the inverse of exp. Fix $a > 0$.

(i) Prove that it is possible to define integers $a_0, a_1, a_2, \ldots$ inductively with $0 \le a_k \le 9$ for $k \ge 1$ such that if x_n and y_n are the rational numbers

$$x_n = a_0 + \frac{a_1}{10} + \cdots + \frac{a_{n-1}}{10^{n-1}} + \frac{a_n}{10^n} \quad \text{and} \quad y_n = a_0 + \frac{a_1}{10} + \cdots$$
$$+ \frac{a_{n-1}}{10^{n-1}} + \frac{a_n + 1}{10^n},$$

then $e^{x_n} \le a < e^{y_n}$.
(ii) Prove that both sequences $\{x_n\}$ and $\{y_n\}$ converge to the same value, call it L. Show that $e^L = a$, where e^L is defined by means of the previous problem. Of course, L is just the logarithm of a as defined in this section.

15. (**The Euler–Mascheroni constant II**) In this problem we prove that the limit $\gamma := \lim \gamma_n$ exists, where $\gamma_n = 1 + \frac{1}{2} + \cdots + \frac{1}{n} - \log n$, following [43].

(i) For $n \ge 2$, define $a_n = \gamma_n - \frac{1}{n}$. Prove that

$$a_n = \log\left(\frac{e^1}{2/1} \cdot \frac{e^{1/2}}{3/2} \cdots \frac{e^{1/(n-1)}}{n/(n-1)}\right).$$

(ii) Using the inequalities in (3.26), prove

$$1 < \frac{e^{1/n}}{(n+1)/n} \quad \text{and} \quad \frac{e^{1/n}}{(n+1)/n} < e^{\frac{1}{n(n+1)}}. \qquad (4.31)$$

(iii) Using the inequalities in (4.31), prove that the sequence $\{a_n\}$, where $n \geq 2$, is strictly increasing such that $0 < a_n < 1$ for all $n \geq 2$.

(iv) Prove that $\lim \gamma_n$ exists; hence the Euler–Mascheroni constant exists.

16. **(The Euler–Mascheroni constant III)** Following [15], we prove that $\gamma :=$ $\lim \gamma_n$ exists, where $\gamma_n = 1 + \frac{1}{2} + \cdots + \frac{1}{n} - \log n$. For each $k \in \mathbb{N}$, define $a_k := e\left(1 + \frac{1}{k}\right)^{-k}$, so that $e = a_k\left(1 + \frac{1}{k}\right)^k$.

(i) For each $k \in \mathbb{N}$, prove that $\frac{1}{k} = \log(a_k^{1/k}) + \log(k+1) - \log k$.

(ii) Prove that

$$1 + \frac{1}{2} + \cdots + \frac{1}{n} - \log(n+1) = \log\left(a_1 \, a_2^{1/2} \, a_3^{1/3} \cdots a_n^{1/n}\right).$$

Remark: In (iii) and (iv), the inequalities in (3.26) will be of use.

(iii) Prove that the sequence $\left\{\log\left(a_1 \, a_2^{1/2} \cdots a_n^{1/n}\right)\right\}$ is nondecreasing.

(iv) Prove that

$$\log\left(a_1 \, a_2^{1/2} \cdots a_n^{1/n}\right) < \log\left(1 + \frac{1}{1}\right) + \frac{1}{2}\log\left(1 + \frac{1}{2}\right) + \cdots + \frac{1}{n}\log\left(1 + \frac{1}{n}\right)$$
$$< \frac{1}{1} + \frac{1}{2} \cdot \frac{1}{2} + \cdots + \frac{1}{n} \cdot \frac{1}{n}.$$

Suggestion: Begin by showing that $a_k < 1 + 1/k$ for each k.

(v) Since the series of the reciprocals of the squares of the natural numbers converges, the sequence $\left\{\log\left(a_1 \, a_2^{1/2} \cdots a_n^{1/n}\right)\right\}$ is bounded and hence converges. Prove that $\{\gamma_n\}$ converges, and hence the Euler–Mascheroni constant exists.

4.8 ★ Proofs that $\sum 1/p$ Diverges

We know that the harmonic series $\sum 1/n$ diverges. In other words, partial sums of reciprocals of the natural numbers increase beyond any positive number. However, if we pick only the squares of natural numbers and sum their reciprocals, then we get the convergent sum $\sum 1/n^2$; here's what we've discussed so far (Fig. 4.35):

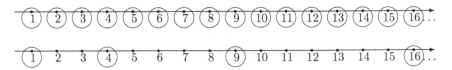

Fig. 4.35 *Top* Considering all natural numbers, the sum of their reciprocals diverges. *Bottom* Considering only the perfect *squares*, the sum of their reciprocals converges

Of course, if we pick perfect cubes, or other higher powers, and sum their reciprocals, we also get convergent series. One may ask: What if we pick the primes

$$\overrightarrow{1 \;\textcircled{2}\; \textcircled{3}\; 4\; \textcircled{5}\; 6\; \textcircled{7}\; 8\; 9\; 10\; \textcircled{11}\; 12\; \textcircled{13}\; 14\; 15\; 16 \dots}$$

and sum their reciprocals:

$$\sum \frac{1}{p} = \frac{1}{2} + \frac{1}{3} + \frac{1}{5} + \frac{1}{7} + \frac{1}{11} + \frac{1}{13} + \frac{1}{17} + \cdots \;;$$

do we get a convergent sum? Since there are arbitrarily large gaps between primes (see Problem 1 on p. 64), one may conjecture that $\sum 1/p$ converges. However, we shall prove that $\sum 1/p$ diverges! The proofs we present are found in [22, 65, 140, 176] (cf. [177]). Other proofs can be found in the exercises. An expository article giving other proofs related to this fascinating divergent sum can be found in [249]. See also [51, 165].

4.8.1 Proof I: Proof by Multiplication and Rearrangement [22, 65]

Suppose, for the sake of contradiction, that $\sum 1/p$ converges. Then we can fix a prime number m such that $\sum_{p>m} 1/p \leq 1/2$. Let $2 < 3 < \cdots < m$ be the list of all prime numbers up to m. For $N > m$, let P_N be the set of natural numbers between 2 and N all of whose prime factors are less than or equal to m, and let Q_N be the set of natural numbers between 2 and N all of whose prime factors are greater than m. Explicitly,

$$
\begin{aligned}
k \in P_N &\iff 2 \leq k \leq N \text{ and } k = 2^i 3^j \cdots m^k, \quad \text{some } i, j, \ldots, k \geq 0, \\
\ell \in Q_N &\iff 2 \leq \ell \leq N \text{ and } \ell = p\, q \cdots r, \quad p, q, \ldots, r > m \text{ are prime.}
\end{aligned}
\tag{4.32}
$$

In the product $p\, q \cdots r$ appearing in the second line, prime numbers may be repeated. Observe that every integer n with $2 \leq n \leq N$ that is not in P_N or Q_N must have prime factors that are both less than or equal to m and greater than m, and hence can be factored in the form $n = k\,\ell$, where $k \in P_N$ and $\ell \in Q_N$. Thus, the finite sum

$$\sum_{k\in P_N}\frac{1}{k} + \sum_{\ell\in Q_N}\frac{1}{\ell} + \Big(\sum_{k\in P_N}\frac{1}{k}\Big)\Big(\sum_{\ell\in Q_N}\frac{1}{\ell}\Big) = \sum_{k\in P_N}\frac{1}{k} + \sum_{\ell\in Q_N}\frac{1}{\ell} + \sum_{k\in P_N,\ell\in Q_N}\frac{1}{k\,\ell},$$

contains every number of the form $1/n$, where $2 \le n \le N$. Of course, the resulting sum contains many other numbers too. In particular,

$$\sum_{k\in P_N}\frac{1}{k} + \sum_{\ell\in Q_N}\frac{1}{\ell} + \Big(\sum_{k\in P_N}\frac{1}{k}\Big)\Big(\sum_{\ell\in Q_N}\frac{1}{\ell}\Big) \ge \sum_{n=2}^{N}\frac{1}{n}.$$

We shall prove that the finite sums on the left remain bounded as $N \to \infty$, which contradicts the fact that the harmonic series diverges. This contradiction shows that our original assumption, that $\sum 1/p$ converges, was nonsense.

To see that $\sum_{P_N} 1/k$ converges, note that each geometric series $\sum_{j=1}^{\infty} 1/p^j = \sum_{j=1}^{\infty}(1/p)^j$ converges (since $1/p < 1$ for all primes p). Moreover, for every prime p, we have

$$\sum_{j=1}^{\infty}\frac{1}{p^j} = \frac{1/p}{1-1/p} = \frac{1}{p-1} \le 1.$$

It follows that for every $N \in \mathbb{N}$, we have

$$\Big(\sum_{i=1}^{N}\frac{1}{2^i}\Big)\Big(\sum_{j=1}^{N}\frac{1}{3^j}\Big)\cdots\Big(\sum_{k=1}^{N}\frac{1}{m^k}\Big) \le 1.$$

Pulling out the summations (that is, using the distributive law many times), we obtain

$$\sum_{i=1}^{N}\sum_{j=1}^{N}\cdots\sum_{k=1}^{N}\frac{1}{2^i \cdot 3^j \cdots m^k} \le 1.$$

The left-hand side of this inequality contains all fractions $1/k$, where $k \in P_N$ (see the definition of P_N in (4.32)). It follows that $\sum_{P_N} 1/k$ is bounded above by 1. Thus, $\lim_{N\to\infty}\sum_{P_N} 1/k$ is finite.

We now prove that $\lim_{N\to\infty}\sum_{Q_N} 1/\ell$ is finite. To do so, recall that we have fixed a prime m such that $\sum_{p>m} 1/p \le 1/2$. For $N > m$, let

$$\alpha_N = \sum_{m<p<N}\frac{1}{p}.$$

Then $\alpha_N \le 1/2$. Also, observe that

$$(\alpha_N)^2 = \Big(\sum_{m<p<N}\frac{1}{p}\Big)\Big(\sum_{m<q<N}\frac{1}{q}\Big) = \sum_{m<p,q<N}\frac{1}{p\,q},$$

where the sum is over all primes p, q with $m < p, q < N$, and

$$(\alpha_N)^3 = \sum_{m<p,q,r<N} \frac{1}{p\,q\,r},$$

where the sum is over all primes p, q, r with $m < p, q, r < N$. We can continue this procedure, showing that $(\alpha_N)^j$ is the sum $\sum 1/(p\,q\cdots r)$, where the sum is over all j-tuples of primes $p, q, \ldots, r$ such that $m < p, q, \ldots, r < N$. By the definition of Q_N, in (4.32), it follows that

$$\sum_{\ell \in Q_N} \frac{1}{\ell} \leq \sum_{j=1}^{\infty} (\alpha_N)^j,$$

which is bounded by $\sum_{j=1}^{\infty} (1/2)^j = 1$. Hence, the limit $\lim_{N\to\infty} \sum_{Q_N} 1/\ell$ is finite, and we have reached a contradiction.

4.8.2   An Elementary Number Theory Fact

Our next proof depends on the idea of square-free integers. A positive integer is said to be **square-free** if no squared prime divides it; that is, if a prime occurs in its prime factorization, then it occurs with multiplicity one. For instance, 1 is square-free, because no squared prime divides it, $10 = 2 \cdot 5$ is square-free, but $24 = 2^3 \cdot 3 = 2^2 \cdot 2 \cdot 3$ is not square-free.

We claim that every positive integer can be written uniquely as the product of a square and a square-free integer. Indeed, let $n \in \mathbb{N}$ and let k be the largest natural number such that k^2 divides n. Then n/k^2 must be square-free, for if n/k^2 is divided by a squared prime p^2, then $pk > k$ divides n, which is not possible by definition of k. Thus, every positive integer n can be uniquely written as $n = k^2$ if n is a perfect square, or

$$n = k^2 \cdot p\,q\cdots r, \tag{4.33}$$

where $k \geq 1$ and where $p, q, \ldots, r$ are distinct primes less than or equal to n. Using the fact that every positive integer can be uniquely written as the product of a square and a square-free integer, we shall prove that $\sum 1/p$ diverges.

4.8.3   Proof II: Proof by Comparison [176, 177]

The trick is to understand, for every natural number $N \geq 3$, the product

$$\prod_{p<N} \left(1 + \frac{1}{p}\right),$$

where the product is over all primes less than N. In other words, if $2 < 3 < \cdots < m$ are all the primes less than N, then

$$\prod_{p<N}\left(1+\frac{1}{p}\right)=\left(1+\frac{1}{2}\right)\left(1+\frac{1}{3}\right)\cdots\left(1+\frac{1}{m}\right).$$

For example, if $N = 5$, then

$$\prod_{p<5}\left(1+\frac{1}{p}\right)=\left(1+\frac{1}{2}\right)\left(1+\frac{1}{3}\right)=1+\frac{1}{2}+\frac{1}{3}+\frac{1}{2\cdot3}.$$

If $N = 6$, then

$$\prod_{p<6}\left(1+\frac{1}{p}\right)=\left(1+\frac{1}{2}\right)\left(1+\frac{1}{3}\right)\left(1+\frac{1}{5}\right)$$

$$=1+\frac{1}{2}+\frac{1}{3}+\frac{1}{5}+\frac{1}{2\cdot3}+\frac{1}{2\cdot5}+\frac{1}{3\cdot5}+\frac{1}{2\cdot3\cdot5}.$$

By induction we can write

$$\prod_{p<N}\left(1+\frac{1}{p}\right)=1+\sum_{p<N}\frac{1}{p}+\sum_{p,q<N}\frac{1}{p\cdot q}+\cdots+\sum_{p,q,\ldots,r<N}\frac{1}{p\cdot q\cdots r}, \quad (4.34)$$

where the kth sum on the right is the sum over over all reciprocals of the form $\frac{1}{p_1\cdot p_2\cdots p_k}$ with $p_1,\ldots,p_k$ distinct primes less than N. To use our fact (4.33), let $N\geq3$ and multiply both sides of (4.34) by $\sum_{k<N}1/k^2$, obtaining

$$\prod_{p<N}\left(1+\frac{1}{p}\right)\cdot\sum_{k<N}\frac{1}{k^2}=\sum_{k<N}\frac{1}{k^2}+\sum_{k<N}\sum_{p<N}\frac{1}{k^2p}$$

$$+\sum_{k<N}\sum_{p,q<N}\frac{1}{k^2\cdot p\cdot q}+\cdots+\sum_{k<N}\sum_{p,q,\ldots,r<N}\frac{1}{k^2\cdot p\cdot q\cdots r}.$$

By our discussion on square-free numbers around (4.33), the right-hand side contains every number of the form $1/n$, where $n < N$ (as well as many other numbers). In particular,

$$\prod_{p<N}\left(1+\frac{1}{p}\right)\cdot\sum_{k<N}\frac{1}{k^2}\geq\sum_{n<N}\frac{1}{n}. \quad (4.35)$$

From this inequality, we shall prove that $\sum 1/p$ diverges. Indeed, we know that $\sum_{k=1}^{\infty}1/k^2$ converges, while $\sum_{n=1}^{\infty}1/n$ diverges, so it follows that

$$\lim_{N\to\infty}\prod_{p<N}\left(1+\frac{1}{p}\right)=\infty.$$

To relate this product to the sum $\sum 1/p$, note that

$$e^x = 1 + x + \frac{x^2}{2!} + \frac{x^3}{3!} + \cdots \geq 1 + x$$

for $x \geq 0$. In fact, this inequality holds for all $x \in \mathbb{R}$ by Theorem 4.30 on p. 301. Hence,

$$\prod_{p<N} \left(1 + \frac{1}{p}\right) \leq \prod_{p<N} \exp(1/p) = \exp\left(\sum_{p<N} \frac{1}{p}\right).$$

Since the left-hand side increases without bound as $N \to \infty$, so must the sum $\sum_{p<N} 1/p$. This ends **Proof II**; see Problem 2 for a related proof.

4.8.4   Proof III: Another Proof by Comparison [140]

The trick now is to understand, for every natural number $N \geq 3$, the product

$$\prod_{p<N} \left(1 - \frac{1}{p}\right)^{-1},$$

where the product is over all primes less than N. By the sum formula for a geometric series, we have

$$\left(1 - \frac{1}{p}\right)^{-1} = \sum_{n=0}^{\infty} \frac{1}{p^n}.$$

Now let $N \geq 3$ and let $2 < 3 < \cdots < m$ be all the primes less than N. It follows that

$$\prod_{p<N} \left(1 - \frac{1}{p}\right)^{-1} = \left(1 - \frac{1}{2}\right)^{-1} \left(1 - \frac{1}{3}\right)^{-1} \cdots \left(1 - \frac{1}{m}\right)^{-1}$$

$$= \left(\sum_{i=0}^{\infty} \frac{1}{2^i}\right)\left(\sum_{j=0}^{\infty} \frac{1}{3^j}\right) \cdots \left(\sum_{k=0}^{\infty} \frac{1}{m^k}\right)$$

$$\geq \left(\sum_{i=0}^{N} \frac{1}{2^i}\right)\left(\sum_{j=0}^{N} \frac{1}{3^j}\right) \cdots \left(\sum_{k=0}^{N} \frac{1}{m^k}\right)$$

$$= \sum_{i=1}^{N} \sum_{j=0}^{N} \cdots \sum_{k=0}^{N} \frac{1}{2^i \cdot 3^j \cdots m^k}. \tag{4.36}$$

Since every natural number $n < N$ can be written in the form

$$n = 2^i \, 3^j \cdots m^k$$

for some nonnegative integers $i, j, \ldots, k$, the sum (4.36) contains all the numbers $\frac{1}{1}, \frac{1}{2}, \frac{1}{3}, \frac{1}{4}, \ldots, \frac{1}{N-1}$ (and of course, many more numbers too). Thus,[8]

$$\prod_{p<N}\left(1 - \frac{1}{p}\right)^{-1} \geq \sum_{n=1}^{N-1} \frac{1}{n}.$$

Rewriting the left-hand side, we get

$$\prod_{p<N} \frac{p}{p-1} \geq \sum_{n=1}^{N-1} \frac{1}{n}. \tag{4.37}$$

Now recall from (4.29) on p. 310 that for every natural number n, we have

$$\log(n+1) - \log n < \frac{1}{n}. \tag{4.38}$$

Therefore, taking logarithms of both sides of (4.37), we get

$$\log\left(\sum_{n=1}^{N-1} \frac{1}{n}\right) \leq \log\left(\prod_{p<N} \frac{p}{p-1}\right)$$

$$= \sum_{p<N}\left(\log p - \log(p-1)\right)$$

$$\leq \sum_{p<N} \frac{1}{p-1} \leq \sum_{p<N} \frac{2}{p},$$

where we used that $p \leq 2(p-1)$ (this is because $n \leq 2(n-1)$ for all natural numbers $n > 1$). Since $\sum_{n=1}^{N-1} 1/n \to \infty$ as $N \to \infty$, $\log\left(\sum_{n=1}^{N-1} 1/n\right) \to \infty$ as $N \to \infty$ as well, so the sum $\sum 1/p$ must diverge.

▶ **Exercises 4.8**

1. Let $s_n = 1/2 + 1/3 + \cdots + 1/p_n$ (where p_n is the nth prime) be the nth partial sum of $\sum 1/p$. We know that $s_n \to \infty$ as $n \to \infty$. However, it turns out that $s_n \to \infty$ avoiding all integers! (That is, for all n, we have $s_n \notin \mathbb{Z}$.) Prove this. Suggestion: Multiply s_n by $2 \cdot 3 \cdots p_{n-1}$.

2. Proof that II can be slightly modified to avoid using the square-free fact. Derive the inequality (4.35) (which, as shown in the main text, implies that $\sum 1/p$ diverges) by proving that for every prime p,

---

[8]Following an idea of Euler [68], the identity (4.37) was used by J.J. Sylvester (1814–1897) in 1888 to prove the number of primes is infinite [238]. In fact, since the harmonic series diverges, the left-hand side of (4.37) must become unbounded as $N \to \infty$; consequently, there must be infinitely many primes!

$$\left(1 + \frac{1}{p}\right) \cdot \sum_{k=0}^{n} \frac{1}{p^{2k}} = \sum_{k=0}^{2n+1} \frac{1}{p^k}.$$

3. Here is another proof that is similar to Proof III in which we replace the inequality (4.38) with the following argument.

   (i) Prove that

   $$\frac{1}{1 - x/2} \le e^x \quad \text{for all } 0 \le x \le 1. \tag{4.39}$$

   Suggestion: Prove that $e^{-x} \le 1 - x/2$ using the series expansion for e^{-x}.
   (ii) Taking logarithms of (4.39), prove that for every prime number p, we have

   $$-\log\left(1 - \frac{1}{p}\right) = -\log\left(1 - \frac{2/p}{2}\right) \le \frac{2}{p}.$$

   (iii) Prove that

   $$\frac{1}{2} \sum_{p<N} \log\left(\frac{p}{p-1}\right) \le \sum_{p<N} \frac{1}{p}.$$

   (iv) Finally, use (4.37) to prove that $\sum 1/p$ diverges.

4. (Cf. [66]) Here's one more proof that $\sum 1/p$ diverges. Assume, to get a contradiction, that $\sum 1/p$ converges. Then we can fix a natural number N such that $\sum_{p>N} 1/p \le 1/2$; derive a contradiction as follows.

   (i) For every $x \in \mathbb{N}$, let A_x be the set of all integers $1 \le n \le x$ such that $n = 1$ or n can be factored into primes all of which are $\le N$. Given $n \in A_x$, we can write $n = k^2 m$, where m is square-free. Prove that $k \le \sqrt{x}$. From this, deduce that

   $$\#A_x \le C \sqrt{x},$$

   where $\#A_x$ denotes the number of elements in the set A_x, and C is a constant (you can take C to equal the number of square-free integers $m \le N$).
   (ii) Given $x \in \mathbb{N}$ and a prime p, prove that the number of integers $1 \le n \le x$ divisible by p is no more than x/p.
   (iii) Given $x \in \mathbb{N}$, prove that $x - \#A_x$ equals the number of integers $1 \le n \le x$ that are divisible by some prime $p > N$. From this fact and Part (ii), together with our assumption that $\sum_{p>N} 1/p \le 1/2$, prove that

   $$x - \#A_x \le \frac{x}{2}.$$

   (iv) Using (iii) and the inequality $\#A_x \le C \sqrt{x}$ that you proved in Part (i), derive a contradiction.

4.9   Defining the Trig Functions and π, and Which is Larger, π^e or e^π?

In high school we learned about sine and cosine using geometric intuition based on either triangles or the unit circle; here's the triangle version:

$$\cos\theta = \frac{\text{adjacent}}{\text{hypotonus}}, \quad \sin\theta = \frac{\text{opposite}}{\text{hypotonus}}, \quad \tan\theta = \frac{\text{opposite}}{\text{adjacent}}$$

(For this point of view, see the interesting paper [226].) In this section we introduce these function from a purely analytic framework, and we prove that these functions have all the properties you learned in high school. In high school we also learned about the number[9] π, again using geometric intuition. In this section we *define* π rigorously using analysis without any geometry. However, we do prove that π has all the geometric properties you think it does. At the end of this section we prove that the above triangle formulas are indeed true.

4.9.1   The Trigonometric and Hyperbolic Functions

Cosine and sine are the functions $\cos : \mathbb{C} \longrightarrow \mathbb{C}$ and $\sin : \mathbb{C} \longrightarrow \mathbb{C}$ defined by the equations

$$\cos z = \frac{e^{iz} + e^{-iz}}{2}, \quad \sin z = \frac{e^{iz} - e^{-iz}}{2i}.$$

In particular, both cosine and sine are continuous functions, being linear combinations of the continuous functions $e^{iz} = \exp(iz)$ and $e^{-iz} = \exp(-iz)$. From these formulas, we see that $\cos 0 = 1$ and $\sin 0 = 0$; other "well-known" values of sine and cosine are discussed in the problems. Multiplying the equation for $\sin z$ by i, we get

$$i\sin z = \frac{e^{iz} - e^{-iz}}{2},$$

and then adding to $\cos z$, the terms with e^{-iz} cancel, and we get $\cos z + i \sin z = e^{iz}$. This equation is the famous **Euler's identity**:

$$e^{iz} = \cos z + i \sin z. \quad \textbf{(Euler's identity)}$$

This formula provides a very easy proof of **de Moivre's formula**, named after its discoverer Abraham de Moivre (1667–1754),

---

[9]"Cosine, secant, tangent, sine, 3.14159; integral, radical, $u\,dv$, slipstick, sliderule, MIT!" MIT cheer.

$$(\cos z + i \sin z)^n = \cos nz + i \sin nz, \quad z \in \mathbb{C}, \qquad \textbf{(de Moivre's formula)},$$

which is given much attention in elementary mathematics and is usually stated only when $z = \theta$, a real variable. Here is the one-line proof:

$$(\cos z + i \sin z)^n = \left(e^{iz}\right)^n = e^{inz} = \cos nz + i \sin nz.$$

In the following theorem, we adopt the standard notation of writing $\sin^2 z$ for $(\sin z)^2$, etc.[10] Here are some well-known trigonometric identities that you memorized in high school and power series you derived in calculus,[11] but now we prove these facts from the basic definitions and even for complex variables.

Basic properties of cosine and sine

> **Theorem 4.35** *Cosine and sine are continuous functions on $\mathbb{C}$. In particular, restricting to real values, they define continuous functions on $\mathbb{R}$. Moreover, for all complex numbers z and w, they satisfy*
>
> *(1)* $\cos(-z) = \cos z$, $\sin(-z) = -\sin z$,
> *(2)* $\cos^2 z + \sin^2 z = 1$ *(Pythagorean identity)*,
> *(3)* *Addition formulas:*
>
> $$\cos(z + w) = \cos z \cos w - \sin z \sin w, \quad \sin(z + w) = \sin z \cos w + \cos z \sin w,$$
>
> *(4)* *Double angle formulas:*
>
> $$\cos(2z) = \cos^2 z - \sin^2 z = 2\cos^2 z - 1 = 1 - 2\sin^2 z,$$
> $$\sin(2z) = 2 \cos z \sin z.$$
>
> *(5)* *Trigonometric series: The following series converge absolutely:*
>
> $$\cos z = \sum_{n=0}^{\infty} (-1)^n \frac{z^{2n}}{(2n)!}, \qquad \sin z = \sum_{n=0}^{\infty} (-1)^n \frac{z^{2n+1}}{(2n+1)!}. \qquad (4.40)$$

---

[10]"$\sin^2 \phi$ is odious to me, even though Laplace made use of it; should it be feared that $\sin^2 \phi$ might become ambiguous, which would perhaps never occur, or at most very rarely when speaking of $\sin(\phi^2)$, well then, let us write $(\sin \phi)^2$, but not $\sin^2 \phi$, which by analogy should signify $\sin(\sin \phi)$." Carl Friedrich Gauss (1777–1855).

[11]In elementary calculus, these series are usually derived via Taylor series and are usually attributed to Isaac Newton (1643–1727), who derived them in his paper "De Methodis Serierum et Fluxionum" (Method of series and fluxions), written in 1671. However, it is interesting to know that these series were first discovered *hundreds* of years earlier by Madhava of Sangamagramma (1350–1425), a mathematician from the Kerala state in southern India!

Proof We shall leave some of this proof to the reader. Note that *(1)* follows directly from the definition of cosine and sine. Consider the addition formula:

$$\cos z \cos w - \sin z \sin w = \left(\frac{e^{iz} + e^{-iz}}{2}\right)\left(\frac{e^{iw} + e^{-iw}}{2}\right)$$

$$- \left(\frac{e^{iz} - e^{-iz}}{2i}\right)\left(\frac{e^{iw} - e^{-iw}}{2i}\right)$$

$$= \frac{1}{4}\left\{e^{i(z+w)} + e^{i(z-w)} + e^{-i(z-w)} + e^{-i(z+w)}\right.$$

$$\left. + e^{i(z+w)} - e^{i(z-w)} - e^{-i(z-w)} + e^{-i(z+w)}\right\}$$

$$= \frac{e^{i(z+w)} + e^{-i(z+w)}}{2} = \cos(z + w).$$

Taking $w = -z$ and using *(1)*, we get the Pythagorean identity:

$$1 = \cos 0 = \cos(z - z) = \cos z \cos(-z) - \sin z \sin(-z) = \cos^2 z + \sin^2 z.$$

We leave the double angle formulas to the reader. To prove *(5)*, we use the infinite series for the exponential to compute

$$e^{iz} + e^{-iz} = \sum_{n=0}^{\infty} \frac{i^n z^n}{n!} + \sum_{n=0}^{\infty} \frac{(-1)^n i^n z^n}{n!}.$$

The terms when n is odd cancel, so

$$\cos z = \frac{e^{iz} + e^{-iz}}{2} = \sum_{n=0}^{\infty} \frac{i^{2n} z^{2n}}{(2n)!} = \sum_{n=0}^{\infty} (-1)^n \frac{z^{2n}}{(2n)!},$$

where we used the fact that $i^{2n} = (i^2)^n = (-1)^n$. This series converges absolutely, since it is the sum of two absolutely convergent series. The series expansion for $\sin z$ is proved in a similar manner. ∎

From the series expansion for sine, it is straightforward to *prove* the following limit from elementary calculus (but now for complex numbers):

$$\lim_{z \to 0} \frac{\sin z}{z} = 1;$$

see Problem 3. Of course, from the identities in Theorem 4.35, one can derive other identities such as the so-called *half-angle formulas*:

$$\cos^2 z = \frac{1 + \cos 2z}{2}, \quad \sin^2 z = \frac{1 - \cos 2z}{2}.$$

The other trigonometric functions are defined in terms of sine and cosine in the usual manner:

$$\tan z = \frac{\sin z}{\cos z}, \qquad \cot z = \frac{1}{\tan z} = \frac{\cos z}{\sin z}$$

$$\sec z = \frac{1}{\cos z}, \qquad \csc z = \frac{1}{\sin z},$$

and are called the **tangent, cotangent, secant,** and **cosecant,** respectively. Note that these functions are defined only for those complex z for which the expressions make sense, e.g., $\tan z$ is defined only for those z such that $\cos z \neq 0$. The extra trig functions satisfy the same identities that you learned in high school, for example, for all complex numbers z, w, we have

$$\tan(z + w) = \frac{\tan z + \tan w}{1 - \tan z \tan w}, \qquad (4.41)$$

for those z, w such that the denominator is not zero. Setting $z = w$, we see that

$$\tan 2z = \frac{2 \tan z}{1 - \tan^2 z}.$$

In Problem 4 we ask you to prove (4.41) and other identities.

Before baking our π, we quickly define the hyperbolic functions. For every complex number z, we define

$$\cosh z = \frac{e^z + e^{-z}}{2}, \qquad \sinh z = \frac{e^z - e^{-z}}{2};$$

these are called the **hyperbolic cosine** and **hyperbolic sine,** respectively. There are hyperbolic tangents, secants, etc., defined in the obvious manner. Observe that by definition, $\cosh z = \cos iz$ and $\sinh z = -i \sin iz$, so after substituting iz for z in the series for cos and sin, we obtain

$$\cosh z = \sum_{n=0}^{\infty} \frac{z^{2n}}{(2n)!}, \qquad \sinh z = \sum_{n=0}^{\infty} (-1)^n \frac{z^{2n+1}}{(2n+1)!}.$$

These functions are intimately related to the trig functions and share many of the same properties, as shown in Problem 8.

4.9.2 The Number π

Substituting $z = x \in \mathbb{R}$ into the series (4.40), we obtain the following formulas learned in elementary calculus:

$$\cos x = \sum_{n=0}^{\infty} (-1)^n \frac{x^{2n}}{(2n)!}, \qquad \sin x = \sum_{n=0}^{\infty} (-1)^n \frac{x^{2n+1}}{(2n+1)!}.$$

In particular, cos, sin : $\mathbb{R} \longrightarrow \mathbb{R}$. In the following lemma and theorem we shall consider these real-valued functions instead of the more general complex versions. The following lemma is the key result needed to define π.

Lemma 4.36 *Sine and cosine have the following properties on* $[0, 2]$:

(1) Sine is nonnegative on $[0, 2]$ *and positive on* $(0, 2]$.
(2) Cosine on $[0, 2]$ *is strictly decreasing with* $\cos 0 = 1$ *and* $\cos 2 < 0$.

Proof It's clear that $\sin 0 = 0$. Since

$$\sin x = \sum_{n=0}^{\infty} (-1)^n \frac{x^{2n+1}}{(2n+1)!} = x\left(1 - \frac{x^2}{2 \cdot 3}\right) + \frac{x^5}{5!}\left(1 - \frac{x^2}{6 \cdot 7}\right) + \cdots$$

and one can check that each term in parentheses is positive for $0 < x < 2$, we have $\sin x > 0$ for all $0 < x < 2$. This proves *(1)*.

Since

$$\cos x = 1 - \frac{x^2}{2!} + \frac{x^4}{4!} - \frac{x^6}{6!} + \cdots,$$

we have

$$\cos 2 = 1 - \frac{2^2}{2!} + \frac{2^4}{4!} - \left(\frac{2^6}{6!} - \frac{2^8}{8!}\right) - \left(\frac{2^{10}}{10!} - \frac{2^{12}}{12!}\right) - \cdots.$$

All the terms in parentheses are positive, because for $k \geq 2$, we have

$$\frac{2^k}{k!} - \frac{2^{k+2}}{(k+2)!} = \frac{2^k}{k!}\left(1 - \frac{4}{(k+1)(k+2)}\right) > 0.$$

Therefore,

$$\cos 2 < 1 - \frac{2^2}{2!} + \frac{2^4}{4!} = -\frac{1}{3} < 0.$$

We now show that cosine is strictly decreasing on $[0, 2]$. Since cosine is continuous, by Theorem 4.28 on p. 297, if we show that cosine is one-to-one on $[0, 2]$, then we can conclude that cosine is strictly monotone on $[0, 2]$; then $\cos 0 = 1$ and $\cos 2 < 0$ tell us that cosine must be strictly decreasing. Suppose that $0 \leq x \leq y \leq 2$ and $\cos x = \cos y$; we shall prove that $x = y$. We already know that sine is nonnegative on $[0, 2]$, so the identity

$$\sin^2 x = 1 - \cos^2 x = 1 - \cos^2 y = \sin^2 y$$

implies that $\sin x = \sin y$. Therefore,

$$\sin(y - x) = \sin y \cos x - \cos y \sin x = \sin x \cos x - \cos x \sin x = 0,$$

and using that $0 \le y - x \le 2$ and *(1)*, we get $y - x = 0$. Hence, $x = y$, so cosine is one-to-one on $[0, 2]$, and our proof is complete.   ∎

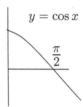

Fig. 4.36 The function $\cos : [0, 2] \longrightarrow \mathbb{R}$ crosses the x-axis at a unique point $0 < c < 2$; we define $\pi := 2c$, so that $c = \pi/2$

We now define the real number π, which is illustrated in Fig. 4.36.

Definition of π

Theorem 4.37 *There exists a unique real number, denoted by the Greek letter π, having the following two properties:*

(1) $3 < \pi < 4$,
(2) $\cos(\pi/2) = 0$.

Moreover, $\cos x > 0$ *for* $0 < x < \pi/2$.

Proof By our lemma, we know that $\cos : [0, 2] \longrightarrow \mathbb{R}$ is strictly decreasing with $\cos 0 = 1$ and $\cos 2 < 0$, so by the intermediate value theorem and the fact that cosine is strictly decreasing, there is a unique point $0 < c < 2$ such that $\cos c = 0$. Define $\pi := 2c$, that is, $c = \pi/2$. Then $0 < c < 2$ implies that $0 < \pi < 4$, and since $\cos$ is strictly decreasing on $[0, 2]$, we have $\cos x > 0$ for $0 < x < \pi/2$ and $\cos x < 0$ for $\pi/2 < x \le 2$. To see that in fact, $3 < \pi < 4$, we just need to show that $\cos(3/2) > 0$; this implies that $3/2 < \pi/2 < 2$ and therefore $3 < \pi < 4$. Plugging $x = 3/2$ into the formula for $\cos x$, we get

$$\cos \frac{3}{2} = \left(1 - \frac{3^2}{2^2\,2!}\right) + \left(\frac{3^4}{2^4\,4!} - \frac{3^6}{2^6\,6!}\right) + \left(\frac{3^8}{2^8\,8!} - \frac{3^{10}}{2^{10}\,10!}\right) + \cdots.$$

As the reader can check, apart from the first term in parentheses, $1 - 3^2/(2^2 2!)$, all the parentheses work out to be positive numbers. In particular, dropping all the terms after the second parentheses, we obtain

$$\cos\frac{3}{2} > \left(1 - \frac{3^2}{2^2\,2!}\right) + \left(\frac{3^4}{2^4\,4!} - \frac{3^6}{2^6\,6!}\right).$$

After a few minutes of calculations, we find that the right-hand side equals $359/$ $(2^{10} \cdot 5)$, a positive number. Thus,

$$\cos\frac{3}{2} > 0,$$

and we're done.   ∎

The number $\pi/180$ is called a **degree**. Thus, $\pi/2 = 90 \cdot \pi/180$ is the same as 90 degrees, which we write as $90°$; $\pi = 180 \cdot \pi/180$ is the same as $180°$; etc.

4.9.3 Properties of π

As we already stated, the approach we have taken to introducing π has been completely analytical without reference to triangles or circles, but surely the π we have defined and the π you have grown up with must be the same. We now show that the π we have defined is not an imposter, but indeed does have all the properties of the π that you have grown to love.

We first state some of the well-known trig identities involving π that you learned in high school, but now we prove them for complex variables.

Theorem 4.38 *We have*

$$\cos(\pi/2) = 0, \quad \cos(\pi) = -1, \quad \cos(3\pi/2) = 0, \quad \cos(2\pi) = 1$$
$$\sin(\pi/2) = 1, \quad \sin(\pi) = 0, \quad \sin(3\pi/2) = -1, \quad \sin(2\pi) = 0.$$

Moreover, for every complex number z, we have the following addition formulas:

$$\cos\left(z + \frac{\pi}{2}\right) = -\sin z, \quad \sin\left(z + \frac{\pi}{2}\right) = \cos z,$$
$$\cos(z + \pi) = -\cos z, \quad \sin(z + \pi) = -\sin z,$$
$$\cos(z + 2\pi) = \cos z, \quad \sin(z + 2\pi) = \sin z.$$

Proof We know that $\cos(\pi/2) = 0$ and, by *(1)* of Lemma 4.36, $\sin(\pi/2) > 0$. Therefore, since

$$\sin^2(\pi/2) = 1 - \cos^2(\pi/2) = 1,$$

we must have $\sin(\pi/2) = 1$. The double angle formulas now imply that

$$\cos(\pi) = \cos^2(\pi/2) - \sin^2(\pi/2) = -1, \quad \sin(\pi) = 2\cos(\pi/2)\sin(\pi/2) = 0,$$

and by another application of the double angle formulas, we get

$$\cos(2\pi) = 1, \quad \sin(2\pi) = 0.$$

The facts just proved plus the addition formulas for cosine and sine in Property *(3)* of Theorem 4.35 imply the last six formulas in the statement of this theorem; for example,

$$\cos\left(z + \frac{\pi}{2}\right) = \cos z \cos \frac{\pi}{2} - \sin z \sin \frac{\pi}{2} = -\sin z,$$

and the other formulas are proved similarly. Finally, substituting $z = \pi$ into

$$\cos\left(z + \frac{\pi}{2}\right) = -\sin z, \quad \sin\left(z + \frac{\pi}{2}\right) = \cos z,$$

prove that $\cos(3\pi/2) = 0$ and $\sin(3\pi/2) = -1$.   ■

The last two formulas in Theorem 4.38 (plus an induction argument) imply that cosine and sine are **periodic** (with period 2π) in the sense that for every $n \in \mathbb{Z}$,

$$\cos(z + 2\pi n) = \cos z, \quad \sin(z + 2\pi n) = \sin z. \tag{4.42}$$

Now, substituting $z = \pi$ into $e^{iz} = \cos z + i\sin z$ and using that $\cos \pi = -1$ and $\sin \pi = 0$, we get $e^{i\pi} = -1$, or by bringing -1 to the left, we get perhaps the most important equation in all of mathematics (at least to some mathematicians!)[12]:

$$\boxed{e^{i\pi} + 1 = 0.}$$

In one shot, this single equation contains the five "most important" constants in mathematics: 0, the additive identity; 1, the multiplicative identity; i, the imaginary unit; and the constants e, the base of the exponential function; and π, the fundamental constant of geometry.

We now study the graphs of $\cos x$ and $\sin x$ for $x \in \mathbb{R}$. By 2π-periodicity (see (4.42)), all we need to know are their graphs on the interval $[0, 2\pi]$ (Fig. 4.37). With this in mind, we understand that the following theorem states that the graphs of cosine and sine go "up and down" on $[0, 2\pi]$, and hence on all of $\mathbb{R}$, as you think they should.

---

[12][After proving a formula equivalent to Euler's formula $e^{i\pi} = -1$ in a lecture] "Gentlemen, that is surely true, it is absolutely paradoxical; we cannot understand it, and we don't know what it means. But we have proved it, and therefore we know it is the truth." Benjamin Peirce (1809–1880). Quoted in E. Kasner and J. Newman [119].

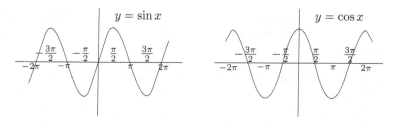

Fig. 4.37 Theorem 4.39 says that on the interval $[0, 2\pi]$, the graphs of cosine and sine oscillate as you learned in high school. Using that cosine and sine are 2π-periodic, we obtain the graph of cosine and sine on $[-2\pi, 0]$

Oscillation theorem

> **Theorem 4.39** *On the interval* $[0, 2\pi]$, *the following monotonicity properties of* cos *and* sin *hold:*
>
> *(1) Cosine decreases from* 1 *to* -1 *on* $[0, \pi]$ *and increases from* -1 *to* 1 *on* $[\pi, 2\pi]$.
> *(2) Sine increases from* 0 *to* 1 *on* $[0, \pi/2]$, *decreases from* 1 *to* -1 *on* $[\pi/2, 3\pi/2]$, *and then increases from* -1 *to* 0 *on* $[3\pi/2, 2\pi]$.

Proof From Lemma 4.36 we know that cosine is strictly decreasing from 1 to 0 on $[0, \pi/2]$, and from this same lemma, we know that sine is positive on $(0, \pi/2)$. Therefore, by the Pythagorean identity,

$$\sin x = \sqrt{1 - \cos^2 x}$$

on $[0, \pi/2]$. Since cosine is positive and strictly decreasing on $[0, \pi/2]$, this formula implies that sine is strictly increasing on $[0, \pi/2]$. Replacing z by $x - \pi/2$ in the identities

$$\cos\left(z + \frac{\pi}{2}\right) = -\sin z, \quad \sin\left(z + \frac{\pi}{2}\right) = \cos z$$

found in Theorem 4.38 gives the formulas

$$\cos x = -\sin\left(x - \frac{\pi}{2}\right), \quad \sin x = \cos\left(x - \frac{\pi}{2}\right).$$

The first formula here plus the fact that sine is increasing on $[0, \pi/2]$ shows that cosine is decreasing on $[\pi/2, \pi]$, while the second formula plus the fact that cosine is decreasing on $[0, \pi/2]$ shows that sine is also decreasing on $[\pi/2, \pi]$. Finally, the formulas

$$\cos x = -\cos(x - \pi), \quad \sin x = -\sin(x - \pi),$$

also obtained as a consequence of Theorem 4.38, and the monotone properties already established for cosine and sine on $[0, \pi]$ imply the rest of the monotone properties in *(1)* and *(2)* of cosine and sine on $[\pi, 2\pi]$. ∎

In geometric terms, the following theorem states that points on the unit circle can be identified with numbers (which we usually call "angles") in the interval $[0, 2\pi)$, through the function $f(\theta) = (\cos \theta, \sin \theta)$. (However, because we like complex notation, we shall write $(\cos \theta, \sin \theta)$ as the complex number $\cos \theta + i \sin \theta = e^{i\theta}$ in the theorem.)

π and the unit circle

Theorem 4.40 *For a real number θ, define*

$$f(\theta) := e^{i\theta} = \cos \theta + i \sin \theta.$$

Then $f : \mathbb{R} \longrightarrow \mathbb{C}$ is a continuous function and has range equal to the unit circle

$$\mathbb{S}^1 := \{(a, b) \in \mathbb{R}^2 \, ; \, a^2 + b^2 = 1\} = \{z \in \mathbb{C} \, ; \, |z| = 1\}.$$

Moreover, $f(\theta) = f(\phi)$ if and only if $\theta - \phi$ is an integer multiple of 2π. Finally, for each $z \in \mathbb{S}^1$, there exists a unique θ with $0 \le \theta < 2\pi$ such that $f(\theta) = z$.

Proof Since the exponential function is continuous, so is the function f, and by the Pythagorean identity, $\cos^2 \theta + \sin^2 \theta = 1$, so we also know that f maps to the unit circle. Given z in the unit circle, we can write $z = a + ib$, where $a^2 + b^2 = 1$. We prove that there exists a unique $0 \le \theta < 2\pi$ such that $f(\theta) = z$, that is, such that $\cos \theta = a$ and $\sin \theta = b$. Now either $b \ge 0$ or $b < 0$. Assume that $b \ge 0$; the case $b < 0$ is proved in a similar way. Since, according to Theorem 4.39, $\sin \theta < 0$ for all $\pi < \theta < 2\pi$, and we are assuming $b \ge 0$, there is no θ with $\pi < \theta < 2\pi$ such that $f(\theta) = z$. Hence, we just have to show that there is a unique $\theta \in [0, \pi]$ such that $f(\theta) = z$. Since $a^2 + b^2 = 1$, we have $-1 \le a \le 1$ and $0 \le b \le 1$. Since cosine strictly decreases from 1 to -1 on $[0, \pi]$, by the intermediate value theorem there is a unique value $\theta \in [0, \pi]$ such that $\cos \theta = a$. The identity

$$\sin^2 \theta = 1 - \cos^2 \theta = 1 - a^2 = b^2,$$

and the fact that $\sin \theta \ge 0$, because $0 \le \theta \le \pi$, imply that $b = \sin \theta$.

Let θ and ϕ be real numbers such that $f(\theta) = f(\phi)$; we shall prove that θ and ϕ differ by an integer multiple of 2π. Let n be the unique integer such that (see the Archimedean property on p. 97)

$$n \le \frac{\theta - \phi}{2\pi} < n + 1.$$

Multiplying everything by 2π and subtracting $2\pi n$, we obtain

$$0 \leq \theta - \phi - 2\pi n < 2\pi.$$

By periodicity (see (4.42)),

$$f(\theta - \phi - 2\pi n) = f(\theta - \phi) = e^{i(\theta - \phi)} = e^{i\theta}e^{-i\phi} = f(\theta)/f(\phi) = 1.$$

Since $\theta - \phi - 2\pi n$ is in the interval $[0, 2\pi)$ and $f(0) = 1$ also, by the uniqueness we proved in the previous paragraph, we conclude that $\theta - \phi - 2\pi n = 0$. This completes the proof of the theorem. ∎

We now solve trigonometric equations. Notice that Property *(2)* of the following theorem shows that cosine vanishes exactly at $\pi/2$ and all its π translates, and *(3)* shows that sine vanishes exactly at integer multiples of π, well-known facts from high school! However, we consider complex variables instead of just real variables.

> **Theorem 4.41** *For complex numbers z and w,*
>
> *(1)* $e^z = e^w$ *if and only if $z = w + 2\pi i n$ for some integer n.*
> *(2)* $\cos z = 0$ *if and only if $z = n\pi + \pi/2$ for some integer n.*
> *(3)* $\sin z = 0$ *if and only if $z = n\pi$ for some integer n.*

Proof The "if" statements follow from Theorem 4.38, so we are left to prove the "only if" statements. Suppose that $e^z = e^w$. Then $e^{z-w} = 1$. Hence, it suffices to prove that $e^z = 1$ implies that z is an integer multiple of $2\pi i$. Let $z = x + iy$ for real numbers x and y. Then,

$$1 = |e^{x+iy}| = |e^x e^{iy}| = e^x.$$

Since the exponential function on the real line is one-to-one, it follows that $x = 0$. Now the equation $1 = e^z = e^{iy}$ implies, by Theorem 4.40, that y must be an integer multiple of 2π. Hence, $z = x + iy = iy$ is an integer multiple of $2\pi i$.

Assume that $\sin z = 0$. Then by definition of $\sin z$, we have $e^{iz} = e^{-iz}$. By *(1)*, we have $iz = -iz + 2\pi i n$ for some integer n. Solving for z, we get $z = \pi n$. Finally, the identity

$$\sin\left(z + \frac{\pi}{2}\right) = \cos z$$

and the result already proved for sine show that $\cos z = 0$ implies that $z = n\pi + \pi/2$ for some integer n. ∎

As a corollary of this theorem we see that the domain of both $\tan z = \sin z/\cos z$ and $\sec z = 1/\cos z$ consists of all complex numbers except odd integer multiples of $\pi/2$.

4.9.4   Which Is Larger, π^e or e^π?

Of course, one can simply check using a calculator that e^π is greater. Here's a mathematical proof following [210]. First recall that $1 + x < e^x$ for every positive real x. Hence, since powers preserve inequalities, for every $x, y > 0$, we obtain

$$\left(1 + \frac{x}{y}\right)^y < (e^{x/y})^y = e^x.$$

In Example 3.49 on p. 222, we noted that $e < 3$. Since $3 < \pi$, we have $\pi - e > 0$. Now substituting $x = \pi - e > 0$ and $y = e$ in the above equation, we get

$$\left(1 + \frac{\pi - e}{e}\right)^e = \left(\frac{\pi}{e}\right)^e < e^{\pi - e},$$

which, after multiplying by e^e, gives the inequality $\pi^e < e^\pi$.

By the way, speaking about e^π, Charles Hermite (1822–1901) made the fascinating discovery that for many values of n, $e^{\pi\sqrt{n}}$ is an **"almost integer"** [47, p. 80]. For example, if you go to a calculator, you'll find that when $n = 1$, e^π is not almost an integer, but $e^\pi - \pi$ is:

$$\boxed{e^\pi - \pi \approx 20.}$$

In fact, $e^\pi - \pi = 19.999099979\ldots$. When $n = 163$, we get the incredible approximation

$$\boxed{e^{\pi\sqrt{163}} = 262537412640768743.9999999999992\ldots} \tag{4.43}$$

Check out $e^{\pi\sqrt{58}}$. Isn't it amazing how e and π show up in the strangest places?

4.9.5   Plane Geometry and Polar Representation of Complex Numbers

Given a nonzero complex number z, we can write $z = r\omega$, where $r = |z|$ and $\omega = z/|z|$. Notice that $|\omega| = 1$, so from our knowledge of π and the unit circle (Theorem 4.40), we know that there is a unique $0 \le \theta < 2\pi$ such that

$$\frac{z}{|z|} = e^{i\theta} = \cos\theta + i\sin\theta.$$

Therefore,

$$z = re^{i\theta} = r\big(\cos\theta + i\sin\theta\big),$$

which is called the **polar representation** of z. We can relate this representation to the familiar *polar coordinates* on $\mathbb{R}^2$ as follows. Recall that $\mathbb{C}$ is really just $\mathbb{R}^2$. Let $z = x + iy$, which recall is the same as $z = (x, y)$, where $i = (0, 1)$. Then

$$r = |z| = \sqrt{x^2 + y^2}$$

is just the familiar radial distance of (x, y) to the origin. Equating the real and imaginary parts of the equation $\cos\theta + i\sin\theta = z/|z|$, we get the two equations

$$\cos\theta = \frac{x}{r} = \frac{x}{\sqrt{x^2 + y^2}} \quad \text{and} \quad \sin\theta = \frac{y}{r} = \frac{y}{\sqrt{x^2 + y^2}}. \qquad (4.44)$$

Summarizing: The equation $(x, y) = z = re^{i\theta} = r\left(\cos\theta + i\sin\theta\right)$ is equivalent to

$$x = r\cos\theta \quad \text{and} \quad y = r\sin\theta.$$

We call (r, θ) the **polar coordinates** of the point $z = (x, y)$. When z is drawn as a point in $\mathbb{R}^2$, r represents the distance of z to the origin, and θ represents (or rather is *by definition*) the **angle** that z makes with the positive real axis:

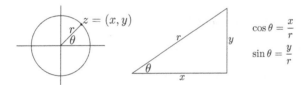

Fig. 4.38 The familiar concept of angle

The number r is often called the **modulus**, and θ the **argument**, of z. In elementary calculus, one usually studies polar coordinates without introducing complex numbers. However, we prefer the complex number approach and in particular, the single notation $z = re^{i\theta}$ instead of the pair notation $x = r\cos\theta$ and $y = r\sin\theta$. We have taken $0 \le \theta < 2\pi$, but it will be very convenient to allow θ to represent *any* real number. In this case, $z = re^{i\theta}$ is not attached to a unique choice of θ, but by our knowledge of π and the unit circle, we know that any two such values of θ differ by an integer multiple of 2π. Thus, the polar coordinates (r, θ) and $(r, \theta + 2\pi n)$ represent the same point for every integer n.

Using polar representations, complex multiplication has a simple geometric interpretation. Let $z = re^{i\theta}$ and $w = \rho e^{i\varphi}$. Then,

$$zw = re^{i\theta} \cdot \rho e^{i\varphi} = r\rho e^{i(\theta + \varphi)}.$$

Thus, to find zw, all we do is add the angles and multiply the moduli (Fig. 4.39):

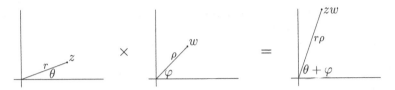

Fig. 4.39 "Arguments add" and "moduli multiply." That is, zw is the vector obtained by rotating w by the argument of z and scaling by the modulus of z

For example, if $r < 1$, it would be instructive to draw the sequence $z, z^2, z^3, z^4, \ldots$; you should get a spiral converging toward the origin. If $r > 1$, the points spiral away from the origin! In any case, after all our work, let us give a . . .

Summary of this section: We have seen that

All that you thought about trigonometry is true!

In particular, from (4.44) and adding the formula $\tan\theta = \sin\theta/\cos\theta = y/x$, from Fig. 4.38 we see that

$$\cos\theta = \frac{\text{adjacent}}{\text{hypotenuse}}, \quad \sin\theta = \frac{\text{opposite}}{\text{hypotenuse}}, \quad \tan\theta = \frac{\text{opposite}}{\text{adjacent}},$$

just as you learned in high school!

▶ **Exercises 4.9**

1. Here are some values of the trigonometric functions.
   (a) Find $\sin i$, $\cos i$, and $\tan(1 + i)$ (in terms of e and i).
   (b) Using trig identities (no triangles allowed!), prove $\sin(\pi/4) = \cos(\pi/4) = 1/\sqrt{2}$.
   (c) Find $\sin(\pi/8)$ and $\cos(\pi/8)$.
   (d) Prove that $\sin(\pi/6) = \cos(\pi/3)$, and $\sin(\pi/3) = \cos(\pi/6)$. Suggestion: Note that $\pi/3 = \pi/2 - \pi/6$.
   (e) Prove $\sin(\pi/6) = \cos(\pi/3) = 1/2$ and $\sin(\pi/3) = \cos(\pi/6) = \sqrt{3}/2$.

2. In this problem we find a very close estimate of π. Prove that for $0 < x < 2$, we have
$$\cos x < 1 - \frac{x^2}{2} + \frac{x^4}{24}.$$

   Use this fact to prove that $3/2 < \pi/2 < \sqrt{6 - 2\sqrt{3}}$, which implies that $3 < \pi < 2\sqrt{6 - 2\sqrt{3}} \approx 3.185$. We'll get a much better estimate in Section 4.12.

3. Using the series representations (4.40) for $\sin z$ and $\cos z$, find the limits
$$\lim_{z\to 0}\frac{\sin z}{z}, \quad \lim_{z\to 0}\frac{\sin z - z}{z^3}, \quad \lim_{z\to 0}\frac{\cos z - 1 + z^2/2}{z^3}, \quad \lim_{z\to 0}\frac{\cos z - 1 + z^2/2}{z^4}.$$

4. Prove some of the following identities.

   (a) For $z, w \in \mathbb{C}$,

   $$2 \sin z \sin w = \cos(z - w) - \cos(z + w),$$
   $$2 \cos z \cos w = \cos(z - w) + \cos(z + w),$$
   $$2 \sin z \cos w = \sin(z + w) + \sin(z - w),$$
   $$\tan(z + w) = \frac{\tan z + \tan w}{1 - \tan z \tan w},$$
   $$1 + \tan^2 z = \sec^2 z, \quad \cot^2 z + 1 = \csc^2 z.$$

   Use the definitions of cosine and sine in terms of exponential functions.

   (b) If $x \in \mathbb{R}$, then for every natural number n,

   $$\cos nx = \sum_{k=0}^{\lfloor n/2 \rfloor} (-1)^k \binom{n}{2k} \cos^{n-2k} x \, \sin^{2k} x,$$

   $$\sin nx = \sum_{k=0}^{\lfloor (n-1)/2 \rfloor} (-1)^k \binom{n}{2k+1} \cos^{n-2k-1} x \, \sin^{2k+1} x,$$

   where $\lfloor t \rfloor$ is the greatest integer less than or equal to $t \in \mathbb{R}$. Suggestion: Expand the left-hand side of de Moivre's formula using the binomial theorem.

   (c) Prove that

   $$\sin^2 \frac{\pi}{5} = \frac{5 - \sqrt{5}}{8}, \quad \cos^2 \frac{\pi}{5} = \frac{3 + \sqrt{5}}{8}, \quad \cos \frac{\pi}{5} = \frac{1 + \sqrt{5}}{4}.$$

   Suggestion: What if you consider $x = \pi/5$ and $n = 5$ in the equation for $\sin nx$ in Part (b)?

5. Prove that for $0 \le r < 1$ and $\theta \in \mathbb{R}$,

   $$\sum_{n=0}^{\infty} r^n \cos(n\theta) = \frac{1 - r \cos \theta}{1 - 2r \cos \theta + r^2}, \quad \sum_{n=1}^{\infty} r^n \sin(n\theta) = \frac{r \sin \theta}{1 - 2r \cos \theta + r^2}.$$

   Suggestion: Let $z = re^{i\theta}$ in the geometric series $\sum_{n=0}^{\infty} z^n$.

6. For every real number β such that $e < \beta$, prove that $\beta^e < e^\beta$.

7. Here's a very neat problem posed by D.J. Newman [168].

(i) As n increases through the natural numbers, prove that[13]

$$\lim_{n \to \infty} n \sin(2\pi\, e\, n!) = 2\pi.$$

Suggestion: Multiply $e = 1 + \frac{1}{2!} + \frac{1}{3!} + \cdots + \frac{1}{n!} + \frac{1}{(n+1)!} + \cdots$ by $2\pi n!$.

(ii) Prove, using (i), that e is irrational.

8. (**Hyperbolic functions**) In this problem we study the hyperbolic functions.

(a) Show that

$$\sinh(z + w) = \sinh z \cosh w + \cosh z \sinh w,$$
$$\cosh(z + w) = \cosh z \cosh w + \sinh z \sinh w,$$
$$\sinh(2z) = 2 \cosh z \sinh z, \quad \cosh^2 z - \sinh^2 z = 1.$$

(b) If $z = x + iy$, prove that

$$\sinh z = \sinh x \cos y + i \cosh x \sin y, \quad \cosh z = \cosh x \cos y + i \sinh x \sin y$$
$$|\sinh z|^2 = \sinh^2 x + \sin^2 y, \qquad |\cosh z|^2 = \sinh^2 x + \cos^2 y.$$

Find all $z \in \mathbb{C}$ such that $\sinh z$ is real. Do the same for $\cosh z$. Find all the zeros of $\sinh z$ and $\cosh z$.

(c) Prove that if $z = x + iy$, then

$$\sin z = \sin x \cosh y + i \cos x \sinh y, \quad \cos z = \cos x \cosh y - i \sin x \sinh y.$$

Find all $z \in \mathbb{C}$ such that $\sin z$ is real. Do the same for $\cos z$.

9. Here is an interesting geometric problem. Let $z \in \mathbb{C}$ and let $G(n, r)$ denote a regular n-gon ($n \geq 3$) of radius r centered at the origin of $\mathbb{C}$. Is there a simple formula for the sum of the squares of the distances from z to the vertices of $G(n, r)$? Using complex numbers, this problem is not too difficult to solve. Proceed as follows.

(i) Show that $0 = \sum_{k=1}^{n} e^{2\pi i k/n} = e^{2\pi i/n} + \left(e^{2\pi i/n}\right)^2 + \cdots + \left(e^{2\pi i/n}\right)^n$.

(ii) Show that

$$\sum_{k=1}^{n} \left| z - r e^{2\pi i k/n} \right|^2 = n(|z|^2 + r^2).$$

How does this formula solve our problem?

---

[13] Typing N[n sin(2 pi e (n!))] in WolframAlpha to numerically compute $n \sin(2\pi e n!)$, for $n \leq 170$, you will find that $n \sin(2\pi e n!)$ seems to approach 2π as n increases (press the button *More digits* for n large). However, for all $n \geq 171$, you will get $n \sin(2\pi e n!) = 0$, which is wrong!

10. In this problem we consider "Thomae-like" functions. Prove that the following functions are continuous at the irrationals and discontinuous at the rationals.

(a) Define $f : \mathbb{R} \longrightarrow \mathbb{R}$ by

$$f(x) = \begin{cases} \sin(1/q) & \text{if } x \in \mathbb{Q} \text{ and } x = p/q \text{ in lowest terms and } q > 0, \\ 0 & \text{if } x \text{ is irrational.} \end{cases}$$

(b) Define $g : (0, \infty) \longrightarrow \mathbb{R}$ by

$$g(x) = \begin{cases} p\sin(1/q) & \text{if } x \in \mathbb{Q} \text{ and } x = p/q \text{ in lowest terms and } q > 0, \\ x & \text{if } x \text{ is irrational.} \end{cases}$$

11. (**Definition of** π) In this problem we give an alternative proof that $\cos$: $[0, 2] \longrightarrow \mathbb{R}$ is one-to-one; as shown in the proof of Theorem 4.37, this is all we need to define π as $2\times$ the unique zero of cosine in $[0, 2]$.

(i) From the series expansion for $\cos x$, prove that $\cos x > 0$ for $x \in [0, 1]$.
(ii) Let $x, y \in [0, 2]$ and assume $\cos x = \cos y$; we shall prove that $x = y$. Using a trigonometric identity, prove that $\cos(x/2) = \cos(y/2)$.
(iii) Prove for all $n \in \mathbb{N}$, $\cos(x/2^n) = \cos(y/2^n)$.
(iv) Prove that $\lim_{z \to 0}(1 - \cos z)/z^2 = 1/2$.
(v) Using (iii) and (iv), prove $x = y$.

12. (**Definition of** π) In this problem we give another way to define π. (See [224, p. 160] for another proof.) Assume only the following properties of $\cos, \sin$: $\mathbb{R} \longrightarrow \mathbb{R}$:

(a) $\cos^2 x + \sin^2 x = 1$ for all $x \in \mathbb{R}$.
(b) $\cos$ and $\sin$ are continuous, $\cos(0) = 1$, and $\sin x > 0$ for $x > 0$ sufficiently small.
(c) $\sin(x \pm y) = \sin x \cos y \pm \cos x \sin y$ for all $x, y \in \mathbb{R}$.

Using only properties of cosine and sine derivable from (a), (b), and (c), we shall prove that the set $A := \{x \geq 0 \,;\, \cos x = 0\}$ is not empty. In particular, we can define π by the formula

$$\pi := 2 \cdot \inf A.$$

Assume, by way of contradiction, that $A = \varnothing$. Proceed as follows.

(i) First establish the following identity: For every $x, y \in \mathbb{R}$,

$$\sin y - \sin x = 2 \cos \frac{x+y}{2} \sin \frac{y-x}{2}.$$

(ii) Show that $\cos x > 0$ for all $x \geq 0$. (Remember, we are assuming $A = \varnothing$.)

(iii) Show that $\sin : [0, \infty) \longrightarrow \mathbb{R}$ is strictly increasing, and use this to show that $\cos : [0, \infty) \longrightarrow \mathbb{R}$ is strictly decreasing. Suggestion: Show that if $0 \leq x < y$ and x and y are sufficiently close, say within some $\delta > 0$, then $\sin x < \sin y$. For arbitrary $0 \leq x < y$, partition the interval $[x, y]$ into points $x = x_0 < x_1 < x_2 < \cdots < x_N = y$, where for each k, x_k and x_{k+1} are within δ.

(iv) Show that $L := \lim_{x \to \infty} \cos x$ exists and $\lim_{x \to \infty} \sin x = \sqrt{1 - L^2}$.

(v) Prove that $\sin 2x = 2 \cos x \sin x$ for all $x \in \mathbb{R}$ and then prove that $L = \frac{1}{2}$.

(vi) Using the identity in (i), prove that for every $y \in \mathbb{R}$, we have $\sin y = 0$. Derive a contradiction.

(vii) Using the identity in (i), prove that for every $y \in \mathbb{R}$ we have $\sin y = 0$. Derive a contradiction. Thus, the assumption that $A = \varnothing$ must have been false, and hence π is well defined.

4.10   ★ Three Proofs of the Fundamental Theorem of Algebra (FTA)

In elementary calculus you were exposed to the *method of partial fractions* to integrate rational functions, in which you had to factor polynomials. The necessity to factor polynomials for the method of partial fractions played a large role in the race to prove the fundamental theorem of algebra; see [63] for more on this history, especially Euler's part in this theorem. Carl Friedrich Gauss (1777–1855) is usually credited as the first to prove the fundamental theorem of algebra, as part of his doctoral thesis (1799) entitled "A new proof of the theorem that every integral rational algebraic function[14] can be decomposed into real factors of the first or second degree" (see, e.g., [35, p. 499]). We present three independent and different guises of one of the more elementary and popular *topological* proofs of the theorem, which is essentially a proof due to Jean Argand (1768–1822) published in 1806 with new versions in 1814/1815.

4.10.1   Proof I of the FTA [199]

This proof could have actually been presented immediately after Section 4.4, but we have chosen to save the proof till now because it fits so well with roots of complex numbers, on which we'll touch in Section 4.10.3.

Given $n \in \mathbb{N}$ and $z \in \mathbb{C}$, a complex number ξ is called an *nth root* of z if $\xi^n = z$. A natural question is: Does every $z \in \mathbb{C}$ have an nth root? This question is easily answered in special cases. For example, if $z = 0$, then for every n, $\xi = 0$ is an nth

---

[14]In plain English, a polynomial with real coefficients. A translation of Gauss's thesis can be found at http://www.fsc.edu/library/archives/manuscripts/gauss.cfm. Gauss's proof was actually incorrect, but he published a correct version in 1816.

root of z, and it is the only nth root (since nonzero complex numbers raised to the nth power cannot equal zero). If $n = 1$, then for every z, $\xi = z$ is a first root of z. If $n = 2$ and if z is a real positive number, then we know that z has a (real) square root, and if z is a negative real number, then $\xi = i\sqrt{-z}$ is a square root of z. If $z = a + ib$, where $b \neq 0$, then the numbers

$$\xi = \pm\left(\sqrt{\frac{|z| + a}{2}} + i\frac{b}{|b|}\sqrt{\frac{|z| - a}{2}}\right)$$

are square roots of z, as the reader can easily verify; see Problem 9. What about roots of higher order than two? In the following lemma we prove that every complex number has an nth root for every $n \in \mathbb{N}$. In Section 4.10.3 we'll give another proof of this lemma using facts about exponential and trigonometric functions developed in the previous sections. However, the following proof is interesting because it is completely elementary in that it avoids any reference to these functions.

Lemma 4.42 *Every complex number has an nth root for all $n \in \mathbb{N}$.*

Proof Let $z \in \mathbb{C}$, which we may assume is nonzero. We shall prove that z has an nth root using strong induction. We already know that z has nth roots for $n = 1, 2$. Let $n > 2$ and assume that z has roots of all orders less than n; we shall prove that z has an nth root.

Suppose first that n is even, say $n = 2m$ for some natural number $m > 2$. Then we are looking for a complex number ξ such that $\xi^{2m} = z$. By our discussion before this lemma, we know that there is a number η such that $\eta^2 = z$, and since $m < n$, by the induction hypothesis we know that there is a number ξ such that $\xi^m = \eta$. Then

$$\xi^n = \xi^{2m} = (\xi^m)^2 = \eta^2 = z,$$

and we've found an nth root of z.

Suppose now that n is odd. If z is a nonnegative real number, then we know that z has a real nth root, so we henceforth assume that z is not a nonnegative real number. Let η be a complex number such that $\eta^2 = z$. Then for $x \in \mathbb{R}$, consider the polynomial $p(x)$ given by taking the imaginary part of $\eta(x - i)^n$:

$$p(x) := \text{Im}\left[\eta(x - i)^n\right] = \frac{1}{2i}\left[\eta(x - i)^n - \overline{\eta}(x + i)^n\right],$$

where we used Property *(4)* of Theorem 2.49 on p. 133 that $\text{Im } w = \frac{1}{2i}(w - \overline{w})$ for every complex number w. Expanding $(x - i)^n$ and $(x + i)^n$ using the binomial theorem, one can show that

$$p(x) = \text{Im}(\eta)\, x^n + \text{lower order terms in} \, x.$$

Since η is not real, the coefficient in front of x^n is nonzero, so $p(x)$ is an nth-degree polynomial in x with real coefficients. In Problem 3 on p. 280 we noted that all odd-degree real-valued polynomials have a real root, so there is some $c \in \mathbb{R}$ with $p(c) = 0$. For this c, we have

$$\eta(c - i)^n - \overline{\eta}(c + i)^n = 0.$$

After a little manipulation, and using that $\eta^2 = z$, we get

$$\frac{(c + i)^n}{(c - i)^n} = \frac{\eta}{\overline{\eta}} = \frac{\eta^2}{|\eta|^2} = \frac{z}{|z|} \implies |z| \frac{(c + i)^n}{(c - i)^n} = z.$$

It follows that $\xi = \sqrt[n]{|z|} \frac{c+i}{c-i}$ satisfies $\xi^n = z$, and our proof is now complete. ∎

We now present our first proof of the celebrated fundamental theorem of algebra. The following proof is very elementary in the sense that looking through the proof, we see that the nontrivial results we use are kept at a minimum:

(1) The Bolzano–Weierstrass theorem (see p. 179).
(2) Every nonzero complex number has a kth root.

As mentioned already, the following proof goes back to Jean Argand's (1768–1822) 1814/1815 proof, versions of which can be found in [74, 204, 239], or (one of my favorites) [198].

The fundamental theorem of algebra, Proof I

Theorem 4.43 *Every complex polynomial of positive degree has at least one complex root.*

Proof Let $p(z) = a_n z^n + a_{n-1} z^{n-1} + \cdots + a_1 z + a_0$ be a polynomial with complex coefficients, $n \geq 1$ with $a_n \neq 0$. We prove this theorem in four steps.

Step 1: We begin by proving a simple, but important, inequality. Since

$$|p(z)| = |a_n z^n + \cdots + a_0| = |z|^n \left| a_n + \frac{a_{n-1}}{z} + \frac{a_{n-2}}{z^2} + \cdots + \frac{a_1}{z^{n-1}} + \frac{a_0}{z^n} \right|,$$

for $|z|$ sufficiently large the absolute value of the sum of all the terms to the right of a_n can be made less than, say $|a_n|/2$. It follows that

$$|p(z)| \geq \frac{|a_n|}{2} \cdot |z|^n, \quad \text{for } |z| \text{ sufficiently large.} \tag{4.45}$$

Step 2: We now prove that there exists a point $c \in \mathbb{C}$ such that $|p(c)| \leq |p(z)|$ for all $z \in \mathbb{C}$; in other words, we are claiming that $|p(z)|$ achieves its minimum value on $\mathbb{C}$. The proof of this involves the Bolzano–Weierstrass theorem. Define

$$m := \inf A, \qquad A := \{|p(z)| \, ; \, z \in \mathbb{C}\}.$$

This infimum certainly exists, since A is nonempty and bounded below by zero. Since m is the greatest lower bound of A, for each $k \in \mathbb{N}$, $m+1/k$ is no longer a lower bound, so there is a point $z_k \in \mathbb{C}$ such that $m \leq |p(z_k)| < m + 1/k$. By (4.45), the sequence $\{z_k\}$ must be bounded, so by the Bolzano–Weierstrass theorem, this sequence has a convergent subsequence $\{w_k\}$. If c is the limit of this subsequence, then by continuity of polynomials, $|p(w_k)| \to |p(c)|$, and since $m \leq |p(z_k)| < m + 1/k$ for all k, by the squeeze theorem we must have $|p(c)| = m$.

Step 3: The rest of the proof involves showing that the minimum m must be zero, which shows that $p(c) = 0$, and so c is a root of $p(z)$. To do so, we introduce an auxiliary polynomial $q(z)$ as follows. Let us suppose, for the sake of contradiction, that $p(c) \neq 0$. Define $q(z) := p(z + c)/p(c)$. Then $|q(z)|$ has a minimum at the point $z = 0$, the minimum being $|q(0)| = |1| = 1$. Since $q(0) = 1$, we can write

$$q(z) = b_n z^n + \cdots + 1 = b_n z^n + \cdots + b_k z^k + 1, \qquad (4.46)$$

where k is the smallest natural number such that $b_k \neq 0$. In our next step we shall prove that 1 is in fact not the minimum of $|q(z)|$, which gives a contradiction.

Step 4: By our lemma, $-1/b_k$ has a kth root a, so that $a^k = -1/b_k$. Then $|q(az)|$ also has a minimum at $z = 0$, and

$$q(az) = 1 + b_k(az)^k + \cdots = 1 - z^k + \cdots,$$

where $\cdots$ represents terms of higher degree than k. Thus, we can write

$$q(az) = 1 - z^k + z^{k+1} r(z),$$

where $r(z)$ is a polynomial of degree at most $n - (k + 1)$. Let $z = x$, a real number with $0 < x < 1$, be so small that $x\,|r(x)| < 1$. Then,

$$
\begin{aligned}
|q(ax)| = |1 - x^k + x^{k+1} r(x)| &\leq |1 - x^k| + x^{k+1}|r(x)| \\
&< 1 - x^k + x^k \cdot 1 = 1 = |q(0)| \implies |q(ax)| < |q(0)|.
\end{aligned}
$$

Therefore, $|q(0)|$ is not the minimum of $|q(z)|$. Hence our assumption that $p(c) \neq 0$ must have been false, and our proof is complete.   ∎

We remark that the other two proofs of the FTA in this section (basically) differ from this proof only at the first line in **Step 4**, in how we claim that there is a complex number a with $a^k = -1/b_k$.

As a consequence of the fundamental theorem of algebra, we can prove the well-known fact that every polynomial can be factored. Let $p(z)$ be a polynomial of positive degree n with complex coefficients and let c_1 be a root of p, which we know exists by the FTA. Then from Lemma 2.60 on p. 142, we can write

$$p(z) = (z - c_1)\, q_1(z),$$

where $q_1(z)$ is a polynomial of degree $n - 1$ in both z and c_1. By the FTA, q_1 has a root, call it c_2. Then from Lemma 2.60, we can write $q_1(z) = (z - c_2)\, q_2(z)$, where q_2 has degree $n - 2$, and substituting q_1 into the formula for p, we obtain

$$p(z) = (z - c_1)(z - c_2)\, q_2(z).$$

Proceeding a total of $n - 2$ more times in this fashion, we eventually arrive at

$$p(z) = (z - c_1)(z - c_2) \cdots (z - c_n)\, q_n,$$

where q_n is a polynomial of degree zero, that is, a necessarily nonzero constant. It follows that $c_1, \ldots, c_n$ are roots of $p(z)$. Moreover, these numbers are the only roots, for if

$$0 = p(c) = (c - c_1)(c - c_2) \cdots (c - c_n)\, q_n,$$

then c must equal one of the c_k, since a product of complex numbers is zero if and only if one of the factors is zero. Summarizing, we have proved the following.

Corollary 4.44 *If $p(z)$ is a polynomial of positive degree n, then p has exactly n complex roots $c_1, \ldots, c_n$ counting multiplicities, and for some $a \in \mathbb{C}$, we can write*

$$p(z) = a\,(z - c_1)(z - c_2) \cdots (z - c_n).$$

4.10.2   Proof II of the FTA [179]

Our second proof of the FTA is almost exactly the same as the first, but we substitute the result in Lemma 4.46 for the argument at the beginning of **Step 4** in the above proof. We start with the following lemma, which is "obvious" by thoroughly studying Fig. 4.40.

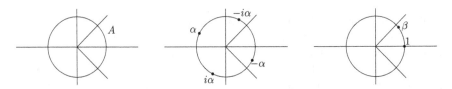

Fig. 4.40 *Left A* is the set of $z = e^{i\theta}$ with $-\pi/4 \le \theta \le \pi/4$. *Middle* for every α with $|\alpha| = 1$, at least one of α, $i\alpha$, $-i\alpha$, and $-\alpha$ lies in A, in this case, $-\alpha$ does. *Right* if $|\beta| = 1$ and $\beta \in A$, then the distance between β and 1 is less than 1

We leave the proof of the following lemma to Problem 6.

Lemma 4.45 *Let α, β be complex numbers on the unit circle with $\beta \in A$, where A is as shown in Fig. 4.40.*

(i) At least one of the numbers α, $i\alpha$, $-i\alpha$, and $-\alpha$ lies in A.
(ii) The distance between β and 1 is less than 1.

Lemma 4.46 *Let ℓ be an odd natural number, $\zeta = (1 + i)/\sqrt{2}$, and let α be a complex number of length 1. Then there is a natural number ν such that*

$$|\alpha \zeta^{2\nu\ell} - 1| < 1.$$

Proof Observe that

$$\zeta^2 = \frac{(1+i)(1+i)}{2} = \frac{1 + 2i + i^2}{2} = i.$$

Therefore, $\alpha \zeta^{2\nu\ell} - 1$ simplifies to $\alpha i^{\nu\ell} - 1$, and we shall use this latter expression for the rest of the proof. Since ℓ is odd, say $\ell = 2j + 1$, we have

$$i^\ell = i^{2j+1} = (-1)^j i.$$

Suppose that j is even in what follows, so that $i^\ell = i$; there is an almost identical argument in the case that j is odd. It follows that $i^{\nu\ell} = i^\nu$. Since i^ν equals i (for $\nu = 1$), -1 (for $\nu = 2$), $-i$ (for $\nu = 3$), or 1 (for $\nu = 4$), by *(1)* of our lemma we can choose ν such that αi^ν lies in the set A shown in Fig. 4.40. Then by *(2)* of our lemma, for this ν we have $|\alpha i^{\nu\ell} - 1| < 1$. ∎

The fundamental theorem of algebra, Proof II

Theorem 4.47 *Every complex polynomial of positive degree has at least one complex root.*

Proof We proceed by strong induction. Certainly the FTA holds for all polynomials of first degree, and therefore assume that $p(z)$ is a polynomial of degree $n \geq 2$ and suppose that the FTA holds for all polynomials of degree less than n.

Now we proceed, *without changing a single word*, exactly as in Proof I up to **Step 4**, where we substitute the following argument.

Step 4 modified: Recall that the polynomial $q(z)$ in (4.46),

$$q(z) = b_n z^n + \cdots + b_k z^k + 1,$$

has the property that $|q(z)|$ has the minimum value 1. We claim that the k in this expression cannot equal n. To see this, for the sake of contradiction, let us suppose that $k = n$. Then $q(z) = b_n z^n + 1$, and $q(z)$ has the property that

$$|q(z)| = |1 + b_n z^n| = \left| 1 + \frac{b_n}{|b_n|} |b_n| z^n \right| = |\alpha\, w^n - 1|$$

has the minimum value 1, where $\alpha = -b_n/|b_n|$ has unit length and $w = |b_n|^{1/n} z$. We derive a contradiction in three cases: $n > 2$ is even, $n = 2$, and n is odd. If $n > 2$ is even, then we can write $n = 2m$ for a natural number m with $2 \le m < n$. By our induction hypothesis (the FTA holds for all polynomials of degree less than n), there is a number η such that $\eta^m - 1/\alpha = 0$, and there is a number ξ such that $\xi^2 - \eta = 0$. Then

$$\xi^n = \xi^{2m} = (\xi^2)^m = \eta^m = 1/\alpha.$$

Thus, for $w = \xi$, we obtain $|\alpha\, w^n - 1| = 0$, which contradicts the fact that $|\alpha\, w^n - 1|$ is never less than 1. Now suppose that $n = 2$. Then by our lemma with $\ell = 1$, there is a ν such that

$$|\alpha\, \zeta^{2\nu} - 1| < 1,$$

where $\zeta = (1 + i)/\sqrt{2}$. This shows that $w = \zeta^\nu$ satisfies $|\alpha\, w^n - 1| < 1$, again contradicting the fact that $|\alpha\, w^n - 1|$ is never less than 1. Finally, suppose that $n = \ell$ is odd. Then by our lemma, there is a ν such that

$$|\alpha\, \zeta^{2\nu n} - 1| < 1,$$

where $\zeta = (1 + i)/\sqrt{2}$, which shows that $w = \zeta^{2\nu}$ satisfies $|\alpha\, w^n - 1| < 1$, again resulting in a contradiction. Therefore, $k < n$.

Now that we've proved $k < n$, we can use our induction hypothesis to conclude that there is a complex number a such that $a^k + 1/b_k = 0$, that is, $a^k = -1/b_k$. We can now proceed exactly as in **Step 4** of Proof I to finish the proof. ∎

4.10.3 Roots of Complex Numbers

Back in Section 2.7 on p. 89, we learned how to find nth roots of nonnegative real numbers; we now generalize this to *complex* numbers using the polar representation of complex numbers studied in Section 4.9 on p. 323.

Let $n \in \mathbb{N}$ and let w be a complex number. We shall find all nth roots of z using trigonometry. If $z = 0$, then the only ξ that works is $\xi = 0$, since the product of nonzero complex numbers is nonzero. Therefore, we assume henceforth that $z \neq 0$. We can write $z = re^{i\theta}$, where $r > 0$ and $\theta \in \mathbb{R}$, and given a nonzero complex ξ, we can write $\xi = \rho e^{i\phi}$, where $\rho > 0$ and $\phi \in \mathbb{R}$. Then $\xi^n = z$ if and only if

$$\rho^n e^{in\phi} = r e^{i\theta}.$$

Taking the absolute value of both sides, and using that $|e^{in\phi}| = 1 = |e^{i\theta}|$, we get $\rho^n = r$, or $\rho = \sqrt[n]{r}$. Now canceling $\rho^n = r$, we see that

$$e^{in\phi} = e^{i\theta},$$

which holds if and only if $n\phi = \theta + 2\pi m$ for some integer m, or

$$\phi = \frac{\theta}{n} + \frac{2\pi m}{n}, \quad m \in \mathbb{Z}.$$

As the reader can easily check, every number of this form differs by an integer multiple of 2π from one of the following numbers:

$$\frac{\theta}{n}, \quad \frac{\theta}{n} + \frac{2\pi}{n}, \quad \frac{\theta}{n} + \frac{4\pi}{n}, \quad \ldots, \quad \frac{\theta}{n} + \frac{2\pi}{n}(n-1).$$

None of these numbers differs by an integer multiple of 2π, and therefore, by our knowledge of π and the unit circle, all the n numbers

$$e^{i\frac{1}{n}(\theta+2\pi k)}, \quad k = 0, 1, 2, \ldots, n-1$$

are distinct. Thus, there is a total of n solutions ξ to the equation $\xi^n = z$, all of them given in the following theorem.

Existence of complex nth roots

> **Theorem 4.48** *For every $n \in \mathbb{N}$, there are exactly n nth roots of every nonzero complex number $z = re^{i\theta}$; the complete set of roots is given by*
>
> $$\sqrt[n]{r}\, e^{i\frac{1}{n}(\theta+2\pi k)} = \sqrt[n]{r}\left[\cos\frac{1}{n}(\theta + 2\pi k) + i\sin\frac{1}{n}(\theta + 2\pi k)\right], \quad k = 0, 1, 2, \ldots, n-1.$$

There is a very convenient way to write these nth roots, as we now describe. First of all, notice that

$$\sqrt[n]{r}\, e^{i\frac{1}{n}(\theta+2\pi k)} = \sqrt[n]{r}\, e^{i\frac{\theta}{n}} \cdot e^{i\frac{2\pi k}{n}} = \sqrt[n]{r}\, e^{i\frac{\theta}{n}} \cdot \left(e^{i\frac{2\pi}{n}}\right)^k.$$

Therefore, the nth roots of z are given by

$$\sqrt[n]{r}\, e^{i\frac{\theta}{n}} \cdot \omega^k, \qquad k = 0, 1, \ldots, n-1, \quad \text{where } \omega = e^{i\frac{2\pi}{n}} = \cos\frac{2\pi}{n} + i\sin\frac{2\pi}{n}.$$

Of all the n distinct roots, there is one called the **principal nth root**, denoted by $\sqrt[n]{z}$. It is the nth root given by

$$\boxed{\sqrt[n]{z} := \sqrt[n]{r}\, e^{i\frac{\theta}{n}}, \quad \text{where } -\pi < \theta \le \pi.}$$

Note that if $z = x$ is a *positive* real number, then $x = re^{i0}$ with $r = x$. Since $-\pi < 0 \le \pi$, the principal nth root of x is by definition $\sqrt[n]{x}e^{i0/n} = \sqrt[n]{x}$, the usual real nth root of x. Thus, there is no ambiguity in notation between the complex principal nth root of a positive real number and its real nth root.

We now give some examples.

Example 4.35 For our first example, we find the square roots of -1. Since $-1 = e^{i\pi}$, because $\cos\pi + i\sin\pi = -1 + i0$, the square roots of -1 are $e^{i(1/2)\pi}$ and $e^{i(1/2)(\pi+2\pi)} = e^{i3\pi/2}$. Since

$$e^{i\pi/2} = \cos\frac{\pi}{2} + i\sin\frac{\pi}{2} = i,$$

and similarly, $e^{i3\pi/2} = -i$, we get i and $-i$ as the square roots of -1. Note that the principal square root of -1 is i, and so $\sqrt{-1} = i$, just as we learned in high school!

Example 4.36 Next let us find all nth roots of 1 (called the roots of unity). Since $1 = 1\,e^{i0}$, all the n nth roots of 1 are given by

$$1, \omega, \omega^2, \ldots, \omega^{n-1}, \quad \text{where } \omega := e^{i\frac{2\pi}{n}} = \cos\frac{2\pi}{n} + i\sin\frac{2\pi}{n}.$$

Consider $n = 4$. In this case, $\cos\frac{2\pi}{4} + i\sin\frac{2\pi}{4} = i$, $i^2 = -1$, and $i^3 = -i$. Therefore, the fourth roots of unity are

$$1, i, -1, -i.$$

Since

$$\cos\frac{2\pi}{3} + i\sin\frac{2\pi}{3} = -\frac{1}{2} + i\frac{\sqrt{3}}{2},$$

the cube roots of unity are

$$1,\ -\frac{1}{2} + i\frac{\sqrt{3}}{2},\ -\frac{1}{2} - i\frac{\sqrt{3}}{2}.$$

4.10.4   Proof III of the FTA

We now give our third proof.

The fundamental theorem of algebra, Proof III

> **Theorem 4.49** *Every complex polynomial of positive degree has at least one complex root.*

Proof We proceed, *without changing a single word*, exactly as in Proof I up to **Step 4**, where we use the following.

Step 4 modified: At the beginning of **Step 4** in Proof I, we used Lemma 4.42 to conclude that there is a complex a such that $a^k = -1/b_k$. Now we can simply invoke Theorem 4.48 to verify that there is such a number a. Explicitly, we can just write $-1/b_k = re^{i\theta}$ and simply define $a = r^{1/k}e^{i\theta/k}$. In any case, now that we have such an a, we can proceed exactly as in **Step 4** of Proof I to finish the proof. ■

▶ **Exercises 4.10**

1. Let $p(z)$ and $q(z)$ be polynomials of degree at most n.

    (a) If p vanishes at $n + 1$ distinct complex numbers, prove that $p = 0$, the zero polynomial.
    (b) If p and q agree at $n + 1$ distinct complex numbers, prove that $p = q$.
    (c) If $c_1, \ldots, c_n$ (with each root repeated according to multiplicity) are roots of $p(z)$, a polynomial of degree n, prove that $p(z) = a_n(z - c_1)(z - c_2) \cdots (z - c_n)$, where a_n is the coefficient of z^n in the expression for $p(z)$.

2. Find the following roots and state which of the roots represents the principal root.

    (a) Find the cube roots of -1.
    (b) Find the square roots of i.
    (c) Find the cube roots of i.
    (d) Find the square roots of $\sqrt{3} + 3i$.

3. Geometrically (not rigorously) demonstrate that the nth roots, with $n \geq 3$, of a nonzero complex number z are the vertices of a regular polygon.

4. Let $n \in \mathbb{N}$ and let $\omega = e^{i\frac{2\pi}{n}}$. If k is an integer that is not a multiple of n, prove that

    $$1 + \omega^k + \omega^{2k} + \omega^{3k} + \cdots + \omega^{(n-1)k} = 0.$$

5. Prove that every quadratic polynomial $az^2 + bz + c = 0$ with complex coefficients has two complex roots, counting multiplicities, given by

    $$z = \frac{-b \pm \sqrt{b^2 - 4ac}}{2a},$$

where $\sqrt{b^2 - 4ac}$ is the principal square root of $b^2 - 4ac$.

6. Prove Lemma 4.45.

7. We show how the ingenious mathematicians of the past solved the general cubic equation $z^3 + bz^2 + cz + d = 0$ with complex coefficients; for the history, see [95].

   (i) First, replacing z with $z - b/3$, show that our cubic equation transforms into an equation of the form $z^3 + \alpha z + \beta = 0$, where α and β are complex. Thus, we may focus our attention on the equation $z^3 + \alpha z + \beta = 0$.

   (ii) Second, show that the substitution $z = w - \alpha/(3w)$ gives an equation of the form
   $$27(w^3)^2 + 27\beta(w^3) - \alpha^3 = 0,$$
   a quadratic equation in w^3. We can solve this equation for w^3 by Problem 5, and therefore we can solve for w, and therefore we can get $z = w - \alpha/(3w)$!

8. A nice application of the previous problem is finding $\sin(\pi/9)$ and $\cos(\pi/9)$.

   (i) Use de Moivre's formula to prove that
   $$\cos 3x = \cos^3 x - 3\cos x \sin^2 x, \qquad \sin 3x = 3\cos^2 x \sin x - \sin^3 x.$$

   (ii) Choose one of these equations, and using $\cos^2 x + \sin^2 x = 1$, turn the right-hand side into a cubic polynomial in $\cos x$ or $\sin x$.

   (iii) Using the equation you get, find $\sin(\pi/9)$ and $\cos(\pi/9)$.

9. This problem is for the classical mathematicians at heart: We find square roots without using the technology of trigonometric functions.

   (i) Let $z = a + ib$ be a nonzero complex number with $b \neq 0$. Show that $\xi = x + iy$ satisfies $\xi^2 = z$ if and only if $x^2 - y^2 = a$ and $2xy = b$.

   (ii) Prove that $x^2 + y^2 = \sqrt{a^2 + b^2} = |z|$, and then $x^2 = \frac{1}{2}(|z| + a)$ and $y^2 = \frac{1}{2}(|z| - a)$.

   (iii) Finally, deduce that z must equal
   $$\xi = \pm\left(\sqrt{\frac{|z| + a}{2}} + i\frac{b}{|b|}\sqrt{\frac{|z| - a}{2}}\right).$$

10. Prove that if r is a root of a polynomial $p(z) = z^n + a_{n-1}z^{n-1} + \cdots + a_0$, then $|r| \leq \max\left\{1, \sum_{k=0}^{n-1}|a_k|\right\}$.

11. (Cf. [245]) (**Continuous dependence of roots**) In this problem we prove the following useful theorem. Let z_0 be a root of multiplicity m of a polynomial $p(z) = z^n + a_{n-1}z^{n-1} + \cdots + a_0$. Then given $\varepsilon > 0$, there is a $\delta > 0$ such that if $q(z) = z^n + b_{n-1}z^{n-1} + \cdots + b_0$ satisfies $|b_j - a_j| < \delta$ for all $j = 0, \ldots, n - 1$, then $q(z)$ has at least m roots within ε of z_0. You may proceed as follows.

(i) Suppose the theorem is false. Prove that there exist $\varepsilon > 0$ and a sequence $\{q_k\}$ of polynomials $q_k(z) = z^n + b_{k,n-1}z^{n-1} + \cdots + b_{k,0}$ such that q_k has at most $m - 1$ roots within ε of z_0 and for each $j = 0, \ldots, n - 1$, we have $b_{k,j} \to a_j$ as $k \to \infty$.

(ii) Let $r_{k,1}, \ldots, r_{k,n}$ be the n roots of q_k. Let $R_k = (r_{k,1}, \ldots, r_{k,n}) \in \mathbb{C}^n = \mathbb{R}^{2n}$. Prove that the sequence $\{R_k\}$ has a convergent subsequence. Suggestion: Problem 10 is helpful.

(iii) By relabeling the subsequence if necessary, we assume that $\{R_k\}$ itself converges; say $R_k = (r_{k,1}, \ldots, r_{k,n}) \to (r_1, \ldots, r_n)$. Prove that at most $m - 1$ of the r_j can equal z_0.

(iv) Prove that $q_k(z) = (z - r_{k,1})(z - r_{k,2}) \cdots (z - r_{k,n})$, and therefore, for each $z \in \mathbb{C}$, $\lim_{k \to \infty} q_k(z) = (z - r_1)(z - r_2) \cdots (z - r_n)$. On the other hand, using that $b_{k,j} \to a_j$ as $k \to \infty$, prove that for each $z \in \mathbb{C}$, $\lim_{k \to \infty} q_k(z) = p(z)$. Derive a contradiction.

4.11 The Inverse Trigonometric Functions and the Complex Logarithm

In this section we study the inverse trigonometric functions. We then use these functions to derive properties of the polar angle, also called the argument of a complex number. In Section 4.7 we developed the properties of real logarithms, and using the logarithm, we defined complex powers of positive bases. In our current section we shall extend logarithms to include complex logarithms, which are then used to define complex powers with complex bases. Finally, we use the complex logarithm to define complex inverse trigonometric functions.

4.11.1 The Real-Valued Inverse Trigonometric Functions

By the oscillation theorem, Theorem 4.39 on p. 331, we know that

$$\sin : [-\pi/2, \pi/2] \longrightarrow [-1, 1] \quad \text{and} \quad \cos : [0, \pi] \longrightarrow [-1, 1]$$

are both strictly monotone bijective continuous functions, sine being strictly increasing and cosine being strictly decreasing, as in the two left-hand pictures in Fig. 4.41. In particular, by the monotone inverse theorem on p. 297, each of these functions has a strictly monotone inverse, which we denote by

$$\arcsin : [-1, 1] \longrightarrow [-\pi/2, \pi/2] \quad \text{and} \quad \arccos : [-1, 1] \longrightarrow [0, \pi],$$

called the **inverse, or arc, sine**, which is strictly increasing, and **inverse, or arc, cosine**, which is strictly decreasing. To explain the term "arc," consider sine and observe that for $\theta \in [-\pi/2, \pi]$, we have

$$\sin \theta = x \quad \Longleftrightarrow \quad \theta = \arcsin x.$$

Therefore, $\arcsin x$ gives the *arc*, or *angle*, whose sine is x. Now being inverse functions, the arcsine and arccosine functions satisfy

$$\sin(\arcsin x) = x, \ -1 \leq x \leq 1 \quad \text{and} \quad \arcsin(\sin x) = x, \quad -\pi/2 \leq x \leq \pi/2,$$

and

$$\cos(\arccos x) = x, \ -1 \leq x \leq 1 \quad \text{and} \quad \arccos(\cos x) = x, \ 0 \leq x \leq \pi.$$

If $0 \leq \theta \leq \pi$, then $-\pi/2 \leq \pi/2 - \theta \leq \pi/2$, so letting x denote both sides of the identity

$$\cos \theta = \sin \left(\frac{\pi}{2} - \theta \right),$$

we get $\theta = \arccos x$ and $\pi/2 - \theta = \arcsin x$, which further implies that

$$\arccos x = \frac{\pi}{2} - \arcsin x, \quad \text{for all } -1 \leq x \leq 1. \tag{4.47}$$

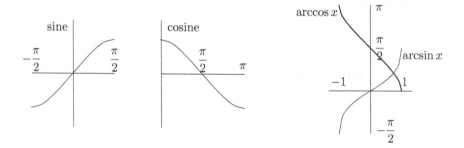

Fig. 4.41 Graphs of sine and cosine and their inverses

We now introduce the inverse tangent function. We first claim that

$$\tan : (-\pi/2, \pi/2) \longrightarrow \mathbb{R}$$

is a strictly increasing bijection; here's the common graph of tangent:

Fig. 4.42   Graphs of tangent and arctangent

Indeed, since

$$\tan x = \frac{\sin x}{\cos x}$$

and sine is strictly increasing on $[0, \pi/2]$ from 0 to 1 and cosine is strictly decreasing on $[0, \pi/2]$ from 1 to 0, we see that tangent is strictly increasing on $[0, \pi/2)$ from 0 to ∞. Using the properties of sine and cosine on $[-\pi/2, 0]$, in a similar manner one can show that tan is strictly decreasing on $(-\pi/2, 0)$ from $-\infty$ to 0. This proves that $\tan : (-\pi/2, \pi/2) \longrightarrow \mathbb{R}$ is a strictly increasing bijection. Therefore, this function has a strictly increasing inverse, which we denote by

$$\arctan : \mathbb{R} \longrightarrow (-\pi/2, \pi/2);$$

it is called the **inverse, or arc, tangent** (Fig. 4.42).

4.11.2   The Argument of a Complex Number

Given a nonzero complex number z, we know that we can write $z = |z|e^{i\theta}$ for some $\theta \in \mathbb{R}$, and all such θ satisfying this equation differ by integer multiples of 2π. Geometrically, θ is interpreted as the angle z makes with the positive real axis when z is drawn as a vector in $\mathbb{R}^2$. Every such angle θ is called an **argument** of z and is denoted by arg z. Thus, we can write

$$z = |z| \, e^{i \arg z}.$$

We remark that arg z is not a function but is referred to as a **multiple-valued function**, since arg z does not represent a single value of θ; however, any two choices for arg z differ by an integer multiple of 2π. If w is another nonzero complex number, written as $w = |w| \, e^{i\phi}$, so that arg $w = \phi$, then

$$zw = \left(|z| \, e^{i\theta}\right)\left(|w| \, e^{i\phi}\right) = |z| \, |w| \, e^{i(\theta+\phi)}.$$

This formula implies that

$$\arg(zw) = \arg z + \arg w,$$

where we interpret this as saying that all choices for these three arguments satisfy this equation up to an integer multiple of 2π. Thus, the *argument of a product is the sum of the arguments*. What other function do you know of that takes products into sums? The logarithm of course. We shall shortly show how arg is involved in the definition of complex logarithms. Similarly, properly interpreted, we have

$$\arg\left(\frac{z}{w}\right) = \arg z - \arg w.$$

Since arg is somewhat ambiguous, being a set of values rather than just a single value, mathematically it would be nice to turn arg into an actual function. To do so, note that given a nonzero complex number z, there is exactly one argument satisfying $-\pi < \arg z \le \pi$; this particular angle is called the **principal argument** of z and is denoted by Arg z. Thus, Arg : $\mathbb{C}\backslash\{0\} \longrightarrow \mathbb{R}$ is characterized by the following properties:

$$\boxed{z = |z|e^{i\,\text{Arg}\,z}, \qquad -\pi < \text{Arg}\,z \le \pi.}$$

Then all arguments of z differ from the principal one by multiples of 2π:

$$\arg z = \text{Arg}\,z + 2\pi n, \qquad n \in \mathbb{Z}.$$

We can find many different formulas for Arg z using the inverse trig functions as follows. Writing z in terms of its real and imaginary parts, $z = x + iy$, and equating this with $|z|e^{i\,\text{Arg}\,z} = |z|\cos(\text{Arg}\,z) + i|z|\sin(\text{Arg}\,z)$, we see that

$$\cos \text{Arg}\,z = \frac{x}{\sqrt{x^2 + y^2}} \quad \text{and} \quad \sin \text{Arg}\,z = \frac{y}{\sqrt{x^2 + y^2}}. \qquad (4.48)$$

By the properties of cosine, we see that

$$-\frac{\pi}{2} < \text{Arg}\,z < \frac{\pi}{2} \quad \Longleftrightarrow \quad x > 0.$$

Since arcsin is the inverse of sin with angles in $(-\pi/2, \pi/2)$, it follows that

$$\text{Arg}\,z = \arcsin\left(\frac{y}{\sqrt{x^2 + y^2}}\right), \qquad x > 0.$$

Perhaps the most common formula for Arg z when $x > 0$ is in terms of arctangent, which is derived by dividing the formulas in (4.48) to get $\tan \text{Arg}\,z = y/x$ and then taking the arctangent of both sides:

$$\text{Arg } z = \arctan \frac{y}{x}, \qquad x > 0.$$

We now derive a formula for Arg z when $y > 0$. By the properties of sine, we see that

$$0 \leq \text{Arg } z \leq \pi \iff y \geq 0 \quad \text{and} \quad -\pi < \text{Arg } z < 0 \iff y < 0.$$

Assuming that $y \geq 0$, that is, $0 \leq \text{Arg } z \leq \pi$, we can take the arccosine of both sides of the first equation in (4.48) to get

$$\text{Arg } z = \arccos\left(\frac{x}{\sqrt{x^2 + y^2}}\right), \qquad y \geq 0.$$

Assume that $y < 0$, that is, $-\pi < \text{Arg } z < 0$. Then $0 < -\text{Arg } z < \pi$, and since $\cos \text{Arg } z = \cos(-\text{Arg } z)$, we get $\cos(-\text{Arg } z) = x/\sqrt{x^2 + y^2}$. Taking the arccosine of both sides, we get

$$\text{Arg } z = -\arccos\left(\frac{x}{\sqrt{x^2 + y^2}}\right), \qquad y \leq 0.$$

Putting together our expressions for Arg z, we obtain the following formulas for the principal argument:

$$\boxed{\text{Arg } z = \arctan \frac{y}{x} \quad \text{if } x > 0,} \tag{4.49}$$

and

$$\boxed{\text{Arg } z = \begin{cases} \arccos\left(\dfrac{x}{\sqrt{x^2 + y^2}}\right) & \text{if } y \geq 0, \\[3mm] -\arccos\left(\dfrac{x}{\sqrt{x^2 + y^2}}\right) & \text{if } y < 0. \end{cases}} \tag{4.50}$$

Using these formulas, we can easily prove the following theorem.

Theorem 4.50 Arg $: \mathbb{C}\backslash(-\infty, 0] \longrightarrow (-\pi, \pi)$ *is continuous.*

Proof Since

$$\mathbb{C}\backslash(-\infty, 0] = \{x + iy \, ; \, x > 0\} \cup \{x + iy \, ; \, y > 0\} \cup \{x + iy \, ; \, y < 0\},$$

all we have to do is prove that Arg is continuous on each of these three sets. But this is easy: The formula (4.49) shows that Arg is continuous when $x > 0$, the first formula in (4.50) shows that Arg is continuous when $y > 0$, and the second formula in (4.50) shows that Arg is continuous when $y < 0$.   ∎

4.11.3   The Complex Logarithm and Powers

Recall from Section 4.7.2 on p. 303 that if $a \in \mathbb{R}$ and $a > 0$, then a real number ξ having the property that

$$e^{\xi} = a$$

is called the logarithm of a; we know that ξ always exists and is unique, since $\exp : \mathbb{R} \longrightarrow (0, \infty)$ is a bijection. Of course, $\xi = \log a$ by definition of log. We now consider *complex* logarithms. We define such logarithms in an analogous way: If $z \in \mathbb{C}$ and $z \neq 0$, then a complex number ξ having the property that

$$e^{\xi} = z$$

is called a **complex logarithm** of z. The reason we assume $z \neq 0$ is that there is no complex ξ such that $e^{\xi} = 0$. We now show that nonzero complex numbers always have logarithms; however, in contrast to the case of real numbers, complex numbers have infinitely many distinct logarithms!

Theorem 4.51   *The complex logarithms of a nonzero complex number z are all of the form*

$$\xi = \log |z| + i\big(\text{Arg } z + 2\pi n\big), \qquad n \in \mathbb{Z}. \tag{4.51}$$

Therefore, all complex logarithms of z have exactly the same real part $\log |z|$, but have imaginary parts that differ from Arg z by integer multiples of 2π.

Proof The idea behind this proof is very simple: We write

$$z = |z| \cdot e^{i \arg z} = e^{\log |z|} \cdot e^{i \arg z} = e^{\log |z| + i \arg z}.$$

Since every argument of z is of the form Arg $z + 2\pi n$ for $n \in \mathbb{Z}$, this equation shows that all the numbers in (4.51) are indeed logarithms. On the other hand, if ξ is a logarithm of z, then $e^{\xi} = z$. Since also $z = e^{\log |z| + i \, \text{Arg } z}$, Theorem 4.41 on p. 515 implies that $\xi = \log |z| + i \, \text{Arg } z + 2\pi i n$ for some $n \in \mathbb{Z}$.   ∎

To isolate one of these infinitely many logarithms, we define the so-called "principal" one. For every nonzero complex number z, we define the **principal (branch of the) logarithm** of z by

$$\boxed{\operatorname{Log} z := \log |z| + i \operatorname{Arg} z.}$$

By Theorem 4.51, *all* logarithms of z are of the form

$$\operatorname{Log} z + 2\pi i \, n, \quad n \in \mathbb{Z}.$$

Note that if $x \in \mathbb{R}$, then $\operatorname{Arg} x = 0$, and therefore,

$$\operatorname{Log} x = \log x,$$

our usual logarithm, so Log is an extension of the real log to complex numbers.

Example 4.37 Observe that since $\operatorname{Arg}(-1) = \pi$ and $\operatorname{Arg} i = \pi/2$ and $\log |-1| = 0 = \log |i|$, since both equal $\log 1$, we have

$$\operatorname{Log}(-1) = i\pi \quad \text{and} \quad \operatorname{Log} i = i\frac{\pi}{2}.$$

The principal logarithm satisfies some of the properties of the real logarithm, but we need to be careful with the addition properties.

Example 4.38 For instance, observe that

$$\operatorname{Log}(-1 \cdot i) = \operatorname{Log}(-i) = \log |-i| + i \operatorname{Arg}(-i) = -i\frac{\pi}{2}.$$

On the other hand,

$$\operatorname{Log}(-1) + \operatorname{Log} i = i\pi + i\frac{\pi}{2} = i\frac{3\pi}{2},$$

so $\operatorname{Log}(-1 \cdot i) \neq \operatorname{Log}(-1) + \operatorname{Log} i$. Another example of this phenomenon is

$$\operatorname{Log}(-i \cdot -i) = \operatorname{Log}(-1) = i\pi,$$

while

$$\operatorname{Log}(-i) + \operatorname{Log}(-i) = -i\frac{\pi}{2} - i\frac{\pi}{2} = -i\pi.$$

However, under certain conditions, Log does satisfy the usual properties.

Theorem 4.52 *Let z and w be complex numbers.*

(1) If $-\pi < \mathrm{Arg}\, z + \mathrm{Arg}\, w \le \pi$, then

$$\mathrm{Log}\, zw = \mathrm{Log}\, z + \mathrm{Log}\, w.$$

(2) If $-\pi < \mathrm{Arg}\, z - \mathrm{Arg}\, w \le \pi$, then

$$\mathrm{Log}\, \frac{z}{w} = \mathrm{Log}\, z - \mathrm{Log}\, w.$$

(3) If $\mathrm{Re}\, z, \mathrm{Re}\, w \ge 0$ with at least one strictly positive, then both (1) and (2) hold.

Proof Suppose that $-\pi < \mathrm{Arg}\, z + \mathrm{Arg}\, w \le \pi$. By definition,

$$\mathrm{Log}\, zw = \log\big(|z|\,|w|\big) + i\big(\mathrm{Arg}\, zw\big) = \log|z| + \log|w| + i\,\mathrm{Arg}\, zw.$$

Since $\arg(zw) = \arg z + \arg w$, $\mathrm{Arg}\, z + \mathrm{Arg}\, w$ is an argument of zw, and since $\mathrm{Arg}(zw)$ is the unique argument of zw in $(-\pi, \pi]$ and $-\pi < \mathrm{Arg}\, z + \mathrm{Arg}\, w \le \pi$, it follows that $\mathrm{Arg}(zw) = \mathrm{Arg}\, z + \mathrm{Arg}\, w$. Thus,

$$\mathrm{Log}\, zw = \log|z| + \log|w| + i\,\mathrm{Arg}\, z + i\,\mathrm{Arg}\, w = \mathrm{Log}\, z + \mathrm{Log}\, w.$$

Property *(2)* is proved in a similar manner. Property *(3)* follows from *(1)* and *(2)*, since if $\mathrm{Re}\, z, \mathrm{Re}\, w \ge 0$ with at least one strictly positive, then as the reader can verify, the hypotheses of both *(1)* and *(2)* are satisfied. ∎

We now use Log to define complex powers of *complex* numbers. Recall from Section 4.7.3 on p. 304 that given a positive real number a and a complex number z, we have $a^z := e^{z \log a}$. Using Log instead of log, we can now define powers for *complex* a. Let a be a nonzero complex number and let z be a complex number. Every number of the form e^{zb}, where b is a complex logarithm of a, is called a **complex power of a to the z**; the choice of principal logarithm defines

$$\boxed{a^z = e^{z\,\mathrm{Log}\, a},}$$

and we call this the **principal value** of a to the power z. As before, a is called the **base** and z is called the **exponent**. Note that if a is a positive real number, then $\mathrm{Log}\, a = \log a$, so

$$a^z = e^{z\,\mathrm{Log}\, a} = e^{z \log a}$$

is the usual complex power of a defined on p. 304. Theorem 4.51 implies the following.

> **Theorem 4.53** *The complex powers of a nonzero complex number a to the power z are all of the form*
>
> $$e^{z\left(\operatorname{Log} a + 2\pi i n\right)}, \quad n \in \mathbb{Z}. \tag{4.52}$$

In general, there are infinitely many complex powers, but in certain cases they actually reduce to a finite number; see Problem 4. Here are some examples.

Example 4.39 Have you ever thought about what i^i equals? In this case, $\operatorname{Log} i = i\pi/2$, so

$$i^i = e^{i \operatorname{Log} i} = e^{i(i\pi/2)} = e^{-\pi/2},$$

a real number! Here is another nice example:

$$(-1)^{1/2} = e^{(1/2)\operatorname{Log}(-1)} = e^{(1/2)i\pi} = \cos\frac{\pi}{2} + i\sin\frac{\pi}{2} = i.$$

Therefore, $(-1)^{1/2} = i$, just as we suspected!

4.11.4 The Complex-Valued Arctangent Function

We now investigate the complex arctangent function; the other complex inverse functions are found in Problem 6. Given a complex number z, in the following theorem we shall find all complex numbers ξ such that

$$\tan\xi = z. \tag{4.53}$$

Of course, if we can find such a ξ, then we would like to call ξ the "inverse tangent of z." However, when this equation does have solutions, it turns out that it has infinitely many.

> **Lemma 4.54** *If $z = \pm i$, then the Eq.(4.53) has no solutions. If $z \neq \pm i$, then*
>
> $$\tan\xi = z \iff e^{2i\xi} = \frac{1 + iz}{1 - iz},$$
>
> *that is, if and only if $\xi = \dfrac{1}{2i} \times$ a complex logarithm of $\dfrac{1 + iz}{1 - iz}$.*

Proof The following statements are equivalent:

$$\tan\xi = z \iff \sin\xi = z\cos\xi \iff e^{i\xi} - e^{-i\xi} = iz(e^{i\xi} + e^{-i\xi})$$
$$\iff (e^{2i\xi} - 1) = iz(e^{2i\xi} + 1) \iff (1 - iz)e^{2i\xi} = 1 + iz.$$

If $z = i$, then this last equation is just $2e^{2i\xi} = 0$, which is impossible, and if $z = -i$, then the last equation is $0 = 2$, again an impossibility. If $z \neq \pm i$, then the last equation is equivalent to

$$e^{2i\xi} = \frac{1 + iz}{1 - iz},$$

which by definition of the complex logarithm just means that $2i\xi$ is a complex logarithm of the number $(1 + iz)/(1 - iz)$.   ∎

We now choose one of the solutions of (4.53), the obvious choice being the one corresponding to the principal logarithm: Given $z \in \mathbb{C}$ with $z \neq \pm i$, we define the **principal inverse, or arc, tangent** of z to be the complex number

$$\boxed{\text{Arctan } z = \frac{1}{2i} \text{ Log } \frac{1 + iz}{1 - iz}.}$$

This defines a function Arctan : $\mathbb{C}\backslash\{\pm i\} \longrightarrow \mathbb{C}$, which does satisfy (4.53):

$$\tan(\text{Arctan } z) = z, \qquad z \in \mathbb{C}, \ z \neq \pm i.$$

You might ask whether Arctan really is an "inverse" of tan. For example, you might ask whether Arctan is a bijection and whether $\text{Arctan } x = \arctan x$ for x real. The answer to the first question is yes if we restrict the domain of Arctan, and the answer to the second question is yes.

Properties of Arctan

Theorem 4.55 *Let*

$$D = \{z \in \mathbb{C}; z \neq iy, \ y \in \mathbb{R}, \ |y| \geq 1\}, \qquad E = \{\xi \in \mathbb{C}; \ |\text{Re } \xi| < \pi/2\}.$$

Then
$$\text{Arctan} : D \longrightarrow E$$

is a continuous bijection from D onto E with inverse $\tan : E \longrightarrow D$, *and when restricted to real values,*

$$\text{Arctan} : \mathbb{R} \longrightarrow (-\pi/2, \pi/2),$$

and this function equals the usual arctangent function $\arctan : \mathbb{R} \longrightarrow (-\pi/2, \pi/2)$.

Proof We begin by showing that $\text{Arctan}(D) \subseteq E$. First of all, by definition of Log, for every $z \in \mathbb{C}$ with $z \neq \pm i$ (not necessarily in D), we have

$$\operatorname{Arctan} z = \frac{1}{2i} \operatorname{Log} \frac{1+iz}{1-iz} = \frac{1}{2i} \left(\log \left| \frac{1+iz}{1-iz} \right| + i \operatorname{Arg} \frac{1+iz}{1-iz} \right)$$

$$= \frac{1}{2} \operatorname{Arg} \frac{1+iz}{1-iz} - \frac{i}{2} \log \left| \frac{1+iz}{1-iz} \right|. \qquad (4.54)$$

Since the principal argument of every complex number lies in $(-\pi, \pi]$, it follows that

$$-\frac{\pi}{2} < \operatorname{Re} \operatorname{Arctan} z \le \frac{\pi}{2}, \quad \text{for all } z \in \mathbb{C}, \ z \neq \pm i.$$

Assume that $\operatorname{Arctan} z \notin E$, which, by the above inequality, is equivalent to

$$2 \operatorname{Re} \operatorname{Arctan} z = \operatorname{Arg} \frac{1+iz}{1-iz} = \pi \quad \Longleftrightarrow \quad \frac{1+iz}{1-iz} \in (-\infty, 0).$$

If $z = x + iy$, then (by multiplying top and bottom of $\frac{1+iz}{1-iz}$ by $1 + i\bar{z}$ and making a short computation) we can write

$$\frac{1+iz}{1-iz} = \frac{1-|z|^2}{|1-iz|^2} + \frac{2x}{|1-iz|^2} i.$$

This formula shows that $(1 + iz)/(1 - iz) \in (-\infty, 0)$ if and only if $x = 0$ and $1 - |z|^2 < 0$, that is, either $x = 0$ and $1 - y^2 < 0$, or $|y| > 1$. Hence,

$$\frac{1+iz}{1-iz} \in (-\infty, 0) \quad \Longleftrightarrow \quad z = iy, \ |y| > 1.$$

In summary, for every $z \in \mathbb{C}$ with $z \neq \pm i$, we have $\operatorname{Arctan} z \notin E \Longleftrightarrow z \notin D$, or

$$\operatorname{Arctan} z \in E \quad \Longleftrightarrow \quad z \in D. \qquad (4.55)$$

Therefore, $\operatorname{Arctan}(D) \subseteq E$.

We now show that $\operatorname{Arctan}(D) = E$, so let $\xi \in E$. Define $z = \tan \xi$. Then according to Lemma 4.54, we have $z \neq \pm i$ and $e^{2i\xi} = \frac{1+iz}{1-iz}$. By definition of E, the real part of ξ satisfies $-\pi/2 < \operatorname{Re} \xi < \pi/2$. Since $\operatorname{Im}(2i\xi) = 2 \operatorname{Re}(\xi)$, we have $-\pi < \operatorname{Im}(2i\xi) < \pi$, and therefore by definition of the principal logarithm,

$$2i\xi = \operatorname{Log} \frac{1+iz}{1-iz}.$$

Hence, by definition of the arctangent, $\xi = \operatorname{Arctan} z$. The complex number z must be in D by (4.55) and the fact that $\xi = \operatorname{Arctan} z \in E$. This shows that $\operatorname{Arctan}(\tan \xi) = \xi$ for all $\xi \in E$, and we already know that $\tan(\operatorname{Arctan} z) = z$ for all $z \in D$ (in fact, for all $z \in \mathbb{C}$ with $z \neq \pm i$), so Arctan is a continuous bijection from D onto E with inverse given by $\tan$.

Finally, it remains to show that Arctan equals arctan when restricted to the real line. To prove this, we just need to prove that Arctan x is real when $x \in \mathbb{R}$. This will imply that Arctan : $\mathbb{R} \longrightarrow (-\pi/2, \pi/2)$ and therefore is just arctan. Now from (4.54), we see that if $x \in \mathbb{R}$, then the imaginary part of Arctan x is

$$\log \left| \frac{1 + ix}{1 - ix} \right| = \log \left| \frac{\sqrt{1 + x^2}}{\sqrt{1 + (-x)^2}} \right| = \log 1 = 0.$$

Thus, Arctan x is real, and our proof is complete.   ∎

Setting $z = 1$, we get Giulio Fagnano's (1682–1766) famous formula (see [13] for more on Giulio Carlo Fagnano dei Toschi):

$$\boxed{\frac{\pi}{4} = \frac{1}{2i} \, \mathrm{Log} \, \frac{1 + i}{1 - i}.}$$

▶ **Exercises 4.11**

1. Find the following logarithms:

$$\mathrm{Log}(1 + i\sqrt{3}), \quad \mathrm{Log}(\sqrt{3} - i), \quad \mathrm{Log}(1 - i)^4,$$

   and find the following powers: 2^i, $(1 + i)^i$, $e^{\mathrm{Log}(3+2i)}$, $i^{\sqrt{3}}$, $(-1)^{2i}$.
2. Prove the following identities:

$$\arctan x + \arctan y = \arctan \left(\frac{x + y}{1 - xy} \right),$$

$$\arctan(x + 1) + \arctan(x - 1) = \arctan \left(\frac{2}{x^2} \right),$$

$$\arctan x + \arctan \frac{1}{x} = \frac{\pi}{2},$$

$$\arcsin x + \arcsin y = \arcsin \left(x\sqrt{1 - y^2} + y\sqrt{1 - x^2} \right),$$

   and give restrictions under which these identities are valid. For example, the first identity holds when $xy \neq 1$ and the left-hand side lies strictly between $-\pi/2$ and $\pi/2$.
3. Find the exact value of the infinite series $\sum_{n=1}^{\infty} \arctan \left(\frac{2}{n^2} \right)$ and prove that the infinite series $\sum_{n=1}^{\infty} \left(\frac{\pi}{2} - \arctan n \right)$ diverges.
4. In this problem we study real powers of complex numbers. Let $a \in \mathbb{C}$ be nonzero.

   (a) Let $n \in \mathbb{N}$ and show that all powers of a to $1/n$ are given by

$$e^{\frac{1}{n} \left(\mathrm{Log} \, a + 2\pi i k \right)}, \quad k = 0, 1, 2, \ldots, n - 1.$$

In addition, show that these values are all the nth roots of a and that the principal nth root of a is the same as the principal value of $a^{1/n}$.

(b) If m/n is a rational number in lowest terms with $n > 0$, show that all powers of a to m/n are given by

$$e^{\frac{m}{n}\left(\operatorname{Log} a + 2\pi i k\right)}, \qquad k = 0, 1, 2, \ldots, n - 1.$$

(c) If x is an irrational number, show that there are infinitely many distinct complex powers of a to the x.

5. Let $a, b, z, w \in \mathbb{C}$ with $a, b \neq 0$ and prove the following:

(a) $1/a^z = a^{-z}$, $a^z \cdot a^w = a^{z+w}$, and $(a^z)^n = a^{zn}$ for all $n \in \mathbb{Z}$.
(b) If $-\pi < \operatorname{Arg} a + \operatorname{Arg} b \leq \pi$, then $(ab)^z = a^z b^z$.
(c) If $-\pi < \operatorname{Arg} a - \operatorname{Arg} b \leq \pi$, then $(a/b)^z = a^z/b^z$.
(d) If $\operatorname{Re} a, \operatorname{Re} b > 0$, then both (b) and (c) hold.
(c) Give examples showing that the conclusions of (b) and (c) are false if the hypotheses are not satisfied.

6. (**Arcsine and arccosine function**) In this problem we define the principal arcsine and arccosine functions. To define the complex arcsine, given $z \in \mathbb{C}$ we want to solve the equation $\sin \xi = z$ for ξ and call ξ the "inverse, or arc, sine of z."

(a) Prove that $\sin \xi = z$ if and only if $(e^{i\xi})^2 - 2iz\,(e^{i\xi}) - 1 = 0$.
(b) Solving this quadratic equation for $e^{i\xi}$ (see Problem 5 on p. 349), prove that $\sin \xi = z$ if and only if

$$\xi \;=\; \frac{1}{i} \times \text{a complex logarithm of } iz \pm \sqrt{1 - z^2}.$$

Because of this formula, we define the **principal inverse, or arc, sine** of z to be the complex number

$$\boxed{\operatorname{Arcsin} z := \frac{1}{i} \, \operatorname{Log}\left(iz + \sqrt{1 - z^2}\right).}$$

Based on the formula (4.47) on p. 352, we define the **principal inverse, or arc, cosine** of z to be the complex number

$$\boxed{\operatorname{Arccos} z := \frac{\pi}{2} - \operatorname{Arcsin} z.}$$

(c) Prove that when restricted to real values, $\operatorname{Arcsin} : [-1, 1] \longrightarrow [-\pi/2, \pi/2]$ and equals the usual arcsine function.
(d) Similarly, prove that when restricted to real values, $\operatorname{Arccos} : [-1, 1] \longrightarrow [0, \pi]$ and equals the usual arccosine function.

7. (**Inverse hyperbolic functions**) We look at the inverse hyperbolic functions.

    (a) Prove that sinh : $\mathbb{R} \longrightarrow \mathbb{R}$ is a strictly increasing bijection. We denote its inverse by arcsinh, the **hyperbolic arcsine**. Show that cosh : $[0, \infty) \longrightarrow [1, \infty)$ is a strictly increasing bijection; the inverse is denoted by arccosh, the **hyperbolic arccosine**.

    (b) Using a similar argument as you did for the arcsine function in Problem 6, prove that $\sinh x = y$ (here, $x, y \in \mathbb{R}$) if and only if $e^{2x} - 2ye^x - 1 = 0$, which holds if and only if $e^x = y \pm \sqrt{y^2 + 1}$. From this, prove that

$$\text{arcsinh}\, x = \log(x + \sqrt{x^2 + 1}).$$

If x is replaced by $z \in \mathbb{C}$ and log by Log, the principal complex logarithm, then this formula is called the **principal hyperbolic arcsine** of z.

    (c) Prove that

$$\text{arccosh}\, x = \log(x + \sqrt{x^2 - 1}).$$

If x is replaced by $z \in \mathbb{C}$ and log by Log, the principal complex logarithm, then this formula is called the **principal inverse hyperbolic cosine** of z.

4.12 ★ The Amazing π and Its Computation from Ancient Times

In the *Measurement of the Circle*, Archimedes (287–212 B.C.), listed three famous propositions involving π. In this section we look at each of these propositions, especially his third one, which uses the first known algorithm to compute π to any desired number of decimal places![15] His basic idea is to approximate a circle by inscribed and circumscribed regular polygons. We begin by looking at a brief history of π.

4.12.1 A Brief (and Very Incomplete) History of π

We begin by giving a short snippet of the history of π with, unfortunately, many details left out. Some of what we choose to put here is based on what will come up later in this book (for example, in the chapter on continued fractions; see Section 8.5 starting on p. 636) or what might be interesting trivia. References include the comprehensive chronicles [212–214], the beautiful books [9, 25, 73, 195], the wonderful websites [180, 182, 221], the short synopsis [200], and (my favorite π papers) [47, 48]. Before

---

[15][On π] "Ten decimal places of are sufficient to give the circumference of the earth to a fraction of an inch, and thirty decimal places would give the circumference of the visible universe to a quantity imperceptible to the most powerful microscope." Simon Newcomb (1835–1909) [153].

discussing this history, assume that the area and circumference of a circle are related to its radius r as follows:

Area of $\bigodot$ of radius $r = \pi r^2$,   Circumference of $\bigodot$ of radius $r = 2\pi r$.

In terms of the diameter $d := 2r$, we have

$$\text{Area of } \bigodot = \pi\frac{d^2}{4}, \quad \text{Circumference of } \bigodot = \pi d, \quad \pi = \frac{\text{circumference}}{\text{diameter}}.$$

(1) (circa 1650 B.C.) The Rhind (or Ahmes) papyrus is the oldest known mathematical text in existence. It is named after the Egyptologist Alexander Henry Rhind (1833–1863), who purchased it in Luxor in 1858, but it was written by a scribe Ahmes (1680 B.C.–1620 B.C.). In this text is written the following rule to find the area of a circle: *Cut $\frac{1}{9}$ off the circle's diameter and construct a square on the remainder.* Thus,

$$\pi\frac{d^2}{4} = \text{area of circle} \approx \text{square of } \left(d - \frac{1}{9}d\right) = \left(d - \frac{1}{9}d\right)^2 = \left(\frac{8}{9}\right)^2 d^2.$$

Canceling d^2 from both extremities, we obtain

$$\pi \approx 4\left(\frac{8}{9}\right)^2 = \left(\frac{4}{3}\right)^4 = 3.160493827\ldots.$$

(2) (circa 1000 B.C.) The Holy Bible in I Kings, Chapter 7, verse 23, and II Chronicles, Chapter 4, verse 2, states:

> And he made a molten sea, ten cubits from the one brim to the other: it was round all about, and his height was five cubits: and a line of thirty cubits did compass it about. I Kings 7:23.

This gives the approximate value (cf. the interesting article [5]):

$$\pi = \frac{\text{circumference}}{\text{diameter}} \approx \frac{30 \text{ cubits}}{10 \text{ cubits}} = 3.$$

Not only the Israelites used 3. Other ancient civilizations used 3 for the value of π for rough purposes (more than good enough for "everyday life"), including the Babylonians, Hindus, and Chinese.

(3) (circa 250 B.C.) Archimedes (287–212) gave the estimate $\pi \approx 22/7 = 3.14285714\ldots$ (correct to two decimal places). We'll thoroughly discuss *Archimedes's method* in a moment.

(4) (circa 500 A.D.) Tsu Chung-Chi (also Zu Chongzhi), of China (429–501) gave the estimate $\pi \approx 355/113 = 3.14159292\ldots$ (correct to six decimal places); he also gave the incredible estimate

$$3.1415926 < \pi < 3.1415927.$$

(5) (circa 1600 A.D.) The Dutch mathematician Adriaan Anthoniszoon (1527–1607) used Archimedes's method to get

$$\frac{333}{106} < \pi < \frac{377}{120}.$$

By taking the average of the numerators and denominators, he found Tsu Chung-Chi's approximation $355/113$.

(6) (1706) The symbol π was first introduced by William Jones (1675–1749) in his beginners' calculus book *Synopsis palmariorum mathesios*, where he published John Machin's (1680–1751) one hundred digit approximation to π; see Section 4.12.5 on p. 373 for more on Machin. The symbol π was popularized and became standard through Leonhard Euler's (1707–1783) famous book *Introduction in Analysing Infinitorum* [69]. The letter π was (reportedly) chosen because it's the first letter of the Greek words "perimeter" and "periphery."

(7) (1761) Johann Lambert (1728–1777) proved that π is irrational.

(8) (1882) Ferdinand von Lindemann (1852–1939) proved that π is transcendental. (See p. 143 for the definition of transcendental.)

(9) (1897) A sad day in the life of π. House bill No. 246, Indiana state legislature, 1897, written by a physician Edwin Goodwin (1828–1902), tried to legally set the value of π to a rational number; see [97, 228] for more about this sad tale. This value would be copyrighted and used in Indiana state math textbooks, and other states would have to pay to use this value! The bill is very convoluted (try to read Goodwin's article [90] and you'll probably get a headache), and (reportedly) the following values of π can be inferred from the bill: $\pi = 9.24$, 3.236, 3.232, and 3.2; it's also implied that $\sqrt{2} = 10/7$. Moreover, Mr. Goodwin claimed that he could trisect an angle, double a cube, and square a circle, which (quoting from the bill) "had been long since given up by scientific bodies as insolvable mysteries and above mans ability to comprehend." These problems "had been long since given up" because they have been proven unsolvable! (See [57, 86] for more on these unsolvable problems, first proved by Pierre Wantzel (1814–1848), and see [61] for other stories of amateur mathematicians claiming to have solved the unsolvable.) This bill passed the house (!), but fortunately, with the help of mathematician C.A. Waldo, of the Indiana Academy of Science, the bill didn't pass in the senate.

Hold on to your seats, because we'll take up our brief history of π again in Section 4.12.5, after a brief intermission.

4.12.2 Archimedes's Three Propositions

The following three propositions are contained in Archimedes's book *Measurement of the Circle* [107]:

(1) The area of a circle is equal to that of a right-angled triangle where the sides including the right angle are respectively equal to the radius and circumference of the circle.
(2) The ratio of the area of a circle to that of a square with side equal to the circle's diameter is close to 11:14.
(3) The ratio of the circumference of any circle to its diameter is less than 3 1/7 but greater than 3 10/71.

Figure 4.43 shows Archimedes's first proposition. Archimedes's second proposition gives the famous estimate $\pi \approx \frac{22}{7}$:

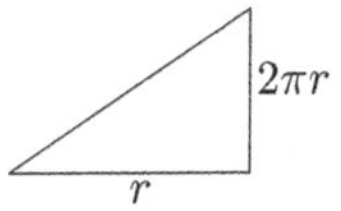

$$2\pi r \qquad \text{area } \triangle = \frac{1}{2}\text{base} \times \text{height} = \frac{1}{2}r \cdot (2\pi r) = \pi r^2$$

Fig. 4.43 Archimedes's first proposition

$$\frac{\text{area of circle}}{\text{area of square}} = \frac{\pi r^2}{(2r)^2} = \frac{\pi}{4} \approx \frac{11}{14} \implies \pi \approx \frac{22}{7}.$$

We now derive Archimedes's third proposition using the same method Archimedes pioneered over two thousand years ago, but we shall employ trigonometric functions! Archimedes's original method used plane geometry to derive his formulas (they didn't have the knowledge of trigonometric functions back then as we do now). However, before doing so, we need a couple of trig facts.

4.12.3 Some Useful Trig Facts

We first consider some useful trig identities.

Lemma 4.56 *We have*

$$\tan z = \frac{\sin(2z)\tan(2z)}{\sin(2z) + \tan(2z)} \quad and \quad 2\sin^2 z = \sin(2z)\tan z.$$

Proof We'll prove the first one and leave the second one to you. Multiplying $\tan z$ by $2\cos z/2\cos z = 1$ and using the double angle formulas $2\cos^2 z = 1 + \cos 2z$ and $\sin(2z) = 2\cos z \sin z$ (see Theorem 4.35 on p. 324), we obtain

$$\tan z = \frac{\sin z}{\cos z} = \frac{2\sin z \cos z}{2\cos^2 z} = \frac{\sin(2z)}{1 + \cos(2z)}.$$

Multiplying the top and bottom by $\tan 2z$, we get

$$\tan z = \frac{\sin(2z)\tan(2z)}{\tan(2z) + \cos(2z)\tan 2z} = \frac{\sin(2z)\tan(2z)}{\tan(2z) + \sin(2z)}. \qquad \blacksquare$$

Next, we consider some useful inequalities.

Lemma 4.57 *For $0 < x < \pi/2$, we have*

$$\sin x < x < \tan x.$$

Proof We first prove that $\sin x < x$ for $0 < x < \pi/2$. We note that the inequality $\sin x < x$ for $0 < x < \pi/2$ automatically implies that $\sin x < x$ holds for all $x > 0$, since x is increasing and $\sin x$ is oscillating. Recalling the infinite series for $\sin x$, the inequality $\sin x < x$, or $-x < -\sin x$, is equivalent to

$$-x < -x + \frac{x^3}{3!} - \frac{x^5}{5!} + \frac{x^7}{7!} - \frac{x^9}{9!} + - \cdots,$$

or after canceling the $-x$'s, this inequality is equivalent to

$$\frac{x^3}{3!}\left(1 - \frac{x^2}{4\cdot 5}\right) + \frac{x^7}{7!}\left(1 - \frac{x^2}{8\cdot 9}\right) + \cdots > 0.$$

For $0 < x < 2$, each of the terms in parentheses is positive. This shows that in particular, this expression is positive for $0 < x < \pi/2$.

We now prove that $x < \tan x$ for $0 < x < \pi/2$. This inequality is equivalent to $x\cos x < \sin x$ for $0 < x < \pi/2$. Substituting the infinite series for cos and sin, the inequality $x\cos x < \sin x$ is equivalent to

$$x - \frac{x^3}{2!} + \frac{x^5}{4!} - \frac{x^7}{6!} + - \cdots < x - \frac{x^3}{3!} + \frac{x^5}{5!} - \frac{x^7}{7!} + - \cdots.$$

Bringing everything to the right, we get an inequality of the form

$$x^3\left(\frac{1}{2!} - \frac{1}{3!}\right) - x^5\left(\frac{1}{4!} - \frac{1}{5!}\right) + x^7\left(\frac{1}{6!} + \frac{1}{7!}\right) - x^9\left(\frac{1}{8!} + \frac{1}{9!}\right) + - \cdots > 0.$$

Combining adjacent terms, the left-hand side is a sum of terms of the form

$$x^{2k-1}\left(\frac{1}{(2k-2)!}-\frac{1}{(2k-1)!}\right)-x^{2k+1}\left(\frac{1}{(2k)!}-\frac{1}{(2k+1)!}\right),\quad k=2,3,4,\cdots.$$

We claim that this term is positive for $0 < x < 3$. This shows that $x\cos x < \sin x$ for $0 < x < 3$, and so in particular, for $0 < x < \pi/2$. Now the above expression is positive if and only if

$$x^2 < \frac{\dfrac{1}{(2k-2)!}-\dfrac{1}{(2k-1)!}}{\dfrac{1}{(2k)!}-\dfrac{1}{(2k+1)!}} = (2k+1)(2k-2),\quad k=2,3,4,\ldots,$$

where we multiplied the top and bottom by $(2k+1)!$. The right-hand side is smallest when $k = 2$, when it equals $5 \cdot 2 = 10$. It follows that these inequalities hold for $0 < x < 3$, and our proof is now complete. ∎

4.12.4 Archimedes's Third Proposition

For a circle of radius r, we have

$$\frac{\text{circumference}}{\text{diameter}} = \frac{2\pi r}{2r} = \pi,$$

which we have to prove is "less than 3 1/7 but greater than 3 10/71." To do so, let us work with a circle of radius $r = 1/2$ and fix a natural number $M \geq 3$. Given $n = 0, 1, 2, 3, \ldots$, we inscribe and circumscribe the circle with regular polygons having $2^n M$ sides, as seen in Fig. 4.44. We denote the perimeter of the inscribed $2^n M$-gon by lowercase p_n and the perimeter of the circumscribed $2^n M$-gon by uppercase P_n. Then geometrically, we can see that

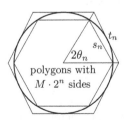

Fig. 4.44 Archimedes inscribed and circumscribed a circle with diameter 1 (radius 1/2) with regular polygons. The sides of these polygons have lengths s_n and t_n, respectively. The central angle of the inscribed and circumscribed $2^n M$-gons equals $2\theta_n$

$$p_n < \pi < P_n, \qquad n = 0, 1, 2, \ldots,$$

and $p_n \to \pi$ and $P_n \to \pi$ as $n \to \infty$; we shall prove these facts analytically in Theorem 4.58. Using plane geometry, Archimedes found iterative formulas for the sequences $\{p_n\}$ and $\{P_n\}$, and using these formulas he proved his third proposition. Recall that everything we thought about trigonometry is true ☺, so we shall use these trig facts to derive Archimedes's famous iterative formulas for the sequences $\{p_n\}$ and $\{P_n\}$. To this end, we let s_n and t_n denote the side lengths of the inscribed and circumscribed polygons, so that

$$P_n = (\#\text{ sides}) \times (\text{length each side}) = 2^n M \cdot t_n$$

and

$$p_n = (\#\text{ sides}) \times (\text{length each side}) = 2^n M \cdot s_n.$$

Let $2\theta_n$ be the central angle of the inscribed and circumscribed $2^n M$-gons as shown in Figs. 4.44 and 4.45 (that is, θ_n is half the central angle).

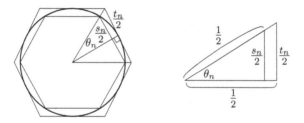

Fig. 4.45 We cut the central angle in half. The *right* picture shows a blowup of the overlapping *triangles* on the *left*

The right picture in Fig. 4.45 gives a blown-up picture of the triangles in the left-hand picture. The outer triangle in the right-hand picture shows that

$$\tan \theta_n = \frac{\text{opposite}}{\text{adjacent}} = \frac{t_n/2}{1/2} \quad \Longrightarrow \quad t_n = \tan \theta_n \quad \Longrightarrow \quad P_n = 2^n M \tan \theta_n.$$

The inner triangle shows that

$$\sin \theta_n = \frac{\text{opposite}}{\text{hypotenuse}} = \frac{s_n/2}{1/2} \quad \Longrightarrow \quad s_n = \sin \theta_n \quad \Longrightarrow \quad p_n = 2^n M \sin \theta_n.$$

Now what's θ_n? Well, since $2\theta_n$ is the central angle of the $2^n M$-gon, we have

$$\text{central angle} = \frac{\text{total angle of circle}}{\#\text{ of sides of regular polygon}} = \frac{2\pi}{2^n M}.$$

Setting this equal to $2\theta_n$, we get

$$\theta_n = \frac{\pi}{2^n M}.$$

In particular,

$$\theta_{n+1} = \frac{\pi}{2^{n+1} M} = \frac{1}{2}\frac{\pi}{2^n M} = \frac{1}{2}\theta_n.$$

Setting z equal to $z/2$ in Lemma 4.56, we see that

$$\tan\left(\frac{1}{2}z\right) = \frac{\sin(z)\tan(z)}{\sin(z) + \tan(z)} \quad \text{and} \quad 2\sin^2\left(\frac{1}{2}z\right) = \sin(z)\tan\left(\frac{1}{2}z\right).$$

Hence,

$$\tan\theta_{n+1} = \tan\left(\frac{1}{2}\theta_n\right) = \frac{\sin(\theta_n)\tan(\theta_n)}{\sin(\theta_n) + \tan(\theta_n)}$$

and

$$2\sin^2(\theta_{n+1}) = 2\sin^2\left(\frac{1}{2}\theta_n\right) = \sin(\theta_n)\tan\left(\frac{1}{2}\theta_n\right) = \sin(\theta_n)\tan(\theta_{n+1}).$$

In particular, recalling that $P_n = 2^n M \tan\theta_n$ and $p_n = 2^n M \sin\theta_n$, we see that

$$P_{n+1} = 2^{n+1} M \tan\theta_{n+1} = 2^{n+1} M \frac{\sin(\theta_n)\tan(\theta_n)}{\sin(\theta_n) + \tan(\theta_n)}$$
$$= 2\frac{2^n M \sin(\theta_n) \cdot 2^n M \tan(\theta_n)}{2^n M \sin(\theta_n) + 2^n M \tan(\theta_n)} = 2\frac{p_n P_n}{p_n + P_n}.$$

Also,

$$2p_{n+1}^2 = 2\left(2^{n+1} M \sin\theta_{n+1}\right)^2 = \left(2^{n+1} M\right)^2 \sin(\theta_n)\tan(\theta_{n+1})$$
$$= 2 \cdot 2^n M \sin(\theta_n)\, 2^{n+1} M \tan(\theta_{n+1}) = 2p_n P_{n+1},$$

or $p_{n+1} = \sqrt{p_n P_{n+1}}$. Finally, since $\theta_0 = \pi/M$ (recall that $\theta_n = \frac{\pi}{2^n M}$), we have $P_0 = M\tan(\frac{\pi}{M})$ and $p_0 = M\sin(\frac{\pi}{M})$. Here's a summary of what we've obtained:

$$\boxed{\begin{array}{l} P_{n+1} = \dfrac{2p_n P_n}{p_n + P_n}, \quad p_{n+1} = \sqrt{p_n P_{n+1}}; \quad \textbf{(Archimedes' algorithm)} \\[2mm] P_0 = M\tan\left(\dfrac{\pi}{M}\right), \quad p_0 = M\sin\left(\dfrac{\pi}{M}\right). \end{array}} \qquad (4.56)$$

These formulas make up the celebrated Archimedes's algorithm. Starting from the values of P_0 and p_0, we can use the iterative definitions for P_{n+1} and p_{n+1} to generate sequences $\{P_n\}$ and $\{p_n\}$ that converge to π, as we now show.

Archimedes' algorithm

Theorem 4.58 *We have*

$$p_n < \pi < P_n, \qquad n = 0, 1, 2, \ldots,$$

and $p_n \to \pi$ and $P_n \to \pi$ as $n \to \infty$.

Proof Note that for every $n = 0, 1, 2, \ldots$, we have $0 < \theta_n = \frac{\pi}{2^n M} < \frac{\pi}{2}$, because $M \geq 3$. Thus, by Lemma 4.57,

$$p_n = 2^n M \sin \theta_n < 2^n M \theta_n < 2^n M \tan \theta_n = P_n.$$

Since $\theta_n = \frac{\pi}{2^n M}$, the middle term is just π, so $p_n < \pi < P_n$ for every $n = 0, 1, 2, \ldots$. Using the limit $\lim_{z \to 0} \sin z / z = 1$, we obtain

$$\lim_{n \to \infty} p_n = \lim_{n \to \infty} 2^n M \sin \theta_n = \lim_{n \to \infty} \pi \frac{\sin \left(\frac{\pi}{2^n M} \right)}{\left(\frac{\pi}{2^n M} \right)} = \pi.$$

Since $\lim_{z \to 0} \cos z = 1$, we have $\lim_{z \to 0} \tan z / z = \lim_{z \to 0} \sin z / (z \cdot \cos z) = 1$, so the same argument we used for p_n shows that $\lim_{n \to \infty} P_n = \pi$. ∎

In Problem 4 you will study how fast p_n and P_n converge to π. Now let's consider a specific example: Let $M = 6$, which is what Archimedes chose! Then,

$$P_0 = 6 \tan \left(\frac{\pi}{6} \right) = 2\sqrt{3} = 3.464101615 \ldots \quad \text{and} \quad p_0 = 6 \sin \left(\frac{\pi}{6} \right) = 3.$$

From these values, we can find P_1 and p_1 from Archimedes's algorithm (4.56):

$$P_1 = \frac{2 p_0 P_0}{p_0 + P_0} = \frac{2 \cdot 3 \cdot 2\sqrt{3}}{3 + 2\sqrt{3}} = 3.159659942 \ldots$$

and

$$p_1 = \sqrt{p_0 P_1} = \sqrt{3 \cdot 3.159659942 \ldots} = 3.105828541 \ldots.$$

Continuing this process (I used a spreadsheet), we can find P_2, p_2, then P_3, p_3, and so forth, arriving at the table

n	p_n	P_n
0	3	3.464101615
1	3.105828541	3.215390309
2	3.132628613	3.159659942
3	3.139350203	3.146086215
4	3.141031951	3.1427146
5	3.141452472	3.14187305
6	3.141557608	3.141662747
7	3.141583892	3.141610177

Archimedes considered $p_4 = 3.14103195\ldots$ and $P_4 = 3.1427146\ldots$. Notice that

$$3\frac{10}{71} = 3.140845070\ldots < p_4 \quad \text{and} \quad P_4 < 3.142857142\ldots = 3\frac{1}{7}.$$

Hence,

$$3\frac{10}{71} < p_4 < \pi < P_4 < 3\frac{1}{7},$$

which proves Archimedes's third proposition. It's interesting to note that Archimedes didn't have computers back then (to find square roots, for instance), or trig functions, or coordinate geometry, or decimal notation, etc., so it's incredible that Archimedes was able to approximate π to such incredible accuracy!

4.12.5 Continuation of Our Brief History of π

Here are (only some!) famous formulas for π (along with the earliest known date of publication I'm aware of) that we'll prove in our journey through our book:

Archimedes $\approx$ 250 B.C.: $\pi = \lim P_n = \lim p_n$, where

$$P_{n+1} = \frac{2p_n P_n}{p_n + P_n}, \quad p_{n+1} = \sqrt{p_n P_{n+1}}; \quad P_0 = M\tan\left(\frac{\pi}{M}\right), \quad p_0 = M\sin\left(\frac{\pi}{M}\right).$$

We remark that Archimedes's algorithm is similar to Borchardt's algorithm (see Problem 1), which is similar to the modern-day AGM method of Eugene Salamin, Richard Brent, and Jonathan and Peter Borwein [31, 32]. This AGM method can generate *billions* of digits of π!

François Viète 1593 (Section 5.1, p. 381):

$$\frac{2}{\pi} = \sqrt{\frac{1}{2}} \cdot \sqrt{\frac{1}{2} + \frac{1}{2}\sqrt{\frac{1}{2}}} \cdot \sqrt{\frac{1}{2} + \frac{1}{2}\sqrt{\frac{1}{2} + \frac{1}{2}\sqrt{\frac{1}{2}}}} \cdots.$$

William Brouncker 1655 (Section 8.2, p. 597):

$$\frac{4}{\pi} = 1 + \cfrac{1^2}{2 + \cfrac{3^2}{2 + \cfrac{5^2}{2 + \cfrac{7^2}{2 + \cdots}}}}.$$

John Wallis 1656 (Section 5.1, p. 381):

$$\frac{\pi}{2} = \prod_{n=1}^{\infty} \frac{2n}{2n-1} \cdot \frac{2n}{2n+1} = \frac{2}{1} \cdot \frac{2}{3} \cdot \frac{4}{3} \cdot \frac{4}{5} \cdot \frac{6}{5} \cdot \frac{6}{7} \cdots.$$

where "$\prod_{n=1}^{\infty}$" means to take the product of the terms substituting $n = 1, 2, 3, \ldots$.
James Gregory, Gottfried Leibniz 1670, Madhava of Sangamagramma ≈ 1400 (Section 5.2, p. 393):

$$\frac{\pi}{4} = 1 - \frac{1}{3} + \frac{1}{5} - \frac{1}{7} + \frac{1}{9} - \frac{1}{11} + - \cdots.$$

John Machin 1706 (Section 6.9, p. 518):

$$\pi = 4 \arctan\left(\frac{1}{5}\right) - \arctan\left(\frac{1}{239}\right) = 4 \sum_{n=0}^{\infty} \frac{(-1)^n}{(2n+1)} \left(\frac{4}{5^{2n+1}} - \frac{1}{239^{2n+1}}\right).$$

Machin calculated 100 digits of π with this formula. William Shanks (1812–1882) is famed for his calculation of π to 707 places in 1873 using Machin's formula. However, only the first 527 places were correct, as discovered by D. Ferguson in 1944 [78] using another Machin-type formula. Ferguson ended up publishing 620 correct places in 1946, which marks the last hand calculation for π ever to so many digits. From that point on, computers have been used to compute π to increasing accuracy. See, for example, Yasumasa Kanada's website http://www.super-computing.org/; he and his coworkers at the University of Tokyo have used Machin-like formulas to compute *trillions* of digits of π. One might ask "why try to find so many digits of π?" Well,

Perhaps in some far distant century they may say, "Strange that those ingenious investigators into the secrets of the number system had so little conception of the fundamental discoveries that would later develop from them!" Derrick N. Lehmer (1867–1938) [270, p. 238].

We now go back to our list of formulas.
Leonhard Euler 1736 (Section 5.2, p. 393):

$$\frac{\pi^2}{6} = \sum_{n=1}^{\infty} \frac{1}{n^2} = 1 + \frac{1}{2^2} + \frac{1}{3^2} + \frac{1}{4^2} + \cdots$$

and (Section 7.6, p. 566):

$$\frac{\pi^2}{6} = \frac{2^2}{2^2 - 1} \cdot \frac{3^2}{3^2 - 1} \cdot \frac{5^2}{5^2 - 1} \cdot \frac{7^2}{7^2 - 1} \cdot \frac{11^2}{11^2 - 1} \cdots.$$

We end our history with questions to ponder: What is the probability that a natural number, chosen at random, is square-free (that is, is not divisible by the square of a prime)? What is the probability that two natural numbers, chosen at random, are relatively (or co) prime (that is, don't have any common prime factors)? The answers, drumroll please (Section 7.6, p. 566):

Probability of being square-free = Probability of being coprime = $\dfrac{6}{\pi^2}$.

▶ **Exercises 4.12**

1. In a letter from Gauss to his teacher Johann Pfaff (1765–1825) around 1800, Gauss asked Pfaff about the following sequences $\{\alpha_n\}$, $\{\beta_n\}$ defined recursively as follows:

$$\alpha_{n+1} = \frac{1}{2}(\alpha_n + \beta_n), \quad \beta_{n+1} = \sqrt{\alpha_{n+1}\beta_n}. \quad \textbf{(Borchardt's algorithm)}$$

   Later, Carl Borchardt (1817–1880) rediscovered this algorithm and since then, this algorithm is called **Borchardt's algorithm** [46]. Prove that Borchardt's algorithm is basically the same as Archimedes's algorithm in the following sense: if you set $\alpha_n := 1/P_n$ and $\beta_n := 1/p_n$ in Archimedes's algorithm, you get Borchardt's algorithm.

2. (**Pfaff's solution I**) Now what if we don't use the starting values $P_0 = M \tan\left(\frac{\pi}{M}\right)$ and $p_0 = M \sin\left(\frac{\pi}{M}\right)$ for Archimedes's algorithm in (4.56), but instead use other starting values? What do the sequences $\{P_n\}$ and $\{p_n\}$ converge to? These questions were answered by Johann Pfaff. Pick starting values P_0 and p_0, and let's assume that $0 \le p_0 < P_0$; the case that $P_0 < p_0$ is handled in the next problem.

   (i) Define

$$\theta = \arccos\left(\frac{p_0}{P_0}\right), \quad r = \frac{p_0 P_0}{\sqrt{P_0^2 - p_0^2}}.$$

   Prove that $P_0 = r \tan\theta$ and $p_0 = r \sin\theta$.

   (ii) Prove by induction that $P_n = 2^n r \tan\left(\frac{\theta}{2^n}\right)$ and $p_n = 2^n r \sin\left(\frac{\theta}{2^n}\right)$.

   (iii) Prove that as $n \to \infty$, both $\{P_n\}$ and $\{p_n\}$ converge to

$$r\theta = \frac{p_0 P_0}{\sqrt{P_0^2 - p_0^2}} \arccos\left(\frac{p_0}{P_0}\right).$$

3. **(Pfaff's solution II)** Now assume that $0 < P_0 < p_0$.

   (i) Define (see Problem 7 on p. 364 for the definition of arccosh)

   $$\theta := \operatorname{arccosh}\left(\frac{p_0}{P_0}\right), \qquad r := \frac{p_0 P_0}{\sqrt{p_0^2 - P_0^2}}.$$

   Prove that $P_0 = r \tanh \theta$ and $p_0 = r \sinh \theta$.
   (ii) Prove by induction that $P_n = 2^n r \tanh\left(\frac{\theta}{2^n}\right)$ and $p_n = 2^n r \sinh\left(\frac{\theta}{2^n}\right)$.
   (iii) Prove that as $n \to \infty$, both $\{P_n\}$ and $\{p_n\}$ converge to

   $$r\theta = \frac{p_0 P_0}{\sqrt{p_0^2 - P_0^2}} \operatorname{arccosh}\left(\frac{p_0}{P_0}\right).$$

4. (Cf. [163, 193]) **(Rate of convergence)**

   (a) Using the formulas $p_n = 2^n M \sin \theta_n$ and $P_n = 2^n M \tan \theta_n$, where $\theta_n = \frac{\pi}{2^n M}$, prove that there are constants $C_1, C_2 > 0$ such that for all n,

   $$|p_n - \pi| \leq \frac{C_1}{4^n} \quad \text{and} \quad |P_n - \pi| \leq \frac{C_2}{4^n}.$$

   Suggestion: For the first estimate, use the expansion $\sin z = z - \frac{z^3}{3!} + \cdots$.
   For the second estimate, notice that $|P_n - \pi| = \frac{1}{\cos \theta_n} |p_n - \pi \cos \theta_n|$.
   (b) Part (a) shows that $\{p_n\}$ and $\{P_n\}$ converge to π very fast, but we can get even faster convergence by looking at the sequence $\{a_n\}$, where $a_n = \frac{1}{3}(2p_n + P_n)$. Prove that there is a constant $C > 0$ such that for all n,

   $$|a_n - \pi| \leq \frac{C}{16^n}.$$

Chapter 5
Some of the Most Beautiful Formulas
in the World I–III

God used beautiful mathematics in creating the world.
Paul Dirac (1902–1984)

In this chapter we present a small sample of *some* of the most beautiful formulas in the world. We begin in Section 5.1, where we present Viète's formula, Wallis's formula, and Euler's sine expansion. Viète's formula, due to François Viète (1540–1603), is the infinite product

$$\frac{2}{\pi} = \sqrt{\frac{1}{2}} \cdot \sqrt{\frac{1}{2} + \frac{1}{2}\sqrt{\frac{1}{2}}} \cdot \sqrt{\frac{1}{2} + \frac{1}{2}\sqrt{\frac{1}{2} + \frac{1}{2}\sqrt{\frac{1}{2}}}} \cdots,$$

published in 1593. This is not only the first recorded infinite product [129, p. 218], it is also the first recorded theoretically *exact* analytical expression for the number π [35, p. 321]. Wallis's formula, named after John Wallis (1616–1703), was the second recorded infinite product [129, p. 219]:

$$\frac{\pi}{2} = \prod_{n=1}^{\infty} \frac{2n}{2n-1} \cdot \frac{2n}{2n+1} = \frac{2}{1} \cdot \frac{2}{3} \cdot \frac{4}{3} \cdot \frac{4}{5} \cdot \frac{6}{5} \cdot \frac{6}{7} \cdots.$$

To explain Euler's sine expansion, recall that if $p(x)$ is a polynomial with roots $r_1, \ldots, r_n$ (repeated according to multiplicity), then we can factor $p(x)$ as $p(x) = a(x - r_1)(x - r_2) \ldots (x - r_n)$, where a is a constant. Assuming that the r_k are nonzero, by factoring out $-r_1, -r_2, \ldots, -r_n$, we can write $p(x)$ as

$$p(x) = b \left(1 - \frac{x}{r_1}\right) \left(1 - \frac{x}{r_2}\right) \cdots \left(1 - \frac{x}{r_n}\right), \tag{5.1}$$

© Paul Loya 2017
P. Loya, *Amazing and Aesthetic Aspects of Analysis*,
https://doi.org/10.1007/978-1-4939-6795-7_5

for a constant b. If we put $x = 0$, we see that $b = p(0)$. Returning now to the sine function, recall that the roots of $\sin x$ are located at

$$0, \pi, -\pi, 2\pi, -2\pi, 3\pi, -3\pi, \ldots,$$

as seen here:

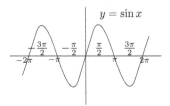

So, if we think of $p(x) := \frac{\sin x}{x} = 1 - \frac{x^2}{3!} + \frac{x^4}{5!} - \cdots$ as an (infinite-degree) polynomial, then $p(x)$ has roots at $\pi, -\pi, 2\pi, -2\pi, \ldots$. Assuming that (5.1) holds for such an infinite polynomial and noting that $p(0) = 1$, we have (recalling that $b = p(0)$)

$$p(x) = p(0)\left(1 - \frac{x}{\pi}\right)\left(1 + \frac{x}{\pi}\right)\left(1 - \frac{x}{2\pi}\right)\left(1 + \frac{x}{2\pi}\right)\left(1 - \frac{x}{3\pi}\right)\left(1 + \frac{x}{3\pi}\right)\cdots$$

$$= \left(1 - \frac{x^2}{\pi^2}\right)\left(1 - \frac{x^2}{2^2\pi^2}\right)\left(1 - \frac{x^2}{3^2\pi^2}\right)\cdots.$$

Replacing $p(x)$ by $\sin x / x$, we obtain the formula

$$\boxed{\sin x = x\left(1 - \frac{x^2}{\pi^2}\right)\left(1 - \frac{x^2}{2^2\pi^2}\right)\left(1 - \frac{x^2}{3^2\pi^2}\right)\left(1 - \frac{x^2}{4^2\pi^2}\right)\left(1 - \frac{x^2}{5^2\pi^2}\right)\cdots,}$$

$$(5.2)$$

which Euler first published in 1735 in his epoch-making paper *De summis serierum reciprocarum* (On the sums of series of reciprocals), which was read in the St. Petersburg Academy on December 5, 1735, and originally published in *Commentarii Academiae Scientiarum Petropolitanae* 7, 1740, and reprinted on pp. 123–134 of *Opera Omnia: Series 1*, Volume 14, pp. 73–86.

In Section 5.2 we study the Basel problem, which asks for the exact value of $\zeta(2) = \sum_{n=1}^{\infty} \frac{1}{n^2}$. Euler, in the same 1735 paper *De summis serierum reciprocarum*, proved that $\zeta(2) = \frac{\pi^2}{6}$:

$$\boxed{\sum_{n=1}^{\infty} \frac{1}{n^2} = 1 + \frac{1}{2^2} + \frac{1}{3^2} + \frac{1}{4^2} + \cdots = \frac{\pi^2}{6}.}$$

Euler actually gave three proofs of this formula in *De summis serierum reciprocarum*, but the third one is the easiest to explain. Here it is. First, we divide both sides of

(5.2) by x; then we formally[1] multiply out the right-hand side. For example, if we multiply out just the first three factors, we get

$$\left(1 - \frac{x^2}{1^2\pi^2}\right)\left(1 - \frac{x^2}{2^2\pi^2}\right)\left(1 - \frac{x^2}{3^2\pi^2}\right) = 1 - \frac{x^2}{\pi^2}\left(\frac{1}{1^2} + \frac{1}{2^2} + \frac{1}{3^2}\right) + x^4 \text{ and } x^6 \text{ terms.}$$

Continuing multiplying, we get, at least formally speaking,

$$\frac{\sin x}{x} = 1 - \frac{x^2}{\pi^2}\left(\frac{1}{1^2} + \frac{1}{2^2} + \frac{1}{3^2} + \cdots\right) + \cdots,$$

where the dots "$\cdots$" involve powers of x of degree at least four or higher. Notice that the coefficient of x^2 on the right involves $\zeta(2)$. Now dividing the infinite series of $\sin x = x - \frac{x^3}{3!} + \cdots$ by x, we conclude that

$$1 - \frac{x^2}{3!} + \cdots = 1 - \frac{x^2}{\pi^2}\zeta(2) + \cdots,$$

where the "$\cdots$" on both sides involve powers of x of degree at least four or higher. Finally, equating powers of x^2, we conclude that

$$\frac{1}{3!} = \frac{\zeta(2)}{\pi^2} \implies \zeta(2) = \frac{\pi^2}{3!} = \frac{\pi^2}{6}.$$

Here is an English translation of Euler's argument from *De summis serierum reciprocarum* (which was originally written in Latin) taken from [20]:

Indeed, it having been put[2] $y = 0$, from which the fundamental equation will turn into this[3]

$$0 = s - \frac{s^3}{1 \cdot 2 \cdot 3} + \frac{s^5}{1 \cdot 2 \cdot 3 \cdot 4 \cdot 5} - \frac{s^7}{1 \cdot 2 \cdot 3 \cdot 4 \cdot 5 \cdot 6 \cdot 7} + \text{etc.}$$

The roots of this equation give all the arcs of which the sine is equal to 0. Moreover, the single smallest root is $s = 0$, whereby the equation divided by s will exhibit all the remaining arcs of which the sine is equal to 0; these arcs will hence be the roots of this equation

$$0 = 1 - \frac{s^2}{1 \cdot 2 \cdot 3} + \frac{s^4}{1 \cdot 2 \cdot 3 \cdot 4 \cdot 5} - \frac{s^6}{1 \cdot 2 \cdot 3 \cdot 4 \cdot 5 \cdot 6 \cdot 7} + \text{etc.}$$

---

[1] "Formal" in mathematics usually refers to "having the form or appearance without the substance or essence," which is the fifth entry for "formal" in Webster's 1828 dictionary. This is very different from the common use of "formal": "according to form; agreeable to established mode; regular; methodical," which is the first entry in Webster's 1828 dictionary. Elaborating on the mathematical use of "formal," it means something like "a symbolic manipulation or expression presented without paying attention to correctness".

[2] Here, Euler set $y = \sin s$.

[3] Instead of writing, e.g., $1 \cdot 2 \cdot 3$, today we would write this as 3!. However, the factorial symbol wasn't invented until 1808 [44, p. 341], by Christian Kramp (1760–1826), more than 70 years after *De summis serierum reciprocarum* was read in the St. Petersburg Academy.

Truly then, those arcs of which the sine is equal to 0 are[4]

$$p, \quad -p, \quad +2p, \quad -2p, \quad 3p, \quad -3p \quad \text{etc.},$$

of which the second of the two of each pair is negative, each of these because the equation indicates for the dimensions of s to be even. Hence the divisors of this equation will be

$$1 - \frac{s}{p}, \quad 1 + \frac{s}{p}, \quad 1 - \frac{s}{2p}, \quad 1 + \frac{s}{2p} \quad \text{etc.},$$

and by the joining of these divisors two by two it will be

$$1 - \frac{s^2}{1 \cdot 2 \cdot 3} + \frac{s^4}{1 \cdot 2 \cdot 3 \cdot 4 \cdot 5} - \frac{s^6}{1 \cdot 2 \cdot 3 \cdot 4 \cdot 5 \cdot 6 \cdot 7} + \text{etc.}$$
$$= \left(1 - \frac{s^2}{p^2}\right)\left(1 - \frac{s^2}{4p^2}\right)\left(1 - \frac{s^2}{9p^2}\right)\left(1 - \frac{s^2}{16p^2}\right) \text{ etc.}$$

It is now clear from the nature of equations for the coefficient[5] of ss that is $\frac{1}{1 \cdot 2 \cdot 3}$ to be equal to

$$\frac{1}{p^2} + \frac{1}{4p^2} + \frac{1}{9p^2} + \frac{1}{16p^2} + \text{etc.}$$

In this last step, Euler says that

$$\frac{1}{1 \cdot 2 \cdot 3} = \frac{1}{p^2} + \frac{1}{4p^2} + \frac{1}{9p^2} + \frac{1}{16p^2} + \text{etc.},$$

which after multiplication by p^2 is exactly the statement that $\zeta(2) = \frac{\pi^2}{6}$. By the way, in this book we give eleven proofs of Euler's formula for $\zeta(2)$. Euler's proof reminds me of a quotation by Charles Hermite (1822–1901):

> There exists, if I am not mistaken, an entire world which is the totality of mathematical truths, to which we have access only with our mind, just as a world of physical reality exists, the one like the other independent of ourselves, both of divine creation. *Quoted in* The Mathematical Intelligencer, *vol. 5, no. 4.*

In Section 5.2 we also prove the Gregory–Leibniz–Madhava series

$$\boxed{\frac{\pi}{4} = 1 - \frac{1}{3} + \frac{1}{5} - \frac{1}{7} + - \cdots .}$$

This formula is usually called **Leibniz's series** after Gottfried Leibniz (1646–1716), because he is usually accredited as the first to mention this formula in print, in 1673, although James Gregory (1638–1675) probably knew about it. However, the great Indian mathematician and astronomer Madhava of Sangamagramma (1350–1425) discovered this formula over 200 years before either Gregory or Leibniz!

---

[4]Here, Euler uses p for π. The notation π for the ratio of the length of a circle to its diameter was introduced in 1706 by William Jones (1675–1749), and around 1736, a year after Euler published *De summis serierum reciprocarum*, Euler seems to have adopted the notation π.

[5]Here, ss means s^2.

Finally, in Section 5.3 we derive Euler's formula for $\zeta(n)$ for all even n.

CHAPTER 5 OBJECTIVES: THE STUDENT WILL BE ABLE TO ...

- Explain the various formulas of Euler, Wallis, Viète, Gregory, Leibniz, Madhava.
- Formally derive Euler's sine expansion and formula for $\pi^2/6$.
- Describe Euler's formulas for $\zeta(n)$ for n even.

5.1 ★ Beautiful Formulas I: Euler, Wallis, and Viète

Historically, Viète's formula was the first infinite product written down, and Wallis's formula was the second [129, pp. 218–219]. In this section we prove these formulas, and we also prove Euler's celebrated sine expansion.

5.1.1 Viète's Formula: The First Analytic Expression For π

François Viète's (1540–1603) formula has a very elementary proof. For every nonzero $z \in \mathbb{C}$, dividing the identity $\sin z = 2 \sin(z/2) \cos(z/2)$ by z, we get

$$\frac{\sin z}{z} = \cos(z/2) \cdot \frac{\sin(z/2)}{z/2}.$$

Replacing z with $z/2$, we get $\sin(z/2)/(z/2) = \cos(z/2^2) \cdot \sin(z/2^2)/(z/2^2)$. Therefore,

$$\frac{\sin z}{z} = \cos(z/2) \cdot \cos(z/2^2) \cdot \frac{\sin(z/2^2)}{z/2^2}.$$

Continuing by induction, we obtain

$$\frac{\sin z}{z} = \cos(z/2) \cdot \cos(z/2^2) \cdots \cos(z/2^n) \cdot \frac{\sin(z/2^n)}{z/2^n}$$

$$= \frac{\sin(z/2^n)}{z/2^n} \cdot \prod_{k=1}^{n} \cos(z/2^k), \tag{5.3}$$

or

$$\prod_{k=1}^{n} \cos(z/2^k) = \frac{z/2^n}{\sin(z/2^n)} \cdot \frac{\sin z}{z}.$$

Since $\lim_{n \to \infty} \frac{z/2^n}{\sin(z/2^n)} = 1$ for every nonzero $z \in \mathbb{C}$, we have

$$\lim_{n\to\infty} \prod_{k=1}^{n} \cos(z/2^k) = \lim_{n\to\infty} \frac{\sin z}{z} \cdot \frac{z/2^n}{\sin(z/2^n)} = \frac{\sin z}{z}.$$

For notational purposes, we can write

$$\frac{\sin z}{z} = \prod_{n=1}^{\infty} \cos(z/2^n) = \cos(z/2) \cdot \cos(z/2^2) \cdot \cos(z/2^4) \cdots \qquad (5.4)$$

and refer to the right-hand side as an **infinite product**, the subject of which we'll thoroughly study in Chapter 7. For the purposes of this chapter, given a sequence $a_1, a_2, a_3, \ldots$, we shall denote by $\prod_{n=1}^{\infty} a_n$ the limit

$$\prod_{n=1}^{\infty} a_n := \lim_{n\to\infty} \prod_{k=1}^{n} a_k = \lim_{n\to\infty} \left(a_1 a_2 \cdots a_n \right),$$

provided that the limit exists. Putting $z = \pi/2$ into (5.4), we get

$$\frac{2}{\pi} = \cos\left(\frac{\pi}{2^2}\right) \cdot \cos\left(\frac{\pi}{2^3}\right) \cdot \cos\left(\frac{\pi}{2^4}\right) \cdot \cos\left(\frac{\pi}{2^5}\right) \cdots = \prod_{n=1}^{\infty} \cos\left(\frac{\pi}{2^{n+1}}\right).$$

We now obtain formulas for $\cos\left(\frac{\pi}{2^{n+1}}\right)$. To do so, note that for every $0 \le \theta \le \pi$, we have

$$\cos\left(\frac{\theta}{2}\right) = \sqrt{\frac{1}{2} + \frac{1}{2}\cos\theta}.$$

(This follows from the double angle formula: $2\cos^2(2z) = 1 + \cos z$.) Thus,

$$\cos\left(\frac{\theta}{2^2}\right) = \sqrt{\frac{1}{2} + \frac{1}{2}\cos\left(\frac{\theta}{2}\right)} = \sqrt{\frac{1}{2} + \frac{1}{2}\sqrt{\frac{1}{2} + \frac{1}{2}\cos\theta}}.$$

Continuing this process (slang for "it can be shown by induction"), we see that

$$\cos\left(\frac{\theta}{2^n}\right) = \sqrt{\frac{1}{2} + \frac{1}{2}\sqrt{\frac{1}{2} + \frac{1}{2}\sqrt{\frac{1}{2} + \cdots + \frac{1}{2}\sqrt{\frac{1}{2} + \frac{1}{2}\cos\theta}}}}, \qquad (5.5)$$

where there are n square roots here. Putting $\theta = \pi/2$, we obtain

$$\cos\left(\frac{\pi}{2^{n+1}}\right) = \sqrt{\frac{1}{2} + \frac{1}{2}\sqrt{\frac{1}{2} + \frac{1}{2}\sqrt{\frac{1}{2} + \cdots + \frac{1}{2}\sqrt{\frac{1}{2}}}}},$$

and hence,

$$\frac{2}{\pi} = \prod_{n=1}^{\infty} \sqrt{\frac{1}{2} + \frac{1}{2}\sqrt{\frac{1}{2} + \frac{1}{2}\sqrt{\frac{1}{2} + \cdots + \frac{1}{2}\sqrt{\frac{1}{2}}}}},$$

where there are n square roots in the nth factor of the infinite product. Writing out the infinite product, we have

$$\frac{2}{\pi} = \sqrt{\frac{1}{2}} \cdot \sqrt{\frac{1}{2} + \frac{1}{2}\sqrt{\frac{1}{2}}} \cdot \sqrt{\frac{1}{2} + \frac{1}{2}\sqrt{\frac{1}{2} + \frac{1}{2}\sqrt{\frac{1}{2}}}} \cdots . \tag{5.6}$$

This formula was given by Viète in 1593.

5.1.2 Expansion of Sine I

Our first proof of Euler's infinite product for sine is based on a neat identity involving tangents that we'll present in Lemma 5.1 below. See Problem 6 for another proof. To begin, we first write, for $z \in \mathbb{C}$,

$$\sin z = \frac{1}{2i}\left(e^{iz} - e^{-iz}\right) = \lim_{n \to \infty} \frac{1}{2i}\left\{\left(1 + \frac{iz}{n}\right)^n - \left(1 - \frac{iz}{n}\right)^n\right\} = \lim_{n \to \infty} F_n(z), \tag{5.7}$$

where F_n is the polynomial of degree n in z given by

$$F_n(z) = \frac{1}{2i}\left\{\left(1 + \frac{iz}{n}\right)^n - \left(1 - \frac{iz}{n}\right)^n\right\}. \tag{5.8}$$

In the following lemma, we factor the polynomial $F_n(z)$ in terms of tangents.

Lemma 5.1 *If $n = 2m + 1$ with $m \in \mathbb{N}$, then we can write*

$$F_n(z) = z \prod_{k=1}^{m} \left(1 - \frac{z^2}{n^2 \tan^2(k\pi/n)}\right).$$

Proof We shall find $n = 2m + 1$ roots of $F_n(z)$. To do so, observe that substituting $z = n \tan \theta$ yields

$$1 + \frac{iz}{n} = 1 + i \tan \theta = 1 + i\frac{\sin \theta}{\cos \theta} = \frac{1}{\cos \theta}\left(\cos \theta + i \sin \theta\right)$$
$$= \sec \theta \, e^{i\theta},$$

and similarly, $1 - iz/n = \sec \theta \, e^{-i\theta}$. Thus,

$$F_n(n \tan \theta) = \frac{1}{2i}\left\{\left(1 + \frac{iz}{n}\right)^n - \left(1 - \frac{iz}{n}\right)^n\right\}\bigg|_{z=n \tan \theta}$$
$$= \frac{1}{2i} \sec^n \theta \left(e^{in\theta} - e^{-in\theta}\right).$$

Since $\sin z = \frac{1}{2i}(e^{iz} - e^{-iz})$, we have

$$F_n(n \tan \theta) = \sec^n \theta \, \sin(n\theta).$$

The sine function vanishes at integer multiples of π, so $F_n(n \tan \theta) = 0$ when $n\theta = k\pi$, for every integer k, that is, when $\theta = k\pi/n$, for every $k \in \mathbb{Z}$. Thus, $F_n(z_k) = 0$ for

$$z_k = n \tan \left(\frac{k\pi}{n}\right) = n \tan \left(\frac{k\pi}{2m+1}\right),$$

where $k \in \mathbb{Z}$, and we recall that $n = 2m + 1$. Since $\tan \theta$ is strictly increasing on the interval $(-\pi/2, \pi/2)$, it follows that

$$z_{-m} < z_{-m+1} < \cdots < z_{-1} < z_0 < z_1 < \cdots < z_{m-1} < z_m.$$

Moreover, since tangent is an odd function, we have $z_{-k} = -z_k$ for each k. In particular, we have found $2m + 1 = n$ distinct roots of $F_n(z)$, so as a consequence of the fundamental theorem of algebra, we can write $F_n(z)$ as a constant times

$$(z - z_0) \cdot \prod_{k=1}^{m}\left\{(z - z_k) \cdot (z - z_{-k})\right\}$$
$$= z \cdot \prod_{k=1}^{m}\left\{(z - z_k) \cdot (z + z_k)\right\} \; (\text{since} z_{-k} = -z_k)$$
$$= z \prod_{k=1}^{m}(z^2 - z_k^2).$$

Factoring out all the $-z_k^2$ terms and gathering them all into one constant, we can conclude that for some constant a,

$$F_n(z) = a z \prod_{k=1}^{m}\left(1 - \frac{z^2}{z_k^2}\right) = a z \prod_{k=1}^{m}\left(1 - \frac{z^2}{n^2 \tan^2(k\pi/n)}\right).$$

Multiplying out the terms in the formula (5.8), we see that $F_n(z) = z$ plus higher powers of z. This implies that $a = 1$ and completes the proof of the lemma. ∎

Using this lemma we can give a formal proof of Euler's sine expansion. We now restrict our attention to real numbers. From (5.7) and Lemma 5.1, we know that for every $x \in \mathbb{R}$,

$$\frac{\sin x}{x} = \lim_{n \to \infty} \frac{F_n(x)}{x} = \lim_{n \to \infty} \prod_{k=1}^{\frac{n-1}{2}} \left(1 - \frac{x^2}{n^2 \tan^2(k\pi/n)} \right), \qquad (5.9)$$

where in the limit we restrict n to *odd* natural numbers. Thus, if we write $n = 2m + 1$, the limit in (5.9) really means

$$\frac{\sin x}{x} = \lim_{m \to \infty} \prod_{k=1}^{m} \left(1 - \frac{x^2}{(2m+1)^2 \tan^2(k\pi/(2m+1))} \right),$$

but we prefer the simpler form in (5.9) with the understanding that n is odd in (5.9). To compute the limit as $n \to \infty$ in (5.9), note that

$$\lim_{n \to \infty} n^2 \tan^2(k\pi/n) = \lim_{n \to \infty} (k\,\pi)^2 \left(\frac{\tan(k\pi/n)}{k\pi/n} \right)^2 = k^2 \pi^2,$$

where we used that $\lim_{z \to 0} \frac{\tan z}{z} = 1$ (which follows from $\lim_{z \to 0} \frac{\sin z}{z} = 1$ and that $\lim_{z \to 0} \cos(z) = \cos(0) = 1$). Hence,

$$\lim_{n \to \infty} \left(1 - \frac{x^2}{n^2 \tan^2(k\pi/n)} \right) = \left(1 - \frac{x^2}{k^2 \pi^2} \right). \qquad (5.10)$$

Thus, formally evaluating the limit in (5.9), we see that

$$\frac{\sin x}{x} = \lim_{n \to \infty} \prod_{k=1}^{\frac{n-1}{2}} \left(1 - \frac{x^2}{n^2 \tan^2(k\pi/n)} \right)$$

$$= \prod_{k=1}^{\infty} \lim_{n \to \infty} \left(1 - \frac{x^2}{n^2 \tan^2(k\pi/n)} \right)$$

$$= \prod_{k=1}^{\infty} \left(1 - \frac{x^2}{k^2 \pi^2} \right),$$

which is Euler's result. Unfortunately, there are two issues with this argument. The first issue is that although by (5.10) the second and third lines are one and the same, we have yet to prove that the infinite product in the last line actually converges! The second issue occurs in switching the limit with the product from the first to second lines:

$$\lim_{n\to\infty} \prod_{k=1}^{\frac{n-1}{2}} \left(1 - \frac{x^2}{n^2 \tan^2(k\pi/n)}\right) = \prod_{k=1}^{\infty} \lim_{n\to\infty} \left(1 - \frac{x^2}{n^2 \tan^2(k\pi/n)}\right). \tag{5.11}$$

See Problem 2 for an example in which such an interchange leads to a wrong answer. On page 547 in Section 7.3 of Chapter 7, we'll encounter Tannery's theorem for infinite products, from which we can easily deduce that (5.11) does indeed hold. However, we'll leave Tannery's theorem for products until Chapter 7, because we can easily justify (5.11) in a very elementary (although a little long-winded) way, which we do in the following theorem.

Euler's theorem

Theorem 5.2 *For every $x \in \mathbb{R}$ we have*

$$\sin x = x \prod_{k=1}^{\infty} \left(1 - \frac{x^2}{\pi^2 k^2}\right)$$

in the sense that the right-hand infinite product converges and equals $\sin x$.

Proof Euler's formula holds for x any integer multiple of π (both sides are zero), so we can fix a real number x not an integer multiple of π. Given $m \in \mathbb{N}$, if we put

$$P_m = \prod_{k=1}^{m} \left(1 - \frac{x^2}{k^2 \pi^2}\right),$$

we need to show that $\sin x / x = \lim_{m\to\infty} P_m$.

Step 1: We begin by understanding $\lim_{n\to\infty} p_n$, where

$$p_n = \prod_{k=1}^{\frac{n-1}{2}} \left(1 - \frac{x^2}{n^2 \tan^2(k\pi/n)}\right),$$

and where we restrict n to *odd* natural numbers.

Let $m \in \mathbb{N}$. Then for every odd n with $m < \frac{n-1}{2}$, we can break up the product p_n from $k = 1$ to m and then from $m + 1$ to $\frac{n-1}{2}$:

$$p_n = \prod_{k=1}^{m} \left(1 - \frac{x^2}{n^2 \tan^2(k\pi/n)}\right) \cdot \prod_{k=m+1}^{\frac{n-1}{2}} \left(1 - \frac{x^2}{n^2 \tan^2(k\pi/n)}\right). \tag{5.12}$$

As $n \to \infty$, the left-hand side, p_n, of (5.12) converges to $\sin x / x$. On the other hand, by the observation (5.10) and the algebra of limits (noting that we have a finite product), we have

$$\lim_{n\to\infty}\prod_{k=1}^{m}\left(1-\frac{x^2}{n^2\tan^2(k\pi/n)}\right)=\prod_{k=1}^{m}\left(1-\frac{x^2}{k^2\pi^2}\right)=P_m,$$

which is not zero, since x is not an integer multiple of π. Thus, taking $n\to\infty$ in (5.12), it follows that the limit

$$Q_m:=\lim_{n\to\infty}\prod_{k=m+1}^{\frac{n-1}{2}}\left(1-\frac{x^2}{n^2\tan^2(k\pi/n)}\right)$$

exists, and

$$\frac{\sin x}{x}=P_mQ_m.$$

Hence, for every $m\in\mathbb{N}$,

$$\left|\frac{\sin x}{x}-P_m\right|=\left|P_mQ_m-P_m\right|=|P_m|\cdot|Q_m-1|.$$

Our goal now is to prove that the right-hand side approaches zero as $m\to\infty$.

Step 2: We claim that $|P_m|$ is bounded. Indeed, observe that for every $m\in\mathbb{N}$,

$$|P_m|=\left|\prod_{k=1}^{m}\left(1-\frac{x^2}{k^2\pi^2}\right)\right|\le\prod_{k=1}^{m}\left(1+\frac{x^2}{k^2\pi^2}\right)$$

$$\le\prod_{k=1}^{m}e^{\frac{x^2}{k^2\pi^2}}\quad\text{(since }1+t\le e^t\text{ for any }t\in\mathbb{R}\text{)}$$

$$=e^{\sum_{k=1}^{m}\frac{x^2}{k^2\pi^2}}\quad\text{(since }e^a\cdot e^b=e^{a+b}\text{)}$$

$$\le e^L,$$

where $L=\sum_{k=1}^{\infty}\frac{x^2}{k^2\pi^2}$, a finite constant, the exact value of which is not important. (Note that $\sum_{k=1}^{\infty}\frac{1}{k^2}$ converges by the p-test with $p=2$.) Hence, for every $m\in\mathbb{N}$,

$$\left|\frac{\sin x}{x}-P_m\right|\le e^L|Q_m-1|.$$

It remains to show that $Q_m\to1$ as $m\to\infty$. To do this, we need some estimates.

Step 3: We find some nice estimates on the quotient $\frac{x^2}{n^2\tan^2(k\pi/n)}$. In Lemma 4.57 on page 368, we proved that

$$\theta<\tan\theta,\quad\text{for }0<\theta<\pi/2.$$

In particular, if $n\in\mathbb{N}$ is odd and $1\le k\le\frac{n-1}{2}$, then

$$\frac{k\pi}{n} < \frac{n-1}{2} \cdot \frac{\pi}{n} < \frac{\pi}{2},$$

so

$$\frac{x^2}{n^2 \tan^2(k\pi/n)} < \frac{x^2}{n^2(k\pi)^2/n^2} = \frac{x^2}{k^2\pi^2}. \tag{5.13}$$

Step 4: We now complete our proof. Let $m \in \mathbb{N}$ with $\frac{x^2}{m^2\pi^2} < 1$. Then for every $n \in \mathbb{N}$ with $m < \frac{n-1}{2}$, by (5.13) we have

$$\frac{x^2}{n^2 \tan^2(k\pi/n)} < \frac{x^2}{k^2\pi^2} < 1 \quad \text{for } k = m+1, m+2, \dots, \frac{n-1}{2}.$$

In particular,

$$0 < \left(1 - \frac{x^2}{n^2 \tan^2(k\pi/n)}\right) < 1 \quad \text{for } k = m+1, m+2, \dots, \frac{n-1}{2}.$$

Hence,

$$\prod_{k=m+1}^{\frac{n-1}{2}} \left(1 - \frac{x^2}{n^2 \tan^2(k\pi/n)}\right) < 1.$$

Taking $n \to \infty$, we conclude that

$$Q_m < 1.$$

In Problem 3 you will prove that for all nonnegative real numbers $a_1, a_2, \dots, a_p \geq 0$, we have

$$1 - (a_1 + a_2 + \cdots + a_p) \leq (1 - a_1)(1 - a_2) \cdots (1 - a_p). \tag{5.14}$$

Using this inequality, it follows that

$$1 - \sum_{k=m+1}^{\frac{n-1}{2}} \frac{x^2}{n^2 \tan^2(k\pi/n)} \leq \prod_{k=m+1}^{\frac{n-1}{2}} \left(1 - \frac{x^2}{n^2 \tan^2(k\pi/n)}\right).$$

By (5.13) we have

$$\sum_{k=m+1}^{\frac{n-1}{2}} \frac{x^2}{n^2 \tan^2(k\pi/n)} \leq \frac{x^2}{\pi^2} \sum_{k=m+1}^{\frac{n-1}{2}} \frac{1}{k^2} \leq s_m,$$

where $s_m = \frac{x^2}{\pi^2} \sum_{k=m+1}^{\infty} \frac{1}{k^2}$. Thus,

$$1 - s_m \le \prod_{k=m+1}^{\frac{n-1}{2}} \left(1 - \frac{x^2}{n^2 \tan^2(k\pi/n)}\right).$$

Taking $n \to \infty$, we conclude that

$$1 - s_m \le Q_m.$$

Combining this with the fact that $Q_m < 1$, we see that

$$1 - s_m \le Q_m < 1.$$

Since $\sum_{k=1}^{\infty} \frac{1}{k^2}$ converges (the p-test with $p = 2$), by the Cauchy criterion for series on page 205, we know that $\lim_{m \to \infty} s_m = 0$. Thus, by the squeeze theorem it follows that $Q_m \to 1$, as desired. ■

We remark that Euler's sine expansion also holds for all complex $z \in \mathbb{C}$ (and not just real $x \in \mathbb{R}$), but we'll wait for Section 7.3, page 547, of Chapter 7 for the proof of the complex version.

5.1.3 Wallis's Formulas

As an application of Euler's sine expansion, we derive John Wallis's (1616–1703) formulas for π.

Wallis's formulas

Theorem 5.3 *We have*

$$\frac{\pi}{2} = \prod_{n=1}^{\infty} \frac{2n}{2n-1} \cdot \frac{2n}{2n+1} = \frac{2}{1} \cdot \frac{2}{3} \cdot \frac{4}{3} \cdot \frac{4}{5} \cdot \frac{6}{5} \cdot \frac{6}{7} \cdots,$$

$$\sqrt{\pi} = \lim_{n \to \infty} \frac{1}{\sqrt{n}} \prod_{k=1}^{n} \frac{2k}{2k-1} = \lim_{n \to \infty} \frac{1}{\sqrt{n}} \cdot \frac{2}{1} \cdot \frac{4}{3} \cdot \frac{6}{5} \cdots \frac{2n}{2n-1}.$$

Proof To obtain the first formula, we set $x = \pi/2$ in Euler's infinite product expansion for sine:

$$\sin x = x \prod_{n=1}^{\infty} \left(1 - \frac{x^2}{\pi^2 n^2}\right) \implies 1 = \frac{\pi}{2} \prod_{n=1}^{\infty} \left(1 - \frac{1}{2^2 n^2}\right).$$

Since $1 - \frac{1}{2^2 n^2} = \frac{2^2 n^2 - 1}{2^2 n^2} = \frac{(2n-1)(2n+1)}{(2n)(2n)}$, we see that

$$\frac{2}{\pi} = \prod_{n=1}^{\infty} \frac{2n-1}{2n} \cdot \frac{2n+1}{2n}.$$

Now taking reciprocals of both sides (you are encouraged to verify that the reciprocal of an infinite product is the product of the reciprocals) we get Wallis's first formula. To obtain the second formula, we write the first formula as

$$\frac{\pi}{2} = \lim_{n \to \infty} \left\{ \left(\frac{2}{1}\right)^2 \cdot \left(\frac{4}{3}\right)^2 \cdots \left(\frac{2n}{2n-1}\right)^2 \cdot \frac{1}{2n+1} \right\}.$$

Then taking square roots we obtain

$$\sqrt{\pi} = \lim_{n \to \infty} \sqrt{\frac{2}{2n+1}} \prod_{k=1}^{n} \frac{2k}{2k-1} = \lim_{n \to \infty} \frac{1}{\sqrt{n}} \frac{1}{\sqrt{1+1/2n}} \prod_{k=1}^{n} \frac{2k}{2k-1}.$$

Using that $1/\sqrt{1+1/2n} \to 1$ as $n \to \infty$ completes our proof. ∎

We now prove a beautiful expression for π due to Jonathan Sondow [232, 259]. To present this formula, we first manipulate Wallis's first formula to

$$\frac{\pi}{2} = \prod_{n=1}^{\infty} \frac{2n}{2n-1} \cdot \frac{2n}{2n+1} = \prod_{n=1}^{\infty} \frac{4n^2}{4n^2-1} = \prod_{n=1}^{\infty} \left(1 + \frac{1}{4n^2-1}\right).$$

Second, using partial fractions, we observe that

$$\sum_{n=1}^{\infty} \frac{1}{4n^2-1} = \frac{1}{2} \sum_{n=1}^{\infty} \left(\frac{1}{2n-1} - \frac{1}{2n+1}\right) = \frac{1}{2} \cdot 1 = \frac{1}{2},$$

since the sum telescopes (see e.g., the telescoping series theorem, Theorem 3.22 on page 200). Dividing these two formulas, we get

$$\pi = \frac{\prod_{n=1}^{\infty} \left(1 + \dfrac{1}{4n^2-1}\right)}{\sum_{n=1}^{\infty} \dfrac{1}{4n^2-1}},$$

quite astonishing! Written out in all its glory, we have

$$\pi = \frac{\left(1 + \dfrac{1}{1 \cdot 3}\right)\left(1 + \dfrac{1}{3 \cdot 5}\right)\left(1 + \dfrac{1}{5 \cdot 7}\right)\left(1 + \dfrac{1}{7 \cdot 9}\right) \cdots}{\dfrac{1}{1 \cdot 3} + \dfrac{1}{3 \cdot 5} + \dfrac{1}{5 \cdot 7} + \dfrac{1}{7 \cdot 9} + \cdots}.$$

▶ Exercises 5.1

1. Here are some Viète–Wallis products from [187, 190].

   (i) From the formulas (5.3) and (5.5) on pages 381 and 382, and Euler's sine expansion, prove that for every $x \in \mathbb{R}$ and $p \in \mathbb{N}$ we have

   $$\frac{\sin x}{x} = \prod_{k=1}^{p} \sqrt{\frac{1}{2} + \frac{1}{2}\sqrt{\frac{1}{2} + \cdots + \frac{1}{2}\sqrt{\frac{1}{2} + \frac{1}{2}\cos x}}} \cdot \prod_{n=1}^{\infty} \left(\frac{2^p \pi n - x}{2^p \pi n} \cdot \frac{2^p \pi n + x}{2^p \pi n} \right),$$

   where there are k square roots in the kth factor of the product $\prod_{k=1}^{p}$.

   (ii) Setting $x = \pi/2$ in (i), show that

   $$\frac{2}{\pi} = \prod_{k=1}^{p} \sqrt{\frac{1}{2} + \frac{1}{2}\sqrt{\frac{1}{2} + \cdots + \frac{1}{2}\sqrt{\frac{1}{2} + \frac{1}{2}}}} \cdot \prod_{n=1}^{\infty} \left(\frac{2^{p+1}n - 1}{2^{p+1}n} \cdot \frac{2^{p+1}n + 1}{2^{p+1}n} \right),$$

   where there are k square roots in the kth factor of the product $\prod_{k=1}^{p}$.

   (iii) Setting $x = \pi/6$ in (i), show that

   $$\frac{3}{\pi} = \prod_{k=1}^{p} \sqrt{\frac{1}{2} + \frac{1}{2}\sqrt{\frac{1}{2} + \cdots + \frac{1}{2}\sqrt{\frac{1}{2} + \frac{1}{2}\left(\frac{\sqrt{3}}{2} \right)}}}$$

   $$\cdot \prod_{n=1}^{\infty} \left(\frac{3 \cdot 2^{p+1}n - 1}{3 \cdot 2^{p+1}n} \cdot \frac{3 \cdot 2^{p+1}n + 1}{3 \cdot 2^{p+1}n} \right),$$

   where there are k square roots in the kth factor of the product $\prod_{k=1}^{p}$.

   (iv) Experiment with two other values of x to derive other Viète–Wallis-type formulas.

2. Suppose that for each $n \in \mathbb{N}$ we are given a finite product

   $$\prod_{k=1}^{a_n} f_k(n),$$

   where $f_k(n)$ is an expression involving k, n, and $a_n \in \mathbb{N}$ is such that $\lim_{n \to \infty} a_n = \infty$. For example, in (5.11) we have $a_n = \frac{n-1}{2}$ and $f_k(n) = \left(1 - \frac{x^2}{n^2 \tan^2(k\pi/n)} \right)$; then (5.11) claims that for this example we have

   $$\lim_{n \to \infty} \prod_{k=1}^{a_n} f_k(n) = \prod_{k=1}^{\infty} \lim_{n \to \infty} f_k(n). \tag{5.15}$$

However, this equality is not always true. Indeed, give an example of an a_n and $f_k(n)$ for which (5.15) does not hold.

3. Prove (5.14) using induction on p.

4. Prove the following splendid formula:

$$\sqrt{\pi} = \lim_{n\to\infty} \frac{(n!)^2 \, 2^{2n}}{(2n)! \, \sqrt{n}}.$$

Suggestion: Wallis's formula is hidden here.

5. (cf. [21]) In this problem we give an elementary proof of the following interesting identity: For every n that is a power of 2 and for every $x \in \mathbb{R}$ we have

$$\sin x = n \sin \left(\frac{x}{n}\right) \cos \left(\frac{x}{n}\right) \prod_{k=1}^{\frac{n}{2}-1} \left(1 - \frac{\sin^2(x/n)}{\sin^2(k\pi/n)}\right). \qquad (5.16)$$

(i) Prove that for every $x \in \mathbb{R}$,

$$\sin x = 2 \sin \left(\frac{x}{2}\right) \sin \left(\frac{\pi+x}{2}\right).$$

(ii) Show that for n equal to a power of 2, we have

$$\sin x = 2^n \sin \left(\frac{x}{n}\right) \sin \left(\frac{\pi+x}{n}\right) \sin \left(\frac{2\pi+x}{n}\right) \cdots$$
$$\cdots \sin \left(\frac{(n-2)\pi+x}{n}\right) \sin \left(\frac{(n-1)\pi+x}{n}\right);$$

note that if $n = 2^1$, we get the formula in (i).

(iii) Show that the formula in (ii) can be written as

$$\sin x = 2^n \sin \left(\frac{x}{n}\right) \sin \left(\frac{\frac{n}{2}\pi+x}{n}\right) \prod_{1\le k<\frac{n}{2}} \sin \left(\frac{k\pi+x}{n}\right) \sin \left(\frac{k\pi-x}{n}\right).$$

(iv) Prove the identity $\sin(\theta+\varphi) \sin(\theta-\varphi) = \sin^2\theta - \sin^2\varphi$ and use this to conclude that the formula in (iii) equals

$$\sin x = 2^n \sin \left(\frac{x}{n}\right) \cos \left(\frac{x}{n}\right) \prod_{1\le k<\frac{n}{2}} \left(\sin^2 \left(\frac{k\pi}{n}\right) - \sin^2 \left(\frac{x}{n}\right)\right).$$

(v) By considering what happens as $x \to 0$ in the formula in (iv), prove that for n a power of 2, we have

$$n = 2^n \prod_{1 \leq k < \frac{n}{2}} \left(\sin^2 \left(\frac{k\pi}{n} \right) \right).$$

Now prove (5.16).

6 (**Expansion of sine II**) We give a second proof of Euler's sine expansion.

(i) Show that taking $n \to \infty$ on both sides of the identity (5.16) from the previous problem gives a *formal* proof of Euler's sine expansion.

(ii) Now using the identity (5.16) and following the ideas found in the proof of Theorem 5.2, give another rigorous proof of Euler's sine expansion.

5.2 ★ Beautiful Formuls II: Euler, Gregory, Leibniz, and Madhava

In this section we present two beautiful formulas involving π: Euler's formula for $\pi^2/6$ and the Gregory–Leibniz–Madhava formula for $\pi/4$. My favorite proofs of these formulas are taken from the article by Josef Hofbauer [110] and are completely "elementary" in the sense that they involve nothing involving derivatives or integrals. They use just a little bit of trigonometric identities and then a dab of some inequalities (or Tannery's theorem if you prefer) to finish them off. However, before presenting Hofbauer's proofs, we present (basically) one of Euler's original proofs of his solution to the Basel problem.

5.2.1 Proof I of Euler's Formula for $\pi^2/6$

The Italian mathematician Pietro Mengoli (1625–1686), in his 1650 book *Novae quadraturae arithmeticae, seu de additione fractionum*, posed the following question: What's the value of the sum

$$\sum_{n=1}^{\infty} \frac{1}{n^2} = 1 + \frac{1}{2^2} + \frac{1}{3^2} + \frac{1}{4^2} + \cdots ?$$

Here's what Mengoli said[6]:

Having concluded with satisfaction my consideration of those arrangements of fractions, I shall move on to those other arrangements that have the unit as numerator, and square numbers as denominators. The work devoted to this consideration has bore some fruit —

---

[6]It took ≈ 5 years to find this passage! The breakthrough came thanks to Emanuele Delucchi, who contacted his sister Rachele Delucchi, who then found Mengoli's book in the library of ETH Zurich, and thanks to Emanuele Delucchi for translating the original Latin.

the question itself still awaiting solution — but it [the work] requires the support of a richer mind, in order to lead to the evaluation of the precise sum of the arrangement [of fractions] that I have set myself as a task.

The task of finding the sum was made popular through Jacob Bernoulli (1654–1705) when he wrote about it in 1689, and was solved by Leonhard Euler (1707–1783) in 1735. Bernoulli was so baffled by the unknown value of the series that he wrote:

If somebody should succeed in finding what till now withstood our efforts and communicate it to us, we shall be much obliged to him. [47, p. 73], [270, p. 345].

Before Euler's solution to this request, known as the *Basel problem* (Bernoulli lived in Basel, Switzerland), this problem had eluded many of the great mathematicians of that day: In 1742, Euler wrote:

Jacob Bernoulli does mention those series, but confesses that, in spite of all his efforts, he could not get through, so that Joh. Bernoulli, de Moivre, and Stirling, great authorities in such matters, were highly surprised when I told them that I had found the sum of $\zeta(2)$, and even of $\zeta(n)$ for n even. [255, pp. 262–263].

(We shall consider $\zeta(n)$ for n even in the next section.) Needless to say, it shocked the mathematical community when Euler found the sum to be $\pi^2/6$; in the introduction to his famous 1735 paper *De summis serierum reciprocarum* (On the sums of series of reciprocals), where he first proves that $\zeta(2) = \pi^2/6$, Euler writes:

So much work has been done on the series $\zeta(n)$ that it seems hardly likely that anything new about them may still turn up ... I, too, in spite of repeated efforts, could achieve nothing more than approximate values for their sums ... Now, however, quite unexpectedly, I have found an elegant formula for $\zeta(2)$, depending on the quadrature of the circle [i.e., upon π] [255, p. 261].

For more on various solutions to the Basel problem, see [49, 118, 209], and for more on Euler, see [10, 128]. On the side is a picture of a Swiss ten-franc banknote honoring Euler.

We already saw Euler's original argument in the introduction to this chapter; we shall now make his argument rigorous. First, we claim that for all nonnegative real numbers $a_1, a_2, \ldots, a_n \geq 0$, we have

$$1 - \sum_{k=1}^{n} a_k \leq \prod_{k=1}^{n}(1 - a_k) \leq 1 - \sum_{k=1}^{n} a_k + \sum_{1 \leq i < j \leq n} a_i\, a_j, \qquad (5.17)$$

where the sum $\sum_{1 \leq i < j \leq n}$ means to sum the $a_i\, a_j$ with $1 \leq i < j \leq n$. You will prove these inequalities in Problem 1. Applying these inequalities to $\prod_{k=1}^{n}\left(1 - \frac{x^2}{k^2\pi^2}\right)$, we obtain

$$1 - \sum_{k=1}^{n} \frac{x^2}{k^2 \pi^2} \leq \prod_{k=1}^{n} \left(1 - \frac{x^2}{k^2 \pi^2}\right) \leq 1 - \sum_{k=1}^{n} \frac{x^2}{k^2 \pi^2} + \sum_{1 \leq i < j \leq n} \frac{x^2}{i^2 \pi^2} \frac{x^2}{j^2 \pi^2}.$$

After some slight simplifications, we can write this as

$$1 - \frac{x^2}{\pi^2} \sum_{k=1}^{n} \frac{1}{k^2} \leq \prod_{k=1}^{n} \left(1 - \frac{x^2}{k^2 \pi^2}\right) \leq 1 - \frac{x^2}{\pi^2} \sum_{k=1}^{n} \frac{1}{k^2} + \frac{x^4}{\pi^4} \sum_{1 \leq i < j \leq n} \frac{1}{i^2 j^2}. \quad (5.18)$$

Let us put

$$\zeta_n(2) = \sum_{k=1}^{n} \frac{1}{k^2} \quad \text{and} \quad \zeta_n(4) = \sum_{k=1}^{n} \frac{1}{k^4},$$

and observe that

$$\zeta_n(2)^2 = \left(\sum_{i=1}^{n} \frac{1}{i^2}\right)\left(\sum_{j=1}^{n} \frac{1}{j^2}\right) = \sum_{i,j=1}^{n} \frac{1}{i^2 j^2}$$

$$= \zeta_n(4) + 2 \sum_{1 \leq i < j \leq n} \frac{1}{i^2 j^2}.$$

Thus, (5.18) can be written as

$$1 - \frac{x^2}{\pi^2} \zeta_n(2) \leq \prod_{k=1}^{n} \left(1 - \frac{x^2}{k^2 \pi^2}\right) \leq 1 - \frac{x^2}{\pi^2} \zeta_n(2) + \frac{x^4}{\pi^4} \frac{\zeta_n(2)^2 - \zeta_n(4)}{2}.$$

Taking $n \to \infty$ and using that $\zeta_n(2) \to \zeta(2)$, $\prod_{k=1}^{n} \left(1 - \frac{x^2}{k^2 \pi^2}\right) \to \frac{\sin x}{x}$, and that $\zeta_n(4) \to \zeta(4)$, we obtain

$$1 - \frac{x^2}{\pi^2} \zeta(2) \leq \frac{\sin x}{x} \leq 1 - \frac{x^2}{\pi^2} \zeta(2) + \frac{x^4}{\pi^4} \frac{\zeta(2)^2 - \zeta(4)}{2}.$$

Replacing $\frac{\sin x}{x}$ by its infinite series, we see that

$$1 - \frac{x^2}{\pi^2} \zeta(2) \leq 1 - \frac{x^2}{3!} + \frac{x^4}{5!} - \frac{x^6}{7!} + \cdots \leq 1 - \frac{x^2}{\pi^2} \zeta(2) + \frac{x^4}{\pi^4} \frac{\zeta(2)^2 - \zeta(4)}{2}.$$

Now subtracting 1 from everything and dividing by x^2, we get

$$-\frac{1}{\pi^2} \zeta(2) \leq -\frac{1}{3!} + \frac{x^2}{5!} - \frac{x^4}{7!} + \cdots \leq -\frac{1}{\pi^2} \zeta(2) + \frac{x^2}{\pi^4} \frac{\zeta(2)^2 - \zeta(4)}{2}.$$

Finally, putting $x = 0$, we conclude that

$$-\frac{1}{\pi^2}\zeta(2) \le -\frac{1}{3!} \le -\frac{1}{\pi^2}\zeta(2).$$

This implies that $\zeta(2) = \frac{\pi^2}{6}$, exactly as Euler stated. See Problem 7 for a derivation of the value of $\zeta(4)$.

5.2.2   Proof II of Euler's Formula for $\pi^2/6$ [110]

We begin with the following identity, valid for $z \in \mathbb{C}$ that are noninteger multiples of π,

$$\frac{1}{\sin^2 z} = \frac{1}{4\sin^2\frac{z}{2}\cos^2\frac{z}{2}} = \frac{1}{4}\left(\frac{1}{\sin^2\frac{z}{2}} + \frac{1}{\cos^2\frac{z}{2}}\right) = \frac{1}{4}\left(\frac{1}{\sin^2\frac{z}{2}} + \frac{1}{\sin^2\left(\frac{\pi-z}{2}\right)}\right),$$

where at the last step we used that $\cos(z) = \sin(\frac{\pi}{2} - z)$. Replacing z with πz, we get for noninteger z,

$$\frac{1}{\sin^2 \pi z} = \frac{1}{2^2}\left(\frac{1}{\sin^2\frac{z\pi}{2}} + \frac{1}{\sin^2\left(\frac{(1-z)\pi}{2}\right)}\right). \tag{5.19}$$

In particular, setting $z = 1/2^2$ gives

$$2 = \frac{1}{2^2}\left(\frac{1}{\sin^2\frac{\pi}{2^3}} + \frac{1}{\sin^2\frac{3\pi}{2^3}}\right),$$

or in summation notation,

$$2 = \frac{1}{2^2}\sum_{k=1}^{2}\frac{1}{\sin^2\frac{(2k-1)\pi}{2^3}}.$$

Applying (5.19) to each term on the right-hand side of this equation (that is, set $z = 1/2^2$, respectively $z = 3/2^2$, in (5.19) to expand $\frac{1}{\sin^2\frac{\pi}{2^3}}$, respectively $\frac{1}{\sin^2\frac{3\pi}{2^3}}$), we obtain

$$2 = \frac{1}{2^2}\left(\frac{1}{2^2}\left[\frac{1}{\sin^2\frac{\pi}{2^4}} + \frac{1}{\sin^2\frac{7\pi}{2^4}}\right] + \frac{1}{2^2}\left[\frac{1}{\sin^2\frac{3\pi}{2^4}} + \frac{1}{\sin^2\frac{5\pi}{2^4}}\right]\right)$$

$$= \frac{1}{2^4}\left(\frac{1}{\sin^2\frac{\pi}{2^4}} + \frac{1}{\sin^2\frac{3\pi}{2^4}} + \frac{1}{\sin^2\frac{5\pi}{2^4}} + \frac{1}{\sin^2\frac{7\pi}{2^4}}\right)$$

$$= \frac{1}{2^4}\sum_{k=1}^{4}\frac{1}{\sin^2\frac{(2k-1)\pi}{2^4}}.$$

Repeatedly applying (5.19) (slang for "use induction"), we arrive at the following.

Lemma 5.4 *For every* $n \in \mathbb{N}$, *we have*

$$2 = \frac{1}{2^{2n}}\sum_{k=1}^{2^n}\frac{1}{\sin^2\frac{(2k-1)\pi}{2^{n+2}}}.$$

To establish Euler's formula, we need one more lemma.

Lemma 5.5 *For* $0 < x < \pi/2$, *we have*

$$-1 + \frac{1}{\sin^2 x} < \frac{1}{x^2} < \frac{1}{\sin^2 x}.$$

Proof For $0 < x < \pi/2$, taking reciprocals in the formula from Lemma 4.57 on page 368,

$$\sin x < x < \tan x,$$

we get $\cot^2 x < x^{-2} < \sin^{-2} x$. Since $\cot^2 x = \cos^2 x / \sin^2 x = \sin^{-2} x - 1$, we conclude that

$$\frac{1}{\sin^2 x} > \frac{1}{x^2} > -1 + \frac{1}{\sin^2 x}, \quad 0 < x < \frac{\pi}{2},$$

which proves the lemma. ∎

Now, observe that for $0 \le k \le 2^n$, we have

$$\frac{(2k-1)\pi}{2^{n+2}} \le \frac{(2(2^n)-1)\pi}{2^{n+2}} = \frac{(2^{n+1}-1)\pi}{2^{n+2}} < \frac{\pi}{2},$$

and therefore, using the identity

$$-1 + \frac{1}{\sin^2 x} < \frac{1}{x^2} < \frac{1}{\sin^2 x}, \quad 0 < x < \frac{\pi}{2},$$

we see that

$$-2^n + \sum_{k=1}^{2^n} \frac{1}{\sin^2 \frac{(2k-1)\pi}{2^{n+2}}} < \sum_{k=1}^{2^n} \frac{1}{\left(\frac{(2k-1)\pi}{2^{n+2}}\right)^2} < \sum_{k=1}^{2^n} \frac{1}{\sin^2 \frac{(2k-1)\pi}{2^{n+2}}}.$$

Multiplying both sides by $1/2^{2n}$ and using Lemma 5.4, we get

$$-\frac{1}{2^n} + 2 < \frac{2^4}{\pi^2} \sum_{k=1}^{2^n} \frac{1}{(2k-1)^2} < 2.$$

Taking $n \to \infty$ and using the squeeze theorem, we conclude that

$$2 \le \frac{2^4}{\pi^2} \sum_{k=1}^{\infty} \frac{1}{(2k-1)^2} \le 2 \quad \Longrightarrow \quad \sum_{k=1}^{\infty} \frac{1}{(2k-1)^2} = \frac{\pi^2}{8}.$$

Finally, summing over the even and odd numbers (see Problem 2a on page 203), we have

$$\sum_{n=1}^{\infty} \frac{1}{n^2} = \sum_{n=1}^{\infty} \frac{1}{(2n-1)^2} + \sum_{n=1}^{\infty} \frac{1}{(2n)^2} = \frac{\pi^2}{8} + \frac{1}{4} \sum_{n=1}^{\infty} \frac{1}{n^2} \qquad (5.20)$$

$$\Longrightarrow \quad \frac{3}{4} \sum_{n=1}^{\infty} \frac{1}{n^2} = \frac{\pi^2}{8}.$$

Solving for $\sum_{n=1}^{\infty} 1/n^2$, we obtain Euler's formula:

$$\frac{\pi^2}{6} = \sum_{n=1}^{\infty} \frac{1}{n^2} = 1 + \frac{1}{2^2} + \frac{1}{3^2} + \frac{1}{4^2} + \cdots.$$

5.2.3   Proof III of Euler's Formula for $\pi^2/6$

In Proof II, we established Euler's formula from Lemma 5.5. This time we apply Tannery's theorem from page 216. The idea is to write the identity in Lemma 5.4 in a form found in Tannery's theorem:

$$2 = \frac{1}{2^{2n}} \sum_{k=1}^{2^n} \frac{1}{\sin^2 \frac{(2k-1)\pi}{2^{n+2}}} = \sum_{k=1}^{2^n} a_k(n), \qquad (5.21)$$

where

$$a_k(n) = \frac{1}{2^{2n}} \frac{1}{\sin^2 \frac{(2k-1)\pi}{2^{n+2}}}.$$

Let us verify the hypotheses of Tannery's theorem. First, since $\lim_{z \to 0} \frac{\sin z}{z} = 1$, we have

$$\lim_{n \to \infty} 2^n \sin \frac{(2k-1)\pi}{2^{n+2}} = \lim_{n \to \infty} \frac{(2k-1)\pi}{2^2} \cdot \frac{\sin \frac{(2k-1)\pi}{2^{n+2}}}{\frac{(2k-1)\pi}{2^{n+2}}} = \frac{(2k-1)\pi}{2^2}.$$

Therefore,

$$\lim_{n \to \infty} a_k(n) = \lim_{n \to \infty} \frac{1}{2^{2n}} \cdot \frac{1}{\sin^2 \frac{(2k-1)\pi}{2^{n+2}}}$$

$$= \lim_{n \to \infty} \frac{1}{\left(2^n \sin \frac{(2k-1)\pi}{2^{n+2}}\right)^2} = \frac{2^4}{\pi^2 (2k-1)^2}.$$

To verify the other hypothesis of Tannery's theorem, we need the following lemma.

Lemma 5.6 *If $R < \pi$, there exists a constant $c > 0$ such that for $z \in \mathbb{C}$ with $|z| \le R$, we have*

$$c|z| \le |\sin z|.$$

Proof Let $R < \pi$ and let B denote the ball of radius R centered at the origin in $\mathbb{C}$. Define $f : B \longrightarrow \mathbb{R}$ by $f(0) = 1$ and $f(z) = |(\sin z)/z|$ for $z \ne 1$. Since $\lim_{z \to 0}(\sin z)/z = 1$, f is continuous on all of B. Note that f is not zero for $z \in B$ (why?). Thus, by Corollary 4.21, there is a constant $c > 0$ such that $c \le f(z)$ for all $z \in B$. Hence, $c|z| \le |\sin z|$ for $z \in B$. ∎

Now observe that for $1 \le k \le 2^n$, we have

$$\frac{(2k-1)\pi}{2^{n+2}} \le \frac{(2(2^n)-1)\pi}{2^{n+2}} = \frac{(2^{n+1}-1)\pi}{2^{n+2}} < \frac{\pi}{2},$$

and therefore, by Lemma 5.6, for some $c > 0$,

$$c \cdot \frac{(2k-1)\pi}{2^{n+2}} \le \sin \frac{(2k-1)\pi}{2^{n+2}} \implies \frac{1}{\sin^2 \frac{(2k-1)\pi}{2^{n+2}}} \le \frac{2^{2n+4}}{c^2 \pi^2 (2k-1)^2}.$$

Multiplying both sides by $1/2^{2n}$, we obtain

$$\frac{1}{2^{2n}} \cdot \frac{1}{\sin^2 \frac{(2k-1)\pi}{2^{n+2}}} \le \frac{2^4}{c^2\pi^2} \cdot \frac{1}{(2k-1)^2} =: M_k.$$

It follows that $|a_k(n)| \le M_k$ for all n, k. Moreover, since the sum $\sum_{k=1}^{\infty} M_k = \sum_{k=1}^{\infty} \frac{2^4}{c^2\pi^2} \cdot \frac{1}{(2k-1)^2}$ converges, we have verified the hypotheses of Tannery's theorem. Hence, taking $n \to \infty$ in (5.21), we get

$$2 = \lim_{n \to \infty} \sum_{k=1}^{2^n} a_k(n) = \sum_{k=1}^{\infty} \lim_{n \to \infty} a_k(n)$$

$$= \sum_{k=1}^{\infty} \frac{2^4}{\pi^2(2k-1)^2} \quad \Longrightarrow \quad \frac{\pi^2}{8} = \sum_{k=1}^{\infty} \frac{1}{(2k-1)^2}.$$

Doing the even–odd trick as we did in (5.20), we know that this formula implies Euler's formula for $\pi^2/6$. See Problem 5 for Proof IV, a classic proof.

5.2.4   Proof I of Gregory–Leibniz–Madhava's Formula for $\pi/4$

Like the proof of Euler's formula, which was based on a trigonometric identity for sines (Lemma 5.4), the proof of Gregory–Leibniz–Madhava's formula,

$$\frac{\pi}{4} = \sum_{n=0}^{\infty} \frac{(-1)^{n-1}}{2n-1} = 1 - \frac{1}{3} + \frac{1}{5} - \frac{1}{7} + \frac{1}{9} - \frac{1}{11} + \cdots,$$

also involves trigonometric identities, but for cotangents. Concerning Leibniz's discovery of this formula, Christiaan Huygens (1629–1695) wrote that "it would be a discovery always to be remembered among mathematicians" [270, p. 316]. In 1676, Isaac Newton (1642–1727) wrote:

> Leibniz's method for obtaining convergent series is certainly very elegant, and it would have sufficiently revealed the genius of its author, even if he had written nothing else. [244, p. 130].

To prove the Gregory–Leibniz–Madhava formula, we begin with the double angle formula

$$2 \cot 2z = 2\frac{\cos 2z}{\sin 2z} = \frac{\cos^2 z - \sin^2 z}{\cos z \sin z} = \cot z - \tan z,$$

from which we see that

$$\cot 2z = \frac{1}{2}\left(\cot z - \tan z\right).$$

Since $\tan z = \cot(\pi/2 - z)$, we find that

$$\cot 2z = \frac{1}{2}\left(\cot z - \cot\left(\frac{\pi}{2} - z\right)\right).$$

Replacing z with $\pi z/2$, we get

$$\cot \pi z = \frac{1}{2}\left(\cot \frac{z\pi}{2} - \cot \frac{(1 - z)\pi}{2}\right), \tag{5.22}$$

which is our fundamental equation. In particular, setting $z = 1/4$, we obtain

$$1 = \frac{1}{2}\left(\cot \frac{\pi}{4 \cdot 2} - \cot \frac{3\pi}{4 \cdot 2}\right) = \frac{1}{2}\sum_{k=1}^{1}\left(\cot \frac{(4k - 3)\pi}{2^3} - \cot \frac{(4k - 1)\pi}{2^3}\right).$$

Applying (5.22) to each term $\cot \frac{\pi}{2^3}$ and $\cot \frac{3\pi}{2^3}$ gives

$$\begin{aligned}
1 &= \frac{1}{2}\left[\frac{1}{2}\left(\cot \frac{\pi}{2^4} - \cot \frac{7\pi}{2^4}\right) - \frac{1}{2}\left(\cot \frac{3\pi}{2^4} - \cot \frac{5\pi}{2^4}\right)\right] \\
&= \frac{1}{2^2}\left[\left(\cot \frac{\pi}{2^4} - \cot \frac{3\pi}{2^4}\right) + \left(\cot \frac{5\pi}{2^4} - \cot \frac{7\pi}{2^4}\right)\right] \\
&= \frac{1}{2^2}\sum_{k=1}^{2}\left(\cot \frac{(4k - 3)\pi}{2^4} - \cot \frac{(4k - 1)\pi}{2^4}\right).
\end{aligned}$$

Repeatedly applying (5.22), one can prove that for every $n \in \mathbb{N}$, we have

$$1 = \frac{1}{2^n}\sum_{k=1}^{2^{n-1}}\left(\cot \frac{(4k - 3)\pi}{2^{n+2}} - \cot \frac{(4k - 1)\pi}{2^{n+2}}\right).$$

(The diligent reader will supply the details!) Thus,

$$1 = \sum_{k=1}^{2^{n-1}} a_k(n), \quad \text{where } a_k(n) = \frac{1}{2^n}\left(\cot \frac{(4k - 3)\pi}{2^{n+2}} - \cot \frac{(4k - 1)\pi}{2^{n+2}}\right). \tag{5.23}$$

This identity implores us to try Tannery's theorem (see page 216)! Let us verify the hypotheses of that theorem. First, to find $\lim_{n\to\infty} a_k(n)$, note that

$$\lim_{z\to 0} z \cot z = \lim_{z\to 0}\left(\frac{z}{\sin z} \cdot \cos z\right) = 1 \cdot \cos(0) = 1.$$

Observe that

$$
a_k(n) = \frac{1}{2^n} \cot \frac{(4k-3)\pi}{2^{n+2}} - \frac{1}{2^n} \cot \frac{(4k-1)\pi}{2^{n+2}}
$$

$$
= \frac{4}{\pi(4k-3)} \cdot \frac{(4k-3)\pi}{2^{n+2}} \cot \frac{(4k-3)\pi}{2^{n+2}} - \frac{4}{\pi(4k-1)} \cdot \frac{(4k-1)\pi}{2^{n+2}} \cot \frac{(4k-1)\pi}{2^{n+2}}.
$$

Taking $n \to \infty$ and using $\lim_{z\to 0} z \cot z = 1$, we obtain

$$
\lim_{n\to\infty} a_k(n) = \frac{4}{\pi}\left(\frac{1}{4k-3} - \frac{1}{4k-1}\right).
$$

We now need to bound $a_k(n)$. Since we know a nice boundedness property of sine from Lemma 5.6, we shall write $a_k(n)$ in terms of sine. To do so, observe that for complex numbers z, w, not integer multiples of π, we have

$$
\cot z - \cot w = \frac{\cos z}{\sin z} - \frac{\cos w}{\sin w} = \frac{\sin w \cos z - \cos w \sin z}{\sin z \sin w}
$$

$$
= \frac{\sin(w-z)}{\sin z \sin w}.
$$

Using this identity in the formula for $a_k(n)$ in (5.23), we obtain

$$
a_k(n) = \frac{1}{2^n} \frac{\sin \frac{\pi}{2^{n+1}}}{\sin \frac{(4k-3)\pi}{2^{n+2}} \cdot \sin \frac{(4k-1)\pi}{2^{n+2}}}.
$$

To find a bound for $a_k(n)$, we shall use the following lemma.

Lemma 5.7 *If* $|z| \le 1$, *then*

$$
|\sin z| \le \frac{6}{5}|z|.
$$

Proof Observe that for $|z| \le 1$, we have $|z|^k \le |z|$ for every $k \in \mathbb{N}$, and

$$
(2n+1)! = (2\cdot 3)\cdot(4\cdot 5)\cdots(2n\cdot(2n+1))
$$

$$
\ge (2\cdot 3)\cdot(2\cdot 3)\cdots(2\cdot 3) = (2\cdot 3)^n = 6^n.
$$

Thus,

$$
|\sin z| = \left|\sum_{n=0}^{\infty}(-1)^n \frac{z^{2n+1}}{(2n+1)!}\right| \le \sum_{n=0}^{\infty} \frac{|z|^{2n+1}}{(2n+1)!}
$$

$$
\le \sum_{n=0}^{\infty} \frac{|z|}{6^n} = |z| \sum_{n=0}^{\infty} \frac{1}{6^n} = \frac{1}{1-(1/6)}|z| = \frac{6}{5}|z|.
$$

■

Since $0 \leq \frac{\pi}{2^{n+1}} \leq 1$ for $n \in \mathbb{N}$ (because $\pi < 4$), by this lemma we have

$$\sin \frac{\pi}{2^{n+1}} \leq \frac{6}{5} \cdot \frac{\pi}{2^{n+1}}. \tag{5.24}$$

Observe that for $1 \leq k \leq 2^{n-1}$ and $0 \leq \ell \leq 4$, we have

$$\frac{(4k - \ell)\pi}{2^{n+2}} \leq \frac{(4(2^{n-1}) - \ell)\pi}{2^{n+2}} = \frac{(2^{n+1} - \ell)\pi}{2^{n+2}} \leq \frac{\pi}{2}.$$

Therefore, by Lemma 5.6, for some constant $c > 0$,

$$c \cdot \frac{(4k - \ell)\pi}{2^{n+2}} \leq \sin \frac{(4k - \ell)\pi}{2^{n+2}} \implies \frac{1}{\sin \frac{(4k-\ell)\pi}{2^{n+2}}} \leq \frac{1}{c} \frac{2^{n+2}}{(4k - \ell)\pi}.$$

Combining this inequality with (5.24), we see that for $1 \leq k \leq 2^{n-1}$, we have

$$\frac{1}{2^n} \frac{\sin \frac{\pi}{2^{n+1}}}{\sin \frac{(4k-3)\pi}{2^{n+2}} \cdot \sin \frac{(4k-1)\pi}{2^{n+2}}} \leq \frac{1}{2^n} \cdot \left(\frac{6}{5} \cdot \frac{\pi}{2^{n+1}}\right) \cdot \left(\frac{1}{c} \frac{2^{n+2}}{(4k - 3)\pi}\right) \cdot \left(\frac{1}{c} \frac{2^{n+2}}{(4k - 1)\pi}\right)$$

$$= \frac{48}{5\pi c^2} \cdot \frac{1}{(4k - 3)(4k - 1)}.$$

It follows that

$$|a_k(n)| \leq \frac{48}{5\pi c^2} \cdot \frac{1}{(4k - 3)(4k - 1)} =: M_k.$$

Since the sum $\sum_{k=1}^{\infty} M_k$ converges, we have verified the hypotheses of Tannery's theorem. Hence, taking $n \to \infty$ in (5.23), we get

$$1 = \lim_{n \to \infty} \sum_{k=1}^{2^{n-1}} a_k(n) = \sum_{k=1}^{\infty} \lim_{n \to \infty} a_k(n)$$

$$= \sum_{k=1}^{\infty} \frac{4}{\pi} \left(\frac{1}{4k - 3} - \frac{1}{4k - 1}\right) \implies \frac{\pi}{4} = \sum_{k=1}^{\infty} \left(\frac{1}{4k - 3} - \frac{1}{4k - 1}\right).$$

The last series is equivalent to Gregory–Leibniz–Madhava's formula, because writing out the series term by term, we obtain

$$\frac{\pi}{4} = \sum_{k=1}^{\infty} \left(\frac{1}{4k - 3} - \frac{1}{4k - 1}\right) = 1 - \frac{1}{3} + \frac{1}{5} - \frac{1}{7} + \frac{1}{9} - \frac{1}{11} + \cdots,$$

which is exactly Gregory–Leibniz–Madhava's formula.

▶ **Exercises 5.2**

1. Prove the formula (5.17) by induction on n.
2. Find the following limit:

$$\lim_{n \to \infty} \left\{ \frac{1}{n^3 \sin\left(\frac{1 \cdot 2}{n^3}\right)} + \frac{1}{n^3 \sin\left(\frac{2 \cdot 3}{n^3}\right)} + \cdots + \frac{1}{n^3 \sin\left(\frac{n \cdot (n+1)}{n^3}\right)} \right\}.$$

3. **(Partial fraction expansion of** $1/\sin^2 x$**, Proof I,** [110]**)** Let $x \in \mathbb{R}$ with x not an integer multiple of π.

   (i) Prove that for all $n \in \mathbb{N}$,

   $$\frac{1}{\sin^2 x} = \frac{1}{2^{2n}} \sum_{k=0}^{2^n - 1} \frac{1}{\sin^2 \frac{x + \pi k}{2^n}}.$$

   (ii) Show that

   $$\frac{1}{\sin^2 x} = \frac{1}{2^{2n}} \sum_{k=-2^{n-1}}^{2^{n-1} - 1} \frac{1}{\sin^2 \frac{x + \pi k}{2^n}}. \tag{5.25}$$

   (iii) Using Lemma 5.5, prove that $\frac{1}{\sin^2 x} = \lim_{n \to \infty} \sum_{k=-n}^{n} \frac{1}{(x + \pi k)^2}$. We usually write this as

   $$\boxed{\frac{1}{\sin^2 x} = \sum_{k \in \mathbb{Z}} \frac{1}{(x + \pi k)^2}.} \tag{5.26}$$

4. **(Partial fraction expansion of** $1/\sin^2 x$**, Proof II)** Give another proof of (5.26) using Tannery's theorem and the formula (5.25).

5. **(Euler's sum for** $\pi^2/6$**, Proof IV)** In this problem we derive Euler's sum via an old argument found in Thomas Bromwich's (1875–1929) book [40, pp. 218–219] (cf. similar ideas found in [6, 132, 191], [267, Problem 145]).

   (i) Recall from Problem 4 on page 336 that for every $n \in \mathbb{N}$ and $x \in \mathbb{R}$,

   $$\sin nx = \sum_{k=0}^{\lfloor (n-1)/2 \rfloor} (-1)^k \binom{n}{2k + 1} \cos^{n-2k-1} x \, \sin^{2k+1} x.$$

   Using this formula, prove that if $\sin x \neq 0$, then

   $$\sin(2n + 1)x = \sin^{2n+1} x \sum_{k=0}^{n} (-1)^k \binom{2n + 1}{2k + 1} (\cot^2 x)^{n-k}.$$

(ii) Prove that if $n \in \mathbb{N}$, then the roots of $\sum_{k=0}^{n}(-1)^k \binom{2n+1}{2k+1} t^{n-k} = 0$ are the n numbers $t = \cot^2 \frac{m\pi}{2n+1}$ where $m = 1, 2, \ldots, n$.

(iii) Prove that if $n \in \mathbb{N}$, then

$$\sum_{k=1}^{n} \cot^2 \frac{k\pi}{2n+1} = \frac{n(2n-1)}{3}. \tag{5.27}$$

Suggestion: Recall that if $p(t)$ is a polynomial of degree n with roots $r_1, \ldots, r_n$, then $p(t) = a(t - r_1)(t - r_2) \cdots (t - r_n)$ for a constant a. What's the coefficient of t^1 if you multiply out $a(t - r_1) \cdots (t - r_n)$?

(iv) From the identity (5.27), derive Euler's sum.

6. (**Euler's sum for $\pi^2/6$, Proof V**) Here's another proof (cf. 110)!

   (i) Use (5.26) in Problem 3 to prove that for all $n \in \mathbb{N}$,

$$\frac{1}{\sin^2 x} = \frac{1}{n^2} \sum_{m=0}^{n-1} \frac{1}{\sin^2 \frac{x+\pi m}{n}}. \tag{5.28}$$

Suggestion: Replace x with $\frac{x+\pi m}{n}$ in (5.26) and sum from $m = 0$ to $n - 1$.

   (ii) Take the $m = 0$ term in (5.28) to the left, replace n by $2n + 1$, and then take $x \to 0$ to derive the identity

$$\sum_{k=1}^{n} \frac{1}{\sin^2 \frac{\pi k}{2n+1}} = \frac{2n(n+1)}{3}. \tag{5.29}$$

   (iii) From the identity (5.29), derive Euler's sum.

7. (**Euler's sum for $\zeta(4)$**) In this problem we prove that $\zeta(4) = \frac{\pi^4}{90}$.

   (i) Prove that for all nonnegative real numbers $a_1, \ldots, a_n$, we have

$$1 - \sum_{k=1}^{n} a_k + \sum_{1 \leq i < j \leq n} a_i a_j - \sum_{1 \leq i < j < k \leq n} a_i a_j a_k \leq \prod_{k=1}^{n}(1 - a_k) \leq 1 - \sum_{k=1}^{n} a_k + \sum_{1 \leq i < j \leq n} a_i a_j.$$

   (ii) Applying the inequalities in (i) to $\prod_{k=1}^{n}\left(1 - \frac{x^2}{k^2\pi^2}\right)$, prove that $\zeta(4) = \pi^4/90$.

8. (**Euler's sum for $\zeta(4)$, again**) Here's another derivation of the value of $\zeta(4)$.

   (i) Prove the identity

$$\frac{1}{\sin^4 z} = \frac{1}{2^4}\left(\frac{1}{\sin^4(z/2)} + \frac{1}{\sin^4((\pi - z)/2)}\right) + \frac{1}{2}\frac{1}{\sin^2 z}.$$

(ii) Prove that for all $n \in \mathbb{N}$,

$$2^2 = \frac{1}{2^{4n}} \sum_{k=1}^{2^n} \frac{1}{\sin^4 \frac{(2k-1)\pi}{2^{n+2}}} + 2^2 \sum_{k=1}^{n} \frac{1}{2^{2k}}.$$

(Note that the summation on the right is a geometric sum.)

(iii) Taking $n \to \infty$ in (ii) and using Tannery's theorem, derive the formula for $\zeta(4)$.

9. (**Euler's sum for** $\zeta(6)$) Here's a derivation of the value for $\zeta(6)$.

(i) Prove the identity

$$\frac{1}{\sin^6 z} = \frac{1}{2^6} \left(\frac{1}{\sin^6(z/2)} + \frac{1}{\sin^6((\pi - z)/2)} \right) + \frac{3}{2^2} \frac{1}{\sin^4 z}.$$

(ii) Prove that for every $n \in \mathbb{N}$, we have

$$2^3 = \frac{1}{2^{6n}} \sum_{k=1}^{2^n} \frac{1}{\sin^6 \frac{(2k-1)\pi}{2^{n+2}}} + 2^3 \sum_{k=1}^{n} \frac{1}{2^{2k}} + 2^4 \sum_{k=1}^{n} \frac{1}{2^{4k}}.$$

(You may use the formula in (ii) of Problem 8.)

(iii) Taking $n \to \infty$ in (ii) and using Tannery's theorem, show that $\frac{\pi^6}{945} = \sum_{n=1}^{\infty} \frac{1}{n^6}$.

5.3 ★ Beautiful Formulas III: Euler's Formula for $\zeta(2k)$

In Euler's famous 1735 paper *De summis serierum reciprocarum*, he found not only $\zeta(2)$ but also $\zeta(n)$ for even n up to $n = 12$, although it is clear from his method that he could, with a lot of work, get the value of $\zeta(n)$ for any even n. Following G.T. Williams [265], we derive Euler's formula for $\zeta(n)$, for all $n \in \mathbb{N}$ even, as a rational multiple of π^n.

5.3.1 Williams's Formula

To find Euler's formula for $\zeta(2k)$, we'll use the following theorem, whose proof is admittedly long, but completely elementary in that it uses only high school arithmetic and basic facts about series. The "hard" part is only understanding the manipulations of some finite multiple summations.

Williams's formula

Theorem 5.8 *For every $k \in \mathbb{N}$ with $k \geq 2$, we have*

$$\left(k + \frac{1}{2}\right) \zeta(2k) = \sum_{\ell=1}^{k-1} \zeta(2\ell)\, \zeta(2k - 2\ell).$$

Proof Fix $k \in \mathbb{N}$ with $k \geq 2$. Then for $N \in \mathbb{N}$, define

$$a_N := \sum_{\ell=1}^{k-1} \left(\sum_{m=1}^{N} \frac{1}{m^{2\ell}}\right)\left(\sum_{n=1}^{N} \frac{1}{n^{2k-2\ell}}\right).$$

By definition of the zeta function, we have

$$\lim_{N\to\infty} a_N = \sum_{\ell=1}^{k-1} \zeta(2\ell)\, \zeta(2k - 2\ell).$$

The plan is to work out a nice formula for a_N, then show that $\lim_{N\to\infty} a_N = \left(k + \frac{1}{2}\right) \zeta(2k)$, which proves the theorem.

Step 1: We begin by making some modifications to the formula for a_N. We first multiply out the terms to the right of a_N and get

$$a_N = \sum_{\ell=1}^{k-1} \sum_{m,n=1}^{N} \frac{1}{m^{2\ell}\, n^{2k-2\ell}},$$

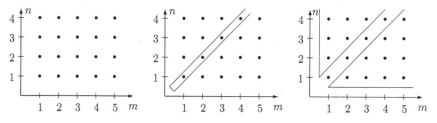

Fig. 5.1 *Left* $\sum_{m,n=1}^{N} a_{mn}$ sums over all the grid points (m, n) with $1 \leq m, n \leq N$. We break this sum up into diagonal and off-diagonal sums. *Middle* $\sum_{n=1}^{N} a_{nn}$ means to sum the a_{mn} with $m = n$. *Right* $\sum_{m\neq n} a_{mn}$ means to sum the a_{mn} for (m, n) off the diagonal

where for simplicity of notation, we write the double summation $\sum_{m=1}^{N} \sum_{n=1}^{N}$ as a single entity $\sum_{m,n=1}^{N}$. By commutativity, we can always switch the order of finite sums, so after noting that $\dfrac{1}{m^{2\ell}\, n^{2k-2\ell}} = \dfrac{1}{n^{2k}}\left(\dfrac{n}{m}\right)^{2\ell}$, we can write

$$a_N = \sum_{m,n=1}^{N} a_{mn} \,,$$

where

$$a_{mn} = \sum_{\ell=1}^{k-1} \frac{1}{n^{2k}}\left(\frac{n}{m}\right)^{2\ell} = \frac{1}{n^{2k}} \sum_{\ell=1}^{k-1} \left(\frac{n}{m}\right)^{2\ell}. \tag{5.30}$$

We now break up the summation $\sum_{m,n=1}^{N} a_{mn}$ into two sums:

$$\sum_{m,n=1}^{N} a_{mn} = \sum_{n=1}^{N} a_{nn} + \sum_{m\neq n}^{N} a_{mn},$$

where the sums on the right are explained in Fig. 5.1. We now work out each of these sums.

Diagonal sum: Recalling the formula (5.30) and noting that $(n/m)^{2\ell} = 1$ for $m = n$, we see that

$$a_{nn} = \frac{1}{n^{2k}} \sum_{\ell=1}^{k-1} 1 = (k-1)\frac{1}{n^{2k}}. \tag{5.31}$$

Off-diagonal sum: Now assume $m \neq n$. Then applying the formula for a geometric sum, $\sum_{\ell=1}^{k-1} r^{\ell} = (r - r^k)/(1 - r)$, with $r = (n/m)^2$ (note that $r \neq 1$, since $m \neq n$), we obtain

$$a_{mn} = \frac{1}{n^{2k}} \sum_{\ell=1}^{k-1} \left(\frac{n}{m}\right)^{2\ell} = \frac{1}{n^{2k}} \cdot \frac{(n/m)^2 - (n/m)^{2k}}{1 - (n/m)^2} = \frac{1}{n^{2k}} \cdot \frac{n^2 - n^{2k}m^{2-2k}}{m^2 - n^2}$$

$$= \frac{1}{n^{2k}} \cdot \frac{n^2}{m^2 - n^2} + \frac{1}{m^{2k}} \cdot \frac{m^2}{n^2 - m^2}.$$

Therefore,

$$\sum_{m\neq n}^{N} a_{mn} = \sum_{m\neq n}^{N} \frac{1}{n^{2k}} \frac{n^2}{m^2 - n^2} + \sum_{m\neq n}^{N} \frac{1}{m^{2k}} \frac{m^2}{n^2 - m^2}. \tag{5.32}$$

If on the right-hand side of (5.32) we switch the letters m and n in the second summation (which we can do, since m and n are just summation indices and we can use whatever letters we want), we get the first summation in (5.32). Hence, the second sum on the right in (5.32) is really twice the first sum. Now combining (5.31) with (5.32), we get

$$a_N = (k-1) \sum_{n=1}^{N} \frac{1}{n^{2k}} + \sum_{m \neq n}^{N} \frac{1}{n^{2k}} \frac{2n^2}{m^2 - n^2}. \qquad (5.33)$$

Step 2: We now find a nice expression for the second sum in (5.33). We first use partial fractions to write

$$\frac{2n}{m^2 - n^2} = \frac{1}{m-n} - \frac{1}{n+m} = -\left(\frac{1}{n-m} + \frac{1}{n+m}\right).$$

Hence,

$$\frac{1}{n^{2k}} \frac{2n^2}{m^2 - n^2} = -\frac{1}{n^{2k-1}} \left(\frac{1}{n-m} + \frac{1}{n+m}\right).$$

Next, we write the summation $\sum_{m \neq n}^{N}$ as shown in Fig. 5.2:

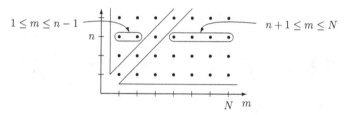

Fig. 5.2 We break up $\sum_{m \neq n}^{N}$ as $\sum_{n=1}^{N} \left(\sum_{m=1}^{n-1} + \sum_{m=n+1}^{N}\right)$; that is, for each $n = 1, \ldots, N$, we sum along the nth horizontal row, from $m = 1$ to $m = n - 1$, skipping $m = n$, then from $m = n + 1$ to $m = N$

$$\sum_{m \neq n}^{N} \frac{1}{n^{2k}} \frac{2n^2}{m^2 - n^2} = -\sum_{n=1}^{N} \frac{1}{n^{2k-1}} \left(\sum_{m=1}^{n-1} + \sum_{m=n+1}^{N}\right) \left(\frac{1}{n-m} + \frac{1}{n+m}\right). \qquad (5.34)$$

Observe that

$$\left(\sum_{m=1}^{n-1} + \sum_{m=n+1}^{N}\right) \frac{1}{n-m}$$

$$= \frac{1}{n-1} + \frac{1}{n-2} + \cdots + \frac{1}{2} + \frac{1}{1} - \frac{1}{1} - \frac{1}{2} - \cdots - \frac{1}{N-n}. \qquad (5.35)$$

On the other hand,

$$\left(\sum_{m=1}^{n-1} + \sum_{m=n+1}^{N} \right) \frac{1}{n+m}$$

$$= \frac{1}{n+1} + \frac{1}{n+2} + \cdots + \frac{1}{2n-1} + \frac{1}{2n+1} + \cdots + \frac{1}{n+N}.$$
$$(5.36)$$

Observe that if we combine the positive terms in (5.35) with all the terms in (5.36), and we add in $1/n$ and $1/(2n)$, we get the sum $1/1 + 1/2 + \cdots + 1/(n+N)$. Thus,

$$\frac{1}{n} + \frac{1}{2n} + \left(\sum_{m=1}^{n-1} + \sum_{m=n+1}^{N} \right) \left(\frac{1}{n-m} + \frac{1}{n+m} \right)$$

$$= \frac{1}{1} + \frac{1}{2} + \cdots + \frac{1}{n+N} - \frac{1}{1} - \frac{1}{2} - \cdots - \frac{1}{N-n}.$$

Canceling like terms, we conclude that

$$\left(\sum_{m=1}^{n-1} + \sum_{m=n+1}^{N} \right) \left(\frac{1}{n-m} + \frac{1}{n+m} \right) = -\frac{3}{2n} + \sum_{m=N-n+1}^{N+n} \frac{1}{m}.$$

Thus, by (5.34), we have

$$2 \sum_{\substack{n=1 \\ m \ne n}}^{N} \frac{n^{2-2k}}{m^2 - n^2} = \sum_{n=1}^{N} \frac{1}{n^{2k-1}} \left(\frac{3}{2n} - \sum_{m=N-n+1}^{N-n} \frac{1}{m} \right)$$

$$= \frac{3}{2} \sum_{n=1}^{N} \frac{1}{n^{2k}} - \sum_{n=1}^{N} \left(\frac{1}{n^{2k-1}} \sum_{m=N-n+1}^{N+n} \frac{1}{m} \right).$$

Plugging this into the formula (5.33) for a_N, we obtain

$$a_N = \left(k + \frac{1}{2} \right) \sum_{n=1}^{N} \frac{1}{n^{2k}} - \sum_{n=1}^{N} \left(\frac{1}{n^{2k-1}} \sum_{m=N-n+1}^{N+n} \frac{1}{m} \right).$$

Therefore, $\lim a_N = (k + 1/2)\zeta(2k)$, provided we can show that

$$0 = \lim_{N \to \infty} \sum_{n=1}^{N} \left(\frac{1}{n^{2k-1}} \sum_{m=N-n+1}^{N+n} \frac{1}{m} \right).$$
$$(5.37)$$

Step 3: Our proof is done once we establish (5.37). To do so, observe that for $N - n + 1 \le m \le N + n$, we have $1/m \le 1/(N-n+1)$. Thus,

$$\sum_{m=N-n+1}^{N+n} \frac{1}{m} \le \sum_{m=N-n+1}^{N+n} \frac{1}{N-n+1} = \frac{2n}{N-n+1}.$$

Since $k \ge 2$, we have $1/n^{2k-1} \le 1/n^3$, and thus

$$\sum_{n=1}^{N} \left(\frac{1}{n^{2k-1}} \sum_{m=N-n+1}^{N+n} \frac{1}{m} \right) \le 2 \sum_{n=1}^{N} \left(\frac{n}{n^3} \frac{1}{N-n+1} \right) = 2 \sum_{n=1}^{N} \frac{1}{n^2(N-n+1)}.$$

The following limit proves (5.37):

$$\lim_{N \to \infty} \sum_{n=1}^{N} \frac{1}{n^2(N-n+1)} = 0. \tag{5.38}$$

This limit is easily proved using Tannery's theorem, which we leave for your enjoyment in Problem 1, but we can prove it in an elementary way. First, using partial fractions, a bit of algebra shows that

$$\frac{1}{n^2(N-n+1)} = \frac{1}{(N+1)} \frac{1}{n^2} + \frac{1}{(N+1)^2} \left(\frac{1}{n} + \frac{1}{N-n+1} \right).$$

The sum in parentheses on the far right is $\le 1/1 + 1/1 = 2$. Thus,

$$\frac{1}{n^2(N-n+1)} \le \frac{1}{N+1} \cdot \frac{1}{n^2} + \frac{2}{(N+1)^2}.$$

Hence,

$$\sum_{n=1}^{N} \frac{1}{n^2(N-n+1)} \le \frac{1}{N+1} \sum_{n=1}^{N} \frac{1}{n^2} + \sum_{n=1}^{N} \frac{2}{(N+1)^2}$$

$$\le \frac{\pi^2/6}{N+1} + \frac{2N}{(N+1)^2}.$$

Taking $N \to \infty$ proves (5.38) and completes our proof. ∎

In particular, setting $k = 2$, we see that $\frac{5}{2}\zeta(4) = \zeta(2)^2$. Thus, $\zeta(4) = \frac{2}{5}\frac{\pi^4}{36} = \frac{\pi^4}{90}$. Taking $k = 3$, we get

$$\frac{7}{2}\zeta(6) = \zeta(2)\zeta(4) + \zeta(4)\zeta(2) = 2\zeta(2)\zeta(4) = 2 \cdot \frac{\pi^2}{6} \cdot \frac{\pi^4}{90},$$

which, after doing the algebra, becomes $\zeta(6) = \pi^6/945$. Thus,

$$\frac{\pi^4}{90} = \sum_{n=1}^{\infty} \frac{1}{n^4} \ , \qquad \frac{\pi^6}{945} = \sum_{n=1}^{\infty} \frac{1}{n^6}.$$

We can also derive explicit formulas for $\zeta(2k)$ for all $k \in \mathbb{N}$.

5.3.2 Euler's Formula for $\zeta(2k)$

To derive Euler's formula, we first define a sequence $C_1, C_2, C_3, \ldots$ by $C_1 = \frac{1}{12}$, and for $k \geq 2$, we define

$$C_k = -\frac{1}{2k+1} \sum_{\ell=1}^{k-1} C_\ell C_{k-\ell}. \tag{5.39}$$

The first few values of C_k are

$$C_1 = \frac{1}{12}, \quad C_2 = -\frac{1}{720}, \quad C_3 = \frac{1}{30240}, \quad C_4 = -\frac{1}{1209600}.$$

The numbers C_k are rational numbers (easily proved by induction) and are related to the **Bernoulli numbers,** to be covered in Section 6.7 on page 501. But it's not necessary to know this.[7] We are now ready to prove . . .

Euler's formulæ

Theorem 5.9 *For every $k \in \mathbb{N}$, we have*

$$\sum_{n=1}^{\infty} \frac{1}{n^{2k}} = (-1)^{k-1} \frac{(2\pi)^{2k} C_k}{2} \ ; \quad or, \ \zeta(2k) = (-1)^{k-1} \frac{(2\pi)^{2k} C_k}{2}. \tag{5.40}$$

Proof When $k = 1$, we have

$$(-1)^{k-1} \frac{(2\pi)^{2k} C_k}{2} = \frac{(2\pi)^2 (1/12)}{2} = \frac{\pi^2}{6} = \zeta(2),$$

so our theorem holds when $k = 1$. Let $k \geq 2$ and assume that our theorem holds for all natural numbers up to and including $k - 1$; we shall prove that it holds for k. Using Williams's formula and the induction hypothesis, we see that

---

[7]Explicitly, $C_k = B_{2k}/(2k)!$, but this formula is not needed.

$$\left(k + \frac{1}{2}\right) \zeta(2k) = \sum_{\ell=1}^{k-1} \zeta(2\ell)\,\zeta(2k - 2\ell)$$

$$= \sum_{\ell=1}^{k-1} \left((-1)^{\ell-1} \frac{(2\pi)^{2\ell} C_\ell}{2}\right) \left((-1)^{k-\ell-1} \frac{(2\pi)^{2k-2\ell} C_{k-2}}{2}\right)$$

$$= \sum_{\ell=1}^{k-1} \left((-1)^{k-2} \frac{(2\pi)^{2k} C_\ell C_{k-\ell}}{4}\right)$$

$$= (-1)^{k-2} \frac{(2\pi)^{2k}}{4} \sum_{\ell=1}^{k-1} C_\ell C_{k-\ell}$$

$$= (-1)^{k-1} \frac{(2\pi)^{2k}}{4} (2k + 1) C_k.$$

Dividing everything by $(k + 1/2) = (1/2)(2k + 1)$ and using the formula (5.39) for C_k proves our result for k. ∎

As a side note, we remark that (5.40) shows that $\zeta(2k)$ is a rational number times π^{2k}; in particular, since π is transcendental (see, for example, [146, 174, 175]), it follows that $\zeta(n)$ is transcendental for n even. One may ask whether there are similar expressions like (5.40) for sums of the reciprocals of the *odd* powers (e.g., $\zeta(3) = \sum_{n=1}^{\infty} \frac{1}{n^3}$). Unfortunately, there are no known formulas! Moreover, it is not even known whether $\zeta(k)$ is transcendental for k odd, and in fact, of all odd numbers, only $\zeta(3)$ is known without a doubt to be irrational; this was proven by Roger Apéry (1916–1994) in 1979 (see [28, 248])!

▶ **Exercises 5.3**

1. Prove (5.38) using Tannery's theorem.
2. (Cf. [125]) Let $H_n = \sum_{m=1}^{n} \frac{1}{m}$, the nth partial sum of the harmonic series. In this problem we prove the equalities

$$\zeta(3) = \sum_{n=1}^{\infty} \frac{1}{n^3} = \sum_{n=1}^{\infty} \frac{H_n}{(n+1)^2} = \frac{1}{2} \sum_{n=1}^{\infty} \frac{H_n}{n^2}. \tag{5.41}$$

(i) Prove that for $N \in \mathbb{N}$,

$$\sum_{m,n=1}^{N} \frac{1}{mn(m+n)} = \sum_{m=1}^{N} \frac{H_m}{m^2} = \sum_{m=1}^{N} \frac{1}{m^3} + \sum_{n=1}^{N-1} \frac{H_k}{(k+1)^2},$$

where the notation $\sum_{m,n=1}^{N}$ is as in the proof of Williams's theorem. Suggestion: For the first equality, use that $\frac{1}{mn(m+n)} = \frac{1}{m^2}\left(\frac{1}{n} - \frac{1}{m+n}\right)$.

(ii) Now prove that for $N \in \mathbb{N}$,

$$\sum_{m,n=1}^{N} \frac{1}{mn(m+n)} = 2 \sum_{m=1}^{N} \sum_{n=1}^{N} \frac{1}{m(m+n)^2}.$$

Suggestion: Use that $\frac{1}{mn(m+n)} = \frac{1}{m(m+n)^2} + \frac{1}{n(m+n)^2}$.

(iii) In Part (ii), instead of using n as the inner summation variable on the right-hand side, change to $k = m + n - 1$, and in doing so, prove that

$$\sum_{m,n=1}^{N} \frac{1}{mn(m+n)} = 2 \sum_{m=1}^{N} \sum_{k=m}^{N} \frac{1}{m(k+1)^2} + b_N, \quad \text{where } b_N = 2 \sum_{m=1}^{N} \sum_{k=N+1}^{m+N-1} \frac{1}{m(k+1)^2}.$$

(iv) Show that $\sum_{m=1}^{N} \sum_{k=m}^{N} \frac{1}{m(k+1)^2} = \sum_{k=1}^{N} \frac{H_k}{(k+1)^2}$ and that $b_N \to 0$ as $N \to \infty$. Now prove (5.41).

3. (**Euler's sum for** $\pi^2/6$, **Proof VI**) In this problem we prove Euler's formula for $\pi^2/6$ by *carefully* squaring Gregory–Leibniz–Madhava's formula for $\pi/4$; thus, taking Gregory–Leibniz–Madhava's formula as given, we derive Euler's formula.[8] The proof is very much in the same spirit as the proof of Williams's formula; see page 526 in Section 6.10 for another, more systematic, proof.

(i) Given $N \in \mathbb{N}$, prove that

$$\left(\sum_{m=0}^{N} \frac{(-1)^m}{(2m+1)} \right) \left(\sum_{n=0}^{N} \frac{(-1)^n}{(2n+1)} \right) = \sum_{n=0}^{N} \frac{1}{(2n+1)^2} + \sum_{m \neq n}^{N} \frac{(-1)^{m+n}}{(2m+1)(2n+1)},$$

where the notation $\sum_{m \neq n}^{N}$ is as in the proof of Williams's theorem.

(ii) For $m \neq n$, prove that[9]

$$\frac{1}{(2m+1)(2n+1)} = \frac{\frac{2m+1}{2n+1} - \frac{2n+1}{2m+1}}{(2m+1)^2 - (2n+1)^2}$$

$$= \frac{2m+1}{2n+1} \cdot \frac{1}{(2m+1)^2 - (2n+1)^2} - \frac{2n+1}{2m+1} \cdot \frac{1}{(2m+1)^2 - (2n+1)^2}.$$

(iii) Prove that

$$\sum_{m \neq n}^{N} \frac{(-1)^{m+n}}{(2m+1)(2n+1)} = 2 \sum_{m \neq n}^{N} \frac{(-1)^{m+n}}{2n+1} \cdot \frac{2m+1}{(2m+1)^2 - (2n+1)^2}$$

$$= 2 \sum_{n=0}^{N} \frac{(-1)^n}{2n+1} \left(\sum_{m=0}^{n-1} + \sum_{m=n+1}^{N} \right) (-1)^m \frac{2m+1}{(2m+1)^2 - (2n+1)^2}.$$

---

[8] Actually, this works in reverse: We can just as well take Euler's formula as given, and then derive Gregory–Leibniz–Madhava's formula!

[9] Alternatively, one can prove that $\frac{1}{(2m+1)(2n+1)} = \frac{1}{2(m-n)(2n+1)} + \frac{1}{2(n-m)(2m+1)}$ and use this decomposition in what follows. However, the decomposition of $\frac{1}{(2m+1)(2n+1)}$ as presented might be helpful if you do "Williams's other formula" in Problem 4.

(iv) Prove that

$$4\left(\sum_{m=0}^{n-1} + \sum_{m=n+1}^{N}\right)(-1)^m \frac{2m+1}{(2m+1)^2 - (2n+1)^2} = -\frac{(-1)^n}{2n+1} + (-1)^N \sum_{m=N-n+1}^{N+n+1} \frac{1}{m}.$$

Suggestion: Note that $4\frac{2m+1}{(2m+1)^2-(2n+1)^2} = \frac{2m+1}{(m+n+1)(m-n)} = \frac{1}{m-n} + \frac{1}{m+n+1}$.

(v) Prove that

$$\left(\sum_{m=0}^{N} \frac{(-1)^m}{(2m+1)}\right)\left(\sum_{n=0}^{N} \frac{(-1)^n}{(2n+1)}\right) = b_N + \frac{1}{2}\sum_{n=0}^{N} \frac{1}{(2n+1)^2},$$

where $b_N = \frac{1}{2}\sum_{n=0}^{N} \frac{(-1)^{N+n}}{(2n+1)}\left(\sum_{m=N-n+1}^{N+n+1} \frac{1}{m}\right).$

(vi) Prove that $b_N \to 0$ as $N \to \infty$, and conclude that $(\pi/4)^2 = \frac{1}{2}\sum_{n=0}^{\infty} \frac{1}{(2n+1)^2}$. Finally, derive Euler's formula for $\pi^2/6$.

4. (**Williams's other formula**) For each $k \in \mathbb{N}$, define

$$\xi(k) = \sum_{n=0}^{\infty}(-1)^n \frac{1}{(2n+1)^k}.$$

For example, by Gregory–Leibniz–Madhava's formula we know that $\xi(1) = \pi/4$. Prove that for every $k \in \mathbb{N}$ with $k \geq 2$, we have

$$\left(k - \frac{1}{2}\right)\sum_{n=0}^{\infty} \frac{1}{(2n+1)^{2k}} = \sum_{\ell=0}^{k-1} \xi(2\ell+1)\,\xi(2k - 2\ell - 1).$$

Suggestion: Imitate the proof of Williams's formula. You will see that ideas from Problem 3 will also be useful.

5. (Cf. [36, 126, 265]) Let $H_n = \sum_{m=1}^{n} \frac{1}{m}$, the nth partial sum of the harmonic series. In this problem we prove that for every $k \in \mathbb{N}$ with $k \geq 2$, we have

$$(k+2)\,\zeta(k+1) = \sum_{\ell=1}^{k-2} \zeta(k-\ell)\,\zeta(\ell+1) + 2\sum_{n=1}^{\infty} \frac{H_n}{n^k}, \tag{5.42}$$

a formula due to Euler (no surprise!). The proof is very similar to the proof of Williams's formula, with some twists of course. You may proceed as follows.

(i) For $N \in \mathbb{N}$, define

$$a_N = \sum_{\ell=1}^{k-2}\left(\sum_{m=1}^{N}\frac{1}{m^{k-\ell}}\right)\left(\sum_{n=1}^{N}\frac{1}{n^{\ell+1}}\right) = \sum_{m,n=1}^{N}\sum_{\ell=1}^{k-2}\frac{1}{m^{k-\ell}\,n^{\ell+1}}.$$

Summing the geometric series $\sum_{\ell=1}^{k-2}\frac{1}{m^{k-\ell}n^{\ell+1}} = \frac{1}{m^k n}\sum_{\ell=1}^{k-2}(m/n)^{\ell}$, prove that

$$a_N = (k-2)\sum_{n=1}^{N}\frac{1}{n^{k+1}} + 2\sum_{m\neq n}^{N}\frac{1}{n^{k-1}\,m\,(m-n)},$$

where the notation $\sum_{m\neq n}^{N}$ is as in the proof of Williams's theorem.

(ii)   Prove that

$$\sum_{m\neq n}^{N}\frac{1}{n^{k-1}\,m\,(m-n)} = \sum_{n=1}^{N}\frac{1}{n^k}\left(\sum_{m=1}^{n-1}+\sum_{m=n+1}^{N}{}'\right)\left(\frac{1}{m-n}-\frac{1}{m}\right).$$

(iii)   Prove that

$$\left(\sum_{m=1}^{n-1}+\sum_{m=n+1}^{N}\right)\left(\frac{1}{m-n}-\frac{1}{m}\right) = \frac{2}{n} - H_n - \sum_{m=N-n+1}^{N}\frac{1}{m}.$$

(iv)   Prove that

$$a_N = (k+2)\sum_{n=1}^{N}\frac{1}{n^{k+1}} - 2\sum_{n=1}^{N}\frac{H_n}{n^k} - b_N, \quad\text{where } b_N = 2\sum_{n=1}^{N}\frac{1}{n^k}\left(\sum_{m=N-n+1}^{N}\frac{1}{m}\right).$$

(v)   Prove that $b_N \to 0$ as $N \to \infty$, and conclude that (5.42) holds.

6.   (Cf. [125, 126]) Here are a couple of applications of (5.42). First, use (5.42) to give a quick proof of (5.41). Second, prove that

$$\boxed{\frac{\pi^4}{72} = \sum_{n=1}^{\infty}\frac{1}{n^3}\left(1+\frac{1}{2}+\frac{1}{3}+\cdots+\frac{1}{n}\right).}$$

Part II
Extracurricular Activities

Chapter 6
Advanced Theory of Infinite Series

Ut non-finitam Seriem finita cöercet,
Summula, & in nullo limite limes adest:
Sic modico immensi vestigia Numinis haerent
Corpore, & angusto limite limes abest.
Cernere in immenso parvum, dic, quanta voluptas!
In parvo immensum cernere, quanta, Deum.

Even as the finite encloses an infinite series
And in the unlimited limits appear,
So the soul of immensity dwells in minutia
And in the narrowest limits no limit in here.
What joy to discern the minute in infinity!
The vast to perceive in the small, what divinity!
Jacob Bernoulli (1654–1705) Ars Conjectandi. [231, p. 271]

This chapter is about going in depth into the theory and application of infinite series. One infinite series that will come up again and again in this chapter and the next chapter as well is the Riemann zeta function

$$\zeta(z) = \sum_{n=1}^{\infty} \frac{1}{n^z},$$

introduced in Section 4.7 on p. 308. Among many other things, in this chapter we'll see how to write some well-known constants in terms of the Riemann zeta function; e.g., we'll derive the following neat formula for our friend log 2 (Section 6.5),

$$\log 2 = \sum_{n=2}^{\infty} \frac{1}{2^n} \zeta(n),$$

another formula for our friend the Euler–Mascheroni constant (Section 6.8),

© Paul Loya 2017
P. Loya, *Amazing and Aesthetic Aspects of Analysis,*
https://doi.org/10.1007/978-1-4939-6795-7_6

$$\gamma = \sum_{n=2}^{\infty} \frac{(-1)^n}{n} \zeta(n),$$

and two more formulas involving our most delicious friend π (see Sections 6.9 and 6.10),

$$\pi = \sum_{n=2}^{\infty} \frac{3^n - 1}{4^n} \zeta(n+1) \quad , \quad \frac{\pi^2}{6} = \zeta(2) = \sum_{n=1}^{\infty} \frac{1}{n^2} = 1 + \frac{1}{2^2} + \frac{1}{3^2} + \frac{1}{4^2} + \cdots.$$

We'll also rederive Gregory–Leibniz–Madhava's formula (Section 6.9),

$$\frac{\pi}{4} = 1 - \frac{1}{3} + \frac{1}{5} - \frac{1}{7} + \frac{1}{9} - \frac{1}{11} + - \cdots,$$

and we derive Machin's formula, which started the "decimal place race" of computing π (Section 6.9):

$$\pi = 4 \arctan\left(\frac{1}{5}\right) - \arctan\left(\frac{1}{239}\right) = 4 \sum_{n=0}^{\infty} \frac{(-1)^n}{(2n+1)} \left(\frac{4}{5^{2n+1}} - \frac{1}{239^{2n+1}}\right).$$

Leibniz's formula for $\pi/4$ is an example of an *alternating series*. We study these types of series in Section 6.1. In Sections 6.2 and 6.3 we look at the ratio and root tests, which you are probably familiar with from elementary calculus. In Section 6.4 we look at power series and prove some pretty powerful properties of power series. The formulas for $\log 2$ and γ, and the formula $\pi = \sum_{n=2}^{\infty} \frac{3^n-1}{4^n} \zeta(n+1)$ displayed above, are proved using a famous theorem called the *Cauchy double series theorem*. This theorem, and double series in general, are the subject of Section 6.5. In Section 6.6 we investigate rearranging series (that is, mixing up the order of their terms). In elementary calculus, you probably never saw the power series representations of tangent and secant. This is because those series are somewhat sophisticated, mathematically speaking. In Section 6.7 we shall derive the power series representations

$$\tan z = \sum_{n=1}^{\infty} (-1)^{n-1} \frac{2^{2n}(2^{2n} - 1) B_{2n}}{(2n)!} z^{2n-1}$$

and

$$\sec z = \sum_{n=0}^{\infty} (-1)^n \frac{E_{2n}}{(2n)!} z^{2n}.$$

Here, the B_{2n} are called *Bernoulli numbers*, and the E_{2n} are called *Euler numbers*, which are certain numbers having extraordinary properties. Although you've probably never seen the tangent and secant power series, you might have seen the logarithmic, binomial, and arctangent series:

$$\log(1+z) = \sum_{n=1}^{\infty} \frac{(-1)^{n-1}}{n} z^n \ , \ (1+z)^{\alpha} = \sum_{n=0}^{\infty} \binom{\alpha}{n} z^n \ , \ \arctan z = \sum_{n=0}^{\infty} (-1)^n \frac{z^{2n+1}}{2n+1}.$$

You most likely used calculus (derivatives and integrals) to derive these formulas. In Section 6.8 we shall derive these formulas without any calculus. Finally, in Sections 6.9 and 6.10 we derive many incredible and awe-inspiring formulas involving π. In particular, we again look at the Basel problem.

CHAPTER 6 OBJECTIVES: THE STUDENT WILL BE ABLE TO ...

- Determine the convergence, and radius and interval of convergence, for an infinite series and power series, respectively, using various tests, including the Dirichlet, Abel, ratio, and root tests.
- Apply Cauchy's double series theorem and know how it relates to rearrangement, and multiplication and composition of power series.
- Identify series formulas for the various elementary functions (logarithm, binomial, arctangent, etc.) and for π.

6.1 Summation by Parts, Bounded Variation, and Alternating Series

In elementary calculus, you studied "integration by parts," a formula I'm sure you used quite often in trying to integrate tricky integrals. In this section we study a discrete version of the integration by parts formula called "summation by parts," which is used to sum tricky summations! Summation by parts has broad applications, including finding sums of powers of integers and deriving some famous convergence tests for series, the Dirichlet and Abel tests.

6.1.1 Summation by Parts and Abel's Lemma

From calculus we learned the integration by parts formula

$$\int_a^b f'(x)\,g(x)\,dx + \int_a^b f(x)\,g'(x)\,dx = f(b)\,g(b) - f(a)\,g(a).$$

If we replace the integral with a sum and f and g by sequences, we get the famous summation by parts formula:

Summation by parts

Theorem 6.1 *For complex sequences $\{a_n\}$ and $\{b_n\}$, we have*

$$\sum_{k=m}^{n}(a_{k+1} - a_k)b_{k+1} + \sum_{k=m}^{n}a_k(b_{k+1} - b_k) = a_{n+1}b_{n+1} - a_m b_m.$$

Proof Combining the two terms on the left, we obtain

$$\sum_{k=m}^{n}\left[b_{k+1}a_{k+1} - b_{k+1}a_k + a_k b_{k+1} - a_k b_k\right] = \sum_{k=m}^{n}\left(b_{k+1}a_{k+1} - a_k b_k\right).$$

This is a telescoping sum, and it simplifies to $a_{n+1}b_{n+1} - a_m b_m$ after all the cancellations. ∎

As a corollary, we get Abel's lemma, named after Niels Abel[1] (1802–1829).

Abel's lemma

Corollary 6.2 *Let $\{a_n\}$ and $\{b_n\}$ be complex sequences and let s_n denote the nth partial sum of the series corresponding to the sequence $\{a_n\}$. Then for every $m < n$ we have*

$$\sum_{k=m+1}^{n}a_k b_k = s_n b_n - s_m b_m - \sum_{k=m}^{n-1}s_k(b_{k+1} - b_k).$$

Proof Applying the summation by parts formula to the sequences $\{s_n\}$ and $\{b_n\}$, we obtain

$$\sum_{k=m}^{n-1}(s_{k+1} - s_k)b_{k+1} + \sum_{k=m}^{n-1}s_k(b_{k+1} - b_k) = s_n b_n - s_m b_m.$$

Since $s_{k+1} - s_k = a_{k+1}$, we conclude that

$$\sum_{k=m}^{n-1}a_{k+1}b_{k+1} + \sum_{k=m}^{n-1}s_k(b_{k+1} - b_k) = s_n b_n - s_m b_m.$$

Replacing k with $k - 1$ in the first sum and bringing the second sum to the right, we get our result. ∎

---

[1] "Abel has left mathematicians enough to keep them busy for 500 years." Charles Hermite (1822–1901), in *Calculus Gems* [225].

6.1.2 Sums of Powers of Integers

We shall apply the summation by parts formula

$$\sum_{k=1}^{n}(a_{k+1} - a_k)b_{k+1} + \sum_{k=1}^{n} a_k(b_{k+1} - b_k) = a_{n+1}b_{n+1} - a_1 b_1$$

to find sums of powers of integers (cf. [84, 272]). See the exercises for more applications.

Example 6.1 Let $a_k = k$ and $b_k = k - 1$. Then each of the differences $a_{k+1} - a_k$ and $b_{k+1} - b_k$ equals 1, so by summation by parts, we have

$$\sum_{k=1}^{n}(1)(k) + \sum_{k=1}^{n}(k)(1) = (n + 1)(n).$$

This equality can be simplified to $2\sum_{k=1}^{n} k = n(n + 1)$. Thus, we obtain the well-known result

$$1 + 2 + \cdots + n = \frac{n(n + 1)}{2}.$$

Example 6.2 Now let $a_k = k^2$ and $b_k = k - 1$. In this case, $a_{k+1} - a_k = (k + 1)^2 - k^2 = 2k + 1$ and $b_{k+1} - b_k = 1$, so by the summation by parts formula, we have

$$\sum_{k=1}^{n}(2k + 1)(k) + \sum_{k=1}^{n}(k^2)(1) = (n + 1)^2 n.$$

Simplifying a bit, we get

$$3\sum_{k=1}^{n} k^2 + \sum_{k=1}^{n} k = (n + 1)^3 n.$$

Since $\sum_{k=1}^{n} k = n(n + 1)/2$ from the previous example, after some algebra we end up with the well-known result

$$\sum_{k=1}^{n} k^2 = 1^2 + 2^2 + \cdots + n^2 = \frac{n(n + 1)(2n + 1)}{6}.$$

Example 6.3 For our final result, let $a_k = k^2$ and $b_k = (k - 1)^2$. Then $a_{k+1} - a_k = (k + 1)^2 - k^2 = 2k + 1$ and $b_{k+1} - b_k = 2k - 1$, so by the summation by parts formula,

$$\sum_{k=1}^{n}(2k+1)(k^2) + \sum_{k=1}^{n}(k^2)(2k-1) = (n+1)^2 \cdot n^2.$$

After some work simplifying the left-hand side and using the formula for the sum of squares, we get

$$1^3 + 2^3 + \cdots + n^3 = \frac{n^2(n+1)^2}{4}.$$

6.1.3   Sequences of Bounded Variation and Dirichlet's Test

A sequence $\{a_n\}$ of complex numbers is said to be of **bounded variation** if

$$\sum_{n=1}^{\infty} |a_{n+1} - a_n| < \infty.$$

If we plot the points $a_1, a_2, a_3, \ldots$ in the complex plane, then $|a_{n+1} - a_n|$ is the distance between a_n and a_{n+1}. Thus, $\sum_{n=1}^{\infty} |a_{n+1} - a_n|$ is the total length of the polygonal curve formed by the points $a_1, a_2, a_3, \ldots$, as seen here:

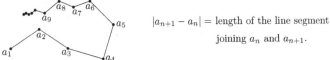

$|a_{n+1} - a_n| =$ length of the line segment

joining a_n and a_{n+1}.

Bounded variation just means that the polygonal curve has finite length. Here are some facts concerning sequences of bounded variation.

Proposition 6.3 *Sequences of bounded variation converge. Examples of sequences of bounded variation include bounded monotone sequences of real numbers and contractive sequences of complex numbers.*

Proof Let $\{a_n\}$ be of bounded variation. Given $m < n$, we can write $a_n - a_m$ as a telescoping sum:

$$a_n - a_m = (a_{m+1} - a_m) + (a_{m+2} - a_{m+1}) + \cdots$$

$$+ (a_{n-1} - a_{n-2}) + (a_n - a_{n-1}) = \sum_{k=m}^{n}(a_{k+1} - a_k).$$

Hence, by the triangle inequality,

$$|a_n - a_m| \le \sum_{k=m}^{n} |a_{k+1} - a_k|.$$

By assumption, the sum $\sum_{k=1}^{\infty} |a_{k+1} - a_k|$ converges, so the sum on the right-hand side of this inequality can be made arbitrarily small as $m, n \to \infty$ (Cauchy's criterion for series on p. 206). Thus, $\{a_n\}$ is Cauchy and hence converges.

Now let $\{a_n\}$ be a bounded nondecreasing sequence. We shall prove that this sequence is of bounded variation; the proof for a nonincreasing sequence is similar. In this case, we have $a_n \le a_{n+1}$ for each n, so for each n,

$$\sum_{k=1}^{n} |a_{k+1} - a_k| = \sum_{k=1}^{n} (a_{k+1} - a_k) = (a_2 - a_1) + (a_3 - a_2)$$

$$+ \cdots + (a_n - a_{n-1}) + (a_{n+1} - a_n) = a_{n+1} - a_1,$$

since the sum telescoped. The sequence $\{a_n\}$ is by assumption bounded, so it follows that the partial sums of the infinite series $\sum_{n=1}^{\infty} |a_{n+1} - a_n|$ are bounded, and hence the series must converge by the nonnegative series test (Theorem 3.19). That contractive sequences are of bounded variation is part of Problem 6. ∎

Example 6.4 The converse, that every convergent sequence is of bounded variation, is false. Consider the sequence given by $a_n = (-1)^{n-1}/n$, $n = 1, 2, 3, \ldots$. This sequence jumps back and forth, as seen here:

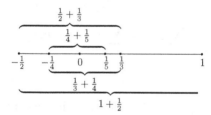

In this case, the polygonal curve has length

$$\sum_{n=1}^{\infty} \left(\frac{1}{n} + \frac{1}{n+1} \right),$$

which is infinite.

Here's a useful test named after Lejeune Dirichlet (1805–1859).

Dirichlet's test

Theorem 6.4 *Suppose that the partial sums of the series $\sum a_n$ are uniformly bounded (although the series $\sum a_n$ may not converge). Then for every sequence $\{b_n\}$ that is of bounded variation and converges to zero, the series $\sum a_n b_n$ converges. In particular, the series $\sum a_n b_n$ converges if $\{b_n\}$ is a monotone sequence of real numbers approaching zero.*

Proof Setting $m = 1$ in Abel's lemma, we have

$$\sum_{k=1}^{n} a_k b_k = s_n b_n - \sum_{k=1}^{n-1} s_k (b_{k+1} - b_k). \tag{6.1}$$

Now we are given two facts: The first is that the partial sums $\{s_n\}$ are bounded, say by a constant C, and the second is that the sequence $\{b_n\}$ is of bounded variation and converges to zero. Since $\{s_n\}$ is bounded and $b_n \to 0$, it follows that $s_n b_n \to 0$. Since $|s_n| \leq C$ for all n and $\{b_n\}$ is of bounded variation, we have

$$\sum_{k=1}^{\infty} |s_k(b_{k+1} - b_k)| \leq C \sum_{k=1}^{\infty} |b_{k+1} - b_k| < \infty.$$

Therefore, $\sum_{k=1}^{\infty} s_k(b_{k+1} - b_k)$ converges absolutely. In particular, taking $n \to \infty$ in (6.1), it follows that the sum $\sum a_k b_k$ converges. Moreover, we actually have found a formula for the sum:

$$\sum_{n=1}^{\infty} a_n b_n = \sum_{n=1}^{\infty} s_n (b_{n+1} - b_n). \tag{6.2}$$

∎

Example 6.5 For each $x \in (0, 2\pi)$, we determine the convergence of the series

$$\sum_{n=1}^{\infty} \frac{e^{inx}}{n}.$$

To do so, we let $a_n = e^{inx}$ and $b_n = 1/n$. Since $\{1/n\}$ is a monotone sequence converging to zero, by Dirichlet's test, if we can prove that the partial sums of $\sum e^{inx}$ are bounded, then $\sum_{n=1}^{\infty} \frac{e^{inx}}{n}$ converges. To establish this boundedness, we observe that

$$\sum_{n=1}^{m} e^{inx} = e^{ix} \frac{1 - e^{imx}}{1 - e^{inx}},$$

where we summed $\sum_{n=1}^{m} (e^{ix})^n$ via the geometric progression (2.3) on p. 40. Hence,

$$\left| \sum_{n=1}^{m} e^{inx} \right| \leq \left| \frac{1 - e^{imx}}{1 - e^{inx}} \right| \leq \frac{1 + |e^{imx}|}{|1 - e^{inx}|} = \frac{2}{|1 - e^{ix}|}.$$

Since $1 - e^{ix} = e^{ix/2}(e^{-ix/2} - e^{ix/2}) = -2ie^{ix/2}\sin(x/2)$, we see that

$$|1 - e^{ix}| = 2|\sin(x/2)| \implies \left| \sum_{n=1}^{m} e^{inx} \right| \leq \frac{1}{\sin(x/2)}.$$

Thus, for each $x \in (0, 2\pi)$, by Dirichlet's test, given a sequence $\{b_n\}$ of bounded variation that converges to zero, the sum $\sum_{n=1}^{\infty} b_n e^{inx}$ converges. In particular, $\sum_{n=1}^{\infty} \frac{e^{inx}}{n}$ converges. Taking real and imaginary parts shows that for every $x \in (0, 2\pi)$,

$$\sum_{n=1}^{\infty} \frac{\cos nx}{n} \quad \text{and} \quad \sum_{n=1}^{\infty} \frac{\sin nx}{n} \quad \text{converge}.$$

More generally, this argument shows that $\sum_{n=1}^{\infty} \frac{e^{inx}}{n^p}$ converges for every $p > 0$.

Before going on to other tests, it might be interesting to note that we could have determined the convergence of the series $\sum_{n=1}^{\infty} \frac{\cos nx}{n}$ immediately after learning trig functions, without having to know anything about Dirichlet's test. The trick here is to use some trig identities and write

$$\cos nx = \frac{\sin(n + 1/2)x - \sin(n - 1/2)x}{2\sin(x/2)},$$

or

$$\cos nx = c_{n+1} - c_n \quad \text{where} \quad c_n = \frac{\sin(n - 1/2)x}{2\sin(x/2)}.$$

Whenever you have a sequence that is "telescoping like," i.e., is the difference of adjacent terms of another sequence, good things should happen. Indeed,

$$\sum_{n=1}^{m} \frac{\cos nx}{n} = \sum_{n=1}^{m} \frac{c_{n+1} - c_n}{n} = \left(\frac{c_2 - c_1}{1} + \frac{c_3 - c_2}{2} + \frac{c_4 - c_3}{3} \right.$$

$$\left. + \cdots + \frac{c_{m+1} - c_m}{m} \right).$$

Gathering like terms, we obtain

$$\sum_{n=1}^{m} \frac{\cos nx}{n} = -c_1 + \frac{c_{m+1}}{m} + \sum_{n=1}^{m-1} c_{n+1}\left(\frac{1}{n} - \frac{1}{n+1}\right)$$

$$= -c_1 + \frac{c_{m+1}}{m} + \sum_{n=1}^{m-1} c_{n+1}\left(\frac{1}{n(n+1)}\right).$$

Replacing all the c_n by their formulas in terms of sine, and using that $c_1 = 1/2$, we obtain

$$\frac{1}{2} + \sum_{n=1}^{m} \frac{\cos nx}{n} = \frac{\sin(m+1/2)x}{2m\sin(x/2)} + \sum_{n=2}^{m}\left(\frac{\sin(n+1/2)x}{2\sin(x/2)} \cdot \frac{1}{n(n+1)}\right). \quad (6.3)$$

Since the sine is always bounded by 1 and $\sum 1/n(n+1)$ converges, it follows that as $m \to \infty$, the first term on the right of (6.3) tends to zero, while the summation on the right of (6.3) converges; in particular, the series in question converges, and we get the following formula:

$$\frac{1}{2} + \sum_{n=1}^{\infty} \frac{\cos nx}{n} = \frac{1}{2\sin(x/2)} \sum_{n=1}^{\infty} \frac{\sin(n+1/2)x}{n(n+1)} \ , \quad x \in (0, 2\pi).$$

In fact, this formula is *exactly* what you get from formula (6.2) in the proof of Dirichlet's test, where $a_n = \cos nx$ and $b_n = 1/n$. In Example 6.42 on p. 515, we'll show that $\sum_{n=1}^{\infty} \frac{\cos nx}{n} = -\log(2\sin(x/2))$.

6.1.4   Alternating Series, Decimal Places, log 2, and e Is Irrational

Since the partial sums of the (divergent) series $\sum (-1)^{n-1}$ are bounded, as a direct consequence of Dirichlet's test, we immediately get the alternating series test.

Alternating series test

Theorem 6.5 *If $\{a_n\}$ is a sequence of bounded variation that converges to zero, then $\sum(-1)^{n-1}a_n$ converges. In particular, if $\{a_n\}$ is a monotone sequence of real numbers approaching zero, then $\sum(-1)^{n-1}a_n$ converges.*

Example 6.6 The **alternating harmonic series**

$$\sum_{n=1}^{\infty}(-1)^{n-1}\frac{1}{n} = 1 - \frac{1}{2} + \frac{1}{3} - \frac{1}{4} + \frac{1}{5} - \frac{1}{6} + - \cdots$$

converges. Of course, we already knew this, and we also know that the value of the alternating harmonic series equals $\log 2$ (see p. 310 in Section 4.7).

We now come to a useful theorem for approximation purposes (cf. Problem 5).

Alternating series error estimate

> **Corollary 6.6** *If* $\{a_n\}$ *is a monotone sequence of real numbers approaching zero, and if* s *denotes the sum* $\sum(-1)^{n-1}a_n$ *and* s_n *denotes the nth partial sum, then*
>
> $$|s - s_n| \leq |a_{n+1}|.$$

Proof To establish the error estimate, we assume that $a_n \geq 0$ for each n, in which case we have $a_1 \geq a_2 \geq a_3 \geq a_4 \geq \cdots \geq 0$. (The case $a_n \leq 0$ is similar or can be derived from the present case by multiplying by -1.) Let's consider how $s = \sum_{n=1}^{\infty}(-1)^{n-1}a_n$ is approximated by the s_n. Observe that $s_1 = a_1$ increases from $s_0 = 0$ by the amount a_1; $s_2 = a_1 - a_2 = s_1 - a_2$ decreases from s_1 by the amount a_2; $s_3 = a_1 - a_2 + a_3 = s_2 + a_3$ increases from s_2 by the amount a_3; and so on; see Fig. 6.1 for a picture of what's going on here. Studying this figure also shows why $|s - s_n| \leq a_{n+1}$ holds. For this reason, we shall leave the exact proof details to the diligent and interested reader! ∎

Given $d \in \mathbb{N}$, what is a good definition for a real number a to "equal 0 to d decimal places"? If we write a as a decimal, taking the finite decimal expansion in case a has two expansions, we certainly want

$$|a| = 0.\underbrace{00\ldots0}_{d\text{ zeros}}a_{d+1}a_{d+2}\ldots. \tag{6.4}$$

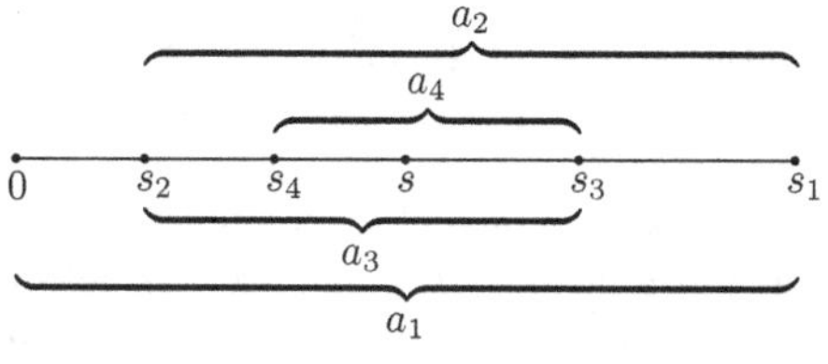

Fig. 6.1 The partial sums $\{s_n\}$ jump forward and backward by the amounts given by the a_n. This picture also shows that $|s - s_1| \leq a_2$, $|s - s_2| \leq a_3$, $|s - s_3| \leq a_4, \ldots$

However, we actually want more. Indeed, if $a_{d+1} \geq 5$, it's common practice to "round up" the preceding 0, so if we truncate a to the first d digits after the decimal point, then we would express a as $0.00\ldots01$, where there are $d - 1$ zeros after the decimal point. On the other hand, if $a_{d+1} < 5$, then we would express a as $0.00\ldots00$, where there are n zeros after the decimal point. This latter case is obviously what we desire. For this reason, we say that a **equals** 0 **to** d **decimal places** if there are d zeros after the decimal point and $a_{d+1} < 5$. It's easy to check that this is equivalent to

$$|a| < 0.\underbrace{00\ldots0}_{d \text{ zeros}}5 = 5 \times 10^{-d-1}.$$

We say that two real numbers x and y are **equal to d decimal places** if $x - y$ equals zero to d decimal places.

Example 6.7 Let's find n such that the nth partial sum of $\log 2 = \sum_{n=1}^{\infty} \frac{(-1)^{n-1}}{n}$ approximates $\log 2$ to two decimal places. By our discussion above, we need n such that

$$|\log 2 - s_n| < 0.005.$$

By Corollary 6.6, we can make this inequality hold by choosing n such that

$$|a_{n+1}| = \frac{1}{n+1} < 0.005 \quad \Longrightarrow \quad 500 < n+1 \quad \Longrightarrow \quad n = 500 \text{ works.}$$

With about five hours of pencil and paper work (and ten coffee breaks ☺), we find that $s_{500} = \sum_{n=1}^{500} \frac{(-1)^n}{n} = 0.69$ to two decimal places. Thus, $\log 2 = 0.69$ to two decimal places. A lot of work just to get two decimal places! (In fact, we need *at least* $n = 100$ to be within two decimal places; see Problem 5.)

Example 6.8 (*Irrationality of e, Proof II*) Another nice application of the alternating series error estimate (or rather its proof) is a simple proof that e is irrational; cf. [7, 192]. Indeed, suppose to the contrary that $e = m/n$, where $m, n \in \mathbb{N}$. Then we can write

$$\frac{n}{m} = e^{-1} = \sum_{k=0}^{\infty} \frac{(-1)^k}{k!} \quad \Longrightarrow \quad \frac{n}{m} - \sum_{k=0}^{m} \frac{(-1)^k}{k!} = \sum_{k=m+1}^{\infty} \frac{(-1)^k}{k!}.$$

Multiplying both sides by $m!$, we obtain

$$n(m-1)! - \sum_{k=0}^{m} (-1)^k \frac{m!}{k!} = \sum_{k=m+1}^{\infty} \frac{(-1)^k m!}{k!} = (-1)^{m+1} \sum_{k=1}^{\infty} \frac{(-1)^{k-1} m!}{(m+k)!}. \quad (6.5)$$

For $0 \le k \le m$, $m!/k!$ is an integer (this is because $m! = 1 \cdot 2 \cdots k \cdot (k+1) \cdots m$ contains a factor of $k!$), therefore the left-hand side of (6.5) is an integer. Hence, if $s = \sum_{k=1}^{\infty} (-1)^{k-1} a_k$, where $a_k = \frac{m!}{(m+k)!}$, then s is also an integer. Thus, as seen in Fig. 6.1, we have

$$0 < s < a_1 = \frac{1}{m+1}.$$

Now recall that $m \in \mathbb{N}$, so $1/(m+1) \le 1/2$. Thus, s is an integer strictly between 0 and $1/2$, an obvious contradiction!

6.1.5 Abel's Test for Series

Now let's modify the sum $\sum_{n=1}^{\infty} \frac{e^{inx}}{n}$, say to the slightly more complicated version

$$\sum_{n=1}^{\infty} \left(1 + \frac{1}{n}\right)^n \frac{e^{inx}}{n}.$$

If we try to determine the convergence of this series using Dirichlet's test, we'll have to do some work, but if we're feeling a little lazy, we can use the following theorem, whose proof uses an $\varepsilon/3$-**trick**.

Abel's test for series

> **Theorem 6.7** *Suppose that $\sum a_n$ converges. Then for every sequence $\{b_n\}$ of bounded variation, the series $\sum a_n b_n$ converges.*

Proof We shall apply Abel's lemma to establish that the sequence of partial sums for $\sum a_n b_n$ forms a Cauchy sequence, which implies that $\sum a_n b_n$ converges. For $m < n$, by Abel's lemma, we have

$$\sum_{k=m+1}^{n} a_k b_k = s_n b_n - s_m b_m - \sum_{k=m}^{n-1} s_k (b_{k+1} - b_k), \tag{6.6}$$

where s_n is the nth partial sum of the series $\sum a_n$. Concentrating on the far right summation in (6.6), if we add and subtract $s := \sum a_n$ to s_k, we get

$$\sum_{k=m}^{n-1} s_k (b_{k+1} - b_k) = \sum_{k=m}^{n-1} (s_k - s)(b_{k+1} - b_k) + s \sum_{k=m}^{n-1} (b_{k+1} - b_k)$$

$$= \sum_{k=m}^{n-1} (s_k - s)(b_{k+1} - b_k) + s b_n - s b_m,$$

since the sum telescoped. Replacing this into (6.6), we obtain

$$\sum_{k=m+1}^{n} a_k b_k = (s_n - s) b_n - (s_m - s) b_m - \sum_{k=m}^{n-1} (s_k - s)(b_{k+1} - b_k).$$

Let $\varepsilon > 0$. Since $\{b_n\}$ is of bounded variation, this sequence converges by Proposition 6.3, so in particular, it is bounded, and therefore, since $s_n \to s$, we have $(s_n - s) b_n \to 0$ and $(s_m - s) b_m \to 0$. Thus, we can choose N such that for $n, m > N$, we have $|(s_n - s) b_n| < \varepsilon/3$, $|(s_m - s) b_m| < \varepsilon/3$, and $|s_n - s| < \varepsilon/3$. Thus, for $N < m < n$, we have

$$\left| \sum_{k=m+1}^{n} a_k b_k \right| \le |(s_n - s)b_n| + |(s_m - s)b_m| + \sum_{k=m}^{n-1} |(s_k - s)(b_{k+1} - b_k)|$$

$$< \frac{\varepsilon}{3} + \frac{\varepsilon}{3} + \frac{\varepsilon}{3} \sum_{k=m}^{n-1} |b_{k+1} - b_k|.$$

Finally, since $\sum |b_{k+1} - b_k|$ converges, by the Cauchy criterion for series, the sum $\sum_{k=m}^{n-1} |b_{k+1} - b_k|$ can be made less than 1 for N chosen larger if necessary. Thus, for $N < m < n$, we have $|\sum_{k=m+1}^{n} a_k b_k| < \varepsilon$. This completes our proof. ∎

Example 6.9 Returning to our discussion above, we can write

$$\sum_{n=1}^{\infty} \left(1 + \frac{1}{n}\right)^n \frac{e^{inx}}{n} = \sum a_n b_n,$$

where $a_n = \frac{e^{inx}}{n}$ and $b_n = (1 + \frac{1}{n})^n$. Since we already know that $\sum_{n=1}^{\infty} a_n$ converges and that $\{b_n\}$ is nondecreasing and bounded above (by e; see p. 180 in Section 3.3) and therefore is of bounded variation, Abel's test shows that the series $\sum a_n b_n$ converges.

▶ **Exercises 6.1**

1. Derive the following formula for the sum of fourth powers:

$$1^4 + 2^4 + \cdots + n^4 = \frac{n^5}{5} + \frac{n^4}{2} + \frac{n^3}{3} - \frac{n}{30}.$$

2. (Cf. [84]) In this problem we use summation by parts to derive neat identities for the Fibonacci numbers. The Fibonacci sequence was defined in Problem 9 on p. 47. Using summation by parts, derive the formulas

   (a) $F_1 + F_2 + F_3 + \cdots + F_n = F_{n+2} - 1$, (b) $F_1^2 + F_2^2 + F_3^2 + \cdots + F_n^2 = F_n F_{n+1}$,

   (c) $F_1 + F_3 + F_5 + \cdots + F_{2n-1} = F_{2n}$, (d) $1 + F_2 + F_4 + F_6 + \cdots + F_{2n} = F_{2n+1}$.

3. (Cf. [83]) In this problem we relate limits of arithmetic means to summation by parts.

   (a) Let $\{a_n\}$, $\{b_n\}$ be sequences of complex numbers. Assume that $b_n \to 0$ and $\frac{1}{n} \sum_{k=1}^{n} k |b_{k+1} - b_k| \to 0$ as $n \to \infty$, and that for some constant C, we have $\left| \frac{1}{n} \sum_{k=1}^{n} a_k \right| \le C$ for all n. Prove that $\frac{1}{n} \sum_{k=1}^{n} a_k b_k \to 0$ as $n \to \infty$.
   (b) Prove that $\frac{1}{n}\left(\sqrt{1} - \sqrt{2} + \sqrt{3} - \sqrt{4} + \cdots + (-1)^{n-1}\sqrt{n}\right) \to 0$ as $n \to \infty$.

4. Determine the convergence or divergence of the following series:

   (a) $\frac{1}{1} + \frac{1}{2} + \frac{1}{3} - \frac{1}{4} - \frac{1}{5} + \frac{1}{6} + \frac{1}{7} - - + + \cdots$, (b) $\sum_{n=1}^{\infty} (-1)^n (\sqrt{n+1} - \sqrt{n})$,

(c) $\sum_{n=2}^{\infty} \dfrac{\cos nx}{\log n}$, (d) $\dfrac{1}{2 \cdot 1} - \dfrac{1}{2 \cdot 2} + \dfrac{1}{3 \cdot 3} - \dfrac{1}{3 \cdot 4} + \dfrac{1}{4 \cdot 5} - \dfrac{1}{4 \cdot 6} + - \cdots$,

(e) $\sum_{n=2}^{\infty} \dfrac{(-1)^{n-1}}{n} \log \dfrac{2n+1}{n}$, (f) $\sum_{n=2}^{\infty} \cos nx \, \sin\left(\dfrac{x}{n}\right) \ (x \in \mathbb{R})$, (g) $\sum_{n=2}^{\infty} (-1)^{n-1} \dfrac{\log n}{n}$.

5. (**More alternating series error estimates**) Let $\{a_n\}$ be a nonincreasing sequence of real numbers with $a_n \to 0$. Let $s = \sum_{n=1}^{\infty} (-1)^{n-1} a_n$ and let $s_n = \sum_{k=1}^{\infty} (-1)^{k-1} a_k$.

   (i)   Prove that for all $n \in \mathbb{N}$, $|s - s_n| = \sum_{k=n+1}^{\infty} (a_k - a_{k+1})$.
   (ii) For each n, define $b_n = a_n - a_{n+1}$ and assume that $\{b_n\}$ is nonincreasing. Prove that $|s - s_{n+1}| \le |s - s_n|$ for all n; that is, the error is nonincreasing.
   (iii) With the assumptions in (ii), prove that $a_{n+1} \le 2|s - s_n|$ for all n.
   (iv) Let s_n denote the nth partial sum of the alternating harmonic series. How large *must* n be in order that we have $|\log 2 - s_n| < 0.005$?

6. Here are some problems on bounded variation sequences.

   (a) Prove that every contractive sequence is of bounded variation.
   (b) Give an example of a convergent sequence, different from the one in Example 6.4, that is not of bounded variation.
   (c) (**Jordan decomposition**) If $\{a_n\}$ is a sequence of real numbers of bounded variation, prove that there are nonnegative, nondecreasing, convergent sequences $\{b_n\}$ and $\{c_n\}$ such that $a_n = b_n - c_n$ for all n. Suggestion: Let $b_n = \sum_{k=1}^{n} |a_{n+1} - a_n|$ and $c_n = b_n - a_n$.

7. Let $a_1, a_2, a_3, \ldots$ be a sequence of distinct natural numbers such that for all n, $s_n = a_1 + a_2 + \cdots + a_n \ge$ sum of the first n natural numbers $= n(n+1)/2$. (This holds, in particular, if $a_1, a_2, a_3, \ldots$ are distinct natural numbers.) Prove that

$$\sum_{k=1}^{n} \frac{1}{k} \le \sum_{k=1}^{n} \frac{a_k}{k^2}.$$

In particular, $\sum_{k=1}^{\infty} a_k / k^2$ diverges to infinity at least as fast as the harmonic series.

6.2 Lim Infs/Sups, Ratio/Roots, and Power Series

It is a fact of life that most sequences simply do not converge. In this section we introduce limit infimums and supremums, which *always* exist, either as real numbers or as $\pm\infty$. We also study their basic properties. We need these limits to study the ratio and root tests. You've probably seen these tests before in elementary calculus, but in this section we'll look at them in a slightly more sophisticated way.

6.2.1   Limit Infimums and Supremums

For an arbitrary sequence $\{a_n\}$ of real numbers we know that $\lim a_n$ may not exist, such as the sequence seen here:

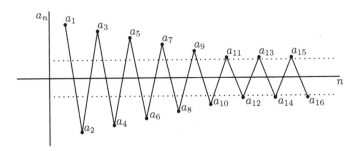

Fig. 6.2 For the oscillating sequence $\{a_n\}$, the *upper dashed line* represents $\limsup a_n$, and the *lower dashed line* represents $\liminf a_n$

However, being mathematicians, we shouldn't let this stop us, and in this sub-section, we define "limits" for an arbitrary sequence. It turns out that there are two notions of "limit" that show up often; one is the limit supremum (limsup) of $\{a_n\}$, which represents the "greatest" limiting value the a_n could possibly have, and the second is the limit infimum (liminf) of $\{a_n\}$, which represents the "least" limiting value that the a_n could possibly have. Here's a number line picture of these ideas for a sequence $\{a_n\}$ that oscillates left and right:

Figure 6.2 shows another picture of this same sequence.

We now make "greatest" limiting value and "least" limiting value precise. Let $a_1, a_2, a_3, \ldots$ be a sequence of real numbers bounded from above. Let us put

$$s_n = \sup_{k \geq n} a_k = \sup\{a_n, a_{n+1}, a_{n+2}, a_{n+3}, \ldots\}.$$

Note that

$$s_{n+1} = \sup\{a_{n+1}, a_{n+2}, \ldots\} \leq \sup\{a_n, a_{n+1}, a_{n+2}, \ldots\} = s_n.$$

Indeed, s_n is an upper bound for $\{a_n, a_{n+1}, a_{n+2}, \ldots\}$ and hence an upper bound for $\{a_{n+1}, a_{n+2}, \ldots\}$. Therefore, s_{n+1}, being the least such upper bound, must satisfy

$s_{n+1} \leq s_n$. Thus, $s_1 \geq s_2 \geq \cdots \geq s_n \geq s_{n+1} \geq \cdots$ is a nonincreasing sequence. In particular, being a monotone sequence, the limit $\lim s_n$ is defined as either a real number or (properly divergent to) $-\infty$. We define

$$\limsup a_n := \lim s_n = \lim_{n \to \infty} \left(\sup\{a_n, a_{n+1}, a_{n+2}, \ldots\} \right).$$

This limit, which again is either a real number or $-\infty$, is called the **limit supremum** or **lim sup** of the sequence $\{a_n\}$. This name fits, since $\limsup a_n$ is exactly that, a limit of supremums. If $\{a_n\}$ is not bounded from above, then we define

$$\limsup a_n := \infty \quad \text{if } \{a_n\} \text{ is not bounded from above.}$$

We define an **extended real number** as a real number or the symbols $\infty = +\infty$, $-\infty$. Then it is worth mentioning that lim sups *always* exist as an extended real number, unlike regular limits which may not exist. For the picture in Fig. 6.2, notice that

$$s_1 = \sup\{a_1, a_2, a_3, \ldots\} = a_1,$$
$$s_2 = \sup\{a_2, a_3, a_4, \ldots\} = a_3,$$
$$s_3 = \sup\{a_3, a_4, a_5, \ldots\} = a_3,$$

and so on. Thus, the sequence $s_1, s_2, s_3, \ldots$ picks out the odd-indexed terms of the sequence $a_1, a_2, \ldots$ and $\limsup a_n = \lim s_n$ is the value given by the upper dashed line in Fig. 6.2; here's a picture:

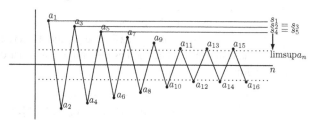

Here are some other examples.

Example 6.10 We shall compute $\limsup a_n$, where $a_n = \frac{1}{n}$. According to the definition of lim sup, we first have to find s_n:

$$s_n := \sup\{a_n, a_{n+1}, a_{n+2}, \ldots\} = \sup\left\{\frac{1}{n}, \frac{1}{n+1}, \frac{1}{n+2}, \frac{1}{n+3}, \ldots\right\} = \frac{1}{n}.$$

Second, we take the limit of the sequence $\{s_n\}$:

$$\limsup a_n := \lim_{n \to \infty} s_n = \lim_{n \to \infty} \frac{1}{n} = 0.$$

Notice that $\lim a_n$ also exists and $\lim a_n = 0$, the same as the lim sup. We'll come back to this observation in Example 6.12 below.

Example 6.11 Consider the sequence $\{(-1)^n\}$. In this case, we know that $\lim(-1)^n$ does not exist. To find $\limsup(-1)^n$, we first compute s_n:

$$s_n = \sup\{(-1)^n, (-1)^{n+1}, (-1)^{n+2}, \ldots\} = \sup\{+1, -1\} = 1,$$

where we used that the set $\{(-1)^n, (-1)^{n+1}, (-1)^{n+2}, \ldots\}$ is just a set consisting of the numbers $+1$ and -1. Hence,

$$\limsup(-1)^n := \lim s_n = \lim 1 = 1.$$

We can also define a corresponding $\liminf a_n$, which is a limit of infimums. To do so, assume for the moment that our generic sequence $\{a_n\}$ is bounded from below. Consider the sequence $\{\iota_n\}$ where

$$\iota_n = \inf_{k \geq n} a_k = \inf\{a_n, a_{n+1}, a_{n+2}, a_{n+3}, \ldots\}.$$

Note that

$$\iota_n = \inf\{a_n, a_{n+2}, \ldots\} \leq \inf\{a_{n+1}, a_{n+2}, \ldots\} = \iota_{n+1},$$

since the set $\{a_n, a_{n+2}, \ldots\}$ on the left of $\leq$ contains the set $\{a_{n+1}, a_{n+2}, \ldots\}$. Thus, $\iota_1 \leq \iota_2 \leq \cdots \leq \iota_n \leq \iota_{n+1} \leq \cdots$ is an nondecreasing sequence. In particular, being a monotone sequence, the limit $\lim \iota_n$ is defined as either a real number or (properly divergent to) ∞. We define

$$\boxed{\liminf a_n := \lim \iota_n = \lim_{n \to \infty} \Big(\inf\{a_n, a_{n+1}, a_{n+2}, \ldots\}\Big),}$$

which exists either as a real number or $+\infty$, is called the **limit infimum** or **lim inf** of $\{a_n\}$. If $\{a_n\}$ is not bounded from below, then we define

$$\boxed{\liminf a_n := -\infty \quad \text{if } \{a_n\} \text{ is not bounded from below.}}$$

Again, as with lim sups, lim infs always exist as extended real numbers. For the picture in Fig. 6.2, here is the corresponding sequence $\iota_1, \iota_2, \iota_3, \ldots$:

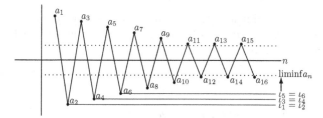

Thus, the sequence $\iota_1, \iota_2, \iota_3, \ldots$ picks out the even-indexed terms of the sequence $a_1, a_2, \ldots$, and $\lim \inf a_n = \lim \iota_n$ is the value given by the lower dashed line. Here are some worked examples.

Example 6.12 We shall compute $\lim \inf a_n$, where $a_n = \frac{1}{n}$. According to the definition of lim inf, we first have to find ι_n:

$$\iota_n := \inf\{a_n, a_{n+1}, a_{n+2}, \ldots\} = \inf\left\{\frac{1}{n}, \frac{1}{n+1}, \frac{1}{n+2}, \frac{1}{n+3}, \ldots\right\} = 0.$$

Second, we take the limit of ι_n:

$$\lim \inf a_n := \lim_{n \to \infty} \iota_n = \lim_{n \to \infty} 0 = 0.$$

Notice that $\lim a_n$ also exists and $\lim a_n = 0$, the same as $\lim \inf a_n$, which is the same as $\lim \sup a_n$, as we saw in Example 6.10. We are thus led to make the following conjecture: If $\lim a_n$ exists, then $\lim \sup a_n = \lim \inf a_n = \lim a_n$; this conjecture is indeed true, as we'll see in Property *(2)* of Theorem 6.8.

Example 6.13 If $a_n = (-1)^n$, then

$$\inf\{a_n, a_{n+1}, a_{n+2}, \ldots\} = \sup\{(-1)^n, (-1)^{n+1}, (-1)^{n+2}, \ldots\} = \inf\{+1, -1\} = -1.$$

Hence,

$$\lim \inf(-1)^n := \lim -1 = -1.$$

The following theorem contains the main properties of limit infimums and supremums that we shall need in the sequel.

Properties of lim inf/sup

> **Theorem 6.8** *If $\{a_n\}$ and $\{b_n\}$ are sequences of real numbers, then*
>
> *(1)* $\limsup a_n = -\liminf(-a_n)$ *and* $\liminf a_n = -\limsup(-a_n)$.
> *(2)* $\lim a_n$ *is defined, as a real number or* $\pm\infty$ *if and only if* $\limsup a_n = \liminf a_n$, *in which case*
>
> $$\lim a_n = \limsup a_n = \liminf a_n.$$
>
> *(3)* *If $a_n \le b_n$ for all n sufficiently large, then*
>
> $$\liminf a_n \le \liminf b_n \quad and \quad \limsup a_n \le \limsup b_n.$$
>
> *(4)* *The following inequality properties hold:*
>
> *(a)* $\limsup a_n < a \implies$ *there is an N such that $n > N \implies a_n < a$.*
> *(b)* $\limsup a_n > a \implies$ *there exist infinitely many n such that $a_n > a$.*
> *(c)* $\liminf a_n < a \implies$ *there exist infinitely many n such that $a_n < a$.*
> *(d)* $\liminf a_n > a \implies$ *there is an N such that $n > N \implies a_n > a$.*

Proof To prove *(1)*, assume first that $\{a_n\}$ is not bounded from above; then $\{-a_n\}$ is not bounded from below. Hence, $\limsup a_n := \infty$ and $\liminf(-a_n) := -\infty$, which implies *(1)* in this case. Assume now that $\{a_n\}$ is bounded above. Recall from Lemma 2.30 on p. 92 that given a nonempty subset $A \subseteq \mathbb{R}$ bounded above, we have $\sup A = -\inf(-A)$. Hence,

$$\sup\{a_n, a_{n+1}, a_{n+2}, \dots\} = -\inf\{-a_n, -a_{n+1}, -a_{n+2}, -a_{n+3}, \dots\}.$$

Taking $n \to \infty$ on both sides, we get $\limsup a_n = -\liminf(-a_n)$.

We now prove *(2)*. Suppose first that $\lim a_n$ converges to a real number L. Then given $\varepsilon > 0$, there exists an N such that

$$L - \varepsilon \le a_k \le L + \varepsilon, \quad \text{for all } k > N,$$

which implies that for $n > N$,

$$L - \varepsilon \le \inf_{k \ge n} a_k \le \sup_{k \ge n} a_k \le L + \varepsilon.$$

Taking $n \to \infty$ implies that

$$L - \varepsilon \le \liminf a_n \le \limsup a_n \le L + \varepsilon.$$

Since $\varepsilon > 0$ was arbitrary, it follows that $\limsup a_n = L = \liminf a_n$. Reversing these steps, we leave you to show that if $\limsup a_n = L = \liminf a_n$, then $\{a_n\}$

converges to L. We now consider *(2)* in the case that $\lim a_n = +\infty$; the case that the limit is $-\infty$ is proved similarly. Then given a real number $M > 0$, there exists an N such that

$$n > N \implies M \leq a_n.$$

This implies that

$$M \leq \inf_{k \geq n} a_k \leq \sup_{k \geq n} a_k.$$

Taking $n \to \infty$, we obtain

$$M \leq \liminf a_n \leq \limsup a_n.$$

Since $M > 0$ was arbitrary, it follows that $\limsup a_n = +\infty = \liminf a_n$. Reversing these steps, we leave you to show that if $\limsup a_n = +\infty = \liminf a_n$, then $a_n \to +\infty$.

To prove *(3)*, note that if $\{a_n\}$ is not bounded from below, then $\liminf a_n := -\infty$, so $\liminf a_n \leq \liminf b_n$ automatically; thus, we may assume that $\{a_n\}$ is bounded from below. In this case, observe that $a_n \leq b_n$ for all n sufficiently large implies that for n sufficiently large,

$$\inf\{a_n, a_{n+1}, a_{n+2}, \ldots\} \leq \inf\{b_n, b_{n+1}, b_{n+2}, b_{n+3}, \ldots\}.$$

Taking $n \to \infty$ and using that limits preserve inequalities now proves *(3)*. The proof that $\limsup a_n \leq \limsup b_n$ is similar.

Because this proof is dragging on ☺, we'll prove only *(a)* and *(b)* of *(4)*, leaving *(c)*, *(d)* to the reader. Assume that $\limsup a_n < a$, that is,

$$\lim_{n \to \infty} \left(\sup\{a_n, a_{n+1}, a_{n+2}, \ldots\} \right) < a.$$

By definition of limit, it follows that for some N, we have

$$n > N \implies \sup\{a_n, a_{n+1}, a_{n+2}, \ldots\} < a,$$

that is, the least upper bound of $\{a_n, a_{n+1}, a_{n+2}, \ldots\}$ is strictly less than a, so we must have we have $a_n < a$ for all $n > N$. Instead of proving *(b)* directly, we prove its contrapositive: If there are at most finitely many n such that $a_n > a$, then $\limsup a_n \leq a$. Indeed, if there are only finitely many n such that $a_n > a$, then for some N we have $a_n \leq a$ for all $n > N$ (where N larger than the largest natural number k satisfying $a_k > a$). Hence, for $n > N$,

$$\sup\{a_n, a_{n+1}, a_{n+2}, \ldots\} \leq a.$$

Taking $n \to \infty$, we get $\limsup a_n \leq a$. ∎

6.2.2 Ratio/Root Tests and The Exponential and ζ-Functions, Again

You undoubtedly studied the ratio/root tests back in elementary calculus. To review them, consider a geometric series $\sum a_n$, where $a_n = a^n$ for some $a \in \mathbb{C}$. Observe that

$$\left| \frac{a_{n+1}}{a_n} \right| = |a| \quad \text{and} \quad |a_n|^{1/n} = |a|,$$

and by the geometric series test,

$$\sum a_n \text{ converges } \iff |a| < 1.$$

Thus, the geometric series converges/diverges according as the ratio $|a_{n+1}/a_n|$ or the root $|a_n|^{1/n}$ is strictly less than/greater than or equal to 1. The ratio and root tests generalize these results to arbitrary series where the exact ratios $|a_{n+1}/a_n|$ and roots $|a_n|^{1/n}$ are replaced by limit sups and infs. Before stating these generalizations, we first consider the following important lemma.

Lemma 6.9 *If $\{a_n\}$ is a sequence of nonzero complex numbers, then*

$$\liminf \left| \frac{a_{n+1}}{a_n} \right| \leq \liminf |a_n|^{1/n} \leq \limsup |a_n|^{1/n} \leq \limsup \left| \frac{a_{n+1}}{a_n} \right|.$$

Proof The middle inequality is automatic (because inf's are $\leq$ sup's), so we just need to prove the left and right inequalities. Consider the left one; the right one is analogous and is left to the reader. If $\liminf |a_{n+1}/a_n| = -\infty$, then there is nothing to prove, so we may assume that $\liminf |a_{n+1}/a_n| \neq -\infty$. Given $b < \liminf |a_{n+1}/a_n|$, we shall prove that $b < \liminf |a_n|^{1/n}$. This proves the left side in our desired inequalities, for if on the contrary, we had $\liminf |a_n|^{1/n} < \liminf |a_{n+1}/a_n|$, then choosing $b = \liminf |a_n|^{1/n}$, we would have

$$\liminf |a_n|^{1/n} < \liminf |a_n|^{1/n},$$

a contradiction. So, let $b < \liminf |a_{n+1}/a_n|$. Choose a such that $b < a < \liminf |a_{n+1}/a_n|$. Then by Property *4 (d)* in Theorem 6.8, for some N, we have

$$n > N \quad \Longrightarrow \quad \left| \frac{a_{n+1}}{a_n} \right| > a.$$

Fix $m > N$ and let $n > m > N$. Then we can write

$$|a_n| = \left| \frac{a_n}{a_{n-1}} \right| \cdot \left| \frac{a_{n-1}}{a_{n-2}} \right| \cdots \left| \frac{a_{m+1}}{a_m} \right| \cdot |a_m|.$$

There are $n - m$ quotients in this equality, each greater than a, so

$$|a_n| > a \cdot a \cdots a \cdot |a_m| = a^{n-m} \cdot |a_m|,$$

which implies that

$$a^{1-m/n} \cdot |a_m|^{1/n} < |a_n|^{1/n}. \tag{6.7}$$

As $n \to \infty$, the left-hand side of (6.7) approaches $a^1 \cdot |a_m|^0 = a$. Thus, since $\liminf = \lim$ when the limit exists and $\liminf$'s preserve inequalities (Properties *(2)* and *(3)* in Theorem 6.8), taking the limit infimums of both sides of (6.7), we obtain

$$a = \liminf \left(a^{1-m/n} \cdot |a_m|^{1/n} \right) \leq \liminf |a_n|^{1/n}.$$

Since $b < a$, we have $b < \liminf |a_n|^{1/n}$, and our proof is complete. ∎

We now state a "souped-up" version[2] of the elementary calculus root test.

Cauchy's root test

Theorem 6.10 *A series* $\sum a_n$ *converges absolutely or diverges according as*

$$\limsup |a_n|^{1/n} < 1 \quad or \quad \limsup |a_n|^{1/n} > 1.$$

Proof Suppose first that $\limsup |a_n|^{1/n} < 1$. Then we can choose $0 < a < 1$ such that $\limsup |a_n|^{1/n} < a$, which, by Property *4 (a)* of Theorem 6.8, implies that for some N,

$$n > N \quad \Longrightarrow \quad |a_n|^{1/n} < a,$$

that is,

$$n > N \quad \Longrightarrow \quad |a_n| < a^n.$$

The geometric series $\sum a^n$ converges, since $0 < a < 1$, and thus by the comparison test, the sum $\sum |a_n|$ also converges, and hence $\sum a_n$ converges as well.

Assume that $\limsup |a_n|^{1/n} > 1$. Then by Property *4 (b)* of Theorem 6.8, there are infinitely many n such that $|a_n|^{1/n} > 1$; that is,

there are infinitely many n such that $|a_n| > 1$.

By the nth term test on p. 196, the series $\sum a_n$ does not converge. ∎

---

[2]In elementary calculus, the ratio and root tests are usually stated with regular limits and not with $\limsup$'s.

It is important to remark that in the case $\limsup |a_n|^{1/n} = 1$, this test does not give information as to convergence.

Example 6.14 The series $\sum 1/n$ diverges, and $\limsup |1/n|^{1/n} = \lim 1/n^{1/n} = 1$ (see Example 3.7 on p. 159 for the proof that $\lim n^{1/n} = 1$). On the other hand, $\sum 1/n^2$ converges, and $\limsup |1/n^2|^{1/n} = \lim (1/n^{1/n})^2 = 1$ as well. Thus, $\limsup |a_n|^{1/n} = 1$ is not enough to determine the convergence of a series.

As with the root test, in elementary calculus you learned the ratio test most likely without proof, and accepting by faith this test as correct, you probably used it to great effect to solve many problems.[3] Here's a "souped-up" version of the elementary calculus ratio test.

d'Alembert's ratio test

> **Theorem 6.11** *A series $\sum a_n$, with a_n nonzero for n sufficiently large, converges absolutely or diverges according as*
>
> $$\limsup \left| \frac{a_{n+1}}{a_n} \right| < 1 \quad or \quad \liminf \left| \frac{a_{n+1}}{a_n} \right| > 1.$$

Proof If we set $L := \limsup |a_n|^{1/n}$, then by Lemma 6.9, we have

$$\liminf \left| \frac{a_{n+1}}{a_n} \right| \leq L \leq \limsup \left| \frac{a_{n+1}}{a_n} \right|. \tag{6.8}$$

Therefore, if $\limsup \left| \frac{a_{n+1}}{a_n} \right| < 1$, then $L < 1$ too, so $\sum a_n$ converges absolutely by the root test. On the other hand, if $\liminf \left| \frac{a_{n+1}}{a_n} \right| > 1$, then $L > 1$ too, so $\sum a_n$ diverges by the root test. ∎

We remark that in the other case, that is, $\liminf \left| \frac{a_{n+1}}{a_n} \right| \leq 1 \leq \limsup \left| \frac{a_{n+1}}{a_n} \right|$, this test does not give information as to convergence. Indeed, the same divergent and convergent examples used for the root test, $\sum 1/n$ and $\sum 1/n^2$, have the property that $\liminf \left| \frac{a_{n+1}}{a_n} \right| = 1 = \limsup \left| \frac{a_{n+1}}{a_n} \right|$.

Note that if $\limsup |a_n|^{1/n} = 1$, that is, the root test fails (to give a decisive answer), then setting $L = 1$ in (6.8), we see that the ratio test also fails. Thus,

$$\text{root test fails} \implies \text{ratio test fails.} \tag{6.9}$$

Therefore, if the root test fails one cannot hope to appeal to the ratio test.

Let's now consider some examples.

---

[3]"Allez en avant, et la foi vous viendra [push on and faith will catch up with you]." Advice to those who questioned the calculus by Jean d'Alembert (1717–1783) [154].

Example 6.15 First, our old friend

$$\exp(z) := \sum_{n=1}^{\infty} \frac{z^n}{n!},$$

which we already knows converges, but for the fun of it, let's apply the ratio test. Observe that

$$\left|\frac{a_{n+1}}{a_n}\right| = \left|\frac{\frac{z^{n+1}}{(n+1)!}}{\frac{z^n}{n!}}\right| = |z| \cdot \frac{n!}{(n+1)!} = \frac{|z|}{n+1}.$$

Hence,

$$\lim \left|\frac{a_{n+1}}{a_n}\right| = 0 < 1.$$

Thus, the exponential function $\exp(z)$ converges absolutely for all $z \in \mathbb{C}$. This proof was a little easier than the one in Section 3.7, but then again, back then we didn't have the up-to-date technology of the ratio test that we have now. Here's an example that fails.

Example 6.16 Consider the Riemann zeta function

$$\zeta(z) = \sum_{n=1}^{\infty} \frac{1}{n^z}, \qquad \operatorname{Re} z > 1.$$

If $z = x + iy$ is separated into its real and imaginary parts, then

$$\left|a_n\right|^{1/n} = \left|\frac{1}{n^z}\right|^{1/n} = \left(\frac{1}{n^x}\right)^{1/n} = \left(\frac{1}{n^{1/n}}\right)^x.$$

Since $\lim n^{1/n} = 1$, it follows that

$$\lim \left|a_n\right|^{1/n} = 1,$$

so the root test fails to give information, which also implies that the ratio test fails as well. Of course, using the comparison test as we did in the proof of Theorem 4.34, we already know that $\zeta(z)$ converges for all $z \in \mathbb{C}$ with $\operatorname{Re} z > 1$.

It's easy to find examples of series for which the ratio test fails but the root test succeeds.

Example 6.17 A general class of examples that foil the ratio test is (see Problem 4)

$$a + b + a^2 + b^2 + a^3 + b^3 + a^4 + b^4 + \cdots , \quad 0 < b < a < 1; \qquad (6.10)$$

here, the odd terms are given by $a_{2n-1} = a^n$, and the even terms are given by $a_{2n} = b^n$. For concreteness, let us consider the series

$$\frac{1}{2} + \frac{1}{3} + \left(\frac{1}{2}\right)^2 + \left(\frac{1}{3}\right)^2 + \left(\frac{1}{2}\right)^3 + \left(\frac{1}{3}\right)^3 + \left(\frac{1}{2}\right)^4 + \left(\frac{1}{3}\right)^4 + \cdots .$$

Since

$$\left|\frac{a_{2n}}{a_{2n-1}}\right| = \left|\frac{(1/3)^n}{(1/2)^n}\right| = \left(\frac{2}{3}\right)^n$$

and

$$\left|\frac{a_{2n+1}}{a_{2n}}\right| = \left|\frac{(1/2)^{n+1}}{(1/3)^n}\right| = \left(\frac{3}{2}\right)^n \cdot \frac{1}{2},$$

it follows that $\liminf |a_{n+1}/a_n| = 0 < 1 < \infty = \limsup |a_{n+1}/a_n|$, so the ratio test does not give information. On the other hand, since

$$|a_{2n-1}|^{1/(2n-1)} = \left((1/2)^n\right)^{1/(2n-1)} = \left(\frac{1}{2}\right)^{\frac{n}{2n-1}}$$

and

$$|a_{2n}|^{1/(2n)} = \left((1/3)^{n-1}\right)^{1/(2n)} = \left(\frac{1}{3}\right)^{\frac{n-1}{2n}},$$

we leave it as an exercise for you to show that $\limsup |a_n|^{1/n} = (1/2)^{1/2}$. Since $(1/2)^{1/2} < 1$, the series converges by the root test.

Thus, in contrast to (6.9),

$$\text{ratio test fails} \quad \not\Longrightarrow \quad \text{root test fails.}$$

However, in the following lemma we show that if the ratio test fails such that the *true* limit $\lim |\frac{a_{n+1}}{a_n}| = 1$, then the root test fails as well.

Lemma 6.12 *If $|\frac{a_{n+1}}{a_n}| \to L$ with $L \in \mathbb{R}$ or $L = \pm\infty$, then $|a_n|^{1/n} \to L$.*

Proof By Lemma 6.9, we know that

$$\liminf \left|\frac{a_{n+1}}{a_n}\right| \le \liminf |a_n|^{1/n} \le \limsup |a_n|^{1/n} \le \limsup \left|\frac{a_{n+1}}{a_n}\right|.$$

By Theorem 6.8, a limit exists if and only if the lim inf and the lim sup have the same limit, so the outside quantities in these inequalities equal L. It follows that $\liminf |a_n|^{1/n} = \limsup |a_n|^{1/n} = L$ as well, and hence $\lim |a_n|^{1/n} = L$. ∎

Let's do one last example:

Example 6.18 Consider the series

$$1 + \sum_{n=1}^{\infty} \frac{1 \cdot 3 \cdot 5 \cdots (2n-1)}{2 \cdot 4 \cdot 6 \cdots (2n)\,(2n+1)}. \tag{6.11}$$

Applying the ratio test, we have

$$\frac{a_{n+1}}{a_n} = \frac{(2n+1)(2n+1)}{(2n+2)(2n+3)} = \frac{4n^2 + 4n + 1}{4n^2 + 10n + 6} = \frac{1 + \dfrac{1}{n} + \dfrac{1}{4n^2}}{1 + \dfrac{5}{2n} + \dfrac{3}{2n^2}}. \tag{6.12}$$

Therefore, $\lim \left| \frac{a_{n+1}}{a_n} \right| = 1$, so the ratio and root tests give no information! What can we do? We'll see that Raabe's test in Section 6.3 will show that (6.11) converges.

6.2.3 Power Series

Our old friend

$$\exp(z) := \sum_{n=0}^{\infty} \frac{z^n}{n!}$$

is an example of a **power series**, by which we mean a series of the form

$$\sum_{n=0}^{\infty} a_n z^n, \quad \text{where } z \in \mathbb{C}, \quad \text{or} \quad \sum_{n=0}^{\infty} a_n x^n, \quad \text{where } x \in \mathbb{R},$$

where $a_n \in \mathbb{C}$ for all n (in particular, the a_n may be real). However, we shall focus on power series of the complex variable z, *although essentially everything we mention works for real variables x.*

Example 6.19 Besides the exponential function, other familiar examples of power series include the trigonometric series, $\sin z = \sum_{n=0}^{\infty} (-1)^n z^{2n+1}/(2n+1)!$, $\cos z = \sum_{n=0}^{\infty} (-1)^n z^{2n}/(2n)!$.

The convergence of power series is quite easy to analyze. First, $\sum_{n=0}^{\infty} a_n z^n = a_0 + a_1 z + a_2 z^2 + \cdots$ certainly converges if $z = 0$. For $|z| > 0$ we can use the root test: Observe that (see Problem 8 for a proof that we can take out $|z|$)

$$\limsup \left| a_n z^n \right|^{1/n} = \limsup \left(|z|\,|a_n|^{1/n} \right) = |z| \limsup |a_n|^{1/n}.$$

Therefore, $\sum a_n z^n$ converges (absolutely) or diverges according as

$$|z| \cdot \limsup |a_n|^{1/n} < 1 \quad \text{or} \quad |z| \cdot \limsup |a_n|^{1/n} > 1.$$

Therefore, if we define $0 \leq R \leq \infty$ by

$$R := \frac{1}{\limsup |a_n|^{1/n}},$$ (6.13)

where by convention, we put $R := +\infty$ when $\limsup |a_n|^{1/n} = 0$ and $R := 0$ when $\limsup |a_n|^{1/n} = +\infty$, then it follows that $\sum a_n z^n$ converges (absolutely) or diverges according to $|z| < R$ or $|z| > R$; when $|z| = R$, anything can happen. It is quite fitting to call R the **radius of convergence**, as seen here (Fig. 6.3).

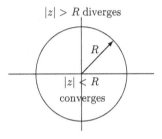

Fig. 6.3 $\sum a_n z^n$ converges (absolutely) or diverges according as $|z| < R$ or $|z| > R$

Let us summarize our findings in the following theorem, named after Cauchy (whom we've already met many times) and Jacques Hadamard (1865–1963).[4]

Cauchy–Hadamard theorem

Theorem 6.13 *If R is the radius of convergence of the power series $\sum a_n z^n$, then the series is absolutely convergent for $|z| < R$ and is divergent for $|z| > R$.*

One final remark. Suppose that the a_n are nonzero for n sufficiently large and suppose that $\lim |\frac{a_n}{a_{n+1}}|$ exists. Then by Lemma 6.12, we have

$$R = \lim \left| \frac{a_n}{a_{n+1}} \right|.$$ (6.14)

This formula for the radius of convergence might, in some cases, be easier to work with than the formula involving $|a_n|^{1/n}$.

▶ **Exercises 6.2**

1. Find the lim infs/sups of the sequence $\{a_n\}$, where a_n is given by

   (a) $\dfrac{2 + (-1)^n}{4}$, (b) $(-1)^n \left(1 - \dfrac{1}{n}\right)$, (c) $2^{(-1)^n}$, (d) $2^{n(-1)^n}$, (e) $\left(1 + \dfrac{(-1)^n}{2}\right)^n$.

---

[4]"The shortest path between two truths in the real domain passes through the complex domain." Jacques Hadamard (1865–1963). Quoted in *The Mathematical Intelligencer* 13 (1991).

(f) If $\{r_n\}$ is a list of all rationals in $(0, 1)$, prove that $\liminf r_n = 0$ and $\limsup r_n = 1$.

2. Investigate the following series for convergence (in (c), $z \in \mathbb{C}$):

(a) $\displaystyle\sum_{n=1}^{\infty} \frac{(n+1)(n+2)\cdots(n+n)}{n^n}$, (b) $\displaystyle\sum_{n=1}^{\infty} \frac{(n+1)^n}{n!}$, (c) $\displaystyle\sum_{n=1}^{\infty} \frac{n^z}{n!}$, (d) $\displaystyle\sum_{n=1}^{\infty} \frac{1}{2^{n+(-1)^n}}$.

3. Find the radius of convergence for the following series:

(a) $\displaystyle\sum_{n=1}^{\infty} \frac{(n+1)^n}{n^{n+1}} z^n$, (b) $\displaystyle\sum_{n=1}^{\infty} \left(\frac{n}{n+1}\right)^n z^n$, (c) $\displaystyle\sum_{n=1}^{\infty} \frac{(2n)!}{(n!)^2} z^n$, (d) $\displaystyle\sum_{n=1}^{\infty} \frac{z^n}{n^p}$,

where in the last sum, $p \in \mathbb{R}$. If $z = x \in \mathbb{R}$, for each series, state all $x \in \mathbb{R}$ such that the series converges. For (c), your answer should depend on p.

4. (a) Investigate the series (6.10) for convergence using both the ratio and the root tests.

   (b) Here is another class of examples:

$$1 + a + b^2 + a^3 + b^4 + a^5 + b^6 + \cdots \quad , \quad 0 < a < b < 1.$$

   Show that the ratio test fails but the root test works.

5. Lemma 6.12 is very useful in finding certain limits that aren't obvious at first glance. Using this lemma, derive the following limits:

(a) $\displaystyle\lim \frac{n}{(n!)^{1/n}} = e$, (b) $\displaystyle\lim \frac{n+1}{(n!)^{1/n}} = e$, (c) $\displaystyle\lim \frac{n}{[(n+1)(n+2)\cdots(n+n)]^{1/n}} = \frac{e}{4}$,

and for $a, b \in \mathbb{R}$ with $a > 0$ and $a + b > 0$,

$$(d) \quad \lim \frac{n}{[(a+b)(2a+b)\cdots(na+b)]^{1/n}} = \frac{e}{a}.$$

Suggestion: For (a), let $a_n = n^n/n!$. Prove that $\lim \frac{a_{n+1}}{a_n} = e$ and hence $\lim a_n^{1/n} = e$ as well. As a side remark, recall that (a) is called (the "weak") Stirling's formula, which we introduced in Eq. (3.27) on p. 181 and proved in Problem 6 on p. 183.

6. In this problem we investigate the interesting power series $\sum_{n=1}^{\infty} \frac{n!}{n^n} z^n$, where $z \in \mathbb{C}$.

   (a) Prove that this series has radius of convergence $R = e$.

   (b) If $|z| = e$, then the ratio and root tests both fail. However, if $|z| = e$, then prove that the infinite series diverges.

   (c) Investigate the convergence/divergence of $\sum_{n=1}^{\infty} \frac{n^n}{n!} z^n$, where $z \in \mathbb{C}$.

7. In this problem we investigate the interesting power series

$$F(z) := \sum_{n=0}^{\infty} F_{n+1} z^n = F_1 + F_2 z + F_3 z^2 + \cdots ,$$

where $\{F_n\}$ is the Fibonacci sequence defined in Problem 9 on p. 47. In that problem you proved that $F_n = \frac{1}{\sqrt{5}}[\Phi^n - (-\Phi)^{-n}]$, where $\Phi = \frac{1+\sqrt{5}}{2}$, the golden ratio.

(i) Prove that $F(z)$ has radius of convergence equal to Φ^{-1}.

(ii) Prove that for all z with $|z| < \Phi^{-1}$, we have $F(z) = \frac{1}{1-z-z^2}$. Suggestion: Show that $(1 - z - z^2)F(z) = 1$. By the way, given a sequence $\{a_n\}_{n=0}^{\infty}$, the power series $\sum_{n=0}^{\infty} a_n z^n$ is called the **generating function** of the sequence $\{a_n\}$. Thus, the generating function for $\{F_{n+1}\}$ has the closed form $1/(1 - z - z^2)$. For more on generating functions, see the free book [264]. Also, if you're interested in a magic trick you can do with the formula $F(z) = 1/(1 - z - z^2)$, see [188].

8. Here are some lim inf/sup problems. Let $\{a_n\}$, $\{b_n\}$ be sequences of real numbers.

(a) Prove that if $c > 0$, then $\liminf(ca_n) = c \liminf a_n$ and $\limsup(ca_n) = c \limsup a_n$. Here, we take the "obvious" conventions: $c \cdot \pm\infty = \pm\infty$.

(b) Prove that if $c < 0$, then $\liminf(ca_n) = c \limsup a_n$ and $\limsup(ca_n) = c \liminf a_n$.

(c) If $\{a_n\}$, $\{b_n\}$ are bounded, prove that $\liminf a_n + \liminf b_n \le \liminf(a_n + b_n)$.

(d) If $\{a_n\}$, $\{b_n\}$ are bounded, prove that $\limsup(a_n + b_n) \le \limsup a_n + \limsup b_n$.

9. If $a_n \to L$, where L is a positive real number, prove that $\limsup(a_n \cdot b_n) = L \limsup b_n$ and $\liminf(a_n \cdot b_n) = L \liminf b_n$. Here are some steps if you want them:

(i) Show that you can get the lim inf statement from the lim sup statement; hence we can focus on the lim sup statement. We shall prove that $\limsup(a_n b_n) \le L \limsup b_n$ and $L \limsup b_n \le \limsup(a_n b_n)$.

(ii) Show that the inequality $\limsup(a_n b_n) \le L \limsup b_n$ follows if the following statement holds: If $\limsup b_n < b$, then $\limsup(a_n b_n) < L b$.

(iii) Now prove that if $\limsup b_n < b$, then $\limsup(a_n b_n) < L b$. Suggestion: If $\limsup b_n < b$, then choose a such that $\limsup b_n < a < b$. Using Property 4 (a) of Theorem 6.8 and the definition of $L = \lim a_n > 0$, prove that there is an N such that $n > N$ implies $b_n < a$ and $a_n > 0$. Conclude that for $n > N$, $a_n b_n < a a_n$. Finally, take lim sups of both sides of $a_n b_n < a a_n$.

(iv) Show that the inequality $L \limsup b_n \le \limsup(a_n b_n)$ follows if the following statement holds: If $\limsup(a_n b_n) < L b$, then $\limsup b_n < b$; then prove this statement.

10. Let $\{a_n\}$ be a sequence of real numbers. We prove that there are monotone subsequences of $\{a_n\}$ that converge to $\liminf a_n$ and $\limsup a_n$. Proceed as follows:

(i) Using the monotone subsequence theorem on p. 178, show that it suffices to prove that there are subsequences converging to $\liminf a_n$ and $\limsup a_n$.

(ii) Show that it suffices to prove that there is a subsequence converging to $\liminf a_n$.

(iii) If $\liminf a_n = \pm\infty$, prove that there is a subsequence converging to $\liminf a_n$.

(iv) Now assume that $\liminf a_n = \lim_{n\to\infty}\left(\inf\{a_n, a_{n+1}, \ldots\}\right) \in \mathbb{R}$. By the definition of limit, show that there is an n such that $a - 1 < \inf\{a_n, a_{n+1}, \ldots\}$ $< a + 1$. Show that we can choose an n_1 such that $a - 1 < a_{n_1} < a + 1$. Then show there is an $n_2 > n_1$ such that $a - \frac{1}{2} < a_{n_2} < a + \frac{1}{2}$. Continue this process.

6.3 A Potpourri of Ratio-Type Tests and "Big $\mathcal{O}$" Notation

The ratio and root tests are indeed very powerful tests, but they sometimes fail to determine convergence or divergence. For example, in the previous section we saw that the ratio and root tests failed for the series

$$1 + \sum_{n=1}^{\infty} \frac{1 \cdot 3 \cdot 5 \cdots (2n-1)}{2 \cdot 4 \cdot 6 \cdots (2n)\,(2n+1)}.$$

In this section we'll develop new technologies that are able to detect the convergence of this and other series for which the ratio and root tests fail to give information.

6.3.1 Kummer's Test

Kummer's test, named after Ernst Kummer (1810–1893), is really the comparison test in disguised form (see [208]), but it is incredibly powerful, since it can used to derive a potpourri of ratio-type tests that work when the ratio test fails. To motivate Kummer's test, let $\{a_n\}$ be a sequence of positive numbers. Then from the "telescoping comparison test" (see Problem 8 on p. 214) we know that $\sum a_n$ converges if and only if there exist a constant $c > 0$ and a sequence $\{x_n\}$ of positive numbers such that for all n sufficiently large,

$$a_n \leq c\,(x_n - x_{n+1}).$$

It was Kummer's insight to realize that we can always write $x_n = a_n\,b_n$ for some $b_n > 0$ (namely $b_n = x_n/a_n$). Thus, $\sum a_n$ converges if and only if there exist a constant $c > 0$ and a sequence $\{b_n\}$ of positive numbers such that for all n sufficiently large, $a_n \leq c\,(a_n\,b_n - a_{n+1}\,b_{n+1})$, or after division by a_n,

$$\frac{1}{c} \leq b_n - \frac{a_{n+1}}{a_n}\,b_{n+1}. \tag{6.15}$$

This is equivalent to

$$\liminf \left(b_n - \frac{a_{n+1}}{a_n} b_{n+1} \right) > 0.$$

Indeed, this is certainly implied by (6.15). On the other hand, assuming that the limit infimum is positive, by Property *4 (d)* of Theorem 6.8, given $a > 0$ less than this limit infimum, we have (6.15) with $c = 1/a$. Thus, we have proved Part *(i)* of the following theorem.[5]

Kummer's test

Theorem 6.14 *Let $\{a_n\}$ be a sequence of positive numbers. Then*

(i) $\sum a_n$ converges if there is a sequence $\{b_n\}$ of positive numbers such that

$$\liminf \left(b_n - \frac{a_{n+1}}{a_n} b_{n+1} \right) > 0.$$

(ii) $\sum a_n$ diverges if there is a sequence $\{b_n\}$ of positive numbers such that $\sum 1/b_n$ diverges, and

$$\limsup \left(b_n - \frac{a_{n+1}}{a_n} b_{n+1} \right) < 0.$$

Proof Suppose there is a positive sequence $\{b_n\}$ with $\limsup(b_n - (a_{n+1}/a_n) b_{n+1}) < 0$. Then by Property *4 (a)* of Theorem 6.8, there is an N such that

$$n > N \quad \Longrightarrow \quad b_n - \frac{a_{n+1}}{a_n} b_{n+1} < 0, \quad \text{that is,} \quad a_n b_n < a_{n+1} b_{n+1}.$$

Thus, for $n > N$, $a_n b_n$ is strictly increasing. In particular, fixing $m > N$, for all $n > m$, we have $C < a_n b_n$, where $C = a_m b_m$ is a constant independent of n. Thus, for all $n > m$, we have $C/b_n < a_n$, and since the sum $\sum 1/b_n$ diverges, the comparison test implies that $\sum a_n$ diverges too. ∎

Note that d'Alembert's ratio test is just Kummer's test with $b_n = 1$ for each n.

6.3.2   Raabe's Test and "Big $\mathcal{O}$" Notation

The following test, attributed to Joseph Ludwig Raabe (1801–1859), follows from Kummer's test with $b_n = n - 1$, as the reader can check.

---

[5]Parts *(i)* and *(ii)* are usually stated with limit infimums and supremums of the sequence $b_n a_n/a_{n+1} - b_{n+1}$. However, I like using $b_n - (a_{n+1}/a_n) b_{n+1}$, because it fits perfectly with the motivation and hence is easy to remember and derive.

Raabe's test

> **Theorem 6.15** *A series $\sum a_n$ of positive terms converges or diverges according as*
> $$\liminf \; n\left(1 - \frac{a_{n+1}}{a_n}\right) > 1 \quad or \quad \limsup \; n\left(1 - \frac{a_{n+1}}{a_n}\right) < 1.$$

In order to effectively apply Raabe's test, it is useful to first introduce some very handy notation. For a nonnegative function g, when we write $f = \mathcal{O}(g)$ ("**big O**" of g), we simply mean that $|f| \leq Cg$ for some constant C. In words,

> $\mathcal{O}$ means "... is, in absolute value, less than or equal to a constant times ...."

This big $\mathcal{O}$ notation was introduced by Paul Bachmann (1837–1920) but became well known through Edmund Landau (1877–1938) [257].

Example 6.20 For $x \geq 0$, we have
$$\frac{x^2}{1+x} = \mathcal{O}(x^2),$$
because $x^2/(1+x) \leq x^2$ for $x \geq 0$. Thus, for $x \geq 0$,
$$\frac{1}{1+x} = 1 - x + \frac{x^2}{1+x} \quad \Longrightarrow \quad \frac{1}{1+x} = 1 - x + \mathcal{O}(x^2). \tag{6.16}$$

In this section, we are mostly interested in using the big $\mathcal{O}$ notation in dealing with natural numbers.

Example 6.21 For $n \in \mathbb{N}$,
$$\frac{5}{2n} + \frac{3}{2n^2} = \mathcal{O}\left(\frac{1}{n}\right), \tag{6.17}$$
because $\frac{5}{2n} + \frac{3}{2n^2} \leq \frac{5}{2n} + \frac{3}{2n} = \frac{C}{n}$, where $C = 4$.

Three important properties of the big $\mathcal{O}$ notation are as follows:

(1) If $f = \mathcal{O}(ag)$ with $a \geq 0$, then $f = \mathcal{O}(g)$.

   If $f_1 = \mathcal{O}(g_1)$ and $f_2 = \mathcal{O}(g_2)$, then

(2) $f_1 f_2 = \mathcal{O}(g_1 g_2)$;
(3) $f_1 + f_2 = \mathcal{O}(g_1 + g_2)$.

To prove these properties, observe that if $|f| \leq C(ag)$, then $|f| \leq C'g$, where $C' = aC$, and that $|f_1| \leq C_1 g_1$ and $|f_2| \leq C_2 g_2$ imply
$$|f_1 f_2| \leq \left(C_1 C_2\right) g_1 g_2 \quad \text{and} \quad |f_1 + f_2| \leq (C_1 + C_2)(g_1 + g_2),$$

whence our three properties.

Example 6.22 In view of (6.17), we have $\mathcal{O}\left[\left(\frac{5}{2n}+\frac{3}{2n^2}\right)^2\right] = \mathcal{O}\left(\frac{1}{n}\cdot\frac{1}{n}\right) = \mathcal{O}\left(\frac{1}{n^2}\right)$.
Therefore, using (the right-hand part of) (6.16), we obtain

$$\frac{1}{1+\left(\dfrac{5}{2n}+\dfrac{3}{2n^2}\right)} = 1 - \frac{5}{2n} - \frac{3}{2n^2} + \mathcal{O}\left[\left(\frac{5}{2n}+\frac{3}{2n^2}\right)^2\right]$$

$$= 1 - \frac{5}{2n} + \mathcal{O}\left(\frac{1}{n^2}\right) + \mathcal{O}\left(\frac{1}{n^2}\right)$$

$$= 1 - \frac{5}{2n} + \mathcal{O}\left(\frac{1}{n^2}\right),$$

since $\mathcal{O}(2/n^2) = \mathcal{O}(1/n^2)$.

Here we can see the very "big" advantage of using the big $\mathcal{O}$ notation: it hides a lot of complicated junk information. For example, the left-hand side of the equation in Example 6.22 is exactly equal to (see the left-hand part of (6.16))

$$\frac{1}{1+\left(\dfrac{5}{2n}+\dfrac{3}{2n^2}\right)} = 1 - \frac{5}{2n} + \left[-\frac{3}{2n^2} + \frac{\left(\frac{5}{2n}+\frac{3}{2n^2}\right)^2}{1+\frac{5}{2n}+\frac{3}{2n^2}}\right].$$

So, the big $\mathcal{O}$ notation allows us to summarize the complicated material on the right as the very simple $\mathcal{O}\left(\frac{1}{n^2}\right)$.

Example 6.23 Consider our "mystery" series

$$1 + \sum_{n=1}^{\infty} \frac{1\cdot 3\cdot 5\cdots(2n-1)}{2\cdot 4\cdot 6\cdots(2n)\,(2n+1)},$$

already considered in (6.11). We saw that the ratio and root tests failed for this series; however, it turns out that Raabe's test works. To see this, let a_n denote the nth term in the "mystery" series. Then from the ratio (6.12) found on p. 445, we see that

$$\frac{a_{n+1}}{a_n} = \frac{1+\dfrac{1}{n}+\dfrac{1}{4n^2}}{1+\dfrac{5}{2n}+\dfrac{3}{2n^2}} = \left(1+\frac{1}{n}+\frac{1}{4n^2}\right)\left(1-\frac{5}{2n}+\mathcal{O}\left(\frac{1}{n^2}\right)\right)$$

$$= \left(1+\frac{1}{n}+\mathcal{O}\left(\frac{1}{n^2}\right)\right)\left(1-\frac{5}{2n}+\mathcal{O}\left(\frac{1}{n^2}\right)\right).$$

Multiplying out the right-hand side, using the properties of big $\mathcal{O}$, we get

$$\frac{a_{n+1}}{a_n} = 1 + \frac{1}{n} - \frac{5}{2n} + \mathcal{O}\left(\frac{1}{n^2}\right),$$

or

$$\frac{a_{n+1}}{a_n} = 1 - \frac{3}{2n} + \mathcal{O}\left(\frac{1}{n^2}\right).$$

Hence,

$$n\left(1 - \frac{a_{n+1}}{a_n}\right) = \frac{3}{2} + \mathcal{O}\left(\frac{1}{n}\right) \implies \lim n\left(1 - \frac{a_{n+1}}{a_n}\right) = \frac{3}{2} > 1,$$

so by Raabe's test, the "mystery" sum converges.[6]

6.3.3   De Morgan and Bertrand's Test

We next study a test named after Augustus De Morgan (1806–1871) and Joseph Bertrand (1822–1900). For this test, we let $b_n = (n-1)\log(n-1)$ in Kummer's test.

De Morgan and Bertrand's test

> **Theorem 6.16** *Let $\{a_n\}$ be a sequence of positive numbers and define α_n by the equation*
>
> $$\frac{a_{n+1}}{a_n} = 1 - \frac{1}{n} - \frac{\alpha_n}{n\log n}.$$
>
> *Then $\sum a_n$ converges or diverges according as $\liminf \alpha_n > 1$ or $\limsup \alpha_n < 1$.*

Proof If we let $b_n = (n-1)\log(n-1)$ in Kummer's test, then

$$\kappa_n = b_n - \frac{a_{n+1}}{a_n}b_{n+1} = (n-1)\log(n-1) - \left(1 - \frac{1}{n} - \frac{\alpha_n}{n\log n}\right)n\log n$$

$$= \alpha_n + n\Big[\log(n-1) - \log n\Big] + \Big[\log n - \log(n-1)\Big].$$

Since

$$(n+1)\Big[\log n - \log(n+1)\Big] = \log\left(1 - \frac{1}{n+1}\right)^{n+1} \to \log\left(e^{-1}\right) = -1$$

---

[6]It turns out that the "mystery" sum equals $\pi/2$; see [146] for a proof.

and

$$\log n - \log(n-1) = \log \frac{n}{n-1} \to \log 1 = 0,$$

we have

$$\liminf \kappa_n = \liminf \alpha_n - 1 \quad \text{and} \quad \limsup \kappa_n = \limsup \alpha_n - 1.$$

Invoking Kummer's test now completes the proof.   ∎

6.3.4 Gauss's Test

We end our potpourri of tests with Gauss's test.

Gauss's test

Theorem 6.17 *Let $\{a_n\}$ be a sequence of positive numbers and suppose that we can write*

$$\frac{a_{n+1}}{a_n} = 1 - \frac{\xi}{n} + \mathcal{O}\left(\frac{1}{n^p}\right),$$

where ξ is a constant and $p > 1$. Then $\sum a_n$ converges or diverges according as $\xi \leq 1$ or $\xi > 1$.

Proof The hypotheses imply that

$$n\left(1 - \frac{a_{n+1}}{a_n}\right) = \xi + n\,\mathcal{O}\left(\frac{1}{n^p}\right) = \xi + \mathcal{O}\left(\frac{1}{n^{p-1}}\right) \to \xi$$

as $n \to \infty$, where we used that $p - 1 > 0$. Thus, Raabe's test shows that the series $\sum a_n$ converges for $\xi > 1$ and diverges for $\xi < 1$. For the case $\xi = 1$, write

$$\frac{a_{n+1}}{a_n} = 1 - \frac{1}{n} - f_n,$$

where $\{f_n\}$ is a sequence such that $f_n = \mathcal{O}\left(\frac{1}{n^p}\right)$. Then we can further rewrite

$$\frac{a_{n+1}}{a_n} = 1 - \frac{1}{n} - \frac{\alpha_n}{n \log n},$$

where $\alpha_n = f_n\, n \log n$. If we let $p = 1 + \delta$, where $\delta > 0$. Then

$$\alpha_n = f_n\, n \log n = \mathcal{O}\left(\frac{1}{n^{1+\delta}}\right) n \log n = \mathcal{O}\left(\frac{\log n}{n^\delta}\right).$$

By Problem 9 on p. 312, we know that $(\log n)/n^{\delta} \to 0$ as $n \to \infty$, so $\lim \alpha_n = 0$. Thus, De Morgan and Bertrand's test shows that the series $\sum a_n$ diverges.   ■

Example 6.24 Gauss's test originated with Gauss's study of the hypergeometric series:

$$1 + \frac{\alpha \cdot \beta}{1 \cdot \gamma} + \frac{\alpha(\alpha - 1) \cdot \beta(\beta - 1)}{2! \cdot \gamma(\gamma + 1)} + \frac{\alpha(\alpha - 1)(\alpha - 2) \cdot \beta(\beta - 1)(\beta - 2)}{3! \cdot \gamma(\gamma + 1)(\gamma + 2)} + \cdots ,$$

where α, β, γ are positive real numbers. We can write this as $\sum a_n$, where

$$a_n = \frac{\alpha(\alpha - 1)(\alpha - 2) \cdots (\alpha - n + 1) \cdot \beta(\beta - 1)(\beta - 2) \cdot (\beta - n + 1)}{n! \cdot \gamma(\gamma + 1)(\gamma + 2) \cdots (\gamma + n - 1)}.$$

Hence, for $n \geq 1$ we have

$$\frac{a_{n+1}}{a_n} = \frac{(\alpha + n)(\beta + n)}{(n + 1)(\gamma + n)} = \frac{n^2 + (\alpha + \beta)n + \alpha\beta}{n^2 + (\gamma + 1)n + \gamma} = \frac{1 + \dfrac{\alpha + \beta}{n} + \dfrac{\alpha\beta}{n^2}}{1 + \dfrac{\gamma + 1}{n} + \dfrac{\gamma}{n^2}}.$$

Using the handy formula (6.16) on p. 451,

$$\frac{1}{1 + x} = 1 - x + \frac{x^2}{1 + x},$$

we see that (after some algebra)

$$\frac{a_{n+1}}{a_n} = \left(1 + \frac{\alpha + \beta}{n} + \frac{\alpha\beta}{n^2}\right)\left[1 - \frac{\gamma + 1}{n} - \frac{\gamma}{n^2} + \mathcal{O}\left(\frac{1}{n^2}\right)\right]$$

$$= 1 - \frac{\gamma + 1 - \alpha - \beta}{n} + \mathcal{O}\left(\frac{1}{n^2}\right).$$

Thus, the hypergeometric series converges if $\gamma > \alpha + \beta$ and diverges if $\gamma \leq \alpha + \beta$.

► **Exercises 6.3**

1. Determine whether each of the following series converges.

(a) $\displaystyle\sum_{n=1}^{\infty} \frac{1 \cdot 3 \cdot 5 \cdots (2n - 1)}{2^n(n + 1)!}$, (b) $\displaystyle\sum_{n=1}^{\infty} \frac{3 \cdot 6 \cdot 9 \cdots (3n)}{7 \cdot 10 \cdot 13 \cdots (3n + 4)}$,

(c) $\displaystyle\sum_{n=1}^{\infty} \frac{1 \cdot 3 \cdot 5 \cdots (2n - 1)}{2 \cdot 4 \cdot 6 \cdots (2n)}$, (d) $\displaystyle\sum_{n=1}^{\infty} \frac{2 \cdot 4 \cdot 6 \cdots (2n + 2)}{1 \cdot 3 \cdot 5 \cdots (2n - 1)(2n)}$.

For $\alpha, \beta \neq 0, -1, -2, \ldots,$

$$(e) \sum_{n=1}^{\infty} \frac{\alpha(\alpha+1)(\alpha+2)\cdots(\alpha+n-1)}{n!},$$

$$(f) \sum_{n=1}^{\infty} \frac{\alpha(\alpha+1)(\alpha+2)\cdots(\alpha+n-1)}{\beta(\beta+1)(\beta+2)\cdots(\beta+n-1)}.$$

If $\alpha, \beta, \gamma, \kappa, \lambda \neq 0, -1, -2, \ldots$, then prove that the following monster series

$$(g) \sum_{n=1}^{\infty} \frac{\alpha(\alpha+1)\cdots(\alpha+n-1)\beta(\beta+1)\cdots(\beta+n-1)\gamma(\gamma+1)\cdots(\gamma+n-1)}{n!\,\kappa(\kappa+1)\cdots(\kappa+n-1)\lambda(\lambda+1)\cdots(\lambda+n-1)}$$

converges for $\kappa + \lambda - \alpha - \beta - \gamma > 0$.

2. Using Raabe's test, prove that $\sum 1/n^p$ converges for $p > 1$ and diverges for $p < 1$.

3. (**Logarithmic test**) We prove a useful test called the **logarithmic test**: If $\sum a_n$ is a series of positive terms, then this series converges or diverges according as

$$\liminf \left(n \log \frac{a_n}{a_{n+1}} \right) > 1 \quad \text{or} \quad \limsup \left(n \log \frac{a_n}{a_{n+1}} \right) < 1.$$

To prove this, proceed as follows.

(i) Suppose first that $\liminf \left(n \log \frac{a_n}{a_{n+1}} \right) > 1$. Show that there exist $a > 1$ and N such that

$$n > N \implies a < n \log \frac{a_n}{a_{n+1}} \implies \frac{a_{n+1}}{a_n} < e^{-a/n}.$$

(ii) Using $\left(1 + \frac{1}{n} \right)^n < e$ from (3.26) on p. 180, the p-test, and the limit comparison test (see Problem 9 on p. 215), prove that $\sum a_n$ converges.

(iii) Similarly, prove that if $\limsup \left(n \log \frac{a_n}{a_{n+1}} \right) < 1$, then $\sum a_n$ diverges.

(iv) Using the logarithmic test, determine the convergence/divergence of

$$\sum_{n=1}^{\infty} \frac{n!}{n^n} \quad \text{and} \quad \sum_{n=1}^{\infty} \frac{n^n}{n!}.$$

6.4 Pretty Powerful Properties of Power Series

The title of this section speaks for itself! As stated already, we focus on power series of a complex variable z, but all the results stated in this section have corresponding statements for power series of a real variable x.

6.4.1 Continuity and the Exponential Function (Again)

We first prove that power series are always continuous (within their radius of convergence).

> **Lemma 6.18** *If a power series $\sum_{n=0}^{\infty} a_n z^n$ has radius of convergence R, then the series $\sum_{n=1}^{\infty} n\, a_n z^{n-1}$ also has radius of convergence R.*

Proof (See Problem 3 for a proof of this lemma using the radius of convergence formula (6.13) on p. 446; we shall give a more hands-on proof.) For $z \neq 0$, $\sum_{n=1}^{\infty} n\, a_n z^{n-1}$ converges if and only if $z \cdot \sum_{n=1}^{\infty} n\, a_n z^{n-1} = \sum_{n=1}^{\infty} n\, a_n z^n$ converges, so we just have to show that $\sum_{n=1}^{\infty} n\, a_n z^n$ has radius of convergence R. Since $|a_n| \leq n|a_n|$, if $\sum_{n=1}^{\infty} n\, |a_n|\, |z|^n$ converges, then $\sum_{n=1}^{\infty} |a_n|\, |z|^n$ also converges by comparison. Hence, the radius of convergence of the series $\sum_{n=1}^{\infty} n\, a_n z^n$ can't be larger than R. To prove that the radius of convergence is at least R, fix z with $|z| < R$; we need to prove that $\sum_{n=1}^{\infty} n\, |a_n|\, |z|^n$ converges. To this end, fix ρ with $|z| < \rho < R$ and note that $\sum_{n=1}^{\infty} n\, (|z|/\rho)^n$ converges, by, e.g., the root test:

$$\lim \left| n \left(\frac{|z|}{\rho} \right)^n \right|^{1/n} = \lim \left(n^{1/n} \cdot \frac{|z|}{\rho} \right) = \frac{|z|}{\rho} < 1.$$

Hence, by the nth term test, $n\, (|z|/\rho)^n \to 0$ as $n \to \infty$. In particular, $n\, (|z|/\rho)^n \leq M$ for some constant M. Next, observe that

$$n\, |a_n|\, |z|^n = n\, |a_n|\, \rho^n \cdot \left(\frac{|z|}{\rho} \right)^n \leq M \cdot |a_n|\, \rho^n.$$

Since $\sum_{n=1}^{\infty} |a_n|\, \rho^n$ converges (because $\rho < R$, the radius of convergence of the series $\sum_{n=0}^{\infty} a_n z^n$), by the comparison test it follows that $\sum n\, |a_n|\, |z|^n$ also converges. This completes our proof. ∎

Continuity theorem for power series

> **Theorem 6.19** *A power series is continuous within its radius of convergence.*

Proof Let $f(z) = \sum_{n=0}^{\infty} a_n z^n$ have radius of convergence R; we need to show that $f(z)$ is continuous at each point $c \in \mathbb{C}$ with $|c| < R$. So, let us fix such a c. Since

$$z^n - c^n = (z - c)\, q_n(z), \quad \text{where } q_n(z) = z^{n-1} + z^{n-2} c + \cdots + z\, c^{n-2} + c^{n-1},$$

which is proved by multiplying out $(z - c)\, q_n(z)$, we can write

$$f(z) - f(c) = \sum_{n=1}^{\infty} a_n(z^n - c^n) = (z - c) \sum_{n=0}^{\infty} a_n q_n(z).$$

To bound the sum $\sum_{n=0}^{\infty} a_n q_n(z)$ we proceed as follows. Fix r such that $|c| < r < R$. Then for $|z - c| < r - |c|$, we have

$$|z| \leq |z - c| + |c| < r - |c| + |c| = r.$$

Thus, since $|c| < r$, for $|z - c| < r - |c|$, we see that

$$|q_n(z)| \leq \underbrace{r^{n-1} + r^{n-2}r + \cdots + r\,r^{n-2} + r^{n-1}}_{n \text{ terms}} = nr^{n-1}.$$

By our lemma, $\sum_{n=1}^{\infty} n\,|a_n|\,r^{n-1}$ converges, so if $C := \sum_{n=1}^{\infty} n\,|a_n|\,r^{n-1}$, then

$$|f(z) - f(c)| \leq |z - c| \sum_{n=1}^{\infty} |a_n|\,|q_n(z)| \leq |z - c| \sum_{n=1}^{\infty} |a_n|\,nr^{n-1} = C|z - c|.$$

This implies that $\lim_{z \to c} f(z) = f(c)$; that is, f is continuous at $z = c$.   ■

6.4.2   Abel's Limit Theorem

Abel's limit theorem has to do with the following question. Let $f(x) = \sum_{n=0}^{\infty} a_n x^n$ have radius of convergence R; this implies, in particular, that $f(x)$ is defined for all $-R < x < R$ and, by Theorem 6.19, is continuous on the interval $(-R, R)$. Let us suppose that $f(R) = \sum_{n=0}^{\infty} a_n R^n$ converges. In particular, $f(x)$ is defined for all $-R < x \leq R$. Question: Is f continuous on the interval $(-R, R]$, that is, is it true that

$$\lim_{x \to R-} f(x) = f(R)? \tag{6.18}$$

The answer to this question is yes, and it follows from Niels Henrik Abel's limit theorem below (Theorem 6.21). The version that we state, which involves some beautiful geometry, is due to Otto Stolz (1842–1905). For a nonzero complex number z_0, the following figure defines a "Stolz angle" (Fig. 6.4)

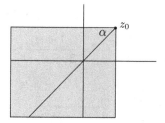

Fig. 6.4 A **Stolz angle** with vertex z_0 is an angular sector with total angle $< \pi$ and vertex at z_0 that is symmetric about, and contains, the ray starting at z_0 passing through the origin. The angle α shown is half the total angle, so $0 \le \alpha < \pi/2$

Of course, in this figure we drew only part of the Stolz angle; the sector is of infinite extent (it continues forever to the west, south, and southwest). Here's an important property of a Stolz angle.

Lemma 6.20 *For every Stolz angle with vertex $z_0 \ne 0$, there are constants $\rho, C > 0$ such that for all z in the Stolz angle with $|z - z_0| < \rho$, we have*

$$\frac{|z_0 - z|}{|z_0| - |z|} \le C.$$

Proof By Problem 7, it's enough to prove this lemma for $z_0 = 1$. Thus, we henceforth assume that $z_0 = 1$; in this case, the Stolz angle looks like

We need to find constants $\rho, C > 0$ such that for all z in the Stolz angle with $|1 - z| < \rho$, we have

$$\frac{|1 - z|}{1 - |z|} \le C.$$

The proof is easy if we write a complex number z as $z = 1 - r\,e^{i\theta}$, where $r = |1 - z|$ is the distance between z and 1, and θ is the angle z makes with the *negative* horizontal. When we write complex numbers in this way, the Stolz angle with vertex at 1 and total angle 2α is given by

$$\{1 - r\,e^{i\theta} ; r \ge 0 \text{ and } -\alpha \le \theta \le \alpha\}.$$

(Do you see why the Stolz angle has this form?) Here are some not-so-difficult-to-prove facts, which we leave you to verify (see Fig. 6.5): For $z = 1 - re^{i\theta}$,

(1) $|z| < 1$ if and only if $r < 2\cos\theta$, and $|z| = 1$ if and only if $r = 2\cos\theta$.

(2) We have

$$\frac{|1 - z|}{1 - |z|} = \frac{1 + |z|}{2\cos\theta - r}.$$

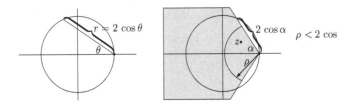

Fig. 6.5 *Left* $z = 1 - r\,e^{i\theta}$ is on the unit circle if and only if $r = 2\cos\theta$. *Right* Here, $2\cos\alpha$ is the length of the segment from 1 to the point where the Stolz angle intersects the unit circle (this follows from the description of the unit circle on the *left*)

The proofs of (1) and (2) just use the form $z = 1 - r\,e^{i\theta} = (1 - r\cos\theta) - ir\sin\theta$ and some algebra. Fix ρ with $0 < \rho < 2\cos\alpha$, as seen in the right picture in Fig. 6.5, and let z be a complex number inside the Stolz angle such that $|1 - z| < \rho$; such a z is shown on the right in Fig. 6.5. Then we can write $z = 1 - r\,e^{i\theta}$, where $-\alpha \le \theta \le \alpha$ and $r < \rho$. It follows that $2\cos\theta - r \ge 2\cos\alpha - \rho$. Since $\rho < 2\cos\alpha$, we have $2\cos\theta - r > 0$, so by (1), $|z| < 1$. Hence by (2), we have

$$\frac{|1 - z|}{1 - |z|} = \frac{1 + |z|}{2\cos\theta - r} < \frac{1 + 1}{2\cos\alpha - \rho} = C,$$

where $C = 2/(2\cos\alpha - \rho)$. This completes our proof. ∎

We can now prove the famous Abel's limit theorem, which in particular implies the result (6.18).

Abel's limit theorem

Theorem 6.21 *Let $f(z) = \sum_{n=0}^{\infty} a_n z^n$ have radius of convergence R and let $z_0 \in \mathbb{C}$ with $|z_0| = R$, where the series $f(z_0) = \sum_{n=0}^{\infty} a_n z_0^n$ converges. Then*

$$\lim_{z \to z_0} f(z) = f(z_0),$$

where $z \to z_0$ on the left inside any given Stolz angle with vertex z_0.

Proof As in the previous proof, we may assume that[7] $z_0 = 1$. Moreover, by adding the constant $-f(1)$ to $f(z)$ if necessary, we also assume that $f(1) = 0$. With

---

[7] Or just prove this theorem for the function $g(z) = f(z_0 z)$ as $z \to 1$ to get the result for f.

these assumptions, if we put $s_n = a_0 + a_1 + \cdots + a_n$, then $0 = f(1) = \sum_{n=0}^{\infty} a_n = \lim s_n$. Let z be inside the unit circle. By Abel's lemma (see Corollary 6.2 on p. 421),

$$\sum_{k=0}^{n} a_k z^k = s_n z^n - \sum_{k=0}^{n-1} s_k (z^{k+1} - z^k),$$

that is,

$$\sum_{k=0}^{n} a_k z^k = (1 - z) \sum_{k=0}^{n} s_k z^k + s_n z^n.$$

Since $s_n \to 0$ and $|z| < 1$, $s_n z^n \to 0$. Therefore, taking $n \to \infty$, we obtain

$$f(z) = \sum_{n=0}^{\infty} a_n z^n = (1 - z) \sum_{n=0}^{\infty} s_n z^n,$$

which implies that

$$|f(z)| \le |1 - z| \sum_{n=0}^{\infty} |s_n| |z|^n.$$

We now fix a Stolz angle with vertex 1. By the previous lemma, there are $\rho, C > 0$ such that for all z in the Stolz angle with $|z - 1| < \rho$, we have $|1 - z|/(1 - |z|) \le C$. Let $\varepsilon > 0$. Since $s_n \to 0$, we can choose $N \in \mathbb{N}$ such that $n > N \implies |s_n| < \varepsilon/(2C)$. Put $K = \sum_{n=0}^{N} |s_n|$. Observe that for all z in the Stolz angle with $|z| < 1$ and $|z - 1| < \rho$, we have

$$|f(z)| = |1 - z| \sum_{n=0}^{N} |s_n| |z|^n + |1 - z| \sum_{n=N+1}^{\infty} |s_n| |z|^n$$

$$\le |1 - z| \sum_{n=0}^{N} |s_n| \cdot 1^n + |1 - z| \sum_{n=N+1}^{\infty} \frac{\varepsilon}{2C} |z|^n$$

$$\le K|1 - z| + \frac{\varepsilon}{2C} |1 - z| \sum_{n=0}^{\infty} |z|^n$$

$$= K|1 - z| + \frac{\varepsilon}{2C} \frac{|1 - z|}{1 - |z|} \le K|1 - z| + \frac{\varepsilon}{2}.$$

Thus, with $\delta := \varepsilon/(2K)$, we have

$$|z - 1| < \delta \text{ with } |z| < 1 \text{ in a Stolz angle with vertex 1} \implies |f(z)| < \varepsilon.$$

This completes our proof. ∎

6.4.3   The Identity Theorem

The identity theorem is a very useful property of power series. It says that if two power series are identical at "sufficiently many" points, then in fact, the power series are identical everywhere! Here's the idea:

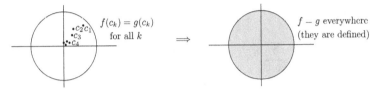

Identity theorem

Theorem 6.22 *Let* $f(z) = \sum a_n z^n$ *and* $g(z) = \sum b_n z^n$ *have positive radii of convergence (not a priori the same) and suppose that* $f(c_k) = g(c_k)$ *for some nonzero sequence* $c_k \to 0$. *Then the power series* $f(z)$ *and* $g(z)$ *are actually identical; that is* $a_n = b_n$ *for every* $n = 0, 1, 2, 3, \ldots$.

Proof We begin by proving that for each $m = 0, 1, 2, \ldots$, the series

$$f_m(z) := \sum_{n=m}^{\infty} a_n z^{n-m} = a_m + a_{m+1} z + a_{m+2} z^2 + a_{m+3} z^3 + \cdots$$

has the same radius of convergence as f. Indeed, since we can write $f_m(z) = z^{-m} \sum_{n=m}^{\infty} a_n z^n$ for $z \neq 0$, the power series $f_m(z)$ converges if and only if $\sum_{n=m}^{\infty} a_n z^n$ converges, which in turn converges if and only if $f(z)$ converges. It follows that $f_m(z)$ and $f(z)$ have the same radius of convergence; in particular, by the continuity theorem for power series, $f_m(z)$ is continuous at 0. Similarly, for each $m = 0, 1, 2, \ldots$, $g_m(z) := \sum_{n=m}^{\infty} b_n z^{n-m}$ has the same radius of convergence as $g(z)$; in particular, $g_m(z)$ is continuous at 0. These continuity facts concerning f_m and g_m are the important facts that will be used below.

Now to our proof. We are given that

$$a_0 + a_1 c_k + a_2 c_k^2 + \cdots = b_0 + b_1 c_k + b_2 c_k^2 + \cdots, \quad \text{that is, } f(c_k) = g(c_k) \tag{6.19}$$

for all k. In particular, taking $k \to \infty$ in the equality $f(c_k) = g(c_k)$, using that $c_k \to 0$ and that f and g are continuous at 0, we obtain $f(0) = g(0)$, or $a_0 = b_0$. Canceling $a_0 = b_0$ and dividing by $c_k \neq 0$ in (6.19), we obtain

$$a_1 + a_2 c_k + a_3 c_k^2 + \cdots = b_1 + b_2 c_k + b_3 c_k^2 + \cdots \quad \text{that is, } f_1(c_k) = g_1(c_k) \tag{6.20}$$

for all k. Taking $k \to \infty$ and using that $c_k \to 0$ and that f_1 and g_1 are continuous at 0, we obtain $f_1(0) = g_1(0)$, or $a_1 = b_1$. Canceling $a_1 = b_1$ and dividing by $c_k \neq 0$ in (6.20), we obtain

$$a_2 + a_3 c_k + a_4 c_k^2 + \cdots = b_2 + b_3 c_k + b_4 c_k^2 + \cdots \quad \text{that is,} \quad f_2(c_k) = g_2(c_k)$$

for all k. Taking $k \to \infty$, using that $c_k \to 0$ and that f_2 and g_2 are continuous at 0, we obtain $f_2(0) = g_2(0)$, or $a_2 = b_2$. Continuing by induction, we get $a_n = b_n$ for all $n = 0, 1, 2, \ldots$, which is exactly what we wanted to prove. ∎

Corollary 6.23 *If $f(z) = \sum a_n z^n$ and $g(z) = \sum b_n z^n$ have positive radii of convergence and $f(x) = g(x)$ for all $x \in \mathbb{R}$ with $|x| < \varepsilon$ for some $\varepsilon > 0$, then $a_n = b_n$ for every n; that is, f and g are actually the same power series.*

Proof To prove this, observe that since $f(x) = g(x)$ for all $x \in \mathbb{R}$ such that $|x| < \varepsilon$, then $f(c_k) = g(c_k)$ for all k sufficiently large, where $c_k = 1/k$; the identity theorem now implies $a_n = b_n$ for every n. ∎

Using the identity theorem, we can deduce certain properties of series.

Example 6.25 Suppose that $f(z) = \sum a_n z^n$ is an **odd function** in the sense that $f(-z) = -f(z)$ for all z within its radius of convergence. In terms of power series, the identity $f(-z) = -f(z)$ is

$$\sum a_n (-1)^n z^n = \sum -a_n z^n.$$

By the identity theorem, we must have $(-1)^n a_n = -a_n$ for each n. Thus, for n even we must have $a_n = -a_n$ or $a_n = 0$, and for n odd, we must have $-a_n = -a_n$, a tautology. In conclusion, we see that f is odd if and only if all coefficients of even powers vanish:

$$f(z) = \sum_{n=0}^{\infty} a_{2n+1} z^{2n+1};$$

that is, f is odd if and only if f has only odd powers in its series expansion.

▶ **Exercises 6.4**

1. Prove that a power series $f(z) = \sum a_n z^n$ is an **even function**, in the sense that $f(-z) = f(z)$ for all z within its radius of convergence, if and only if f has only even powers in its expansion, that is, f takes the form $f(z) = \sum_{n=0}^{\infty} a_{2n} z^{2n}$.
2. Recall that the binomial coefficient is $\binom{n}{k} = \frac{n!}{k!(n-k)!}$ for $0 \leq k \leq n$. Prove the not-obvious-at-first-sight result

$$\binom{m+n}{k} = \sum_{j=0}^{k} \binom{m}{j} \binom{n}{k-j}.$$

Suggestion: Apply the binomial formula to $(1 + z)^{m+n}$, which equals $(1 + z)^m \cdot (1 + z)^n$. Prove that

$$\binom{2n}{n} = \sum_{k=0}^{n} \binom{n}{k}^2.$$

3. Prove Lemma 6.18 using the radius of convergence formula (6.13) on p. 446. You will need Problem 9 on p. 448.

4. (**Abel summability**) We say that a series $\sum a_n$ is **Abel summable** to L if the power series $f(x) := \sum a_n x^n$ is defined for all $x \in [0, 1)$ and $\lim_{x \to 1^-} f(x) = L$.

   (a) Prove that if $\sum a_n$ converges to $L \in \mathbb{C}$, then $\sum a_n$ is also Abel summable to L.

   (b) Derive the following amazing formulas (properly interpreted!):

   $$\boxed{\begin{aligned} 1 - 1 + 1 - 1 + 1 - 1 + - \cdots &=_a \frac{1}{2}, \\ 1 + 2 - 3 + 4 - 5 + 6 - 7 + - \cdots &=_a \frac{1}{4}, \end{aligned}}$$

   where $=_a$ means "is Abel summable to." You will need Problem 6 on p. 204.

5. We continue the previous problem on Abel summability. Let $a_0, a_1, a_2, \ldots$ be a positive nonincreasing sequence tending to zero (in particular, $\sum (-1)^{n-1} a_n$ converges by the alternating series test). Define $b_n := a_0 + a_1 + \cdots + a_n$. We shall prove the neat formula

$$b_0 - b_1 + b_2 - b_3 + b_4 - b_5 + - \cdots =_a \frac{1}{2} \sum_{n=0}^{\infty} (-1)^n a_n,$$

where $=_a$ mean "is Abel summable to."

   (i) Let $f(x) = \sum_{n=0}^{\infty} (-1)^n b_n x^n$. Prove that f has radius of convergence 1. Suggestion: Use the ratio test.

   (ii) Let

$$f_n(x) = \sum_{k=0}^{n} (-1)^k b_k x^k = a_0 - (a_0 + a_1)x + (a_0 + a_1 + a_2)x^2 - \cdots$$
$$+ (-1)^n (a_0 + a_1 + \cdots + a_n) x^n$$

   be the nth partial sum of $f(x)$. Prove that

$$f_n(x) = \frac{1}{1+x}\left(a_0 - a_1 x + a_2 x^2 - a_3 x^3 + \cdots + (-1)^n a_n x^n\right)$$
$$+ (-1)^n \frac{x^{n+1}}{1+x}\left(a_0 + a_2 + a_3 + \cdots + a_n\right).$$

(iii) Prove that[8]

$$f(x) = \frac{1}{1+x}\sum_{n=0}^{\infty}(-1)^n a_n x^n.$$

Finally, from this formula prove the desired result.

(iv) Establish the remarkable formula

$$\boxed{\; 1 - \left(1 + \frac{1}{2}\right) + \left(1 + \frac{1}{2} + \frac{1}{3}\right) - \left(1 + \frac{1}{2} + \frac{1}{3} + \frac{1}{4}\right) + - \cdots =_a \frac{1}{2}\log 2. \;}$$

6. Let $\sum a_n$ be a divergent series of positive real numbers and suppose that $f(z) = \sum a_n z^n$ has radius of convergence 1. Prove that $\lim_{x \to 1^-} f(x) = +\infty$.

7. (**More on Stolz angles**) Let $z_0 \in \mathbb{C}$ be nonzero and write $z_0 = r_0\, e^{i\theta_0}$ in polar coordinates, where $r_0 > 0$ and $\theta_0 \in \mathbb{R}$. Denote the Stolz angle as seen in Fig. 6.4 by $S(z_0, \alpha)$, where $0 \le \alpha < \pi/2$, and which we can write as

$$S(z_0, \alpha) = \{z_0 - r\, e^{i\theta}\,;\, r \ge 0 \text{ and } \theta_0 - \alpha \le \theta \le \theta_0 + \alpha\}.$$

Find a one-to-one correspondence between $S(z_0, \alpha)$ and $S(1, \alpha)$ such that if $z \in S(z_0, \alpha)$ corresponds to $w \in S(1, \alpha)$, then

$$\frac{|z_0 - z|}{|z_0| - |z|} = \frac{|1 - w|}{1 - |w|}.$$

This formula explains why we could assume $z_0 = 1$ in Lemma 6.20.

6.5 Cauchy's Double Series Theorem and A ζ-Function Identity

After studying single integrals in elementary calculus, you probably took a course in which you studied double integrals. In a similar way, now that we have a thorough background in "single infinite series," we now move to the topic of "double infinite series." The main result of this section is Cauchy's double series theorem, Theorem 6.26, which we'll use quite often in the sequel. If you did Problem 9 on p. 225, you already saw a version of Cauchy's theorem derived from Tannery's theorem

---

[8] We could prove this identity more quickly using Cauchy's double series theorem from Section 6.5.

(however, we won't assume Tannery's theorem for this section). The books [157, Chapter 3] and [40, Chapter 5,] have lots of material on double series.

6.5.1 Double Series: Basic Definitions

Recall that a complex sequence is really just a function $a : \mathbb{N} \to \mathbb{C}$, where we usually denote $a(n)$ by a_n. By analogy, we define a **double sequence** of complex numbers as a function $a : \mathbb{N} \times \mathbb{N} \longrightarrow \mathbb{C}$. We usually denote $a(m, n)$ by a_{mn} and the corresponding double sequence by $\{a_{mn}\}$.

Recall that if $\{a_n\}$ is a sequence of complex numbers, then we say that $\sum a_n$ converges if the sequence $\{s_n\}$ converges, where $s_n := \sum_{k=1}^{n} a_k$. By analogy, we define a **double series** of complex numbers as follows. Let $\{a_{mn}\}$ be a double sequence of complex numbers and let

$$s_{mn} := \sum_{i=1}^{m} \sum_{j=1}^{n} a_{ij},$$

which we call the m, n**th partial sum** of $\sum a_{mn}$. Figure 6.6 shows a picture of s_{mn}, where we imagine putting the a_{ij} in an infinite matrix and summing them within an $m \times n$ rectangle. We say that the double series $\sum a_{mn}$ **converges** if the double sequence $\{s_{mn}\}$ of partial sums converges in the following sense: There is a complex number L having the property that given $\varepsilon > 0$, there is a real number N such that

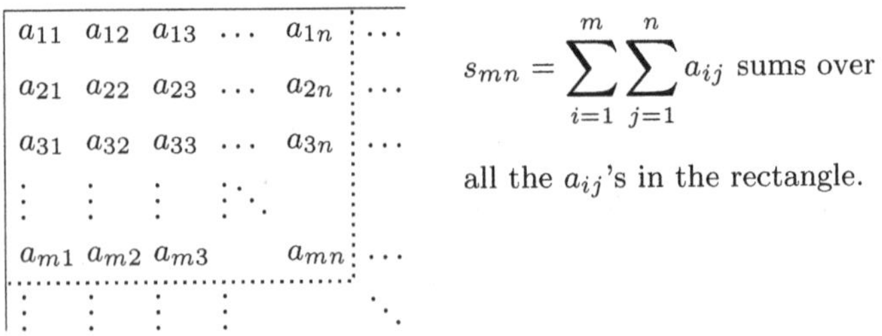

$$
\begin{array}{ccc}
a_{11} & a_{12} & a_{13} & \cdots & a_{1n} & \cdots \\
a_{21} & a_{22} & a_{23} & \cdots & a_{2n} & \cdots \\
a_{31} & a_{32} & a_{33} & \cdots & a_{3n} & \cdots \\
\vdots & \vdots & \vdots & \ddots \\
a_{m1} & a_{m2} & a_{m3} & & a_{mn} & \cdots \\
\end{array}
$$

$s_{mn} = \displaystyle\sum_{i=1}^{m} \sum_{j=1}^{n} a_{ij}$ sums over

all the a_{ij}'s in the rectangle.

Fig. 6.6 The partial sum s_{mn} equals the sum of the a_{ij} in the $m \times n$ rectangle. The sum $\sum a_{mn}$ is the limit of these sums as the rectangles become arbitrarily large; that is, as $m, n \to \infty$

$$m, n > N \implies |L - s_{mn}| < \varepsilon.$$

If this is the case, we define[9]

$$\sum a_{mn} := L.$$

---

[9]It's easy to check that L, if it exists, is unique.

Geometrically, this means that as we take larger and larger rectangles as shown in Fig. 6.6, the sums of the entries of the rectangles get closer and closer to L. If a double series does not converge, we say that it **diverges**.

Example 6.26 Let $p, q > 1$ and consider the double series $\sum 1/(m^p n^q)$. Observe that

$$s_{mn} = \sum_{i=1}^{m} \sum_{j=1}^{n} \frac{1}{i^p j^q} = \sum_{i=1}^{m} \sum_{j=1}^{n} \frac{1}{i^p} \cdot \frac{1}{j^q} = \left(\sum_{i=1}^{m} \frac{1}{i^p} \right) \left(\sum_{j=1}^{n} \frac{1}{j^q} \right) = S_m \cdot T_n,$$

where

$$S_m = \sum_{i=1}^{m} \frac{1}{i^p} \quad \text{and} \quad T_n = \sum_{j=1}^{n} \frac{1}{j^q}$$

are the mth and nth partial sums of the p and q series, respectively. Since $p, q > 1$, we know that the p and q series converge, with values $\zeta(p)$ and $\zeta(q)$, the zeta function evaluated at p and q, respectively. Thus, we have

$$\lim_{m \to \infty} S_m = \zeta(p) \quad \text{and} \quad \lim_{n \to \infty} T_n = \zeta(q).$$

Since $s_{mn} = S_m T_n$, it's not difficult to prove, and we ask you to do so in Problem 1, that $\{s_{mn}\}$ converges to $\zeta(p)\, \zeta(q)$. Thus,

$$\zeta(p)\, \zeta(q) = \sum 1/(m^p n^q).$$

Example 6.27 We can generalize the previous example as follows. Let $\sum a_n$ and $\sum b_n$ be convergent infinite series. Then in Problem 1 you will prove that the double series $\sum a_m b_n$ converges, and

$$\boxed{\left(\sum a_m \right) \left(\sum b_n \right) = \sum a_m b_n.}$$

Note that by the double series $\sum a_m b_n$ on the right, we mean the double series $\sum a_{mn}$, where $a_{mn} = a_m b_n$ for each m, n.

6.5.2 Iterated Double Series and Pringsheim's Theorem

You might recall from calculus that immediately after discussing "double integrals," you studied "iterated integrals." We can do the same for double series and study "iterated series." Thus, we can study the series

$$\sum_{m=1}^{\infty}\sum_{n=1}^{\infty} a_{mn} \quad \text{and} \quad \sum_{n=1}^{\infty}\sum_{m=1}^{\infty} a_{mn} . \tag{6.21}$$

The **iterated series**, or **summations**, here are defined by performing the *inside* infinite summation first, then the *outside* summation next. To explain this carefully, consider the left-hand sum in (6.21). First, *for every* m, we require the infinite series

$$\sum_{n=1}^{\infty} a_{mn} = a_{m1} + a_{m2} + a_{m3} + a_{m4} + \cdots \quad \text{to converge.}$$

This sum depends on m, so let's give it a name, say r_m. Geometrically, for each m, the number r_m is the sum of the a_{mn} in the mth *row*, as seen in the left-hand picture here:

a_{11}	a_{12}	a_{13}	a_{14}	$\cdots$	r_1
a_{21}	a_{22}	a_{23}	a_{24}	$\cdots$	r_2
a_{31}	a_{32}	a_{33}	a_{34}	$\cdots$	r_3
a_{41}	a_{42}	a_{43}	a_{44}	$\cdots$	r_4
$\vdots$	$\vdots$	$\vdots$	$\vdots$	$\ddots$	

a_{11}	a_{12}	a_{13}	a_{14}	$\cdots$
a_{21}	a_{22}	a_{23}	a_{24}	$\cdots$
a_{31}	a_{32}	a_{33}	a_{34}	$\cdots$
a_{41}	a_{42}	a_{43}	a_{44}	$\cdots$
$\vdots$	$\vdots$	$\vdots$	$\vdots$	$\ddots$
c_1	c_2	c_3	c_4	

Fig. 6.7 In the first array we are "summing by rows," and in the second array we are "summing by columns"

We then declare

$$\sum_{m=1}^{\infty}\sum_{n=1}^{\infty} a_{mn} := \sum_{m=1}^{\infty} r_m , \quad \text{provided that the right infinite series converges.}$$

Another way to express this is to use parentheses:

$$\sum_{m=1}^{\infty}\sum_{n=1}^{\infty} a_{mn} = \sum_{m=1}^{\infty}\left(\sum_{n=1}^{\infty} a_{mn}\right),$$

where, provided that the summations converge, for every m we sum the *inside* series over n, and then we sum the *outside* series over m. Similarly, the right-hand iterated sum in (6.21) is defined by

$$\sum_{n=1}^{\infty}\sum_{m=1}^{\infty} a_{mn} := \sum_{n=1}^{\infty}\left(\sum_{m=1}^{\infty} a_{mn}\right),$$

where, provided that the summations converge, for every n we sum the *inside* series over m, and then we sum the *outside* series over n. Geometrically, we can view this series as "summing by columns," for the inside series is exactly the sum of the entries in the nth column shown on the right in Fig. 6.7, and if we call this sum c_n, then the iterated series is the sum over the c_n's.

Another way to think of iterated *series* is as iterated *limits* of the partial sums. That is, we claim that

$$\sum_{m=1}^{\infty} \sum_{n=1}^{\infty} a_{mn} = \lim_{m \to \infty} \left(\lim_{n \to \infty} s_{mn} \right) \text{ and } \sum_{n=1}^{\infty} \sum_{m=1}^{\infty} a_{mn} = \lim_{n \to \infty} \left(\lim_{m \to \infty} s_{mn} \right),$$

provided that the limits involved actually converge. Indeed, recall that

$$s_{mn} = \sum_{i=1}^{m} \sum_{j=1}^{n} a_{ij} .$$

If we take $n \to \infty$, we get

$$\lim_{n \to \infty} s_{mn} = \sum_{i=1}^{m} \sum_{j=1}^{\infty} a_{ij} ,$$

provided that the infinite series in j converges for every i. The right-hand side is the sum of the first m rows $r_1, r_2, \ldots, r_m$ in Fig. 6.7. Now taking $m \to \infty$, we get, provided that the limits exist,

$$\lim_{m \to \infty} \left(\lim_{n \to \infty} s_{mn} \right) = \sum_{i=1}^{\infty} \sum_{j=1}^{\infty} a_{ij} ,$$

as claimed (except for the change of the letters i and j to m and n, respectively).

Now that we have gone over double series and iterated series, we should ask ourselves whether these various notions are equivalent; for example, is it true that

$$\sum a_{mn} = \sum_{m=1}^{\infty} \sum_{n=1}^{\infty} a_{mn} = \sum_{n=1}^{\infty} \sum_{m=1}^{\infty} a_{mn} ?$$

It turns out that there may be no relationships between these series.

Example 6.28 Consider the a_{mn} shown in Fig. 6.8. The left picture in Fig. 6.8 shows an $m \times n$ rectangle with $m < n$ (the horizontally longer rectangle) and also a rectangle with $m \geq n$ (the vertically longer rectangle). In particular, we see that $s_{mn} = 0$ if $m < n$, while $s_{mn} = -1$ if $m \geq n$. It follows that the double sum $\sum a_{mn}$ does not converge. On the other hand, from Fig. 6.8, we see that

1	−1	0	0	0	...
0	1	−1	0	0	...
0	0	1	−1	0	...
0	0	0	1	−1	...
0	0	0	0	1	...

1	−1	0	0	0	...
0	1	−1	0	0	...
0	0	1	−1	0	...
0	0	0	1	−1	...
0	0	0	0	1	...

1	−1	0	0	0	...
0	1	−1	0	0	...
0	0	1	−1	0	...
0	0	0	1	−1	...
0	0	0	0	1	...

Fig. 6.8 *Left* The double series $\sum a_{mn}$ does not exist. The double sum $\sum a_{mn}$ does not converge. *Middle* The sum by rows converges to 0. *Right* The sum by columns converges to 1

$$\sum_{m=1}^{\infty}\sum_{n=1}^{\infty} a_{mn} = \text{sum by rows} = 0 + 0 + 0 + 0 + \cdots = 0,$$

while

$$\sum_{n=1}^{\infty}\sum_{m=1}^{\infty} a_{mn} = \text{sum by columns} = 1 + 0 + 0 + 0 + \cdots = 1.$$

Example 6.29 Consider the a_{mn} shown in Fig. 6.9. Since $s_{mn} = 0$ for all $m, n \geq 2$, it follows that $\sum a_{mn}$ converges and equals 0. However, the sums by rows and by columns do not exist.

0	0	1	2	3	4	...
0	0	−1	−2	−3	−4	...
1	−1	0	0	0	0	...
2	−2	0	0	0	0	...
3	−3	0	0	0	0	...
4	−4	0	0	0	0	...

0	0	1	2	3	4	...
0	0	−1	−2	−3	−4	...
1	−1	0	0	0	0	...
2	−2	0	0	0	0	...
3	−3	0	0	0	0	...
4	−4	0	0	0	0	...

Fig. 6.9 *Left* We have $s_{mn} = 0$ for all $m, n \geq 2$. *Right* The sums of the entries of the *first column* and the *second column* do not converge. Therefore, the sum by columns does not exist. Similarly, the sums of the entries of the *first row* and *second row* do not converge either. Thus, the sum by rows does not exist

As these examples show, the convergence of the double series may not imply that each iterated series converges, and conversely, if an iterated series converges, the double series may not. *However*, by the following theorem of Alfred Pringsheim (1850–1941) (cf. [40, p. 79]), if a double series converges and if it happens that all the rows converge and all the columns converge, then in fact both iterated series converge and equal the double series.

Pringsheim's theorem for series

Theorem 6.24 *Let $\{a_{mn}\}$ be a double sequence and assume:*

(1) The double series $\sum a_{mn}$ is convergent.
(2) For every m, the sum of the mth row, $\sum_{n=1}^{\infty} a_{mn}$, converges.
(3) For every n, the sum of the nth column, $\sum_{m=1}^{\infty} a_{mn}$, converges.

Then both iterated series converge, and we have the equality

$$\sum a_{mn} = \sum_{m=1}^{\infty} \sum_{n=1}^{\infty} a_{mn} = \sum_{n=1}^{\infty} \sum_{m=1}^{\infty} a_{mn} . \tag{6.22}$$

Proof Let $\varepsilon > 0$. Then by the definition of convergence for a double series, for $L = \sum a_{mn}$, there is an N such that for all $m, n > N$, we have

$$|L - s_{mn}| < \varepsilon/2 .$$

Taking $n \to \infty$, we get, for $m > N$, $|L - \lim_{n \to \infty} s_{mn}| \leq \varepsilon/2$. Thus,

$$m > N \implies \left| L - \lim_{n \to \infty} s_{mn} \right| < \varepsilon .$$

By the definition of convergence, this means that

$$\lim_{m \to \infty} \left(\lim_{n \to \infty} s_{mn} \right) = L ,$$

which is to say,

$$\sum_{m=1}^{\infty} \sum_{n=1}^{\infty} a_{mn} = L .$$

A similar argument establishes the equality for the other iterated series. ∎

Later we shall study the most useful theorem on iterated series, Cauchy's double series theorem on p. 477, which states that (6.22) always holds for absolutely convergent series. Here, a double series $\sum a_{mn}$ is said to **converge absolutely** if the double series of absolute values $\sum |a_{mn}|$ converges. However, before presenting Cauchy's theorem, we first generalize summing by rows and columns to "summing by curves."

6.5.3 *"Summing by Curves"*

Before we present the *sum by curves theorem* (Theorem 6.25 below), it might be helpful to give a couple of examples of this theorem to help in understanding what it says. Let $\{a_{mn}\}$ be a double sequence.

Example 6.30 Let

$$S_k = \{(m, n)\,;\ 1 \le m \le k\,,\ 1 \le n \le k\},$$

which represents a $k \times k$ square of numbers; here's a 4 × 4 block:

$$\begin{vmatrix} a_{11} & a_{12} & a_{13} & a_{14} & \cdots \\ a_{21} & a_{22} & a_{23} & a_{24} & \cdots \\ a_{31} & a_{32} & a_{33} & a_{34} & \cdots \\ a_{41} & a_{42} & a_{43} & a_{44} & \cdots \\ \vdots & \vdots & \vdots & \vdots & \ddots \end{vmatrix}$$

The left-hand picture in Fig. 6.10 shows $1 \times 1, 2 \times 2, 3 \times 3,$ and 4×4 examples. We denote by $\sum_{(m,n)\in S_k} a_{mn}$ the sum of those a_{mn} within the $k \times k$ square S_k. Explicitly,

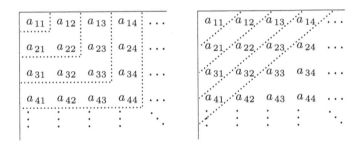

Fig. 6.10 "Summing by squares" and "summing by triangles"

$$\sum_{(m,n)\in S_k} a_{mn} = \sum_{m=1}^{k}\sum_{n=1}^{k} a_{mn}.$$

It is natural to refer to the limit (provided it exists)

$$\lim_{k\to\infty} \sum_{(m,n)\in S_k} a_{mn} = \lim_{k\to\infty} \sum_{m=1}^{k}\sum_{n=1}^{k} a_{mn}$$

as "summing by squares," since as we already noted, $\sum_{(m,n)\in S_k} a_{mn}$ involves summing the a_{mn} within a $k \times k$ square.

Example 6.31 Now let

$$S_k = T_1 \cup \cdots \cup T_k \quad , \quad \text{where } T_\ell = \{(m,n) \, ; \, m+n = \ell+1\}.$$

Notice that $T_\ell = \{(m,n) \, ; \, m+n = \ell+1\} = \{(1,\ell), (2, \ell-1), \ldots, (\ell, 1)\}$ represents the ℓth diagonal in the right-hand picture in Fig. 6.10; for instance, $T_3 = \{(1,3), (2,2), (3,1)\}$ is the third diagonal in Fig. 6.10. Then

$$\sum_{(m,n)\in S_k} a_{mn} = \sum_{\ell=1}^{k} \sum_{(m,n)\in T_\ell} a_{mn}$$

is the sum of the a_{mn} that are within the triangle consisting of the first k diagonals. It is natural to refer to the limit (provided it exists)

$$\lim_{k\to\infty} \sum_{(m,n)\in S_k} a_{mn} = \lim_{k\to\infty} \sum_{\ell=1}^{k} \sum_{(m,n)\in T_\ell} a_{mn}$$

as "summing by triangles." Using that $T_\ell = \{(1,\ell), (2, \ell-1), \ldots, (\ell, 1)\}$, we can express the summation by triangles as

$$\sum_{k=1}^{\infty} \left(a_{1,k} + a_{2,k-1} + \cdots + a_{k,1} \right).$$

More generally, we can "sum by curves" as long as the curves increasingly fill up the array like the squares or triangles shown in Fig. 6.10. More precisely, let $\mathscr{C}$ be a collection of nondecreasing finite sets $S_1 \subseteq S_2 \subseteq S_3 \subseteq \cdots \subseteq \mathbb{N} \times \mathbb{N}$ having the property that for every pair m, n there is a k such that

$$\{1, 2, \ldots, m\} \times \{1, 2, \ldots, n\} \subseteq S_k \subseteq S_{k+1} \subseteq S_{k+2} \subseteq \cdots . \tag{6.23}$$

This condition says that the sets S_k grow so as to encompass any given rectangle. For geometric purposes, one can think of the sets S_k as "curves" (like the boundaries of squares and triangles). Here's an example "curve" S_k (Fig. 6.11):

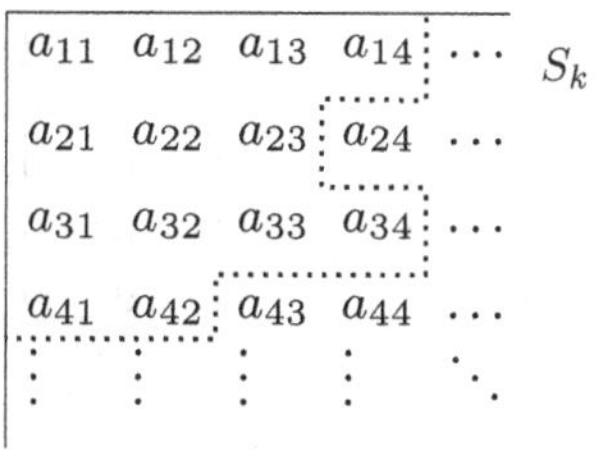

Fig. 6.11 A "curve." These curves grow larger and larger so as to enclose any given rectangle

For each $k \in \mathbb{N}$, let s_k be the sum of the a_{mn} for (m, n) inside the "curve" S_k:

$$s_k := \sum_{(m,n) \in S_k} a_{mn}.$$

We say that $\sum a_{mn}$ is **summable by the curves** $\mathscr{C}$ if $\lim_{k \to \infty} s_k$ exists, and we denote this limit by $\sum_{\mathscr{C}} a_{mn}$.

Example 6.32 Consider the a_{mn} shown in Fig. 6.12. From the left picture in Fig. 6.12, we see that $s_{mn} = 0$ for all $m, n \geq 2$. Thus, $\sum a_{mn}$ converges and equals 0. Consider the "triangles"

$$S_k = T_1 \cup \cdots \cup T_k , \quad \text{where } T_\ell = \{(m, n) ; \ m + n = \ell + 1\}.$$

Observe that for $k \geq 2$, we have

$$s_k = \sum_{(m,n) \in S_k} a_{mn} = 0 + 2 + 4 + 6 + 8 + \cdots + 2(k - 2).$$

Hence, the "sum by triangles" does not converge.

Example 6.33 Consider the a_{mn} shown in Fig. 6.13. From the left picture in Fig. 6.8, one can verify that $s_{mn} = m - n$ for all m, n; for example, the picture shows a 4×6

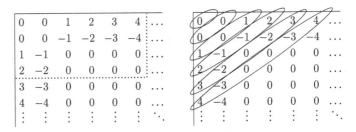

Fig. 6.12 "Summing by triangles." *Left* For every $m, n \geq 2$, the partial sum s_{mn} equals zero *Right* The sum by triangles equals $2 + 4 + 6 + 8 + \cdots$, which does not converge

Fig. 6.13 "Summing by triangles." *Left* We have $s_{mn} = m - n$. *Right* The sum by triangles equals $0 + 0 + 0 + 0 + \cdots = 0$

rectangle, where $s_{4,6} = -2$. It follows that $\sum a_{mn}$ does not converge. On the other hand, consider the "triangles" as we had in our previous example. Observe that for $k \geq 3$, we have

$$s_k = \sum_{(m,n)\in S_k} a_{mn} = 0 + 0 + 0 + 0 + \cdots + 0 = 0.$$

Hence, the "sum by triangles" converges to zero.

These examples show that in general, there is no relationship between convergence of the double series and convergence of a sum by curves. The following important theorem says that if the double series *converges absolutely*, then there is a relationship: the double series and sum by curves both converge and are equal.

Sum by curves theorem

> **Theorem 6.25** *If a double series $\sum a_{mn}$ of complex numbers is absolutely convergent, then $\sum a_{mn}$ itself converges, and moreover, given any collection of "curves" $\mathscr{C}$ satisfying (6.23), the sum by curves $\sum_{\mathscr{C}} a_{mn}$ converges, and*
>
> $$\sum a_{mn} = \sum_{\mathscr{C}} a_{mn}.$$

Proof Fix a collection $\mathscr{C}$ satisfying (6.23). For each k, let $s_k = \sum_{(m,n)\in S_k} a_{mn}$. We will show that $\{s_k\}$ is Cauchy and therefore converges; then we prove that $\sum a_{mn}$ converges and $\sum a_{mn} = \lim s_k$.

Step 1: We first prove a useful inequality. Let $A, A', B, B' \subseteq \mathbb{N} \times \mathbb{N}$ with $B' \subseteq B \subseteq A \subseteq A'$. Observe that

$$\left| \sum_{(i,j)\in A} a_{ij} - \sum_{(i,j)\in B} a_{ij} \right| = \left| \sum_{(i,j)\in A\setminus B} a_{ij} \right| \leq \sum_{(i,j)\in A\setminus B} |a_{ij}|$$

$$= \sum_{(i,j)\in A} |a_{ij}| - \sum_{(i,j)\in B} |a_{ij}|$$

$$\leq \sum_{(i,j)\in A'} |a_{ij}| - \sum_{(i,j)\in B'} |a_{ij}|.$$

Step 2: To prove that $\{s_k\}$ is Cauchy, let $\varepsilon > 0$ be given. By assumption, $\sum |a_{mn}|$ converges, so if L denotes its limit and t_{mn} its m, nth partial sum, we can choose N such that

$$m, n > N \implies |L - t_{mn}| < \frac{\varepsilon}{2}. \tag{6.24}$$

Fix $n > N$. Then by the property (6.23), there is an $N' \in \mathbb{N}$ such that

$$\{1, 2, \ldots, n\} \times \{1, 2, \ldots, n\} \subseteq S_{N'} \subseteq S_{N'+1} \subseteq S_{N'+2} \subseteq \cdots.$$

Now let k, ℓ be integers with $N' < \ell < k$, and then choose m large enough that $m > n$ and $S_k \subseteq \{1, 2, \ldots, m\} \times \{1, 2, \ldots, m\}$. Then,

$$\{1, 2, \ldots, n\} \times \{1, 2, \ldots, n\} \subseteq S_\ell \subseteq S_k \subseteq \{1, 2, \ldots, m\} \times \{1, 2, \ldots, m\}.$$

It follows from **Step 1** that

$$|s_k - s_\ell| \leq \sum_{i=1}^{m} \sum_{j=1}^{m} |a_{ij}| - \sum_{i=1}^{n} \sum_{j=1}^{n} |a_{ij}|$$

$$= t_{mm} - t_{nn}$$

$$= (t_{mm} - L) + (L - t_{nn}) < \frac{\varepsilon}{2} + \frac{\varepsilon}{2} = \varepsilon,$$

where we used (6.24). Hence, $|s_k - s_\ell| < \varepsilon$, which gives the Cauchy condition.

Step 3: We now show that $\sum a_{mn}$ converges with sum equal to $s := \lim s_k$. Let $\varepsilon > 0$ be given and choose N such that (6.24) holds with $\varepsilon/2$ replaced with $\varepsilon/3$. Fix natural numbers $m, n > N$. By the property (6.23) and the fact that $s_k \to s$, we can choose $k > N$ such that

$$\{1, 2, \ldots, m\} \times \{1, 2, \ldots, n\} \subseteq S_k \quad \text{and} \quad |s_k - s| < \varepsilon/3.$$

Now choose $m' \in \mathbb{N}$ such that $S_k \subseteq \{1, 2, \ldots, m'\} \times \{1, 2, \ldots, m'\}$. By **Step 1**,

$$|s_k - s_{mn}| \leq \sum_{i=1}^{m'} \sum_{j=1}^{m'} |a_{ij}| - \sum_{i=1}^{m} \sum_{j=1}^{n} |a_{ij}|$$

$$= t_{m'm'} - t_{mn}$$

$$= (t_{m'm'} - L) + (L - t_{mn}) < \frac{\varepsilon}{3} + \frac{\varepsilon}{3} = \frac{2\varepsilon}{3},$$

where we used the property (6.24) (with $\varepsilon/2$ replaced with $\varepsilon/3$). Finally, recalling that $|s_k - s| < \varepsilon/3$, by the triangle inequality, we have

$$|s_{mn} - s| \leq |s_{mn} - s_k| + |s_k - s| < \frac{2\varepsilon}{3} + \frac{\varepsilon}{3} = \varepsilon.$$

This proves that $\sum a_{mn} = s$ and completes our proof.   ∎

We recommend that the reader look at Problem 12 for a related result.

6.5.4   Cauchy's Double Series Theorem

We now come to the most important theorem of this section.

Cauchy's double series theorem

> **Theorem 6.26** *For a double series $\sum a_{mn}$ of complex numbers, the following are equivalent statements:*
>
> *(a) The series $\sum a_{mn}$ is absolutely convergent.*
> *(b) $\sum_{m=1}^{\infty} \sum_{n=1}^{\infty} |a_{mn}|$ converges.*
> *(c) $\sum_{n=1}^{\infty} \sum_{m=1}^{\infty} |a_{mn}|$ converges.*
>
> *In each of these cases, we have*
>
> $$\sum_{m=1}^{\infty} \sum_{n=1}^{\infty} a_{mn} = \sum_{n=1}^{\infty} \sum_{m=1}^{\infty} a_{mn} = \sum a_{mn}, \qquad (6.25)$$
>
> *in the sense that both iterated sums converge and equal the sum of the series.*

Proof Note that the sum by curves part in (6.25) follows from the sum by curves theorem, so can omit proving that part in what follows. We proceed in three steps.

Step 1: Assume first that the sum $\sum a_{mn}$ converges absolutely; we shall prove that both iterated sums $\sum_{m=1}^{\infty} \sum_{n=1}^{\infty} |a_{mn}|$, $\sum_{n=1}^{\infty} \sum_{m=1}^{\infty} |a_{mn}|$ converge. Since $\sum |a_{mn}|$ converges, setting $s := \sum |a_{mn}|$ and denoting by s_{mn} the m, nth partial sum, by the definition of convergence we can choose N such that

$$m, n > N \implies |s - s_{mn}| < 1 \implies s_{mn} < s + 1. \qquad (6.26)$$

Given $p \in \mathbb{N}$, choose $m \geq p$ such that $m > N$ and let $n > N$. Then in view of (6.26), we have

$$\sum_{j=1}^{n} |a_{pj}| \leq \sum_{i=1}^{m} \sum_{j=1}^{n} |a_{ij}| = s_{mn} < s + 1. \qquad (6.27)$$

Therefore, the partial sums of $\sum_{j=1}^{\infty} |a_{pj}|$ are bounded above by a fixed constant and hence (by the nonnegative series test, Theorem 3.19 on p. 197) for every $p \in \mathbb{N}$, the sum $\sum_{j=1}^{\infty} |a_{pj}|$ converges. Similarly, for each $q \in \mathbb{N}$, the sum $\sum_{i=1}^{\infty} |a_{iq}|$ converges. Therefore, by Pringsheim's theorem for series, both iterated series $\sum_{m=1}^{\infty} \sum_{n=1}^{\infty} |a_{mn}|$, $\sum_{n=1}^{\infty} \sum_{m=1}^{\infty} |a_{mn}|$ converge (and equal $\sum |a_{mn}|$).

Step 2: Assuming that $\sum a_{mn}$ converges absolutely, we now establish the equality (6.25). Indeed, by the sum by curves theorem, we know that $\sum a_{mn}$ converges, and we showed in **Step 1** that for each $p, q \in \mathbb{N}$, the sums $\sum_{n=1}^{\infty} |a_{pn}|$ and $\sum_{m=1}^{\infty} |a_{mq}|$ exist. This implies that for each $p, q \in \mathbb{N}$, $\sum_{n=1}^{\infty} a_{pn}$ and $\sum_{m=1}^{\infty} a_{mq}$ converge. Now (6.25) follows from Pringsheim's theorem.

Step 3: Now assume that

$$\sum_{m=1}^{\infty} \sum_{n=1}^{\infty} |a_{mn}| = t < \infty.$$

We will show that $\sum a_{mn}$ is absolutely convergent; a similar proof shows that if $\sum_{n=1}^{\infty} \sum_{m=1}^{\infty} |a_{mn}| < \infty$, then $\sum a_{mn}$ is absolutely convergent. Let $\varepsilon > 0$. Then the fact that $\sum_{i=1}^{\infty} \left(\sum_{j=1}^{\infty} |a_{ij}| \right) < \infty$ implies, by the Cauchy criterion for series, that there is an N such that

$$k > m > N \implies \sum_{i=m+1}^{k} \left(\sum_{j=1}^{\infty} |a_{ij}| \right) < \frac{\varepsilon}{2}.$$

Let $m, n > N$. Then for every $k > m$, we have

$$\left| \sum_{i=1}^{k} \sum_{j=1}^{\infty} |a_{ij}| - \sum_{i=1}^{m} \sum_{j=1}^{n} |a_{ij}| \right| = \sum_{i=m+1}^{k} \sum_{j=1}^{\infty} |a_{ij}| < \frac{\varepsilon}{2}.$$

Taking $k \to \infty$ shows that for all $m, n > N$,

$$\left| t - \sum_{i=1}^{m} \sum_{j=1}^{n} |a_{ij}| \right| \leq \frac{\varepsilon}{2} < \varepsilon,$$

which proves that $\sum |a_{mn}|$ converges and completes the proof of our result. ∎

Now for some applications.

Example 6.34 For an application of Cauchy's theorem and the sum by curves theorem, we look at the double sum $\sum z^{m+n}$ for $|z| < 1$. For such z, this sum converges absolutely, because

$$\sum_{m=0}^{\infty} \sum_{n=0}^{\infty} |z|^{m+n} = \sum_{m=0}^{\infty} |z|^m \cdot \frac{1}{1 - |z|} = \frac{1}{(1 - |z|)^2} < \infty,$$

where we used the geometric series test (twice): If $|r| < 1$, then $\sum_{k=0}^{\infty} r^k = \frac{1}{1-r}$. So $\sum z^{m+n}$ converges absolutely by Cauchy's double series theorem, and

$$\sum z^{m+n} = \sum_{m=0}^{\infty} \sum_{n=0}^{\infty} z^{m+n} = \sum_{m=0}^{\infty} z^m \cdot \frac{1}{1 - z} = \frac{1}{(1 - z)^2}.$$

On the other hand, by our sum by curves theorem, we can find $\sum z^{m+n}$ by summing over curves; we shall choose to sum over triangles. Thus, if we set

$$S_k = T_0 \cup T_1 \cup T_2 \cup \cdots \cup T_k \quad , \quad \text{where } T_\ell = \{(m, n)\,;\, m + n = \ell\,,\, m, n \geq 0\},$$

then

$$\sum z^{m+n} = \lim_{k \to \infty} \sum_{(m,n) \in S_k} z^{m+n} = \lim_{k \to \infty} \sum_{\ell=0}^{k} \sum_{(m,n) \in T_\ell} z^{m+n}.$$

Since $T_\ell = \{(m, n) \,;\, m + n = \ell\} = \{(0, \ell), (1, \ell - 1), \ldots, (\ell, 0)\}$, we have

$$\sum_{(m,n) \in T_\ell} z^{m+n} = z^{0+\ell} + z^{1+(\ell-1)} + z^{2+(\ell-2)} + \cdots + z^{\ell+0} = (\ell + 1)z^\ell.$$

Thus, $\sum z^{m+n} = \sum_{k=0}^{\infty}(k + 1)z^k$. However, we already proved that $\sum z^{m+n} = 1/(1 - z)^2$, so

$$\frac{1}{(1 - z)^2} = \sum_{n=1}^{\infty} nz^{n-1}. \tag{6.28}$$

See Problem 5 for an easier proof of (6.28) using Cauchy's double series theorem.

Example 6.35 Another very neat application of Cauchy's double series theorem is to derive nonobvious identities. Fix $z \in \mathbb{C}$ with $|z| < 1$, and consider the series

$$\sum_{n=1}^{\infty} \frac{z^n}{1 + z^{2n}} = \frac{z}{1 + z^2} + \frac{z^2}{1 + z^4} + \frac{z^3}{1 + z^6} + \cdots ;$$

we'll see why this series converges in a moment. Observe that (since $|z| < 1$)

$$\frac{1}{1 + z^{2n}} = \sum_{m=0}^{\infty}(-1)^m z^{2mn},$$

by the familiar geometric series test with $r = -z^{2n}$: $\sum_{k=0}^{\infty} r^k = \frac{1}{1-r}$ for $|r| < 1$. Therefore,

$$\sum_{n=1}^{\infty} \frac{z^n}{1 + z^{2n}} = \sum_{n=1}^{\infty} z^n \cdot \sum_{m=0}^{\infty}(-1)^m z^{2mn} = \sum_{n=1}^{\infty} \sum_{m=0}^{\infty}(-1)^m z^{(2m+1)n}.$$

We claim that the double sum $\sum(-1)^m z^{(2m+1)n}$ converges absolutely. To prove this, observe that

$$\sum_{n=1}^{\infty} \sum_{m=0}^{\infty} |z|^{(2m+1)n} = \sum_{n=1}^{\infty} |z|^n \sum_{m=0}^{\infty} |z|^{2nm} = \sum_{n=1}^{\infty} \frac{|z|^n}{1 - |z|^{2n}}.$$

Since $\frac{1}{1-|z|^{2n}} \le \frac{1}{1-|z|}$ (this is because $|z|^{2n} \le |z|$ for $|z| < 1$), we have

$$\frac{|z|^n}{1 - |z|^{2n}} \le \frac{1}{1 - |z|} \cdot |z|^n.$$

Since $\sum |z|^n$ converges, by the comparison theorem, $\sum_{n=1}^{\infty} \frac{|z|^n}{1-|z|^{2n}}$ converges too. Hence, Cauchy's double series theorem applies, and

$$\sum_{n=1}^{\infty}\sum_{m=0}^{\infty}(-1)^m z^{(2m+1)n} = \sum_{m=0}^{\infty}\sum_{n=1}^{\infty}(-1)^m z^{(2m+1)n}$$

$$= \sum_{m=0}^{\infty}(-1)^m \sum_{n=1}^{\infty} z^{(2m+1)n}$$

$$= \sum_{m=0}^{\infty}(-1)^m \frac{z^{2m+1}}{1-z^{2m+1}}.$$

Thus,

$$\sum_{n=1}^{\infty} \frac{z^n}{1+z^{2n}} = \sum_{m=0}^{\infty}(-1)^m \frac{z^{2m+1}}{1-z^{2m+1}}.$$

Therefore, we have derived the following striking identity between even and odd powers of z:

$$\frac{z}{1+z^2} + \frac{z^2}{1+z^4} + \frac{z^3}{1+z^6} + \cdots = \frac{z}{1-z} - \frac{z^3}{1-z^3} + \frac{z^5}{1-z^5} - + \cdots.$$

There are more beautiful series like this found in the exercises (see Problem 6, or better yet, Problem 8). We just touch on one more because it's so nice.

6.5.5 A Neat ζ-Function Identity

Recall that the ζ-function is defined by $\zeta(z) = \sum_{n=1}^{\infty} \frac{1}{n^z}$, which converges absolutely for $z \in \mathbb{C}$ with $\operatorname{Re} z > 1$. Here's a beautiful theorem from Flajolet and Vardi [82, 250].

Theorem 6.27 If $f(z) = \sum_{n=2}^{\infty} a_n z^n$ and $\sum_{n=2}^{\infty} |a_n|$ converges, then

$$\sum_{n=1}^{\infty} f\left(\frac{1}{n}\right) = \sum_{n=2}^{\infty} a_n \zeta(n).$$

Proof We first write

$$\sum_{n=1}^{\infty} f\left(\frac{1}{n}\right) = \sum_{n=1}^{\infty}\sum_{m=2}^{\infty} a_m \frac{1}{n^m}.$$

Now if we set $C = \sum_{m=2}^{\infty} |a_m| < \infty$, then

$$\sum_{n=1}^{\infty} \sum_{m=2}^{\infty} \left| a_m \frac{1}{n^m} \right| \le \sum_{n=1}^{\infty} \sum_{m=2}^{\infty} |a_m| \frac{1}{n^2} \le C \sum_{n=1}^{\infty} \frac{1}{n^2} < \infty.$$

Hence, by Cauchy's double series theorem, we can switch the order of summation:

$$\sum_{n=1}^{\infty} f\left(\frac{1}{n}\right) = \sum_{n=1}^{\infty} \sum_{m=2}^{\infty} a_m \frac{1}{n^m} = \sum_{m=2}^{\infty} a_m \sum_{n=1}^{\infty} \frac{1}{n^m} = \sum_{m=2}^{\infty} a_m \, \zeta(m),$$

which completes our proof.   ∎

Using this theorem, we can derive the pretty formula (see Problem 10):

$$\boxed{\log 2 = \sum_{n=2}^{\infty} \frac{1}{2^n} \, \zeta(n).}   \tag{6.29}$$

Not only is this formula pretty, it converges to $\log 2$ much faster than the usual series $\sum_{n=1}^{\infty} \frac{(-1)^{n-1}}{n}$ (from which (6.29) is derived by the help of Theorem 6.27); see [82, 250] for a discussion of such convergence issues.

▶ **Exercises 6.5**

1. **(Products of infinite series)** If $\sum_{m=1}^{\infty} a_m$ and $\sum_{n=1}^{\infty} b_n$ are convergent infinite series, with sums A and B respectively, prove that the double series[10] $\sum a_m b_n$ and the iterated series $\sum_{m=1}^{\infty} \sum_{n=1}^{\infty} a_m b_n$ and $\sum_{n=1}^{\infty} \sum_{m=1}^{\infty} a_m b_n$ converge to AB; that is, prove that

$$\left(\sum_{m=1}^{\infty} a_m \right) \left(\sum_{n=1}^{\infty} b_n \right) = \sum a_m b_n = \sum_{m=1}^{\infty} \sum_{n=1}^{\infty} a_m b_n = \sum_{n=1}^{\infty} \sum_{m=1}^{\infty} a_m b_n.$$

2. **(Comparison test)** Let $\{a_{mn}\}$ and $\{b_{mn}\}$ be double sequences such that $0 \le a_{mn} \le b_{mn}$ for all m, n and such that the double series $\sum b_{mn}$ converges. Prove that $\sum a_{mn}$ also converges. Use the comparison test to prove that the double series $\sum 1/(m^4 + n^4)$ converges.

3. Determine the convergence, iterated convergence, and absolute convergence for the double series

(a) $\displaystyle\sum_{m,n \ge 1} \frac{(-1)^{mn}}{mn}$,   (b) $\displaystyle\sum_{m,n \ge 1} \frac{(-1)^n}{(m + n^p)(m + n^p - 1)}$, $p > 1$,   (c) $\displaystyle\sum_{m \ge 2, n \ge 1} \frac{1}{m^n}$.

Suggestion: For (b), show that $\sum_{m=1}^{\infty} \frac{1}{(m+n^p)(m+n^p-1)}$ telescopes.

---

[10]By the double series $\sum a_m b_n$, we mean $\sum a_{mn}$, where $a_{mn} = a_m b_n$ for each m, n.

4. (*mn*-**term test for double series**) Show that if $\sum a_{mn}$ converges, then $a_{mn} \to$ 0 as $m, n \to \infty$. Suggestion: First verify that $a_{mn} = s_{mn} - s_{m-1,n} - s_{m,n-1} + s_{m-1,n-1}$.

5. Let $z \in \mathbb{C}$ with $|z| < 1$. For $(m, n) \in \mathbb{N} \times \mathbb{N}$, define $a_{mn} = z^n$ if $m \le n$ and define $a_{mn} = 0$ otherwise. Using Cauchy's double series theorem on $\sum a_{mn}$, prove (6.28). Using (6.28), find $\sum_{n=1}^{\infty} \frac{n}{2^n}$ (cf. Problem 3 on p. 203).

6. Let $|z| < 1$. Using Cauchy's double series theorem, derive the identities

(a) $\dfrac{z}{1+z^2} + \dfrac{z^3}{1+z^6} + \dfrac{z^5}{1+z^{10}} + \cdots = \dfrac{z}{1-z^2} - \dfrac{z^3}{1-z^6} + \dfrac{z^5}{1-z^{10}} - + \cdots$,

(b) $\dfrac{z}{1+z^2} - \dfrac{z^2}{1+z^4} + \dfrac{z^3}{1+z^6} - + \cdots = \dfrac{z}{1+z} - \dfrac{z^3}{1+z^3} + \dfrac{z^5}{1+z^5} - + \cdots$,

(c) $\dfrac{z}{1+z} - \dfrac{2z^2}{1+z^2} + \dfrac{3z^3}{1+z^3} - + \cdots = \dfrac{z}{(1+z)^2} - \dfrac{z^2}{(1+z^2)^2} + \dfrac{z^3}{(1+z^3)^2} - + \cdots$.

Suggestion: For (c), you need the formula $1/(1 - z)^2 = \sum_{n=1}^{\infty} n z^{n-1}$ found in (6.28).

7. Here's a neat formula for $\zeta(k)$ found in [39]: For every $k \in \mathbb{N}$ with $k \ge 3$, we have

$$\boxed{\zeta(k) = \sum_{\ell=1}^{k-2} \sum_{m=1}^{\infty} \sum_{n=1}^{\infty} \frac{1}{m^\ell (m + n)^{k-\ell}}.}$$

To prove this, you may proceed as follows.

(i) Show that

$$\sum_{\ell=1}^{k-2} \frac{1}{m^\ell (m + n)^{k-\ell}} = \frac{1}{(m + n)^k} \sum_{\ell=1}^{k-2} \left(\frac{m + n}{m} \right)^\ell = \frac{1}{m^{k-2} n (m + n)} - \frac{1}{n (m + n)^{k-1}}.$$

(ii) Use (i) to show that

$$\sum_{\ell=1}^{k-2} \sum_{m=1}^{\infty} \sum_{n=1}^{\infty} \frac{1}{m^\ell (m + n)^{k-\ell}} = \sum_{m=1}^{\infty} \sum_{n=1}^{\infty} \frac{1}{m^{k-2} n (m + n)} - \sum_{m=1}^{\infty} \sum_{n=1}^{\infty} \frac{1}{n (m + n)^{k-1}}.$$

Make sure you justify each step; in particular, why does each sum converge?

(iii) Use the partial fractions expansion $\frac{1}{n(m+n)} = \frac{1}{n} - \frac{1}{m+n}$ to show that

$$\sum_{m=1}^{\infty} \sum_{n=1}^{\infty} \frac{1}{m^{k-2} n (m + n)} = \sum_{m=1}^{\infty} \frac{1}{m^{k-1}} \sum_{n=1}^{m} \frac{1}{n}.$$

(iv) Replace the summation variable n with $\ell = m + n$ in $\sum_{m=1}^{\infty} \sum_{n=1}^{\infty} \frac{1}{n(m+n)^{k-1}}$ to get a new sum in terms of m and ℓ; then use Cauchy's double series theorem to change the order of summation. Finally, prove the desired result.

8. (**Number theory series**) Here are some pretty formulas involving number theory!

   (a) For $n \in \mathbb{N}$, let $\tau(n)$ denote the number of positive divisors of n (that is, the number of positive integers that divide n). For example, $\tau(1) = 1$ and $\tau(4) = 3$ (because precisely 1, 2, 4 divide 4). Prove that

$$\sum_{n=1}^{\infty} \frac{z^n}{1-z^n} = \sum_{n=1}^{\infty} \tau(n)z^n, \quad |z| < 1. \tag{6.30}$$

   Suggestion: Write $1/(1-z^n) = \sum_{m=0}^{\infty} z^{mn} = \sum_{m=1}^{\infty} z^{n(m-1)}$; then prove that the left-hand side of (6.30) equals $\sum z^{mn}$. Finally, use the sum by curves theorem with the set S_k given by $S_k = T_1 \cup \cdots \cup T_k$, where $T_k = \{(m, n) \in \mathbb{N} \times \mathbb{N} ; \ m \cdot n = k\}$.

   (b) For $n \in \mathbb{N}$, let $\sigma(n)$ denote the sum of the positive divisors of n. (For example, $\sigma(1) = 1$ and $\sigma(4) = 1 + 2 + 4 = 7$). Prove that

$$\sum_{n=1}^{\infty} \frac{z^n}{(1-z^n)^2} = \sum_{n=1}^{\infty} \sigma(n)z^n, \quad |z| < 1.$$

9. Let $f(z) = \sum_{n=1}^{\infty} a_n z^n$ and $g(z) = \sum_{n=1}^{\infty} b_n z^n$ be power series with positive radii of convergence. Show that for $z \in \mathbb{C}$ sufficiently near the origin, we have

$$\sum_{n=1}^{\infty} b_n f(z^n) = \sum_{n=1}^{\infty} a_n g(z^n).$$

   From this formula, derive the following formulas:

$$\sum_{n=1}^{\infty} f(z^n) = \sum_{n=1}^{\infty} \frac{a_n z^n}{1-z^n}, \quad \sum_{n=1}^{\infty} (-1)^{n-1} f(z^n) = \sum_{n=1}^{\infty} \frac{a_n z^n}{1+z^n},$$

   and my favorite:

$$\boxed{\sum_{n=1}^{\infty} \frac{f(z^n)}{n!} = \sum_{n=1}^{\infty} a_n e^{z^n}.}$$

10. In this problem we derive (6.29).

    (i) Prove that $\log 2 = \sum_{n=1}^{\infty} \frac{1}{2n(2n-1)} = \sum_{n=1}^{\infty} f\left(\frac{1}{n}\right)$, where $f(z) = \frac{z^2}{2(2-z)}$.

    (ii) Show that $f(z) = \sum_{n=2}^{\infty} \frac{z^n}{2^n}$, and from this and Theorem 6.27, prove (6.29).

11. (Cf. [82, 250]) Prove the following extension of Theorem 6.27: If $f(z) = \sum_{n=2}^{\infty} a_n z^n$ and for some $N \in \mathbb{N}$, $\sum_{n=2}^{\infty} \frac{|a_n|}{N^n}$ converges, then

484 6 Advanced Theory of Infinite Series

$$\sum_{n=N}^{\infty} f\left(\frac{1}{n}\right) = \sum_{n=2}^{\infty} a_n \left\{ \zeta(n) - \left(1 + \frac{1}{2^n} + \cdots + \frac{1}{(N-1)^n}\right) \right\},$$

where the sum $\left(1 + \frac{1}{2^n} + \cdots + \frac{1}{(N-1)^n}\right)$ is (by convention) zero if $N = 1$.

12. (**Arbitrary rearrangements of double series**) Let $f : \mathbb{N} \to \mathbb{N} \times \mathbb{N}$ be a bijective function and set $\nu_n = f(n) \in \mathbb{N} \times \mathbb{N}$; therefore $\nu_1, \nu_2, \nu_3, \ldots$ is a list of all elements of $\mathbb{N} \times \mathbb{N}$. For a double series $\sum a_{mn}$ of complex numbers, prove that $\sum_{n=1}^{\infty} a_{\nu_n}$ is absolutely convergent if and only if $\sum a_{mn}$ is absolutely convergent, in which case $\sum_{n=1}^{\infty} a_{\nu_n} = \sum a_{mn}$.

6.6 Rearrangements and Multiplication of Power Series

We already know that the associative law holds for infinite series. That is, we can group the terms of an infinite series in any way we wish and the resulting series still converges with the same sum (Theorem 3.21 on p. 199). A natural question that you may ask is whether the commutative law holds for infinite series. That is, suppose that $s = a_1 + a_2 + a_3 + \cdots$ exists. Can we commute the a_n in any way we wish and still get the same sum? For instance, is it true that

$$s = a_1 + a_2 + a_4 + a_3 + a_6 + a_8 + a_5 + a_{10} + a_{12} + \cdots?$$

For general series, the answer is, perhaps shocking at first (but in agreement with the order-dependent summation cases presented before in this book), no!

6.6.1 Rearrangements

A sequence $\nu_1, \nu_2, \nu_3, \ldots$ of natural numbers such that every natural number occurs exactly once in this list is called a **rearrangement** of the natural numbers.

Example 6.36 $1, 2, 4, 3, 6, 8, 5, 10, 12, \ldots$, where we follow every odd number by two adjacent even numbers, is a rearrangement.

A **rearrangement** of a series $\sum_{n=1}^{\infty} a_n$ is a series $\sum_{n=1}^{\infty} a_{\nu_n}$, where $\{\nu_n\}$ is a rearrangement of $\mathbb{N}$.

Example 6.37 Let us rearrange the alternating harmonic series

$$\log 2 = \sum_{n=1}^{\infty} (-1)^{n-1} \frac{1}{n} = 1 - \frac{1}{2} + \frac{1}{3} - \frac{1}{4} + \frac{1}{5} - \frac{1}{6} + \frac{1}{7} - \frac{1}{8} + - \cdots$$

using the rearrangement $1, 2, 4, 3, 6, 8, 5, 10, 12, \ldots$ we've already mentioned:

$$s = 1 - \frac{1}{2} - \frac{1}{4} + \frac{1}{3} - \frac{1}{6} - \frac{1}{8} + \frac{1}{5} - \frac{1}{10} - \frac{1}{12} + --$$

$$\cdots + \frac{1}{2k - 1} - \frac{1}{4k - 2} - \frac{1}{4k} + \cdots,$$

provided of course that this sum converges. Here, the bottom three terms represent the general formula for the kth triplet of a positive term followed by two negative ones. To see that this sum converges, let s_n denote its nth partial sum. Then we can write $n = 3k + \ell$, where ℓ is 0, 1, or 2, and so

$$s_n = 1 - \frac{1}{2} - \frac{1}{4} + \frac{1}{3} - \frac{1}{6} - \frac{1}{8} + - - \cdots + \frac{1}{2k - 1} - \frac{1}{4k - 2} - \frac{1}{4k} + r_n,$$

where r_n consists of the next ℓ (=0, 1, 2) terms of the series for s_n. Note that $r_n \to 0$ as $n \to \infty$. Now observe that

$$s_n = \left(1 - \frac{1}{2}\right) - \frac{1}{4} + \left(\frac{1}{3} - \frac{1}{6}\right) - \frac{1}{8} + - - \cdots + \left(\frac{1}{2k - 1} - \frac{1}{4k - 2}\right) - \frac{1}{4k} + r_n$$

$$= \frac{1}{2} - \frac{1}{4} + \frac{1}{6} - \frac{1}{8} + - \cdots + \frac{1}{4k - 2} - \frac{1}{4k} + r_n$$

$$= \frac{1}{2}\left(1 - \frac{1}{2} + \frac{1}{3} - \frac{1}{4} + - \cdots + \frac{1}{2k - 1} - \frac{1}{2k}\right) + r_n.$$

Taking $n \to \infty$, we see that

$$s = \frac{1}{2} \log 2.$$

Thus, the rearrangement converges to half the sum of the original series!

6.6.2 Riemann's Rearrangement Theorem

The previous example showed that rearrangements of series can have different sums from that of the original series. In fact, it turns out that a convergent series can be rearranged to get a different value if and only if the series is not absolutely convergent. The "only if" portion is proved in Theorem 6.29 to come, and the "if" portion is proved in the following theorem.

Riemann's rearrangement theorem

Theorem 6.28 *If a series $\sum a_n$ of real numbers converges, but not absolutely, then there are rearrangements of the series that can be made to converge to $\pm\infty$ or any real number whatsoever.*

Proof We shall prove that there are rearrangements of the series that converge to any real number whatsoever; following the argument for this case, you should be able to handle the $\pm\infty$ cases yourself.

Step 1: We first show that each of the two series corresponding to the positive and negative terms in $\sum a_n$ diverges. Let $b_1, b_2, b_3, \dots$ denote the terms in the sequence $\{a_n\}$ that are nonnegative, in the order in which they occur, and let $c_1, c_2, c_3, \dots$ denote the absolute values of the terms in $\{a_n\}$ that are negative, again in the order in which they occur. We claim that both series $\sum b_n$ and $\sum c_n$ diverge. To see this, observe that

$$\sum_{k=1}^{n} a_k = \sum_i b_i - \sum_j c_j, \tag{6.31}$$

where the right-hand sums are over the natural numbers i, j such that b_i and c_j occur in the left-hand sum. The left-hand side converges as $n \to \infty$ by assumption, so if either sum $\sum_{n=1}^{\infty} b_n$ or $\sum_{n=1}^{\infty} c_n$ of nonnegative numbers converges, then the equality (6.31) would imply that the other sum converges. But this would then imply that

$$\sum_{k=1}^{n} |a_k| = \sum_i b_i + \sum_j c_j$$

converges as $n \to \infty$, which is false by assumption. Hence, both sums $\sum b_n$ and $\sum c_n$ diverge.

Step 2: We now rearrange. Let $\xi \in \mathbb{R}$. We shall produce a rearrangement

$$b_1 + \cdots + b_{m_1} - c_1 - \cdots - c_{n_1} + b_{m_1+1} + \cdots + b_{m_2} \tag{6.32}$$
$$- c_{n_1+1} - \cdots - c_{n_2} + b_{m_2+1} + \cdots + b_{m_3} - c_{n_2+1} - \cdots$$

such that its partial sums converge to ξ. The idea is simple to explain; consider, for example, this area picture of summation:

We know that the areas on both sides are infinite. The first step is to add enough b_i's to get an area A_1 slightly greater than ξ (which is possible, since the b_i's "add to infinity"). Then we subtract enough c_j's from A_1 to get a net area A_2 slightly less than ξ. Next, we add more b_i's to get an area A_3 slightly greater than ξ. Continuing, we are able to add the areas in such a way to equal ξ in the limit!

To make this idea rigorous, let $\{\beta_n\}$ and $\{\gamma_n\}$ denote the partial sums for $\sum b_n$ and $\sum c_n$, respectively. Since $\beta_n \to \infty$, for n sufficiently large, $\beta_n > \xi$. We define m_1 as the smallest natural number such that

$$\beta_{m_1} > \xi.$$

Here's a picture (Fig. 6.14):

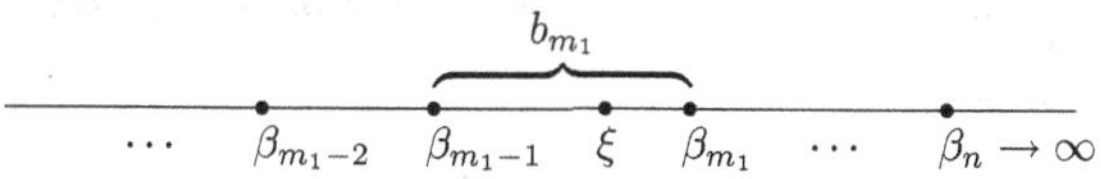

Fig. 6.14 We have
$\beta_{m_1-1} \leq \xi < \beta_{m_1}$

Since for every n we have $\beta_n = \beta_{n-1} + b_n$ (from the definition of partial sum), it follows that β_{m_1} differs from ξ by at most b_{m_1}. Since $\gamma_n \to \infty$, for n sufficiently large, $\beta_{m_1} - \gamma_n < \xi$. We define n_1 to be the smallest natural number such that

$$\beta_{m_1} - \gamma_{n_1} < \xi.$$

Note that the left-hand side differs from ξ by at most c_{n_1} (do you see why?). Now define m_2 as the smallest natural number greater than m_1 such that

$$\beta_{m_2} - \gamma_{n_1} > \xi.$$

As before, such a number exists because $\beta_n \to \infty$, and the left-hand side differs from ξ by at most b_{m_2}. We define the number n_2 as the smallest natural number greater than n_1 such that

$$\beta_{m_2} - \gamma_{n_2} < \xi,$$

where the left-hand side differs from ξ by at most c_{n_2}. Continuing this process, we produce sequences $m_1 < m_2 < m_3 < \cdots$ and $n_1 < n_2 < n_3 < \cdots$ such that for every k,

$$\beta_{m_k} - \gamma_{n_{k-1}} > \xi,$$

where the left-hand side differs from ξ by at most b_{m_k}, and

$$\beta_{m_k} - \gamma_{n_k} < \xi,$$

where the left-hand side differs from ξ by at most c_{n_k}. This produces the rearrangement (6.32). By assumption, $\sum a_n$ converges, so by the nth term test, it follows that $b_{m_k}, c_{n_k} \to 0$ as $k \to \infty$. This implies, by the way the β_{m_k} and γ_{n_k} were chosen, that the rearrangement (6.32) converges to ξ, as the reader can check. This completes our proof. ∎

We now prove that a convergent series can be rearranged to get a different value only if the series is not absolutely convergent. Actually, we shall prove the contrapositive: If a series is absolutely convergent, then every rearrangement has the same value as the original sum. This is a consequence of the following theorem.

Dirichlet's theorem

> **Theorem 6.29** *All rearrangements of an absolutely convergent series of complex numbers converge with the same sum as the original series.*

Proof Let $\sum a_n$ converge absolutely. We shall prove that every rearrangement of this series converges to the same value as the sum itself. To see this, let $\nu_1, \nu_2, \nu_3, \ldots$ be a rearrangement of the natural numbers and define

$$a_{mn} = \begin{cases} a_m & \text{if } m = \nu_n, \\ 0 & \text{else.} \end{cases}$$

Then by definition of a_{mn}, we have

$$a_m = \sum_{n=1}^{\infty} a_{mn} \quad \text{and} \quad a_{\nu_n} = \sum_{m=1}^{\infty} a_{mn}.$$

Moreover,

$$\sum_{m=1}^{\infty} \sum_{n=1}^{\infty} |a_{mn}| = \sum_{m=1}^{\infty} |a_m| < \infty,$$

so by Cauchy's double series theorem,

$$\sum_{m=1}^{\infty} a_m = \sum_{m=1}^{\infty} \sum_{n=1}^{\infty} a_{mn} = \sum_{n=1}^{\infty} \sum_{m=1}^{\infty} a_{mn} = \sum_{n=1}^{\infty} a_{\nu_n}.$$

■

We now move to the important topic of multiplication of series.

6.6.3 Multiplication of Power Series and Infinite Series

If we have two convergent infinite series $\sum_{m=0}^{\infty} a_m$ and $\sum_{n=0}^{\infty} b_n$, then we know from Problem 1 on p. 481 from Exercises 6.5 on double series that

$$\left(\sum_{m=0}^{\infty} a_m \right) \cdot \left(\sum_{n=0}^{\infty} b_n \right) = \sum_{m,n} a_m b_n. \tag{6.33}$$

Here, the right-hand side is the double series of the double sequence $a_m b_n$, seen here:

$$
\begin{array}{llllll}
a_0 b_0 & a_0 b_1 & a_0 b_2 & a_0 b_3 & \cdots \\[4pt]
a_1 b_0 & a_1 b_1 & a_1 b_2 & a_1 b_3 & \cdots \\[4pt]
a_2 b_0 & a_2 b_1 & a_2 b_2 & a_2 b_3 & \cdots \\[4pt]
a_3 b_0 & a_3 b_1 & a_3 b_2 & a_3 b_3 & \cdots \\[4pt]
\vdots & \vdots & \vdots & \vdots & \vdots & \ddots
\end{array}
$$

The equality (6.33) says that the double series $\sum_{m,n} a_m b_n$ converges and its sum equals $(\sum a_m) \cdot (\sum b_n)$. We learned in Section 6.5 that we could try to sum a double series "by curves." The sum may not agree with the original double series, but we'll get to that later! In this particular instance, it turns out that "summing by triangles" is the most natural. Let's see why.

Example 6.38 Consider the two power series $f(z) = \sum_{n=0}^{\infty} a_n z^n$ and $g(z) = \sum_{n=0}^{\infty} b_n z^n$, and let's multiply them:

$$
f(z)\, g(z) = (a_0 + a_1 z + a_2 z^2 + \cdots)(b_0 + b_1 z + b_2 z^2 + \cdots),
$$

where the dots "$\cdots$" here and below represents z^3 and higher-power terms. We now use the distributive law in a *formal*[11] way. We first multiply the first term a_0 by each term of $g(z)$,

$$
\left(a_0 + a_1 z + a_2 z^2 + \cdots \right)\left(b_0 + b_1 z + b_2 z^2 + \cdots \right).
$$

We do the same with the second term $a_1 z$,

$$
\left(a_0 + a_1 z + a_2 z^2 + \cdots \right)\left(b_0 + b_1 z + b_2 z^2 + \cdots \right),
$$

and again with the third term,

$$
\left(a_0 + a_1 z + a_2 z^2 + \cdots \right)\left(b_0 + b_1 z + b_2 z^2 + \cdots \right).
$$

If we add all the terms obtained through multiplication and omit writing any z^3 terms or higher powers, we obtain

$$
\begin{aligned}
f(z)\, g(z) = \; & a_0 b_0 + a_0 b_1 z + a_0 b_2 z^2 + \cdots \\
& + a_1 b_0 z + a_1 b_2 z^2 + \cdots \\
& + a_2 b_0 z^2 + \cdots .
\end{aligned}
$$

---

[11] Recall that as in most of mathematics, "formal" refers to "a symbolic manipulation or expression presented without paying attention to correctness.".

Collecting like powers of z, we get

$$f(z)\, g(z) = a_0 b_0 + (a_0 b_1 + a_1 b_0)\, z + (a_0 b_2 + a_1 b_1 + a_2 b_0)\, z^2 + \cdots .$$

In other words,

$$f(z)\, g(z) = c_0 + c_1 z + c_2 z^2 + \cdots ,$$

where

$$c_0 = a_0 b_0 , \quad c_1 = a_0 b_1 + a_1 b_0 , \quad c_2 = a_0 b_2 + a_1 b_1 + a_2 b_0 , \quad \ldots .$$

In general, for $f(z) = \sum_{n=0}^{\infty} a_n z^n$ and $g(z) = \sum_{n=0}^{\infty} b_n z^n$, we have the following result, which is called the **Cauchy product** of the power series:

Cauchy product formula : $f(z)\, g(z) = \displaystyle\sum_{n=0}^{\infty} c_n z^n$,   where

$$c_n = \sum_{k=0}^{n} a_k b_{n-k} = a_0 b_n + a_1 b_{n-1} + \cdots + a_n b_0 .$$

To summarize: The Cauchy product is exactly what you would expect if you formally multiplied the series and collected like powers of z.

In Corollary 6.31 we'll see that this formal computation is correct! In particular, assuming for the moment that the Cauchy product formula is correct, at $z = 1$ we get (again, only formally!)

$$\begin{aligned}
\left(a_0 + a_1 + a_2 + a_3 + \cdots\right)&\left(b_0 + b_1 + b_2 + b_3 + \cdots\right) = \\
&a_0 b_0 + (a_0 b_1 + a_1 b_0) + (a_0 b_2 + a_1 b_1 + a_2 b_0) \\
&+ (a_0 b_3 + a_1 b_2 + a_2 b_1 + a_3 b_0) + \cdots .
\end{aligned}$$

These thoughts suggest the following definition. Given two series $\sum_{n=0}^{\infty} a_n$ and $\sum_{n=0}^{\infty} b_n$, we define their **Cauchy product** as the series $\sum_{n=0}^{\infty} c_n$, where

$$c_n = a_0 b_n + a_1 b_{n-1} + \cdots + a_n b_0 = \sum_{k=0}^{n} a_k b_{n-k} .$$

Thus, if we put the products $a_m b_n$ in an infinite array as in Fig. 6.15, then c_n is just the sum of the nth diagonal, where we call the first diagonal $a_0 b_0$ the zeroth diagonal. In other words, the Cauchy product is just the double series $\sum a_m b_n$ "summed by triangles."

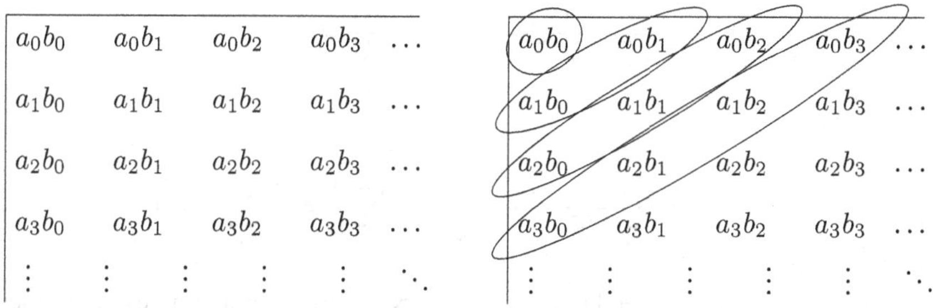

$$
\begin{array}{llllll}
a_0b_0 & a_0b_1 & a_0b_2 & a_0b_3 & \cdots \\
a_1b_0 & a_1b_1 & a_1b_2 & a_1b_3 & \cdots \\
a_2b_0 & a_2b_1 & a_2b_2 & a_2b_3 & \cdots \\
a_3b_0 & a_3b_1 & a_3b_2 & a_3b_3 & \cdots \\
\vdots & \vdots & \vdots & \vdots & \vdots & \ddots
\end{array}
$$

Fig. 6.15 The Cauchy product is just summing the series $\sum a_m b_n$ "by triangles"!

A natural question to ask is this: If $\sum_{n=0}^{\infty} a_n$ and $\sum_{n=0}^{\infty} b_n$ converge, then is it true that the Cauchy product $\sum c_n$ converges, and

$$
\left(\sum_{n=0}^{\infty} a_n \right) \left(\sum_{n=0}^{\infty} b_n \right) = \sum_{n=0}^{\infty} c_n \ ?
$$

Based on Examples 6.32 and 6.33 starting on p. 474, the answer is an obvious no, since a double series and a sum by curves are not necessarily related. Here's an explicit example due to Cauchy.

Example 6.39 Consider the product $(\sum_{n=1}^{\infty} \frac{(-1)^{n-1}}{\sqrt{n}})(\sum_{n=1}^{\infty} \frac{(-1)^{n-1}}{\sqrt{n}})$ (Fig. 6.16).

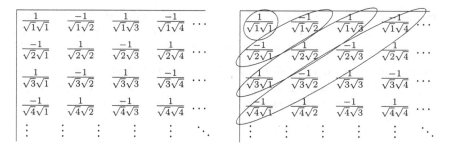

Fig. 6.16 Cauchy's example. The "sum by triangles" (that is, the Cauchy product) does not converge!

Observe that

$$
c_n = (-1)^{n-1} \left(\frac{1}{\sqrt{n}\,\sqrt{1}} + \frac{1}{\sqrt{n-1}\,\sqrt{2}} + \frac{1}{\sqrt{n-2}\,\sqrt{2}} + \cdots + \frac{1}{\sqrt{2}\,\sqrt{n-1}} + \frac{1}{\sqrt{1}\,\sqrt{n}} \right)
$$

$$
= (-1)^{n-1} \sum_{k=1}^{n} \frac{1}{\sqrt{n+1-k}\,\sqrt{k}}.
$$

Since for $1 \leq k \leq n$, we have

$$\frac{1}{\sqrt{n+1-k\sqrt{k}}} \geq \frac{1}{\sqrt{n}\sqrt{n}} = \frac{1}{n},$$

we see that

$$(-1)^n c_n = \sum_{k=1}^{n} \frac{1}{\sqrt{n+1-k}\sqrt{k}} \geq \sum_{k=1}^{n} \frac{1}{n} = \frac{1}{n}\sum_{k=1}^{n} 1 = 1.$$

Thus, the terms c_n do not tend to zero as $n \to \infty$, so by the nth term test, the series $\sum_{n=0}^{\infty} c_n$ does not converge.

The problem with this example is that the series $\sum \frac{(-1)^{n-1}}{\sqrt{n}}$ does not converge absolutely. For absolutely convergent series, we have no worries.

Cauchy's multiplication theorem

Theorem 6.30 *If two series $\sum a_n = A$ and $\sum b_n = B$ converge absolutely, then the double series $\sum_{m,n} a_m b_n$ can be summed using any curves; in particular, the Cauchy product (or sum by triangles) converges with sum equal to AB.*

Proof Since

$$\sum_{m=0}^{\infty}\sum_{n=0}^{\infty} |a_m b_n| = \sum_{m=0}^{\infty} |a_m| \sum_{n=0}^{\infty} |b_n| = \left(\sum_{m=0}^{\infty} |a_m|\right)\left(\sum_{n=0}^{\infty} |b_n|\right) < \infty,$$

by Cauchy's double series theorem, the double series $\sum a_m b_n$ converges absolutely. Hence, by the sum by curves theorem on p. 475, the double series $\sum a_m b_n$ equals its sum by triangles, which is just the Cauchy product. ∎

Example 6.40 Back in Theorem 3.30 on p. 219, using our faithful Tannery's theorem, we proved the formula $\exp(z)\exp(w) = \exp(z+w)$ for $z, w \in \mathbb{C}$. Using Cauchy's multiplication theorem, we can give an alternative and quick proof: Forming the Cauchy product, we have

$$\exp(z)\exp(w) = \left(\sum_{n=0}^{\infty} \frac{z^n}{n!}\right) \cdot \left(\sum_{n=0}^{\infty} \frac{w^n}{n!}\right)$$

$$= \sum_{n=0}^{\infty} \left(\sum_{k=0}^{n} \frac{z^k}{k!} \cdot \frac{w^{n-k}}{(n-k)!}\right)$$

$$= \sum_{n=0}^{\infty} \frac{1}{n!}\left(\sum_{k=0}^{n} \frac{n!}{k!(n-k)!} z^k w^{n-k}\right)$$

$$= \sum_{n=0}^{\infty} \frac{1}{n!}\left(\sum_{k=0}^{n} \binom{n}{k} z^k w^{n-k}\right) = \sum_{n=0}^{\infty} \frac{1}{n!}(z+w)^n = \exp(z+w),$$

where we used the binomial theorem for $(z + w)^n$ going from the third to fourth lines.

Returning to our motivating Example 6.38, for the case of power series, Theorem 6.30 takes the following form.

Cauchy product of power series

Corollary 6.31 *If $f(z) = \sum_{n=0}^{\infty} a_n z^n$ and $g(z) = \sum_{n=0}^{\infty} b_n z^n$ are power series with positive radii of convergence, then their product $f(z)g(z)$ is a power series, and moreover, the product power series is obtained by formally multiplying the power series and combining like powers of z.*

6.6.4 Mertens and Abel Multiplication Theorems

Requiring absolute convergence is rather restrictive, so for the rest of this section we go over two theorems that relax this condition. The first theorem is due to Franz Mertens (1840–1927), and it requires only one of the two series to be absolutely convergent.

Mertens's multiplication theorem

Theorem 6.32 *If at least one of two convergent series $\sum a_n = A$ and $\sum b_n = B$ is absolutely convergent, then their Cauchy product converges with sum equal to AB.*

Proof Because our notation is symmetric in A and B, we may assume that $\sum a_n$ is absolutely convergent. Now consider the partial sums of the Cauchy product:

$$
\begin{aligned}
C_n &= c_0 + c_1 + \cdots + c_n \\
&= a_0 b_0 + (a_0 b_1 + a_1 b_0) + \cdots + (a_0 b_n + a_1 b_{n-1} + \cdots + a_n b_0) \\
&= a_0(b_0 + \cdots + b_n) + a_1(b_0 + \cdots + b_{n-1}) + \cdots + a_n b_0. \quad (6.34)
\end{aligned}
$$

We need to show that C_n tends to AB as $n \to \infty$. If A_n denotes the nth partial sum of $\sum a_n$, and B_n that of $\sum b_n$, then from (6.34), we have

$$
C_n = a_0 B_n + a_1 B_{n-1} + \cdots + a_n B_0.
$$

If we set $B_k = B + \beta_k$, then $\beta_k \to 0$, and we can write

$$
\begin{aligned}
C_n &= a_0(B + \beta_n) + a_1(B + \beta_{n-1}) + \cdots + a_n(B + \beta_0) \\
&= A_n B + (a_0 \beta_n + a_1 \beta_{n-1} + \cdots + a_n \beta_0).
\end{aligned}
$$

Since $A_n \to A$, the first part of this sum converges to AB. Thus, we just need to show that the term in parentheses tends to zero as $n \to \infty$. To see this, let $\varepsilon > 0$ be given. Putting $\alpha = \sum |a_n|$ and using that $\beta_n \to 0$, we can choose a natural number N such that for all $n > N$, we have $|\beta_n| < \varepsilon/(2\alpha)$. Also, since $\beta_n \to 0$, we can choose a constant C such that $|\beta_n| \leq C$ for every n. Then for $n > N$,

$$
\begin{aligned}
|a_0\beta_n &+ a_1\beta_{n-1} + \cdots + a_n\beta_0| - |a_0\beta_n + a_1\beta_{n-1} + \cdots + a_{n-N+1}\beta_{N+1} \\
&\qquad\qquad\qquad\qquad\qquad + a_{n-N}\beta_N + \cdots + a_n\beta_0| \\
&\leq |a_0\beta_n + a_1\beta_{n-1} + \cdots + a_{n-N+1}\beta_{N+1}| + |a_{n-N}\beta_N + \cdots + a_n\beta_0| \\
&< \left(|a_0| + |a_1| + \cdots + |a_{n-N+1}|\right) \cdot \frac{\varepsilon}{2\alpha} + \left(|a_{n-N}| + \cdots + |a_n|\right) \cdot C \\
&\leq \alpha \cdot \frac{\varepsilon}{2\alpha} + C\left(|a_{n-N}| + \cdots + |a_n|\right) \\
&= \frac{\varepsilon}{2} + C\left(|a_{n-N}| + \cdots + |a_n|\right).
\end{aligned}
$$

Since $\sum |a_n| < \infty$, by the Cauchy criterion for series, we can choose $N' > N$ such that

$$
n > N' \quad\Longrightarrow\quad |a_{n-N}| + \cdots + |a_n| < \frac{\varepsilon}{2C}.
$$

Then for $n > N'$, we see that

$$
|a_0\beta_n + a_1\beta_{n-1} + \cdots + a_n\beta_0| < \frac{\varepsilon}{2} + \frac{\varepsilon}{2} = \varepsilon.
$$

Since $\varepsilon > 0$ was arbitrary, this completes the proof of the theorem. ∎

The last theorem of this section does not require any absolute convergence assumptions; it requires only that the Cauchy product actually converge. (In the Cauchy and Mertens theorems, the convergence of the Cauchy product was a conclusion of the theorems.)

Abel's multiplication theorem

Theorem 6.33 *If the Cauchy product of two convergent series $\sum a_n = A$ and $\sum b_n = B$ converges, then the Cauchy product has the value AB.*

Proof In my opinion, the slickest proof of this theorem is Abel's original, proved in 1826 [129, p. 321] using his limit theorem, Theorem 6.21 on p. 460. Let

$$
f(z) = \sum a_n z^n, \quad g(z) = \sum b_n z^n, \quad h(z) = \sum c_n z^n,
$$

where $c_n = a_0 b_n + \cdots + a_n b_0$. These power series converge at $z = 1$, so they must have radii of convergence at least 1. In particular, each series converges absolutely for $|z| < 1$, and for these values of z, we know that (by either the Cauchy or Mertens multiplication theorem)

$$h(z) = f(z) \cdot g(z).$$

Since each of the sums $\sum a_n$, $\sum b_n$, and $\sum c_n$ converges, by Abel's limit theorem, the functions f, g, and h converge to A, B, and $C = \sum c_n$, respectively, as $z = x \to 1$ from the left. Thus,

$$C = \lim_{x \to 1-} h(x) = \lim_{x \to 1-} \left(f(x) \cdot g(x) \right) = A \cdot B. \qquad \blacksquare$$

Example 6.41 For example, let us square the nonabsolutely convergent series $\log 2 = \sum_{n=1}^{\infty} \frac{(-1)^{n-1}}{n}$. It turns out that it will be convenient to write $\log 2$ in two ways: $\log 2 = \sum_{n=1}^{\infty} a_n$, where $a_0 = 0$ and $a_n = \frac{(-1)^{n-1}}{n}$ for $n = 1, 2, \ldots$, and $\log 2 = \sum_{n=0}^{\infty} b_n$, where $b_n = \frac{(-1)^n}{n+1}$. Their Cauchy product is given by $c_0 = a_0 b_0 = 0$, and for $n = 1, 2, \ldots$, we have

$$c_n = \sum_{k=0}^{n} a_k b_{n-k} = \sum_{k=1}^{n} \frac{(-1)^{k-1}(-1)^{n-k}}{k(n+1-k)} = (-1)^{n-1} \alpha_n,$$

where $\alpha_n = \sum_{k=1}^{n} \frac{1}{k(n+1-k)}$. In order to use Abel's multiplication theorem, we need to check that $\sum_{n=0}^{\infty} c_n = \sum_{n=1}^{\infty} (-1)^{n-1} \alpha_n$ converges. By the alternating series test, this sum converges if we can prove that $\{\alpha_n\}$ is nonincreasing and converges to zero. To prove these statements, observe that

$$\frac{1}{k(n-k+1)} = \frac{1}{n+1}\left(\frac{1}{k} + \frac{1}{n-k+1}\right).$$

Therefore,

$$\alpha_n = \frac{1}{1} \cdot \frac{1}{n} + \frac{1}{2} \cdot \frac{1}{n-1} + \frac{1}{3} \cdot \frac{1}{n-2} + \cdots + \frac{1}{n} \cdot \frac{1}{1}$$

$$= \frac{1}{n+1}\left[\left(1 + \frac{1}{n}\right) + \left(\frac{1}{2} + \frac{1}{n-1}\right) + \left(\frac{1}{3} + \frac{1}{n-2}\right) + \cdots + \left(\frac{1}{n} + \frac{1}{1}\right)\right].$$

In the brackets are two copies of $1 + \frac{1}{2} + \cdots + \frac{1}{n}$. Thus,

$$\alpha_n = \frac{2}{n+1} H_n, \qquad \text{where } H_n := 1 + \frac{1}{2} + \frac{1}{3} + \cdots + \frac{1}{n}.$$

It is common to use the notation H_n for the nth partial sum of the harmonic series. Now, by our work on p. 309 on the Euler–Mascheroni constant, we found that $\gamma_n := H_n - \log n$ is bounded above by 1, so

$$\alpha_n = \frac{2}{n+1}(\gamma_n + \log n) \le \frac{2}{n+1} + 2\frac{\log n}{n+1}$$

$$= \frac{2}{n+1} + 2 \cdot \frac{n}{n+1} \cdot \frac{1}{n} \log n$$

$$= \frac{2}{n+1} + 2 \cdot \frac{n}{n+1} \cdot \log(n^{1/n}) \to 0 + 2 \cdot 1 \cdot \log 1 = 0$$

as $n \to \infty$. Thus, $\alpha_n \to 0$. Moreover,

$$\alpha_n - \alpha_{n+1} = \frac{2}{n+1}H_n - \frac{2}{n+2}H_{n+1} = \frac{2}{n+1}H_n - \frac{2}{n+2}\left(H_n + \frac{1}{n+1}\right)$$

$$= \left(\frac{2}{n+1} - \frac{2}{n+2}\right)H_n - \frac{2}{(n+1)(n+2)}$$

$$= \frac{2}{(n+1)(n+2)}H_n - \frac{2}{(n+1)(n+2)}$$

$$= \frac{2}{(n+1)(n+2)}(H_n - 1) \ge 0.$$

Thus, $\alpha_n \ge \alpha_{n+1}$, so $\sum c_n = \sum (-1)^{n-1}\alpha_n$ converges. Abel's multiplication theorem now implies $(\log 2)^2 = \sum (-1)^{n-1}\alpha_n$. Dividing by 2, we have proved the following pretty formula:

$$\frac{1}{2}\left(\log 2\right)^2 = \sum_{n=1}^{\infty} \frac{(-1)^{n-1}}{n+1}H_n$$

$$= \sum_{n=1}^{\infty} \frac{(-1)^{n-1}}{n+1}\left(1 + \frac{1}{2} + \cdots + \frac{1}{n}\right).$$

▶ **Exercises 6.6**

1. Here are some alternating series problems:

   (a) Prove that

   $$\frac{1}{1} + \frac{1}{3} - \frac{1}{2} + \frac{1}{5} + \frac{1}{7} - \frac{1}{4} + \cdots + \frac{1}{4k-3} + \frac{1}{4k-1} - \frac{1}{2k} + \cdots = \frac{3}{2}\log 2.$$

   That is, we rearrange the alternating harmonic series so that two positive terms are followed by one negative one, otherwise keeping the ordering the same. Suggestion: Observe that

   $$\frac{1}{2}\log 2 = \frac{1}{2} - \frac{1}{4} + \frac{1}{6} - \frac{1}{8} + \frac{1}{9} - \frac{1}{10} + \cdots$$

   $$= 0 + \frac{1}{2} + 0 - \frac{1}{4} + 0 + \frac{1}{6} + 0 - \frac{1}{8} + \cdots.$$

Add this term by term to the series for log 2.

(b) Prove that

$$\frac{1}{1} + \frac{1}{3} + \frac{1}{5} + \frac{1}{7} - \frac{1}{2} + \cdots + \frac{1}{8k-7} + \frac{1}{8k-5} + \frac{1}{8k-3}$$
$$+ \frac{1}{8k-1} - \frac{1}{2k} + \cdots = \frac{3}{2} \log 2;$$

that is, we rearrange the alternating harmonic series so that four positive terms are followed by one negative one, otherwise keeping the ordering the same.

(c) What's wrong with the following argument?

$$1 - \frac{1}{2} + \frac{1}{3} - \frac{1}{4} + \frac{1}{5} - \frac{1}{6} + \cdots = \left(1 + \frac{1}{2} + \frac{1}{3} + \frac{1}{4} + \frac{1}{5} + \frac{1}{6} + \cdots\right)$$
$$- 2\left(\frac{1}{2} + 0 + \frac{1}{4} + 0 + \frac{1}{6} + \cdots\right)$$
$$= \left(1 + \frac{1}{2} + \frac{1}{3} + \frac{1}{4} + \frac{1}{5} + \frac{1}{6} + \cdots\right)$$
$$- \left(1 + \frac{1}{2} + \frac{1}{3} + \frac{1}{4} + \frac{1}{5} + \frac{1}{6} + \cdots\right) = 0.$$

2. Let $f(z) = \sum_{n=0}^{\infty} a_n z^n$ be absolutely convergent for $|z| < 1$. Prove that for $|z| < 1$, we have

$$\frac{f(z)}{1-z} = \sum_{n=0}^{\infty} (a_0 + a_1 + a_2 + \cdots + a_n) z^n.$$

3. Using the previous problem, prove that for $z \in \mathbb{C}$ with $|z| < 1$,

$$\frac{1}{(1-z)^2} = \sum_{n=0}^{\infty} (n+1) z^n; \quad \text{that is, } \left(\sum_{n=0}^{\infty} z^n\right) \cdot \left(\sum_{n=0}^{\infty} z^n\right) = \sum_{n=0}^{\infty} (n+1) z^n.$$

Using this formula, derive the following neat-looking formula: For $z \in \mathbb{C}$ with $|z| < 1$,

$$\left(\sum_{n=0}^{\infty} \cos n\theta \, z^n\right) \cdot \left(\sum_{n=0}^{\infty} \sin n\theta \, z^n\right) = \frac{1}{2} \sum_{n=0}^{\infty} (n+1) \sin n\theta \, z^n. \qquad (6.35)$$

Suggestion: Put $z = e^{i\theta} x$ with x real into the formula $\left(\sum_{n=0}^{\infty} z^n\right) \cdot \left(\sum_{n=0}^{\infty} z^n\right) = \sum_{n=0}^{\infty} (n+1) z^n$, then equate imaginary parts of both sides; this proves (6.35) for $z = x$ real and $|x| < 1$. Why does (6.35) hold for $z \in \mathbb{C}$ with $|z| < 1$?

4. Derive the following beautiful formula: For $|z| < 1$,

$$\Big(\sum_{n=1}^{\infty} \frac{\cos n\theta}{n} z^n\Big) \cdot \Big(\sum_{n=1}^{\infty} \frac{\sin n\theta}{n} z^n\Big) = \frac{1}{2} \sum_{n=2}^{\infty} \frac{H_n \sin n\theta}{n} z^n.$$

5. In this problem we prove the following fact: Let $f(z) = \sum_{n=0}^{\infty} a_n z^n$ be a power series with radius of convergence $R > 0$ and let $\alpha \in \mathbb{C}$ with $|\alpha| < R$. Then we can write

$$f(z) = \sum_{n=0}^{\infty} b_n (z - \alpha)^n,$$

where this series converges absolutely for $|z - \alpha| < R - |\alpha|$.

   (i) Show that

$$f(z) = \sum_{n=0}^{\infty} \sum_{m=0}^{n} a_n \binom{n}{m} \alpha^{n-m} (z - \alpha)^m. \qquad (6.36)$$

   (ii) Prove that

$$\sum_{n=0}^{\infty} \sum_{m=0}^{n} |a_n| \binom{n}{m} |\alpha|^{n-m} |z - \alpha|^m = \sum_{n=0}^{\infty} |a_n| \big(|z - \alpha| + |\alpha|\big)^m < \infty.$$

   (iii) Prove the result by verifying that you can change the order of summation in (6.36).

6. If $f(z) = \sum_{n=0}^{\infty} a_n z^n$ is a power series with a positive radius of convergence, prove that for all $m \in \mathbb{N}$,

$$f(z)^m = \sum_{n=0}^{\infty} c_n z^n,$$

where the coefficients are given by

$$c_n = \sum_{k_1 + k_2 + \cdots + k_m = m} a_{k_1} a_{k_2} \cdots a_{k_m}.$$

Here, the summation is over all products $a_{k_1} a_{k_2} \cdots a_{k_m}$, where $(k_1, k_2, \ldots, k_m)$ belongs to the set $\{(k_1, k_2, \ldots, k_m)\,;\, k_1, \ldots, k_m \geq 0,\ k_1 + \cdots + k_m = m\}$.

6.7   Composition of Power Series and Bernoulli and Euler Numbers

We've kept you in suspense long enough concerning the extraordinary Bernoulli and Euler numbers, so in this section we finally get to these fascinating numbers.

6.7.1 Composition and Division of Power Series

The Bernoulli and Euler numbers come up when one divides power series, so before we do anything, we need to understand division of power series, and to understand that, we first need to consider the composition of power series. The following theorem basically says that the composition of power series is again a power series.

Power series composition theorem

> **Theorem 6.34** *If $f(z)$ and $g(z)$ are power series, then the composition $f(g(z))$ can be written as a power series that converges at least for those values of z satisfying*
>
> $$\sum_{n=0}^{\infty} |a_n z^n| < \text{the radius of convergence of } f,$$
>
> *where $g(z) = \sum_{n=0}^{\infty} a_n z^n$. Moreover, the composition $f(g(z))$ is exactly the power series obtained by formally plugging the series $g(z)$ into the power series $f(z)$ and collecting like powers of z.*

Proof Thus, $f(g(z))$ is claimed to be a power series for those z in the shaded domain shown here:

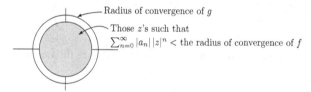

The composition $f(g(z))$ can converge for more values of z; see Problem 1. The idea of the proof is exactly what the last statement of this theorem says. First, for each m we can "formally multiply" $g(z)^m = g(z) \cdot g(z) \cdots g(z)$ (m factors of $g(z)$), and write $g(z)^m$ as a power series

$$g(z)^m = \left(\sum_{n=0}^{\infty} a_n z^n \right)^m = \sum_{n=0}^{\infty} \alpha_{mn} z^n,$$

for some coefficients α_{mn}. Actually, by the Cauchy multiplication theorem applied $m - 1$ times (see Corollary 6.31 on p. 493), this series is valid for each z inside the radius of convergence of g. Now let $f(z) = \sum_{m=0}^{\infty} b_m z^m$. Then provided that $g(z)$ is inside the radius of convergence of f, we have

$$f(g(z)) = \sum_{m=0}^{\infty} b_m g(z)^m = \sum_{m=0}^{\infty} \sum_{n=0}^{\infty} b_m \alpha_{mn} z^n.$$

We now "collect like powers of z," which means to interchange the order of summation in $f(g(z))$:

$$f(g(z)) = \sum_{n=0}^{\infty} \sum_{m=0}^{\infty} b_m \alpha_{mn} \, z^n = \sum_{n=0}^{\infty} \beta_n \, z^n, \qquad \text{where} \quad \beta_n = \sum_{m=0}^{\infty} b_m \alpha_{mn},$$

provided that the interchange is valid. Once we verify that the interchange is valid, our proof is complete. To prove that it was valid, fix z such that

$$\xi := \sum_{n=0}^{\infty} |a_n z^n| = \sum_{n=0}^{\infty} |a_n| \, |z|^n < \text{the radius of convergence of } f.$$

In particular, since $f(\xi) = \sum_{m=0}^{\infty} b_m \xi^m$ is absolutely convergent,

$$\sum_{m=0}^{\infty} |b_m| \, \xi^m < \infty. \tag{6.37}$$

Now according to Cauchy's double series theorem, interchanging the order of summation in $f(g(z))$ was indeed valid if we can show that

$$\sum_{m=0}^{\infty} \sum_{n=0}^{\infty} |b_m \alpha_{mn} z^n| = \sum_{m=0}^{\infty} |b_m| \left(\sum_{n=0}^{\infty} |\alpha_{mn}| \, |z|^n \right) < \infty. \tag{6.38}$$

To prove this, we claim that the inner summation satisfies the inequality

$$\textbf{Claim:} \qquad \sum_{n=0}^{\infty} |\alpha_{mn}| \, |z|^n \le \xi^m. \tag{6.39}$$

Once we prove this claim, then putting this inequality into (6.38) and using (6.37) completes the proof of this theorem. To prove the claim (6.39), consider the case $m = 2$. Recall that the coefficients α_{2n} are obtained by the Cauchy product:

$$g(z)^2 = \left(\sum_{n=0}^{\infty} a_n \, z^n \right)^2 = \sum_{n=0}^{\infty} \alpha_{2n} \, z^n \quad \text{where} \quad \alpha_{2n} = \sum_{k=0}^{n} a_k a_{n-k}.$$

Thus, $|\alpha_{2n}| \le \sum_{k=0}^{n} |a_k| \, |a_{n-k}|$. On the other hand, we can express ξ^2 via the Cauchy product:

$$\xi^2 = \left(\sum_{n=0}^{\infty} |a_n| \, |z|^n \right)^2 = \sum_{n=0}^{\infty} c_n \, |z|^n \quad \text{where} \quad c_n = \sum_{k=0}^{n} |a_k| \, |a_{n-k}|.$$

By the triangle inequality, we have $|\alpha_{2n}| \le c_n$, so

$$\sum_{n=0}^{\infty} |\alpha_{2n}| \, |z|^n \le \sum_{n=0}^{\infty} c_n \, |z|^n = \xi^2,$$

which proves (6.39) for $m = 2$. An induction argument shows that (6.39) holds for all m. This completes the proof of the claim and hence our theorem. ∎

We already know that the product of two power series is again a power series. As a consequence of the following theorem, we get the same statement for division.

Power series division theorem

Theorem 6.35 *If $f(z)$ and $g(z)$ are power series with positive radii of convergence and with $g(0) \ne 0$, then $f(z)/g(z)$ is also a power series with positive radius of convergence.*

Proof Since $f(z)/g(z) = f(z) \cdot (1/g(z))$ and we know that the product of two power series is a power series, all we have to do is show that $1/g(z)$ is a power series. To this end, let $g(z) = \sum_{n=0}^{\infty} a_n z^n$ and define

$$\tilde{g}(z) = \frac{1}{a_0} g(z) - 1 = \sum_{n=1}^{\infty} \alpha_n z^n,$$

where $\alpha_n = \frac{a_n}{a_0}$ and where we recall that $a_0 = g(0) \ne 0$. Then $\tilde{g}$ has a positive radius of convergence and $\tilde{g}(0) = 0$. Now let $h(z) := \frac{1}{a_0(1+z)}$, which can be written as a geometric series with radius of convergence 1. Note that for $|z|$ small, $\sum_{n=1}^{\infty} |\alpha_n| \, |z|^n < 1$ (why?), and thus by the previous theorem, for such z,

$$\frac{1}{g(z)} = \frac{1}{a_0(1 + \tilde{g}(z))} = h(\tilde{g}(z))$$

has a power series expansion with a positive radius of convergence. ∎

6.7.2 Bernoulli Numbers

See [54, 129, 220], or [92] for more information on Bernoulli numbers. Since

$$\frac{e^z - 1}{z} = \frac{1}{z} \cdot \sum_{n=1}^{\infty} \frac{1}{n!} z^n = \sum_{n=1}^{\infty} \frac{1}{n!} z^{n-1} = \sum_{n=0}^{\infty} \frac{1}{(n+1)!} z^n$$

has a power series expansion and equals 1 at $z = 0$, by our division of power series theorem, its reciprocal $z/(e^z - 1)$ also has a power series expansion near $z = 0$. It is customary to denote its coefficients by $B_n/n!$, in which case we can write

$$\boxed{\frac{z}{e^z - 1} = \sum_{n=0}^{\infty} \frac{B_n}{n!} z^n}, \qquad (6.40)$$

where the series has a positive radius of convergence. The numbers B_n are called the **Bernoulli numbers** after Jacob Bernoulli (1654–1705), who discovered them while searching for formulas involving powers of integers; see Problems 4 and 5. We can find a remarkable symbolic equation for these Bernoulli numbers as follows. First, we multiply both sides of (6.40) by $(e^z - 1)/z$ and use Cauchy's multiplication theorem to get

$$1 = \left(\sum_{n=0}^{\infty} \frac{B_n}{n!} z^n \right) \cdot \left(\sum_{n=0}^{\infty} \frac{1}{(n+1)!} z^n \right) = \sum_{n=0}^{\infty} \sum_{k=0}^{n} \left(\frac{B_k}{k!} \cdot \frac{1}{(n-k+1)!} \right) z^n.$$

By the identity theorem on p. 462, the $n = 0$ term on the right must equal 1, while all other terms must vanish. The $n = 0$ term on the right is just B_0, so $B_0 = 1$, and for $n > 1$, we must have $\sum_{k=0}^{n} \frac{B_k}{k!} \cdot \frac{1}{(n+1-k)!} = 0$. Multiplying this by $(n + 1)!$, we get

$$0 = \sum_{k=0}^{n} \frac{B_k}{k!} \cdot \frac{(n+1)!}{(n+1-k)!} = \sum_{k=0}^{n} \frac{(n+1)!}{k!(n+1-k)!} \cdot B_k = \sum_{k=0}^{n} \binom{n+1}{k} B_k,$$

and adding $B_{n+1} = \binom{n+1}{n+1} B_{n+1}$ to both sides of this equation, we get

$$B_{n+1} = \sum_{k=0}^{n+1} \binom{n+1}{k} B_k.$$

The right-hand side might look familiar from the binomial formula. Recall from the binomial formula that for every complex number a, we have

$$(a + 1)^{n+1} = \sum_{k=0}^{n+1} \binom{n+1}{k} a^k \cdot 1^{n-k} = \sum_{k=0}^{n+1} \binom{n+1}{k} a^k.$$

Notice that the right-hand side of this expression is exactly the right-hand side of the previous equation if we put $a = B$ and make the superscript k into a subscript k. Thus, if we use the notation

$$\boxed{\doteq \text{ to mean "equals after making superscripts into subscripts,"}}$$

then we can write

$$\boxed{B^{n+1} \doteq (B+1)^{n+1} \quad , \quad n = 1, 2, 3, \ldots \quad \text{with } B_0 = 1.} \tag{6.41}$$

Using the identity (6.41), one can in principle find all the Bernoulli numbers. For example, when $n = 1$, we see that

$$B^2 \doteq (B+1)^2 = B^2 + 2B^1 + 1 \quad \Longrightarrow \quad 0 = 2B_1 + 1 \quad \Longrightarrow \quad B_1 = -\frac{1}{2}.$$

When $n = 2$, we see that

$$B^3 \doteq (B+1)^3 = B^3 + 3B^2 + 3B^1 + 1 \Longrightarrow 0 = 3B_2 + 3B_1 + 1 \Longrightarrow B_2 = \frac{1}{6}.$$

Here is a partial list through B_{14}:

$$B_0 = 1, \quad B_1 = -\frac{1}{2}, \quad B_2 = \frac{1}{6}, \quad B_3 = 0,$$

$$B_4 = -\frac{1}{30}, \quad B_5 = B_7 = B_9 = B_{11} = B_{13} = B_{15} = 0,$$

$$B_6 = \frac{1}{42}, \quad B_8 = -\frac{1}{30}, \quad B_{10} = \frac{5}{66}, \quad B_{12} = -\frac{691}{2730}, \quad B_{14} = \frac{7}{6}.$$

These numbers are rational, but besides this fact, there is no known regular pattern to which these numbers conform. However, we can easily deduce that all Bernoulli numbers with odd index except for B_1 are zero. Indeed, we can rewrite (6.40) as

$$\frac{z}{e^z - 1} + \frac{z}{2} = 1 + \sum_{n=2}^{\infty} \frac{B_n}{n!} z^n. \tag{6.42}$$

The fractions on the left-hand side can be combined into one fraction

$$\frac{z}{e^z - 1} + \frac{z}{2} = \frac{z(e^z + 1)}{2(e^z - 1)} = \frac{z(e^{z/2} + e^{-z/2})}{2(e^{z/2} - e^{-z/2})}, \tag{6.43}$$

which is an even function of z. Thus, (see Problem 1 on p. 463)

$$B_{2n+1} = 0, \quad n = 1, 2, 3, \ldots. \tag{6.44}$$

Other properties are given in the exercises (see, e.g., Problem 4).

6.7.3 Trigonometric Functions

We already know the power series expansions for $\sin z$ and $\cos z$. It turns out that other trigonometric functions have power series expansions involving Bernoulli numbers! For example, to find the expansion for $\cot z$, we replace z with $2iz$ in (6.42) and (6.43) to get

$$\frac{iz(e^{iz} + e^{-iz})}{(e^{iz} - e^{-iz})} = 1 + \sum_{n=2}^{\infty} \frac{B_n}{n!} (2iz)^n = 1 + \sum_{n=1}^{\infty} \frac{B_{2n}}{(2n)!} (-1)^n (2z)^{2n},$$

where we used that $B_3, B_5, B_7, \ldots$ all vanish in order to sum only over all even Bernoulli numbers. Since $\cot z = \cos z / \sin z$, using the definition of $\cos z$ and $\sin z$ in terms of $e^{\pm iz}$, we see that the left-hand side is exactly $z \cot z$. Thus, we have derived the formula

$$\boxed{z \cot z = \sum_{n=0}^{\infty} (-1)^n \frac{2^{2n} B_{2n}}{(2n)!} z^{2n}.}$$

From this formula, we can get the expansion for $\tan z$ using the identity

$$2 \cot(2z) = 2\frac{\cos 2z}{\sin 2z} = 2\frac{\cos^2 z - \sin^2 z}{2 \sin z \cos z} = \cot z - \tan z.$$

Hence,

$$\tan z = \cot z - 2\cot(2z) = \sum_{n=0}^{\infty} (-1)^n \frac{2^{2n} B_{2n}}{(2n)!} z^{2n} - 2\sum_{n=0}^{\infty} (-1)^n \frac{2^{2n} B_{2n}}{(2n)!} 2^{2n} z^{2n},$$

which, after combining the terms on the right, takes the form

$$\boxed{\tan z = \sum_{n=1}^{\infty} (-1)^{n-1} \frac{2^{2n}(2^{2n} - 1) B_{2n}}{(2n)!} z^{2n-1}.}$$

In Problem 2, we derive the power series expansion of $\csc z$. In conclusion, we have power series expansions for $\sin z$, $\cos z$, $\tan z$, $\cot z$, $\csc z$. What about $\sec z$?

6.7.4 The Euler Numbers

It turns out that the expansion for $\sec z$ involves the Euler numbers, which are defined similarly to the Bernoulli numbers. By the division of power series theorem, the function $2e^z/(e^{2z} + 1)$ has a power series expansion near zero. It is customary to denote its coefficients by $E_n/n!$, so

$$\frac{2e^z}{e^{2z}+1} = \sum_{n=0}^{\infty} \frac{E_n}{n!} z^n \qquad (6.45)$$

where the series has a positive radius of convergence. The numbers E_n are called the **Euler numbers**. We can get the missing expansion for $\sec z$ as follows. First, observe that

$$\sum_{n=0}^{\infty} \frac{E_n}{n!} z^n = \frac{2e^z}{e^{2z}+1} = \frac{2}{e^z + e^{-z}} = \frac{1}{\cosh z} = \operatorname{sech} z,$$

where $\operatorname{sech} z := 1/\cosh z$ is the hyperbolic secant. As one can check, $\operatorname{sech} z$ is an even function (that is, $\operatorname{sech}(-z) = \operatorname{sech} z$), so it follows that all E_n with n odd vanish. Hence,

$$\operatorname{sech} z = \sum_{n=0}^{\infty} \frac{E_{2n}}{(2n)!} z^{2n} . \qquad (6.46)$$

In particular, putting iz for z in (6.46) and using that $\cosh(iz) = \cos z$, we get the missing expansion for $\sec z$:

$$\sec z = \sum_{n=0}^{\infty} (-1)^n \frac{E_{2n}}{(2n)!} z^{2n} .$$

Just as with the Bernoulli numbers, we can derive a symbolic equation for the Euler numbers. To do so, we multiply (6.46) by $\cosh z = \sum_{n=0}^{\infty} \frac{1}{(2n)!} z^{2n}$ and use Cauchy's multiplication theorem to get

$$1 = \left(\sum_{n=0}^{\infty} \frac{E_{2n}}{(2n)!} z^{2n} \right) \cdot \left(\sum_{n=0}^{\infty} \frac{1}{(2n)!} z^{2n} \right) = \sum_{n=0}^{\infty} \sum_{k=0}^{n} \left(\frac{E_{2k}}{(2k)!} \cdot \frac{1}{(2n-2k)!} \right) z^{2n}.$$

By the identity theorem, the $n = 0$ term on the right must equal 1, while all other terms must vanish. The $n = 0$ term on the right is just E_0, so $E_0 = 1$, and for $n > 1$, we must have $\sum_{k=0}^{n} \frac{E_{2k}}{(2k)!} \cdot \frac{1}{(2n-2k)!} = 0$. Multiplying this by $(2n)!$, we get

$$0 = \sum_{k=0}^{n} \frac{E_{2k}}{(2k)!} \cdot \frac{(2n)!}{(2n-2k)!} = \sum_{k=0}^{n} \frac{(2n)!}{(2k)!(2n-2k)!} \cdot E_{2k}. \qquad (6.47)$$

Now, from the binomial formula, for every complex number a, we have

$$(a+1)^{2n} + (a-1)^{2n} = \sum_{k=0}^{2n} \frac{(2n)!}{k!(2n-k)!} a^k + \sum_{k=0}^{2n} \frac{(2n)!}{k!(2n-k)!} a^k (-1)^{2n-k}$$

$$= \sum_{k=0}^{2n} \frac{(2n)!}{k!(2n-k)!} a^k + \sum_{k=0}^{2n} \frac{(2n)!}{k!(2n-k)!} a^k (-1)^k$$

$$= \sum_{k=0}^{2n} \frac{(2n)!}{(2k)!(2n-2k)!} a^{2k},$$

since all the odd terms cancel. Notice that the right-hand side of this expression is exactly the right-hand side of (6.47) if put $a = E$ and we make the superscript $2k$ into a subscript $2k$. Thus,

$$\boxed{(E+1)^{2n} + (E-1)^{2n} \doteq 0 \ , \ n = 1, 2, \ldots \text{ with } E_0 = 1 \text{ and } E_{\text{odd}} = 0.}$$ (6.48)

Using the identity (6.48), one can in principle find all the Euler numbers. For example, when $n = 1$, we see that

$$(E^2 + 2E^1 + 1) + (E^2 - 2E^1 + 1) \doteq 0 \implies 2E_2 + 2 = 0 \implies E_2 = -1.$$

Here is a partial list through E_{12}:

$$E_0 = 1, \quad E_1 = E_2 = E_3 = \cdots = 0 \ (E_{\text{odd}} = 0), \quad E_2 = -1, \quad E_4 = 5$$
$$E_6 = -61, \quad E_8 = 1385, \quad E_{10} = -50,521, \quad E_{12} = 2,702,765, \quad \ldots .$$

These numbers are all integers, but besides this fact, there is no known regular pattern to which these numbers conform.

▶ **Exercises 6.7**

1. Let $a, b \in \mathbb{C}$ and let $f(z) = 1/(a+z)$ and $g(z) = 1/(b-z)$, which have radii of convergence $|a|$ and $|b|$, respectively. (i) Find the radius of convergence of $f(g(z))$. (ii) Let $a > 0$ and prove that as $a \to 0$, the radius of convergence of $f(g(z))$ goes to infinity.

2. Recall that $\csc z = 1/\sin z$. Prove that $\csc z = \cot z + \tan(z/2)$, and from this identity deduce that

$$\boxed{z \csc z = \sum_{n=0}^{\infty} (-1)^{n-1} \frac{(2^{2n} - 2) B_{2n}}{(2n)!} z^{2n}.}$$

3. (a) Let $f(z) = \sum a_n z^n$ and $g(z) = \sum b_n z^n$ with $b_0 \neq 0$ be power series with positive radii of convergence. Show that $f(z)/g(z) = \sum c_n z^n$, where $\{c_n\}$ is the sequence defined recursively as follows:

$$c_0 = \frac{a_0}{b_0} \quad , \quad b_0 c_n = a_n - \sum_{k=1}^{n} b_k \, c_{n-k}.$$

(b) Use Part (a) to find the first few coefficients of the expansion for $\tan z = \sin z / \cos z$.

4. (Cf. [129, p. 526], which is reproduced in [178]) In this and the next problem we give an elegant application of the theory of Bernoulli numbers to find the sum of the first kth powers of integers, Bernoulli's original motivation for his numbers.

(i) For $n \in \mathbb{N}$, derive the formula

$$1 + e^z + e^{2z} + \cdots + e^{nz} = \frac{z}{e^z - 1} \cdot \frac{e^{(n+1)z} - 1}{z}.$$

(ii) Writing each side of the identity in (i) as a power series (on the right, you need to use the Cauchy product), derive the formula

$$1^k + 2^k + \cdots + n^k = \sum_{j=0}^{k} \binom{k}{j} B_j \frac{(n+1)^{k+1-j}}{k+1-j}, \quad k = 1, 2, \ldots. \qquad (6.49)$$

Plug in $k = 1, 2, 3$ to derive some pretty formulas!

5. Here's another proof of (6.49) that is aesthetically appealing.

(i) Prove that for every complex number a and all natural numbers k, n,

$$(n + 1 + a)^{k+1} - (n + a)^{k+1} = \sum_{j=1}^{k+1} \binom{k+1}{j} n^{k+1-j} \left((a+1)^j - a^j \right).$$

(ii) Prove that

$$1^k + 2^k + \cdots + n^k \doteq \frac{1}{k+1} \left\{ (n + 1 + B)^{k+1} - B^{k+1} \right\}.$$

Suggestion: Look for a telescoping sum and recall that $(B + 1)^j \doteq B^j$ for $j \geq 2$.

6. The nth Bernoulli polynomial $B_n(t)$, where $t \in \mathbb{C}$, is by definition $n!$ times the coefficient of z^n in the power series expansion in z of the function $f(z, t) := ze^{zt}/(e^z - 1)$; that is,

$$\frac{z \, e^{zt}}{e^z - 1} = \sum_{n=0}^{\infty} \frac{B_n(t)}{n!} z^n. \qquad (6.50)$$

Here we are using the fact that the function e^{zt} has a power series in z and so does $z/(e^z - 1)$, so their product $f(z, t)$ is also a power series in z.

(a) Prove that $B_n(t) = \sum_{k=0}^n \binom{n}{k} B_k t^{n-k}$, where the B_k are the Bernoulli numbers. Thus, the first few Bernoulli polynomials are

$$B_0(t) = 1, \quad B_1(t) = t - \frac{1}{2}, \quad B_2(t) = t^2 - t + \frac{1}{6}, \quad B_3(t) = t^3 - \frac{3}{2}t^2 + \frac{1}{2}t.$$

(b) Prove that $B_n(0) = B_n$ for $n = 0, 1, \ldots$ and that $B_n(0) = B_n(1) = B_n$ for $n \neq 1$. Suggestion: Show that $f(z, 1) = z + f(z, 0)$.

(c) Prove that $B_n(t + 1) - B_n(t) = nt^{n-1}$ for $n = 0, 1, 2, \ldots$. Suggestion: Show that $f(z, t + 1) - f(z, t) = ze^{zt}$.

(d) Prove that $B_{2n+1}(0) = 0$ for $n = 1, 2, \ldots$ and that $B_{2n+1}(1/2) = 0$ for $n = 0, 1, \ldots$.

6.8 The Logarithmic, Binomial, Arctangent Series, and γ

From elementary calculus, you might have seen the logarithmic, binomial, and arctangent series (discovered by Nikolaus Mercator (1620–1687), Isaac Newton (1643–1727), and Madhava of Sangamagramma (1350–1425), respectively):

$$\log(1 + x) = \sum_{n=1}^{\infty} \frac{(-1)^{n-1}}{n} x^n, \quad (1 + x)^\alpha = \sum_{n=0}^{\infty} \binom{\alpha}{n} x^n, \quad \arctan x = \sum_{n=0}^{\infty} (-1)^n \frac{x^{2n+1}}{2n + 1},$$

where $\alpha \in \mathbb{R}$. (Below, we'll discuss the meaning of $\binom{\alpha}{n}$.) I can bet that you used calculus (derivatives and integrals) to derive these formulas. In this section we'll derive even more general complex versions of these formulas without derivatives!

6.8.1 The Binomial Coefficients

From our familiar binomial theorem, we know that for every $z \in \mathbb{C}$ and $k \in \mathbb{N}$, we have $(1 + z)^k = \sum_{n=0}^k \binom{k}{n} z^n$, where $\binom{k}{0} := 1$ and for $n = 1, 2, \ldots, k$,

$$\binom{k}{n} := \frac{k!}{n!(k-n)!} = \frac{1 \cdot 2 \cdots k}{n! \cdot 1 \cdot 2 \cdots (k-n)} = \frac{k(k-1) \cdots (k-n+1)}{n!}. \quad (6.51)$$

The formula $(1 + z)^k = \sum_{n=0}^k \binom{k}{n} z^n$ trivially holds when $k = 0$ too. Another way to express this formula is

$$(1+z)^k = 1 + \sum_{n=1}^{k} \frac{k(k-1)\cdots(k-n+1)}{n!} z^n.$$

With this motivation, given a complex number α, we define the **binomial coefficient** $\binom{\alpha}{n}$ for every nonnegative integer n as follows: $\binom{\alpha}{0} = 1$ and for $n > 0$,

$$\binom{\alpha}{n} = \frac{\alpha(\alpha-1)\cdots(\alpha-n+1)}{n!}. \tag{6.52}$$

Note that if $\alpha = 0, 1, 2, \ldots$, then we see that all $\binom{\alpha}{n}$ vanish for $n \geq \alpha + 1$, and $\binom{\alpha}{n}$ is exactly the usual binomial coefficient (6.51). In the following lemma, we derive an identity that will be useful later.

Lemma 6.36 *For every $\alpha, \beta \in \mathbb{C}$, we have*

$$\binom{\alpha+\beta}{n} = \sum_{k=0}^{n} \binom{\alpha}{k}\binom{\beta}{n-k}, \qquad n = 0, 1, 2, \ldots.$$

Proof Throughout this proof, we put $\mathbb{N}_0 := \{0, 1, 2, 3, \ldots\}$.

Step 1: First of all, our lemma holds when both α and β are in $\mathbb{N}_0$. Indeed, if $\alpha = p, \beta = q$ are in $\mathbb{N}_0$, then expressing both sides of the identity $(1+z)^{p+q} = (1+z)^p(1+z)^q$ using the binomial formula, we obtain

$$\sum_{n=0}^{p+q} \binom{p+q}{n} z^n = \left(\sum_{k=0}^{p} \binom{p}{k} z^k\right) \cdot \left(\sum_{k=0}^{q} \binom{q}{k} z^k\right)$$

$$= \sum_{n=0}^{p+q} \left(\sum_{k=0}^{n} \binom{p}{k}\binom{q}{n-k}\right) z^n,$$

where at the last step we formed the Cauchy product of $(1+z)^p(1+z)^q$. By the identity theorem we must have

$$\binom{p+q}{n} = \sum_{k=0}^{n} \binom{p}{k}\binom{q}{n-k}, \qquad \text{for all } p, q, n \in \mathbb{N}_0.$$

Step 2: Assume now that $\beta = q \in \mathbb{N}_0$, $n \in \mathbb{N}_0$, and define $f : \mathbb{C} \longrightarrow \mathbb{C}$ by

$$f(z) := \binom{z+q}{n} - \sum_{k=0}^{n} \binom{z}{k}\binom{q}{n-k}.$$

In view of the definition (6.52) of the binomial coefficient, it follows that $f(z)$ is a polynomial in z of degree at most n. Moreover, by **Step 1** we know that $f(p) = 0$ for all $p \in \mathbb{N}_0$. In particular, the polynomial $f(z)$ has more than n roots. Therefore, $f(z)$ must be the zero polynomial, so in particular, given $\alpha \in \mathbb{C}$, we have $f(\alpha) = 0$; that is,

$$\binom{\alpha + q}{n} = \sum_{k=0}^{n} \binom{\alpha}{k} \binom{q}{n-k}, \quad \text{for all } \alpha \in \mathbb{C}, q, n \in \mathbb{N}_0.$$

Step 3: Let $\alpha \in \mathbb{C}$, $n \in \mathbb{N}_0$, and define $g : \mathbb{C} \longrightarrow \mathbb{C}$ by

$$g(z) := \binom{\alpha + z}{n} - \sum_{k=0}^{n} \binom{\alpha}{k} \binom{z}{n-k}.$$

As with the function $f(z)$ in **Step 2**, $g(z)$ is a polynomial in z of degree at most n. Also, by **Step 2** we know that $g(q) = 0$ for all $q \in \mathbb{N}_0$, and consequently, $g(z)$ must be the zero polynomial. In particular, given $\beta \in \mathbb{C}$, we have $g(\beta) = 0$, which completes our proof.   ∎

6.8.2 The Complex Logarithm and Binomial Series

In Theorem 6.38 we shall derive (along with a power series for Log) the **binomial series**:

$$(1+z)^\alpha = \sum_{n=0}^{\infty} \binom{\alpha}{n} z^n = 1 + \alpha z + \frac{\alpha(\alpha-1)}{1!} z^2 + \cdots, \quad |z| < 1. \quad (6.53)$$

Let us define $f(\alpha, z) := \sum_{n=0}^{\infty} \binom{\alpha}{n} z^n$. Our goal is to show that $f(\alpha, z) = (1+z)^\alpha$ for all $\alpha \in \mathbb{C}$ and $|z| < 1$, where

$$(1+z)^\alpha := \exp(\alpha \operatorname{Log}(1+z)),$$

with Log the principal logarithm of the complex number $1 + z$. If $\alpha = k = 0, 1, 2, \ldots$, then we already know that all the $\binom{k}{n}$ vanish for $n \geq k + 1$, and these binomial coefficients are the usual ones, so $f(k, z)$ converges with sum $f(k, z) = (1+z)^k$. To see that $f(\alpha, z)$ converges for all other α, assume that $\alpha \in \mathbb{C}$ is not a nonnegative integer. Then setting $a_n = \binom{\alpha}{n}$, we have

$$\left| \frac{a_n}{a_{n+1}} \right| = \left| \frac{\alpha(\alpha-1)\cdots(\alpha-n+1)}{n!} \cdot \frac{(n+1)!}{\alpha(\alpha-1)\cdots(\alpha-n)} \right| = \frac{n+1}{|\alpha-n|},$$

which approaches 1 as $n \to \infty$. Thus, the radius of convergence of $f(\alpha, z)$ is 1 by the radius of convergence formula (6.14) on p. 446. In conclusion, $f(\alpha, z)$ is convergent for all $\alpha \in \mathbb{C}$ and $|z| < 1$.

We now prove the real versions of the logarithm series and the binomial series (6.53); see Theorem 6.38 below for the more general complex version. It is worth emphasizing that we do not use the advanced technology of the differential and integral calculus to derive these formulas!

Lemma 6.37 *For all $x \in \mathbb{R}$ with $|x| < 1$, we have*

$$\log(1 + x) = \sum_{n=1}^{\infty} \frac{(-1)^{n-1}}{n} x^n,$$

and for all $\alpha \in \mathbb{C}$ and $x \in \mathbb{R}$ with $|x| < 1$, we have

$$(1 + x)^\alpha = \sum_{n=0}^{\infty} \binom{\alpha}{n} x^n = 1 + \alpha x + \frac{\alpha(\alpha - 1)}{1!} x^2 + \cdots.$$

Proof We prove this lemma in three steps.

Step 1: We first show that $f(r, x) = (1 + x)^r$ for all $r = p/q \in \mathbb{Q}$, where $p, q \in \mathbb{N}$ with q odd and $x \in \mathbb{R}$ with $|x| < 1$. To see this, observe that for every $z \in \mathbb{C}$ with $|z| < 1$, taking the Cauchy product of $f(\alpha, z)$ and $f(\beta, z)$ and using our lemma, we obtain

$$f(\alpha, z) \cdot f(\beta, z) = \sum_{n=0}^{\infty} \left(\sum_{j=0}^{n} \binom{\alpha}{j} \binom{\beta}{n-j} \right) z^n = \sum_{n=0}^{\infty} \binom{\alpha + \beta}{n} z^n = f(\alpha + \beta, z).$$

By induction it easily follows that

$$f(\alpha_1, z) \cdot f(\alpha_2, z) \cdots f(\alpha_k, z) = f(\alpha_1 + \alpha_2 + \cdots + \alpha_k, z).$$

Using this identity, we obtain

$$f(1/q, z)^q = \underbrace{f(1/q, z) \cdots f(1/q, z)}_{q\text{ times}} = f(\underbrace{1/q + \cdots + 1/q}_{q\text{ times}}, z) = f(1, z) = 1 + z.$$

Now put $z = x \in \mathbb{R}$ with $|x| < 1$ and let $q \in \mathbb{N}$ be odd. Then $f(1/q, x)^q = 1 + x$, so taking qth roots, we get $f(1/q, x) = (1 + x)^{1/q}$. Here we used that every real number has a unique qth root, which holds because q is odd; for q even we could conclude only that $f(1/q, x) = \pm(1 + x)^{1/q}$ (unless we checked that $f(1/q, x)$ is positive, and then we would get $f(1/q, x) = (1 + x)^{1/q}$). Thus, for $r = p/q$ with $p \in \mathbb{N}$,

$$f(r, x) = f(p/q, x) = f(\underbrace{1/q + \cdots + 1/q}_{p \text{ times}}, x) = \underbrace{f(1/q, x) \cdots f(1/q, x)}_{p \text{ times}}$$

$$= f(1/q, x)^p = (1 + x)^{p/q} = (1 + x)^r.$$

Step 2: Second, we prove that for every $z \in \mathbb{C}$ with $|z| < 1$, $f(\alpha, z)$ can be written as a power series in α that converges for all $\alpha \in \mathbb{C}$:

$$f(\alpha, z) = 1 + \sum_{m=1}^{\infty} a_m(z) \, \alpha^m.$$

In particular, since we know that power series are continuous, $f(\alpha, z)$ is a continuous function of $\alpha \in \mathbb{C}$. Here, the coefficients $a_m(z)$ depend on z (which we'll see are power series in z), and we'll show that

$$a_1(z) = \sum_{n=1}^{\infty} \frac{(-1)^{n-1}}{n} z^n. \tag{6.54}$$

To prove these statements, note that for $n \geq 1$, $\alpha(\alpha - 1) \cdots (\alpha - n + 1)$ is a polynomial of degree n in α, so for $n \geq 1$ we can write

$$\binom{\alpha}{n} = \frac{\alpha(\alpha - 1) \cdots (\alpha - n + 1)}{n!} = \sum_{m=1}^{n} a_{mn} \, \alpha^m, \tag{6.55}$$

for some coefficients a_{mn}. Defining $a_{mn} = 0$ for $m = n + 1, n + 2, n + 3, \ldots$, we can write $\binom{\alpha}{n} = \sum_{m=0}^{\infty} a_{mn} \, \alpha^m$. Hence,

$$f(z, \alpha) = 1 + \sum_{n=1}^{\infty} \binom{\alpha}{n} z^n = 1 + \sum_{n=1}^{\infty} \left(\sum_{m=1}^{\infty} a_{mn} \, \alpha^m \right) z^n. \tag{6.56}$$

To make this a power series in α, we need to switch the order of summation, which we can do by Cauchy's double series theorem if we can demonstrate absolute convergence by showing that

$$\sum_{n=1}^{\infty} \sum_{m=1}^{\infty} \left| a_{mn} \, \alpha^m z^n \right| = \sum_{n=1}^{\infty} \sum_{m=1}^{\infty} |a_{mn}| \, |\alpha|^m \, |z|^n < \infty.$$

To verify this, we first observe that for all $\alpha \in \mathbb{C}$, we have

$$\frac{\alpha(\alpha+1)(\alpha+2)\cdots(\alpha+n-1)}{n!} = \sum_{m=1}^{n} b_{mn}\,\alpha^m, \qquad (6.57)$$

where the b_{mn} are nonnegative real numbers. This is certainly plausible, because each of the numbers $1, 2, \ldots, n-1$ on the left comes with a positive sign; in any case, this statement can be verified by induction, for instance, so we leave it to the reader. We secondly observe that replacing α with $-\alpha$ in (6.55), we get

$$\sum_{m=1}^{n} a_{mn}\,(-1)^m \alpha^m = \frac{-\alpha(-\alpha-1)\cdots(-\alpha-n+1)}{n!}$$

$$= (-1)^n \frac{\alpha(\alpha+1)\cdots(\alpha+n-1)}{n!} = \sum_{m=1}^{n}(-1)^n b_{mn}\,\alpha^m.$$

By the identity theorem, we have $a_{mn}(-1)^m = (-1)^n b_{mn}$. In particular, $|a_{mn}| = b_{mn}$, since $b_{mn} > 0$, and therefore, in view of (6.57), we see that

$$\sum_{m=0}^{\infty} |a_{mn}|\,|\alpha|^m = \sum_{m=0}^{n} |a_{mn}|\,|\alpha|^m = \sum_{m=0}^{n} b_{mn}\,|\alpha|^m = \frac{|\alpha|(|\alpha|+1)\cdots(|\alpha|+n-1)}{n!}.$$

Therefore,

$$\sum_{n=1}^{\infty}\sum_{m=1}^{\infty} |a_{mn}|\,|\alpha|^m\,|z|^n = \sum_{n=1}^{\infty} \frac{|\alpha|(|\alpha|+1)\cdots(|\alpha|+n-1)}{n!}\,|z|^n.$$

Using the now very familiar ratio test, it's easily checked that since $|z| < 1$, the series on the right converges. Thus, we can iterate sums in (6.56) and conclude that

$$f(\alpha, z) = 1 + \sum_{n=1}^{\infty}\left(\sum_{m=1}^{\infty} a_{mn}\,\alpha^m\right) z^n = 1 + \sum_{m=1}^{\infty}\left(\sum_{n=1}^{\infty} a_{mn}\,z^n\right)\alpha^m.$$

Thus, $f(\alpha, z)$ is indeed a power series in α. To prove (6.54), we need to find the coefficient of α^1 in (6.55), which we see is given by

$$a_{1n} = \text{coefficient of } \alpha \text{ in } \frac{\alpha(\alpha-1)(\alpha-2)\cdots(\alpha-n+1)}{n!}$$

$$= \frac{(-1)(-2)(-3)\cdots(-n+1)}{n!} = (-1)^{n-1}\frac{(n-1)!}{n!} = \frac{(-1)^{n-1}}{n}.$$

Therefore,

$$a_1(z) = \sum_{n=1}^{\infty} a_{1n}\,z^n = \sum_{n=1}^{\infty} \frac{(-1)^{n-1}}{n}\,z^n,$$

just as we stated in (6.54). This completes **Step 2**.

Step 3: We are finally ready to prove our theorem. Let $x \in \mathbb{R}$ with $|x| < 1$. By **Step 2**, we know that for every $\alpha \in \mathbb{C}$,

$$f(\alpha, x) = 1 + \sum_{m=1}^{\infty} a_m(x)\, \alpha^m$$

is a power series in α. However,

$$(1 + x)^{\alpha} = \exp(\alpha \log(1 + x)) = \sum_{n=0}^{\infty} \frac{1}{n!} \log(1 + x)^n \cdot \alpha^n$$

is also a power series in $\alpha \in \mathbb{C}$. By **Step 1**, $f(\alpha, x) = (1 + x)^{\alpha}$ for all $\alpha \in \mathbb{Q}$ with $\alpha > 0$ having odd denominators. The identity theorem applies to this situation (why?), so we must have $f(\alpha, x) = (1 + x)^{\alpha}$ for all $\alpha \in \mathbb{C}$. Also by the identity theorem, the coefficients of α^n must be identical; in particular, the coefficients of α^1 are identical: $a_1(x) = \log(1 + x)$. Now (6.54) implies the series for $\log(1 + x)$. ∎

Using this lemma and the identity theorem, we are ready to generalize these formulas for real x to formulas for complex z.

The complex logarithm and binomial series

Theorem 6.38 *We have*

$$\mathrm{Log}(1 + z) = \sum_{n=1}^{\infty} \frac{(-1)^{n-1}}{n} z^n, \quad |z| \le 1,\ z \ne -1,$$

and given $\alpha \in \mathbb{C}$, we have

$$(1 + z)^{\alpha} = \sum_{n=0}^{\infty} \binom{\alpha}{n} z^n = 1 + \alpha z + \frac{\alpha(\alpha - 1)}{1!} z^2 + \cdots, \quad |z| < 1.$$

Proof We prove this theorem first for $\mathrm{Log}(1 + z)$, then for $(1 + z)^{\alpha}$.

Step 1: We shall prove that $\mathrm{Log}(1 + z) = \sum_{n=1}^{\infty} \frac{(-1)^{n-1}}{n} z^n$ holds for $|z| < 1$. To this end, define $f(z) = \sum_{n=1}^{\infty} \frac{(-1)^{n-1}}{n} z^n$; we will prove that $\mathrm{Log}(1 + z) = f(z)$ for $|z| < 1$. First, one can check that the radius of convergence of $f(z)$ is 1, so by our power series composition theorem, we know that $\exp(f(z))$ can be written as a power series:

$$\exp(f(z)) = \sum_{n=0}^{\infty} a_n z^n, \quad |z| < 1.$$

Restricting to real values of z, by our lemma we know that $f(x) = \log(1 + x)$, so

$$\sum_{n=0}^{\infty} a_n x^n = \exp(f(x)) = \exp(\log(1 + x)) = 1 + x.$$

By the identity theorem for power series, we must have $a_0 = 1$, $a_1 = 1$, and all other $a_n = 0$. Thus, $\exp(f(z)) = 1 + z$. Since $\exp(\mathrm{Log}(1 + z)) = 1 + z$ as well, we have

$$\exp(f(z)) = \exp(\mathrm{Log}(1 + z)) \quad \text{for all } z \in \mathbb{C} \text{ with } |z| < 1.$$

By Part (1) of Theorem 4.41 on p. 333, it follows that there is an integer k such that $f(z) = \mathrm{Log}(1 + z) + 2\pi i k$ for $|z| < 1$ (try to think about why this statement holds). If we set $z = 0$ and note that $f(0) = 0$ and $\mathrm{Log}(1) = 0$, we get $k = 0$. This completes **Step 1**.

Step 2: We now prove that $\mathrm{Log}(1 + z) = \sum_{n=1}^{\infty} \frac{(-1)^{n-1}}{n} z^n$ holds for $|z| = 1$ with $z \neq -1$ (note that for $z = -1$, both sides of this equality are not defined). Let $z \in \mathbb{C}$ with $|z| = 1$ and $z \neq -1$. Then we can write $z = -e^{ix}$ with $x \in (0, 2\pi)$. Recall from Example 6.5 on p. 426 that for every $x \in (0, 2\pi)$, the series $\sum_{n=1}^{\infty} \frac{e^{inx}}{n}$ converges. Since

$$-\sum_{n=1}^{\infty} \frac{(-1)^{n-1}}{n} z^n = \sum_{n=1}^{\infty} \frac{(-1)^n(-e^{ix})^n}{n} = \sum_{n=1}^{\infty} \frac{e^{inx}}{n}, \tag{6.58}$$

it follows that $\sum_{n=1}^{\infty} \frac{(-1)^{n-1}}{n} z^n$ converges for $|z| = 1$ with $z \neq -1$. We can now use Abel's theorem (Theorem 6.21 on p. 460): taking $w \to z$ through the straight line from 0 to z, it follows that

$$\sum_{n=1}^{\infty} \frac{(-1)^{n-1}}{n} z^n = \lim_{w \to z} \sum_{n=1}^{\infty} \frac{(-1)^{n-1}}{n} w^n = \lim_{w \to z} \mathrm{Log}(1 + w) = \mathrm{Log}(1 + z),$$

$$\tag{6.59}$$

where at the second equality we used **Step 1** and for the third equality we used that $\mathrm{Log}(1 + w)$ is continuous for $|w| \leq 1$ with $w \neq -1$.

Step 3: Let $\alpha \in \mathbb{C}$. To prove the binomial series, we note that by the power series composition theorem, $(1 + z)^{\alpha} = \exp(\alpha \mathrm{Log}(1 + z))$, being the composition of $\exp(z)$ and $\alpha \mathrm{Log}(1 + z)$, both of which are power series in z, is also a power series in z, convergent for $|z| < 1$. Let $f(\alpha, z) = \sum_{n=0}^{\infty} \binom{\alpha}{n} z^n$ for $z \in \mathbb{C}$ with $|z| < 1$. Then restricting to real $z = x \in \mathbb{R}$ with $|x| < 1$, by our lemma we know that $(1 + x)^{\alpha} = f(\alpha, x)$. Hence, by the identity theorem, we must have $(1 + z)^{\alpha} = f(\alpha, z)$ for all $z \in \mathbb{C}$ with $|z| < 1$. This proves the binomial series. ∎

For every $z \in \mathbb{C}$ with $|z| < 1$, we have $\mathrm{Log}\big((1 + z)/(1 - z)\big) = \mathrm{Log}(1 + z) - \mathrm{Log}(1 - z)$. Using this fact together with Theorem 6.38, in Problem 1 you will prove the interesting formula

$$\frac{1}{2} \operatorname{Log}\left(\frac{1+z}{1-z}\right) = \sum_{n=0}^{\infty} \frac{z^{2n+1}}{2n+1}. \qquad (6.60)$$

The next example contains two other interesting formulas.

Example 6.42 In the proof of Theorem 6.38 we used that for $x \in (0, 2\pi)$, the series $\sum_{n=1}^{\infty} \frac{e^{inx}}{n} = \sum_{n=1}^{\infty} \frac{\cos nx}{n} + i \sum_{n=1}^{\infty} \frac{\sin nx}{n}$ converges. We shall prove that

$$\sum_{n=1}^{\infty} \frac{\cos nx}{n} = -\log\left(2\sin(x/2)\right) \quad \text{and} \quad \sum_{n=1}^{\infty} \frac{\sin nx}{n} = \frac{\pi - x}{2} \quad \text{for } x \in (0, 2\pi).$$

To prove this, let $x \in (0, 2\pi)$ and note that from Eqs. (6.58) and (6.59) above,

$$\sum_{n=1}^{\infty} \frac{\cos nx}{n} + i \sum_{n=1}^{\infty} \frac{\sin nx}{n} = -\operatorname{Log}(1 - e^{ix}).$$

Since

$$1 - e^{ix} = e^{ix/2}(e^{-ix/2} - e^{ix/2}) = -2ie^{ix/2}\sin(x/2) = 2\sin(x/2)e^{ix/2 - i\pi/2},$$

by definition of Log, we have

$$\operatorname{Log}(1 - e^{ix}) = \log\left(2\sin(x/2)\right) + i\frac{x - \pi}{2}.$$

Our two interesting formulas now follow.

6.8.3   Gregory–Madhava Series and Formulas for γ

Recall from Section 4.11 that

$$\operatorname{Arctan} z = \frac{1}{2i} \operatorname{Log} \frac{1 + iz}{1 - iz}.$$

Using (6.60), we get the famous formula first discovered by Madhava of Sangama-gramma (1350–1425) around 1400 and rediscovered over 200 years later in Europe by James Gregory (1638–1675), who found it in 1671! In fact, the mathematicians of Kerala in southern India discovered not only the arctangent series, they also discovered the infinite series for sine and cosine, but their results were written up in Sanskrit and not brought to Europe until the 1800s. For more history on this fascinating topic, see the articles [120, 203], and the website [184].

Theorem 6.39 *For every complex number z with $|z| < 1$, we have*

$$\text{Arctan } z = \sum_{n=0}^{\infty} (-1)^n \frac{z^{2n+1}}{2n+1}, \qquad \textit{Gregory–Madhava's series.}$$

This series is commonly known as **Gregory's arctangent series**, but we shall call it the **Gregory–Madhava arctangent series** because of Madhava's contribution to this series. Setting $z = x$, a real variable, we obtain the usual formula learned in elementary calculus:

$$\arctan x = \sum_{n=0}^{\infty} (-1)^n \frac{x^{2n+1}}{2n+1}.$$

In Problem 6 we prove the following stunning formulas for the Euler–Mascheroni constant γ in terms of the Riemann ζ-function $\zeta(z)$:

$$\begin{aligned}
\gamma &= \sum_{n=2}^{\infty} \frac{(-1)^n}{n} \zeta(n) \\
&= 1 - \sum_{n=2}^{\infty} \frac{1}{n} \left(\zeta(n) - 1 \right) \\
&= \frac{3}{2} - \log 2 - \sum_{n=2}^{\infty} \frac{(-1)^n}{n} (n-1)\left(\zeta(n) - 1 \right).
\end{aligned} \tag{6.61}$$

The first two formulas are due to Euler and the last one to Philippe Flajolet and Ilan Vardi (see [217, pp. 4, 5], [82]).

▶ **Exercises 6.8**

1. Fill in the details in the proof of formula (6.60).
2. Derive the following remarkably pretty formulas:

$$2(\text{Arctan } z)^2 = \sum_{n=0}^{\infty} \frac{(-1)^n}{2n+2} \left(1 + \frac{1}{3} + \frac{1}{5} + \cdots + \frac{1}{2n+1} \right) z^{2n+2},$$

and

$$\frac{1}{2}(\text{Log}(1+z))^2 = \sum_{n=0}^{\infty} \frac{(-1)^n}{n+2} \left(1 + \frac{1}{2} + \frac{1}{3} + \cdots + \frac{1}{n+1} \right) z^{n+2},$$

both valid for $|z| < 1$.

3. Before looking at the next section, prove that

$$\arctan x = \sum_{n=0}^{\infty} (-1)^n \frac{x^{2n+1}}{2n+1} \quad \text{and} \quad \log(1+x) = \sum_{n=0}^{\infty} \frac{(-1)^{n-1}}{n} x^n$$

are valid for $-1 < x \le 1$. Suggestion: I know you are Abel to do this! Note that setting $x = 1$ in the formulas, we obtain

$$\frac{\pi}{4} = 1 - \frac{1}{3} + \frac{1}{5} - \frac{1}{7} + -\cdots \quad \text{and} \quad \log 2 = 1 - \frac{1}{2} + \frac{1}{3} - \frac{1}{4} + \cdots.$$

4. For $\alpha \in \mathbb{R}$, prove that $\sum_{n=0}^{\infty} \binom{\alpha}{n}$ converges if and only if $\alpha \le 0$ or $\alpha \in \mathbb{N}$, in which case

$$2^\alpha = \sum_{n=0}^{\infty} \binom{\alpha}{n}.$$

Suggestion: To prove convergence, use Gauss's test.

5. Prove the exquisite formulas

$$(a) \quad \sum_{n=1}^{\infty} \frac{1}{n} \frac{z^n}{1-z^n} = \sum_{n=1}^{\infty} \mathrm{Log} \frac{1}{1-z^n}, \quad |z| < 1,$$

$$(b) \quad \sum_{n=1}^{\infty} \frac{(-1)^{n-1}}{n} \frac{z^n}{1-z^n} = \sum_{n=1}^{\infty} \mathrm{Log}(1+z^n), \quad |z| < 1.$$

Suggestion: Cauchy's double series theorem.

6. In this problem, we prove the stunning formulas in (6.61).

   (i) Using the first formula for γ in Problem 7a on p. 312, prove that $\gamma = \sum_{n=1}^{\infty} f\left(\frac{1}{n}\right)$, where $f(z) = \sum_{n=2}^{\infty} \frac{(-1)^n}{n} z^n$.

   (ii) Prove that $\gamma = 1 - \log 2 + \sum_{n=2}^{\infty} \frac{(-1)^n}{n}(\zeta(n) - 1)$ using (i) and Problem 6.5 on p. 484. Show that this formula is equivalent to the first formula in (6.61).

   (iii) Using the second and third formulas in Problem 7a on p. 312, derive the second and third formulas in (6.61).

6.9 ★ π, Euler, Fibonacci, Leibniz, Madhava, and Machin

In this section, we continue our fascinating study of formulas for π that we initiated in Section 4.12 starting on p. 364. In particular, we derive (using a very different method from the one presented in Section 5.2 on p. 399) Gregory–Leibniz–Madhava's formula for $\pi/4$, formulas for π discovered by Euler involving the arctangent function and even the Fibonacci numbers, and finally, we look at Machin's formula for π, versions of which have been used to compute trillions of digits of π by Yasumasa Kanada and his coworkers at the University of Tokyo.[12] For other formulas for $\pi/4$ in terms of arctangents, see the articles [105, 142]. For more on computing π, see [12] and the famous *New Yorker* article "The Mountains of Pi" [196], on David and Gregory Chudnovsky. For interesting historical facets on π in general, see [9, 47, 48]. The website [221] has tons of information.

6.9.1 Gregory–Leibniz–Madhava's Formula for $\pi/4$, Proof II

Recall Gregory–Madhava's formula for real values:

$$\arctan x = \sum_{n=0}^{\infty} (-1)^{n-1} \frac{x^{2n-1}}{2n-1}.$$

By the alternating series theorem, we know that $\sum_{n=0}^{\infty} (-1)^{n-1}/(2n-1)$ converges, therefore by Abel's limit theorem (Theorem 6.21), we know that

$$\frac{\pi}{4} = \lim_{x \to 1-} \arctan x = \sum_{n=0}^{\infty} (-1)^{n-1} \frac{1}{2n-1} = 1 - \frac{1}{3} + \frac{1}{5} - \frac{1}{7} + - \cdots.$$

Therefore, we obtain another derivation of

$$\boxed{\frac{\pi}{4} = 1 - \frac{1}{3} + \frac{1}{5} - \frac{1}{7} + - \cdots, \quad \textbf{Gregory–Leibniz–Madhava's series}.}$$

Madhava of Sangamagramma (1350–1425) was the first to discover this formula, over 200 years before James Gregory (1638–1675) and Gottfried Leibniz (1646–1716) were even born! Note that the Gregory–Leibniz–Madhava series is really just a special case of Gregory–Madhava's formula for $\arctan x$ (just set $x = 1$), which, recall, was discovered in 1671 by Gregory and much earlier by Madhava. Leibniz discovered the formula for $\pi/4$ (using geometric arguments) around 1673. Although there is no published record of Gregory noting the formula for $\pi/4$ (he published few of his mathematical results and he died at only 37 years old), it would be hard to believe that he didn't know about the formula for $\pi/4$. For more history, including Nilakantha Somayaji's (1444–1544) contribution, see [37, 120, 184, 203].

---

[12] "The value of π has engaged the attention of many mathematicians and calculators from the time of Archimedes to the present day, and has been computed from so many different formulae, that a complete account of its calculation would almost amount to a history of mathematics." James Glaisher (1848–1928) [89].

Example 6.43 Suppose we approximate $\pi/4$ by Gregory–Leibniz–Madhava's series to within, say, the reasonable amount of seven decimal places. Then denoting the nth partial sum of Gregory–Leibniz–Madhava's series by s_n, according to Problem 5 on p. 433, if we have seven decimal places, then

$$\frac{1}{2(2n+1)} \leq \left| \frac{\pi}{4} - s_n \right| < 0.00000005 = 5 \times 10^{-8},$$

which implies that $2n + 1 > 10^7$, or $n \geq 5,000,000$. Thus, to approximate $\pi/4$ to within seven decimal places by the partial sums of the Gregory–Leibniz–Madhava series, we are required to use at least *five million* terms! Thus, although Gregory–Leibniz–Madhava's series is beautiful, it is quite useless for computing π.

Example 6.44 From Gregory–Leibniz–Madhava's formula, we can easily derive the breathtaking formula (see Problem 4)

$$\boxed{\pi = \sum_{n=2}^{\infty} \frac{3^n - 1}{4^n} \zeta(n+1),}$$ (6.62)

due to Philippe Flajolet and Ilan Vardi (see [218, p. 1], [82, 250]).

6.9.2 Euler's Arctangent Formula and the Fibonacci Numbers

In 1738, Euler derived a very pretty two-angle arctangent expression for π:

$$\frac{\pi}{4} = \arctan \frac{1}{2} + \arctan \frac{1}{3}.$$ (6.63)

This formula is very easy to derive. We start off with the addition formula for tangent (see (4.41), but now considering real variables)

$$\frac{\tan \theta + \tan \phi}{1 - \tan \theta \tan \phi} = \tan(\theta + \phi),$$ (6.64)

where it is assumed that $1 - \tan \theta \tan \phi \neq 0$. Let $x = \tan \theta$ and $y = \tan \phi$ and assume that $-\pi/2 < \theta + \phi < \pi/2$. Then taking arctangents of both sides of the above equation, we obtain

$$\arctan \left(\frac{x+y}{1-xy} \right) = \theta + \phi,$$

or after putting the right-hand side in terms of x, y, we get

$$\arctan\left(\frac{x+y}{1-xy}\right) = \arctan x + \arctan y. \tag{6.65}$$

Setting $x = 1/2$ and $y = 1/3$, we find that

$$\arctan 1 = \arctan\frac{1}{2} + \arctan\frac{1}{3}.$$

This equality is just (6.63).

In Problem 9 on p. 47 we studied the **Fibonacci sequence**, named after Leonardo Fibonacci (1170–1250): $F_0 = 0$, $F_1 = 1$, and $F_n = F_{n-1} + F_{n-2}$ for all $n \geq 2$, and you proved that for every n,

$$F_n = \frac{1}{\sqrt{5}}\left[\Phi^n - (-\Phi)^{-n}\right], \quad \text{where} \quad \Phi = \frac{1+\sqrt{5}}{2}. \tag{6.66}$$

Observe that (6.63) can be written as

$$\frac{\pi}{4} = \arctan\left(\frac{1}{F_3}\right) + \arctan\left(\frac{1}{F_4}\right).$$

We can use (6.65) and the definition of the Fibonacci numbers to rewrite this as

$$\frac{\pi}{4} = \arctan\left(\frac{1}{F_3}\right) + \arctan\left(\frac{1}{F_5}\right) + \arctan\left(\frac{1}{F_6}\right).$$

Continuing, we can prove the following fascinating formula for $\pi/4$ in terms of the Fibonacci numbers due to Derrick H. Lehmer (1905–1991) [141] (see Problem 2 and [143]):

$$\boxed{\frac{\pi}{4} = \sum_{n=1}^{\infty} \arctan\left(\frac{1}{F_{2n+1}}\right).} \tag{6.67}$$

Also, in Problem 3 you will prove the following series due to Dario Castellanos [47]:

$$\boxed{\frac{\pi}{\sqrt{5}} = \sum_{n=0}^{\infty} \frac{(-1)^n F_{2n+1} 2^{2n+3}}{(2n+1)(3+\sqrt{5})^{2n+1}}.} \tag{6.68}$$

6.9.3 Machin's Arctangent Formula for π

In 1706, John Machin (1680–1752) derived a fairly rapidly convergent series for π. To arrive at his expansion, consider the smallest positive angle α whose tangent is $1/5$:

$$\tan \alpha = \frac{1}{5} \quad (\text{that is, } \alpha := \arctan(1/5)).$$

Now setting $\theta = \phi = \alpha$ in (6.64), we obtain

$$\tan 2\alpha = \frac{2 \tan \alpha}{1 - \tan^2 \alpha} = \frac{2/5}{1 - 1/25} = \frac{5}{12},$$

so

$$\tan 4\alpha = \frac{2 \tan 2\alpha}{1 - \tan^2 2\alpha} = \frac{5/6}{1 - 25/144} = \frac{120}{119},$$

which is just slightly above 1. Hence, $4\alpha - \pi/4$ is positive, and moreover,

$$\tan \left(4\alpha - \frac{\pi}{4} \right) = \frac{\tan 4\alpha + \tan \pi/4}{1 + \tan 4\alpha \, \tan \pi/4} = \frac{1/119}{1 + 120/119} = \frac{1}{239}.$$

Taking the inverse tangent of both sides and solving for $\frac{\pi}{4}$, we get

$$\frac{\pi}{4} = 4 \tan^{-1} \frac{1}{5} - \tan^{-1} \frac{1}{239}.$$

Substituting $1/5$ and $1/239$ into the Gregory–Madhava series for the inverse tangent, we arrive at Machin's formula for π:

Machin's formula

Theorem 6.40 *We have*

$$\pi = 16 \sum_{n=0}^{\infty} \frac{(-1)^n}{(2n + 1)5^{2n+1}} - 4 \sum_{n=0}^{\infty} \frac{(-1)^n}{(2n + 1) \, 239^{2n+1}}.$$

Example 6.45 Machin's formula gives many decimal places of π without much effort. Let s_n denote the nth partial sum of $s := 16 \sum_{n=0}^{\infty} \frac{(-1)^n}{(2n+1)5^{2n+1}}$ and t_n that of $t := 4 \sum_{n=0}^{\infty} \frac{(-1)^n}{(2n+1) \, 239^{2n+1}}$. Then $\pi = s - t$, and by the alternating series error estimate,

$$|s - s_3| \le \frac{16}{9 \cdot 5^9} \approx 9.102 \times 10^{-7}$$

and

$$|t - t_0| \le \frac{4}{3 \cdot (239)^3} \approx 10^{-7}.$$

Therefore,

$$|\pi - (s_3 - t_0)| = |(s - t) - (s_3 - t_0)| \leq |s - s_3| + |t - t_0| < 5 \times 10^{-6}.$$

A manageable computation (even without a calculator!) shows that $s_3 - t_0 = 3.14159\ldots$. Therefore, $\pi = 3.14159$ to five decimal places!

▶ **Exercises 6.9**

1. From Gregory–Madhava's series, derive the following pretty series:

$$\boxed{\frac{\pi}{2\sqrt{3}} = 1 - \frac{1}{3 \cdot 3} + \frac{1}{5 \cdot 3^2} - \frac{1}{7 \cdot 3^3} + \frac{1}{9 \cdot 3^4} - + \cdots.}$$

Suggestion: Consider $\arctan(1/\sqrt{3}) = \pi/6$. How many terms of this series do you need to approximate $\pi/(2\sqrt{3})$ to within seven decimal places? *History Bite:* Abraham Sharp (1651–1742) used this formula in 1669 to compute π to 72 decimal places, and Thomas de Lagny (1660–1734) used this formula in 1717 to compute π to 126 decimal places (with a mistake in the 113th place) [47].

2. In this problem we prove (6.67).

   (i) Prove that $\arctan \frac{1}{3} = \arctan \frac{1}{5} + \arctan \frac{1}{8}$, and use this to prove that

   $$\frac{\pi}{4} = \arctan \frac{1}{2} + \arctan \frac{1}{5} + \arctan \frac{1}{8}.$$

   Prove that $\arctan \frac{1}{8} = \arctan \frac{1}{13} + \arctan \frac{1}{21}$, and use this to prove that

   $$\frac{\pi}{4} = \arctan \frac{1}{2} + \arctan \frac{1}{5} + \arctan \frac{1}{13} + \arctan \frac{1}{21}.$$

   From here you can now see the appearance of Fibonacci numbers.

   (ii) To continue this by induction, prove that for every natural number n,

   $$F_{2n} = \frac{F_{2n+1} F_{2n+2} - 1}{F_{2n+3}}.$$

   Suggestion: Can you use (6.66)?
   (iii) Using the formula in (ii), prove that

   $$\arctan\left(\frac{1}{F_{2n}}\right) = \arctan\left(\frac{1}{F_{2n+1}}\right) + \arctan\left(\frac{1}{F_{2n+2}}\right).$$

   Now derive (6.67).

3. In this problem we prove (6.68).

   (i) Using (6.65), prove that

$$\tan^{-1} \frac{\sqrt{5}\,x}{1-x^2} = \tan^{-1}\left(\frac{1+\sqrt{5}}{2}\right)x - \tan^{-1}\left(\frac{1-\sqrt{5}}{2}\right)x.$$

(ii) Now prove that

$$\tan^{-1} \frac{\sqrt{5}\,x}{1-x^2} = \sum_{n=0}^{\infty} \frac{(-1)^n F_{2n+1}\, x^{2n+1}}{5^n\,(2n+1)}.$$

(iii) Finally, derive the formula (6.68).

4. In this problem, we prove the breathtaking formula (6.62).

(i) Prove that

$$\frac{\pi}{4} = \sum_{n=1}^{\infty}\left(\frac{1}{4n-3} - \frac{1}{4n-1}\right) = \sum_{n=1}^{\infty} f\left(\frac{1}{n}\right),$$

where $f(z) = \frac{z}{4-3z} - \frac{z}{4-z}$.

(ii) Use Theorem 6.27 on p. 480 to derive our breathtaking formula.

6.10 ★ Another Proof that $\pi^2/6 = \sum_{n=1}^{\infty} 1/n^2$ (The Basel Problem)

Assuming only Gregory–Leibniz–Madhava's series, $\frac{\pi}{4} = \sum_{n=0}^{\infty} \frac{(-1)^n}{2n+1}$, we give our seventh proof of the fact that[13]

$$\boxed{\frac{\pi^2}{6} = \sum_{n=1}^{\infty} \frac{1}{n^2} = 1 + \frac{1}{2^2} + \frac{1}{3^2} + \frac{1}{4^2} + \cdots.}$$

The main ideas of the proof we are about to give are attributed to Nicolaus Bernoulli (1687–1759), and the proof "may be regarded as the most elementary of all known proofs, since it borrows nothing from the theory of functions except the Leibniz series" [129, p. 324].

6.10.1 Cauchy's Arithmetic Mean Theorem

Before giving our seventh proof of Euler's sum, we prove the following theorem (attributed to Cauchy [129, p. 72]).

---

[13]This proof is a systematized version of the sixth proof, Problem 3 on p. 414.

Cauchy's arithmetic mean theorem

Theorem 6.41 *If a sequence $a_1, a_2, a_3, \ldots$ converges to L, then the sequence of arithmetic means (or averages)*

$$m_n := \frac{1}{n}\Big(a_1 + a_2 + \cdots + a_n\Big), \quad n = 1, 2, 3, \ldots,$$

also converges to L. Moreover, if the sequence $\{a_n\}$ is nonincreasing, then so is the sequence of arithmetic means $\{m_n\}$.

Proof To show that $m_n \to L$, we need to show that

$$m_n - L = \frac{1}{n}\Big((a_1 - L) + (a_2 - L) + \cdots + (a_n - L)\Big)$$

tends to zero as $n \to \infty$. Let $\varepsilon > 0$ and fix a natural number $N \in \mathbb{N}$ such that for all $n > N$, we have $|a_n - L| < \varepsilon/2$. Then for $n > N$, we can write

$$|m_n - L| \le \frac{1}{n}\Big(|(a_1 - L) + \cdots + (a_N - L)|\Big) + \frac{1}{n}\Big(|(a_{N+1} - L) + \cdots + (a_n - L)|\Big)$$

$$\le \frac{1}{n}\Big(|(a_1 - L) + \cdots + (a_N - L)|\Big) + \frac{1}{n}\Big(\frac{\varepsilon}{2} + \cdots + \frac{\varepsilon}{2}\Big)$$

$$= \frac{1}{n}\Big(|(a_1 - L) + \cdots + (a_N - L)|\Big) + \frac{n - N}{n} \cdot \frac{\varepsilon}{2}$$

$$\le \frac{1}{n}\Big(|(a_1 - L) + \cdots + (a_N - L)|\Big) + \frac{\varepsilon}{2}.$$

By choosing n larger, we can make $\frac{1}{n}\Big(|(a_1 - L) + \cdots + (a_N - L)|\Big)$ also less than $\varepsilon/2$, which shows that $|m_n - L| < \varepsilon$ for n sufficiently large. This shows that $m_n \to L$.

Assume now that $\{a_n\}$ is nonincreasing. We need to prove that $\{m_n\}$ is also nonincreasing; that is, for each n,

$$\frac{1}{n+1}\Big(a_1 + \cdots + a_n + a_{n+1}\Big) \le \frac{1}{n}\Big(a_1 + \cdots + a_n\Big),$$

or, after multiplying both sides by $n(n + 1)$,

$$n\Big(a_1 + \cdots + a_n\Big) + n a_{n+1} \le n\Big(a_1 + \cdots + a_n\Big) + \Big(a_1 + \cdots + a_n\Big).$$

Canceling, we conclude that the sequence $\{m_n\}$ is nonincreasing if and only if

$$n a_{n+1} = \underbrace{a_{n+1} + a_{n+1} + \cdots a_{n+1}}_{n \text{ times}} \le a_1 + a_2 + \cdots + a_n.$$

But this inequality certainly holds, since $a_{n+1} \leq a_k$ for $k = 1, 2, \ldots, n$. This completes the proof.   ∎

There is a related theorem for geometric means found in Problem 2, which can be used to derive the following neat formula:

$$
e = \lim_{n \to \infty} \left[\left(\frac{2}{1}\right)^1 \left(\frac{3}{2}\right)^2 \left(\frac{4}{3}\right)^3 \cdots \left(\frac{n+1}{n}\right)^n \right]^{1/n}.
\tag{6.69}
$$

6.10.2 Proof VII of Euler's Formula for $\pi^2/6$

First we shall apply Abel's multiplication theorem to Gregory–Leibniz–Madhava's series:

$$
\left(\frac{\pi}{4}\right)^2 = \left(\sum_{n=0}^{\infty} (-1)^n \frac{1}{2n+1} \right) \cdot \left(\sum_{n=0}^{\infty} (-1)^n \frac{1}{2n+1} \right).
$$

To do so, we first form the nth term in the Cauchy product:

$$
c_n = \sum_{k=0}^{n} (-1)^k \frac{1}{2k+1} \cdot (-1)^{n-k} \frac{1}{2n-2k+1} = (-1)^n \sum_{k=0}^{n} \frac{1}{(2k+1)(2n-2k+1)}.
$$

Using partial fractions, one can check that

$$
\frac{1}{(2k+1)(2n-2k+1)} = \frac{1}{2(n+1)} \left(\frac{1}{2k+1} + \frac{1}{2n-2k+1} \right),
$$

which implies that

$$
c_n = \frac{(-1)^n}{2(n+1)} \left(\sum_{k=0}^{n} \frac{1}{2k+1} + \sum_{k=0}^{n} \frac{1}{2n-2k+1} \right) = \frac{(-1)^n}{n+1} \sum_{k=0}^{n} \frac{1}{2k+1},
$$

since $\sum_{k=0}^{n} \frac{1}{2n-2k+1} = \sum_{k=0}^{n} \frac{1}{2k+1}$. Thus, by Abel's theorem, we can write

$$
\left(\frac{\pi}{4}\right)^2 = \sum_{n=0}^{\infty} (-1)^n m_n, \quad \text{where} \quad m_n = \frac{1}{n+1} \left(1 + \frac{1}{3} + \cdots + \frac{1}{2n+1} \right),
$$

provided that the series converges! To see that this series converges, note that m_n is exactly the arithmetic mean, or average, of the numbers $1, 1/3, \ldots, 1/(2n+1)$. Since $1/(2n+1) \to 0$ monotonically, Cauchy's arithmetic mean theorem shows that these averages also tend to zero monotonically. In particular, by the alternating series theorem, $\sum_{n=0}^{\infty} (-1)^n m_n$ converges, so by Abel's multiplication theorem, we get (not quite $\pi^2/6$, but pretty nonetheless)

$$\boxed{\frac{\pi^2}{16} = \sum_{n=0}^{\infty} (-1)^n \frac{1}{n+1} \left(1 + \frac{1}{3} + \cdots + \frac{1}{2n+1} \right).}$$ (6.70)

We evaluate the right-hand side using the following theorem (whose proof is technical, so you can skip it if you like).

Theorem 6.42 *Let $\{a_n\}$ be a nonincreasing sequence of positive numbers such that $\sum a_n^2$ converges. Then both series*

$$s := \sum_{n=0}^{\infty} (-1)^n a_n \quad \text{and} \quad \delta_k := \sum_{n=0}^{\infty} a_n a_{n+k}, \quad k = 1, 2, 3, \ldots$$

converge. Moreover, $\Delta := \sum_{k=1}^{\infty} (-1)^{k-1} \delta_k$ also converges, and we have the formula

$$\sum_{n=0}^{\infty} a_n^2 = s^2 + 2\Delta.$$

Proof Figure 6.17 shows why this theorem is "obvious." The proof, however, is another story, since the series we are dealing with are not all absolutely convergent. In any case, here goes the rather long proof.

a_{00}	a_{01}	a_{02}	a_{03}	a_{04}	$\cdots$
a_{10}	a_{11}	a_{12}	a_{13}	a_{14}	$\cdots$
a_{20}	a_{21}	a_{22}	a_{23}	a_{24}	$\cdots$
a_{30}	a_{31}	a_{32}	a_{33}	a_{34}	$\cdots$
a_{40}	a_{41}	a_{42}	a_{43}	a_{44}	$\cdots$
$\vdots$	$\vdots$	$\vdots$	$\vdots$	$\vdots$	$\ddots$

a_0^2	$-a_0 a_1$	$a_0 a_2$	$-a_0 a_3$	$a_0 a_4$	$\cdots$
$-a_1 a_0$	a_1^2	$-a_1 a_2$	$a_1 a_3$	$-a_1 a_4$	$\cdots$
$a_2 a_0$	$-a_2 a_1$	a_2^2	$-a_2 a_3$	$a_2 a_4$	$\cdots$
$-a_3 a_0$	$a_3 a_1$	$-a_3 a_2$	a_3^2	$-a_3 a_4$	$\cdots$
$a_4 a_0$	$-a_4 a_1$	$a_4 a_2$	$-a_4 a_3$	a_4^2	$\cdots$
$\vdots$	$\vdots$	$\vdots$	$\vdots$	$\vdots$	$\ddots$

Fig. 6.17 Observe that $s^2 = \sum_{m=0}^{\infty} \sum_{n=0}^{\infty} a_{mn}$, where $a_{mn} = (-1)^{m+n} a_m a_n$. Putting the a_{mn} in the infinite array on the left gives the right array. The identity $s^2 = \sum_{n=0}^{\infty} a_n^2 - 2\Delta$ is really just the sum of the diagonal terms plus the sum of the off-diagonal terms!

Since $\sum a_n^2$ converges, we must have $a_n \to 0$, which implies that $\sum (-1)^n a_n$ converges by the alternating series test. By monotonicity, $a_n a_{n+k} \le a_n \cdot a_n = a_n^2$, and since $\sum a_n^2$ converges, by comparison so does each series $\delta_k = \sum_{n=0}^{\infty} a_n a_{n+k}$. Also by monotonicity,

$$\delta_{k+1} = \sum_{n=0}^{\infty} a_n a_{n+k+1} \leq \sum_{n=0}^{\infty} a_n a_{n+k} = \delta_k,$$

so by the alternating series test, the sum Δ converges if $\delta_k \to 0$. To prove that this holds, let $\varepsilon > 0$ and choose N (by invoking the Cauchy criterion for series) such that $a_{N+1}^2 + a_{N+2}^2 + \cdots < \varepsilon/2$. Then, since the sequence $\{a_n\}$ is nondecreasing, we can write

$$\begin{aligned}
\delta_k &= \sum_{n=0}^{\infty} a_n a_{n+k} \\
&= \left(a_0 a_k + \cdots + a_N a_{N+k} \right) + \left(a_{N+1} a_{N+1+k} + a_{N+2} a_{N+2+k} + \cdots \right) \\
&\leq \left(a_0 a_k + \cdots + a_N a_k \right) + \left(a_{N+1}^2 + a_{N+2}^2 + a_{N+3}^2 + \cdots \right) \\
&< a_k \cdot \left(a_0 + \cdots + a_N \right) + \frac{\varepsilon}{2}.
\end{aligned}$$

As $a_k \to 0$ we can make the first term less than $\varepsilon/2$ for all k large enough. Thus, $\delta_k < \varepsilon$ for all k sufficiently large. This proves that $\delta_k \to 0$ and hence $\Delta = \sum_{k=1}^{\infty} (-1)^{k-1} \delta_k$ converges. Finally, we need to prove the equality

$$\sum_{n=0}^{\infty} a_n^2 = s^2 + 2\Delta = s^2 + 2 \sum_{k=1}^{\infty} (-1)^{k-1} \delta_k.$$

To prove this, let s_n denote the nth partial sum of the series $s = \sum_{n=0}^{\infty} (-1)^n a_n$. We have

$$s_n^2 = \left(\sum_{k=0}^{n} (-1)^k a_k \right)^2 = \sum_{k=0}^{n} \sum_{\ell=0}^{n} (-1)^{k+\ell} a_k a_\ell.$$

We can break up the double sum on the right as a sum over (k, ℓ) such that $k = \ell$, $k < \ell$, and $\ell < k$:

$$\sum_{k=0}^{n} \sum_{\ell=0}^{n} (-1)^{k+\ell} a_k a_\ell = \sum_{k=\ell} (-1)^{k+\ell} a_k a_\ell + \sum_{k<\ell} (-1)^{k+\ell} a_k a_\ell + \sum_{\ell<k} (-1)^{k+\ell} a_k a_\ell,$$

where the smallest k and ℓ can be is 0 and the largest is n. The first sum is just $\sum_{k=0}^{n} a_k^2$, and by symmetry in k and ℓ, the last two sums are the same, so

$$s_n^2 = \sum_{k=0}^{n} a_k^2 + 2 \sum_{0 \leq k < \ell \leq n} (-1)^{k+\ell} a_k a_\ell.$$

In the second sum, $0 \le k < \ell \le n$, so we can write $\ell = k + j$, where $1 \le j \le n$ and $0 \le k \le n - j$. Hence,

$$\sum_{1 \le k < \ell \le n} (-1)^{k+\ell} a_k\, a_\ell = \sum_{j=1}^{n} \sum_{k=0}^{n-j} (-1)^{k+(k+j)} a_k\, a_{k+j} = \sum_{j=1}^{n} \sum_{k=0}^{n-j} (-1)^{j} a_k\, a_{k+j}.$$

In summary, we have

$$s_n^2 = \sum_{k=0}^{n} a_k^2 + 2 \sum_{j=1}^{n} (-1)^{j} \left(\sum_{k=0}^{n-j} a_k\, a_{k+j} \right).$$

Let Δ_n denote the nth partial sum of $\Delta = \sum_{j=1}^{\infty} (-1)^{j-1} \delta_j$; we need to show that $s_n^2 + 2\Delta_n \to \sum_{k=0}^{\infty} a_k^2$ as $n \to \infty$. To this end, observe that

$$s_n^2 + 2\Delta_n = \sum_{k=0}^{n} a_k^2 + 2 \sum_{j=1}^{n} (-1)^{j} \left(\sum_{k=0}^{n-j} a_k\, a_{k+j} \right) + 2 \sum_{j=1}^{n} (-1)^{j-1} \delta_j$$

$$= \sum_{k=0}^{n} a_k^2 + 2 \sum_{j=1}^{n} (-1)^{j} \left(\sum_{k=0}^{n-j} a_k\, a_{k+j} - \delta_j \right).$$

Recalling that $\delta_j = \sum_{k=0}^{\infty} a_k a_{k+j}$, we can write $s_n^2 + 2\Delta_n$ as

$$s_n^2 + 2\Delta_n = \sum_{k=0}^{n} a_k^2 + 2 \sum_{j=1}^{n} (-1)^{j} \alpha_j,$$

where

$$\alpha_j := \sum_{k=n-j+1}^{\infty} a_k a_{k+j} = a_{n-j+1} a_{n+1} + a_{n-j+2} a_{n+2} + a_{n-j+3} a_{n+3} + \cdots.$$

Since the sequence $\{a_n\}$ is nonincreasing, it follows that the sequence $\{\alpha_j\}$ is nondecreasing:

$$\alpha_j = a_{n-j+1} a_{n+1} + a_{n-j+2} a_{n+2} + \cdots \le a_{n-j} a_{n+1} + a_{n-j+1} a_{n+2} + \cdots = \alpha_{j+1}.$$

Now assuming that n is even, we have

$$\frac{1}{2}\left|s_n^2 + 2\Delta_n - \sum_{k=0}^{n} a_k^2\right| = \left|(-\alpha_1 + \alpha_2) + (-\alpha_3 + \alpha_4) + \cdots + (-\alpha_{n-1} + \alpha_n)\right|$$

$$= (-\alpha_1 + \alpha_2) + (-\alpha_3 + \alpha_4) + \cdots + (-\alpha_{n-1} + \alpha_n)$$

$$= -\alpha_1 - (\alpha_3 - \alpha_2) - (\alpha_5 - \alpha_4) - \cdots - (\alpha_{n-1} - \alpha_{n-2}) + \alpha_n$$

$$\leq \alpha_n = a_1 a_{n+1} + a_2 a_{n+2} + \cdots = \delta_n - a_0 a_n,$$

where we used the fact that the terms in parentheses are all nonnegative, because the α_j are nondecreasing. Using a very similar argument, we get

$$\frac{1}{2}\left|s_n^2 + 2\Delta_n - \sum_{k=0}^{n} a_k^2\right| \leq \delta_n - a_0 a_n \qquad (6.71)$$

for n odd. Therefore, (6.71) holds for all n. We already know that $\delta_n \to 0$ and $a_n \to 0$, so (6.71) shows that the left-hand side tends to zero as $n \to \infty$. This completes the proof of the theorem. ∎

Finally, we are ready to prove Euler's formula for $\pi^2/6$. To do so, we apply the preceding theorem to the sequence $a_n = 1/(2n + 1)$. In this case,

$$\delta_k = \sum_{n=0}^{\infty} a_n a_{n+k} = \sum_{n=0}^{\infty} \frac{1}{(2n + 1)(2n + 2k + 1)}.$$

Writing in partial fractions,

$$\frac{1}{(2n + 1)(2n + 2k + 1)} = \frac{1}{2k}\left\{\frac{1}{2n + 1} - \frac{1}{2n + 2k + 1}\right\},$$

we get (after some cancellations)

$$\delta_k = \frac{1}{2k}\sum_{n=0}^{\infty}\left\{\frac{1}{2n + 1} - \frac{1}{2n + 2k + 1}\right\} = \frac{1}{2k}\left(1 + \frac{1}{3} + \cdots + \frac{1}{2k - 1}\right).$$

Hence, the equality $\sum_{n=0}^{\infty} a_n^2 = s^2 + 2\Delta$ takes the form

$$\sum_{n=0}^{\infty} \frac{1}{(2n + 1)^2} = \left(\frac{\pi}{4}\right)^2 + \sum_{k=1}^{\infty}(-1)^{k-1}\frac{1}{k}\left(1 + \frac{1}{3} + \cdots \frac{1}{2k - 1}\right).$$

However, see (6.70), we already proved that the Cauchy product of Gregory–Leibniz–Madhava's series with itself is given by the sum on the right. Thus,

$$\sum_{n=0}^{\infty} \frac{1}{(2n + 1)^2} = \left(\frac{\pi}{4}\right)^2 + \left(\frac{\pi}{4}\right)^2 = \frac{\pi^2}{8}. \qquad (6.72)$$

Finally, summing over the even and odd denominators, we have

$$\sum_{n=1}^{\infty} \frac{1}{n^2} = \sum_{n=0}^{\infty} \frac{1}{(2n+1)^2} + \sum_{n=1}^{\infty} \frac{1}{(2n)^2} = \frac{\pi^2}{8} + \frac{1}{4} \sum_{n=1}^{\infty} \frac{1}{n^2},$$

and solving for $\sum_{n=1}^{\infty} 1/n^2$, we obtain Euler's formula: $\frac{\pi^2}{6} = \sum_{n=1}^{\infty} \frac{1}{n^2}$.

▶ **Exercises 6.10**

1. Find the following limits:

$$(a) \ \lim \frac{1 + 2^{1/2} + 3^{1/3} + \cdots + n^{1/n}}{n},$$

$$(b) \ \lim \frac{\left(1 + \frac{1}{1}\right)^1 + \left(1 + \frac{1}{2}\right)^2 + \left(1 + \frac{1}{3}\right)^3 + \cdots + \left(1 + \frac{1}{n}\right)^n}{n}.$$

2. If a sequence $a_1, a_2, a_3, \ldots$ of positive numbers converges to $L > 0$, prove that the sequence of geometric means $(a_1 a_2 \cdots a_n)^{1/n}$ also converges to L. Suggestion: Take logs of the geometric means. Using this result, prove (6.69). Using (6.69), prove that

$$e = \lim \frac{n}{(n!)^{1/n}}.$$

3. Prove the following generalization of Cauchy's arithmetic mean theorem: If a sequence $\{a_n\}$ converges to a and a sequence $\{b_n\}$ converges to b, then the sequence

$$\frac{1}{n}\left(a_1 b_n + a_2 b_{n-1} + \cdots + a_{n-1} b_2 + a_n b_1\right)$$

converges to ab.

Chapter 7
More on the Infinite: Products and Partial Fractions

Reason's last step is the recognition that there are an infinite number of things which are beyond it.
Blaise Pascal (1623–1662), Pensées. 1670.

This chapter is devoted entirely to the theory and application of infinite products, and as a bonus prize we also talk about partial fractions. In Sections 7.1 and 7.2 we present the basics of infinite products. In fact, we already saw infinite products when we studied François Viète's infinite product expression for π in Sections 4.12 and 5.1 (see p. 380). Now hold on to your seats, because the rest of the chapter is full of surprises!

We begin with the following Viète-like formula for $\log 2$, which is due to Philipp von Seidel (1821–1896) and published in 1871 [222]:

$$\log 2 = \frac{2}{1 + \sqrt{2}} \cdot \frac{2}{1 + \sqrt{\sqrt{2}}} \cdot \frac{2}{1 + \sqrt{\sqrt{\sqrt{2}}}} \cdot \frac{2}{1 + \sqrt{\sqrt{\sqrt{\sqrt{2}}}}} \cdots.$$

In Section 7.3, we prove Euler's sine formula, but now for complex arguments:

$$\sin \pi z = \pi z \left(1 - \frac{z^2}{1^2}\right)\left(1 - \frac{z^2}{2^2}\right)\left(1 - \frac{z^2}{3^2}\right)\left(1 - \frac{z^2}{4^2}\right)\left(1 - \frac{z^2}{5^2}\right) \cdots.$$

In Section 7.4, we look at partial fraction expansions of the trig functions. Recall from elementary calculus that if $p(z)$ is a polynomial with distinct roots $r_1, \ldots, r_n$, then factoring $p(z)$ as $p(z) = a(z - r_1)(z - r_2) \cdots (z - r_n)$, we can write

$$\frac{1}{p(z)} = \frac{1}{a(z - r_1)(z - r_2) \cdots (z - r_n)} = \frac{a_1}{z - r_1} + \frac{a_2}{z - r_2} + \cdots + \frac{a_n}{z - r_n}$$

© Paul Loya 2017
P. Loya, *Amazing and Aesthetic Aspects of Analysis*,
https://doi.org/10.1007/978-1-4939-6795-7_7

for some constants $a_1, \ldots, a_n$. You probably studied this in the "partial fraction method of integration" section in your elementary calculus course. We may formally write Euler's sine expansion as

$$\sin \pi z = az(z-1)(z+1)(z-2)(z+2)(z-3)(z+3) \cdots.$$

Euler thought that we should be able to apply the partial fraction decomposition to $1/\sin \pi z$:

$$\frac{1}{\sin \pi z} = \frac{a_1}{z} + \frac{a_2}{z-1} + \frac{a_3}{z+1} + \frac{a_4}{z-2} + \frac{a_5}{z+2} + \cdots.$$

In Section 7.4, we'll prove that this can be done, where the a_n follow in the pattern

$$\frac{\pi}{\sin \pi z} = \frac{1}{z} - \frac{1}{z-1} - \frac{1}{z+1} + \frac{1}{z-2} + \frac{1}{z+2} - \frac{1}{z-3} - \frac{1}{z+3} + \cdots.$$

Combining the adjacent factors $\frac{1}{z-n} + \frac{1}{z+n} = \frac{2z}{z^2-n^2}$, we get Euler's celebrated partial fraction expansion for sine:

$$\frac{\pi}{\sin \pi z} = \frac{1}{z} + \sum_{n=1}^{\infty} (-1)^n \frac{2z}{n^2 - z^2}.$$

We'll also derive partial fraction expansions for the other trig functions. In Section 7.5, we give more proofs of Euler's sum for $\pi^2/6$ using the infinite products and partial fractions we found in Sections 7.3 and 7.4. In Section 7.6, we prove one of the most famous formulas for the Riemann zeta function, namely writing it as an infinite product involving only the *prime* numbers:

$$\zeta(z) = \frac{2^z}{2^z - 1} \cdot \frac{3^z}{3^z - 1} \cdot \frac{5^z}{5^z - 1} \cdot \frac{7^z}{7^z - 1} \cdot \frac{11^z}{11^z - 1} \cdots.$$

In particular, setting $z = 2$, we get the following expression for $\pi^2/6$:

$$\frac{\pi^2}{6} = \prod \frac{p^2}{p^2 - 1} = \frac{2^2}{2^2 - 1} \cdot \frac{3^2}{3^2 - 1} \cdot \frac{5^2}{5^2 - 1} \cdots.$$

As a bonus prize, we see how π is related to questions from probability. Finally, in Section 5.3, we derive some awe-inspiring beautiful formulas (too many to list at this moment!). Here are a couple of my favorite formulas of all time:

$$\frac{\pi}{4} = \frac{3}{4} \cdot \frac{5}{4} \cdot \frac{7}{8} \cdot \frac{11}{12} \cdot \frac{13}{12} \cdot \frac{17}{16} \cdot \frac{19}{20} \cdot \frac{23}{24} \cdots.$$

The numerators of the fractions on the right are the odd prime numbers, and the denominators are even numbers divisible by four and differing from the numerators by one. The next one is also a beaut:

$$\frac{\pi}{2} = \frac{3}{2} \cdot \frac{5}{6} \cdot \frac{7}{6} \cdot \frac{11}{10} \cdot \frac{13}{14} \cdot \frac{17}{18} \cdot \frac{19}{18} \cdot \frac{23}{22} \cdots .$$

The numerators of the fractions are the odd prime numbers, and the denominators are even numbers not divisible by four and differing from the numerators by one.

CHAPTER 7 OBJECTIVES: THE STUDENT WILL BE ABLE TO ...

- Determine (absolute) convergence for an infinite product.
- Explain the infinite products and partial fractions of the trig functions.
- Describe Euler's formulas for powers of π and their relationship to Riemann's zeta function.

7.1 Introduction to Infinite Products

We start our journey through infinite products taking careful steps to define what these phenomenal products are.

7.1.1 Basic Definitions and Examples

Let $\{b_n\}$ be a sequence of complex numbers. Our goal is to define the infinite product

$$\prod_{n=1}^{\infty} b_n = b_1 \cdot b_2 \cdot b_3 \cdots .$$

Here, the capital $\prod$ means to take the *products* of the numbers after it. We say that the infinite product $\prod_{n=1}^{\infty} b_n$ **converges** if there exists an $m \in \mathbb{N}$ such that the b_n are nonzero for all $n \geq m$, and the limit of the **partial products** $\prod_{k=m}^{n} b_k = b_m \cdot b_{m+1} \cdots b_n$,

$$\lim_{n \to \infty} \prod_{k=m}^{n} b_k = \lim_{n \to \infty} \left(b_m \cdot b_{m+1} \cdots b_n \right), \tag{7.1}$$

converges to a *nonzero* complex value, say p. In this case, we define

$$\prod_{n=1}^{\infty} b_n := b_1 \cdot b_2 \cdots b_{m-1} \cdot p.$$

This definition is of course independent of the m chosen such that the b_n are nonzero for all $n \geq m$. The infinite product $\prod_{n=1}^{\infty} b_n$ **diverges** if it doesn't converge; that is, either there are infinitely many zero b_n, or the limit (7.1) either diverges or converges to zero. In this latter case, we say that the infinite product **diverges to zero**. Just as sequences and series can start at any integer, products can also start at any integer: $\prod_{n=k}^{\infty} b_n$, with straightforward modifications of the definition.

Example 7.1 Consider the "harmonic product" $\prod_{n=2}^{\infty}(1 - 1/n)$. Observe that

$$\prod_{k=2}^{n} \left(1 - \frac{1}{k}\right) = \left(1 - \frac{1}{2}\right)\left(1 - \frac{1}{3}\right) \cdots \left(1 - \frac{1}{n}\right) = \frac{1}{2} \cdot \frac{2}{3} \cdot \frac{3}{4} \cdots \frac{n-1}{n}.$$

Canceling each denominator with the next numerator, we obtain

$$\prod_{k=2}^{n} \left(1 - \frac{1}{k}\right) = \frac{1}{n} \to 0.$$

Thus, the harmonic product diverges to zero.

Example 7.2 On the other hand, the product $\prod_{n=2}^{\infty}(1 - 1/n^2)$ converges, because

$$\prod_{k=2}^{n} \left(1 - \frac{1}{k^2}\right) = \prod_{k=2}^{n} \frac{k^2 - 1}{k^2} = \prod_{k=2}^{n} \frac{(k-1)(k+1)}{k \cdot k}$$

$$= \frac{1 \cdot 3}{2 \cdot 2} \cdot \frac{2 \cdot 4}{3 \cdot 3} \cdot \frac{3 \cdot 5}{4 \cdot 4} \cdot \frac{4 \cdot 6}{5 \cdot 5} \cdots \frac{(n-1)(n+1)}{n \cdot n} = \frac{n+1}{2n} \to \frac{1}{2} \neq 0.$$

Therefore,

$$\prod_{n=2}^{\infty} \left(1 - \frac{1}{n^2}\right) = \frac{1}{2}.$$

Note that the infinite product $\prod_{n=1}^{\infty}(1 - 1/n^2)$ also converges, but in this case,

$$\prod_{n=1}^{\infty} \left(1 - \frac{1}{n^2}\right) := \left(1 - \frac{1}{1^2}\right) \cdot \lim_{n \to \infty} \prod_{k=2}^{n} \left(1 - \frac{1}{k^2}\right) = 0 \cdot \frac{1}{2} = 0.$$

Proposition 7.1 *If an infinite product converges, then its factors tend to one. Also, a convergent infinite product has the value 0 if and only if it has a zero factor.*

Proof The second statement is automatic from the definition of convergence. If none of the b_n vanish for $n \geq m$ and $p_n = b_m \cdot b_{m+1} \cdots b_n$, then assuming $p_n \to p$, where p is a nonzero number, we have

$$b_n = \frac{b_m \cdot b_{m+1} \cdots b_{n-1} \cdot b_n}{b_m \cdot b_{m+1} \cdots b_{n-1}} = \frac{p_n}{p_{n-1}} \to \frac{p}{p} = 1. \qquad \blacksquare$$

Because the factors of a convergent infinite product always tend to one, we henceforth write b_n as $1 + a_n$, so the infinite product takes the form

$$\prod (1 + a_n);$$

then convergence of this infinite product implies that $a_n \to 0$.

7.1.2 Infinite Products and Series: The Nonnegative Case

The following theorem states that the convergence of a product $\prod (1 + a_n)$ with all the a_n *nonnegative real numbers* is aligned with that of the series $\sum a_n$.

Theorem 7.2 *An infinite product* $\prod (1 + a_n)$ *with nonnegative terms* a_n *converges if and only if the series* $\sum a_n$ *converges.*

Proof Let the partial products and partial sums be denoted by

$$p_n = \prod_{k=1}^{n} (1 + a_k) \quad \text{and} \quad s_n = \sum_{k=1}^{n} a_k.$$

Since all the a_k are nonnegative, both sequences $\{p_n\}$ and $\{s_n\}$ are nondecreasing, so they converge if and only if they are bounded. Since $1 \le 1 + x \le e^x$ for every real number x (see Theorem 4.30 on p. 301), it follows that

$$1 \le p_n = \prod_{k=1}^{n} (1 + a_k) \le \prod_{k=1}^{n} e^{a_k} = e^{\sum_{k=1}^{n} a_k} = e^{s_n}.$$

This equation shows that if the sequence $\{s_n\}$ is bounded, then the sequence $\{p_n\}$ is also bounded, and hence converges. Its limit must be ≥ 1, so in particular, it is not zero. On the other hand,

$$p_n = (1 + a_1)(1 + a_2) \cdots (1 + a_n) \ge 1 + a_1 + a_2 + \cdots + a_n = 1 + s_n,$$

since the left-hand side, when multiplied out, contains the sum $1 + a_1 + a_2 + \cdots + a_n$ (and a lot of other nonnegative terms too). This shows that if the sequence $\{p_n\}$ is bounded, then the sequence $\{s_n\}$ is also bounded. $\blacksquare$

See Problem 4 for the case in which the a_n are negative.

Example 7.3 Thus, as a consequence of this theorem, the product

$$\prod \left(1 + \frac{1}{n^p}\right)$$

converges for $p > 1$ and diverges for $p \leq 1$.

7.1.3 Infinite Products for log 2 and e

I found the following gem in [219]. Define a sequence $\{e_n\}$ by $e_1 = 1$ and $e_{n+1} = (n+1)(e_n + 1)$ for $n = 1, 2, 3, \ldots$; e.g.,

$$e_1 = 1, \ e_2 = 4, \ e_3 = 15, \ e_4 = 64, \ e_5 = 325, \ e_6 = 1956, \ldots.$$

Then

$$e = \prod_{n=1}^{\infty} \frac{e_n + 1}{e_n} = \frac{2}{1} \cdot \frac{5}{4} \cdot \frac{16}{15} \cdot \frac{65}{64} \cdot \frac{326}{325} \cdot \frac{1957}{1956} \cdots. \tag{7.2}$$

You will be asked to prove this in Problem 7.

We now prove Philipp von Seidel's (1821–1896) formula for log 2:

$$\log 2 = \frac{2}{1 + \sqrt{2}} \cdot \frac{2}{1 + \sqrt{\sqrt{2}}} \cdot \frac{2}{1 + \sqrt{\sqrt{\sqrt{2}}}} \cdot \frac{2}{1 + \sqrt{\sqrt{\sqrt{\sqrt{2}}}}} \cdots.$$

To prove this, we follow the proof of Viète's formula in Section 5.1.1 on p. 381 using hyperbolic functions instead of trigonometric functions. Let $x \in \mathbb{R}$ be nonzero. Then dividing the identity $\sinh x = 2 \cosh(x/2) \sinh(x/2)$ (see Problem 8 on p. 338) by x, we get

$$\frac{\sinh x}{x} = \cosh(x/2) \cdot \frac{\sinh(x/2)}{x/2}.$$

Replacing x with $x/2$, we get $\sinh(x/2)/(x/2) = \cosh(x/2^2) \cdot \sinh(x/2^2)/(x/2^2)$. Therefore,

$$\frac{\sinh x}{x} = \cosh(x/2) \cdot \cosh(x/2^2) \cdot \frac{\sinh(x/2^2)}{x/2^2}.$$

Continuing by induction, we obtain for every $n \in \mathbb{N}$,

$$\frac{\sinh x}{x} = \left(\prod_{k=1}^{n} \cosh(x/2^k)\right) \cdot \frac{\sinh(x/2^n)}{x/2^n}.$$

Since $\lim_{z\to 0}\frac{\sinh z}{z} = 1$ (why?), we have $\lim_{n\to\infty}\frac{\sinh(x/2^n)}{x/2^n} = 1$, so taking $n \to \infty$, it follows that

$$\frac{x}{\sinh x} = \lim_{n\to\infty}\prod_{k=1}^{n}\frac{1}{\cosh(x/2^k)}.$$

Expressing the hyperbolic sine and cosine in terms of the exponential function, we can rewrite this limit as

$$\frac{2x}{e^x - e^{-x}} = \lim_{n\to\infty}\prod_{k=1}^{n}\frac{2}{e^{x/2^k} + e^{-x/2^k}} = \lim_{n\to\infty}\left(\prod_{k=1}^{n}e^{x/2^k}\frac{2}{1 + e^{x/2^{k-1}}}\right)$$

$$= \lim_{n\to\infty}\left(e^{x\sum_{k=1}^{n}1/2^k}\prod_{k=1}^{n}\frac{2}{1 + e^{x/2^{k-1}}}\right).$$

Since $\lim_{n\to\infty}\sum_{k=1}^{n}\frac{1}{2^k} = 1$ (this is just the geometric series $\sum_{k=1}^{\infty}\frac{1}{2^k}$), and

$$\prod_{k=1}^{n}\frac{2}{1 + e^{x/2^{k-1}}} = \frac{2}{1 + e^x}\prod_{k=1}^{n-1}\frac{2}{1 + e^{x/2^k}},$$

we see that

$$\frac{2x}{e^x - e^{-x}} = \frac{2e^x}{1 + e^x}\lim_{n\to\infty}\prod_{k=1}^{n}\frac{2}{1 + e^{x/2^k}}.$$

Finally, putting $x = \log\theta$, or $\theta = e^x$, into this equation and doing a small amount of algebra, we get, by the definition of infinite products, the following beautiful infinite product expansion for $\frac{\log\theta}{\theta-1}$:

$$\boxed{\frac{\log\theta}{\theta - 1} = \prod_{k=1}^{\infty}\frac{2}{1 + \theta^{1/2^k}} = \frac{2}{1 + \sqrt{\theta}}\cdot\frac{2}{1 + \sqrt{\sqrt{\theta}}}\cdot\frac{2}{1 + \sqrt{\sqrt{\sqrt{\theta}}}}\cdots\textit{Seidel's formula.}}$$

In particular, taking $\theta = 2$, we get Seidel's infinite product formula for $\log 2$.

▶ **Exercises 7.1**

1. Prove that

$$(a)\ \prod_{n=2}^{\infty}\left(1 + \frac{1}{n^2 - 1}\right) = 2, \qquad (b)\ \prod_{n=3}^{\infty}\left(1 - \frac{2}{n(n-1)}\right) = \frac{1}{3},$$

$$(c)\ \prod_{n=2}^{\infty}\left(1 + \frac{2}{n^2 + n - 2}\right) = 3, \qquad (d)\ \prod_{n=2}^{\infty}\left(1 + \frac{(-1)^n}{n}\right) = 1.$$

2. Prove that for every $z \in \mathbb{C}$ with $|z| < 1$,

$$\prod_{n=0}^{\infty} \left(1 + z^{2^n}\right) = \frac{1}{1-z}.$$

For example, $\prod_{n=0}^{\infty} \left(1 + \left(\frac{1}{2}\right)^{2^n}\right) = 2$. Suggestion: Derive, e.g., by induction, a formula for $p_n = \prod_{k=0}^{n}(1 + z^{2^k})$ as a geometric sum as in Problem 3e on p. 45.

3. For which $x \in \mathbb{R}$, do the following products converge and diverge?

$$\text{(a)} \ \prod_{n=1}^{\infty} \left(1 + \sin^2 \left(\frac{x}{n}\right)\right), \quad \text{(b)} \ \prod_{n=1}^{\infty} \left(\frac{1 + x^2 + x^{2n}}{1 + x^{2n}}\right)$$

$$\text{(c)} \ \prod_{n=1}^{\infty} \left(1 + \frac{x^{4n}}{\log(1 + x^{2n})}\right), \ x \neq 0.$$

4. In this problem, we prove that an infinite product $\prod(1 - a_n)$ with $0 \le a_n < 1$ converges if and only if the series $\sum a_n$ converges.

   (i) Let $p_n = \prod_{k=1}^{n}(1 - a_k)$ and $s_n = \sum_{k=1}^{n} a_k$. Show that $p_n \le e^{-s_n}$. Conclude that if $\sum a_n$ diverges, then $\prod(1 - a_n)$ also diverges (in this case, diverges to zero).

   (ii) Suppose now that $\sum a_n$ converges. Then we can choose m such that $a_m + a_{m+1} + \cdots < 1/2$. Prove by induction that

   $$(1 - a_m)(1 - a_{m+1}) \cdots (1 - a_n) \ge 1 - (a_m + a_{m+1} + \cdots + a_n)$$

   for $n = m, m + 1, m + 2, \ldots$. Conclude that $p_n/p_m \ge 1/2$ for all $n \ge m$, and from this, prove that $\prod(1 - a_n)$ converges.

   (iii) For what p is $\prod_{n=2}^{\infty} \left(1 - \frac{1}{n^p}\right)$ convergent, and for what p is it divergent?

5. In this problem we prove that the limit $\lim_{n \to \infty} n \cdot \prod_{k=2}^{n}(2 - e^{1/n})$ exists. Proceed as follows.

   (i) Prove that the infinite product $\prod_{n=1}^{\infty}(2 - e^{1/n})$ diverges. (Use Problem 4.)
   (ii) Prove that $\prod_{n=2}^{\infty}[(2 - e^{1/n})/(1 - 1/n)]$ converges.
   (iii) Finally, prove that the desired limit exists.

6. In this problem we derive relationships between series and products. Let $\{a_n\}$ be a sequence of complex numbers.

   (a) Prove that for $n \ge 2$,

   $$\prod_{k=1}^{n}(1 + a_k) = 1 + a_1 + \sum_{k=2}^{n}(1 + a_1) \cdots (1 + a_{k-1})a_k.$$

So if $a_n \neq -1$ for all n, then the product $\prod_{n=1}^{\infty}(1 + a_n)$ converges if and only if $1 + a_1 + \sum_{k=2}^{\infty}(1 + a_1)\cdots(1 + a_{k-1})a_k$ converges to a nonzero value, in which case the infinite product and infinite series have the same value.

(b) Assume that $a_1 + \cdots + a_k \neq 0$ for every k. Prove that for $n \geq 2$,

$$\sum_{k=1}^{n} a_k = a_1 \prod_{k=2}^{n}\left(1 + \frac{a_k}{a_1 + a_2 + \cdots + a_{k-1}}\right).$$

Thus, $\sum_{n=1}^{\infty} a_n$ converges if and only if $a_1 \prod_{n=2}^{\infty}\left(1 + \frac{a_n}{a_1 + a_2 + \cdots + a_{n-1}}\right)$ either converges or diverges to zero, in which case they have the same value.

(c) Using (b) and the sum $\sum_{n=1}^{\infty} \frac{1}{(n+a-1)(n+a)} = \frac{1}{a}$ from Eq. (3.36) on p. 200, prove that

$$\prod_{n=2}^{\infty}\left(1 + \frac{a}{(n+a)(n-1)}\right) = a + 1.$$

7. In this problem we prove (7.2).

(i) Let $s_n = \sum_{k=0}^{n} \frac{1}{k!}$. Prove that $e_n = n! \, s_{n-1}$ for $n = 1, 2, \ldots$.

(ii) Show that $s_n/s_{n-1} = (e_n + 1)/e_n$.

(iii) Show that $s_n = \prod_{k=1}^{n} \frac{e_k+1}{e_k}$ and then complete the proof. Suggestion: Note that we can write $s_n = (s_1/s_0) \cdot (s_2/s_1) \cdots (s_n/s_{n-1})$.

7.2 Absolute Convergence for Infinite Products

Way back in Section 3.6, we introduced absolute convergence for infinite series, and since then we have experienced how incredibly useful this notion is. In this section we continue our study of the basic properties of infinite products by introducing the notion of absolute convergence for infinite products. We also present a general convergence test that is able to test the convergence of an infinite product in terms of a corresponding series of logarithms.

7.2.1 Absolute Convergence for Infinite Products

An infinite product $\prod(1 + a_n)$ of complex numbers is said to **converge absolutely** if $\prod(1 + |a_n|)$ converges. By Theorem 7.2, $\prod(1 + |a_n|)$ converges if and only if $\sum |a_n|$ converges. Therefore, $\prod(1 + a_n)$ converges absolutely if and only if the infinite series $\sum a_n$ converges absolutely. We know that if an infinite series is absolutely convergent, then the series itself converges; is this the same for infinite products? The answer is yes, but before proving this, we first need the following lemma.

Lemma 7.3 *Let* $\{p_k\}_{k=m}^{\infty}$*, where* $m \in \mathbb{N}$*, be a sequence of complex numbers.*

(a) $\{p_k\}$ *converges if and only if the infinite series* $\sum_{k=m+1}^{\infty}(p_k - p_{k-1})$ *converges, in which case*

$$\lim_{k \to \infty} p_k = p_m + \sum_{k=m+1}^{\infty}(p_k - p_{k-1}).$$

(b) *If* $\{a_j\}_{j=m}^{\infty}$ *is a sequence of complex numbers and* $p_k = \prod_{j=m}^{k}(1 + a_j)$*, then*

$$|p_k - p_{k-1}| \le |a_k| \, e^{\sum_{j=m}^{k-1}|a_j|}.$$

Proof The identity in *(a)* is really the telescoping series theorem, Theorem 3.22 on p. 200, but let us prove *(a)* directly. To do so, we note that for $k \ge m$, we have

$$p_k = p_m + \sum_{j=m+1}^{k}(p_j - p_{j-1}), \tag{7.3}$$

since the sum on the right telescopes. It follows that $\lim p_k$ exists if and only if $\lim_{k \to \infty} \sum_{j=m+1}^{k}(p_j - p_{j-1})$ exists; in other words, if and only if the infinite series $\sum_{j=m+1}^{\infty}(p_j - p_{j-1})$ converges. In case of convergence, the limit equality in *(a)* follows from taking $k \to \infty$ in (7.3).

To prove *(b)*, observe that

$$p_k - p_{k-1} = \prod_{j=m}^{k}(1 + a_j) - \prod_{j=m}^{k-1}(1 + a_j)$$

$$= (1 + a_k)\prod_{j=m}^{k-1}(1 + a_j) - \prod_{j=m}^{k-1}(1 + a_j) = a_k \prod_{j=m}^{k-1}(1 + a_j).$$

Therefore, $|p_k - p_{k-1}| \le |a_k| \prod_{j=m}^{k-1}(1 + |a_j|)$. Since $1 + x \le e^x$ for all real numbers x and $e^{|a_1|} \cdots e^{|a_{k-1}|} = e^{|a_1| + \cdots + |a_{k-1}|}$ (law of exponents), we have

$$|p_k - p_{k-1}| \le |a_k| \prod_{j=m}^{k-1} e^{|a_j|} = |a_k| \, e^{\sum_{j=m}^{k-1}|a_j|}.$$

This completes our proof.   ∎

> **Theorem 7.4** *Every absolutely convergent infinite product converges.*

Proof Let $\prod(1 + a_n)$ be absolutely convergent, which is equivalent to the series $\sum |a_n|$ converging; we need to prove that $\prod(1 + a_n)$ converges in the usual sense. Since $\sum |a_n|$ converges, by the Cauchy criterion for series we can choose m such that $\sum_{n=m}^{\infty} |a_n| < \frac{1}{2}$. In particular, $|a_k| < 1$ for $k \geq m$, so $1 + a_k$ is nonzero for $k \geq m$. For $n \geq m$, let $p_n = \prod_{k=m}^{n}(1 + a_k)$. From *(a)* in Lemma 7.3, we know that $\lim p_n$ exists if and only if the infinite series $\sum_{k=m+1}^{\infty}(p_k - p_{k-1})$ converges. To prove that this series converges, note that by *(b)* in Lemma 7.3, for $k > m$ we have

$$|p_k - p_{k-1}| \leq |a_k| \, e^{\sum_{j=m}^{k-1} |a_j|} \quad \Longrightarrow \quad |p_k - p_{k-1}| \leq C \, |a_k|,$$

with $C = e^{1/2}$, where we recall that $\sum_{j=m}^{\infty} |a_j| < \frac{1}{2}$. In particular, since $\sum |a_k|$ converges, by the comparison test, the series $\sum_{k=m+1}^{\infty} |p_k - p_{k-1}|$ converges, and hence $\sum_{k=m+1}^{\infty}(p_k - p_{k-1})$ also converges.

To prove that $\prod(1 + a_n)$ converges, it remains to prove that $\lim p_n \neq 0$. To this end, we claim that for each $n \geq m$, we have

$$|p_n| = \prod_{k=m}^{n} |1 + a_k| \geq 1 - \sum_{k=m}^{n} |a_k| . \tag{7.4}$$

Since $\sum_{n=m}^{\infty} |a_n| < \frac{1}{2}$, it then follows that $\lim p_n \neq 0$. We prove (7.4) by induction on $n = m, m + 1, m + 2, \ldots$. To check the base case, $|1 + a_m| \geq 1 - |a_m|$, observe that for every complex number z,

$$1 = |1 + z - z| \leq |1 + z| + |z| \quad \Longrightarrow \quad |1 + z| \geq 1 - |z|, \tag{7.5}$$

which in particular proves the base case. Assume that our result holds for $n \geq m$; we prove it for $n + 1$. Observe that

$$\prod_{k=m}^{n+1} |1 + a_k| = \prod_{k=m}^{n} |1 + a_k| \cdot |1 + a_{n+1}|$$

$$\geq \left(1 - \sum_{k=m}^{n} |a_k| \right)(1 - |a_{n+1}|) \quad \text{(induction hypothesis and (7.5))}$$

$$= 1 - \sum_{k=m}^{n} |a_k| - |a_{n+1}| + \sum_{k=m}^{n} |a_k| |a_{n+1}|$$

$$\geq 1 - \sum_{k=m}^{n+1} |a_k|,$$

which is exactly the $n + 1$ case. ∎

Just as for infinite series, the converse of this theorem is not true. For example, the infinite product $\prod_{n=2}^{\infty} \left(1 + \frac{(-1)^n}{n}\right)$ converges (and equals 1; see Problem 1 on p. 539), but this product is not absolutely convergent.

7.2.2   Infinite Products and Series: The General Case

Suppose that $a_1, a_2, a_3, \ldots$ are nonnegative real numbers and observe that since logarithms take products to sums, we have

$$\log\left(\prod_{k=1}^{n}(1 + a_k)\right) = \sum_{k=1}^{n} \log(1 + a_k).$$

Exponentiating both sides, we get

$$\prod_{k=1}^{n}(1 + a_k) = \exp\left(\sum_{k=1}^{n} \log(1 + a_k)\right).$$

Using these formulas, it's not difficult to show that

$$\prod_{n=1}^{\infty}(1 + a_n) \text{ converges} \quad \Longleftrightarrow \quad \sum_{n=1}^{\infty} \log(1 + a_n) \text{ converges}.$$

Moreover, in the case of convergence and $L = \sum_{n=1}^{\infty} \log(1 + a_n)$, then $\prod_{n=1}^{\infty}(1 + a_n) = e^L$. The following theorem is an extension of these remarks to general complex infinite products.

Theorem 7.5 *An infinite product $\prod(1 + a_n)$ of complex numbers converges if and only if $a_n \to 0$ and the series*

$$\sum_{n=m+1}^{\infty} \text{Log}(1 + a_n),$$

starting from a suitable index $m + 1$, converges. Moreover, if L is the sum of the series, then
$$\prod(1 + a_n) = (1 + a_1) \cdots (1 + a_m) e^L.$$

Proof First of all, we remark that the statement "starting from a suitable index $m + 1$" concerning the sum of logarithms is put there to make sure that none of the terms $1 + a_n$ is zero (otherwise, $\text{Log}(1 + a_n)$ is undefined). By Proposition 7.1, if the product $\prod(1 + a_n)$ converges, then we must have $a_n \to 0$. For this reason, we may assume that $a_n \to 0$; in particular, we can fix m such that $n > m$ implies $|a_n| < 1$.

Let $b_n = 1 + a_n$. We need to prove that the infinite product $\prod b_n$ converges if and only if the series $\sum_{n=m+1}^{\infty} \text{Log}\, b_n$ converges; moreover, if L denotes the sum of the series, we need to prove that

$$\prod b_n = b_1 \cdots b_m\, e^L. \tag{7.6}$$

For $n > m$, let the partial products and partial sums be denoted by

$$p_n = \prod_{k=m+1}^{n} b_k \quad \text{and} \quad s_n = \sum_{k=m+1}^{n} \text{Log}\, b_k\,.$$

By definition of infinite product, we have

$$\prod b_n = b_1 \cdots b_m \cdot \left(\lim p_n \right),$$

provided that $\lim p_n$ exists and is nonzero. Also, since $\exp(\text{Log}\, z) = z$ for every nonzero complex number z, by the law of exponents we have

$$\exp(s_n) = p_n. \tag{7.7}$$

It follows that if $\{s_n\}$ converges, then $\{p_n\}$ also converges. Moreover, if $\lim s_n = L$, then (7.7) shows that $\lim p_n = e^L$, which is nonzero, and also implies the formula (7.6).

Conversely, suppose that $\{p_n\}$ converges to a nonzero complex number; we shall prove that $\{s_n\}$ also converges. Observe that the formula (7.7) implies that for $n > m$, we have

$$s_n - \text{Log}\, p - \text{Log}(p_n/p) = 2\pi i k_n$$

for some integer k_n (since the left-hand side exponentiates to 1). We already know that $p_n \to p$, so it follows that $\text{Log}(p_n/p)$ converges. Thus, to prove that $\{s_n\}$ converges, all we need to do is prove that $\{k_n\}$ converges. To this end, observe that

$$s_n - s_{n-1} = \left(\sum_{k=m+1}^{n} \text{Log}\, b_k \right) - \left(\sum_{k=m+1}^{n-1} \text{Log}\, b_k \right) = \text{Log}\, b_n.$$

Thus,

$$\text{Log}\, b_n = s_n - s_{n-1} = \text{Log}(p_n/p) - \text{Log}(p_{n-1}/p) + 2\pi i (k_n - k_{n-1}).$$

By assumption, $a_n \to 0$, so $b_n = 1 + a_n \to 1$, and we also know that $p_n/p \to 1$. Therefore, taking $n \to \infty$ in the previous displayed line shows that

$$k_n - k_{n-1} \to 0.$$

Now, $k_n - k_{n-1}$ is an integer, so it can approach 0 only if k_n and k_{n-1} are the same integer, say k, for all n sufficiently large. It follows that $\{k_n\}$ converges to k, and our proof is complete.   ∎

▶ **Exercises 7.2**

1. For what $z \in \mathbb{C}$ are the following products absolutely convergent?

$$(a)\; \prod_{n=1}^{\infty} \left(1 + z^n \right) \;,\quad (b)\; \prod_{n=1}^{\infty} \left(1 + \left(\frac{nz}{1+n} \right)^n \right),$$

$$(c)\; \prod_{n=1}^{\infty} \left(1 + \sin^2 \left(\frac{z}{n} \right) \right) \;,\quad (d)\; \prod_{n=2}^{\infty} \left(1 + \frac{z^n}{n \log n} \right) \;,\quad (e)\; \prod_{n=1}^{\infty} \frac{\sin(z/n)}{z/n}.$$

2. Here is a nice convergence test: Suppose that $\sum a_n^2$ converges. Then $\prod (1 + a_n)$ converges if and only if the series $\sum a_n$ converges. You may proceed as follows.

   (i) Since $\sum a_n^2$ converges, we know that $a_n \to 0$, so we may henceforth assume that $|a_n|^2 < \frac{1}{2}$ for all n. Prove that

   $$\left| \text{Log}(1 + a_n) - a_n \right| \le |a_n|^2.$$

   Suggestion: Use the power series expansion for $\text{Log}(1 + z)$.
   (ii) Prove that $\sum (\text{Log}(1 + a_n) - a_n)$ is absolutely convergent.
   (iii) Prove that $\sum a_n$ converges if and only if $\sum \text{Log}(1 + a_n)$ converges, and from this, prove the desired result.
   (iv) Does the product $\prod_{n=2}^{\infty} \left(1 + \frac{(-1)^n}{n} \right)$ converge? What about the product

   $$\left(1 + \frac{1}{2} \right)\left(1 + \frac{1}{3} \right)\left(1 - \frac{1}{4} \right)\left(1 + \frac{1}{5} \right)\left(1 + \frac{1}{6} \right)\left(1 - \frac{1}{7} \right)\left(1 + \frac{1}{8} \right)\left(1 + \frac{1}{9} \right) \cdots?$$

3. Let $\{a_n\}$ be a sequence of real numbers and assume that $\sum a_n$ converges but $\sum a_n^2$ diverges. In this problem we shall prove that $\prod (1 + a_n)$ diverges.

   (i) Prove there is a constant $C > 0$ such that for all $x \in \mathbb{R}$ with $|x| \le 1/2$, we have

   $$x - \log(1 + x) \ge Cx^2.$$

(ii) Since $\sum a_n$ converges, we know that $a_n \to 0$, so we may assume that $|a_n| \leq 1/2$ for all n. Using (i), prove that $\sum \log(1 + a_n)$ diverges, and hence $\prod(1 + a_n)$ diverges.

(iii) Does $\prod(1 + \frac{(-1)^{n-1}}{\sqrt{n}})$ converge or diverge?

4. Using the formulas from Problem 5 on p. 518, prove that for $|z| < 1$,

$$\prod_{n=1}^{\infty}(1 - z^n) = \exp\left(-\sum_{n=1}^{\infty}\frac{1}{n}\frac{z^n}{1 - z^n}\right) ,$$

$$\prod_{n=1}^{\infty}(1 + z^n) = \exp\left(\sum_{n=1}^{\infty}\frac{(-1)^{n-1}}{n}\frac{z^n}{1 - z^n}\right).$$

5. In this problem we prove that $\prod(1 + a_n)$ is *absolutely convergent* if and only if the series $\sum_{n=m+1}^{\infty} \mathrm{Log}(1 + a_n)$, starting from a suitable index $m + 1$, is absolutely convergent. Proceed as follows.

(i) Prove that for every complex number z with $|z| \leq 1/2$, we have

$$\frac{1}{2}|z| \leq |\mathrm{Log}(1 + z)| \leq \frac{3}{2}|z|. \tag{7.8}$$

Suggestion: Look at the power series expansion for $\frac{\mathrm{Log}(1+z)}{z} - 1$, and using this power series, prove that for $|z| \leq 1/2$, we have $\left|\frac{\mathrm{Log}(1+z)}{z} - 1\right| \leq \frac{1}{2}$.

(ii) Now use (7.8) to prove the desired result.

7.3 Euler and Tannery: Product Expansions Galore

The goal of this section is to learn Tannery's theorem for products and use it to give two proofs of a complex version of Euler's celebrated sine product expansion.

Euler's product for sine

> **Theorem 7.6** *For every complex number z, we have*
>
> $$\sin \pi z = \pi z \prod_{n=1}^{\infty}\left(1 - \frac{z^2}{n^2}\right).$$

Two proofs of this result, when z is real, are on pp. 386 and 393.

7.3.1 Tannery's Theorem for Products

See Problem 6 for another (much shorter) proof of the following theorem, one that uses complex logarithms.

Tannery's theorem for infinite products

> **Theorem 7.7** *For each natural number n, let $\prod_{k=1}^{m_n}(1 + a_k(n))$ be a finite product such that $m_n \to \infty$ as $n \to \infty$. If for each k, $\lim_{n\to\infty} a_k(n)$ exists and there is a convergent series $\sum_{k=1}^{\infty} M_k$ of nonnegative real numbers such that $|a_k(n)| \leq M_k$ for all $n \in \mathbb{N}$ and $1 \leq k \leq m_n$, then*
>
> $$\lim_{n\to\infty} \prod_{k=1}^{m_n}(1 + a_k(n)) = \prod_{k=1}^{\infty} \lim_{n\to\infty}(1 + a_k(n));$$
>
> *that is, the limits and products on both sides converge and are equal.*

Proof First of all, we remark that the infinite product on the right converges. Indeed, if we put $a_k := \lim_{n\to\infty} a_k(n)$, which exists by assumption, then taking $n \to \infty$ in the inequality $|a_k(n)| \leq M_k$, we have $|a_k| \leq M_k$ as well. Since $\sum_{k=1}^{\infty} M_k$ is assumed convergent, by the comparison test, $\sum_{k=1}^{\infty} a_k$ converges absolutely, and hence by Theorem 7.4, the infinite product $\prod_{k=1}^{\infty}(1 + a_k)$ converges.

Now to our proof. Since $\sum M_k$ converges, $M_k \to 0$, so we can choose $m > 1$ such that for all $k \geq m$, we have $M_k < 1$. This implies that $|a_k| < 1$ for $k \geq m$, so $1 + a_k$ is nonzero for $k \geq m$. For n large enough that $m_n > m$, write

$$\prod_{k=1}^{m_n}(1 + a_k(n)) = q(n) \cdot \prod_{k=m}^{m_n}(1 + a_k(n)), \text{ where } q(n) = \prod_{k=1}^{m-1}(1 + a_k(n)).$$

Since $q(n)$ is a finite product, $q(n) \to \prod_{k=1}^{m-1}(1 + a_k)$ as $n \to \infty$; therefore we just have to prove that

$$\lim_{n\to\infty} \prod_{k=m}^{m_n}(1 + a_k(n)) = \prod_{k=m}^{\infty}(1 + a_k).$$

Consider the partial products

$$p_k(n) = \prod_{j=m}^{k}(1 + a_j(n)) \text{ and } p_k = \prod_{j=m}^{k}(1 + a_j).$$

Since these are finite products and $a_j = \lim_{n\to\infty} a_j(n)$, by the algebra of limits we have $\lim_{n\to\infty} p_k(n) = p_k$. Now observe that

$$\prod_{j=m}^{m_n} (1 + a_j(n)) = p_{m_n}(n) = p_m(n) + \sum_{k=m+1}^{m_n} (p_k(n) - p_{k-1}(n)),$$

since the right-hand side telescopes to $p_{m_n}(n)$, and by the limit identity in *(a)* of Lemma 7.3 on p. 542, we know that

$$\prod_{j=m}^{\infty} (1 + a_j) = p_m + \sum_{k=m+1}^{\infty} (p_k - p_{k-1}),$$

since $\prod_{j=m}^{\infty}(1 + a_j) := \lim_{k \to \infty} p_k$. Also, by Part *(b)* of Lemma 7.3, we have

$$|p_k(n) - p_{k-1}(n)| \leq |a_k(n)| \, e^{\sum_{j=m}^{k-1} |a_j(n)|} \leq M_k \, e^{\sum_{j=m}^{k-1} M_j} \leq C M_k,$$

where $C = e^{\sum_{j=m}^{\infty} M_j}$. Since $\sum_{k=m+1}^{\infty} C M_k$ converges, by Tannery's theorem for series on p. 215 we have

$$\lim_{n \to \infty} \sum_{k=m+1}^{m_n} (p_k(n) - p_{k-1}(n)) = \sum_{k=m+1}^{\infty} \lim_{n \to \infty} (p_k(n) - p_{k-1}(n))$$

$$= \sum_{k=m+1}^{\infty} (p_k - p_{k-1}).$$

Therefore,

$$\lim_{n \to \infty} \prod_{j=m}^{m_n} (1 + a_j(n)) = \lim_{n \to \infty} \left(p_m(n) + \sum_{k=m+1}^{m_n} (p_k(n) - p_{k-1}(n)) \right)$$

$$= p_m + \lim_{n \to \infty} \sum_{k=m+1}^{m_n} (p_k(n) - p_{k-1}(n))$$

$$= p_m + \sum_{k=m+1}^{\infty} (p_k - p_{k-1}) = \prod_{j=m}^{\infty} (1 + a_j),$$

which is what we wanted to prove. ∎

7.3.2 Expansion of Sine III

(Cf. [40, p. 294]). Our third proof of Euler's infinite product for sine is a Tannery's theorem version of the proof found in Section 5.1. First, recall from Lemma 5.1 on p. 383 that for every $z \in \mathbb{C}$, we have

$$\sin z = \lim_{n \to \infty} F_n(z),$$

where $n = 2m + 1$ is odd and

$$F_n(z) = z \prod_{k=1}^{m} \left(1 - \frac{z^2}{n^2 \tan^2(k\pi/n)} \right).$$

Thus,

$$\sin z = \lim_{m \to \infty} \left\{ z \prod_{k=1}^{m} \left(1 - \frac{z^2}{n^2 \tan^2(k\pi/n)} \right) \right\}$$

$$= \lim_{m \to \infty} z \prod_{k=1}^{m} (1 + a_k(m)),$$

where $a_k(m) := -\frac{z^2}{n^2 \tan^2(k\pi/n)}$ with $n = 2m + 1$. Second, since $\lim_{z \to 0} \frac{\tan z}{z} = \lim_{z \to 0} \frac{\sin z}{z} \cdot \frac{1}{\cos z} = 1$, we see that

$$\lim_{m \to \infty} a_k(m) = \lim_{m \to \infty} -\frac{z^2}{(2m + 1)^2 \tan^2(k\pi/(2m + 1))}$$

$$= \lim_{m \to \infty} -\frac{z^2}{k^2 \pi^2 \left(\frac{\tan(k\pi/(2m+1))}{k\pi/(2m+1)} \right)^2} = -\frac{z^2}{k^2 \pi^2}.$$

Third, in Lemma 4.57 on p. 368, we proved that

$$x < \tan x, \quad \text{for } 0 < x < \pi/2. \tag{7.9}$$

In particular, for every $z \in \mathbb{C}$, if $n = 2m + 1$ and $1 \leq k \leq m$, then

$$\left| \frac{z^2}{n^2 \tan^2(k\pi/n)} \right| \leq \frac{|z|^2}{n^2(k\pi)^2/n^2} = \frac{|z|^2}{k^2 \pi^2}.$$

Thus, $|a_k(m)| \leq M_k$, where $M_k = \frac{|z|^2}{k^2 \pi^2}$. Finally, since the sum $\sum_{k=1}^{\infty} M_k$ converges, by Tannery's theorem for infinite products, we have

$$\sin z = \lim_{m \to \infty} z \prod_{k=1}^{m} (1 + a_k(m)) = z \prod_{k=1}^{\infty} \lim_{m \to \infty} (1 + a_k(m)) = z \prod_{k=1}^{\infty} \left(1 - \frac{z^2}{k^2 \pi^2} \right).$$

After replacing z by πz, we get Euler's infinite product expansion for $\sin \pi z$. This completes the proof of Theorem 7.6. In particular, we see that

$$\pi i \prod_{k=1}^{\infty}\left(1+\frac{1}{k^2}\right) = \pi i \prod_{k=1}^{\infty}\left(1-\frac{i^2}{k^2}\right) = \sin \pi i = \frac{e^{-\pi} - e^{\pi}}{2i}.$$

Thus, we have derived the very pretty formula

$$\boxed{\frac{e^{\pi} - e^{-\pi}}{2\pi} = \prod_{n=1}^{\infty}\left(1+\frac{1}{n^2}\right).}$$

Recall from Section 7.1 how easy it was to find that $\prod_{n=1}^{\infty}\left(1-\frac{1}{n^2}\right) = 1/2$, but replacing $-1/n^2$ with $+1/n^2$ is a whole different story!

7.3.3 Expansion of Sine IV

Our fourth proof of Euler's infinite product for sine is based on the following neat identity involving sines instead of tangents!

Lemma 7.8 *If* $n = 2m + 1$ *with* $m \in \mathbb{N}$, *then for every* $z \in \mathbb{C}$,

$$\sin nz = n \sin z \prod_{k=1}^{m}\left(1 - \frac{\sin^2 z}{\sin^2(k\pi/n)}\right).$$

Proof Lemma 2.27 on p. 86 shows that for each $k \in \mathbb{N}$, $2\cos kz$ is a polynomial in $2\cos z$ of degree k (with integer coefficients, although this fact is not important for this lemma). Technically speaking, Lemma 2.27 was proved under the assumption that z is real, but the proof used only the angle addition formula for cosine, which holds for complex variables as well. In any case, dividing by 2, it follows that $\cos kz = Q_k(\cos z)$, where Q_k is a polynomial of degree k. In particular, replacing z by $2z$ and using that $\cos 2z = 1 - 2\sin^2 z$, we see that

$$\cos 2kz = R_k(\sin^2 z),$$

where $R_k(\sin^2 z) = Q_k(1 - 2\sin^2 z)$ is a polynomial of degree k in $\sin^2 z$. Now using the addition formulas for sine, we get, for each $k \in \mathbb{N}$,

$$\sin(2k+1)z - \sin(2k-1)z = 2\sin z \cdot \cos(2kz) = 2\sin z \cdot R_k(\sin^2 z). \quad (7.10)$$

We claim that for every $m = 0, 1, 2, \ldots$, we have

$$\sin(2m+1)z = \sin z \cdot P_m(\sin^2 z), \quad (7.11)$$

where P_m is a polynomial of degree m. Indeed, we can write $\sin(2m + 1)z$ as the telescoping sum

$$\sin(2m + 1)z = \sin z + \sum_{k=1}^{m} \left[\sin(2k + 1)z - \sin(2k - 1)z \right].$$

By (7.10), each difference $\sin(2k + 1)z - \sin(2k - 1)z$ can be written as $\sin z$ times a polynomial of degree k in $\sin^2 z$. Factoring out the common factor $\sin z$, we get (7.11). Now observe that $\sin(2m + 1)z$ is zero when $z = z_k$ with $z_k = k\pi/(2m + 1)$, where $k = 1, 2, \ldots, m$. Also observe that since $0 < z_1 < z_2 < \cdots < z_m < \pi/2$, the m values $\sin z_k$ are distinct positive values. Hence, according to (7.11), $P_m(w) = 0$ at the m distinct values $w = \sin^2 z_k$, $k = 1, 2, \ldots, m$. Thus, as a consequence of the fundamental theorem of algebra, the polynomial $P_m(w)$ can be factored into a constant times

$$(w - z_1)(w - z_2) \cdots (w - z_m) = \prod_{k=1}^{m} \left(w - \sin^2 \left(\frac{k\pi}{2m + 1} \right) \right).$$

Factoring out the $\sin^2$ terms and putting $n = 2m + 1$, we get

$$P_m(w) = a \prod_{k=1}^{m} \left(1 - \frac{w}{\sin^2(k\pi/n)} \right),$$

for some constant a. Setting $w = \sin^2 z$, we obtain

$$\sin(2m + 1)z = \sin z \cdot P_m(\sin^2 z) = a \sin z \cdot \prod_{k=1}^{m} \left(1 - \frac{\sin^2 z}{\sin^2(k\pi/n)} \right).$$

Since $\sin(2m + 1)z / \sin z$ has limit equal to $2m + 1$ as $z \to 0$, it follows that $a = 2m + 1$. This completes the proof of the lemma. ∎

We are now ready to give our fourth proof of Euler's infinite product for sine. To this end, we let $n \geq 3$ be odd, and we replace z by z/n in Lemma 7.8 to get

$$\sin z = n \sin(z/n) \prod_{k=1}^{m} \left(1 - \frac{\sin^2(z/n)}{\sin^2(k\pi/n)} \right),$$

where $n = 2m + 1$. Since

$$\lim_{m \to \infty} (2m + 1) \sin(z/(2m + 1)) = \lim_{m \to \infty} z \frac{\sin(z/(2m + 1))}{z/(2m + 1)} = z,$$

we have

$$\sin z = z \lim_{m \to \infty} \prod_{k=1}^{m} \left(1 - \frac{\sin^2(z/n)}{\sin^2(k\pi/n)} \right) = z \lim_{m \to \infty} \prod_{k=1}^{m} (1 + a_k(m)),$$

where $a_k(m) := -\frac{\sin^2(z/n)}{\sin^2(k\pi/n)}$ with $n = 2m + 1$. Since we are taking $m \to \infty$, we can always make sure that $n = 2m + 1 > |z|$, which we henceforth assume. Now recall from Lemmas 5.6 and 5.7 that there is a constant $c > 0$ such that for every $z \in \mathbb{C}$ with (say) $|z| \leq \frac{\pi}{2}$, we have $c\,|z| \leq |\sin z|$, and for every $w \in \mathbb{C}$ with $|w| \leq 1$, we have $|\sin w| \leq \frac{6}{5}|w|$. It follows that for all $k = 1, 2, \ldots, m$,

$$\left| \frac{\sin^2(z/n)}{\sin^2(k\pi/n)} \right| \leq \frac{(6/5|z/n|)^2}{c^2(k\pi/n)^2} = \frac{36|z|^2}{25c^2\pi^2} \cdot \frac{1}{k^2} =: M_k.$$

Since the sum $\sum_{k=1}^{\infty} M_k$ converges, and

$$\lim_{m \to \infty} a_k(m) = - \lim_{m \to \infty} \frac{\sin^2(z/(2m + 1))}{\sin^2(k\pi/(2m + 1))}$$

$$= - \lim_{m \to \infty} \frac{z^2}{k^2\pi^2} \cdot \frac{\left(\frac{\sin(z/(2m+1))}{z/(2m+1)} \right)^2}{\left(\frac{\sin(k\pi/(2m+1))}{k\pi/(2m+1)} \right)^2} = -\frac{z^2}{k^2\pi^2},$$

Tannery's theorem for infinite products implies that

$$\sin z = z \lim_{m \to \infty} \prod_{k=1}^{m} (1 + a_k(m)) = z \prod_{k=1}^{\infty} \lim_{m \to \infty} (1 + a_k(m)) = z \prod_{k=1}^{\infty} \left(1 - \frac{z^2}{k^2\pi^2} \right).$$

Finally, replacing z by πz completes Proof IV of Euler's product formula.

7.3.4 Euler's Cosine Expansion

We can derive an infinite product expansion for the cosine function easily from the sine expansion. In fact, using the double angle formula for sine, we get

$$\cos \pi z = \frac{\sin 2\pi z}{2 \sin \pi z} = \frac{2\pi z \cdot \prod_{n=1}^{\infty} \left(1 - \frac{4z^2}{n^2} \right)}{2\pi z \cdot \prod_{n=1}^{\infty} \left(1 - \frac{z^2}{n^2} \right)} = \frac{\prod_{n=1}^{\infty} \left(1 - \frac{4z^2}{n^2} \right)}{\prod_{n=1}^{\infty} \left(1 - \frac{z^2}{n^2} \right)}.$$

The top product can be split into a product of terms with even denominators and a product of terms with odd denominators:

$$\prod_{n=1}^{\infty}\left(1 - \frac{4z^2}{(2n-1)^2}\right)\prod_{n=1}^{\infty}\left(1 - \frac{4z^2}{(2n)^2}\right) = \prod_{n=1}^{\infty}\left(1 - \frac{4z^2}{(2n-1)^2}\right)\prod_{n=1}^{\infty}\left(1 - \frac{z^2}{n^2}\right),$$

from which we get (see Problem 3 for three more proofs)

$$\cos \pi z = \prod_{n=1}^{\infty}\left(1 - \frac{4z^2}{(2n-1)^2}\right).$$

▶ **Exercises 7.3**

1. Put $z = 1/4$ into the cosine expansion to derive the following elegant product for $\sqrt{2}$:

$$\sqrt{2} = \frac{2}{1}\cdot\frac{2}{3}\cdot\frac{6}{5}\cdot\frac{6}{7}\cdot\frac{10}{9}\cdot\frac{10}{11}\cdots.$$

Compare this with Wallis's formula:

$$\frac{\pi}{2} = \frac{2}{1}\cdot\frac{2}{3}\cdot\frac{4}{3}\cdot\frac{4}{5}\cdot\frac{6}{5}\cdot\frac{6}{7}\cdot\frac{8}{7}\cdot\frac{8}{9}\cdot\frac{10}{9}\cdot\frac{10}{11}\cdots.$$

Thus, the product for $\sqrt{2}$ is obtained from Wallis's formula for $\pi/2$ by removing the factors with numerators that are multiples of 4.

2. (**Infinite products for hyperbolic functions**) Prove that

$$\sinh \pi z = \pi z \prod_{k=1}^{\infty}\left(1 + \frac{z^2}{k^2}\right) \quad \text{and} \quad \cosh \pi z = \prod_{n=1}^{\infty}\left(1 + \frac{z^2}{(2n-1)^2}\right).$$

3. (**Euler's infinite product for $\cos \pi z$**) Here are three more proofs!

   (a) Replace z by $-z + 1/2$ in the sine product to derive the cosine product. Suggestion: Begin by showing that

   $$\left(1 - \frac{(-z+\frac{1}{2})^2}{n^2}\right) = \left(1 - \frac{1}{4n^2}\right)\cdot\left(1 + \frac{2z}{2n-1}\right)\left(1 - \frac{2z}{2n+1}\right).$$

   (b) For our second proof, show that for n even, we can write

   $$\cos z = \prod_{k=1}^{n-1}\left(1 - \frac{\sin^2(z/n)}{\sin^2(k\pi/2n)}\right), \quad k = 1, 3, 5, \ldots, n-1.$$

   Using Tannery's theorem, deduce the cosine expansion.

   (c) Write $\cos z = \lim_{n\to\infty} G_n(z)$, where

$$G_n(z) = \frac{1}{2}\left\{\left(1 + \frac{iz}{n}\right)^n + \left(1 - \frac{iz}{n}\right)^n\right\}.$$

Prove that if $n = 2m$ with $m \in \mathbb{N}$, then

$$G_n(z) = \prod_{k=1}^{m}\left(1 - \frac{z^2}{n^2 \tan^2((2k-1)\pi/(2n))}\right).$$

Using Tannery's theorem, deduce the cosine expansion.

4. Prove that

$$1 - \sin z = \left(1 - \frac{2z}{1}\right)^2\left(1 + \frac{2z}{3}\right)^2\left(1 - \frac{2z}{5}\right)^2\left(1 + \frac{2z}{7}\right)^2\cdots.$$

Suggestion: First show that $1 - \sin z = 2\sin^2(\frac{\pi}{4} - \frac{z}{2})$.

5. Find the following limits.

(a) $\displaystyle\lim_{n\to\infty}\left\{\left(1 - \frac{1}{4n^2\log\left(1 + \left(\frac{2}{2n}\right)^2\right)}\right)\cdot\left(1 - \frac{1}{4n^2\log\left(1 + \left(\frac{3}{2n}\right)^2\right)}\right)\cdot\right.$

$$\left(1 - \frac{1}{4n^2\log\left(1 + \left(\frac{4}{2n}\right)^2\right)}\right)\cdots\left.\left(1 - \frac{1}{4n^2\log\left(1 + \left(\frac{n}{2n}\right)^2\right)}\right)\right\},$$

(b) $\displaystyle\lim_{n\to\infty}\left\{\left(1 + \frac{1}{4n^2\sin\left(\frac{4\cdot 1^2 - 1}{4n^2 - 1}\right)}\right)\cdot\left(1 + \frac{1}{4n^2\sin\left(\frac{4\cdot 2^2 - 1}{4n^2 - 1}\right)}\right)\cdots\left(1 + \frac{1}{4n^2\sin\left(\frac{4\cdot n^2 - 1}{4n^2 - 1}\right)}\right)\right\}.$

6. (**Another proof of Tannery's theorem**) Here we prove Tannery's theorem for products using complex logarithms. Assume the hypotheses and notation of Theorem 7.7. Since $\sum M_k$ converges, $M_k \to 0$, so we can choose m such that for all $k \geq m$, we have $M_k < 1/2$. Then as in the proof of Theorem 7.7, we just have to show that

$$\lim_{n\to\infty}\prod_{k=m}^{m_n}(1 + a_k(n)) = \prod_{k=m}^{\infty}(1 + a_k). \tag{7.12}$$

(i) Show that Tannery's theorem for series implies that

$$\lim_{n\to\infty}\sum_{k=m}^{m_n}\text{Log}(1 + a_k(n)) = \sum_{k=m}^{\infty}\text{Log}(1 + a_k).$$

Suggestion: Use the inequality (7.8) in Problem 5 on p. 547.

(ii) From (i), deduce (7.12).

7. (**Tannery's theorem II**) For each natural number n, let $\prod_{k=1}^{\infty}(1 + a_k(n))$ be a convergent infinite product. If for each k, $\lim_{n \to \infty} a_k(n)$ exists, and there is a series $\sum_{k=1}^{\infty} M_k$ of nonnegative real numbers such that $|a_k(n)| \leq M_k$ for all k, n, prove that

$$\lim_{n \to \infty} \prod_{k=1}^{\infty}(1 + a_k(n)) = \prod_{k=1}^{\infty} \lim_{n \to \infty}(1 + a_k(n));$$

that is, both sides are well defined (the limits and products converge) and are equal.

7.4 Partial Fraction Expansions of the Trigonometric Functions

The goal of this section is to prove partial fraction expansions for the trig functions. For example, we'll prove ...

Euler's partial fraction($\frac{\pi}{\sin \pi z}$)

Theorem 7.9 *We have*

$$\frac{\pi}{\sin \pi z} = \frac{1}{z} + \sum_{n=1}^{\infty}(-1)^n \frac{2z}{z^2 - n^2} \quad \text{for all } z \in \mathbb{C} \setminus \mathbb{Z}.$$

We also get our third proof of the Gregory–Leibniz–Madhava formula for $\pi/4$.

7.4.1 Partial Fraction Expansion of the Cotangent

We shall prove the following theorem, from which we'll derive the sine expansion stated above.

Euler's partial fraction($\pi z \cot \pi z$)

Theorem 7.10 *We have*

$$\pi z \cot \pi z = 1 + \sum_{n=1}^{\infty} \frac{2z^2}{z^2 - n^2} \quad \text{for all } z \in \mathbb{C} \setminus \mathbb{Z}.$$

Our proof of Euler's expansion of the cotangent is based on the following lemma.

> **Lemma 7.11** *For every noninteger complex number z and $n \in \mathbb{N}$, we have*
>
> $$\pi z \cot \pi z = \frac{\pi z}{2^n} \cot \frac{\pi z}{2^n} + \sum_{k=1}^{2^{n-1}-1} \frac{\pi z}{2^n} \left(\cot \frac{\pi(z+k)}{2^n} + \cot \frac{\pi(z-k)}{2^n} \right) - \frac{\pi z}{2^n} \tan \frac{\pi z}{2^n}.$$

Proof Using the double angle formula

$$2 \cot 2z = 2 \frac{\cos 2z}{\sin 2z} = \frac{\cos^2 z - \sin^2 z}{\cos z \sin z} = \cot z - \tan z,$$

we see that

$$\cot 2z = \frac{1}{2} \left(\cot z - \tan z \right).$$

Replacing z with $\pi z/2$, we get

$$\cot \pi z = \frac{1}{2} \left(\cot \frac{\pi z}{2} - \tan \frac{\pi z}{2} \right). \tag{7.13}$$

Multiplying this equality by πz proves our lemma for $n = 1$. In order to proceed by induction, we note that since $\tan z = -\cot(z \pm \pi/2)$, we have

$$\cot \pi z = \frac{1}{2} \left(\cot \frac{\pi z}{2} + \cot \frac{\pi(z \pm 1)}{2} \right). \tag{7.14}$$

This is the main formula on which induction may be applied to prove our lemma. For instance, let's take the case $n = 2$. Considering the positive sign in the second cotangent, we obtain

$$\cot \pi z = \frac{1}{2} \left(\cot \frac{\pi z}{2} + \cot \frac{\pi(z+1)}{2} \right).$$

Applying (7.14) to each cotangent on the right of this equation, using the plus sign for the first and the minus sign for the second, we get

$$\cot \pi z = \frac{1}{2^2} \left\{ \left(\cot \frac{\pi z}{2^2} + \cot \frac{\pi(\frac{z}{2}+1)}{2} \right) + \left(\cot \frac{\pi(z+1)}{2^2} + \cot \frac{\pi(\frac{z+1}{2}-1)}{2} \right) \right\}$$

$$= \frac{1}{2^2} \left\{ \cot \frac{\pi z}{2^2} + \cot \frac{\pi(z+2)}{2^2} + \cot \frac{\pi(z+1)}{2^2} + \cot \frac{\pi(z-1)}{2^2} \right\}.$$

Bringing the second cotangent on the right to the end, we see that

$$\cot \pi z = \frac{1}{2^2} \cot \frac{\pi z}{2^2} + \frac{1}{2^2} \left\{ \cot \frac{\pi(z+1)}{2^2} + \cot \frac{\pi(z-1)}{2^2} \right\} + \frac{1}{2^2} \cot \left(\frac{\pi z}{2^2} + \frac{\pi}{2} \right).$$

The last term is exactly $-1/(2^2) \tan \pi z/2^2$, so our lemma is proved for $n = 2$. Continuing by induction proves our lemma for general n. ∎

Fix a noninteger z; we shall prove Euler's expansion for the cotangent. Note that $\lim_{n \to \infty} \frac{\pi z}{2^n} \tan(\frac{\pi z}{2^n}) = 0 \cdot \tan 0 = 0$, and since

$$\lim_{w \to 0} w \cot w = \lim_{w \to 0} \frac{w}{\sin w} \cdot \cos w = 1 \cdot 1 = 1, \tag{7.15}$$

we have $\lim_{n \to \infty} \frac{\pi z}{2^n} \cot \frac{\pi z}{2^n} = 1$. Therefore, taking $n \to \infty$ in the formula from the preceding lemma, Lemma 7.11, we conclude that

$$\pi z \cot \pi z = 1 + \lim_{n \to \infty} \left\{ \sum_{k=1}^{2^{n-1}-1} \frac{\pi z}{2^n} \left(\cot \frac{\pi(z+k)}{2^n} + \cot \frac{\pi(z-k)}{2^n} \right) \right\}$$

$$= 1 + \lim_{n \to \infty} \sum_{k=1}^{2^{n-1}-1} a_k(n),$$

where

$$a_k(n) = \frac{\pi z}{2^n} \left(\cot \frac{\pi(z+k)}{2^n} + \cot \frac{\pi(z-k)}{2^n} \right).$$

We shall apply Tannery's theorem to this sum. To this end, observe that from (7.15),

$$\lim_{n \to \infty} \frac{\pi z}{2^n} \cot \frac{\pi(z+k)}{2^n} = \frac{z}{z+k} \lim_{n \to \infty} \frac{\pi(z+k)}{2^n} \cot \frac{\pi(z+k)}{2^n} = \frac{z}{z+k},$$

and in a similar way,

$$\lim_{n \to \infty} \frac{\pi z}{2^n} \cot \frac{\pi(z-k)}{2^n} = \frac{z}{z-k}.$$

Thus,

$$\lim_{n \to \infty} a_k(n) = \frac{z}{z+k} + \frac{z}{z-k} = \frac{2z^2}{z^2 - k^2}.$$

Hence, Tannery's theorem gives Euler's cotangent expansion

$$\pi z \cot \pi z = 1 + \sum_{k=1}^{\infty} \frac{2z^2}{z^2 - k^2},$$

provided that we can show that $|a_k(n)| \le M_k$, where the constants M_k satisfy $\sum M_k < \infty$. To bound each $a_k(n)$, we use the formula

$$\cot z + \cot w = \frac{\sin(z + w)}{\sin z \, \sin w},$$

which you should be able to prove. It follows that

$$a_k(n) = \frac{\pi z}{2^n} \cdot \frac{\sin \frac{\pi z}{2^{n-1}}}{\sin \frac{\pi(z+k)}{2^n} \cdot \sin \frac{\pi(z-k)}{2^n}}.$$

Choose $N \in \mathbb{N}$ such that for all $n > N$, we have $|\pi z/2^n| \le 1/4$. Then for $n > N$, $|\pi z/2^{n-1}| \le 1/2$, so according to Lemma 5.7 on p. 402,

$$\left| \sin \frac{\pi z}{2^{n-1}} \right| \le \frac{6}{5} \cdot \frac{\pi \, |z|}{2^{n-1}}.$$

Also, for $n > N$ and $1 \le k \le 2^{n-1}$, we have

$$\left| \frac{\pi(z \pm k)}{2^n} \right| \le \frac{\pi \, |z|}{2^n} + \frac{\pi \, 2^{n-1}}{2^n} \le \frac{\pi}{4} + \frac{\pi}{2} = \frac{3\pi}{4}.$$

Thus, by Lemma 5.6 on p. 399, for some $c > 0$,

$$c \left| \frac{\pi(z \pm k)}{2^n} \right| \le \left| \sin \frac{\pi(z \pm k)}{2^n} \right|.$$

Hence, for $n > N$ and $1 \le k \le 2^{n-1}$, we have

$$\begin{aligned}
|a_k(n)| &\le \frac{\pi \, |z|}{2^n} \cdot \left(\frac{6}{5} \cdot \frac{\pi \, |z|}{2^{n-1}} \right) \cdot \left(\frac{1}{c} \frac{2^n}{|\pi(z+k)|} \right) \cdot \left(\frac{1}{c} \frac{2^n}{|\pi(z-k)|} \right) \\
&= \frac{12 \, |z|^2}{5 \, c^2} \cdot \frac{1}{|z^2 - k^2|} =: M_k.
\end{aligned}$$

We leave you to prove that $\sum_{k=1}^{\infty} M_k$ converges, which justifies the assumptions of Tannery's theorem II in this case, and hence completes the proof of Euler's cotangent expansion.

7.4.2 Partial Fraction Expansions of the Other Trig Functions

We shall leave most of the details to the exercises. Using the formula (see (7.13))

$$\pi \tan \frac{\pi z}{2} = \pi \cot \frac{\pi z}{2} - 2\pi \cot \pi z.$$

and substituting in the partial fraction expansion of the cotangent gives, as the diligent reader will show in Problem 1, for $z \in \mathbb{C}$ not an odd integer,

$$\boxed{\pi \tan \frac{\pi z}{2} = \sum_{n=0}^{\infty} \frac{4z}{(2n+1)^2 - z^2}.}$$ (7.16)

To derive the partial fraction expansion for $\frac{\pi}{\sin \pi z}$ mentioned at the very beginning of this section, we first derive the identity

$$\frac{1}{\sin z} = \cot z + \tan \frac{z}{2}.$$

To see this, observe that

$$\begin{aligned}
\cot z + \tan \frac{z}{2} &= \frac{\cos z}{\sin z} + \frac{\sin(z/2)}{\cos(z/2)} = \frac{\cos z \cos(z/2) + \sin z \sin(z/2)}{\sin z \cos(z/2)} \\
&= \frac{\cos(z - (z/2))}{\sin z \cos(z/2)} = \frac{\cos(z/2)}{\sin z \cos(z/2)} = \frac{1}{\sin z}.
\end{aligned}$$

This identity, together with the partial fraction expansions of the tangent and cotangent and a little algebra, which the extremely diligent reader will supply in Problem 1, implies that for noninteger $z \in \mathbb{C}$,

$$\boxed{\frac{\pi}{\sin \pi z} = \frac{1}{z} + \sum_{n=1}^{\infty} (-1)^n \frac{2z}{z^2 - n^2}.}$$ (7.17)

Finally, the incredibly awesome diligent reader ☺ will supply the details for the following secant expansion: For $z \in \mathbb{C}$ not an odd integer,

$$\boxed{\frac{\pi}{\cos \pi z} = 4 \sum_{n=0}^{\infty} (-1)^n \frac{2n+1}{(2n+1)^2 - 4z^2}.}$$ (7.18)

Finally, see Problem 2 for Proof III of Gregory–Leibniz–Madhava.

▶ **Exercises 7.4**

1. Fill in the details for the proofs of (7.16) and (7.17). For (7.18), first show that

$$\frac{\pi}{\sin \pi z} = \frac{1}{z} - \left(\frac{1}{z-1} + \frac{1}{z+1} \right) + \left(\frac{1}{z-2} + \frac{1}{z+2} \right) - \cdots.$$

Replacing z with $\frac{1}{2} - z$ and doing some algebra, derive the expansion (7.18).

2. (**Gregory–Leibniz–Madhava's formula for** $\pi/4$, **Proof III**) Derive the
   Gregory–Leibniz–Madhava series $\frac{\pi}{4} = \sum_{n=1}^{\infty} \frac{(-1)^{n-1}}{2n-1} = 1 - \frac{1}{3} + \frac{1}{4} - \frac{1}{5} + \cdots$ by
   evaluating the partial fractions for $\pi z \cot \pi z$, $\pi/\sin \pi z$, and $\pi/\cos \pi z$, at certain
   values of z.

3. Derive the following formulas for π:

$$\pi = z \tan\left(\frac{\pi}{z}\right) \cdot \left[1 - \frac{1}{z-1} + \frac{1}{z+1} - \frac{1}{2z-1} + \frac{1}{2z+1} - + \cdots\right]$$

and

$$\pi = z \sin\left(\frac{\pi}{z}\right) \cdot \left[1 + \frac{1}{z-1} - \frac{1}{z+1} - \frac{1}{2z-1} + \frac{1}{2z+1} + - - + + \cdots\right].$$

Put $z = 3, 4, 6$ to derive some pretty formulas.

7.5   ★ More Proofs that $\pi^2/6 = \sum_{n=1}^{\infty} 1/n^2$

In this section, we continue our discussions from Sections 5.2 and 6.10, concerning
the Basel problem of determining the sum of the reciprocals of the squares. A good
reference for this material is [118], and for more on Euler, see [10].

7.5.1   Proof VIII of Euler's Formula For $\pi^2/6$

(Cf. [47, p. 74].) One can consider this proof a "logarithmic" version of Euler's
original (third) proof of the formula for $\pi^2/6$, which we explained in the introduction
to Chapter 5 on p. 378. As with Euler, we begin with Euler's sine expansion restricted
to $0 \leq x < 1$:

$$\frac{\sin \pi x}{\pi x} = \prod_{n=1}^{\infty} \left(1 - \frac{x^2}{n^2}\right).$$

However, in contrast to Euler, we take logarithms of both sides:

$$\log\left(\frac{\sin \pi x}{\pi x}\right) = \log\left(\lim_{m \to \infty} \prod_{n=1}^{m} \left(1 - \frac{x^2}{n^2}\right)\right) = \lim_{m \to \infty} \log\left(\prod_{n=1}^{m} \left(1 - \frac{x^2}{n^2}\right)\right)$$

$$= \lim_{m \to \infty} \sum_{n=1}^{m} \log\left(1 - \frac{x^2}{n^2}\right),$$

where in the second equality we can pull out the limit because $\log$ is continuous, and
at the last step we used that logarithms take products to sums. Thus, we have shown

that

$$\log\left(\frac{\sin \pi x}{\pi x}\right) = \sum_{n=1}^{\infty} \log\left(1 - \frac{x^2}{n^2}\right), \quad 0 \le x < 1.$$

Recalling that $\log(1 + t) = \sum_{m=1}^{\infty} \frac{(-1)^{m-1}}{m} t^m$, we see that

$$\log(1 - t) = -\sum_{m=1}^{\infty} \frac{1}{m} t^m,$$

so replacing t by x^2/n^2, we obtain

$$\log\left(\frac{\sin \pi x}{\pi x}\right) = -\sum_{n=1}^{\infty}\sum_{m=1}^{\infty} \frac{1}{m} \frac{x^{2m}}{n^{2m}}, \quad 0 \le x < 1.$$

Since $0 \le x < 1$, the geometric series $\sum_{m=1}^{\infty} |x|^{2m}$ converges, so in view of the inequality $\frac{1}{mn^{2m}} \le \frac{1}{n^2}$, we have

$$\sum_{n=1}^{\infty}\sum_{m=1}^{\infty} \left| \frac{1}{m} \frac{x^{2m}}{n^{2m}} \right| \le \sum_{n=1}^{\infty}\sum_{m=1}^{\infty} \frac{|x|^{2m}}{n^2} = \zeta(2) \sum_{m=1}^{\infty} |x|^{2m} < \infty.$$

Hence, by Cauchy's double series theorem on p. 477, we can iterate sums and write

$$-\log\left(\frac{\sin \pi x}{\pi x}\right) = \sum_{m=1}^{\infty}\left(\sum_{n=1}^{\infty} \frac{1}{n^{2m}}\right) \frac{x^{2m}}{m} \tag{7.19}$$

$$= x^2 \sum_{n=1}^{\infty} \frac{1}{n^2} + \frac{x^4}{2} \sum_{n=1}^{\infty} \frac{1}{n^4} + \frac{x^6}{3} \sum_{n=1}^{\infty} \frac{1}{n^6} + \cdots.$$

On the other hand, since

$$\frac{\sin \pi x}{\pi x} = 1 - g(x), \quad \text{where} \quad g(x) = \frac{\pi^2 x^2}{3!} - \frac{\pi^4 x^4}{5!} + \frac{\pi^6 x^6}{7!} - \cdots,$$

we have

$$-\log\left(\frac{\sin \pi x}{\pi x}\right) = -\log\left(1 - g(x)\right)$$

$$= g(x) + \frac{1}{2} g(x)^2 + \frac{1}{3} g(x)^3 + \cdots$$

$$= \frac{\pi^2}{3!} x^2 + \left(-\frac{\pi^4}{5!} + \frac{\pi^4}{2 \cdot (3!)^2}\right) x^4 + \left(\frac{\pi^6}{7!} - \frac{\pi^6}{3! \cdot 5!} + \frac{\pi^6}{3 \cdot (3!)^3}\right) x^6 + \cdots,$$

$$\tag{7.20}$$

where we formally multiplied $g(x)^2$, $g(x)^3$, and so forth, collecting like powers of x, which is valid by the power series composition theorem on p. 499. Equating this with (7.19), we obtain

$$\frac{\pi^2}{3!}x^2 + \left(-\frac{\pi^4}{5!} + \frac{\pi^4}{2 \cdot (3!)^2}\right)x^4 + \left(\frac{\pi^6}{7!} - \frac{\pi^6}{3! \cdot 5!} + \frac{\pi^6}{3 \cdot (3!)^3}\right)x^6 + \cdots$$
$$= x^2 \sum_{n=1}^{\infty} \frac{1}{n^2} + \frac{x^4}{2} \sum_{n=1}^{\infty} \frac{1}{n^4} + \frac{x^6}{3} \sum_{n=1}^{\infty} \frac{1}{n^6} + \cdots,$$

or after simplification,

$$\frac{\pi^2}{6}x^2 + \frac{\pi^4}{180}x^4 + \frac{\pi^6}{2835}x^6 + \cdots = x^2 \sum_{n=1}^{\infty} \frac{1}{n^2} + \frac{x^4}{2} \sum_{n=1}^{\infty} \frac{1}{n^4} + \frac{x^6}{3} \sum_{n=1}^{\infty} \frac{1}{n^6} + \cdots.$$
$$(7.21)$$

By the identity theorem, the coefficients of x^k must be identical. Thus, comparing the x^2 terms, we get Euler's formula

$$\frac{\pi^2}{6} = \sum_{n=1}^{\infty} \frac{1}{n^2},$$

comparing the x^4 terms, we get

$$\boxed{\frac{\pi^4}{90} = \sum_{n=1}^{\infty} \frac{1}{n^4},} \qquad (7.22)$$

and finally, comparing the x^6 terms, we get

$$\boxed{\frac{\pi^6}{945} = \sum_{n=1}^{\infty} \frac{1}{n^6}.} \qquad (7.23)$$

Now what if we took more terms in (7.19) and (7.20), say to x^{2k}? Could we then find a formula for $\sum 1/n^{2k}$? The answer is certainly in the affirmative, but the work required to get a formula is rather intimidating; see Problem 1 for a formula when $k = 4$. Of course, in Section 5.3 (Theorem 5.9 on p. 412) we found formulas for $\zeta(2k)$ for all k. In Section 7.7 we will again find formulas for $\zeta(2k)$ for all k.

7.5.2   Proof IX

(Cf. [49, 132].) For this proof, we start with Lemma 7.8 on p. 551, which states that if $n = 2m + 1$ with $m \in \mathbb{N}$, then

$$\sin nz = n \sin z \prod_{k=1}^{m} \left(1 - \frac{\sin^2 z}{\sin^2(k\pi/n)} \right). \qquad (7.24)$$

We fix an m; later we shall take $m \to \infty$. We now substitute the expansion

$$\sin nz = nz - \frac{n^3 z^3}{3!} + \frac{n^5 z^5}{5!} - + \cdots$$

into the left-hand side of (7.24), and the expansions

$$\sin z = z - \frac{z^3}{3!} + \frac{z^5}{5!} - + \cdots$$

and

$$\sin^2 z = \frac{1}{2}(1 - \cos 2z) = z^2 - \frac{2}{3}z^4 + - \cdots$$

into the right-hand side of (7.24). Then multiplying everything out and simplifying, we obtain (after a lot of algebra)

$$nz - \frac{n^3 z^3}{3!} + - \cdots = nz + \left(-\frac{n}{6} - n \sum_{k=1}^{m} \frac{1}{\sin^2(k\pi/n)} \right) z^3 + \cdots.$$

Comparing the z^3 terms, by the identity theorem we conclude that

$$-\frac{n^3}{6} = -\frac{n}{6} - n \sum_{k=1}^{m} \frac{1}{\sin^2(k\pi/n)},$$

which can be written in the form

$$\frac{1}{6} - \sum_{k=1}^{m} \frac{1}{n^2 \sin^2(k\pi/n)} = \frac{1}{6n^2}. \qquad (7.25)$$

To establish Euler's formula, we apply Tannery's theorem to this sum. First, it follows from Lemma 5.6 on p. 399 that for some positive constant c,

$$c\,|z| \le |\sin z| \quad \text{for } |z| \le \pi/2.$$

For $0 \le k \le m = (n-1)/2$, we have $k\pi/n < \pi/2$, so for such k,

$$c \cdot \frac{k\pi}{n} \leq \sin \frac{k\pi}{n},$$

which gives

$$\frac{1}{n^2} \cdot \frac{1}{\sin^2(k\pi/n)} \leq \frac{1}{n^2} \cdot \frac{n^2}{(c\pi)^2 k^2} = \frac{1}{c^2\pi^2} \cdot \frac{1}{k^2}.$$

By the p-test, we know that the sum $\sum_{k=1}^{\infty} \frac{1}{c^2\pi^2} \cdot \frac{1}{k^2}$ converges. Second, since $n \sin(x/n) \to x$ as $n \to \infty$, we have

$$\lim_{n \to \infty} \frac{1}{n^2 \sin^2(k\pi/n)} = \frac{1}{k^2\pi^2}.$$

Thus, taking $m \to \infty$ in (7.25), Tannery's theorem gives

$$\frac{1}{6} - \sum_{k=1}^{\infty} \frac{1}{k^2\pi^2} = 0,$$

which is equivalent to Euler's formula. See Problem 7.5 for a proof that uses (7.25) but doesn't use Tannery's theorem.

▶ Exercises 7.5

1. Find the sum $\sum_{n=1}^{\infty} \frac{1}{n^8}$ using Euler's method, that is, in the same manner as we derived (7.22) and (7.23).

2. (Cf. [49, 132]) (**Euler's sum, Proof X**) Instead of using Tannery's theorem to derive Euler's formula from (7.25), we can proceed as follows.

   (i) Fix $M \in \mathbb{N}$ and let $m > M$. Using (7.25), prove that for $n = 2m + 1$,

   $$\frac{1}{6} - \sum_{k=1}^{M} \frac{1}{n^2 \sin^2(k\pi/n)} = \frac{1}{n^2} + \sum_{k=M+1}^{m} \frac{1}{n^2 \sin^2(k\pi/n)}.$$

   (ii) Using that $c x \leq \sin x$ for $0 \leq x \leq \pi/2$ with $c > 0$, prove that

   $$0 \leq \frac{1}{6} - \sum_{k=1}^{M} \frac{1}{n^2 \sin^2(k\pi/n)} \leq \frac{1}{n^2} + \frac{1}{c^2\pi^2} \sum_{k=M+1}^{\infty} \frac{1}{k^2}.$$

   (iii) Finally, letting $m \to \infty$ (so that $n = 2m + 1 \to \infty$ as well) and then letting $M \to \infty$, establish Euler's formula.

3. (Cf. [56]) Let $S \subseteq \mathbb{N}$ denote the set of square-free natural numbers; see Section 4.8.2 on p. 318 for a review of square-free numbers.

   (i) Let $N \in \mathbb{N}$ and prove that

$$\sum_{n<N} \frac{1}{n^2} \le \left(\sum_{k<N} \frac{1}{k^4}\right)\left(\sum_{n \in S, \, n<N} \frac{1}{n^2}\right) \le \sum_{n=1}^{\infty} \frac{1}{n^2}.$$

(ii) If $\sum_{n \in S} \frac{1}{n^2} := \lim_{N \to \infty} \sum_{n \in S, \, n<N} \frac{1}{n^2}$, using (i), prove that

$$\sum_{n \in S} \frac{1}{n^2} = \frac{15}{\pi^2}.$$

4. (Cf. [56]) Let $A \subseteq \mathbb{N}$ denote the set of natural numbers that are not perfect squares. With $\sum_{n \in A} \frac{1}{n^2} := \lim_{N \to \infty} \sum_{n \in A, \, n<N} \frac{1}{n^2}$, prove that

$$\sum_{n \in A} \frac{1}{n^2} = \frac{\pi^2}{90}(15 - \pi^2).$$

7.6 ★ Riemann's Remarkable ζ-Function, Probability, and $\pi^2/6$

We have already seen the Riemann zeta function at work in many examples. In this section we're going to look at some of its relationships with number theory; this will give just a hint as to the great importance of the zeta function in mathematics. As a consolation prize to our discussion on Riemann's ζ-function we'll find an incredible connection between probability theory and $\pi^2/6$.

7.6.1 The Riemann Zeta Function and Number Theory

We begin with the following theorem proved by Euler that connects $\zeta(z) = \sum_{n=1}^{\infty} \frac{1}{n^z}$ to prime numbers. See Problem 1 for a proof using the ubiquitous Tannery's theorem!

Euler and Riemann

Theorem 7.12 *For all $z \in \mathbb{C}$ with* $\operatorname{Re} z > 1$, *we have*

$$\zeta(z) = \prod \left(1 - \frac{1}{p^z}\right)^{-1} = \prod \frac{p^z}{p^z - 1},$$

where the infinite product is over all prime numbers $p \in \mathbb{N}$.

Proof We give two proofs.

Proof I: Let $r > 1$ be arbitrary and let $\operatorname{Re} z \geq r$. Let $2 < N \in \mathbb{N}$ and let $2 < 3 < \cdots$ $< m < N$ be all the primes less than N. Observe that

$$\prod_{p<N}\left(1-\frac{1}{p^z}\right)^{-1} = \left(1-\frac{1}{2^z}\right)^{-1}\left(1-\frac{1}{3^z}\right)^{-1}\cdots\left(1-\frac{1}{m^z}\right)^{-1}$$

$$= \left(\sum_{i=0}^{\infty}\frac{1}{2^{iz}}\right)\left(\sum_{j=0}^{\infty}\frac{1}{3^{jz}}\right)\cdots\left(\sum_{k=0}^{\infty}\frac{1}{m^{kz}}\right)$$

$$= \lim_{N\to\infty}\left(\sum_{i=0}^{N}\frac{1}{2^{iz}}\right)\left(\sum_{j=0}^{N}\frac{1}{3^{jz}}\right)\cdots\left(\sum_{k=0}^{N}\frac{1}{m^{kz}}\right)$$

$$= \lim_{N\to\infty}\left(\sum_{i=0}^{N}\sum_{j=0}^{N}\cdots\sum_{k=0}^{N}\frac{1}{2^{iz}3^{jz}\cdots m^{kz}}\right). \qquad (7.26)$$

By unique factorization, every natural number $n < N$ can be written as $n = 2^i\, 3^j \cdots m^k$ for some nonnegative integers $i, j, \ldots, k$; for such a natural number, we have

$$n^z = \left(2^i\, 3^j \cdots m^k\right)^z = 2^{iz}\, 3^{jz} \cdots m^{kz}.$$

It follows that

$$\sum_{i=0}^{N}\sum_{j=0}^{N}\cdots\sum_{k=0}^{N}\frac{1}{2^{iz}3^{jz}\cdots m^{kz}}$$

contains the numbers $1, \frac{1}{2^z}, \frac{1}{3^z}, \frac{1}{4^z}, \frac{1}{5^z}, \ldots, \frac{1}{(N-1)^z}$, along with some other numbers $\frac{1}{n^z}$ with $n \geq N$ having prime factors $2, 3, \ldots, m$. Thus,

$$\sum_{i=0}^{N}\sum_{j=0}^{N}\cdots\sum_{k=0}^{N}\frac{1}{2^{iz}3^{jz}\cdots m^{kz}} = \sum_{n=1}^{N-1}\frac{1}{n^z} + \sum\frac{1}{n_k^z}, \qquad (7.27)$$

where $N \leq n_1 < n_2 < n_3 < \cdots$ is a finite subsequence of natural numbers with prime factors in $\{2, 3, \ldots, m\}$. Observe that

$$\left|\sum\frac{1}{n_k^z}\right| \leq \sum_{n=N}^{\infty}\left|\frac{1}{n^z}\right| \leq \sum_{n=N}^{\infty}\frac{1}{n^r},$$

since $\operatorname{Re} z \geq r$. By the p-test (with $p = r > 1$), $\sum\frac{1}{n^r}$ converges, so by Cauchy's criterion for series, $\lim_{N\to\infty}\sum_{n=N}^{\infty}\frac{1}{n^r} = 0$. From (7.27) we conclude that

$$\lim_{N \to \infty} \left(\sum_{i=0}^{N} \sum_{j=0}^{N} \cdots \sum_{k=0}^{N} \frac{1}{2^{iz} 3^{jz} \cdots m^{kz}} \right) = \sum_{n=1}^{\infty} \frac{1}{n^z} = \zeta(z),$$

and then from (7.26) we get

$$\lim_{N \to \infty} \prod_{p < N} \left(1 - \frac{1}{p^z} \right)^{-1} = \zeta(z).$$

Proof II: Here's Euler's beautiful proof using a *sieving method* made famous by Eratosthenes of Cyrene (276 B.C.–194 B.C.). First we get rid of all the numbers in $\zeta(z)$ that have factors of 2: Observe that

$$\frac{1}{2^z} \zeta(z) = \frac{1}{2^z} \sum_{n=1}^{\infty} \frac{1}{n^z} = \sum_{n=1}^{\infty} \frac{1}{(2n)^z},$$

and therefore,

$$\left(1 - \frac{1}{2^z} \right) \zeta(z) = \sum_{n=1}^{\infty} \frac{1}{n^z} - \sum_{n=1}^{\infty} \frac{1}{(2n)^z} = \sum_{n\,;\,2 \nmid n} \frac{1}{n^z}.$$

Next, we get rid of all the numbers in $\left(1 - \frac{1}{2^z} \right) \zeta(z)$ that have factors of 3: Observe that

$$\frac{1}{3^z} \left(1 - \frac{1}{2^z} \right) \zeta(z) = \frac{1}{3^z} \sum_{n\,;\,2 \nmid n} \frac{1}{n^z} = \sum_{n\,;\,2 \nmid n} \frac{1}{(3n)^z},$$

and therefore,

$$\left(1 - \frac{1}{3^z} \right) \left(1 - \frac{1}{2^z} \right) \zeta(z) = \left(1 - \frac{1}{2^z} \right) \zeta(z) - \frac{1}{3^z} \left(1 - \frac{1}{2^z} \right) \zeta(z)$$

$$= \sum_{n\,;\,2 \nmid n} \frac{1}{n^z} - \sum_{n\,;\,2 \nmid n} \frac{1}{(3n)^z}$$

$$= \sum_{n\,;\,2,3 \nmid n} \frac{1}{n^z}.$$

Repeating this argument, we get, for every prime q:

$$\left\{ \prod_{p \text{ prime} \leq q} \left(1 - \frac{1}{p^z} \right) \right\} \zeta(z) = \sum_{n\,;\,2,3,\ldots,q \nmid n} \frac{1}{n^z},$$

where the sum is over all $n \in \mathbb{N}$ that are not divisible by the primes from 2 to q. Therefore, choosing $r > 1$ such that $|z| > r$, we have

$$\left| \left\{ \prod_{p \text{ prime} \leq q} \left(1 - \frac{1}{p^z} \right) \right\} \zeta(z) - 1 \right| = \left| \sum_{n \,;\, n \neq 1 \,\&\, 2,3,\dots,q \nmid n} \frac{1}{n^z} \right|$$

$$\leq \sum_{n \,;\, n \neq 1 \,\&\, 2,3,\dots,q \nmid n} \frac{1}{n^r} \leq \sum_{n=q}^{\infty} \frac{1}{n^r}.$$

By Cauchy's criterion for series, $\lim_{q \to \infty} \sum_{n=q}^{\infty} \frac{1}{n^r} = 0$, so we conclude that

$$\left\{ \prod_{p \text{ prime}} \left(1 - \frac{1}{p^z} \right) \right\} \zeta(z) = 1,$$

which is equivalent to Euler's product formula. ∎

In particular, since we know that $\zeta(2) = \pi^2/6$, we have

$$\boxed{\frac{\pi^2}{6} = \prod \frac{p^2}{p^2 - 1} = \frac{2^2}{2^2 - 1} \cdot \frac{3^2}{3^2 - 1} \cdot \frac{5^2}{5^2 - 1} \cdots.}$$

Our next connection of the zeta function with number theory involves the following strange but interesting, function, defined for $n \in \mathbb{N}$:

$$\mu(n) = \begin{cases} 1 & \text{if } n = 1 \\ (-1)^k & \text{if } n = p_1 \, p_2 \cdots p_k \text{ is a product } k \text{ distinct prime numbers,} \\ 0 & \text{otherwise.} \end{cases}$$

This function is called the **Möbius function** after August Möbius (1790–1868), who introduced the function in 1831. Some of its values are

$$\mu(1) = 1 \,,\; \mu(2) = -1 \,,\; \mu(3) = -1 \,,\; \mu(4) = 0 \,,\; \mu(5) = -1 \,,\; \mu(6) = 1 \,,\; \dots.$$

Theorem 7.13 *For all $z \in \mathbb{C}$ with $\operatorname{Re} z > 1$, we have*

$$\boxed{\frac{1}{\zeta(z)} = \prod \left(1 - \frac{1}{p^z} \right) = \sum_{n=1}^{\infty} \frac{\mu(n)}{n^z}.}$$

Proof Let $r > 1$ be arbitrary and let $\text{Re } z \geq r$. Let $2 < N \in \mathbb{N}$ and let $2 < 3 < \cdots < m < N$ be all the primes less than N. Then observe that the product

$$\prod_{n < N} \left(1 - \frac{1}{p^z} \right) = \left(1 + \frac{-1}{2^z} \right) \left(1 + \frac{-1}{3^z} \right) \left(1 + \frac{-1}{5^z} \right) \cdots \left(1 + \frac{-1}{m^z} \right),$$

when multiplied out, contains 1 and all numbers of the form

$$\left(\frac{-1}{p_1^z} \right) \cdot \left(\frac{-1}{p_2^z} \right) \cdot \left(\frac{-1}{p_3^z} \right) \cdots \left(\frac{-1}{p_k^z} \right) = \frac{(-1)^k}{p_1^z p_2^z \cdots p_k^z} = \frac{(-1)^k}{n^z}, \quad n = p_1 p_2 \cdots p_k,$$

where $p_1 < p_2 < \cdots < p_k < N$ are distinct primes. In particular, $\prod_{n < N} \left(1 - \frac{1}{p^z} \right)$ contains the numbers $\frac{\mu(n)}{n^z}$ for $n = 1, 2, \ldots, N - 1$ (along with all other numbers $\frac{\mu(n)}{n^z}$ with $n \geq N$ having prime factors $2, 3, \ldots, m$), so

$$\left| \sum_{n=1}^{\infty} \frac{\mu(n)}{n^z} - \prod_{p < N} \left(1 - \frac{1}{p^z} \right) \right| \leq \sum_{n=N}^{\infty} \left| \frac{\mu(n)}{n^z} \right| \leq \sum_{n=N}^{\infty} \frac{1}{n^r},$$

since $\text{Re } z \geq r$. By the p-test (with $p = r > 1$), $\sum \frac{1}{n^r}$ converges, so the right-hand side tends to zero as $N \to \infty$. This completes our proof. ∎

See the exercises for other neat connections of $\zeta(z)$ with number theory.

7.6.2   The Eta Function

A function related to the zeta function is the "alternating zeta function" or **Dirichlet eta function**:

$$\eta(z) := \sum_{n=1}^{\infty} \frac{(-1)^{n-1}}{n^z}.$$

In the next section we find the values of the eta function at positive even integers. To do so, we write the eta function in terms of the zeta function as follows.

Theorem 7.14 *We have*

$$\eta(z) = (1 - 2^{1-z})\zeta(z), \quad \text{Re } z > 1.$$

Proof Splitting into sums over even and odd values of n, we get

$$\sum_{n=1}^{\infty} \frac{(-1)^{n-1}}{n^z} = -\sum_{n=1}^{\infty} \frac{1}{(2n)^z} + \sum_{n=1}^{\infty} \frac{1}{(2n-1)^z}$$

$$= -\sum_{n=1}^{\infty} \frac{1}{2^z}\frac{1}{n^z} + \sum_{n=1}^{\infty} \frac{1}{(2n-1)^z}$$

$$= -2^{-z}\zeta(z) + \sum_{n=1}^{\infty} \frac{1}{(2n-1)^z}.$$

On the other hand, breaking the zeta function into sums of even and odd numbers, we get

$$\zeta(z) = \sum_{n=1}^{\infty} \frac{1}{n^z} = \sum_{n=1}^{\infty} \frac{1}{(2n)^z} + \sum_{n=1}^{\infty} \frac{1}{(2n-1)^z} = 2^{-z}\zeta(z) + \sum_{n=1}^{\infty} \frac{1}{(2n-1)^z}.$$

Substituting this expression into the previous one, we see that

$$\sum_{n=1}^{\infty} \frac{(-1)^{n-1}}{n^z} = -2^{-z}\zeta(z) + \zeta(z) - 2^{-z}\zeta(z),$$

which is equivalent to the expression that we desired to prove. ∎

We now consider a shocking connection between probability theory, prime numbers, divisibility, and $\pi^2/6$ (cf. [2, 116]).[1] **Question:** What is the probability that a natural number, chosen at random, is square-free? Answer (drum roll please): $6/\pi^2$, a result that follows from work of Dirichlet in 1849 [131, p. 324], [102, p. 272]. Here's another **Question:** What is the probability that any two natural numbers, chosen at random, are relatively prime? Answer (drum roll please): $6/\pi^2$, first proved by Leopold Gegenbauer (1849–1903) [102, p. 272], who proved it in 1885 (Fig. 7.1).

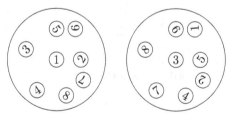

Fig. 7.1 We have two bags, each containing all the natural numbers. We reach into the first bag and grab a number; what is the probability that the number is square-free? We throw that number back into the first bag, and then we draw a number from each bag; what is the probability the two numbers drawn are relatively prime?

---

[1] Such shocking connections in science perhaps made Albert Einstein (1879–1955) state that "the scientist's religious feeling takes the form of a rapturous amazement at the harmony of natural law, which reveals an intelligence of such superiority that, compared with it, all the systematic thinking and acting of human beings is an utterly insignificant reflection" [112].

7.6.3 Elementary Probability Theory

You will prove these results with complete rigor in Problems 11 and 10. However, we are going to derive them intuitively—*not* rigorously (!)—based on some basic probability ideas that should be "obvious" (or at least believable) to you; see [75, 76, 247] for standard books on probability in case you want the hardcore theory. We just need the basics. We denote the probability, or chance, that an event A happens by $P(A)$. If we are conducting a random experiment with equally likely elementary outcomes, then the classic definition is

$$P(A) = \frac{\text{number of outcomes in } A}{\text{total number of possibilities}}. \tag{7.28}$$

For example, consider a classroom with ten students, m men and w women (so that $m + w = 10$). You put the names of the ten students in a bag and pull out a name. The probability of randomly "choosing a man" $(=M)$ is

$$P(M) = \frac{\text{number of men}}{\text{total number of possibilities}} = \frac{m}{10}.$$

Similarly, the probability of randomly choosing a woman is $w/10$. We next need to discuss complementary events. If A^c is the event that A does not happen, then

$$P(A^c) = 1 - P(A). \tag{7.29}$$

For instance, according to (7.29), the probability of "*not* choosing a man," M^c, should be $P(M^c) = 1 - P(M) = 1 - m/10$. But this is certainly true, because "*not* choosing a man" is the same as "choosing a woman" W, so recalling that $m + w = 10$, we have

$$P(M^c) = P(W) = \frac{w}{10} = \frac{10 - m}{10} = 1 - \frac{m}{10}.$$

Finally, we need to discuss independence. Whenever an event A is *unrelated* to an event B (such events are called **independent**), we have the fundamental relation

$$P(A \text{ and } B) = P(A) \cdot P(B).$$

For example, let's say that we have two classrooms of ten students each, the first one with m_1 men and w_1 women, and the second one with m_2 men and w_2 women. Let us randomly choose a pair of names, one from the first classroom and the other from the second. What is the probability of "choosing a man from the first classroom" $=A$ *and* "choosing a woman from the second classroom" $=B$? Certainly A and B don't depend on each other, so by our formula above, we should have

$$P(A \text{ and } B) = P(A) \cdot P(B) = \frac{m_1}{10} \cdot \frac{w_2}{10} = \frac{m_1 w_2}{100}.$$

To see that this is indeed true, note that the number of ways to pair a man in classroom 1 with a woman in classroom 2 is $m_1 \cdot w_2$, and the total number of possible pairs of people is $10^2 = 100$. Thus,

$$P(A \text{ and } B) = \frac{\text{number of men-women pairs}}{\text{total number of possible pairs of people}} = \frac{m_1 \cdot w_2}{100},$$

in agreement with our previous calculation. We remark that for any number of events $A_1, A_2, \ldots$ that are unrelated to each other, we have the generalized result

$$P(A_1 \text{ and } A_2 \text{ and } \cdots) = P(A_1) \cdot P(A_2) \cdots. \tag{7.30}$$

7.6.4 Probability and $\pi^2/6$

To begin discussing our two incredible and shocking problems, we first look at the following question: Given a natural number k, what is the probability, or chance, that a randomly chosen natural number is divisible by k? Since the definition (7.28) involves finite quantities, we can't use this definition as it stands. We can instead use the following modified version:

$$P(A) = \lim_{n \to \infty} \frac{\text{number of occurrences of } A \text{ among } n \text{ possibilities}}{n}. \tag{7.31}$$

Using this formula, in Problem 8, you will prove that the probability that a randomly chosen natural number is divisible by k is $1/k$; of course, this is "obvious".

However, instead of using (7.31), we shall employ the following heuristic trick (which works to give the correct answer). Choose an "extremely large" natural number N, and consider the very large sample of numbers

$$1, 2, 3, 4, 5, 6, \ldots, Nk.$$

There are exactly N numbers in this list that are divisible by k, namely the N numbers $k, 2k, 3k, \ldots, Nk$, and no others, and there is a total of Nk numbers in this list (Fig. 7.2). Thus, the probability that a natural number n, randomly chosen among the large sample, is divisible by k is exactly the probability that n is one of the N

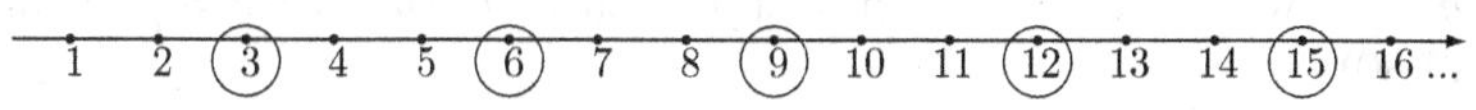

Fig. 7.2 Take for example $k = 3$. Then every third number is divisible by 3. Thus, there should be a one out of three, or $1/3$, probability that a randomly chosen number is divisible by 3

numbers $k, 2k, 3k, \ldots, Nk$, so

$$P(k \text{ divides } n) = \frac{\text{number of occurrences of divisibility}}{\text{total number of possibilities listed}} = \frac{N}{Nk} = \frac{1}{k}. \qquad (7.32)$$

For instance, the probability that a randomly chosen natural number is divisible by 1 is 1, which makes sense. The probability that a randomly chosen natural number is divisible by 2 is 1/2; in other words, the probability that a randomly chosen natural number is even is 1/2, which also makes sense.

We are now ready to solve our two problems. **Question**: What is the probability that a natural number, chosen at random, is square-free? Consider a randomly chosen $n \in \mathbb{N}$. Then n is square-free just means that $p^2 \nmid n$ (p^2 does not divide n) for all primes p. Thus,

$$P(n \text{ is square free}) = P((2^2 \nmid n) \text{ and } (3^2 \nmid n) \text{ and } (5^2 \nmid n) \text{ and } (7^2 \nmid n) \text{ and } \cdots).$$

Since n was randomly chosen, the events $2^2 \nmid n$, $3^2 \nmid n$, $5^2 \nmid n$, etc. are unrelated, so by (7.30),

$$P(n \text{ is square free}) = P(2^2 \nmid n) \cdot P(3^2 \nmid n) \cdot P(5^2 \nmid n) \cdot P(7^2 \nmid n) \cdots.$$

To see what the right-hand side is, we use (7.29) and (7.32) to write

$$P(p^2 \nmid n) = 1 - P(p^2 \text{ divides } n) = 1 - \frac{1}{p^2}.$$

Thus,

$$P(n \text{ is square free}) = \prod_{p \text{ prime}} P(p^2 \nmid n) = \prod_{p \text{ prime}} \left(1 - \frac{1}{p^2}\right) = \frac{1}{\zeta(2)} = \frac{6}{\pi^2},$$

and our first question is answered!

Question: What is the probability that two given numbers, chosen at random, are relatively prime (or coprime)? Consider randomly chosen $m, n \in \mathbb{N}$. Then m and n are **relatively prime**, or **coprime**, just means that m and n have no common factors (except 1), which means[2] that $p \nmid$ both m, n for all prime numbers p. Thus,

$P(m, n \text{ are relatively prime})$

$$= P((2 \nmid \text{ both } m, n) \text{ and } (3 \nmid \text{ both } m, n) \text{ and } (5 \nmid \text{ both } m, n) \text{ and } \cdots).$$

Since m and n were randomly chosen, that $p \nmid$ both m, n is unrelated to $q \nmid$ both m, n, so by (7.30),

---

[2]Explicitly, "$p \nmid$ both m, n" means "it's *not* the case that $p|m$ and $p|n$".

$$P(m, n \text{ are relatively prime}) = \prod_{p \text{ prime}} P(p \nmid \text{ both } m, n).$$

To see what the right-hand side is, we use (7.29), (7.30), and (7.32) to write

$$P(p \nmid \text{ both } m, n) = 1 - P(p \text{ divides both } m, n)$$
$$= 1 - P(p \text{ divides } m \text{ and } p \text{ divides } n)$$
$$= 1 - P(p \text{ divides } m) \cdot P(p \text{ divides } n) = 1 - \frac{1}{p} \cdot \frac{1}{p} = 1 - \frac{1}{p^2}.$$

Thus,

$$P(m, n \text{ are relatively prime}) = \prod_{p \text{ prime}} P(p \nmid \text{ both } m, n) = \prod_{p \text{ prime}} \left(1 - \frac{1}{p^2}\right) = \frac{6}{\pi^2},$$

and our second question is answered!

▶ **Exercises 7.6**

1. ($\zeta(z)$ **product formula, Proof III**) We prove Theorem 7.12 using the good old Tannery's theorem for products.

   (i) Let $r > 1$ be arbitrary and let $\operatorname{Re} z \geq r$. Prove that

   $$\left| \prod_{p < N} \frac{p^z - (1/p^z)^N}{p^z - 1} - \sum_{n=1}^{\infty} \frac{1}{n^z} \right| \leq \sum_{n=N}^{\infty} \frac{1}{n^r}.$$

   Suggestion: $\frac{p^z - (1/p^z)^N}{p^z - 1} = \frac{1 - (1/p^z)^{N+1}}{1 - 1/p^z} = 1 + 1/p^z + 1/p^{2z} + \cdots + 1/p^{Nz}$.

   (ii) Write $\frac{p^z - (1/p^z)^N}{p^z - 1} = 1 + \frac{1 - (1/p^z)^N}{p^z - 1}$. Show that

   $$\left| \frac{1 - (1/p^z)^N}{p^z - 1} \right| \leq \frac{2}{p^r - 1} \leq \frac{4}{p^r}$$

   and $\sum 4/p^r$ converges. Now prove Theorem 7.12 using Tannery's theorem for products.

2. Prove that for $z \in \mathbb{C}$ with $\operatorname{Re} z > 1$,

$$\boxed{\frac{\zeta(z)}{\zeta(2z)} = \sum_{n=1}^{\infty} \frac{|\mu(n)|}{n^z}.}$$

Suggestion: Show that $\frac{\zeta(z)}{\zeta(2z)} = \prod \left(1 + \frac{1}{p^z}\right)$ and copy the proof of Theorem 7.13.

3. (**Möbius inversion formula**) In this problem we prove the Möbius inversion formula.

(i) Given $n \in \mathbb{N}$ with $n > 1$, let $p_1, \ldots, p_k$ be the distinct prime factors of n. For $1 \leq i \leq k$, let

$$A_i = \{m \in \mathbb{N}\,;\; m = \text{a product of exactly } i \text{ distinct prime factors of } n\}.$$

Show that

$$\sum_{d|n} \mu(d) = 1 + \sum_{i=1}^{k} \sum_{m \in A_i} \mu(m),$$

where $\sum_{d|n} \mu(d)$ means to sum over all $d \in \mathbb{N}$ such that $d|n$. Next, show that

$$\sum_{m \in A_i} \mu(m) = (-1)^i \binom{k}{i}.$$

(ii) For every $n \in \mathbb{N}$, prove that

$$\sum_{d|n} \mu(d) = \begin{cases} 1 & \text{if } n = 1, \\ 0 & \text{if } n > 1. \end{cases}$$

(iii) Let $f : (0, \infty) \to \mathbb{R}$ be a function such that $f(x) = 0$ for $x < 1$. Define

$$g(x) = \sum_{n=1}^{\infty} f\left(\frac{x}{n}\right).$$

Note that $g(x) = 0$ for $x < 1$ and that this infinite series is really a finite sum, since $f(x) = 0$ for $x < 1$; specifically, choosing $N \in \mathbb{N}$ with $N \geq \lfloor x \rfloor$ (the greatest integer $\leq x$), we have $g(x) = \sum_{n=1}^{N} f(x/n)$. Prove that

$$f(x) = \sum_{n=1}^{\infty} \mu(n)\, g\left(\frac{x}{n}\right) \qquad \textbf{(Möbius inversion formula).}$$

As before, this sum is really a finite summation. Suggestion: If you've not gotten anywhere after some time, let $S = \{(k, n) \in \mathbb{N} \times \mathbb{N}\,;\; n|k\}$ and consider the sum

$$\sum_{(k,n) \in S} \mu(n)\, f\left(\frac{x}{k}\right).$$

Write this sum as $\sum_{k=1}^{\infty} \sum_{n\,;\,n|k} \mu(n)\, f(x/k)$, then as $\sum_{n=1}^{\infty} \sum_{k\,;\,n|k} \mu(n)\, f(x/k)$, and simplify each iterated sum.

4. (**Liouville's function**) Define, for $n \in \mathbb{N}$,

$$\lambda(n) = \begin{cases} 1 & \text{if } n = 1, \\ 1 & \text{if the number of prime factors of } n, \text{ counted with repetitions, is even,} \\ -1 & \text{if the number of prime factors of } n, \text{ counted with repetitions, is odd.} \end{cases}$$

This function is called **Liouville's function** after Joseph Liouville (1809–1882). Prove that for $z \in \mathbb{C}$ with $\text{Re } z > 1$,

$$\boxed{\frac{\zeta(2z)}{\zeta(z)} = \sum_{n=1}^{\infty} \frac{\lambda(n)}{n^z}.}$$

Suggestion: Show that $\frac{\zeta(2z)}{\zeta(z)} = \prod \left(1 + \frac{1}{p^z}\right)^{-1}$.

5. For $n \in \mathbb{N}$, let $\tau(n)$ denote the number of positive divisors of n (that is, the number of positive integers that divide n). Prove that for $z \in \mathbb{C}$ with $\text{Re } z > 1$,

$$\boxed{\zeta(z)^2 = \sum_{n=1}^{\infty} \frac{\tau(n)}{n^z}.}$$

Suggestion: By absolute convergence, we can write $\zeta(z)^2 = \sum_{m,n} 1/(m \cdot n)^z$, where this double series can be summed in any way we wish. Use Theorem 6.25 on p. 475, the sum by curves theorem, with the set S_k given by $S_k = T_1 \cup \cdots \cup T_k$ where $T_k = \{(m, n) \in \mathbb{N} \times \mathbb{N} ; m \cdot n = k\}$.

6. Let $\zeta(z, a) := \sum_{n=0}^{\infty} (n + a)^{-z}$ for $z \in \mathbb{C}$ with $\text{Re } z > 1$ and $a > 0$; this function is called the **Hurwitz zeta function** after Adolf Hurwitz (1859–1919). Prove that

$$\sum_{m=1}^{k} \zeta\left(z, \frac{m}{k}\right) = k^z \zeta(z).$$

7. In this problem, we find useful bounds and limits for $\zeta(x)$ with $x > 1$ real.

(a) For $\eta(x)$ the Dirichlet eta function, prove that $1 - \frac{1}{2^x} < \eta(x) < 1$.

(b) Prove that

$$\frac{1 - 2^{-x}}{1 - 2^{1-x}} < \zeta(x) < \frac{1}{1 - 2^{1-x}}.$$

(c) Prove the following limits: $\zeta(x) \to 1$ as $x \to \infty$, $\zeta(x) \to \infty$ as $x \to 1^+$, and $(x - 1)\zeta(x) \to 1$ as $x \to 1^+$.

8. Using the definition (7.31), prove that given a natural number k, the probability that a randomly chosen natural number is divisible by k is $1/k$. Suggestion:

Among the n natural numbers $1, 2, 3, \ldots, n$, show that q_n numbers are divisible by k, where q_n is the quotient of n divided by k. Then find $\lim_{n \to \infty} q_n/n$.

9. (Cf. [24, 116]) Let $k \in \mathbb{N}$ with $k \geq 2$. We say that a natural number n is kth-**power-free** if $p^k \nmid n$ for all primes p. What is the probability that a natural number, chosen at random, is kth-power-free? What is the probability that k natural numbers, chosen at random, are relatively prime (have not common factors except 1)?

10. (**Square-free numbers**) Define $S : (0, \infty) \to \mathbb{R}$ by

$$S(x) = \#\{k \in \mathbb{N} \,;\, 1 \leq k \leq x \text{ and } k \text{ is square-free}\};$$

note that $S(x) = 0$ for $x < 1$. We shall prove that

$$\lim_{n \to \infty} \frac{S(n)}{n} = \frac{6}{\pi^2}.$$

Do you see why this formula makes precise the statement "The probability that a randomly chosen natural number is square-free equals $6/\pi^2$"?

(i) For every real number $x > 0$ and $n \in \mathbb{N}$, define

$$A(x, n) = \{k \in \mathbb{N} \,;\, 1 \leq k \leq x \text{ and } n^2 \text{ is the largest square that divides } k\}.$$

Note that $A(x, n) = \varnothing$ for $n^2 > x$. Prove that $A(x, 1)$ consists of all square-free numbers $\leq x$, and also prove that

$$\{k \in \mathbb{N} \,;\, 1 \leq k \leq x\} = \bigcup_{n=1}^{\infty} A(x, n).$$

(ii) Show that there is a bijection between $A(x, n)$ and $A(x/n^2, 1)$.

(iii) Show that for every $x > 0$, we have

$$\lfloor x \rfloor = \sum_{n=1}^{\infty} S\left(\frac{x}{n^2}\right).$$

Using the Möbius inversion formula from Problem 3, conclude that

$$S(x) = \sum_{n=1}^{\infty} \mu(n) \left\lfloor \frac{x}{n^2} \right\rfloor.$$

(iv) Finally, prove that $\lim_{x \to \infty} S(x)/x = 6/\pi^2$, which proves our result.

11. (**Relatively prime numbers**; for different proofs, see [131, p. 337] and [102, p. 268]) Define $R : (0, \infty) \to \mathbb{R}$ by

$R(x) = \#\{(k, \ell) \in \mathbb{N} \; ; \; 1 \leq k, \ell \leq x$ and k and ℓ are relatively prime$\}$;

note that $R(x) = 0$ for $x < 1$. We shall prove that

$$\lim_{n \to \infty} \frac{R(n)}{n^2} = \frac{6}{\pi^2}.$$

Do you see why this formula makes precise the statement "The probability that two randomly chosen natural numbers are relatively prime equals $6/\pi^2$"?

(i) For every real number $x > 0$ and $n \in \mathbb{N}$, define

$A(x, n) = \{(k, \ell) \in \mathbb{N} \times \mathbb{N} \; ; \; 1 \leq k, \ell \leq x$ and n is the largest divisor of both k and $\ell\}$.

Note that $A(x, n) = \varnothing$ for $n > x$. Prove that $A(x, 1)$ consists of all pairs (k, ℓ) of relatively prime natural numbers that are $\leq x$, and also prove that

$$\{(k, \ell) \in \mathbb{N} \times \mathbb{N} \; ; \; 1 \leq k, \ell \leq x\} = \bigcup_{n=1}^{\infty} A(x, n).$$

(ii) Show that there is a bijection between $A(x, n)$ and $A(x/n, 1)$.
(iii) Show that for every $x > 0$, we have

$$\lfloor x \rfloor^2 = \sum_{n=1}^{\infty} R\left(\frac{x}{n}\right).$$

Using the Möbius inversion formula from Problem 3, conclude that

$$R(x) = \sum_{n=1}^{\infty} \mu(n) \left\lfloor \frac{x}{n} \right\rfloor^2.$$

(iv) Finally, prove that $\lim_{x \to \infty} R(x)/x^2 = 6/\pi^2$, which proves our result.

7.7 ★ Some of the Most Beautiful Formulas in the World IV

Hold on to your seats, for you're about to be taken on another journey through a beautiful world of mathematical formulas! In this section we derive many formulas found in Euler's wonderful book *Introduction to analysis of the infinite* [69]; his second book [70] is also great. We also give our tenth proof of Euler's formula for $\pi^2/6$ and our fourth proof of Gregory–Leibniz–Madhava's formula for $\pi/4$.

7.7.1   Bernoulli Numbers and Evaluating Sums/Products

We start our onslaught of beautiful formulas with a formula for $\zeta(2k) = \sum_{n=1}^{\infty} \frac{1}{n^{2k}}$
in terms of Bernoulli numbers; this complements the formulas in Section 5.3, when
we didn't know about Bernoulli numbers. To find such a formula, we begin with the
partial fraction expansion of the cotangent from Section 7.4:

$$\pi z \cot \pi z = 1 + \sum_{n=1}^{\infty} \frac{2z^2}{z^2 - n^2} = 1 - 2 \sum_{n=1}^{\infty} \frac{z^2}{n^2 - z^2}.$$

Next, we apply Cauchy's double series theorem to this sum. Let $z \in \mathbb{C}$ with $|z| < 1$
and observe that

$$\frac{z^2}{n^2 - z^2} = \frac{z^2/n^2}{1 - z^2/n^2} = \sum_{k=1}^{\infty} \left(\frac{z^2}{n^2}\right)^k,$$

where we used the geometric series formula $\sum_{k=1}^{\infty} r^k = \frac{r}{1-r}$ for $|r| < 1$. Therefore,

$$\pi z \cot \pi z = 1 - 2 \sum_{n=1}^{\infty} \sum_{k=1}^{\infty} \frac{z^{2k}}{n^{2k}}.$$

Since $|z| < 1$, the geometric series $\sum_{k=1}^{\infty} |z|^{2k}$ converges, so using that $1/n^{2k} \leq 1/n^2$,
we have

$$\sum_{n=1}^{\infty} \sum_{k=1}^{\infty} \left|\frac{z^{2k}}{n^{2k}}\right| \leq \sum_{n=1}^{\infty} \sum_{k=1}^{\infty} \frac{|z|^{2k}}{n^2} = \zeta(2) \sum_{k=1}^{\infty} |z|^{2k} < \infty.$$

Therefore, by Cauchy's double series theorem, for $|z| < 1$ we have

$$\pi z \cot \pi z = 1 - 2 \sum_{k=1}^{\infty} \sum_{n=1}^{\infty} \left(\frac{z^2}{n^2}\right)^k = 1 - 2 \sum_{k=1}^{\infty} \left(\sum_{n=1}^{\infty} \frac{1}{n^{2k}}\right) z^{2k}. \tag{7.33}$$

On the other hand, we recall from p. 498 in Section 6.7 that

$$z \cot z = \sum_{k=0}^{\infty} (-1)^k \frac{2^{2k} B_{2k}}{(2k)!} z^{2k} \quad \text{(for } |z| \text{ small)},$$

where the B_{2k} are the Bernoulli numbers. Replacing z with πz, we get

$$\pi z \cot \pi z = 1 + \sum_{k=1}^{\infty} (-1)^k \frac{2^{2k} B_{2k}}{(2k)!} \pi^{2k} z^{2k}.$$

Comparing this equation with (7.33) and using the identity theorem, Theorem 6.22 on p. 462, we see that

$$-2 \sum_{n=1}^{\infty} \frac{1}{n^{2k}} = (-1)^k \frac{2^{2k} B_{2k}}{(2k)!} \pi^{2k}, \quad k = 1, 2, 3, \ldots.$$

Rewriting this slightly, we obtain Euler's famous result: For $k = 1, 2, 3, \ldots$,

$$\boxed{\sum_{n=1}^{\infty} \frac{1}{n^{2k}} = (-1)^{k-1} \frac{(2\pi)^{2k} B_{2k}}{2(2k)!} \; ; \quad \text{that is,} \; \zeta(2k) = (-1)^{k-1} \frac{(2\pi)^{2k} B_{2k}}{2(2k)!}.}$$

$$(7.34)$$

Using the known values of the Bernoulli numbers found in Section 6.7, setting $k = 1, 2, 3$, we get, in particular, our eleventh proof of Euler's formula for $\pi^2/6$:

$$\frac{\pi^2}{6} = \sum_{n=1}^{\infty} \frac{1}{n^2} \quad \textbf{(Euler's sum, Proof XI)} \quad , \quad \frac{\pi^4}{90} = \sum_{n=1}^{\infty} \frac{1}{n^4} \quad , \quad \frac{\pi^6}{945} = \sum_{n=1}^{\infty} \frac{1}{n^6}.$$

Using (7.34), we can derive many other pretty formulas. First, recall that the eta function is defined by $\eta(z) = \sum_{n=1}^{\infty} \frac{(-1)^{n-1}}{n^z}$. In Theorem 7.14 on p. 570 we proved that

$$\eta(z) = (1 - 2^{1-z})\zeta(z), \quad \operatorname{Re} z > 1.$$

In particular, setting $z = 2k$, we find that for $k = 1, 2, 3, \ldots$,

$$\boxed{\eta(2k) = \sum_{n=1}^{\infty} \frac{(-1)^{n-1}}{n^{2k}} = (-1)^{k-1}\left(1 - 2^{1-2k}\right) \frac{(2\pi)^{2k} B_{2k}}{2(2k)!};} \quad (7.35)$$

what formulas do you get when you set $k = 1, 2$? Second, recall from Theorem 7.12 on p. 566 that

$$\sum_{n=1}^{\infty} \frac{1}{n^z} = \prod \frac{p^z}{p^z - 1} = \frac{2^z}{2^z - 1} \cdot \frac{3^z}{3^z - 1} \cdot \frac{5^z}{5^z - 1} \cdot \frac{7^z}{7^z - 1} \cdots, \quad (7.36)$$

where the product is over all primes. In particular, setting $z = 2$, we get

$$\boxed{\frac{\pi^2}{6} = \frac{2^2}{2^2 - 1} \cdot \frac{3^2}{3^2 - 1} \cdot \frac{5^2}{5^2 - 1} \cdot \frac{7^2}{7^2 - 1} \cdot \frac{11^2}{11^2 - 1} \cdots,} \quad (7.37)$$

and setting $z = 4$, we get

$$\frac{\pi^4}{90} = \frac{2^4}{2^4 - 1} \cdot \frac{3^4}{3^4 - 1} \cdot \frac{5^4}{5^4 - 1} \cdot \frac{7^4}{7^4 - 1} \cdot \frac{11^4}{11^4 - 1} \cdots .$$

Dividing these two formulas and using that

$$\frac{\dfrac{n^4}{n^4 - 1}}{\dfrac{n^2}{n^2 - 1}} = n^2 \cdot \frac{n^2 - 1}{n^4 - 1} = n^2 \cdot \frac{n^2 - 1}{(n^2 - 1)(n^2 + 1)} = \frac{n^2}{n^2 + 1},$$

we obtain

$$\frac{\pi^2}{15} = \frac{2^2}{2^2 + 1} \cdot \frac{3^2}{3^2 + 1} \cdot \frac{5^2}{5^2 + 1} \cdot \frac{7^2}{7^2 + 1} \cdot \frac{11^2}{11^2 + 1} \cdots . \qquad (7.38)$$

Third, recall from Theorem 7.13 that

$$\frac{1}{\zeta(z)} = \sum_{n=1}^{\infty} \frac{\mu(n)}{n^z},$$

where $\mu(n)$ is the Möbius function. In particular, setting $z = 2$, we find that

$$\frac{6}{\pi^2} = 1 - \frac{1}{2^2} - \frac{1}{3^2} - \frac{1}{5^2} + \frac{1}{6^2} - \frac{1}{7^2} + \frac{1}{10^2} - \frac{1}{11^2} + \cdots ;$$

what formula do you get when you set $z = 4$?

7.7.2 Euler Numbers and Evaluating Sums

We now derive a formula for the *alternating* sum of the odd natural numbers to odd powers:

$$1 - \frac{1}{3^{2k+1}} + \frac{1}{5^{2k+1}} - \frac{1}{7^{2k+1}} + \frac{1}{9^{2k+1}} - + \cdots , \qquad k = 0, 1, 2, 3, \ldots .$$

To this end, let $|z| < 1$ and note from p. 560 that we can write

$$\frac{\pi}{4 \cos \frac{\pi z}{2}} = \frac{1}{1^2 - z^2} - \frac{3}{3^2 - z^2} + \frac{5}{5^2 - z^2} + \cdots = \sum_{n=0}^{\infty} (-1)^n \frac{(2n + 1)}{(2n + 1)^2 - z^2}. \qquad (7.39)$$

Expanding as a geometric series, observe that

$$\frac{(2n+1)}{(2n+1)^2 - z^2} = \frac{1}{(2n+1)} \cdot \frac{1}{1 - \frac{z^2}{(2n+1)^2}} = \sum_{k=0}^{\infty} \frac{z^{2k}}{(2n+1)^{2k+1}}. \tag{7.40}$$

Thus,

$$\frac{\pi}{4\cos\frac{\pi z}{2}} = \sum_{n=0}^{\infty}\sum_{k=0}^{\infty} (-1)^n \frac{z^{2k}}{(2n+1)^{2k+1}}. \tag{7.41}$$

We would like to interchange the order of summation using Cauchy's double series theorem. However, the problem is that when k = 0 the corresponding part of the right-hand side of (7.41) is not absolutely convergent:

$$\sum_{n=0}^{\infty} \left| (-1)^n \frac{z^{2\cdot 0}}{(2n+1)^{2\cdot 0+1}} \right| = \sum_{n=0}^{\infty} \frac{1}{2n+1} \quad \text{is not convergent.}$$

Therefore, we cannot apply Cauchy's double series theorem immediately. However, we can easily get around this impasse by peeling off the $k = 0$ summation first. In (7.41) we separate the $k = 0$ term from the rest of the sum:

$$\frac{\pi}{4\cos\frac{\pi z}{2}} = \sum_{n=0}^{\infty} (-1)^n \frac{1}{2n+1} + \sum_{n=0}^{\infty}\sum_{k=1}^{\infty} (-1)^n \frac{z^{2k}}{(2n+1)^{2k+1}}.$$

Now it's easily checked that the double sum on the right *is* absolutely convergent for $|z| < 1$; indeed, using that $1/(2n+1)^{2k+1} \leq 1/(2n+1)^3$ for $k \geq 1$, we see that

$$\sum_{n=0}^{\infty}\sum_{k=1}^{\infty} \left| (-1)^n \frac{z^{2k}}{(2n+1)^{2k+1}} \right| \leq \sum_{n=0}^{\infty}\sum_{k=1}^{\infty} \frac{|z|^{2k}}{(2n+1)^3}$$

$$= \left(\sum_{n=0}^{\infty} \frac{1}{(2n+1)^3} \right)\left(\sum_{k=1}^{\infty} |z|^{2k} \right) < \infty.$$

Thus, for $|z| < 1$, Cauchy's double series theorem implies

$$\frac{\pi}{4\cos\frac{\pi z}{2}} = \sum_{n=0}^{\infty} (-1)^n \frac{1}{2n+1} + \sum_{k=1}^{\infty}\sum_{n=0}^{\infty} (-1)^n \frac{z^{2k}}{(2n+1)^{2k+1}}.$$

Putting the first sum on the right into the double sum, we arrive at the iterated sum we originally wanted:

$$\frac{\pi}{4\cos\frac{\pi z}{2}} = \sum_{k=0}^{\infty}\sum_{n=0}^{\infty} \left(\frac{(-1)^n}{(2n+1)^{2k+1}} \right) z^{2k}. \tag{7.42}$$

Now recall from p. 505 in Section 6.7 that

$$\frac{1}{\cos z} = \sec z = \sum_{k=0}^{\infty} (-1)^k \frac{E_{2k}}{(2k)!} z^{2k},$$

where the E_{2k} are the Euler numbers. Replacing z with $\pi z/2$ and multiplying by $\pi/4$, we get

$$\frac{\pi}{4 \cos \frac{\pi z}{2}} = \frac{\pi}{4} \sum_{k=0}^{\infty} (-1)^k \frac{E_{2k}}{(2k)!} \left(\frac{\pi}{2}\right)^{2k} z^{2k}.$$

Comparing this equation with (7.42) and using the identity theorem on p. 461, we conclude that for $k = 0, 1, 2, 3, \ldots$,

$$\sum_{n=0}^{\infty} \frac{(-1)^n}{(2n+1)^{2k+1}} = (-1)^k \frac{E_{2k}}{2(2k)!} \left(\frac{\pi}{2}\right)^{2k+1}. \qquad (7.43)$$

In particular, setting $k = 0$ (and recalling that $E_0 = 1$) we get our fourth proof of Gregory–Leibniz–Madhava's formula:

$$\frac{\pi}{4} = 1 - \frac{1}{3} + \frac{1}{5} - \frac{1}{7} + \cdots, \quad \textbf{(Gregory–Leibniz–Madhava, Proof IV)}.$$

What pretty formulas do you get when you set $k = 1, 2$? (Here, you need the Euler numbers calculated in Section 6.7.) We can derive many other pretty formulas from (7.43). To start this onslaught, we first state an "odd version" of Theorem 7.12.

Theorem 7.15 *For every $z \in \mathbb{C}$ with $\mathrm{Re}\, z > 1$, we have*

$$\sum_{n=0}^{\infty} \frac{(-1)^n}{(2n+1)^z} = \frac{3^z}{3^z+1} \cdot \frac{5^z}{5^z-1} \cdot \frac{7^z}{7^z+1} \cdot \frac{11^z}{11^z+1} \cdot \frac{13^z}{13^z-1} \cdots,$$

where the product is over odd primes (all primes except 2) and where the $\pm$ signs in the denominators depend on whether the prime is of the form $4k + 3$ (+ sign) or $4k + 1$ (− sign), where $k = 0, 1, 2, \ldots$.

Since the proof of this theorem is similar to that of Theorem 7.12, we shall leave the proof of this theorem to the interested reader; see Problem 5. In particular, setting $z = 1$, we get

$$\frac{\pi}{4} = \frac{3}{4} \cdot \frac{5}{4} \cdot \frac{7}{8} \cdot \frac{11}{12} \cdot \frac{13}{12} \cdot \frac{17}{16} \cdot \frac{19}{20} \cdot \frac{23}{24} \cdots. \qquad (7.44)$$

The numerators of the fractions on the right are the odd prime numbers, and the denominators are even numbers divisible by four and differing from the numerators by one. In (7.37), we found that

$$\frac{\pi^2}{6} = \frac{2^2}{2^2-1} \cdot \frac{3^2}{3^2-1} \cdot \frac{5^2}{5^2-1} \cdots = \frac{4}{3} \cdot \frac{3 \cdot 3}{2 \cdot 4} \cdot \frac{5 \cdot 5}{4 \cdot 6} \cdot \frac{7 \cdot 7}{6 \cdot 8} \cdot \frac{11 \cdot 11}{10 \cdot 12} \cdot \frac{13 \cdot 13}{12 \cdot 14} \cdots .$$

Dividing this expression by (7.44) and canceling like terms, we obtain

$$\frac{4\pi}{6} = \frac{\pi^2/6}{\pi/4} = \frac{4}{3} \cdot \frac{3}{2} \cdot \frac{5}{6} \cdot \frac{7}{6} \cdot \frac{11}{10} \cdot \frac{13}{14} \cdot \frac{17}{18} \cdots .$$

Multiplying both sides by 3/4, we get another one of Euler's famous formulas:

$$\boxed{\frac{\pi}{2} = \frac{3}{2} \cdot \frac{5}{6} \cdot \frac{7}{6} \cdot \frac{11}{10} \cdot \frac{13}{14} \cdot \frac{17}{18} \cdot \frac{19}{18} \cdot \frac{23}{22} \cdots .} \tag{7.45}$$

The numerators of the fractions are the odd prime numbers, and the denominators are even numbers not divisible by four and differing from the numerators by one. The remarkable Eqs. (7.44) and (7.45) are two of my favorite infinite product expansions for π.

7.7.3 Benoit Cloitre's e and π in a Mirror

In this section we prove an unbelievable fact connecting e and π that is due to Benoit Cloitre [52, 81, 219]. Define sequences $\{a_n\}$ and $\{b_n\}$ by $a_1 = b_1 = 0$, $a_2 = b_2 = 1$, and the rest as the following "mirror images":

$$a_{n+2} = a_{n+1} + \frac{1}{n}a_n,$$

$$b_{n+2} = \frac{1}{n}b_{n+1} + b_n.$$

We shall prove that

$$\boxed{e = \lim_{n \to \infty} \frac{n}{a_n} , \quad \frac{\pi}{2} = \lim_{n \to \infty} \frac{n}{b_n^2}.} \tag{7.46}$$

The sequences $\{a_n\}$ and $\{b_n\}$ and $\{\frac{n}{a_n}\}$ and $\{\frac{n}{b_n^2}\}$ are at a glance not so different, yet they generate very different numbers.

To prove the formula for e, let us define a sequence $\{s_n\}$ by $s_n = a_n/n$. Then $s_1 = a_1/1 = 0$ and $s_2 = a_2/2 = 1/2$. Observe that for $n \geq 2$, we have

$$
\begin{aligned}
s_{n+1} - s_n &= \frac{a_{n+1}}{n+1} - \frac{a_n}{n} = \frac{1}{n+1}\left(a_{n+1} - \frac{n+1}{n}a_n\right) \\
&= \frac{1}{n+1}\left(a_n + \frac{1}{n-1}a_{n-1} - \left(1 + \frac{1}{n}\right)a_n\right) \\
&= \frac{1}{n+1}\left(\frac{1}{n-1}a_{n-1} - \frac{a_n}{n}\right) \\
&= \frac{-1}{n+1}(s_n - s_{n-1}).
\end{aligned}
$$

Using induction, we see that

$$
\begin{aligned}
s_{n+1} - s_n &= \frac{-1}{n+1}(s_n - s_{n-1}) = \frac{-1}{n+1}\cdot\frac{-1}{n}(s_{n-1} - s_{n-2}) \\
&= \frac{-1}{n+1}\cdot\frac{-1}{n}\cdot\frac{-1}{n-1}(s_{n-2} - s_{n-3}) = \cdots \text{ etc.} \\
&= \frac{-1}{n+1}\cdot\frac{-1}{n}\cdot\frac{-1}{n-1}\cdots\frac{-1}{3}(s_2 - s_1) \\
&= \frac{-1}{n+1}\cdot\frac{-1}{n}\cdot\frac{-1}{n-1}\cdots\frac{-1}{3}\cdot\frac{1}{2} = \frac{(-1)^{n-3}}{(n+1)!} = \frac{(-1)^{n+1}}{(n+1)!}.
\end{aligned}
$$

Thus, writing this as a telescoping sum, we obtain

$$
s_n = s_1 + \sum_{k=2}^{n}(s_k - s_{k-1}) = 0 + \sum_{k=2}^{n}\frac{(-1)^k}{k!} = \sum_{k=0}^{n}\frac{(-1)^k}{k!},
$$

which is exactly the nth partial sum for the series expansion of e^{-1}. It follows that $s_n \to e^{-1}$, and so

$$
\lim_{n\to\infty}\frac{n}{a_n} = \lim_{n\to\infty}\frac{1}{s_n} = \frac{1}{e^{-1}} = e,
$$

as we claimed. The limit for π in (7.46) will be left to you (see Problem 2).

▶ **Exercises 7.7**

1. In this problem we derive other neat formulas:

   (a) Dividing (7.38) by $\pi^2/6$, prove that

$$
\boxed{\frac{5}{2} = \frac{2^2+1}{2^2-1}\cdot\frac{3^2+1}{3^2-1}\cdot\frac{5^2+1}{5^2-1}\cdot\frac{7^2+1}{7^2-1}\cdot\frac{11^2+1}{11^2-1}\cdots,}
$$

   quite a neat expression for 2.5.

   (b) Dividing (7.45) by (7.44), prove that

$$2 = \frac{2}{1} \cdot \frac{2}{3} \cdot \frac{4}{3} \cdot \frac{6}{5} \cdot \frac{6}{7} \cdot \frac{8}{9} \cdot \frac{10}{9} \cdot \frac{12}{11} \cdots,$$

quite a neat expression for 2. The fractions on the right are formed as follows: Given an odd prime $3, 5, 7, \ldots$, we take the pair of even numbers immediately above and below the prime, divide them by two, then put the resulting even number as the numerator and the odd number as the denominator.

2. In this problem, we prove the limit for π in (7.46).

   (i) Define $t_n = b_{n+1}/b_n$ for $n = 2, 3, 4, \ldots$. Prove that (for $n = 2, 3, 4, \ldots$), $t_{n+1} = 1/n + 1/t_n$ and then

   $$t_n = \begin{cases} 1 & n \text{ even,} \\ \frac{n}{n-1} & n \text{ odd.} \end{cases}$$

   (ii) Prove that $b_n^2 = t_2^2 \cdot t_3^2 \cdot t_4^2 \cdots t_{n-1}^2$, then using Wallis's formula, derive the limit for π in (7.46).

3. From Problem 7 on p. 576, prove that

   $$\frac{2(2n)! \, (1 - 2^{-2n})}{(2\pi)^{2n} \, (1 - 2^{1-2n})} < |B_{2n}| < \frac{2(2n)!}{(2\pi)^{2n} \, (1 - 2^{1-2n})}.$$

   This estimate shows that the Bernoulli numbers grow very fast as $n \to \infty$.

4. **(Radius of convergence)** In this problem we (finally) find the radii of convergence of

   $$z \cot z = \sum_{n=0}^{\infty} (-1)^n \frac{2^{2n} B_{2n}}{(2n)!} z^{2n}, \quad \tan z = \sum_{n=1}^{\infty} (-1)^{n-1} \frac{2^{2n}(2^{2n} - 1) B_{2n}}{(2n)!} z^{2n-1}.$$

   (a) Let $a_{2n} = (-1)^n \frac{2^{2n} B_{2n}}{(2n)!}$. Prove that

   $$\lim_{n \to \infty} |a_{2n}|^{1/2n} = \lim_{n \to \infty} \frac{1}{\pi} \cdot 2^{1/2n} \cdot \zeta(2n)^{1/2n} = \frac{1}{\pi}.$$

   Conclude that the radius of convergence of $z \cot z$ is π.

   (b) Using a similar argument, show that the radius of convergence of $\tan z$ is $\pi/2$.

5. In this problem, we prove Theorem 7.15.

   (i) Let us call an odd number of "type I" if it is of the form $4k + 1$ for some $k = 0, 1, \ldots$ and of "type II" if it is of the form $4k + 3$ for some $k = 0, 1, \ldots$. Prove that every odd number is either type I or type II.

(ii) Prove that type I × type I = type I, type I × type II = type II, and type II × type II = type I.

(iii) Let $a, b, \ldots, c \in \mathbb{N}$ be odd. Prove that if there is an *odd* number of type II integers among $a, b, \ldots, c$, then $a \cdot b \cdots c$ is of type II; otherwise, $a \cdot b \cdots c$ is type I.

(iv) Show that

$$\sum_{n=0}^{\infty} \frac{(-1)^n}{(2n+1)^z} = \sum_{n=0}^{\infty} \frac{1}{(4n+1)^z} - \sum_{n=0}^{\infty} \frac{1}{(4n+3)^z},$$

a sum of type I and type II natural numbers!

(v) Let $r > 1$ and let $z \in \mathbb{C}$ with $\operatorname{Re} z \geq r > 1$, let $1 < N \in \mathbb{N}$, and let $3 < 5 < \cdots < m < 2N + 1$ be the odd prime numbers less than $2N + 1$. In a similar manner as in the proof of Theorem 7.12, show that

$$\left| \sum_{n=1}^{\infty} \frac{(-1)^n}{(2n+1)^z} - \frac{3^z}{3^z+1} \cdot \frac{5^z}{5^z-1} \cdot \frac{7^z}{7^z+1} \cdot \frac{11^z}{11^z-1} \cdots \frac{m^z}{m^z \pm 1} \right|$$

$$\leq \sum_{n=N}^{\infty} \left| \frac{1}{(2n+1)^z} \right| \leq \sum_{n=N}^{\infty} \frac{1}{(2n+1)^r},$$

where the $+$ signs in the product are for type II odd primes and the $-$ signs for type I odd primes. Now finish the proof of Theorem 7.15.

Chapter 8
Infinite Continued Fractions

From time immemorial, the infinite has stirred men's emotions more than any other question. Hardly any other idea has stimulated the mind so fruitfully ... In a certain sense, mathematical analysis is a symphony of the infinite.
David Hilbert (1862–1943) "On the infinite" [23].

We dabbled a little into the theory of continued fractions (that is, fractions that continue on and on and on ...) way back on p. 192 in the exercises of Section 3.4. In this chapter we concentrate on this fascinating subject. We begin in Section 8.1 by showing that such fractions occur very naturally in long division, and we give their basic definitions. In Section 8.2, we prove some pretty dramatic formulas (this is one reason continued fractions are so fascinating, at least to me). For example, we'll show that $4/\pi$ and π can be written as the continued fractions:

$$\frac{4}{\pi} = 1 + \cfrac{1^2}{2 + \cfrac{3^2}{2 + \cfrac{5^2}{2 + \cfrac{7^2}{2 + \cdots}}}} \quad , \quad \pi = 3 + \cfrac{1^2}{6 + \cfrac{3^2}{6 + \cfrac{5^2}{6 + \cfrac{7^2}{6 + \cdots}}}} .$$

The continued fraction on the left is due to William Brouncker (and is the first continued fraction ever recorded), and the one on the right is due to Euler. If you think that these π formulas are cool, we'll derive the following formulas for e as well:

© Paul Loya 2017
P. Loya, *Amazing and Aesthetic Aspects of Analysis*,
https://doi.org/10.1007/978-1-4939-6795-7_8

$$e = 2 + \cfrac{2}{2 + \cfrac{3}{3 + \cfrac{4}{4 + \cfrac{5}{5 + \ddots}}}} = 1 + \cfrac{1}{0 + \cfrac{1}{1 + \cfrac{1}{1 + \cfrac{1}{2 + \cfrac{1}{1 + \cfrac{1}{1 + \cfrac{1}{4 + \ddots}}}}}}}.$$

We'll prove the formula on the left in Section 8.2, but you'll have to wait for the formula on the right until Section 8.7. In Section 8.3, we discuss elementary properties of continued fractions. In this section we also discuss how a Greek mathematician, Diophantus of Alexandrea (c. 200–284 A.D.), can help you if you're stranded on an island with guys you can't trust and a monkey with a healthy appetite! In Section 8.4 we study the convergence properties of continued fractions.

Recall from our discussion on the amazing number π and its computations from ancient times (see p. 364 in Section 4.12) that throughout the years, the following approximation to π came up: 3, 22/7, 333/106, and 355/113. Did you ever wonder why these particular numbers occur? Also, did you ever wonder why our calendar is constructed the way it is (e.g., why leap years occur)? Finally, did you ever wonder why a piano has twelve keys (within an octave)? In Sections 8.5 and 8.6 you'll find out that these mysteries have to do with continued fractions! In Section 8.8 we study special types of continued fractions having to do with square roots, and in Section 8.9 we learn why Archimedes needed around 8×10^{206544} cattle in order to "have abundant of knowledge in this science [mathematics]"!

In the very last section, Section 8.10, we look at continued fractions and transcendental numbers.

CHAPTER 8 OBJECTIVES: THE STUDENT WILL BE ABLE TO . . .

- Define continued fractions, and state the Wallis–Euler and fundamental recurrence relations.
- Apply the continued fraction convergence theorem (Theorem 8.14 on p. 628).
- Compute the canonical continued fraction of a given real number.
- Understand the relationship between convergents of a simple continued fraction and best approximations, and the relationship between periodic simple continued fractions and quadratic irrationals.
- Solve simple Diophantine equations (of linear and Pell type).

8.1 Introduction to Continued Fractions

In this section we introduce the basics of continued fractions and see how they arise out of high school division and also from solving equations.

8.1.1 Continued Fractions Arise During "Repeated Divisions"

In the following example we "repeatedly divide" (repeatedly use the division algorithm).

Example 8.1 Take, for instance, high school division of 68 into 157. Here, $157 = 2 \cdot 68 + 21$, so $\frac{157}{68} = 2 + \frac{21}{68}$. Inverting the fraction $\frac{21}{68}$, we have

$$\frac{157}{68} = 2 + \frac{1}{\dfrac{68}{21}}.$$

Since we can further divide $\frac{68}{21} = 3 + \frac{5}{21} = 3 + \frac{1}{21/5}$, we can write $\frac{157}{68}$ in the somewhat fancy way

$$\frac{157}{68} = 2 + \cfrac{1}{3 + \cfrac{1}{\dfrac{21}{5}}}.$$

Since $\frac{21}{5} = 4 + \frac{1}{5}$, we can write

$$\frac{157}{68} = 2 + \cfrac{1}{3 + \cfrac{1}{4 + \cfrac{1}{5}}}. \tag{8.1}$$

At this point our repeated division process stops.

The expression on the right in (8.1) is called a **finite simple continued fraction**. There are many ways to denote the right-hand side, but we shall stick with the following two:

$$\langle 2; 3, 4, 5 \rangle \quad \text{or} \quad 2 + \frac{1}{3+} \frac{1}{4+} \frac{1}{5} \quad \text{represent} \quad 2 + \cfrac{1}{3 + \cfrac{1}{4 + \cfrac{1}{5}}}.$$

Thus, continued fractions (that is, fractions that "continue on") arise naturally out of writing rational numbers in a somewhat fancy way by repeated divisions. Of course, 157 and 68 were not special, by repeated divisions one can take *any* two integers a and b with $b \neq 0$ and write a/b as a finite simple continued fraction; see Theorem 8.1 below. Here's an example involving a negative number.

Example 8.2 Consider 157 into -68. Since $-68 = (-1) \cdot 157 + 89$, we have $-\frac{68}{157} = -1 + \frac{89}{157}$. Inverting the fraction $\frac{89}{157}$, we obtain

$$-\frac{68}{157} = -1 + \cfrac{1}{\dfrac{157}{89}}.$$

Since $\frac{157}{89} = 1 + \frac{68}{89} = 1 + \frac{1}{89/68}$, we have

$$-\frac{68}{157} = -1 + \cfrac{1}{1 + \cfrac{1}{\dfrac{89}{68}}}.$$

Repeatedly using the division algorithm, we eventually arrive at

$$-\frac{68}{157} = -1 + \cfrac{1}{1 + \cfrac{1}{1 + \cfrac{1}{3 + \cfrac{1}{4 + \cfrac{1}{5}}}}}.$$

In Section 8.4, we shall prove that every real number, not necessarily rational, can be expressed as a simple (possibly infinite) continued fraction.

8.1.2 Continued Fractions Arise in Solving Equations

Continued fractions also arise naturally in trying to solve equations.

Example 8.3 Let $a, b > 0$ and suppose we want to find the positive solution x to the equation $x^2 + ax - b = 0$. Writing the equation $x^2 + ax - b = 0$ as $x(x + a) = b$ and dividing by $x + a$, we get

$$x = \frac{b}{a + x}.$$

We can replace x in the denominator with $x = b/(a + x)$ to get

$$x = \cfrac{b}{a + \cfrac{b}{a + x}}.$$

Repeating this many times, we can write

$$x = \cfrac{b}{a + \cfrac{b}{a + \cfrac{b}{\ddots \atop a + \cfrac{b}{a + \cfrac{b}{x}}}}}.$$

Repeating this "to infinity," we get

$$\text{``} \; x = \cfrac{b}{a + \cfrac{b}{a + \cfrac{b}{a + \cfrac{b}{a + \ddots}}}}. \; \text{''}$$

For example, given $\alpha > 1$, $x = \sqrt{\alpha} - 1$ is the positive solution to $x^2 + 2x - (\alpha - 1) = 0$, so we can get a pretty continued fraction for any square root using $a = 2$ and $b = \alpha - 1$. For instance, if $\alpha = 2$, we find that

$$\text{``} \; \sqrt{2} = 1 + \cfrac{1}{2 + \cfrac{1}{2 + \cfrac{1}{2 + \cfrac{1}{2 + \ddots}}}}. \; \text{''}$$

Quite a remarkable formula for $\sqrt{2}$! The reason for the quotation marks is that the computation was "formal," although we did prove it rigorously in Problem 8 on p. 191. We shall study such continued fractions for general square roots in Section 8.8.

Here's another neat example:

Example 8.4 Consider the slightly modified formula $x^2 - x - 1 = 0$. Then $\Phi = \frac{1+\sqrt{5}}{2}$, called the **golden ratio**, is the only positive solution. We can rewrite $\Phi^2 - \Phi - 1 = 0$ as $\Phi = 1 + \frac{1}{\Phi}$. Replacing Φ in the denominator with $\Phi = 1 + \frac{1}{\Phi}$, we get

$$\Phi = 1 + \cfrac{1}{1 + \cfrac{1}{\Phi}}.$$

Repeating this substitution process "to infinity," we can write

$$``\ \Phi = 1 + \cfrac{1}{1 + \cfrac{1}{1 + \cfrac{1}{1 + \cfrac{1}{1 + \cdots}}}},\ "$$ (8.2)

quite a beautiful expression, which we proved rigorously in Problem 8 on p. 191. As a side remark, there are many false rumors concerning the golden ratio; see [159] for the rundown.

8.1.3 Basic Definitions for Continued Fractions

A general finite continued fraction can be written as

$$a_0 + \cfrac{b_1}{a_1 + \cfrac{b_2}{a_2 + \cfrac{b_3}{a_3 + \cfrac{\ddots}{a_{n-1} + \cfrac{b_n}{a_n}}}}},$$ (8.3)

where the a_k and b_k are real numbers. Of course, we are implicitly assuming that these fractions are all well defined, e.g., no divisions by zero are allowed. Also, when you simplify this big fraction by combining fractions, *you need to go from the bottom up*. Notice that if $b_m = 0$ for some m, then

$$a_0 + \cfrac{b_1}{a_1 + \cfrac{b_2}{a_2 + \cfrac{b_3}{a_3 + \cfrac{\ddots}{a_{n-1} + \cfrac{b_n}{a_n}}}}} = a_0 + \cfrac{b_1}{a_1 + \cfrac{b_2}{a_2 + \cfrac{\ddots}{a_{m-2} + \cfrac{b_{m-1}}{a_{m-1}}}}},$$ (8.4)

since the $b_m = 0$ will zero out everything below it. The continued fraction is called
simple if all the b_k are 1 and the a_k are integers with a_k positive for $k \geq 1$. Instead
of writing the continued fraction as we did above, which takes up a lot of space, we
shall shorten it to

$$a_0 + \frac{b_1}{a_1+} \frac{b_2}{a_2+} \frac{b_3}{a_3+} \cdots \frac{b_n}{+a_n}.$$

In the case all b_n are equal to 1, the following "bracket notation" is often used:

$$a_0 + \frac{1}{a_1+} \frac{1}{a_2+} \frac{1}{a_3+} \cdots \frac{1}{+a_n} = \langle a_0; a_1, a_2, a_3, \ldots, a_n \rangle.$$

If $a_0 = 0$, some authors write $\langle a_1, a_2, \ldots, a_n \rangle$ instead of $\langle 0; a_1, \ldots a_n \rangle$.

Note that every finite simple continued fraction is a rational number, because it
is made up of additions and divisions of rational numbers (recall that *simple* means
that the a_n are integers) and the rational numbers are closed under such operations.
The converse is also true; see Theorem 8.1. There is an interesting "even–odd fact"
about finite continued fractions. To explain this fact, recall that

$$\frac{157}{68} = 2 + \cfrac{1}{3 + \cfrac{1}{4 + \cfrac{1}{5}}} = \langle 2; 3, 4, 5 \rangle,$$

which has an odd number of terms (three to be exact) after the integer part 2. We can
trivially modify this continued fraction by breaking up 5 as $4 + 1$:

$$\frac{157}{68} = 2 + \cfrac{1}{3 + \cfrac{1}{4 + \cfrac{1}{4 + \cfrac{1}{1}}}} = \langle 2; 3, 4, 4, 1 \rangle,$$

which has an even number of terms after the integer part. This example is typical:
Every finite simple continued fraction can be written with an even or odd number
of terms (by modifying the last term by 1). We summarize these remarks in the
following theorem, whose proof we leave for Problem 2.

Finite simple continued fractions

> **Theorem 8.1** *A real number can be expressed as a finite simple continued frac-
> tion if and only if it is rational, in which case the rational number can be expressed
> as a continued fraction with either an even or an odd number of terms.*

We now discuss infinite continued fractions. Let $\{a_n\}$, $n = 0, 1, 2, \ldots$, and $\{b_n\}$, $n = 1, 2, \ldots$, be sequences of real numbers and suppose that

$$c_n := a_0 + \frac{b_1}{a_1 +} \frac{b_2}{a_2 +} \frac{b_3}{a_3 +} \cdots \frac{b_n}{+ a_n}$$

is defined for all n. We call c_n the n**th convergent**. If the limit, $\lim c_n$, exists, then we say that the **infinite continued fraction**

$$a_0 + \cfrac{b_1}{a_1 + \cfrac{b_2}{a_2 + \cfrac{b_3}{a_3 + \cdots}}} \quad \text{or} \quad a_0 + \frac{b_1}{a_1 +} \frac{b_2}{a_2 +} \frac{b_3}{a_3 +} \cdots \tag{8.5}$$

converges, and we use either of these notations to denote the limiting value $\lim c_n$. In the case that all b_n are equal to 1, in place of (8.5), it's common to use brackets,

$$\langle a_0; a_1, a_2, a_3, \ldots \rangle \quad \text{instead of} \quad a_0 + \frac{1}{a_1 +} \frac{1}{a_2 +} \frac{1}{a_3 +} \cdots.$$

In Section 8.4 we shall prove that every simple continued fraction converges; in particular, we'll give another proof of the formula

$$\Phi = 1 + \frac{1}{1 +} \frac{1}{1 +} \frac{1}{1 +} \cdots.$$

In the case that there is some b_m term that vanishes, then convergence of (8.5) is easy, because (see (8.4)) for all $n \geq m$, we have $c_n = c_{m-1}$. Hence in this case,

$$a_0 + \frac{b_1}{a_1 +} \frac{b_2}{a_2 +} \frac{b_3}{a_3 +} \cdots = a_0 + \frac{b_1}{a_1 +} \frac{b_2}{a_2 +} \frac{b_3}{a_3 +} \cdots \frac{b_{m-1}}{+ a_{m-1}}$$

converges automatically; such a continued fraction is said to **terminate** or be **finite**. However, general convergence issues are not so straightforward. We shall deal with the subtleties of convergence in Section 8.4.

▶ **Exercises 8.1**

1. Expand the following fractions into finite simple continued fractions:

   (a) $\dfrac{7}{11}$, (b) $-\dfrac{11}{7}$, (c) $\dfrac{3}{13}$, (d) $\dfrac{13}{3}$, (e) $-\dfrac{42}{31}$, (f) $\dfrac{31}{42}$.

2. Prove Theorem 8.1. Suggestion: Reviewing the division algorithm (see Theorem 2.16 on p. 60) might help: For $a, b \in \mathbb{Z}$ with $b > 0$, we have $a = qb + r$, where $q, r \in \mathbb{Z}$ with $0 \leq r < b$; if a, b are both nonnegative, then so is q.

3. Let $\xi = a_0 + \dfrac{b_1}{a_1 +} \dfrac{b_2}{a_2 +} \dfrac{b_3}{a_3 +} \cdots \dfrac{b_n}{+ a_n} \neq 0$. Prove that

$$\frac{1}{\xi} = \frac{1}{a_0 +} \frac{b_1}{a_1 +} \frac{b_2}{a_2 +} \frac{b_3}{a_3 +} \cdots \frac{b_n}{+ a_n}.$$

In particular, if $\xi = \langle a_0; a_1, \ldots, a_n \rangle \neq 0$, show that $\frac{1}{\xi} = \langle 0; a_0, a_1, a_2, \ldots, a_n \rangle$.

4. A useful technique in studying continued fraction is the following artifice of writing a continued fraction within a continued fraction. For a continued fraction

$$\xi = a_0 + \frac{b_1}{a_1 +} \frac{b_2}{a_2 +} \frac{b_3}{a_3 +} \cdots + \frac{b_n}{a_n},$$

if $m < n$, prove that

$$\xi = a_0 + \frac{b_1}{a_1 +} \frac{b_2}{a_2 +} \frac{b_3}{a_3 +} \cdots + \frac{b_m}{\eta}, \quad \text{where} \quad \eta = \frac{b_{m+1}}{a_{m+1} +} \cdots + \frac{b_n}{a_n}.$$

8.2 ★ Some of the Most Beautiful Formulas in the World V

Hold on to your seats, for you're about to be taken on another journey through the beautiful world of mathematical formulas!

8.2.1 Transformation of Continued Fractions

It will often be convenient to transform one continued fraction into another one. For example, let ρ_1, ρ_2, ρ_3 be nonzero real numbers and suppose that the finite fraction

$$\xi = a_0 + \cfrac{b_1}{a_1 + \cfrac{b_2}{a_2 + \cfrac{b_3}{a_3}}},$$

where the a_k and b_k are real numbers, is defined. Then multiplying the top and bottom of the fraction by ρ_1, we get

$$\xi = a_0 + \cfrac{\rho_1 b_1}{\rho_1 a_1 + \cfrac{\rho_1 b_2}{a_2 + \cfrac{b_3}{a_3}}}.$$

Multiplying the top and bottom of the fraction with $\rho_1 b_2$ as numerator by ρ_2, and then the last fraction's top and bottom by ρ_3, gives

$$\xi = a_0 + \cfrac{\rho_1 b_1}{\rho_1 a_1 + \cfrac{\rho_1 \rho_2 b_2}{\rho_2 a_2 + \cfrac{\rho_2 b_3}{a_3}}} \quad \text{and} \quad \xi = a_0 + \cfrac{\rho_1 b_1}{\rho_1 a_1 + \cfrac{\rho_1 \rho_2 b_2}{\rho_2 a_2 + \cfrac{\rho_2 \rho_3 b_3}{\rho_3 a_3}}}.$$

In summary,

$$a_0 + \cfrac{b_1}{a_1 +} \cfrac{b_2}{a_2 +} \cfrac{b_3}{a_3} = a_0 + \cfrac{\rho_1 b_1}{\rho_1 a_1 +} \cfrac{\rho_1 \rho_2 b_2}{\rho_2 a_2} + \cfrac{\rho_2 \rho_3 b_3}{\rho_3 a_3}.$$

A simple induction argument proves the following.

Transformation rules

Theorem 8.2 *For real numbers* $a_1, a_2, a_3, \ldots, b_1, b_2, b_3, \ldots,$ *and nonzero constants* $\rho_1, \rho_2, \rho_3, \ldots,$ *we have*

$$a_0 + \cfrac{b_1}{a_1 +} \cfrac{b_2}{a_2 +} \cfrac{b_3}{a_3 +} \cdots + \cfrac{b_n}{a_n} = a_0 + \cfrac{\rho_1 b_1}{\rho_1 a_1 +} \cfrac{\rho_1 \rho_2 b_2}{\rho_2 a_2} + \cfrac{\rho_2 \rho_3 b_3}{\rho_3 a_3} + \cdots + \cfrac{\rho_{n-1} \rho_n b_n}{\rho_n a_n},$$

in the sense that when the left-hand side is defined, so is the right-hand side, and equality holds. In particular, if the limit as $n \to \infty$ *of the left-hand side exists, then the limit of the right-hand side also exists, and*

$$a_0 + \cfrac{b_1}{a_1 +} \cfrac{b_2}{a_2 +} \cdots + \cfrac{b_n}{a_n +} \cdots = a_0 + \cfrac{\rho_1 b_1}{\rho_1 a_1 +} \cfrac{\rho_1 \rho_2 b_2}{\rho_2 a_2} + \cdots + \cfrac{\rho_{n-1} \rho_n b_n}{\rho_n a_n} + \cdots.$$

8.2.2   Two Stupendous Series and Continued Fraction Identities

Let $\alpha_1, \alpha_2, \alpha_3, \ldots$ be real numbers with $\alpha_k \neq 0$ and $\alpha_k \neq \alpha_{k-1}$ for all k. Observe that

$$\frac{1}{\alpha_1} - \frac{1}{\alpha_2} = \frac{\alpha_2 - \alpha_1}{\alpha_1 \alpha_2} = \frac{1}{\frac{\alpha_1 \alpha_2}{\alpha_2 - \alpha_1}}.$$

Since $\dfrac{\alpha_1 \alpha_2}{\alpha_2 - \alpha_1} = \dfrac{\alpha_1 (\alpha_2 - \alpha_1) + \alpha_1^2}{\alpha_2 - \alpha_1} = \alpha_1 + \dfrac{\alpha_1^2}{\alpha_2 - \alpha_1}$, we get

$$\frac{1}{\alpha_1} - \frac{1}{\alpha_2} = \frac{1}{\alpha_1 + \frac{\alpha_1^2}{\alpha_2 - \alpha_1}}.$$

Generalizing this formula, we obtain the following theorem.

Theorem 8.3 *If $\alpha_1, \alpha_2, \alpha_3, \ldots$ are nonzero real numbers with $\alpha_k \neq \alpha_{k-1}$ for all k, then for every $n \in \mathbb{N}$,*

$$\sum_{k=1}^{n} \frac{(-1)^{k-1}}{\alpha_k} = \cfrac{1}{\alpha_1 + \cfrac{\alpha_1^2}{\alpha_2 - \alpha_1 + \cfrac{\alpha_2^2}{\alpha_3 - \alpha_2 + \cfrac{\ddots}{\cfrac{\alpha_{n-1}^2}{\alpha_n - \alpha_{n-1}}}}}}.$$

In particular, taking $n \to \infty$, we conclude that

$$\sum_{k=1}^{\infty} \frac{(-1)^{k-1}}{\alpha_k} = \frac{1}{\alpha_1 +} \frac{\alpha_1^2}{\alpha_2 - \alpha_1 +} \frac{\alpha_2^2}{\alpha_3 - \alpha_2 +} \frac{\alpha_3^2}{\alpha_4 - \alpha_3 +} \cdots \qquad (8.6)$$

as long as either side (and hence both sides) makes sense.

Proof This theorem is certainly true for alternating sums with $n = 1$ terms. Assume that it is true for sums with n terms; we shall prove it holds for sums with $n + 1$ terms. Observe that we can write

$$\sum_{k=1}^{n+1} \frac{(-1)^{k-1}}{\alpha_k} = \frac{1}{\alpha_1} - \frac{1}{\alpha_2} + \cdots + \frac{(-1)^{n-1}}{\alpha_n} + \frac{(-1)^n}{\alpha_{n+1}}$$

$$= \frac{1}{\alpha_1} - \frac{1}{\alpha_2} + \cdots + (-1)^{n-1}\left(\frac{1}{\alpha_n} - \frac{1}{\alpha_{n+1}}\right)$$

$$= \frac{1}{\alpha_1} - \frac{1}{\alpha_2} + \cdots + (-1)^{n-1}\left(\frac{\alpha_{n+1} - \alpha_n}{\alpha_n \alpha_{n+1}}\right)$$

$$= \frac{1}{\alpha_1} - \frac{1}{\alpha_2} + \cdots + (-1)^{n-1}\frac{1}{\frac{\alpha_n \alpha_{n+1}}{\alpha_{n+1} - \alpha_n}}.$$

This is now a sum of n terms. Thus, we can apply the induction hypothesis to conclude that

$$\sum_{k=1}^{n+1} \frac{(-1)^{k-1}}{\alpha_k} = \frac{1}{\alpha_1 +} \frac{\alpha_1^2}{\alpha_2 - \alpha_1 +} \cdots \frac{\alpha_{n-1}^2}{\frac{\alpha_n \alpha_{n+1}}{\alpha_{n+1} - \alpha_n} - \alpha_{n-1}}. \qquad (8.7)$$

Since

$$\frac{\alpha_n \alpha_{n+1}}{\alpha_{n+1} - \alpha_n} - \alpha_{n-1} = \frac{\alpha_n(\alpha_{n+1} - \alpha_n) + \alpha_n^2}{\alpha_{n+1} - \alpha_n} - \alpha_{n-1}$$

$$= \alpha_n - \alpha_{n-1} + \frac{\alpha_n^2}{\alpha_{n+1} - \alpha_n},$$

putting this into (8.7) gives

$$\sum_{k=1}^{n+1} \frac{(-1)^{k-1}}{\alpha_k} = \frac{1}{\alpha_1 +} \frac{\alpha_1^2}{\alpha_2 - \alpha_1 +} \cdots + \frac{\alpha_{n-1}^2}{\alpha_n - \alpha_{n-1} + \frac{\alpha_n^2}{\alpha_{n+1} - \alpha_n}}.$$

This proves our induction step and completes our proof.   ∎

Example 8.5 Since we know that

$$\log 2 = \sum_{k=1}^{\infty} \frac{(-1)^{k-1}}{k} = \frac{1}{1} - \frac{1}{2} + \frac{1}{3} - \frac{1}{4} + \cdots,$$

setting $\alpha_k = k$ in the identity (8.6) in Theorem 8.3, we can write

$$\log 2 = \frac{1}{1+} \frac{1^2}{1+} \frac{2^2}{1+} \frac{3^2}{1+} \cdots,$$

which we can also write as the equally beautiful expression

$$\log 2 = \cfrac{1}{1 + \cfrac{1^2}{1 + \cfrac{2^2}{1 + \cfrac{3^2}{1 + \cfrac{4^2}{1 + \cdots}}}}}.$$

See Problem 1 for a continued fraction formula for $\log(1 + x)$.

Here is another interesting identity. Let $\alpha_1, \alpha_2, \alpha_3, \ldots$ be real numbers, all nonzero and none equal to 1. Then observe that

$$\frac{1}{\alpha_1} - \frac{1}{\alpha_1 \alpha_2} = \frac{\alpha_2 - 1}{\alpha_1 \alpha_2} = \frac{1}{\frac{\alpha_1 \alpha_2}{\alpha_2 - 1}}.$$

Since

$$\frac{\alpha_1 \alpha_2}{\alpha_2 - 1} = \frac{\alpha_1(\alpha_2 - 1) + \alpha_1}{\alpha_2 - 1} = \alpha_1 + \frac{\alpha_1}{\alpha_2 - 1},$$

we get

$$\frac{1}{\alpha_1} - \frac{1}{\alpha_1 \alpha_2} = \frac{1}{\alpha_1 + \frac{\alpha_1}{\alpha_2 - 1}}.$$

We can continue by induction in much the same manner as we did in the proof of Theorem 8.3 to obtain the following result.

Theorem 8.4 *For every real sequence* $\alpha_1, \alpha_2, \alpha_3, \ldots$ *with* $\alpha_k \neq 0, 1$ *for each* k, *we have*

$$\sum_{k=1}^{n} \frac{(-1)^{k-1}}{\alpha_1 \cdots \alpha_k} = \cfrac{1}{\alpha_1 + \cfrac{\alpha_1}{\alpha_2 - 1 + \cfrac{\alpha_2}{\alpha_3 - 1 + \cfrac{\ddots}{\alpha_{n-1} + \cfrac{\alpha_{n-1}}{\alpha_n - 1}}}}}.$$

In particular, taking $n \to \infty$, *we conclude that*

$$\sum_{k=1}^{\infty} \frac{(-1)^{k-1}}{\alpha_1 \cdots \alpha_k} = \frac{1}{\alpha_1 +} \frac{\alpha_1}{\alpha_2 - 1 +} \frac{\alpha_2}{\alpha_3 - 1 +} \cdots \frac{\alpha_{n-1}}{\alpha_n - 1 +} \cdots, \qquad (8.8)$$

as long as either side (and hence both sides) makes sense.

Theorems 8.3 and 8.4 turn series to continued fractions; in Problem 9 we do the same for infinite products.

8.2.3 Continued Fractions for arctan *and* π

We now use the identities just learned to derive some remarkable continued fractions.

Example 8.6 First, since

$$\frac{\pi}{4} = \frac{1}{1} - \frac{1}{3} + \frac{1}{5} - \frac{1}{7} + \cdots,$$

using the limit expression (8.6) in Theorem 8.3:

$$\frac{1}{\alpha_1} - \frac{1}{\alpha_2} + \frac{1}{\alpha_3} - \frac{1}{\alpha_4} + \cdots = \frac{1}{\alpha_1 +} \frac{\alpha_1^2}{\alpha_2 - \alpha_1 +} \frac{\alpha_2^2}{\alpha_3 - \alpha_2 +} \frac{\alpha_3^2}{\alpha_4 - \alpha_3 +} \cdots,$$

we can write

$$\frac{\pi}{4} = \cfrac{1}{1 + \cfrac{1^2}{2 + \cfrac{3^2}{2 + \cfrac{5^2}{2 + \cfrac{7^2}{2 + \cdots}}}}}.$$

Inverting both sides (see Problem 3 on p. 597), we obtain the incredible expansion

$$\frac{4}{\pi} = 1 + \cfrac{1^2}{2 + \cfrac{3^2}{2 + \cfrac{5^2}{2 + \cfrac{7^2}{2 + \cdots}}}}. \tag{8.9}$$

This continued fraction was the very first continued fraction ever recorded, and was written down without proof by William Brouncker (1620–1686), the first president of the Royal Society of London.

Actually, we can derive (8.9) from a related expansion of the arctangent function, which is so neat that we shall derive in two ways, using Theorem 8.3 then using Theorem 8.4.

Example 8.7 Recall that

$$\arctan x = x - \frac{x^3}{3} + \frac{x^5}{5} - \frac{x^7}{7} + \cdots + (-1)^{n-1} \frac{x^{2n-1}}{2n-1} + \cdots.$$

Setting $\alpha_1 = \frac{1}{x}$, $\alpha_2 = \frac{3}{x^3}$, $\alpha_3 = \frac{5}{x^5}$, and in general, $\alpha_n = \frac{2n-1}{x^{2n-1}}$ into the formula (8.6) from Theorem 8.3, we get the somewhat complicated formula

$$\arctan x = \cfrac{1}{\frac{1}{x} +} \frac{\frac{1}{x^2}}{\frac{3}{x^2} - \frac{1}{x} +} \frac{\left(\frac{3}{x^3}\right)^2}{\frac{5}{x^5} - \frac{3}{x^3} +} \cdots + \frac{\left(\frac{2n-3}{x^{2n-3}}\right)^2}{\frac{2n-1}{x^{2n-1}} - \frac{2n-3}{x^{2n-3}} +} \cdots.$$

However, we can simplify this using the transformation rule from Theorem 8.2:

$$\frac{b_1}{a_1 +} \frac{b_2}{a_2 +} \cdots + \frac{b_n}{a_n +} \cdots = \frac{\rho_1 b_1}{\rho_1 a_1 +} \frac{\rho_1 \rho_2 b_2}{\rho_2 a_2 +} \cdots + \frac{\rho_{n-1} \rho_n b_n}{\rho_n a_n +} \cdots.$$

(Here we drop the a_0 term from that theorem.) Let $\rho_1 = x$, $\rho_2 = x^3$, ..., and in general, $\rho_n = x^{2n-1}$. Then,

$$\frac{1}{\frac{1}{x}} + \frac{\frac{1}{x^2}}{\frac{3}{x^3} - \frac{1}{x}} + \frac{\left(\frac{3}{x^3}\right)^2}{\frac{5}{x^5} - \frac{3}{x^3}} + \frac{\left(\frac{5}{x^5}\right)^2}{\frac{7}{x^7} - \frac{5}{x^5}} + \cdots = \frac{x}{1} + \frac{x^2}{3 - x^2} + \frac{3^2 x^2}{5 - 3x^2} + \frac{5^2 x^2}{7 - 5x^2} + \cdots$$

Thus,

$$\arctan x = \frac{x}{1} + \frac{x^2}{3 - x^2} + \frac{3^2 x^2}{5 - 3x^2} + \frac{5^2 x^2}{7 - 5x^2} + \cdots,$$

or in a fancier way:

$$\arctan x = \cfrac{x}{1 + \cfrac{x^2}{(3 - x^2) + \cfrac{3^2 x^2}{(5 - 3x^2) + \cfrac{5^2 x^2}{(7 - 5x^2) + \cdots}}}}. \qquad (8.10)$$

In particular, setting $x = 1$ and inverting, we get Brouncker's formula:

$$\frac{4}{\pi} = 1 + \cfrac{1^2}{2 + \cfrac{3^2}{2 + \cfrac{5^2}{2 + \cfrac{7^2}{2 + \cdots}}}}.$$

Example 8.8 We can also derive (8.10) using Theorem 8.4. Once again we use the formula

$$\arctan x = x - \frac{x^3}{3} + \frac{x^5}{5} - \frac{x^7}{7} + \cdots + (-1)^{n-1} \frac{x^{2n-1}}{2n-1} + \cdots.$$

Setting $\alpha_1 = \frac{1}{x}$, $\alpha_2 = \frac{3}{x^2}$, $\alpha_3 = \frac{5}{3x^2}$, $\alpha_4 = \frac{7}{5x^2}$, $\cdots$, $\alpha_n = \frac{2n-1}{(2n-3)x^2}$ for $n \geq 2$, into the limiting formula (8.8) from Theorem 8.4, we obtain

$$\frac{1}{\alpha_1} - \frac{1}{\alpha_1 \alpha_2} + \frac{1}{\alpha_1 \alpha_2 \alpha_3} - \cdots = \frac{1}{\alpha_1} + \frac{\alpha_1}{\alpha_2 - 1} + \frac{\alpha_2}{\alpha_3 - 1} + \cdots + \frac{\alpha_n}{\alpha_{n+1} - 1} + \cdots$$

we obtain

$$\arctan x = \frac{1}{\frac{1}{x}} + \frac{\frac{1}{x}}{\frac{3}{x^2} - 1} + \frac{\frac{3}{x^2}}{\frac{5}{3x^2} - 1} + \cdots + \frac{\frac{2n-1}{(2n-3)x^2}}{\frac{2n+1}{(2n-1)x^2} - 1} + \cdots.$$

From Theorem 8.2, we know that

$$\frac{b_1}{a_1+}\;\frac{b_2}{a_2+}\;\cdots\;\frac{b_n}{a_n+}\;\cdots=\frac{\rho_1 b_1}{\rho_1 a_1+}\;\frac{\rho_1\rho_2 b_2}{\rho_2 a_2}\;+\cdots+\;\frac{\rho_{n-1}\rho_n b_n}{\rho_n a_n}\;+\;\cdots.$$

In particular, setting $\rho_1 = x$, $\rho_2 = x^2$, $\rho_3 = 3x^2$, $\rho_4 = 5x^2$, and in general, $\rho_n = (2n-3)x^2$ for $n \geq 2$ into the formula for $\arctan x$, we obtain (as you are invited to verify) the exact same expression (8.10)!

Example 8.9 We leave the next two beauts to you! Applying Theorem 8.3 and/or Theorem 8.4 to Euler's sum $\frac{\pi^2}{6} = \frac{1}{1^2} + \frac{1}{2^2} + \frac{1}{3^2} + \cdots$, in Problem 2 we ask you to derive the formula

$$\frac{6}{\pi^2} = 0^2 + 1^2 - \cfrac{1^4}{1^2 + 2^2 - \cfrac{2^4}{2^2 + 3^2 - \cfrac{3^4}{3^2 + 4^2 - \cfrac{4^4}{4^2 + 5^2 - \ddots}}}}, \qquad (8.11)$$

which is, after inversion, the last formula on the front cover of this book.

Example 8.10 In Problem 9 we derive Euler's splendid formula [47, p. 89]:

$$\frac{\pi}{2} = 1 + \cfrac{1}{1 + \cfrac{1 \cdot 2}{1 + \cfrac{2 \cdot 3}{1 + \cfrac{3 \cdot 4}{1 + \ddots}}}}. \qquad (8.12)$$

8.2.4 Another Continued Fraction for π

We now derive another remarkable formula for π, which is due to Euler (according to [47, p. 89]; the proof we give is found in [139]). Consider first the telescoping sum

$$\sum_{n=1}^{\infty}(-1)^{n-1}\left(\frac{1}{n}+\frac{1}{n+1}\right)=\left(\frac{1}{1}+\frac{1}{2}\right)-\left(\frac{1}{2}+\frac{1}{3}\right)+\left(\frac{1}{3}+\frac{1}{4}\right)-+\cdots=1.$$

Since

$$\frac{\pi}{4}=\frac{1}{1}-\frac{1}{3}+\frac{1}{5}-\frac{1}{7}+\cdots=1-\sum_{n=1}^{\infty}\frac{(-1)^{n-1}}{2n+1},$$

multiplying this expression by 4 and using the previous expression, we can write

$$\pi = 4 - 4\sum_{n=1}^{\infty} \frac{(-1)^{n-1}}{2n+1} = 3 + 1 - 4\sum_{n=1}^{\infty} \frac{(-1)^{n-1}}{2n+1}$$

$$= 3 + \sum_{n=1}^{\infty} (-1)^{n-1}\left(\frac{1}{n} + \frac{1}{n+1}\right) - 4\sum_{n=1}^{\infty} \frac{(-1)^{n-1}}{2n+1}$$

$$= 3 + \sum_{n=1}^{\infty} (-1)^{n-1}\left(\frac{1}{n} + \frac{1}{n+1} - \frac{4}{2n+1}\right)$$

$$= 3 + 4\sum_{n=1}^{\infty} \frac{(-1)^{n-1}}{2n(2n+1)(2n+2)},$$

where we combined fractions in going from the third to fourth lines. We now apply the limiting formula (8.6) from Theorem 8.3 with $\alpha_n = 2n(2n+1)(2n+2)$. Observe that

$$\alpha_n - \alpha_{n-1} = 2n(2n+1)(2n+2) - 2(n-1)(2n-1)(2n)$$

$$= 4n\big[(2n+1)(n+1) - (n-1)(2n-1)\big]$$

$$= 4n\big[2n^2 + 2n + n + 1 - (2n^2 - n - 2n + 1)\big] = 4n(6n) = 24n^2.$$

Now putting the α_n into the formula

$$\frac{1}{\alpha_1} - \frac{1}{\alpha_2} + \frac{1}{\alpha_3} - \frac{1}{\alpha_4} + \cdots = \frac{1}{\alpha_1 +} \frac{\alpha_1^2}{\alpha_2 - \alpha_1 +} \frac{\alpha_2^2}{\alpha_3 - \alpha_2 +} \frac{\alpha_3^2}{\alpha_4 - \alpha_3 +} \cdots,$$

we get

$$4\sum_{n=1}^{\infty} \frac{(-1)^{n-1}}{2n(2n+1)(2n+2)} = 4 \cdot \left(\frac{1}{2\cdot3\cdot4 +} \frac{(2\cdot3\cdot4)^2}{24\cdot2^2 +} \frac{(4\cdot5\cdot6)^2}{24\cdot3^2 +} \cdots\right)$$

$$= \frac{1}{2\cdot3 +} \frac{(2\cdot3)^2\cdot4}{24\cdot2^2 +} \frac{(4\cdot5\cdot6)^2}{24\cdot3^2 +} \cdots.$$

Hence,

$$\pi = 3 + \frac{1}{6 +} \frac{(2\cdot3)^2\cdot4}{24\cdot2^2 +} \frac{(4\cdot5\cdot6)^2}{24\cdot3^2 +} \cdots + \frac{(2(n-1)(2n-1)(2n))^2}{24\cdot n^2} + \cdots,$$
$$\tag{8.13}$$

which is beautiful, but we can make this even more beautiful using the transformation rule from Theorem 8.2:

$$\frac{b_1}{a_1 +} \frac{b_2}{a_2 +} \cdots + \frac{b_n}{a_n +} \cdots = \frac{\rho_1 b_1}{\rho_1 a_1 +} \frac{\rho_1 \rho_2 b_2}{\rho_2 a_2 +} \cdots + \frac{\rho_{n-1}\rho_n b_n}{\rho_n a_n +} \cdots.$$

Setting $\rho_1 = 1$ and $\rho_n = \frac{1}{4n^2}$ for $n \geq 2$, and using the appropriate a_n, b_n in (8.13), we find that

$$\rho_{n-1}\rho_n b_n = (2n-1)^2 \quad \text{and} \quad \rho_n a_n = 6.$$

Thus, the formula (8.13) simplifies considerably to

$$\pi = 3 + \frac{1^2}{6+} \frac{3^2}{6+} \frac{5^2}{6+} \frac{7^2}{6+} \cdots + \frac{(2n-1)^2}{6} + \cdots,$$

or in more elegant notation,

$$\pi = 3 + \cfrac{1^2}{6 + \cfrac{3^2}{6 + \cfrac{5^2}{6 + \cfrac{7^2}{6 + \ddots}}}}. \tag{8.14}$$

8.2.5 Continued Fractions and e

For our final beautiful example, we shall compute a continued fraction expansion for e. Since $e^{-1} = \sum_{n=0}^{\infty} \frac{(-1)^n}{n!}$, we have

$$\frac{e-1}{e} = 1 - e^{-1} = \frac{1}{1} - \frac{1}{1 \cdot 2} + \frac{1}{1 \cdot 2 \cdot 3} - \frac{1}{1 \cdot 2 \cdot 3 \cdot 4} + \cdots.$$

Thus, setting $\alpha_k = k$ in the formula (8.8),

$$\frac{1}{\alpha_1} - \frac{1}{\alpha_1 \alpha_2} + \frac{1}{\alpha_1 \alpha_2 \alpha_3} - \cdots = \frac{1}{\alpha_1 +} \frac{\alpha_1}{\alpha_2 - 1 +} \frac{\alpha_2}{\alpha_3 - 1 +} \cdots + \frac{\alpha_{n-1}}{\alpha_n - 1 +} \cdots,$$

we obtain

$$\frac{e-1}{e} = \cfrac{1}{1 + \cfrac{1}{1 + \cfrac{2}{2 + \cfrac{3}{3 + \ddots}}}}.$$

We can make this into an expression for e as follows. First, invert the expression and then subtract 1 from both sides to get

$$\frac{e}{e-1} = 1 + \cfrac{1}{1 + \cfrac{2}{2 + \cfrac{3}{3 + \ddots}}}, \quad \text{then} \quad \frac{1}{e-1} = \cfrac{1}{1 + \cfrac{2}{2 + \cfrac{3}{3 + \ddots}}}.$$

Second, invert again to obtain

$$e - 1 = 1 + \cfrac{2}{2 + \cfrac{3}{3 + \cfrac{4}{4 + \cfrac{5}{5 + \ddots}}}}.$$

Finally, adding 1 to both sides, we get the incredibly beautiful expression

$$e = 2 + \cfrac{2}{2 + \cfrac{3}{3 + \cfrac{4}{4 + \cfrac{5}{5 + \ddots}}}}, \tag{8.15}$$

or in shorthand,

$$e = 2 + \frac{2}{2+} \frac{3}{3+} \frac{4}{4+} \frac{5}{5+} \cdots.$$

In the exercises you will derive other amazing formulas.

▶ Exercises 8.2

1. Recall that $\log(1 + x) = \sum_{n=0}^{\infty} (-1)^n \frac{x^{n+1}}{n+1}$. Using this formula, the formula (8.6) in Theorem 8.3, and also the transformation rule, prove that fabulous formula

$$\log(1 + x) = \cfrac{x}{1 + \cfrac{1^2 x}{(2 - 1x) + \cfrac{2^2 x}{(3 - 2x) + \cfrac{3^2 x}{(4 - 3x) + \ddots}}}}.$$

Plug in $x = 1$ to derive our beautiful formula for $\log 2$.

2. Using Euler's sum $\frac{\pi^2}{6} = \frac{1}{1^2} + \frac{1}{2^2} + \frac{1}{3^2} + \cdots$, give two proofs of the formula (8.11), one using Theorem 8.3 and the other using Theorem 8.4. The transformation rules will come in handy.

3. From the expansion $\frac{\pi}{\sin \pi x} = \frac{1}{x} + \sum_{n=1}^{\infty}(-1)^n \frac{2x}{x^2-n^2}$, derive the beautiful expression

$$\frac{\sin \pi x}{\pi x} = 1 + \frac{-2x^2}{x^2+1+} \frac{(x^2-1^2)^2}{3} + \frac{(x^2-2^2)^2}{5} + \frac{(x^2-3^2)^2}{7} + \frac{(x^2-5^2)^2}{9} + \cdots$$

4. (i) For all real numbers $\{\alpha_k\}$, prove that for every n,

$$\sum_{k=0}^{n} \alpha_k x^k = \alpha_0 + \frac{\alpha_1 x}{1} + \frac{-\frac{\alpha_2}{\alpha_1}x}{1 + \frac{\alpha_2}{\alpha_1}x+} \frac{-\frac{\alpha_3}{\alpha_2}x}{1 + \frac{\alpha_3}{\alpha_2}x+} \cdots + \frac{-\frac{\alpha_n}{\alpha_{n-1}}x}{1 + \frac{\alpha_n}{\alpha_{n-1}}x},$$

provided, of course, that the right-hand side is defined, which we assume holds for every n.

(ii) Deduce that if the infinite series $\sum_{n=0}^{\infty} \alpha_n x^n$ converges, then

$$\sum_{n=0}^{\infty} \alpha_n x^n = \alpha_0 + \frac{\alpha_1 x}{1} + \frac{-\frac{\alpha_2}{\alpha_1}x}{1 + \frac{\alpha_2}{\alpha_1}x+} \frac{-\frac{\alpha_3}{\alpha_2}x}{1 + \frac{\alpha_3}{\alpha_2}x+} \cdots + \frac{-\frac{\alpha_n}{\alpha_{n-1}}x}{1 + \frac{\alpha_n}{\alpha_{n-1}}x+} \cdots.$$

Transforming the continued fraction on the right, prove that

$$\sum_{n=0}^{\infty} \alpha_n x^n = \alpha_0 + \frac{\alpha_1 x}{1} + \frac{-\alpha_2 x}{\alpha_1 + \alpha_2 x+} \frac{-\alpha_1 \alpha_3 x}{\alpha_2 + \alpha_3 x+} \cdots + \frac{-\alpha_{n-2}\alpha_n x}{\alpha_{n-1} + \alpha_n x+} \cdots.$$

5. Writing $\arctan x = x(1 - \frac{y}{3} + \frac{y^2}{5} - \frac{y^3}{7} + \cdots)$, where $y = x^2$, and using the previous problem on $(1 - \frac{y}{3} + \frac{y^2}{5} - \frac{y^3}{7} + \cdots)$, derive the formula (8.10).

6. Let $x, y > 0$. Prove that

$$\sum_{n=0}^{\infty} \frac{(-1)^n}{x + ny} = \frac{1}{x+} \frac{x^2}{y+} \frac{(x+y)^2}{y} + \frac{(x+2y)^2}{y} + \frac{(x+3y)^2}{y} + \cdots.$$

Suggestion: The formula (8.6) in Theorem 8.3 might help.

7. Recall the partial fraction expansion $\pi x \cot \pi x = 1 + 2x^2 \sum_{n=1}^{\infty} \frac{1}{x^2-n^2}$.

(a) Splitting $\frac{2x}{x^2-n^2}$ into two parts, prove that

$$\pi \cot \pi x = \frac{1}{x} - \frac{1}{1-x} + \frac{1}{1+x} - \frac{1}{2-x} + \frac{1}{2+x} - + \cdots.$$

(b) Derive the remarkable formula

$$\pi \cot \pi x = \frac{1}{x} + \frac{x^2}{1-2x} + \frac{(1-x)^2}{2x} + \frac{(1+x)^2}{1-2x} + \frac{(2-x)^2}{2x} + \frac{(2+x)^2}{1-2x} + \cdots$$

Putting $x = 1/4$, give a new proof of Brouncker's formula.

(c) Derive

$$\frac{\tan \pi x}{\pi x} = 1 + \frac{x}{1-2x} + \frac{(1-x)^2}{2x} + \frac{(1+x)^2}{1-2x} + \frac{(2-x)^2}{2x} + \frac{(2+x)^2}{1-2x} + \cdots$$

8. From the expansion $\frac{\pi}{4\cos\frac{\pi x}{2}} = \sum_{n=0}^{\infty} (-1)^n \frac{(2n+1)}{(2n+1)^2 - x^2}$, derive the beautiful expression

$$\frac{\cos\frac{\pi x}{2}}{\frac{\pi}{2}} = x + 1 + \frac{(x+1)^2}{-2\cdot 1} + \frac{(x-1)^2}{-2} + \frac{(x-3)^2}{2\cdot 3} + \frac{(x+3)^2}{2} + \frac{(x+5)^2}{-2\cdot 5} + \frac{(x-5)^2}{-2} + \cdots$$

9. (Cf. [123]) In this problem we turn infinite products into continued fractions.

(a) Let $\alpha_1, \alpha_2, \alpha_3, \ldots$ be a sequence of real numbers with $\alpha_k \neq 0, -1$ for all k. Define sequences $b_1, b_2, b_3, \ldots$ and $a_0, a_1, a_2, \ldots$ by $b_1 = (1 + \alpha_0)\alpha_1$, $a_0 = 1 + \alpha_0, a_1 = 1$, and

$$b_n = -(1 + \alpha_{n-1})\frac{\alpha_n}{\alpha_{n-1}} \quad , \quad \alpha_n = 1 - a_n \text{ for } n = 2, 3, 4, \ldots.$$

Prove (say by induction) that for every $n \in \mathbb{N}$,

$$\prod_{k=0}^{n}(1 + \alpha_k) = a_0 + \frac{b_1}{a_1 +} \frac{b_2}{a_2 +} \cdots \frac{b_n}{+ a_n}.$$

Taking $n \to \infty$, get a formula between infinite products and fractions.

(b) Using that $\frac{\sin \pi x}{\pi x} = \prod_{n=1}^{\infty}\left(1 - \frac{x}{n^2}\right) = (1-x)(1+x)\left(1 - \frac{x}{2}\right)\left(1 + \frac{x}{2}\right)$ $\left(1 - \frac{x}{3}\right)\left(1 + \frac{x}{3}\right) \cdots$ and (a), derive the continued fraction expansion

$$\frac{\sin \pi x}{\pi x} = 1 - \frac{x}{1+} \frac{1\cdot(1-x)}{x} + \frac{1\cdot(1+x)}{1-x} + \frac{2\cdot(2-x)}{x} + \frac{2\cdot(2+x)}{1-x} + \cdots$$

(c) Putting $x = 1/2$, prove (8.12). Putting $x = -1/2$, derive another continued fraction for $\pi/2$.

8.3   Recurrence Relations, Diophantus's Tomb, and Shipwrecked Sailors

In this section we define the Wallis–Euler recurrence relations, which generate sequences of numerators and denominators for convergents of continued fractions. Diophantine equations is the subject of finding integer solutions to polynomial equations. Continued fractions (through the special properties of the Wallis–Euler recurrence relations) turn out to play a very important role in this subject.

8.3.1   Convergents and Recurrence Relations

We call a continued fraction

$$a_0 + \frac{b_1}{a_1 +} \frac{b_2}{a_2 +} \frac{b_3}{a_3 +} \cdots + \frac{b_n}{a_n +} \cdots \qquad (8.16)$$

nonnegative if the a_n, b_n are real numbers with $a_n > 0, b_n \geq 0$ for all $n \geq 1$ (a_0 can be any real number). We shall not spend a lot of time on continued fractions when the a_n and b_n in (8.16), for $n \geq 1$, are arbitrary real numbers; it is only for nonnegative infinite continued fractions that we develop their convergence properties in Section 8.4. However, we shall come across continued fractions in which some of the a_n, b_n are negative; see, for instance, the beautiful expression (8.45) on p. 666 for $\cot x$ (and the following one for $\tan x$) in Section 8.7! We focus on nonnegative continued fractions because their convergents

$$c_n := a_0 + \frac{b_1}{a_1 +} \frac{b_2}{a_2 +} \frac{b_3}{a_3 +} \cdots + \frac{b_n}{a_n}$$

are *always* well defined. For arbitrary a_n, b_n, weird things can happen. For instance, consider the elementary example $\frac{1}{1+} \frac{-1}{1+} \frac{1}{1}$. Let us form its convergents. Note that $c_1 = 1$, which is OK, but

$$c_2 = \frac{1}{1+} \frac{-1}{1} = \frac{1}{1 + \dfrac{-1}{1}} = \frac{1}{1-1} = \frac{1}{0} = ???,$$

which is not OK.[1] However,

---

[1] Actually, in the continued fraction community, we always define $a/0 = \infty$ for $a \neq 0$ so we can get around this division by zero predicament by simply defining it away.

$$c_3 = \frac{1}{1+} \; \frac{-1}{1+} \; \frac{1}{1} = \frac{1}{1 + \dfrac{-1}{1 + \dfrac{1}{1}}} = \frac{1}{1 + \dfrac{-1}{2}} = \frac{1}{\dfrac{1}{2}} = 2,$$

which is OK again! To avoid such craziness, we shall focus on nonnegative continued fractions, *but we emphasize that much of our analysis works in greater generality.*

Let $\{a_n\}_{n=0}^{\infty}$, $\{b_n\}_{n=1}^{\infty}$ be sequences of real numbers with $a_n > 0$, $b_n \geq 0$ for all $n \geq 1$ (there is no restriction on a_0). The following sequences $\{p_n\}$, $\{q_n\}$ are central in the theory of continued fractions:

Wallis–Euler recurrence relations

$$p_n = a_n p_{n-1} + b_n p_{n-2} \quad , \quad q_n = a_n q_{n-1} + b_n q_{n-2} \qquad (8.17)$$
$$p_{-1} = 1 \, , \; p_0 = a_0 \quad , \quad q_{-1} = 0 \, , \; q_0 = 1.$$

We call these recurrence relations the **Wallis–Euler recurrence relations**. You'll see why they're so central in a moment. In particular,

$$p_1 = a_1 p_0 + b_1 p_{-1} = a_1 a_0 + b_1$$
$$q_1 = a_1 q_0 + b_1 q_{-1} = a_1. \qquad (8.18)$$

We remark that $q_n > 0$ for $n = 0, 1, 2, 3, \ldots$. This is easily proved by induction: Certainly, $q_0 = 1, q_1 = a_1 > 0$ (recall that $a_n > 0$ for $n \geq 1$); thus assuming that $q_n > 0$ for $n = 0, \ldots, n-1$, we have (recall that $b_n \geq 0$),

$$q_n = a_n q_{n-1} + b_n q_{n-2} > 0 \cdot 0 + 0 = 0,$$

and our induction is complete. Observe that the zeroth convergent of the continued fraction

$$a_0 + \frac{b_1}{a_1+} \; \frac{b_2}{a_2+} \; \frac{b_3}{a_3+} \; \cdots \; \frac{b_n}{a_n+} \; \cdots$$

is $c_0 = a_0$, which also equals p_0/q_0, since $p_0 = a_0$ and $q_0 = 1$. Also, the first convergent is

$$c_1 = a_0 + \frac{b_1}{a_1} = \frac{a_1 a_0 + b_1}{a_1} = \frac{p_1}{q_1}.$$

The central property of the p_n, q_n is the fact that $c_n = p_n/q_n$ for all n.

Theorem 8.5 *For every positive real number x, we have*

$$a_0 + \frac{b_1}{a_1 +} \frac{b_2}{a_2 +} \frac{b_3}{a_3 +} \cdots + \frac{b_n}{x} = \frac{xp_{n-1} + b_n p_{n-2}}{xq_{n-1} + b_n q_{n-2}}, \quad n = 1, 2, 3, \ldots. \quad (8.19)$$

(Note that the denominator is positive, because $q_n > 0$ for $n \geq 0$.) In particular, setting $x = a_n$ and using the definition of p_n, q_n, we have

$$c_n = a_0 + \frac{b_1}{a_1 +} \frac{b_2}{a_2 +} \frac{b_3}{a_3 +} \cdots + \frac{b_n}{a_n} = \frac{p_n}{q_n}, \quad n = 0, 1, 2, 3, \ldots.$$

Proof We prove (8.19) by induction. If $n = 1$, observe that

$$a_0 + \frac{b_1}{x} = \frac{a_0 x + b_1}{x} = \frac{xp_0 + b_1 p_{-1}}{xq_0 + q_{-1}},$$

since $p_0 = a_0$, $p_{-1} = 1$, $q_0 = 1$, and $q_{-1} = 0$. Assume that (8.19) holds when there are n terms after a_0; we shall prove that it holds for fractions with $n + 1$ terms after a_0. To do so, we write (see Problem 4 on p. 579 for the general technique)

$$a_0 + \frac{b_1}{a_1 +} \frac{b_2}{a_2 +} \cdots + \frac{b_n}{a_n +} \frac{b_{n+1}}{x} = a_0 + \frac{b_1}{a_1 +} \frac{b_2}{a_2 +} \cdots + \frac{b_n}{y},$$

where $y = a_n + \dfrac{b_{n+1}}{x} = \dfrac{xa_n + b_{n+1}}{x}$. By our induction hypothesis, we have

$$\begin{aligned}
a_0 + \frac{b_1}{a_1 +} \frac{b_2}{a_2 +} \cdots + \frac{b_{n+1}}{x} &= \frac{yp_{n-1} + b_n p_{n-2}}{yq_{n-1} + b_n q_{n-2}} \\
&= \frac{\left(\dfrac{xa_n + b_{n+1}}{x}\right) p_{n-1} + b_n p_{n-2}}{\left(\dfrac{xa_n + b_{n+1}}{x}\right) q_{n-1} + b_n q_{n-2}} \\
&= \frac{xa_n p_{n-1} + b_{n+1} p_{n-1} + x b_n p_{n-2}}{xa_n q_{n-1} + b_{n+1} q_{n-1} + x b_n q_{n-2}} \\
&= \frac{x(a_n p_{n-1} + b_n p_{n-2}) + b_{n+1} p_{n-1}}{x(a_n q_{n-1} + b_n q_{n-2}) + b_{n+1} q_{n-1}} \\
&= \frac{xp_n + b_{n+1} p_{n-1}}{xq_n + b_{n+1} q_{n-1}},
\end{aligned}$$

which completes our induction step and finishes our proof. ∎

In the next theorem, we give various useful identities that the p_n, q_n satisfy.

Fundamental recurrence relations

Theorem 8.6 *For all $n \geq 1$, the following identities hold:*

$$p_n q_{n-1} - p_{n-1} q_n = (-1)^{n-1} b_1 b_2 \cdots b_n$$
$$p_n q_{n-2} - p_{n-2} q_n = (-1)^n a_n b_1 b_2 \cdots b_{n-1}$$

and (where the formula for $c_n - c_{n-2}$ is valid only for $n \geq 2$)

$$c_n - c_{n-1} = \frac{(-1)^{n-1} b_1 b_2 \cdots b_n}{q_n q_{n-1}}, \qquad c_n - c_{n-2} = \frac{(-1)^n a_n b_1 b_2 \cdots b_{n-1}}{q_n q_{n-2}}.$$

Proof To prove that $p_n q_{n-1} - p_{n-1} q_n = (-1)^{n-1} b_1 b_2 \cdots b_n$ for $n = 1, 2, \ldots$, we proceed by induction. For $n = 1$, the left-hand side is (see (8.18))

$$p_1 q_0 - p_0 q_1 = (a_1 a_0 + b_1) \cdot 1 - a_0 \cdot a_1 = b_1,$$

which is the right-hand side when $n = 1$. Assume that our equality holds for n; we prove that it holds for $n + 1$. By the Wallis–Euler recurrence relations, we have

$$
\begin{aligned}
p_{n+1} q_n - p_n q_{n+1} &= (a_{n+1} p_n + b_{n+1} p_{n-1}) q_n - p_n (a_{n+1} q_n + b_{n+1} q_{n-1}) \\
&= b_{n+1} p_{n-1} q_n - p_n b_{n+1} q_{n-1} \\
&= -b_{n+1} (p_n q_{n-1} - p_{n-1} q_n) \\
&= -b_{n+1} \cdot (-1)^{n-1} b_1 b_2 \cdots b_n = (-1)^n b_1 b_2 \cdots b_n b_{n+1},
\end{aligned}
$$

which completes our induction step. Next, we use the Wallis–Euler recurrence relations and the equality just proved to obtain

$$
\begin{aligned}
p_n q_{n-2} - p_{n-2} q_n &= (a_n p_{n-1} + b_n p_{n-2}) q_{n-2} - p_{n-2} (a_n q_{n-1} + b_n q_{n-2}) \\
&= a_n p_{n-1} q_{n-2} - p_{n-2} a_n q_{n-1} \\
&= a_n (p_{n-1} q_{n-2} - p_{n-2} q_{n-1}) \\
&= a_n \cdot (-1)^{n-2} b_1 b_2 \cdots b_{n-1} = (-1)^n a_n b_1 b_2 \cdots b_{n-1}.
\end{aligned}
$$

Finally, the equations for $c_n - c_{n-1}$ and $c_n - c_{n-2}$ follow from dividing

$$p_n q_{n-1} - p_{n-1} q_n = (-1)^{n-1} b_1 \cdots b_n \quad \text{and} \quad p_n q_{n-2} - p_{n-2} q_n = (-1)^{n-1} a_n b_1 \cdots b_{n-1}$$

by $q_n q_{n-1}$ and $q_n q_{n-2}$, respectively. ∎

For simple continued fractions, the Wallis–Euler relations (8.17) and (8.18) and the fundamental recurrence relations take the following particularly simple forms.

Simple fundamental recurrence relations

Corollary 8.7 *For simple continued fractions, for all $n \geq 1$, if*

$$p_n = a_n p_{n-1} + p_{n-2} \quad , \quad q_n = a_n q_{n-1} + q_{n-2}$$
$$p_0 = a_0 \, , \ p_1 = a_0 a_1 + 1 \quad , \quad q_0 = 1 \, , \ q_1 = a_1,$$

then

$$c_n = \langle a_0; a_1, a_2, a_3, \ldots, a_n \rangle = \frac{p_n}{q_n} \ \text{ for all } n \geq 0.$$

Also, for every $x > 0$,

$$\langle a_0; a_1, a_2, a_3, \ldots, a_n, x \rangle = \frac{x p_{n-1} + p_{n-2}}{x q_{n-1} + q_{n-2}}, \quad n = 1, 2, 3, \ldots. \qquad (8.20)$$

Moreover, for all $n \geq 1$, the following identities hold:

$$p_n q_{n-1} - p_{n-1} q_n = (-1)^{n-1}$$
$$p_n q_{n-2} - p_{n-2} q_n = (-1)^n a_n$$

and

$$c_n - c_{n-1} = \frac{(-1)^{n-1}}{q_n \, q_{n-1}} \quad , \quad c_n - c_{n-2} = \frac{(-1)^n a_n}{q_n \, q_{n-2}},$$

where $c_n - c_{n-2}$ is valid only for $n \geq 2$.

We also have the following interesting result.

Corollary 8.8 *All the p_n, q_n for a simple continued fraction are relatively prime; that is, $c_n = p_n/q_n$ is automatically in lowest terms.*

Proof The fact that p_n, q_n are in lowest terms follows from the fact that

$$p_n q_{n-1} - p_{n-1} q_n = (-1)^{n-1},$$

so if an integer happens to divide divide both p_n and q_n, then it divides $p_n q_{n-1} - p_{n-1} q_n$ also, so it must divide $(-1)^{n-1} = \pm 1$, which is impossible unless the number was ± 1. ∎

8.3.2 Diophantine Equations, Sailors, Coconuts, and Monkeys

The following puzzle is very fun; for more cool coconut puzzles, see [87, 88, 246, 227], and Problems 5 and 6. See also [229] for the long history of such problems.

Example 8.11 (Going nuts I) Five sailors get shipwrecked on an island where there is one coconut tree and a very slim monkey. The sailors gathered all the coconuts into a gigantic pile and went to sleep. At midnight, one sailor woke up, and because he didn't trust his mates, he divided the coconuts into five equal piles, but with one coconut left over, (Fig. 8.1).

Fig. 8.1 The sailor divides the pile of coconuts into five equal piles with one coconut *left* over. He hides one pile, combines the remaining four piles, and throws the extra coconut to the monkey

He threw the extra one to the monkey, hid his pile, put the remaining coconuts back into a pile, and went to sleep. At one o'clock, the second sailor woke up, and because he was untrusting of his mates, he divided the coconuts into five equal piles, but again with one coconut left over. He threw the extra one to the monkey, hid his pile, put the remaining coconuts back into a pile, and went to sleep. This exact same scenario continued throughout the night with the other three sailors. In the morning, all the sailors woke up, and pretending that nothing had happened in the night, divided the now minuscule pile of coconuts into five equal piles, and they found yet again one coconut left over, which they threw to the now very overweight monkey.

Question: What is the smallest possible number of coconuts in the original pile?

Let x equal the original number of coconuts. Recall that sailor #1 divided x into five parts, but with one left over. This means that if y_1 is the number of coconuts he took, then $x = 5y_1 + 1$ and he left $4 \cdot y_1$ coconuts. If y_2 is the number of coconuts that sailor #2 took, then $4y_1 = 5y_2 + 1$, and he left $4 \cdot y_2$ coconuts. Repeating this argument, if y_3, y_4 and y_5 are the numbers of coconuts sailors #3, #4, and #5, respectively took, then

$$4y_2 = 5y_3 + 1 \quad , \quad 4y_3 = 5y_4 + 1 \quad , \quad 4y_4 = 5y_5 + 1.$$

At the end, the sailors divided the last amount of coconuts $4 \cdot y_5$ into five piles, with one coconut left over. Thus, if y is the number of coconuts in each final pile, then $4y_5 = 5y + 1$. To summarize, we have obtained the equations

$$x = 5y_1 + 1 \quad , \quad 4y_1 = 5y_2 + 1 \quad , \quad 4y_2 = 5y_3 + 1 \quad , \quad 4y_3 = 5y_4 + 1,$$
$$4y_4 = 5y_5 + 1 \quad , \quad 4y_5 = 5y + 1.$$

We can eliminate the variables y_1, y_2, y_3, y_4, y_5, and after some algebra we get

$$1024x - 15625y = 11529.$$

We are searching for nonnegative *integers* x, y solving this equation. We will find such integers after a brief intermission.

The sailor–coconut–monkey puzzles lead us to the subject of Diophantine equations, which are polynomial equations that we wish to solve in integers. We are particularly interested in solving linear equations. Before doing so, it might of interest to know that the subject of Diophantine equations is named after the Greek mathematician Diophantus of Alexandrea (200–284 A.D.). Diophantus is famous for at least two things: his book *Arithmetica*, which studies equations that we now call Diophantine equations in his honor, and for the following riddle, which was supposedly written on his tombstone:

> *This tomb hold Diophantus. Ah, what a marvel! And the tomb tells scientifically the measure of his life. God vouchsafed that he should be a boy for the sixth part of his life; when a twelfth was added, his cheeks acquired a beard; He kindled for him the light of marriage after a seventh, and in the fifth year after his marriage He granted him a son. Alas! late-begotten and miserable child, when he had reached the measure of half his father's life, the chill grave took him. After consoling his grief by this science of numbers for four years, he reached the end of his life.* [171].

Try to find how old Diophantus was when he died using elementary algebra. (Let x equal his age when he died; then you should end up with trying to solve the equation $x = \frac{1}{6}x + \frac{1}{12}x + \frac{1}{7}x + 5 + \frac{1}{2}x + 4$, obtaining $x = 84$.) Here is an easy way to find his age: Unraveling the above fancy language, and picking out two facts, we know that $1/12$th of his life was in youth and $1/7$th was as a bachelor. In particular, his age must divide 7 and 12. The only integer that does this, and which is within the human lifespan, is $7 \cdot 12 = 84$. In particular, he spent $84/6 = 14$ years as a child, $84/12 = 7$ as a youth, $84/7 = 12$ years as a bachelor. He married at $14 + 7 + 12 = 33$, at $33 + 5 = 38$, his son was born, who later died at the age of $84/2 = 42$ years old, when Diophantus was 80. Finally, after 4 years doing the "science of numbers," Diophantus died at the ripe old age of 84.

After taking a moment to wipe away our tears, let us go back to linear Diophantine equations. Let a, $c \in \mathbb{Z}$, $b \in \mathbb{N}$, and suppose we would like to find integer pairs (x, y) such that

$$ax - by = c.$$

Of course, the pairs of real numbers (x, y) satisfying this equation defines a line:

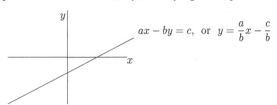

$$ax - by = c, \text{ or } y = \frac{a}{b}x - \frac{c}{b}$$

Thus, we are asking whether pairs of integers lie on this line. This is not always true; for example, the equation $2x - 4y = 1$ has no integer solutions (because for x, y integers, the left-hand side is divisible by 2, while the right side is not). However, if we assume that a and b are relatively prime, then there are always (infinitely many) integer pairs on the line.

Theorem 8.9 *If $a \in \mathbb{Z}$ and $b \in \mathbb{N}$ are relatively prime, then for every $c \in \mathbb{Z}$, the equation*

$$ax - by = c$$

has an infinite number of integer solutions (x, y). Moreover, if (x_0, y_0) is any one integral solution of the equation with $c = 1$, then for general $c \in \mathbb{Z}$, all solutions are of the form

$$x = cx_0 + bt \quad , \quad y = cy_0 + at \quad , \quad \text{for all } t \in \mathbb{Z}.$$

Proof In Problem 8 we ask you to prove this theorem using Problem 5 on p. 66; but we shall use continued fractions just for fun. We first solve the equation $ax - by = 1$. To do so, we write a/b as a finite simple continued fraction: $a/b = \langle a_0; a_1, a_2, \ldots, a_n \rangle$, and by Theorem 8.1 we can choose n to be *odd*. Then a/b is equal to the nth convergent p_n/q_n, which implies that $p_n = a$ and $q_n = b$ since p_n and q_n are relatively prime (Corollary 8.8). Also, by our relations in Corollary 8.7, we know that

$$p_n q_{n-1} - q_n p_{n-1} = (-1)^{n-1} = 1,$$

where we used that n is odd. Thus, $a q_{n-1} - b p_{n-1} = 1$, so

$$\boxed{(x_0, y_0) = (q_{n-1}, p_{n-1})} \tag{8.21}$$

solves $ax_0 - by_0 = 1$. Multiplying $ax_0 - by_0 = 1$ by c, we get

$$a(cx_0) - b(cy_0) = c.$$

Then $ax - by = c$ holds if and only if

$$ax - by = a(cx_0) - b(cy_0) \quad \Longleftrightarrow \quad a(x - cx_0) = b(y - cy_0).$$

This shows that a divides $b(y - cy_0)$, which can be possible if and only if a divides $y - cy_0$, since a and b are relatively prime. Thus, $y - cy_0 = at$ for some $t \in \mathbb{Z}$. Plugging $y - cy_0 = at$ into the equation $a(x - cx_0) = b(y - cy_0)$, we get

$$a(x - cx_0) = b \cdot (at) = abt.$$

Canceling a, we get $x - cx_0 = bt$, and our proof is now complete.   ■

We also remark that the proof of Theorem 8.9, in particular the formula (8.21), also tells us *how* to find (x_0, y_0): Just write a/b as a simple continued fraction with an *odd* number n of terms after the integer part of a/b, and compute the $(n - 1)$st convergent to get $(x_0, y_0) = (q_{n-1}, p_{n-1})$.

Example 8.12 Consider the Diophantine equation

$$157x - 68y = 12.$$

We already know, see Example 8.1 on p. 591, that the continued fraction expansion of $a/b = \frac{157}{68}$ with an odd $n = 3$ is $\frac{157}{68} = \langle 2; 3, 4, 5 \rangle = \langle a_0; a_1, a_2, a_3 \rangle$. Thus,

$$c_2 = 2 + \cfrac{1}{3 + \cfrac{1}{4}} = 2 + \frac{4}{13} = \frac{30}{13}.$$

Therefore, $(13, 30)$ is one solution of $157x - 68y = 1$, which we should check just to be sure:

$$157 \cdot 13 - 68 \cdot 30 = 2041 - 2040 = 1.$$

Since $cx_0 = 12 \cdot 13 = 156$ and $cy_0 = 12 \cdot 30 = 360$, the general solution of $157x - 68y = 12$ is

$$x = 156 + 68t \ , \quad y = 360 + 157t, \qquad t \in \mathbb{Z}.$$

Example 8.13 (*Going nuts II*) To solve the sailor–coconut–monkey puzzle we need integer solutions to

$$1024x - 15625y = 11529.$$

Since $1024 = 2^{10}$ and $15625 = 5^6$ are relatively prime, we can solve the equation by Theorem 8.9. First, we solve $1024x - 15625y = 1$ by writing $1024/15625$ as a continued fraction (this takes some algebra) and forcing n to be odd (in this case $n = 9$)[2]:

$$\frac{1024}{15625} = \langle 0; 15, 3, 1, 6, 2, 1, 3, 2, 1 \rangle.$$

Second, we take the $(n - 1)$st convergent:

$$c_8 = \langle 0; 15, 3, 1, 6, 2, 1, 3, 2 \rangle = \frac{711}{10849}.$$

Thus, $(x_0, y_0) = (10849, 711)$. Since $cx_0 = 11529 \cdot 10849 = 125078121$ and $cy_0 = 11529 \cdot 711 = 8197119$, the integer solutions to $1024x - 15625y = 11529$ are

---

[2]See http://www.mcs.surrey.ac.uk/Personal/R.Knott/Fibonacci/cfCALC.html for a continued fraction calculator.

$$x = 125078121 + 15625t \quad, \quad y = 8197119 + 1024t \quad, \quad t \in \mathbb{Z}. \qquad (8.22)$$

This of course gives us infinitely many solutions. However, we want the smallest *nonnegative* solutions since x and y represent numbers of coconuts; thus, we need

$$x = 125078121 + 15625t \geq 0 \quad \Longrightarrow \quad t \geq -\frac{125078121}{15625} = -8004.999744\ldots,$$

and

$$y = 8197119 + 1024t \geq 0 \quad \Longrightarrow \quad t \geq -\frac{8197119}{1024} = -8004.9990234\ldots.$$

Thus, taking $t = -8004$ in (8.22), we finally arrive at $x = 15621$ and $y = 1023$. In conclusion, the smallest number of coconuts in the original piles is 15621 coconuts.

▶ **Exercises 8.3**

1. Find the general integral solutions of

   (a) $7x - 11y = 1$,   (b) $13x - 3y = 5$,   (c) $13x - 15y = 100$,   (d) $13x + 15y = 100$.

2. If all $a_0, a_1, \ldots, a_n$ are positive (in particular, $p_0 = a_0 > 0$), prove that for $n = 1, 2, \ldots$,

   $$\frac{p_n}{p_{n-1}} = \langle a_n; a_{n-1}, a_{n-2}, \ldots, a_2, a_1, a_0 \rangle \quad \text{and} \quad \frac{q_n}{q_{n-1}} = \langle a_n; a_{n-1}, a_{n-2}, \ldots, a_2, a_1 \rangle.$$

   Suggestion: Observe that $\frac{p_k}{p_{k-1}} = \frac{a_k p_{k-1} + p_{k-2}}{p_{k-1}} = a_k + \frac{1}{\frac{p_{k-1}}{p_{k-2}}}$.

3. In this problem, we relate a special continued fraction to the Fibonacci sequence $\{F_n\}$ (see p. 520 for the definition of the Fibonacci sequence). For $n \in \mathbb{N}$, let p_n/q_n denote the nth convergent of the continued fraction $\langle 1; 1, 1, 1, \ldots \rangle$.

   (a) Prove that $p_n = F_{n+2}$ and $q_n = F_{n+1}$ for all $n = -1, 0, 1, 2, \ldots$.
   (b) Prove that F_n and F_{n+1} are relatively prime and derive the following famous identity, named after Giovanni Cassini (1625–1712) (also called Jean-Dominique Cassini)

   $$F_{n-1}F_{n+1} - F_n^2 = (-1)^n \qquad \textbf{(Cassini's identity)}.$$

4. Imitating the proof of Theorem 8.9, show that a solution of $ax - by = -1$ can be found by writing a/b as a simple continued fraction with an *even* number n of terms after the integer part of a/b and finding the $(n - 1)$th convergent. Apply this method to find a solution of $157x - 68y = -1$ and $7x - 12y = -1$.

5. (**Three sailor coconut problems**) The algebra for these problems is easier.

   (a) Solve the coconut problem in Example 8.11 for the case of three sailors. Thus, during the night each sailor divides the pile into three equal piles with one left over and in the morning, the same occurs.

(b) Solve the three sailor coconut problem when there are no coconuts left over after the final division in the morning. (As before, during the night there was always one coconut left over.) Thus, what is the smallest possible number of coconuts in the original pile given that after the sailors divided the coconuts in the morning, there are no coconuts left over?

6. (**Five and seven sailor coconut problems**) Here are more coconut problems.

(a) Solve the coconut problem assuming the same antics as in the text, except for one thing: there are no coconuts left over for the monkey at the end.

(b) Solve the coconut problem assuming the same antics as in the text except that during the night each sailor divided the pile into five equal piles with *none* left over; however, after he puts the remaining coconuts back into a pile, the monkey (being a thief himself) steals one coconut from the pile (before the next sailor wakes up). In the morning, there is still one coconut left over for the monkey.

(c) Solve the coconut problem when there are seven sailors, otherwise everything is the same as in the text. (Warning: Set aside an evening for long computations!)

7. Let $\alpha = \langle a_0; a_1, a_2, \ldots, a_m \rangle$, $\beta = \langle b_0; b_1, \ldots, b_n \rangle$ with $m, n \geq 0$ and the a_k, b_k integers with $a_m, b_n > 1$ (such finite continued fractions are called **regular**). Prove that if $\alpha = \beta$, then $a_k = b_k$ for all $k = 0, 1, 2, \ldots$. In other words, distinct regular finite simple continued fractions define different rational numbers.

8. Prove Theorem 8.9 using Problem 5 on p. 65.

8.4   Convergence Theorems for Infinite Continued Fractions

In this section we shall investigate the delicate issues surrounding convergence of infinite continued fractions (see Theorem 8.14, the continued fraction convergence theorem); in particular, we prove that *every* simple continued fraction converges. We also show how to write *every* real number as a simple continued fraction via the **canonical continued fraction algorithm**. Finally, we prove that a real number is irrational if and only if its simple continued fraction expansion is infinite.

8.4.1   Monotonicity Properties of Convergents

Let $\{c_n\}$ denote the convergents of a *nonnegative* infinite continued fraction

$$a_0 + \frac{b_1}{a_1+} \, \frac{b_2}{a_2+} \, \frac{b_3}{a_3+} \, \cdots \, \frac{b_n}{a_n+} \, \cdots,$$

where recall that nonnegative means that the a_n, b_n are real numbers with $a_n > 0$, $b_n \geq 0$ for all $n \geq 1$, and there is no restriction on the sign of a_0. The Wallis–Euler recurrence relations (8.17) are

$$p_n = a_n p_{n-1} + b_n p_{n-2} \quad , \quad q_n = a_n q_{n-1} + b_n q_{n-2},$$

where $p_{-1} = 1$, $p_0 = a_0$, $q_{-1} = 0$, $q_0 = 1$. Then (cf. (8.18))

$$p_1 = a_1 p_0 + b_1 p_{-1} = a_1 a_0 + b_1 \quad , \quad q_1 = a_1 q_0 + b_1 q_{-1} = a_1,$$

and all the q_n are positive (see the discussion below (8.18) on p. 611). By Theorem 8.5 we have $c_n = p_n/q_n$ for all n, and by Theorem 8.6, for all $n \geq 1$ we have the fundamental recurrence relations

$$p_n q_{n-1} - p_{n-1} q_n = (-1)^{n-1} b_1 b_2 \cdots b_n,$$
$$p_n q_{n-2} - p_{n-2} q_n = (-1)^n a_n b_1 b_2 \cdots b_{n-1},$$

and

$$c_n - c_{n-1} = \frac{(-1)^{n-1} b_1 b_2 \cdots b_n}{q_n q_{n-1}} \quad , \quad c_n - c_{n-2} = \frac{(-1)^n a_n b_1 b_2 \cdots b_{n-1}}{q_n q_{n-2}},$$

where the formula for $c_n - c_{n-2}$ is valid only for $n \geq 2$. Using these fundamental recurrence relations, we shall prove the following monotonicity properties of the c_n, which is important in the study of the convergence properties of the c_n.

Proposition 8.10 *Let j be even, k be odd, and suppose that $b_1, b_2, \ldots, b_\ell > 0$, where ℓ is the larger of j, k. Then,*

$$c_0 < c_2 < c_4 < \cdots < c_j < c_k < \cdots < c_5 < c_3 < c_1.$$

Proof If n is even and $n \leq \ell + 1$, then the fundamental relation for $c_n - c_{n-2}$ implies

$$c_n - c_{n-2} = \frac{a_n b_1 b_2 \cdots b_{n-1}}{q_n q_{n-2}} > 0 \implies c_{n-2} < c_n.$$

In particular, the convergents with even indices $\leq \ell + 1$ are strictly increasing. Analogously, the convergents with odd indices $\leq \ell + 1$ are strictly decreasing.

Suppose j (which is even) is larger than k. By the fundamental recurrence relation for $c_j - c_{j-1}$, we have

$$c_j - c_{j-1} = -\frac{b_1 b_2 \cdots b_j}{q_j q_{j-1}} < 0 \implies c_j < c_{j-1}.$$

We already proved that the convergents with odd indices are strictly decreasing, so since $j - 1$ is odd (and $k \le j - 1$), it follows that $c_{j-1} \le c_k$. This proves that $c_j < c_k$. Analogously, if k is larger than j, we obtain the same result, $c_j < c_k$. This completes our proof.   ∎

By the monotone criterion for sequences, we have the following.

Corollary 8.11 *The limits of the even and odd convergents exist, and*

$$c_0 < c_2 < c_4 < \cdots < \lim c_{2n} \le \lim c_{2n-1} < \cdots < c_5 < c_3 < c_1.$$

Here's a rational illustration of the monotonicity of convergents.

Example 8.14 Consider the first continued fraction we studied:

$$\frac{157}{68} = 2 + \frac{1}{3+}\frac{1}{4+}\frac{1}{5} = \langle 2; 3, 4, 5 \rangle.$$

Here's a picture of the convergents:

$$c_0 = 2, \ c_1 = \tfrac{7}{3} = 2.333\ldots$$
$$c_2 = \tfrac{30}{13} = 2.30769\ldots, \ c_3 = \tfrac{157}{68} = 2.30882\ldots$$

Note how the convergent $c_2 = 30/13$ is a fairly simple fraction, and it approximates the rather complicated fraction $157/68$ almost perfectly (they differ by $0.0011\ldots$). This is a general phenomenon with convergents; see Section 8.5.

Here's an irrational illustration of the monotonicity of convergents.

Example 8.15 The continued fraction for Φ is (see Example 8.16 on p. 625)

$$\Phi = 1 + \frac{1}{1+}\frac{1}{1+}\frac{1}{1+} \cdots.$$

Here's a picture of the convergents $c_0, \ldots, c_5$:

$$c_0 = 1, \ c_1 = 2, \ c_2 = \tfrac{3}{2} = 1.5$$
$$c_3 = \tfrac{5}{3} = 1.6666\ldots, \ c_4 = \tfrac{8}{5} = 1.6, \ c_5 = \tfrac{13}{8} = 1.625\ldots$$

8.4.2 Convergence Results for Continued Fractions

As a consequence of the previous corollary, it follows that $\lim c_n$ exists if and only if $\lim c_{2n} = \lim c_{2n-1}$, which holds if and only if

$$c_{2n} - c_{2n-1} = \frac{-b_1 b_2 \cdots b_{2n}}{q_{2n} \, q_{2n-1}} \to 0 \quad \text{as } n \to \infty. \tag{8.23}$$

In the following theorem, we give one condition under which this is satisfied.

Theorem 8.12 *Let $\{a_n\}_{n=0}^{\infty}$, $\{b_n\}_{n=1}^{\infty}$ be sequences such that $a_n, b_n > 0$ for $n \geq 1$ and*

$$\sum_{n=1}^{\infty} \frac{a_n a_{n+1}}{b_{n+1}} = \infty.$$

Then (8.23) holds, so the continued fraction $\xi := a_0 + \dfrac{b_1}{a_1+} \dfrac{b_2}{a_2+} \dfrac{b_3}{a_3+} \dfrac{b_4}{a_4+} \cdots$ converges. Moreover, for even j and odd k, we have

$$c_0 < c_2 < c_4 < \cdots < c_j < \cdots < \xi < \cdots < c_k < \cdots < c_5 < c_3 < c_1.$$

Proof Observe that for $n \geq 2$, we have $q_{n-1} = a_{n-1} q_{n-2} + b_{n-1} q_{n-3} \geq a_{n-1} q_{n-2}$, since $b_{n-1}, q_{n-3} \geq 0$. Thus, for $n \geq 2$ we have

$$q_n = a_n q_{n-1} + b_n q_{n-2} \geq a_n \cdot (a_{n-1} q_{n-2}) + b_n q_{n-2} = q_{n-2}(a_n a_{n-1} + b_n),$$

so

$$q_n \geq q_{n-2}(a_n a_{n-1} + b_n).$$

Applying this formula over and over again, we find that for $n \geq 1$,

$$q_{2n} \geq q_{2n-2}(a_{2n} a_{2n-1} + b_{2n})$$
$$\geq q_{2n-4}(a_{2n-2} a_{2n-3} + b_{2n-2}) \cdot (a_{2n} a_{2n-1} + b_{2n})$$
$$\geq \quad \vdots$$
$$\geq q_0 (a_2 a_1 + b_2)(a_4 a_3 + b_4) \cdots (a_{2n} a_{2n-1} + b_{2n}).$$

A similar argument shows that for $n \geq 2$,

$$q_{2n-1} \geq q_1 (a_3 a_2 + b_3)(a_5 a_4 + b_5) \cdots (a_{2n-1} a_{2n-2} + b_{2n-1}).$$

Thus, for $n \geq 2$, we have

$$q_{2n}q_{2n-1} \geq q_0q_1(a_2a_1 + b_2)(a_3a_2 + b_3) \cdots (a_{2n-1}a_{2n-2} + b_{2n-1})(a_{2n}a_{2n-1} + b_{2n}).$$

Factoring out all the b_k, we conclude that

$$q_{2n}q_{2n-1} \geq q_0q_1b_2 \cdots b_{2n} \cdots \left(1 + \frac{a_2a_1}{b_2}\right)\left(1 + \frac{a_3a_2}{b_3}\right) \cdots \left(1 + \frac{a_{2n}a_{2n-1}}{b_{2n}}\right).$$

This shows that

$$\frac{b_1b_2 \cdots b_{2n}}{q_{2n}\,q_{2n-1}} \leq \frac{b_1}{q_0q_1} \cdot \frac{1}{\prod_{k=1}^{2n-1}\left(1 + \frac{a_ka_{k+1}}{b_{k+1}}\right)}. \tag{8.24}$$

Now recall that (see Theorem 7.2 on p. 537) a series $\sum_{k=1}^{\infty} \alpha_k$ of positive real numbers converges if and only if the infinite product $\prod_{k=1}^{\infty}(1 + \alpha_k)$ converges. Thus, since we are given that $\sum_{k=1}^{\infty} \frac{a_ka_{k+1}}{b_{k+1}} = \infty$, we have $\prod_{k=1}^{\infty}\left(1 + \frac{a_ka_{k+1}}{b_{k+1}}\right) = \infty$ as well, so the right-hand side of (8.24) vanishes as $n \to \infty$. The fact that for even j and odd k, we have

$$c_0 < c_2 < c_4 < \cdots < c_j < \cdots < \xi < \cdots < c_k < \cdots < c_5 < c_3 < c_1$$

follows from Corollary 8.11. This completes our proof. ∎

For other convergence theorems, see Problems 6 and 9. An important example for which this theorem applies is to simple continued fractions: For a simple continued fraction $\langle a_0; a_1, a_2, a_3, \ldots \rangle$, all the b_n equal 1, so

$$\sum_{n=1}^{\infty} \frac{a_na_{n+1}}{b_{n+1}} = \sum_{n=1}^{\infty} a_na_{n+1} = \infty,$$

since all the a_n are positive integers.

Corollary 8.13 *Infinite simple continued fractions always converge, and if ξ is the limit of such a fraction, then for even j and odd k, the convergents satisfy*

$$c_0 < c_2 < c_4 < \cdots < c_j < \cdots < \xi < \cdots < c_k < \cdots < c_5 < c_3 < c_1.$$

Example 8.16 In particular, the very special fraction $\Phi := \langle 1; 1, 1, 1, \ldots \rangle$ converges. We already know that it converges to the golden ratio, but let's prove it again. To this end, observe that the nth convergent of Φ is

$$c_n = 1 + \cfrac{1}{1 + \cfrac{1}{1 + \cfrac{\ddots}{1 + \cfrac{1}{1}}}} = 1 + \frac{1}{c_{n-1}}.$$

Thus, if we set $\Phi = \lim c_n$, which we know exists and is larger than $c_0 = 1$, then taking $n \to \infty$ on both sides of $c_n = 1 + \frac{1}{c_{n-1}}$, we get $\Phi = 1 + 1/\Phi$. Thus, $\Phi^2 - \Phi - 1 = 0$. Solving for Φ, taking the root that's greater than 1, we get

$$\Phi = \frac{1 + \sqrt{5}}{2},$$

the golden ratio. We can also get Φ more quickly by noticing that

$$\Phi = 1 + \cfrac{1}{1 + \cfrac{1}{1 + \cfrac{1}{\ddots}}} = 1 + \frac{1}{\Phi} \quad \Longrightarrow \quad \Phi = 1 + \frac{1}{\Phi}.$$

From this equality we get $\Phi^2 - \Phi - 1 = 0$ and then $\Phi = \frac{1+\sqrt{5}}{2}$ as before.

As a side note, unrelated to the present example, we remark that Φ can be used to get a fairly accurate and well-known approximation to π:

$$\boxed{\pi \approx \frac{6}{5}\Phi^2 = 3.1416\dots.}$$

Example 8.17 The continued fraction $\xi := 3 + \frac{4}{6+}\frac{4}{6+}\frac{4}{6+}\frac{4}{6+}\dots$ was studied by Rafael Bombelli (1526–1572) and was one of the first continued fractions ever to be studied. Since $\sum_{n=1}^{\infty} \frac{a_n a_{n+1}}{b_{n+1}} = \sum_{n=1}^{\infty} \frac{6^2}{4} = \infty$, this continued fraction converges. To find the value of ξ, first define the continued fraction $\eta := 6 + \frac{4}{6+}\frac{4}{6+}\frac{4}{6+}\dots$, which also converges and is greater than its 0th convergent, which is 6. Note that $\xi = \eta - 3$ and

$$\eta = 6 + \cfrac{4}{6 + \cfrac{4}{6 + \cfrac{4}{\ddots}}} = 6 + \frac{1}{\eta} \quad \Longrightarrow \quad \eta = 6 + \frac{1}{\eta} \quad \Longrightarrow \quad \eta^2 - 6\eta - 1 = 0.$$

Solving this quadratic equation for η and taking the root larger than 6, we find that $\eta = 3 + \sqrt{13}$. Since $\xi = \eta - 3$, it follows that $\xi = \sqrt{13}$. Isn't this fun!

8.4.3   The Canonical Continued Fraction Algorithm

What if we want to *construct* a simple continued fraction expansion of a real number? We know how to construct such an expansion for rational numbers, so let us review that and see whether the method can be generalized to all real numbers. Recall that to construct the continued fraction expansion of a rational number $\xi = a/b$, where $a \in \mathbb{Z}$ and $b \in \mathbb{N}$, we used the division algorithm (Theorem 2.16 on p. 60): $a = bq + r$ for some integers q, r with $0 \leq r < b$. Dividing both sides of $a = bq + r$ by b, we obtain $\xi = q + r/b$, which we can also write as

$$\xi = q + \frac{1}{\eta},$$

where $\eta = b/r$, provided that $r \neq 0$, and otherwise, $\xi = q$. We then iterate this process over and over again to construct a simple continued fraction representing ξ. The keys to generalize this formula to irrational numbers are the following observations. First, since $0 \leq r < b$, we have $0 \leq r/b < 1$. Thus, the formula $\xi = q + r/b$ implies that $q = \lfloor \xi \rfloor$, the greatest integer $\leq \xi$. Second, solving the above formula for η, we obtain $\eta = 1/(\xi - q)$. To summarize, we have

$$\xi = q + \frac{1}{\eta}, \quad \text{where} \quad q = \lfloor \xi \rfloor \text{ and } \eta = \frac{1}{\xi - q} \text{ if } \xi \neq q. \qquad (8.25)$$

Note that all the quantities involved in this formula, namely $\lfloor \xi \rfloor$ and $1/(\xi - \lfloor \xi \rfloor)$, are defined for irrational numbers as well! The formula (8.25) will be used to construct the simple fraction expansions of real numbers!

Thus, let ξ be an irrational number. First, put $\xi_0 = \xi$ and define $a_0 = \lfloor \xi_0 \rfloor \in \mathbb{Z}$. By definition of the greatest integer function, we have $0 < \xi_0 - a_0 < 1$ (note that $\xi_0 \neq a_0$ since ξ_0 is irrational). Thus, we can write

$$\xi_0 = a_0 + \frac{1}{\xi_1} \quad , \quad \text{where} \quad \xi_1 = \frac{1}{\xi_0 - a_0} > 1.$$

Here's a picture (Fig. 8.2).

$$\xi_0 = k + b \text{ where } k \in \mathbb{Z} \text{ and } 0 < b < 1.$$

Put $a_0 = k$ and $\xi_1 = 1/b$. Then $\xi_0 = a_0 + 1/\xi_1$ where $a_0 \in \mathbb{Z}$ and $\xi_1 > 1$.

Fig. 8.2 We have $\xi_0 = a_0 + \dfrac{1}{\xi_1}$, where a_0 is an integer and $\xi_1 > 1$. We will iterate this process to construct *integers* $a_0, a_1, a_2, \ldots$, with $a_n \geq 1$ for $n \geq 1$, and real numbers $\xi_1, \xi_2, \xi_3, \xi_4, \ldots > 1$

Note that ξ_1 is irrational, because if not, then ξ_0 would be rational, contrary to assumption. Second, we define $a_1 = \lfloor \xi_1 \rfloor \in \mathbb{N}$. Then $0 < \xi_1 - a_1 < 1$, so we can write

$$\xi_1 = a_1 + \frac{1}{\xi_2} \ , \quad \text{where} \quad \xi_2 = \frac{1}{\xi_1 - a_1} > 1.$$

Note that ξ_2 is irrational, and

$$\xi_0 = a_0 + \cfrac{1}{a_1 + \cfrac{1}{\xi_2}}.$$

Third, we define $a_2 = \lfloor \xi_2 \rfloor \in \mathbb{N}$. Then, $0 < \xi_2 - a_2 < 1$, so we can write

$$\xi_2 = a_2 + \frac{1}{\xi_3} \ , \quad \text{where} \quad \xi_3 = \frac{1}{\xi_2 - a_2} > 1.$$

Note that ξ_3 is irrational, and

$$\xi_0 = a_0 + \cfrac{1}{a_1 + \cfrac{1}{a_2 + \cfrac{1}{\xi_3}}}.$$

We can continue this procedure to "infinity," creating a sequence $\{\xi_n\}_{n=0}^{\infty}$ of real numbers with $\xi_n > 0$ for $n \geq 1$, called the **complete quotients** of ξ, and a sequence $\{a_n\}_{n=0}^{\infty}$ of integers with $a_n > 0$ for $n \geq 1$, called the **partial quotients** of ξ, such that

$$\xi_n = a_n + \frac{1}{\xi_{n+1}}, \quad n = 0, 1, 2, 3, \ldots.$$

Putting $\xi_0, \xi_1, \xi_2, \ldots$ together into one formula, we get

$$\xi = \xi_0 = a_0 + \frac{1}{\xi_1} = a_0 + \cfrac{1}{a_1 + \cfrac{1}{\xi_2}} = \cdots " = " a_0 + \cfrac{1}{a_1 + \cfrac{1}{a_2 + \cfrac{1}{a_3 + \cfrac{1}{a_4 + \ddots}}}}.$$

$$(8.26)$$

We put the quotation marks to emphasize that we have actually not proved that ξ is *equal* to the infinite continued fraction on the far right, although it's certainly plausible that they are equal! But as a consequence of Theorem 8.14 below, this equality in fact holds. We remark that by construction, the complete quotients and partial quotients satisfy, for each $n \geq 1$,

$$\xi = \langle a_0; a_1, a_2, \ldots, a_{n-1}, \xi_n \rangle.$$

We also remark that the continued fraction in (8.26) is called the **canonical (simple) continued fraction expansion** of ξ. In Problem 8 you will prove that the canonical simple fraction expansion of a real number is unique.

8.4.4   The Continued Fraction Convergence Theorem

The following theorem is the main result of this section.

Continued fraction convergence theorem

> **Theorem 8.14** *Let $\xi_0, \xi_1, \xi_2, \ldots$ be a sequence of real numbers with $\xi_n > 0$ for $n \geq 1$ and suppose that these numbers are related by*
>
> $$\xi_n = a_n + \frac{b_{n+1}}{\xi_{n+1}} \quad, \quad n = 0, 1, 2, \ldots,$$
>
> *for sequences of real numbers $\{a_n\}_{n=0}^{\infty}, \{b_n\}_{n=1}^{\infty}$ with $a_n, b_n > 0$ for $n \geq 1$ that satisfy $\sum_{n=1}^{\infty} \frac{a_n a_{n+1}}{b_{n+1}} = \infty$. Then ξ_0 is equal to the continued fraction*
>
> $$\xi_0 = a_0 + \frac{b_1}{a_1 +} \frac{b_2}{a_2 +} \frac{b_3}{a_3 +} \frac{b_4}{a_4 +} \frac{b_5}{a_5 +} \cdots.$$
>
> *In particular, for every real number ξ, the canonical continued fraction expansion (8.26) converges to ξ.*

Proof Let $\{c_k = p_k/q_k\}$ denote the convergents of the infinite continued fraction $a_0 + \frac{b_1}{a_1 +} \frac{b_2}{a_2 +} \frac{b_3}{a_3 +} \cdots$. By Theorem 8.12, $\{c_k\}$ converges; we need to prove that $c_k \to \xi_0$.

Now consider the *finite* continued fraction obtained in a similar manner as we did in (8.26) above, by writing out ξ_0 to the nth term:

$$\xi_0 = a_0 + \frac{b_1}{a_1 +} \frac{b_2}{a_2 +} \frac{b_3}{a_3 +} \cdots + \frac{b_{n-1}}{a_{n-1} +} \frac{b_n}{\xi_n}.$$

Let $\{c_k' = p_k'/q_k'\}$ denote the convergents of this finite continued fraction. Then observe that $p_k = p_k'$ and $q_k = q_k'$ for $k \leq n - 1$ and $c_n' = \xi_0$. Therefore, by our fundamental recurrence relations, we have

$$|\xi_0 - c_{n-1}| = |c_n' - c_{n-1}'| = \frac{b_1 b_2 \cdots b_n}{q_n' q_{n-1}'} = \frac{b_1 b_2 \cdots b_n}{q_n' q_{n-1}}.$$

By the Wallis–Euler relations, we have

$$q_n' = \xi_n q_{n-1}' + b_n q_{n-2}' = \left(a_n + \frac{b_{n+1}}{\xi_{n+1}} \right) q_{n-1} + b_n q_{n-2} > a_n q_{n-1} + b_n q_{n-2} = q_n.$$

Hence, $q_n' > q_n$, so

$$|\xi_0 - c_{n-1}| = \frac{b_1 b_2 \cdots b_n}{q_n' q_{n-1}} < \frac{b_1 b_2 \cdots b_n}{q_n q_{n-1}} = |c_n - c_{n-1}|.$$

Since $\{c_k\}$ converges, we have $|c_n - c_{n-1}| \to 0$. This proves that $c_{n-1} \to \xi_0$. Therefore, $c_n \to \xi_0$ as well, and our proof is complete. ∎

Example 8.18 Consider $\xi_0 = \sqrt{3} = 1.73205\ldots$. In this case, $a_0 = \lfloor \xi_0 \rfloor = 1$. Thus,

$$\xi_1 = \frac{1}{\xi_0 - a_0} = \frac{1}{\sqrt{3} - 1} = \frac{1 + \sqrt{3}}{2} = 1.36602\ldots \implies a_1 = \lfloor \xi_1 \rfloor = 1.$$

Therefore,

$$\xi_2 = \frac{1}{\xi_1 - a_1} = \frac{1}{\frac{1 + \sqrt{3}}{2} - 1} = 1 + \sqrt{3} = 2.73205\ldots \implies a_2 = \lfloor \xi_2 \rfloor = 2.$$

Hence,

$$\xi_3 = \frac{1}{\xi_2 - a_2} = \frac{1}{\sqrt{3} - 1} = \frac{1 + \sqrt{3}}{2} = 1.36602\ldots \implies a_3 = \lfloor \xi_3 \rfloor = 1.$$

Here we notice that $\xi_3 = \xi_1$ and $a_3 = a_1$. Therefore,

$$\xi_4 = \frac{1}{\xi_3 - a_3} = \frac{1}{\xi_1 - a_1} = \xi_2 = 1 + \sqrt{3} \implies a_4 = \lfloor \xi_4 \rfloor = \lfloor \xi_2 \rfloor = 2.$$

At this point, we see that we will get the repeating pattern $1, 2, 1, 2, \ldots$, so we conclude that

$$\sqrt{3} = \langle 1; 1, 2, 1, 2, 1, 2, \ldots \rangle = \langle 1; \overline{1, 2} \rangle,$$

where we indicate that the $1, 2$ pattern repeats by putting a bar over $1, 2$.

Example 8.19 Here is a neat example concerning the Fibonacci and Lucas numbers; for other fascinating topics on these numbers, see the website [130]. Let us find the continued fraction expansion of the irrational number $\xi_0 = \Phi/\sqrt{5}$, where Φ is the golden ratio $\Phi = \frac{1+\sqrt{5}}{2}$:

$$\xi_0 = \frac{\Phi}{\sqrt{5}} = \frac{1+\sqrt{5}}{2\sqrt{5}} = 0.72360679\ldots \implies a_0 = \lfloor \xi_0 \rfloor = 0.$$

Thus,

$$\xi_1 = \frac{1}{\xi_0 - a_0} = \frac{1}{\xi_0} = \frac{2\sqrt{5}}{1+\sqrt{5}} = 1.3819660\ldots \implies a_1 = \lfloor \xi_1 \rfloor = 1.$$

Therefore,

$$\xi_2 = \frac{1}{\xi_1 - a_1} = \frac{1}{\dfrac{2\sqrt{5}}{1+\sqrt{5}} - 1} = \frac{1+\sqrt{5}}{\sqrt{5}-1} = 2.6180339\ldots \implies a_2 = \lfloor \xi_2 \rfloor = 2.$$

Hence,

$$\xi_3 = \frac{1}{\xi_2 - a_2} = \frac{1}{\dfrac{1+\sqrt{5}}{\sqrt{5}-1} - 2} = \frac{\sqrt{5}-1}{3-\sqrt{5}} = 1.2360679\ldots \implies a_3 = \lfloor \xi_3 \rfloor = 1.$$

Thus,

$$\xi_4 = \frac{1}{\xi_3 - a_3} = \frac{1}{\dfrac{\sqrt{5}-1}{3-\sqrt{5}} - 1} = \frac{3-\sqrt{5}}{2\sqrt{5}-4} = \frac{1+\sqrt{5}}{2} = 1.6180339\ldots;$$

that is, $\xi_4 = \Phi$, and so $a_4 = \lfloor \xi_4 \rfloor = 1$. Let us do this one more time:

$$\xi_5 = \frac{1}{\xi_4 - a_4} = \frac{1}{\dfrac{1+\sqrt{5}}{2} - 1} = \frac{2}{\sqrt{5}-1} = \frac{1+\sqrt{5}}{2} = \Phi,$$

and so $a_5 = a_4 = 1$. Continuing this process, we will get $\xi_n = \Phi$ and $a_n = 1$ for the rest of the n's. In conclusion, we have

$$\frac{\Phi}{\sqrt{5}} = \langle 0; 1, 2, 1, 1, 1, 1, \ldots \rangle = \langle 0; 1, 2, \overline{1} \rangle.$$

The convergents of this continued fraction are fascinating. Recall that the Fibonacci sequence $\{F_n\}$, named after Leonardo Fibonacci (1170–1250), is defined by $F_0 = 0$, $F_1 = 1$, and $F_n = F_{n-1} + F_{n-2}$ for all $n \geq 2$, which gives the sequence

$$0, 1, 1, 2, 3, 5, 8, 13, 21, 34, 55, 89, 144, \ldots .$$

The **Lucas numbers** $\{L_n\}$, named after François Lucas (1842–1891), are defined by

$$L_0 = 2 \,, \ L_1 = 1 \,, \ \ L_n = L_{n-1} + L_{n-2} \,, \ n = 2, 3, 4, \ldots .$$

They give the sequence

$$2, 1, 3, 4, 7, 11, 18, 29, 47, 76, 123, \ldots .$$

If you work out the convergents of $\frac{\Phi}{\sqrt{5}} = \langle 0; 1, 2, 1, 1, 1, 1, \ldots \rangle$, what you get is the fascinating result

$$\frac{\Phi}{\sqrt{5}} = \langle 0; 1, 2, \overline{1} \rangle \ \text{has convergents}$$

$$\frac{0}{2}, \frac{1}{1}, \frac{2}{3}, \frac{3}{4}, \frac{5}{7}, \frac{8}{11}, \frac{13}{18}, \frac{21}{29}, \frac{34}{47}, \frac{55}{76}, \frac{89}{123}, \ldots = \frac{\text{Fibonacci numbers}}{\text{Lucas numbers}};$$

(8.27)

of course, we do miss the other 1 in the Fibonacci sequence. For more fascinating facts on Fibonacci numbers see Problem 7.

8.4.5 The Numbers π and e

We now discuss the continued fraction expansions for the famous numbers π and e. Consider π first:

$$\xi_0 = \pi = 3.141592653 \ldots \ \implies \ a_0 = \lfloor \xi_0 \rfloor = 3.$$

Thus,

$$\xi_1 = \frac{1}{\pi - 3} = \frac{1}{0.141592653 \ldots} = 7.062513305 \ldots \ \implies \ a_1 = \lfloor \xi_1 \rfloor = 7.$$

Therefore,

$$\xi_2 = \frac{1}{\xi_1 - a_1} = \frac{1}{0.062513305 \ldots} = 15.99659440 \ldots \ \implies \ a_2 = \lfloor \xi_2 \rfloor = 15.$$

Hence,

$$\xi_3 = \frac{1}{\xi_2 - a_2} = \frac{1}{0.996594407 \ldots} = 1.00341723 \ldots \ \implies \ a_3 = \lfloor \xi_3 \rfloor = 1.$$

Let us do this one more time:

$$\xi_4 = \frac{1}{\xi_3 - a_3} = \frac{1}{0.003417231\ldots} = 292.6345908\ldots \implies a_4 = \lfloor \xi_4 \rfloor = 292.$$

Continuing this process (at a cafe and after 314 refills of coffee), we get

$$\pi = \langle 3; 7, 15, 1, 292, 1, 1, 1, 2, 1, 3, 1, 14, 2, 1, 1, 2, 2, 2, 2, 1, 84, 2, 1, \ldots \rangle.$$
(8.28)

Unfortunately (or fortunately, to keep life full of surprises), there is no known pattern that the partial quotients follow! The first few convergents for $\pi = 3.141592653\ldots$ are

$$c_0 = 3 \,, \; c_1 = \frac{22}{7} = 3.142857142\ldots \,, \; c_2 = \frac{333}{106} = 3.141509433\ldots,$$

$$c_4 = \frac{355}{113} = 3.141592920\ldots \,, \; c_5 = \frac{103993}{33102} = 3.141592653\ldots.$$

In stark contrast to π, Euler's number e has a shockingly simple pattern, which we ask you to work out in Problem 2:

$$e = \langle 2, 1, 2, 1, 1, 4, 1, 1, 6, 1, 1, 8, \ldots \rangle$$

We will prove that this pattern continues in Section 8.7! We now discuss the important topic of when a continued fraction represents an irrational number.

8.4.6   Irrational Continued Fractions

Consider the following (cf. [166]).

> **Theorem 8.15** *Let* $\{a_n\}_{n=0}^{\infty}$, $\{b_n\}_{n=1}^{\infty}$ *be sequences of rational numbers such that* $a_n, b_n > 0$ *for* $n \geq 1$, $0 < b_n \leq a_n$ *for all* n *sufficiently large, and* $\sum_{n=1}^{\infty} \frac{a_n a_{n+1}}{b_{n+1}} = \infty$. *Then the real number*
>
> $$\xi = a_0 + \frac{b_1}{a_1 +} \frac{b_2}{a_2 +} \frac{b_3}{a_3 +} \frac{b_4}{a_4 +} \frac{b_5}{a_5 +} \cdots \quad \text{is irrational.}$$

Proof First of all, the continued fraction defining ξ converges by Theorem 8.12. Choose a natural number m such that $0 < b_n \leq a_n$ for all $n \geq m + 1$. Observe that if we define

$$\eta = a_m + \frac{b_{m+1}}{a_{m+1}+} \frac{b_{m+2}}{a_{m+2}+} \frac{b_{m+3}}{a_{m+3}+} \cdots,$$

which also converges by Theorem 8.12, then $\eta > a_m > 0$, and we can write

$$\xi = a_0 + \frac{b_1}{a_1+} \frac{b_2}{a_2+} \frac{b_3}{a_3+} \cdots + \frac{b_m}{\eta}.$$

By Theorem 8.5 on p. 612, we know that

$$\xi = a_0 + \frac{b_1}{a_1+} \frac{b_2}{a_2+} \frac{b_3}{a_3+} \cdots + \frac{b_m}{\eta} = \frac{\eta p_m + b_m p_{m-1}}{\eta q_m + b_m q_{m-1}}.$$

Solving the last equation for η, a bit of algebra gives

$$\xi = \frac{\eta p_m + b_m p_{m-1}}{\eta q_m + b_m q_{m-1}} \iff \eta = \frac{\xi b_m q_{m-1} - b_m p_{m-1}}{p_m - \xi q_m}.$$

Note that since $\eta > a_m$, we have $\xi \neq p_m/q_m$. Since all the a_n, b_n's are rational, it follows that ξ is irrational if and only if η is irrational. Thus, all we have to do is prove that η is irrational. Since a_m is rational, all we have to do is prove that $\frac{b_{m+1}}{a_{m+1}+} \frac{b_{m+2}}{a_{m+2}+} \frac{b_{m+3}}{a_{m+3}+} \cdots$ is irrational, where $0 < b_n \leq a_n$ for all $n \geq m+1$. In conclusion, we might as well assume from the start that

$$\xi = \frac{b_1}{a_1+} \frac{b_2}{a_2+} \frac{b_3}{a_3+} \frac{b_4}{a_4+} \frac{b_5}{a_5+} \cdots,$$

where $0 < b_n \leq a_n$ for all n. We shall do this for the rest of the proof. Assume, by way of contradiction, that ξ *is* rational. Define $\xi_n := \frac{b_n}{a_n+} \frac{b_{n+1}}{a_{n+1}+} \frac{b_{n+2}}{a_{n+2}+} \cdots$. Then for each $n = 1, 2, \ldots$, we have

$$\xi_n = \frac{b_n}{a_n + \xi_{n+1}}. \tag{8.29}$$

By definition, we have $\xi_n > 0$ for all n, and since $0 < b_n \leq a_n$ for all n, it follows that $0 < \xi_n < 1$ for all n. Since $\xi_0 = \xi$ is rational, from (8.29) with $n = 0$ we get that ξ_1 is rational. By induction it follows that ξ_n is rational for all n. Since $0 < \xi_n < 1$ for all n, we can therefore write $\xi_n = s_n/t_n$, where $0 < s_n < t_n$ for all n with s_n and t_n relatively prime integers. Now from the second equality in (8.29) we see that

$$\frac{s_n}{t_n} = \frac{b_n}{a_n + s_{n+1}/t_{n+1}}.$$

Cross multiplying, we arrive at the equality

$$s_n s_{n+1} = (b_n t_n - a_n s_n) t_{n+1}.$$

Thus, $t_{n+1} \mid s_n s_{n+1}$. By assumption, s_{n+1} and t_{n+1} are relatively prime, so t_{n+1} must divide s_n. In particular, $t_{n+1} < s_n$. However, $s_n < t_n$ by assumption, so $t_{n+1} < t_n$. In summary, $\{t_n\}$ is a sequence of positive integers satisfying

$$t_1 > t_2 > t_3 > \cdots > t_n > t_{n+1} > \cdots > 0,$$

which of course is an absurdity, because we would eventually reach zero!   ∎

Example 8.20 (*Irrationality of e, Proof III*) Since we already know that (see (8.15))

$$e = 2 + \frac{2}{2+} \frac{3}{3+} \frac{4}{4+} \frac{5}{5+} \cdots,$$

we certainly have $b_n \le a_n$ for all n, whence e is irrational!

As another application of this theorem, we get the following corollary.

Corollary 8.16 *Every infinite simple continued fraction represents an irrational number. In particular, a real number is irrational if and only if it can be represented by an infinite simple continued fraction.*

Indeed, for a simple continued fraction we have $b_n = 1$ for all n, so $0 < b_n \le a_n$ for all $n \ge 1$.

▶ **Exercises 8.4**

1. (a) Use the simple continued fraction algorithm to the find the expansions of

$$(a)\ \sqrt{2}\ ,\quad (b)\ \frac{1-\sqrt{8}}{2}\ ,\quad (c)\ \sqrt{19}\ ,\quad (d)\ 3.14159\ ,\quad (e)\ \sqrt{7}.$$

   (b) Find the value of the continued fraction expansions

$$(a)\ 4 + \frac{2}{8+} \frac{2}{8+} \frac{2}{8+} \cdots\ ,\quad (b)\ \langle \overline{3} \rangle = \langle 3; 3, 3, 3, 3, 3, \ldots \rangle.$$

   The continued fraction in (a) was studied by Pietro Antonio Cataldi (1548–1626) and is one of the earliest infinite continued fractions on record.

2. In Section 8.7, we will prove the conjectures you make in (a) and (b) below.

   (a) Using a calculator, we find that $e \approx 2.718281828$. Verify that $2.718281828 = \langle 2; 1, 2, 1, 1, 4, 1, 1, 6, \ldots \rangle$. From this, conjecture a formula for a_n, $n = 0, 1, 2, 3, \ldots$, in the canonical continued fraction expansion for e.

   (b) Using a calculator, we find that $\frac{e+1}{e-1} \approx 2.1639534137$. Find a_0, a_1, a_2, a_3 in the canonical continued fraction expansion for 2.1639534137 and conjecture a formula for a_n, $n = 0, 1, 2, 3, \ldots$, in the canonical continued fraction expansion for $\frac{e+1}{e-1}$.

3. Let $n \in \mathbb{N}$. Prove that $\sqrt{n^2 + 1} = \langle n; \overline{2n} \rangle$ using the simple continued fraction algorithm on $\sqrt{n^2 + 1}$. Using the same technique, find the canonical expansion of $\sqrt{n^2 + 2}$. (See Problem 5 below for other proofs.)

4. In this problem we show that every positive real number can be written as two different infinite continued fractions. Let a be a positive real number. Prove that

$$a = 1 + \cfrac{k}{1+} \cfrac{k}{1+} \cfrac{k}{1+} \cfrac{k}{1+} \cdots = \cfrac{\ell}{1+} \cfrac{\ell}{1+} \cfrac{\ell}{1+} \cfrac{\ell}{1+} \cdots,$$

where $k = a^2 - a$ and $\ell = a^2 + a$. Suggestion: Link the limits of continued fractions on the right to the quadratic equations $x^2 - x - k = 0$ and $x^2 + x - \ell = 0$, respectively. Find neat infinite continued fractions for $1, 2$, and 3.

5. Let x be a positive real number and suppose that $x^2 - ax - b = 0$, where a, b are positive. Prove that

$$x = a + \cfrac{b}{a+} \cfrac{b}{a+} \cfrac{b}{a+} \cfrac{b}{a+} \cfrac{b}{a+} \cdots.$$

Using this, prove that for all $\alpha, \beta > 0$,

$$\sqrt{\alpha^2 + \beta} = \alpha + \cfrac{\beta}{2\alpha+} \cfrac{\beta}{2\alpha+} \cfrac{\beta}{2\alpha+} \cfrac{\beta}{2\alpha+} \cdots.$$

6. (a) Prove that a continued fraction $a_0 + \frac{b_1}{a_1+} \frac{b_2}{a_2+} \frac{b_3}{a_3+} \cdots$ converges if and only if

$$c_0 + \sum_{n=1}^{\infty} \frac{(-1)^{n-1} b_1 b_2 \cdots b_n}{q_n q_{n-1}}$$

converges, in which case this sum is exactly $a_0 + \frac{b_1}{a_1+} \frac{b_2}{a_2+} \frac{b_3}{a_3+} \cdots$. In particular, for a simple continued fraction $\xi = \langle a_0; a_1, a_2, a_3, \ldots \rangle$, we have

$$\xi = 1 + \sum_{n=1}^{\infty} \frac{(-1)^{n-1}}{q_n q_{n-1}}.$$

(b) Assume that $\xi = a_0 + \frac{b_1}{a_1+} \frac{b_2}{a_2+} \frac{b_3}{a_3+} \cdots$ converges. Prove that

$$\xi = c_0 + \sum_{n=2}^{\infty} \frac{(-1)^n a_n b_1 b_2 \cdots b_{n-1}}{q_n q_{n-2}}.$$

In particular, for a simple continued fraction $\xi = \langle a_0; a_1, a_2, a_3, \ldots \rangle$, we have

$$\xi = 1 + \sum_{n=2}^{\infty} \frac{(-1)^n a_n}{q_n q_{n-2}}.$$

7. Let $\{c_n\}$ be the convergents of $\Phi = \langle 1; 1, 1, 1, 1, 1, 1, \ldots \rangle$.

   (1) Prove that for $n \geq 1$, we have $\frac{F_{n+1}}{F_n} = c_{n-1}$. (That is, $p_n = F_{n+2}$ and $q_n = F_{n+1}$.) Conclude that

$$\Phi = \lim_{n \to \infty} \frac{F_{n+1}}{F_n},$$

   a beautiful (but nontrivial) fact (also proved in Problem 9 on p. 193).
   (2) Using the previous problem, prove the incredibly beautiful formulas

$$\Phi = \sum_{n=1}^{\infty} \frac{(-1)^{n-1}}{F_n F_{n+1}} \quad \text{and} \quad \Phi^{-1} = \sum_{n=2}^{\infty} \frac{(-1)^n}{F_n F_{n+2}}.$$

8. Let $\alpha = \langle a_0; a_1, a_2, \ldots \rangle$, $\beta = \langle b_0; b_1, b_2, \ldots \rangle$ be infinite simple continued fractions. Prove that if $\alpha = \beta$, then $a_k = b_k$ for all $k = 0, 1, 2, \ldots$, which shows that the canonical simple fraction expansion of an irrational real number is unique. See Problem 7 on p. 619 for the rational case.

9. A continued fraction $a_0 + \frac{1}{a_1} + \frac{1}{a_2} + \frac{1}{a_3} + \frac{1}{a_4} + \cdots$ where the a_n are real numbers with $a_n > 0$ for $n \geq 1$ is said to be **unary**. In this problem we prove that a unary continued fraction converges if and only if $\sum a_n = \infty$. Let $a_0 + \frac{1}{a_1} + \frac{1}{a_2} + \frac{1}{a_3} + \cdots$ be unary.

   (i) Prove that $q_n \leq \prod_{k=1}^{n}(1 + a_k)$.
   (ii) Using the inequality derived in (i), prove that if the unary continued fraction converges, then $\sum a_n = \infty$.
   (iii) Prove that

$$q_{2n} \geq 1 + a_1(a_2 + a_4 + \cdots + a_{2n}) \ , \quad q_{2n-1} \geq a_1 + a_3 + \cdots + a_{2n-1},$$

   where the first inequality holds for $n \geq 1$ and the second for $n \geq 2$.
   (iv) Prove that if $\sum a_n = \infty$, then the unary continued fraction converges.

8.5 Diophantine Approximations and the Mystery of π Solved!

For practical purposes, it is necessary to approximate irrational numbers by rational numbers. Also, if a rational number has a very large denominator, e.g., $\frac{1234567}{121110987654321}$, then it is hard to work with, so for practical purposes it would be nice to have a "good" approximation to such a rational number by a rational number with a more manageable denominator. Diophantine approximations is the subject of finding "good" or even "best" rational approximations to real numbers. Continued fractions play a very important role in this subject. We start our journey by reviewing the mysterious fraction representations of π.

8.5.1 The Mystery of π and Good and Best Approximations

Here we review some approximations to $\pi = 3.14159265\ldots$ that have been discovered throughout the centuries (see p. 634 in Section 4.12 for a thorough study).

(1) 3 in the Holy Bible circa 1000 B.C. by the Hebrews; See Book of I Kings, Chapter 7, verse 23, and Book of II Chronicles, Chapter 4, verse 2:

> And he made a molten sea, ten cubits from the one brim to the other: it was round all about, and his height was five cubits: and a line of thirty cubits did compass it about. I Kings 7:23.

(2) $22/7 = 3.14285714\ldots$ (correct to two decimal places) by Archimedes (287 B.C.–212 B.C.) circa 250 B.C.
(3) $333/106 = 3.14150943\ldots$ (correct to four decimal places), a lower bound found by Adriaan Anthoniszoon (1527–1607) circa 1600 A.D.
(4) $355/113 = 3.14159292\ldots$ (correct to six decimal places) by Tsu Chung-Chi (429–501) circa 500 A.D.

Hmmm… these numbers certainly seem familiar! These numbers are exactly the first four convergents of the continued fraction expansion of π that we worked out on p. 631 in Section 8.4.5! From this example, it seems like approximating real numbers by rational numbers is intimately related to continued fractions; this is indeed the case, as we shall see. To start our adventure in approximations, we start with the concepts of "good" and "best" approximations.

A rational number p/q is called a **good approximation** to a real number ξ if

$$\boxed{\text{for all rational } \frac{a}{b} \neq \frac{p}{q} \text{ with } 1 \leq b \leq q, \text{ we have } \left|\xi - \frac{p}{q}\right| < \left|\xi - \frac{a}{b}\right|;}$$

in other words, we cannot get closer to the real number ξ with a different rational number having a denominator $\leq q$.

For example, let's consider good approximations to π. Here's a picture of rationals between 2 and 4 with denominators ≤ 2 (Fig. 8.3).

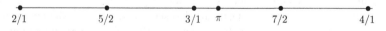

2/1   5/2   3/1 π 7/2   4/1

Fig. 8.3 We would like to approximate π by rationals with denominators ≤ 2

Example 8.21 4/1 is *not* a good approximation to π, because 3/1, which has an equal denominator, is closer to π:

$$\left|\pi - \frac{3}{1}\right| = 0.141592\ldots < \left|\pi - \frac{4}{1}\right| = 0.858407\ldots.$$

Example 8.22 As another example, 7/2 is *not* a good approximation to π because 3/1, which has a smaller denominator than 7/2, is closer to π:

$$\left|\pi - \frac{3}{1}\right| = 0.141592\ldots < \left|\pi - \frac{7}{2}\right| = 0.358407\ldots.$$

This example shows that you wouldn't want to approximate π with 7/2, because you can approximate it with the "simpler" number 3/1, which has a smaller denominator. Consider now fractions with denominators ≤ 4 (Fig. 8.4).

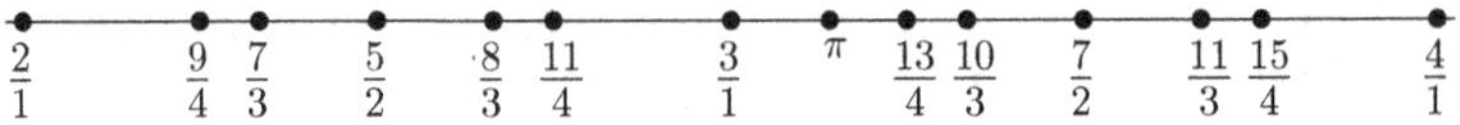

$$\frac{2}{1} \qquad \frac{9}{4}\ \frac{7}{3} \qquad \frac{5}{2} \qquad \frac{8}{3}\ \frac{11}{4} \qquad \frac{3}{1} \qquad \pi \qquad \frac{13}{4}\ \frac{10}{3} \qquad \frac{7}{2} \qquad \frac{11}{3}\ \frac{15}{4} \qquad \frac{4}{1}$$

Fig. 8.4 We would like to approximate π by rationals with denominators ≤ 4

Example 8.23 13/4 is a good approximation to π. This is because

$$\left|\pi - \frac{13}{4}\right| = 0.108407\ldots,$$

and there are no fractions closer to π with denominator 4, and the closest fractions with the smaller denominators 1, 2, and 3 are 3/1, 7/2, and 10/3, which satisfy

$$\left|\pi - \frac{3}{1}\right| = 0.141592\ldots\ , \quad \left|\pi - \frac{7}{2}\right| = 0.358407\ldots\ , \quad \left|\pi - \frac{10}{3}\right| = 0.191740\ldots.$$

Thus,

for all rational $\dfrac{a}{b} \neq \dfrac{13}{4}$ with $1 \leq b \leq 4$, we have $\left|\pi - \dfrac{13}{4}\right| < \left|\pi - \dfrac{a}{b}\right|$.

Now one can argue: Is 13/4 really that great of an approximation to π? For although 3/1 is not as close to π, it is certainly much easier to work with than 13/4 because of the larger denominator 4; moreover, we have $13/4 = 3.25$, so we didn't even gain a single decimal place of accuracy in going from 3.00 to 3.25. These are definitely valid arguments. One can also see the validity of this argument by combining fractions in the inequality in the definition of good approximation: p/q is a good approximation to ξ if

for all rational $\dfrac{a}{b} \neq \dfrac{p}{q}$ with $1 \leq b \leq q$, we have $\dfrac{|q\xi - p|}{q} < \dfrac{|b\xi - a|}{b}$, where we used that $q, b > 0$. Here, we can see that $\frac{|q\xi - p|}{q} < \frac{|b\xi - a|}{b}$ may hold, not because p/q is dramatically much closer to ξ than is a/b, but simply because q is a lot larger than b (as in the case $13/4$ and $3/1$, where 4 is larger than 1). To try to correct this somewhat misleading notion of "good," we introduce the concept of a "best" approximation by clearing the denominators.

A rational number p/q is called a **best approximation** to a real number ξ if[3]

$$\boxed{\text{for all rational } \frac{a}{b} \neq \frac{p}{q} \text{ with } 1 \leq b \leq q, \text{ we have } |q\xi - p| < |b\xi - a|.}$$

This notion has the geometric interpretation shown in Fig. 8.5.

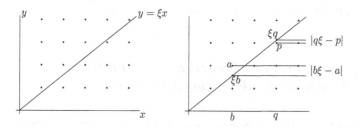

Fig. 8.5 *Left* The line $y = \xi x$ and the integer lattice. *Right* Let p/q be a best approximation to ξ and let $a/b \in \mathbb{Q}$ with $1 \leq b \leq q$. Then the vertical distance between the line at $x = q$ and the pth vertical lattice point (i.e., $|q\xi - p|$) is strictly less than the vertical distance between the line at $x = b$ and the ath vertical lattice point (i.e., $|b\xi - a|$)

Example 8.24 We can see that $p/q = 13/4$ is *not* a best approximation to π, because with $a/b = 3/1$, we have $1 \leq 1 \leq 4$, yet

$$|4 \cdot \pi - 13| = 0.433629\ldots \not< |1 \cdot \pi - 3| = 0.141592\ldots.$$

Thus, $13/4$ is a good approximation to π but is far from a best approximation.

Thus, good $\not\Longrightarrow$ best. However, every best approximation is also good.

Proposition 8.17 *A best approximation is also a good one.*

---

[3] Warning: As a heads up, some authors define good approximation as follows: $\frac{p}{q}$ is a good approximation to ξ if for all rational $\frac{a}{b}$ with $1 \leq b < q$, we have $|\xi - \frac{p}{q}| < |\xi - \frac{a}{b}|$. This definition, although just slightly different from ours, makes some proofs *considerably easier*. Moreover, with this definition, $1\,000\,000/1$ is a good approximation to π (why?)! (In fact, *every* integer, no matter how big, is a good approximation to π.) On the other hand, with the definition we used, the only integer that is a good approximation to π is 3. Also, some authors define best approximation as follows: $\frac{p}{q}$ is a best approximation to ξ if for all rational $\frac{a}{b}$ with $1 \leq b < q$, we have $|q\xi - p| < |b\xi - a|$; with this definition of "best," one can shorten the proof of Theorem 8.20, but then one must live with the fact that $1\,000\,000/1$ is a best approximation to π.

Proof Let p/q be a best approximation to ξ; we shall prove that p/q is a good one too. Let $a/b \neq p/q$ be rational with $1 \leq b \leq q$. Then $|q\xi - p| < |b - \xi a|$, since p/q is a best approximation, and also, $\frac{1}{q} \leq \frac{1}{b}$, since $b \leq q$; hence

$$\left|\xi - \frac{p}{q}\right| = \frac{|q\xi - p|}{q} < \frac{|b\xi - a|}{q} \leq \frac{|b\xi - a|}{b} = \left|\xi - \frac{a}{b}\right| \implies \left|\xi - \frac{p}{q}\right| < \left|\xi - \frac{a}{b}\right|.$$

This shows that p/q is a good approximation.   ∎

In the following subsection, we shall prove the best approximation theorem, Theorem 8.20 on p. 646, which we state here:

> (**Best approximation theorem**) *Every best approximation of a real number (rational or irrational) is a convergent of its canonical continued fraction expansion, and conversely, each of the convergents $c_1, c_2, c_3, \ldots$ is a best approximation.*

8.5.2 Approximations, Convergents, and the "Most Irrational" of All Irrational Numbers

The objective of this subsection is to understand how convergents approximate real numbers. In the following theorem, we show that the convergents of the simple continued fraction of a real number ξ get increasingly closer to ξ. (See Problem 5 for the general case of nonsimple continued fractions.)

Fundamental approximation theorem

> **Theorem 8.18** *If ξ is irrational and $\{c_n = p_n/q_n\}$ are the convergents of its canonical continued fraction, then the following inequalities hold:*
>
> $$\left|\xi - c_n\right| < \frac{1}{q_n q_{n+1}}, \quad \left|\xi - c_{n+1}\right| < \left|\xi - c_n\right|, \quad \left|q_{n+1}\xi - p_{n+1}\right| < \left|q_n\xi - p_n\right|.$$
>
> *If ξ is a rational number and the convergent c_{n+1} is defined (that is, if $\xi \neq c_n$), then these inequalities still hold, with the exception that if $\xi = c_{n+1}$, then the first inequality is replaced with the equality $|\xi - c_n| = \frac{1}{q_n q_{n+1}}$.*

Proof We prove this theorem for ξ irrational; the rational case is proved using a similar argument, which we leave to you if you're interested. The proof of this theorem is very simple. We just need the inequalities (see Corollary 8.13 on 625)

$$c_n < c_{n+2} < \xi < c_{n+1} \quad \text{or} \quad c_{n+1} < \xi < c_{n+2} < c_n, \qquad (8.30)$$

depending on whether n is even or odd, respectively, and the fundamental recurrence relations (see Corollary 8.7 on p. 614)

$$c_{n+1} - c_n = \frac{(-1)^n}{q_n \, q_{n+1}} \quad , \quad c_{n+2} - c_n = \frac{(-1)^n a_{n+2}}{q_n \, q_{n+2}}. \tag{8.31}$$

Now the first inequality of our theorem follows easily:

$$\left| \xi - c_n \right| \overset{\text{by (8.30)}}{<} \left| c_{n+1} - c_n \right| \overset{\text{by (8.31)}}{=} \left| \frac{(-1)^n}{q_n \, q_{n+1}} \right| = \frac{1}{q_n \, q_{n+1}}.$$

We now prove that $\left| q_{n+1} \xi - p_{n+1} \right| < \left| q_n \xi - p_n \right|$. To prove this, we work on the left- and right-hand sides separately. For the left-hand side, we have

$$\left| q_{n+1} \xi - p_{n+1} \right| = q_{n+1} \left| \xi - \frac{p_{n+1}}{q_{n+1}} \right| = q_{n+1} \left| \xi - c_{n+1} \right| < q_{n+1} \left| c_{n+2} - c_{n+1} \right| \text{ by (8.30)}$$

$$= q_{n+1} \frac{1}{q_{n+1} \, q_{n+2}} \quad \text{by (8.31)}$$

$$= \frac{1}{q_{n+2}}.$$

Hence, $\frac{1}{q_{n+2}} > \left| q_{n+1} \xi - p_{n+1} \right|$. Now,

$$\left| q_n \xi - p_n \right| = q_n \left| \xi - \frac{p_n}{q_n} \right| = q_n \left| \xi - c_n \right| > q_n \left| c_{n+2} - c_n \right| \quad \text{by (8.30)}$$

$$= q_n \frac{a_{n+2}}{q_n \, q_{n+2}} \quad \text{by (8.31)}$$

$$= \frac{a_{n+2}}{q_{n+2}} \geq \frac{1}{q_{n+2}} > \left| q_{n+1} \xi - p_{n+1} \right|.$$

This proves our third inequality. Finally, using what we just proved, and that

$$q_{n+1} = a_{n+1} q_n + q_{n-1} \geq q_n + q_{n-1} > q_n \quad \Longrightarrow \quad \frac{1}{q_{n+1}} < \frac{1}{q_n},$$

we see that

$$\left| \xi - c_{n+1} \right| = \left| \xi - \frac{p_{n+1}}{q_{n+1}} \right| = \frac{1}{q_{n+1}} \left| q_{n+1} \xi - p_{n+1} \right|$$

$$< \frac{1}{q_{n+1}} \left| q_n \xi - p_n \right|$$

$$< \frac{1}{q_n} \left| q_n \xi - p_n \right| = \left| \xi - \frac{p_n}{q_n} \right| = \left| \xi - c_n \right|.$$

∎

It is important to use only the *canonical* expansion when ξ is rational. This is because the statement that $\left|q_{n+1}\xi - p_{n+1}\right| < \left|q_n\xi - p_n\right|$ may not *not* be true if we don't use the canonical expansion.

Example 8.25 Consider $5/3$, which has the canonical expansion

$$\frac{5}{3} = \langle 1; 1, 2\rangle = 1 + \cfrac{1}{1 + \cfrac{1}{2}}.$$

We can write this as a noncanonical expansion by breaking up the 2:

$$\xi = \langle 1; 1, 1, 1\rangle = 1 + \cfrac{1}{1 + \cfrac{1}{1 + \cfrac{1}{1}}} = \frac{5}{3}.$$

The convergents for this noncanonical expansion of ξ are $c_0 = 1/1$, $c_2 = 2/1$, $c_3 = 3/2$, and $\xi = c_4 = 5/3$. In this case,

$$\left|q_3\xi - p_3\right| = \left|2 \cdot \frac{5}{3} - 3\right| = \frac{1}{3} = \left|1 \cdot \frac{5}{3} - 2\right| = \left|q_2\xi - p_2\right|,$$

so for this example, $\left|q_2\xi - p_2\right| \not< \left|q_1\xi - p_1\right|$.

We now discuss the "most irrational" of all irrational numbers. From the best approximation theorem (Theorem 8.20, which we'll prove in a moment), we know that the best approximations of a real number ξ are convergents, and from the fundamental approximation theorem 8.18, we have the error estimate

$$\left|\xi - c_n\right| < \frac{1}{q_n q_{n+1}} \quad \Longrightarrow \quad \left|q_n\xi - p_n\right| < \frac{1}{q_{n+1}}. \tag{8.32}$$

This shows you that the larger the q_n are, the better the best approximations are. Since the q_n are determined by the recurrence relation $q_n = a_n q_{n-1} + q_{n-2}$, we see that the larger the a_n are, the larger the q_n are. In summary, ξ can be approximated in a way that is very "good" by rational numbers when it has *large* a_n and a way that is very "bad" by rational numbers when it has *small* a_n.

Example 8.26 Here is a "good" example. Recall from Eq. (8.28) on p. 632 the continued fraction for π:

$$\pi = \langle 3; 7, 15, 1, 292, 1, 1, 1, 2, 1, \ldots\rangle,$$

which has convergents

$$c_0 = 3, \quad c_1 = \frac{22}{7}, \quad c_2 = \frac{333}{106}, \quad c_3 = \frac{355}{113}, \quad c_4 = \frac{103993}{33102}, \ldots$$

Because of the large number $a_4 = 292$, we see from (8.32) that we can approximate π very nicely with c_3; indeed, the left-hand equation in (8.32) implies

$$\left| \pi - c_3 \right| < \frac{1}{q_3 q_4} = \frac{1}{113 \cdot 33102} = 0.000000267\ldots$$

Thus, $c_3 = \frac{355}{113}$ approximates π to within six decimal places! (Just to check, note that $\pi = 3.14159265\ldots$ and $\frac{355}{113} = 3.14159292\ldots$) It's amazing how many decimal places of accuracy we can get with just taking the c_3 convergent!

Example 8.27 (*The "most irrational" number*) Here is a "bad" example. From our discussion after (8.32), we saw that the smaller the a_n are, the worse it can be approximated by rationals. Of course, since 1 is the smallest natural number, we can consider the golden ratio

$$\Phi = \frac{1 + \sqrt{5}}{2} = \langle 1; 1, 1, 1, 1, 1, 1, 1, \ldots \rangle = 1.6180339887\ldots$$

as being the "worst" of all irrational numbers that can be approximated by rational numbers. Indeed, we saw that we could get six decimal places of π by just taking c_3; for Φ we need c_{18}! (Just to check, we find that $c_{17} = \frac{4181}{2584} = 1.6180340557\ldots$, not quite six decimal places, and $c_{18} = \frac{6765}{4181} = 1.618033963\ldots$ got the sixth one. Also notice the large denominator 4181 just to get six decimal places.) Therefore, Φ wins the prize for the "most irrational" number in that it's the "farthest" from the rationals! We continue our discussion on "most irrational" on p. 697 in Section 8.10.3.

We now show that best approximations are exactly convergents; this is one of the most important properties of continued fractions. We first need the following lemma, whose ingenious proof we learned from Beskin's beautiful book [27].

Lemma 8.19 *If p_n/q_n, $n \geq 0$, is a convergent of the canonical continued fraction expansion of a real number ξ and $p/q \neq p_n/q_n$ is a rational number with $q > 0$ and $1 \leq q < q_{n+1}$, then*

$$|q_n \xi - p_n| \leq |q \xi - p|.$$

Moreover, if $n \geq 1$ and $q \leq q_n$, then this inequality is strict.

Proof Let p_n/q_n, $n \geq 0$, be a convergent of the canonical continued fraction expansion of a real number ξ and let $p/q \neq p_n/q_n$ be a rational number with $q > 0$ and $1 \leq q < q_{n+1}$. Note that if ξ happens to be rational, we are implicitly assuming that $\xi \neq p_n/q_n$, so that q_{n+1} is defined.

Step 1: The trick. To prove that $|q_n\xi - p_n| \leq |q\xi - p|$, the trick is to write p and q as linear combinations of $p_n, p_{n+1}, q_n, q_{n+1}$:

$$p = p_n x + p_{n+1} y,$$
$$q = q_n x + q_{n+1} y. \tag{8.33}$$

How do we know such x, y exist? The reason is that we can solve these equations for x and y; after some linear algebra we obtain

$$x = (-1)^n \left(p_{n+1} q - p q_{n+1} \right) , \quad y = (-1)^n \left(p q_n - p_n q \right).$$

(The fact that $p_{n+1} q_n - p_n q_{n+1} = (-1)^n$ was used to simplify the equations for x and y.) These formulas are not needed below except for the important fact that these formulas show that x and y are *integers*. Now, using the formulas in (8.33), we see that

$$q\xi - p = \left(q_n x + q_{n+1} y \right)\xi - p_n x - p_{n+1} y$$
$$= \left(q_n \xi - p_n \right)x + \left(q_{n+1} \xi - p_{n+1} \right)y.$$

Therefore,

$$|q\xi - p| = \left| \left(q_n \xi - p_n \right)x + \left(q_{n+1} \xi - p_{n+1} \right)y \right|. \tag{8.34}$$

Step 2: Our goal is to simplify the right-hand side of (8.34) by understanding the signs of the terms in the absolute values. First of all, we claim that x and y are both nonzero and have opposite signs. In fact, if $x = 0$, then the second formula (8.33) shows that $q = q_{n+1} y$. Since q and q_{n+1} are positive, we must have $y > 0$, and we have $q \geq q_{n+1}$, contradicting that $q < q_{n+1}$. (Note that $y > 0$ is the same thing as saying $y \geq 1$, because y is an integer.) If $y = 0$, then the formulas (8.33) show that $p = p_n x$ and $q = q_n x$, so $p/q = p_n/q_n$, and this contradicts the assumption that $p/q \neq p_n/q_n$. If $x < 0$ and $y < 0$, then the second formula in (8.33) implies that $q < 0$, contradicting that $q > 0$. Finally, if $x > 0$ and $y > 0$, then the second formula in (8.33) implies that

$$q = q_n x + q_{n+1} y > q_{n+1},$$

contradicting that $q < q_{n+1}$. Thus, x and y are indeed both nonzero and have opposite signs. Now by Corollary 8.11 on p. 622 concerning even- and odd-indexed convergents, we know that

$$\xi - \frac{p_n}{q_n} \quad \text{and} \quad \xi - \frac{p_{n+1}}{q_{n+1}}$$

have opposite signs. Therefore, $q_n \xi - p_n$ and $q_{n+1} \xi - p_{n+1}$ have opposite signs, and hence, since x and y also have opposite signs,

$$\left(q_n \xi - p_n \right)x \quad \text{and} \quad \left(q_{n+1} \xi - p_{n+1} \right)y$$

have the same sign. Therefore, in (8.34), we have

$$|q\xi - p| = |q_n\xi - p_n||x| + |q_{n+1}\xi - p_{n+1}||y|. \tag{8.35}$$

Step 3: We now prove our result. Since $x \neq 0$, we have $|x| \geq 1$ (because x is an integer), so by (8.35) we see that

$$|q_n\xi - p_n| \leq |q\xi - p|,$$

as stated in our lemma.

Now assume that $n \geq 1$ and we have $|q_n\xi - p_n| = |q\xi - p|$; we shall prove that $q > q_n$. Note that $|q_n\xi - p_n| = |q\xi - p| \neq 0$, for otherwise, we would have $\xi = p_n/q_n$ and $\xi = p/q$, contradicting that $p/q \neq p_n/q_n$. In particular, recalling that x is nonzero, we see that (8.35) implies that $|x| = 1$. If $x = +1$, then $y < 0$ (because x and y have opposite signs), so $y \leq -1$, since y is an integer, and hence by the second equation in (8.33), we have

$$q = q_n x + q_{n+1}y = q_n + q_{n+1}y \leq q_n - q_{n+1} \leq 0,$$

because $q_n \leq q_{n+1}$. This is impossible, since $q > 0$ by assumption. Hence, $x = -1$. In this case, $y > 0$, and hence $y \geq 1$. Again by the second equation in (8.33) and also by the Wallis–Euler recurrence relations, we have

$$q = q_n x + q_{n+1}y \geq -q_n + q_{n+1} = (a_{n+1} - 1)q_n + q_{n-1}.$$

For $n \geq 1$, we have $a_{n+1} \geq 2$ and $q_{n-1} \geq 1$. Therefore, $q > q_n$. ∎

As an easy consequence of this lemma, it follows that every convergent p_n/q_n with $n \geq 1$ of the canonical continued fraction expansion of a real number ξ must be a best approximation. To see this, observe that if $\xi = p_n/q_n$, then of course p_n/q_n is a best approximation of ξ. So assume that $\xi \neq p_n/q_n$, where $n \geq 1$, and let $p/q \neq p_n/q_n$ with $1 \leq q \leq q_n$. Since $n \geq 1$, we have $q_n < q_{n+1}$, so $1 \leq q < q_{n+1}$ as well. Therefore by Lemma 8.19,

$$|q_n\xi - p_n| < |q\xi - p|.$$

Note that we left out the $n = 0$ case on purpose; the reason is that p_0/q_0 may not be a best approximation!

Example 8.28 Consider $\sqrt{3} = 1.73205080\ldots$. The best integer approximation to $\sqrt{3}$ is 2. In Example 8.18 on p. 629 we found that $\sqrt{3} = \langle 1; \overline{1, 2} \rangle$. Thus, $p_0/q_0 = 1$, which is not a best approximation. However, $p_1/q_1 = 1 + \frac{1}{1} = 2$ is a best approximation.

Best approximation theorem

> **Theorem 8.20** *Every best approximation of a real number (rational or irrational) is a convergent of its canonical continued fraction expansion, and conversely, each of the convergents $c_1, c_2, c_3, \ldots$ is a best approximation.*

Proof Let p/q with $q > 0$ be a best approximation to a real number ξ; we must prove that p/q is a convergent. Let $1 = q_0 \le q_1 < q_2 < \cdots$ be the sequence of denominators for the convergents.

Case 1: Suppose there is a k such that $q_k \le q < q_{k+1}$. By Lemma 8.19, if it were the case that $p/q \ne p_k/q_k$, then we would have $|b\xi - a| \le |q\xi - p|$, where $a = p_k$ and $b = q_k$. Since $b \le q$, this contradicts that p/q is a best approximation to ξ. Therefore, $p/q = p_k/q_k$, so p/q is a convergent of ξ.

Case 2: There is no k such that $q_k \le q < q_{k+1}$. Then ξ must be a rational number (if ξ is irrational, then $q_k \to \infty$ as $k \to \infty$, so **Case 1** always occurs). Hence, $\xi = p_{n+1}/q_{n+1}$ for some $n = -1, 0, 1, \ldots$, and to be outside of **Case 1** we must have $q_{n+1} \le q$. If $p/q \ne p_{n+1}/q_{n+1}$, then by definition of best approximation, we would have

$$|q\xi - p| < |q_{n+1}\xi - p_{n+1}|, \quad \text{or} \quad |q\xi - p| < 0.$$

This is absurd, so $p/q = p_{n+1}/q_{n+1}$ and we're done. ∎

8.5.3 Legendre's Approximation Theorem

This theorem is named after Adrien-Marie Legendre (1752–1833).

Legendre's approximation theorem

> **Theorem 8.21** *Among two consecutive convergents p_n/q_n, p_{n+1}/q_{n+1} with $n \ge 0$ of the canonical continued fraction expansion to a real number (rational or irrational) ξ, one of them satisfies*
>
> $$\left| \xi - \frac{p}{q} \right| < \frac{1}{2q^2}. \tag{8.36}$$
>
> *Conversely, if a rational number p/q satisfies (8.36), then it is a convergent.*

Proof We begin by proving that a rational number satisfying (8.36) must be a convergent; then we show that convergents satisfy (8.36).

Step 1: Assume that p/q satisfies (8.36). To prove that it must be a convergent, we just need to show that it is a best approximation. To this end, assume that $a/b \ne p/q$ with $b > 0$ and that

$$|b\xi - a| \le |q\xi - p|; \tag{8.37}$$

we must show that $q < b$. To prove this, all we do is look at the difference $|p/q - a/b|$. First of all,

$$\left|\frac{p}{q} - \frac{a}{b}\right| = \left|\frac{pb - qa}{bq}\right| \geq \frac{1}{bq},$$

since $|pb - qa|$ is a positive integer (since $a/b \neq p/q$). Secondly,

$$\left|\frac{p}{q} - \frac{a}{b}\right| = \left|\frac{p}{q} - \xi + \xi - \frac{a}{b}\right| \leq \left|\frac{p}{q} - \xi\right| + \left|\xi - \frac{a}{b}\right| < \frac{1}{2q^2} + \frac{|b\xi - a|}{b} \quad \text{(by (8.36))}$$

$$\leq \frac{1}{2q^2} + \frac{|q\xi - p|}{b} \quad \text{(by (8.37))}$$

$$= \frac{1}{2q^2} + \frac{q}{b}\left|\xi - \frac{p}{q}\right|$$

$$< \frac{1}{2q^2} + \frac{q}{2bq^2} \quad \text{(by (8.36))}$$

$$= \frac{b + q}{2bq^2}.$$

Thus,

$$\frac{1}{bq} < \frac{b + q}{2bq^2} \quad \Longrightarrow \quad 2q < b + q \quad \Longrightarrow \quad q < b,$$

just as we wanted to show.

We now show that one of two consecutive convergents satisfies (8.36). Let p_n/q_n and p_{n+1}/q_{n+1}, $n \geq 0$, be two consecutive convergents.

Step 2: Assume first that $q_n = q_{n+1}$. Since $q_{n+1} = a_{n+1}q_n + q_{n-1}$, we see that $q_n = q_{n+1}$ if and only if $n = 0$ (because $q_{n-1} = 0$ if and only if $n = 0$) and $a_1 = 1$, in which case $q_1 = q_0 = 1$, $p_0 = a_0$, and $p_1 = a_0a_1 + 1 = a_0 + 1$. Therefore, $p_0/q_0 = a_0/1$ and $p_1/q_1 = (a_0 + 1)/1$, so we just have to show that

$$|\xi - a_0| < \frac{1}{2} \quad \text{or} \quad |\xi - (a_0 + 1)| < \frac{1}{2}.$$

But one of these must hold, because $a_0 = \lfloor \xi \rfloor$, so

$$a_0 \leq \xi < a_0 + 1.$$

Note that the special situation in which ξ is exactly halfway between a_0 and $a_0 + 1$, that is, $\xi = a_0 + 1/2 = \langle a_0; 2 \rangle$, is not possible under our current assumptions, because in this special situation, $q_1 = 2 \neq 1 = q_0$.

Step 3: Assume now that $q_n \neq q_{n+1}$. Consider two consecutive convergents c_n and c_{n+1}. We know that either

$$c_n < \xi \leq c_{n+1} \quad \text{or} \quad c_{n+1} \leq \xi < c_n,$$

depending on whether n is even or odd. For concreteness, assume that n is even; the odd case is entirely similar. Then from $c_n < \xi \leq c_{n+1}$ and the fundamental recurrence relation $c_{n+1} - c_n = 1/q_n q_{n+1}$, we see that

$$\left|\xi - c_n\right| + \left|c_{n+1} - \xi\right| = (\xi - c_n) + (c_{n+1} - \xi) = c_{n+1} - c_n = \frac{1}{q_n q_{n+1}}.$$

Recall that for real numbers x, y with $x \neq y$, we have $xy \leq (x^2 + y^2)/2$ (just work out $(x - y)^2 > 0$). Hence,

$$\frac{1}{q_n q_{n+1}} < \frac{1}{2q_n^2} + \frac{1}{2q_{n+1}^2},$$

so

$$\left|\xi - c_n\right| + \left|\xi - c_{n+1}\right| < \frac{1}{2q_n^2} + \frac{1}{2q_{n+1}^2}. \tag{8.38}$$

It follows that $\left|\xi - c_n\right| < 1/2q_n^2$ or $\left|\xi - c_{n+1}\right| < 1/2q_{n+1}^2$, for otherwise, (8.38) would fail to hold. This completes our proof.   ∎

► **Exercises 8.5**

1. (a) In this problem we find all the good approximations to $2/7$. First, to see things better, let's write down all the fractions with denominators less than 7 in an area around $2/7$, for example:

$$\frac{0}{1} < \frac{1}{6} < \frac{1}{5} < \frac{1}{4} < \frac{2}{7} < \frac{1}{3} < \frac{2}{5} < \frac{1}{2}.$$

By examining the absolute values $\left|\xi - \frac{a}{b}\right|$ for the fractions listed, show that the good approximations to $2/7$ are $0/1$, $1/2$, $1/3$, $1/4$, and of course, $2/7$.

   (b) Now let's find which of the good approximations are best *without* using the best approximation theorem. To do so, compute the absolute values

$$\left|1 \cdot \frac{2}{7} - 0\right| , \quad \left|2 \cdot \frac{2}{7} - 1\right| , \quad \left|3 \cdot \frac{2}{7} - 1\right| , \quad \left|4 \cdot \frac{2}{7} - 1\right|$$

and from these numbers, determine which of the good approximations are best.

   (c) Using a similar method, find the good and best approximations to $3/7$, $3/5$, $8/5$, and $2/9$.

2. (**Continuation of Lemma** 8.19) Let p_n/q_n, $n \geq 0$, be a convergent of the canonical continued fraction expansion of a real number ξ and let $p/q \neq p_n/q_n$ be a rational number with $q > 0$ and $1 \leq q < q_{n+1}$. Prove that

$$|q_n \xi - p_n| = |q\xi - p|$$

if and only if

$$\xi = \frac{p_{n+1}}{q_{n+1}}, \quad p = p_{n+1} - p_n, \quad \text{and} \quad q = q_{n+1} - q_n.$$

3. Prove that a real number ξ is irrational if and only if there are infinitely many rational numbers p/q satisfying

$$\left| \xi - \frac{p}{q} \right| < \frac{1}{q^2}.$$

4. In this problem we find very beautiful approximations to π.

   (a) Using the canonical continued fraction algorithm, prove that

   $$\pi^4 = 97.40909103400242\ldots = \langle 97, 2, 2, 3, 1, 16539, 1, \ldots \rangle.$$

   (Warning: If your calculator doesn't have enough decimal places of accuracy, you'll probably get a different value for 16539.)

   (b) Compute $c_4 = \frac{2143}{22}$ and therefore, $\pi \approx \left(\frac{2143}{22} \right)^{1/4}$. Note that $\pi = 3.141592653\ldots$, while $(2143/22)^{1/4} = 3.141592652$, quite accurate! This approximation is due to Srinivasa Ramanujan (1887–1920) [26, p. 160].[4] As explained in [258], we can write this approximation in **pandigital** form, that is, using all digits $0, 1, \ldots, 9$ exactly once:

   $$\pi \approx \left(\frac{2143}{22} \right)^{1/4} = \sqrt{\sqrt{0 + 3^4 + \frac{19^2}{78 - 56}}}.$$

   (c) By determining certain convergents of the continued fraction expansions of π^2, π^3, and π^5, derive the equally fascinating results:

   $$\pi \approx \sqrt{10}, \quad \left(\frac{227}{23} \right)^{1/2}, \quad 31^{1/3}, \quad \left(\frac{4930}{159} \right)^{1/3}, \quad 306^{1/5}, \quad \left(\frac{77729}{254} \right)^{1/5}.$$

   The approximation $\pi \approx \sqrt{10} = 3.162\ldots$ was known in Mesopotamia thousands of years before Christ [182]!

5. Let $\xi = a_0 + \dfrac{b_1}{a_1+} \dfrac{b_2}{a_2+} \cdots$, where $a_n \geq 1$ for $n \geq 1$ and $b_n > 0$ for all n, and $\sum_{n=1}^{\infty} \frac{a_n a_{n+1}}{b_{n+1}} = \infty$. If $c_n = a_0 + \dfrac{b_1}{a_1+} \cdots + \dfrac{b_n}{a_n}$ is the nth convergent of ξ, prove that for all $n = 0, 1, 2, \ldots$, we have $|\xi - c_{n+1}| < |\xi - c_n|$ and $|q_{n+1}\xi - p_{n+1}| < |q_n \xi - p_n|$ (cf. Theorem 8.18).

6. (Cf. [215]) (**Pythagorean triples**) Please review Problems 8 and 9 on pages 64 and 65 concerning primitive Pythagorean triples. We ask the following question:

---

[4]"An equation means nothing to me unless it expresses a thought of God." Srinivasa Ramanujan (1887–1920).

Given a right triangle, is there a primitive right triangle similar to it? The answer
is "not always," since e.g., the triangle with sides $(1, 1, \sqrt{2})$ is not similar to any
triangle with integer sides (why?). So we ask: Given a right triangle, is there a
primitive right triangle "nearly" similar to it? The answer is yes, and here's one
way to do it.

(i) Given a right triangle $\triangle$, let θ be one of its acute angles. Prove that if $\tan(\theta/2)$
is rational, then there is a primitive right triangle similar to $\triangle$. Suggestion:
If $\tan(\theta/2) = p/q$, where $p, q \in \mathbb{Z}$ have no common factors with $q > 0$,
prove that $\tan\theta = 2pq/(q^2 - p^2)$. Then recall from Problem 9 on p. 67
that

$$(x, y, z) \text{ is primitive, where } \begin{cases} x = 2pq, \ y = q^2 - p^2, \ z = p^2 + q^2, \text{ or} \\ x = pq, \ y = \frac{q^2 - p^2}{2}, \ z = \frac{p^2 + q^2}{2}, \end{cases}$$

according as p and q have opposite or the same parity.

(ii) Of course in general, $\tan(\theta/2)$ is not rational. To get around this, the idea
is that we can always approximate $\tan(\theta/2)$ by rationals so that $\tan(\theta/2)$ is
"nearly" rational; then by (i) there is a primitive right triangle "nearly" similar to $\triangle$. Of course, continued fractions are perfect instruments to approximate real numbers! Let's apply this idea to the triangle $(1, 1, \sqrt{2})$. In this
case, $\theta = 45°$. Prove that $\tan(\theta/2) = \sqrt{2} - 1$. Prove that the convergents of
the continued fraction expansion of $\sqrt{2} - 1$ are of the form $c_n = u_n/u_{n+1}$,
where $u_n = 2u_{n-1} + u_{n-2}$ ($n \geq 2$) with $u_0 = 0$, $u_1 = 1$. Finally, prove that
(x_n, y_n, z_n), where $x_n = 2u_n u_{n+1}$, $y_n = u_{n+1}^2 - u_n^2$, and $z_n = u_{n+1}^2 + u_n^2$,
forms a sequence of primitive Pythagorean triples such that $x_n/y_n \to \tan\theta$.

8.6 ★ Continued Fractions, Calendars, and Musical Scales

We now do some fun stuff with continued fractions and their applications to calendars
and pianos! In the exercises, you'll see how Christiaan Huygens (1629–1695), a
Dutch physicist, made his model of the solar system (cf. [160]).

8.6.1 Calendars

Calendar making is an amazing subject; see the free book [242] for a fascinating
look at calendars. A year, technically a **tropical year**, is the time it takes from one
vernal equinox to the next.[5] Here, an equinox is basically (there is a more technical

---

[5] A tropical year is close to the amount of time it takes the Earth to make one full revolution around
the Sun; they are about 20 minutes off.

definition) the time when night and day have the same length, and there are two of them: The *vernal equinox* occurs around March 21, the first day of spring, and the *autumnal equinox* occurs around September 23, the first day of fall. A year is approximately 365.24219 days. As you might guess, not being a whole number of days makes it quite difficult to make accurate calendars, and for this reason, the art of calendar making has been around since the beginning. Here are some approximations to a year that you might know about:

(1) 365 days, the ancient Egyptians and others.
(2) $365\frac{1}{4}$ days, Julius Caesar (100–44 B.C.), 46 B.C., giving rise to the **Julian calendar**.
(3) $365\frac{97}{400}$ days, Pope Gregory XIII (1502–1585), 1585, giving rise to the **Gregorian calendar**, the calendar that is now the most widely used calendar.

See Problem 1 for Persian calendars and their link to continued fractions. Let us analyze these calendars more thoroughly. First, the ancient calendar consisting of 365 days. Since a true year is approximately 365.24219 days, an ancient year has

0.24219 *fewer* days than a true year.

Thus, after four years, with an ancient calendar you'll lose approximately

$$4 \times 0.24219 = 0.9687 \text{ days} \approx 1 \text{ day.}$$

After 125 years, with an ancient calendar you'll lose approximately

$$125 \times 0.24219 = 30.27375 \text{ days} \approx 1 \text{ month.}$$

So, instead of having spring around March 21, you'll have it in February! After 500 years, with an ancient calendar you'll lose approximately

$$500 \times 0.24219 = 121.095 \text{ days} \approx 4 \text{ months.}$$

So, instead of having spring around March 21, you'll have it in November! As you can see, this is getting quite ridiculous.

In the Julian calendar, there is an average of $365\frac{1}{4}$ days in a Julian year. The fraction $\frac{1}{4}$ is played out as we all know: We add *one* day in February to the ancient calendar every *four* years, giving us a "leap year," that is, a year with 366 days. Thus, just as we said, a Julian calendar year gives the estimate

$$\frac{4 \times 365 + 1 \text{ days}}{4 \text{ years}} = 365\frac{1}{4} \frac{\text{days}}{\text{year}}.$$

The Julian year has

$$365.25 - 365.24219 = 0.00781 \text{ *more* days than a true year.}$$

So, for instance, after 125 years, with a Julian calendar you'll gain

$$125 \times 0.00781 = 0.97625 \text{ days} \approx 1 \text{ day.}$$

Not bad. After 500 years, with a Julian calendar you'll gain

$$500 \times 0.00781 = 3.905 \text{ days} \approx 4 \text{ days.}$$

Again, not bad! But, still, four days gained is still four days gained.

In the Gregorian calendar, there is an average of $365\frac{97}{400}$ days, that is, we add *ninety-seven* days to the ancient calendar every *four hundred* years. These extra days are added as follows: Every four years we add one extra day, a "leap year" just as in the Julian calendar. However, this gives us 100 extra days in 400 years; so to offset this, we do not have a leap year for the century marks except 400, 800, 1200, 1600, 2000, 2400, ..., multiples of 400. For example, consider the years

$$1604, 1608, \ldots, 1696, 1700, 1704, \ldots, 1796, 1800, 1804, \ldots, 1896,$$
$$1900, 1904, \ldots, 1996, 2000.$$

Each of these years is a leap year except the three years 1700, 1800, and 1900 (but note that the year 2000 was a leap year, as you can verify on your old calendar, since it is a multiple of 400). Hence, in the four hundred years from the end of 1600 to the end of 2000, we added only 97 total days, since we didn't add extra days in 1700, 1800, and 1900. So, just as we said, a Gregorian calendar gives the estimate

$$\frac{400 \times 365 + 97}{400} = 365\frac{97}{400} \frac{\text{days}}{\text{year}}.$$

Since $365\frac{97}{400} = 365.2425$, the Gregorian year has

$$365.2425 - 365.24219 = 0.00031 \text{ more days than a true year.}$$

For instance, after 500 years, with a Gregorian calendar you'll gain

$$500 \times 0.00031 = 0.155 \text{ days} \approx 0 \text{ days!}$$

Now let's link calendars with continued fractions. Here is the continued fraction expansion of the tropical year:

$$365.24219 = \langle 365; 4, 7, 1, 3, 24, 6, 2, 2 \rangle.$$

This has convergents

$$c_0 = 365 \,, \ c_1 = 365\frac{1}{4} \,, \ c_2 = 365\frac{7}{29} \,, \ c_3 = 365\frac{8}{33} \,, \ c_4 = 365\frac{31}{128} \,, \ldots.$$

Here, we see that c_0 is the ancient calendar and c_1 is the Julian calendar, but where is the Gregorian calendar? It's not on this list, but it's almost c_3, since

$$\frac{8}{33} = \frac{8}{33} \cdot \frac{12}{12} = \frac{96}{396} \approx \frac{97}{400}.$$

However, it turns out that $c_3 = 365\frac{8}{33}$ is *exactly* the average number of days in the Persian calendar introduced by the mathematician, astronomer, and poet Omar Khayyam (1048–1131)! See Problem 1 for the modern Persian calendar!

8.6.2 Musical Scales

We now move from calendars to pianos. For more on the interaction between continued fractions and pianos, see [8, 14, 64, 96, 100, 144, 211]. Let's start by giving a short lesson on music based on Euler's letter to a German princess [38] (see also [114]). When a piano wire or guitar string vibrates, it causes the air molecules around it to vibrate, and those air molecules cause neighboring molecules to vibrate, and finally, those molecules bounce against our ears, and we have the sensation of "sound." The rapidness of the vibrations, in number of vibrations per second, is called **frequency**. Let's say that we hear two notes with two different frequencies. In general, these frequencies mix together and don't produce a pleasing sound, but according to Euler, when the *ratio* of their frequencies happens to equal certain ratios of integers, then we hear a pleasant sound![6] Fascinating isn't it? We'll call the ratio of the frequencies an **interval** between the notes or the frequencies. For example, consider two notes, one with frequency f_1 and the other with frequency f_2 such that

$$\frac{f_2}{f_1} = \frac{2}{1} \iff f_2 = 2f_1 \quad \text{(octave)};$$

in other words, the interval between the first and second note is 2, which is to say, f_2 is just twice f_1. This special interval is called an **octave**. It turns out that when two notes an octave apart are played at the same time, they sound beautiful together! Another interval that corresponds to a beautiful sound is called the **fifth**, which occurs when the ratio is $3/2$:

$$\frac{f_2}{f_1} = \frac{3}{2} \iff f_2 = \frac{3}{2}f_1 \quad \text{(fifth)}.$$

Other intervals (which, remember, just refer to ratios) that have names are

---

[6]"The pleasure we obtain from music comes from counting, but counting unconsciously. Music is nothing but unconscious arithmetic." Gottfried Leibniz, in a letter to Christian Goldbach, 27 April 1712 [207].

4/3 (fourth)   9/8 (major tone)   25/24 (chromatic semitone),
5/4 (major third) 10/9 (lesser tone)   81/80 (comma of Didymus),
6/5 (minor thirds) 16/15 (diatonic semitone).

However, it is probably of universal agreement that the octave and the fifth make the prettiest sounds. Ratios such as 7/6, 8/7, 11/10, 12/11, ... don't seem to agree with our ears.

Now let's take a quick look at two facts concerning the piano. We all know what a piano keyboard looks like:

Fig. 8.6 The kth key,
starting from $k = 0$, is
labeled by its frequency f_k

Let us label the (fundamental) frequencies of the piano keys, counting both white and black, by $f_0, f_1, f_2, f_3, \ldots$ starting from the far left key on the keyboard.[7] The first fact is that keys that are twelve keys apart are exactly an octave apart! For instance, f_0 and, jumping twelve keys to the right, f_{12} are an octave apart, f_7 and f_{19} are an octave apart, etc. For this reason, a piano scale really has only twelve basic frequencies, say $f_0, \ldots, f_{11}$, since by doubling these frequencies we get the twelve frequencies above, $f_{12}, \ldots, f_{23}$, and by doubling these we get $f_{24}, \ldots, f_{35}$, etc. The second fact is that a piano is **evenly tempered**, which means that the interval between adjacent keys is constant. Let this constant be c. Then,

$$\frac{f_{n+1}}{f_n} = c \implies f_{n+1} = cf_n$$

for all n. In particular,

$$f_{n+k} = cf_{n+k-1} = c(cf_{n+k-2}) = c^2 f_{n+k-2} = \cdots = c^k f_n. \tag{8.39}$$

Since $f_{n+12} = 2f_n$ (because f_n and f_{n+12} are an octave apart), it follows that with $k = 12$, we get

$$2f_n = c^{12} f_n \implies 2 = c^{12} \implies c = 2^{1/12}.$$

Thus, the interval between adjacent keys is $2^{1/12}$.

A question that might come to mind is this: What is so special about the number twelve for a piano scale? Why not eleven or fifteen? Answer: It has to do with continued fractions! To see why, let us imagine that we have an evenly tempered piano with q basic frequencies, that is, keys that are q apart have frequencies differing

----

[7]A piano wire also gives off **overtones**, but we focus here just on the fundamental frequency. Also, some of what we say here is not quite true for the strings near the ends of the keyboard, because they don't vibrate well due to their stiffness, leading to the phenomenon called **inharmonicity**.

by an octave. Question: Which q's make the best pianos? (Note: We had better come up with $q = 12$ as one of the "best" ones!) By a very similar argument to that given above, we can see that the interval between adjacent keys is $2^{1/q}$. Now we have to ask, What makes a good piano? Well, our piano by design has octaves, but we would also like our piano to have fifths, the other beautiful interval. Let us label the keys of our piano as in Fig. 8.6. Then we would like to have a p such that the interval between every frequency f_n and f_{n+p} is a fifth, that is,

$$\frac{f_{n+p}}{f_n} = \frac{3}{2}.$$

By the formula (8.39), which we can use in the present setup as long as we put $c = 2^{1/q}$, we have $f_{n+p} = (2^{1/q})^p f_n = 2^{p/q} f_n$. Thus, we want

$$2^{p/q} = \frac{3}{2} \quad \Longrightarrow \quad \frac{p}{q} = \frac{\log(3/2)}{\log 2}.$$

This is, unfortunately, impossible, because p/q is rational, yet $\frac{\log(3/2)}{\log 2}$ is irrational (see Theorem 2.29 on p. 87 for a related result)! Thus, it is impossible for our piano (even if $q = 12$ like our everyday piano) to have a fifth. However, hope is not lost, because although our piano can never have a *perfect* fifth, it can certainly have an *approximate* fifth: We just need to find rational approximations to the irrational number $\frac{\log(3/2)}{\log 2}$. This we know how to do using continued fractions! After some work, I found that

$$\frac{\log(3/2)}{\log 2} = \langle 1, 1, 2, 2, 3, 1, \ldots \rangle,$$

which has convergents

$$0, \frac{1}{1}, \frac{1}{2}, \frac{3}{5}, \frac{7}{12}, \frac{24}{41}, \frac{31}{53}, \frac{179}{306}, \ldots.$$

Thus, $1, 2, 5, 12, 41, 53, 306, \ldots$ are the q's that make the "best" pianos. Lo and behold, we see a twelve! In particular, by the best approximation theorem (Theorem 8.20), we know that $7/12$ approximates $\frac{\log(3/2)}{\log 2}$ better than any rational number with a smaller denominator than twelve, which is to say, we cannot find a piano scale with fewer than twelve basic keys that will give a better approximation to a fifth. This is why our everyday piano has twelve keys! What about the other numbers in the list? Supposedly [144], in 40 B.C. King-Fang, a scholar of the Han dynasty, found the fraction $24/41$, although to my knowledge, there has never been an instrument built with a scale of $q = 41$. However, King-Fang also found the fraction $31/53$, and in this case, the $q = 53$ scale was advocated by Nicholas Mercator (1620–1687) circa 1650 and was actually implemented by Robert H.M. Bosanquet (1841–1912) in his instrument *Enharmonic Harmonium* [33]!

We have focused on the interval of a fifth. See Problem 2 for other intervals.

▶ **Exercises 8.6**

1. (**Persian calendars**) The official calendar (the solar Hijri calendar) in Iran and Afghanistan today has an average of $365\frac{683}{2820}$ days per year. The Persian calendar introduced by Omar Khayyam (1048–1131) (the Jalali calendar) has an average of $365\frac{8}{33}$ days per year. Khayyam amazingly calculated the year to be 365.24219858156 days. Find the continued fraction expansion of 365.24219858156, and if $\{c_n\}$ are its convergents, show that c_0 is the ancient calendar, c_1 is the Julian calendar, c_3 is the calendar introduced by Khayyam, and c_7 is the modern Persian calendar!

2. Find the q's that will make a piano with the "best" approximations to a minor third. (Just as we found the q's that will make a piano with the "best" approximations to fifth.) Do you see why many musicians, e.g., Aristoxenus, Kornerup, Ariel, Yasser, who enjoyed minor thirds, liked $q = 19$ musical scales?

3. (**A solar system model**) Christiaan Huygens (1629–1695) made a scale model of the solar system using gears. Huygens chose 206 teeth for the Saturn gear and 7 teeth for the Earth gear. Why? The answer is that in his day, it was thought that it took Saturn 29.43 years (that is, Earth years) to make it once around the sun. Find the continued fraction expansion of 29.43 and explain why Huygens chose 206 and 7. For more on the use of continued fractions to solve gear problems, see [160].

8.7   The Elementary Functions and the Irrationality of $e^{p/q}$

In this section we derive some beautiful and classical continued fraction expansions for $\coth x$, $\tanh x$, and e^x. The book [135, Section 11.7] has a very nice presentation of this material.

8.7.1   A Hypergeometric Function

For a complex number $a \neq 0, -1, -2, \ldots$, the function

$$F(a, z) := 1 + \frac{1}{a}z + \frac{1}{a(a+1)}\frac{z^2}{2!} + \frac{1}{a(a+1)(a+2)}\frac{z^3}{3!} + \cdots, \quad z \in \mathbb{C},$$

is an example of a **hypergeometric function**, and more precisely, it's called the **confluent hypergeometric limit function**. Using the ratio test, it is straightforward to check that $F(a, z)$ converges for all $z \in \mathbb{C}$. For $a \in \mathbb{C}$, the **pochhammer symbol**, introduced by Leo August Pochhammer (1841–1920), is defined by

$$(a)_n := \begin{cases} 1 & n = 0 \\ a(a+1)(a+2)\cdots(a+n-1) & n = 1, 2, 3, \ldots. \end{cases}$$

Thus, we can write the hypergeometric function in shorthand notation

$$F(a, z) = \sum_{n=0}^{\infty} \frac{1}{(a)_n} \frac{z^n}{n!}.$$

Actually, the true hypergeometric function is defined by (which we analyzed in Example 6.24 on p. 455 in the case $z = 1$)

$$F(a, b, c, z) = \sum_{n=0}^{\infty} \frac{(a)_n (b)_n}{(c)_n} \frac{z^n}{n!},$$

but we won't need this function. Many familiar functions can be written in terms of these hypergeometric functions. For instance, consider the following.

Proposition 8.22 *We have*

$$F\left(\frac{1}{2}, \frac{z^2}{4}\right) = \cosh z \quad, \quad z\, F\left(\frac{3}{2}, \frac{z^2}{4}\right) = \sinh z.$$

Proof The proofs of these identities are the same: We simply check that both sides have the same series expansions. For example, let us check the second identity; the identity for cosh is proved similarly. Observe that

$$z\, F\left(\frac{3}{2}, \frac{z^2}{4}\right) = z \cdot \sum_{n=0}^{\infty} \frac{1}{(3/2)_n} \frac{(z^2/2^2)^n}{n!}$$

$$= \sum_{n=0}^{\infty} \frac{1}{(3/2)_n} \frac{z^{2n+1}}{2^{2n}\, n!},$$

and recall that

$$\sinh z = \sum_{n=0}^{\infty} \frac{z^{2n+1}}{(2n+1)!}.$$

Thus, we just have to show that $(3/2)_n\, 2^{2n}\, n! = (2n+1)!$ for each n. Certainly this holds for $n = 0$. For $n \geq 1$, we have

$$(3/2)_n \, 2^{2n} \, n! = \frac{3}{2} \left(\frac{3}{2} + 1 \right) \left(\frac{3}{2} + 2 \right) \cdots \left(\frac{3}{2} + n - 1 \right) \cdot 2^{2n} n!$$

$$= \frac{3}{2} \cdot \frac{5}{2} \cdot \frac{7}{2} \cdots \frac{2n+1}{2} \cdot 2^n \cdot 2^n \cdot 1 \cdot 2 \cdot 3 \cdots n$$

$$= 3 \cdot 5 \cdot 7 \cdots (2n + 1) \cdot 2 \cdot 4 \cdot 6 \cdots 2n!.$$

The last quantity is, after rearrangement, $(2n + 1)!$. This completes our proof. ∎

The hypergeometric function also satisfies an interesting, and useful as we'll see in a moment, recurrence relation.

Proposition 8.23 *The following recurrence relation holds:*

$$F(a, z) = F(a + 1, z) + \frac{z}{a(a + 1)} F(a + 2, z).$$

Proof To prove this identity, we simply check that both sides have the same series expansions. We can write

$$F(a + 1, z) + \frac{z}{a(a+1)} F(a + 2, z) = \sum_{n=0}^{\infty} \frac{1}{(a+1)_n} \frac{z^n}{n!} + \sum_{n=0}^{\infty} \frac{1}{a(a+1)(a+2)_n} \frac{z^{n+1}}{n!}.$$

The constant term on the right is 1, which is also the constant term for $F(a, z)$, and for $n \geq 1$, the coefficient of z^n on the right is

$$\frac{1}{(a+1)_n \, n!} + \frac{1}{a(a+1)(a+2)_{n-1} \, (n-1)!}$$

$$= \frac{1}{(a+1)\cdots(a+n) \, n!} + \frac{1}{a(a+1)(a+2)\cdots(a+n) \, (n-1)!}$$

$$= \frac{1}{(a+1)\cdots(a+n) \, (n-1)!} \cdot \left(\frac{1}{n} + \frac{1}{a} \right)$$

$$= \frac{1}{(a+1)\cdots(a+n) \, (n-1)!} \cdot \left(\frac{a+n}{a \cdot n} \right)$$

$$= \frac{1}{a(a+1)\cdots(a+n-1) \, n(n-1)!} = \frac{1}{(a)_n \, n!}.$$

This is exactly the coefficient of z^n for $F(a, z)$. ∎

This recurrence relation easily proves the following result.

Corollary 8.24 *For real numbers $x > 0$ and $a > -1$, we have the continued fraction expansion*

$$\frac{aF(a, x)}{F(a+1, x)} = a + \frac{x}{a+1+} \frac{x}{a+2+} \frac{x}{a+3+} \frac{x}{a+4+} \frac{x}{a+5+} \cdots$$

Proof For $x > 0$ and $a > -1$, we have $F(a+1, x) > 0$ (by the defining formula for the hypergeometric function), so we can divide by $F(a+1, x)$ in Proposition 8.23, obtaining the recurrence relation

$$\frac{F(a, x)}{F(a+1, x)} = 1 + \frac{x}{a(a+1)} \frac{F(a+2, x)}{F(a+1, x)},$$

which we can write as

$$\frac{aF(a, x)}{F(a+1, x)} = a + \frac{x}{\dfrac{(a+1)F(a+1, x)}{F(a+2, x)}}.$$

Replacing a with $a + n$ with $n = 0, 1, 2, 3, \ldots$, we get

$$\frac{(a+n)F(a+n, x)}{F(a+n+1, x)} = a + n + \frac{x}{\dfrac{(a+n+1)F(a+n+1, x)}{F(a+n+2, x)}}.$$

Thus, if we put

$$\xi_n = \frac{(a+n)F(a+n, x)}{F(a+n+1, x)} \quad , \quad a_n = a + n, \; b_n = x,$$

then

$$\xi_n = a_n + \frac{b_{n+1}}{\xi_{n+1}}, \quad n = 0, 1, 2, 3, \ldots. \tag{8.40}$$

Since

$$\sum_{n=1}^{\infty} \frac{a_n a_{n+1}}{b_n} = \sum_{n=1}^{\infty} \frac{(a+n)(a+n+1)}{x} = \infty,$$

by the continued fraction convergence theorem on p. 627, we know that

$$\frac{aF(a, x)}{F(a+1, x)} = \xi_0 = a + \frac{x}{a+1+} \frac{x}{a+2+} \frac{x}{a+3+} \frac{x}{a+4+} \frac{x}{a+5+} \cdots$$

∎

8.7.2   Continued Fraction Expansion of the Hyperbolic Cotangent

The preceding corollary easily yields the following result.

Theorem 8.25 *For every real number $x \neq 0$, we have*

$$\coth x = \frac{1}{x} + \cfrac{x}{3 + \cfrac{x^2}{5 + \cfrac{x^2}{7 + \cfrac{x^2}{9 + \cdots}}}}.$$

Proof Both $\coth x$ and its proposed continued fraction expansion are odd functions of x, so we may assume $x > 0$. By the previous corollary, we know that for $a > -1$, we have

$$\frac{a F(a, x)}{F(a + 1, x)} = a + \frac{x}{a + 1 +} \; \frac{x}{a + 2 +} \; \frac{x}{a + 3 +} \; \frac{x}{a + 4 +} \; \frac{x}{a + 5 +} \cdots.$$

Since $F\left(1/2, x^2/4\right) = \cosh x$ and $x\, F\left(3/2, x^2/4\right) = \sinh x$ by Proposition 8.22, when we set $a = 1/2$ and replace x with $x^2/4$ in the previous continued fraction, we obtain

$$\frac{x \cosh x}{2 \sinh x} = \frac{x}{2} \coth x = \frac{1}{2} + \frac{x^2/4}{3/2 +} \; \frac{x^2/4}{5/2 +} \; \frac{x^2/4}{7/2 +} \; \frac{x^2/4}{9/2 +} \cdots,$$

or after multiplication by 2 and dividing by x, we get

$$\coth x = \frac{1}{x} + \frac{x/2}{3/2 +} \; \frac{x^2/4}{5/2 +} \; \frac{x^2/4}{7/2 +} \; \frac{x^2/4}{9/2 +} \cdots,$$

Finally, using the transformation rule (Theorem 8.2 on p. 598)

$$a_0 + \frac{b_1}{a_1 +} \; \frac{b_2}{a_2 +} \cdots + \frac{b_n}{a_n +} \cdots = a_0 + \frac{\rho_1 b_1}{\rho_1 a_1 +} \; \frac{\rho_1 \rho_2 b_2}{\rho_2 a_2 +} \cdots + \frac{\rho_{n-1} \rho_n b_n}{\rho_n a_n} + \cdots$$

with $\rho_n = 2$ for all n, we get

$$\coth x = \frac{1}{x} + \frac{x}{3 +} \; \frac{x^2}{5 +} \; \frac{x^2}{7 +} \; \frac{x^2}{9 +} \cdots,$$

exactly what we set out to prove.   ∎

Given x, we certainly have $0 < b_n = x^2 < 2n + 1 = a_n$ for all n sufficiently large, so by Theorem 8.15 on p. 632, it follows that when x is rational, $\coth x$ is irrational, or writing it out, for x rational,

$$\coth x = \frac{e^x + e^{-x}}{e^x - e^{-x}} = \frac{e^{2x} + 1}{e^{2x} - 1}$$

is irrational. It follows that for x rational, e^{2x} must be irrational too, for otherwise $\coth x$ would be rational, contrary to assumption. Replacing x with $x/2$ and calling this r, we get the following neat theorem.

Theorem 8.26 e^r *is irrational for every rational* r.

By the way, as Johann Lambert (1728–1777) originally did back in 1761 [35, p. 463], you can use continued fractions to prove that π is irrational; see [136, 166]. Using the cotangent expansion, we can get the continued fraction expansion for $\tanh x$. To do so, multiply the continued fraction for $\coth x$ by x:

$$x \coth x = b \ , \quad \text{where} \ \ b = 1 + \frac{x^2}{3} + \frac{x^2}{5} + \frac{x^2}{7} + \frac{x^2}{9} + \cdots$$

Thus, $\tanh x = x/b$, or replacing b with its continued fraction, we get

$$\tanh x = \cfrac{x}{1 + \cfrac{x^2}{3 + \cfrac{x^2}{5 + \cfrac{x^2}{7 + \cdots}}}}.$$

We derive one more beautiful expression that we'll need later. As before, we have

$$\coth x = \frac{e^x + e^{-x}}{e^x - e^{-x}} = \frac{e^{2x} + 1}{e^{2x} - 1} = \frac{1}{x} + \frac{x}{3} + \frac{x^2}{5} + \frac{x^2}{7} + \frac{x^2}{9} + \cdots$$

Replacing x with $1/x$, we obtain

$$\frac{e^{2/x} + 1}{e^{2/x} - 1} = x + \frac{1/x}{3} + \frac{1/x^2}{5} + \frac{1/x^2}{7} + \frac{1/x^2}{9} + \cdots$$

Finally, using the now familiar transformation rule, after a little algebra we get

$$\frac{e^{2/x} + 1}{e^{2/x} - 1} = x + \cfrac{1}{3x + \cfrac{1}{5x + \cfrac{1}{7x + \ddots}}}. \tag{8.41}$$

8.7.3   Continued Fraction Expansion of the Exponential

We can now get the famous continued fraction expansion for e^x, which was first discovered by (as you might have guessed) Euler. To start, we observe that

$$\coth(x/2) = \frac{e^{x/2} + e^{-x/2}}{e^{x/2} - e^{-x/2}} = \frac{1 + e^{-x}}{1 - e^{-x}} \implies e^{-x} = \frac{\coth(x/2) - 1}{1 + \coth(x/2)},$$

where we solved the equation on the left for e^{-x}. Thus,

$$e^{-x} = 1 - \frac{2}{1 + \coth(x/2)},$$

so taking reciprocals, we get

$$e^x = \cfrac{1}{1 - \cfrac{2}{1 + \coth(x/2)}}.$$

By Theorem 8.25, we have

$$1 + \coth(x/2) = 1 + \frac{2}{x} + \frac{x/2}{3} + \frac{x^2/4}{5} + \cdots = \frac{x + 2}{x} + \frac{x/2}{3} + \frac{x^2/4}{5} + \frac{x^2/4}{7} + \cdots,$$

so

$$e^x = \frac{1}{1} + \frac{-2}{\frac{x+2}{x}} + \frac{x/2}{3} + \frac{x^2/4}{5} + \frac{x^2/4}{7} + \cdots,$$

or using the transformation rule on p. 598,

$$\frac{b_1}{a_1} + \frac{b_2}{a_2} + \cdots + \frac{b_n}{a_n} + \cdots = \frac{\rho_1 b_1}{\rho_1 a_1} + \frac{\rho_1 \rho_2 b_2}{\rho_2 a_2} + \cdots + \frac{\rho_{n-1} \rho_n b_n}{\rho_n a_n} + \cdots,$$

with $\rho_1 = 1$, $\rho_2 = x$, and $\rho_n = 2$ for all $n \geq 3$, we get

$$e^x = \frac{1}{1} + \frac{-2x}{x+2} + \frac{x^2}{6} + \frac{x^2}{10} + \frac{x^2}{14} + \cdots.$$

This is Euler's celebrated continued fraction expansion for e^x:

Theorem 8.27 *For every real number x, we have*

$$e^x = \cfrac{1}{1 - \cfrac{2x}{x + 2 + \cfrac{x^2}{6 + \cfrac{x^2}{10 + \cfrac{x^2}{14 + \cdots}}}}}.$$

In particular, if we let $x = 1$, we obtain

$$e = \cfrac{1}{1 - \cfrac{2}{3 + \cfrac{1}{6 + \cfrac{1}{10 + \cfrac{1}{14 + \cdots}}}}}.$$

Although beautiful, we can get an even more beautiful continued fraction expansion for e, which is a *simple* continued fraction.

8.7.4 The Simple Continued Fraction Expansion of e

If we expand the decimal number 2.718281828 into a simple continued fraction, we get (see Problem 2 on p. 634)

$$2.718281828 = \langle 2; 1, 2, 1, 1, 4, 1, 1, 6, 1, 1, 8, 1 \rangle.$$

For this reason, we should be able to conjecture that e is the continued fraction

$$e = \langle 2; 1, 2, 1, 1, 4, 1, 1, 6, 1, 1, 8, 1, 1, \ldots \rangle. \tag{8.42}$$

This is true, and it was proved by (as you might have guessed again) Euler. Here,

$$a_0 = 2 \,, \ a_1 = 1 \,, \ a_2 = 2 \,, \ a_3 = 1 \,, \ a_4 = 1 \,, \ a_5 = 4 \,, \ a_6 = 1 \,, \ a_7 = 1,$$

and in general, for all $n \in \mathbb{N}$, $a_{3n-1} = 2n$ and $a_{3n} = a_{3n+1} = 1$. Since

$$2 = 1 + \cfrac{1}{0 + \cfrac{1}{1}},$$

we can write (8.42) in a prettier way that shows the full pattern:

$$\boxed{e = \langle 1; 0, 1, 1, 2, 1, 1, 4, 1, 1, 6, 1, 1, 8, 1, 1, \ldots \rangle,}$$

or in the expanded form

$$e = 1 + \cfrac{1}{0 + \cfrac{1}{1 + \cfrac{1}{1 + \cfrac{1}{2 + \cfrac{1}{1 + \cfrac{1}{1 + \cfrac{1}{4 + \cdots}}}}}}}.$$

To prove this incredible formula, denote the convergents of the right-hand continued fraction in (8.42) by r_k/s_k. Since we have such simple relations $a_{3n-1} = 2n$ and $a_{3n} = a_{3n+1} = 1$ for all $n \in \mathbb{N}$, one might think that it is quite easy to compute formulas for the convergents; this is indeed the case.

Lemma 8.28 *For all $n \geq 2$, we have*

$$r_{3n+1} = 2(2n + 1)r_{3(n-1)+1} + r_{3(n-2)+1}$$
$$s_{3n+1} = 2(2n + 1)s_{3(n-1)+1} + s_{3(n-2)+1}.$$

Proof Both formulas are proved in similar ways, so we shall focus on the formula for r_{3n+1}. First, we apply our Wallis–Euler recursive formulas:

$$r_{3n+1} = r_{3n} + r_{3n-1} = \left(r_{3n-1} + r_{3n-2}\right) + r_{3n-1} = 2r_{3n-1} + r_{3n-2}.$$

We again apply the Wallis–Euler recursive formula to r_{3n-1}:

$$r_{3n+1} = 2\left(2nr_{3n-2} + r_{3n-3}\right) + r_{3n-2}$$

$$= \left(2(2n) + 1\right)r_{3n-2} + 2r_{3n-3}$$

$$= \left(2(2n) + 1\right)r_{3n-2} + r_{3n-3} + r_{3n-3}. \tag{8.43}$$

Again applying the Wallis–Euler recursive formula to the last term, we get

$$r_{3n+1} = \left(2(2n) + 1\right)r_{3n-2} + r_{3n-3} + \left(r_{3n-4} + r_{3n-5}\right)$$

$$= \left(2(2n) + 1\right)r_{3n-2} + \left(r_{3n-3} + r_{3n-4}\right) + r_{3n-5}.$$

Since $r_{3n-2} = r_{3n-3} + r_{3n-4}$ by our Wallis–Euler recursive formulas, we finally get

$$r_{3n+1} = \left(2(2n) + 1\right)r_{3n-2} + r_{3n-2} + r_{3n-5}$$

$$= \left(2(2n) + 2\right)r_{3n-2} + r_{3n-5}$$

$$= 2\left((2n) + 1\right)r_{3(n-1)+1} + r_{3(n-2)+1}.$$

∎

Now putting $x = 1$ in (8.41), let us look at

$$\frac{e+1}{e-1} = \langle 2; 6, 10, 14, 18, \ldots \rangle$$

that is, if the right-hand side is $\langle \alpha_0; \alpha_1, \ldots \rangle$, then $\alpha_n = 2(2n + 1)$ for all $n = 0, 1, 2, \ldots$. If p_n/q_n are the convergents of this continued fraction, then we see that

$$p_n = 2(2n + 1)p_{n-1} + p_{n-2} \quad \text{and} \quad q_n = 2(2n + 1)q_{n-1} + q_{n-2}, \tag{8.44}$$

which are similar to the relations in our lemma! Thus, it is not one bit surprising that the r_{3n+1} and s_{3n+1} are related to the p_n and q_n. The exact relation is given in the following lemma.

Lemma 8.29 *For all $n = 0, 1, 2, \ldots$, we have*

$$r_{3n+1} = p_n + q_n \quad \text{and} \quad s_{3n+1} = p_n - q_n.$$

Proof As with the previous lemma, we shall prove only the formula for r_{3n+1}. The proof is really easy: For $n = 0, 1, 2, \ldots$, define

$$u_n = r_{3n+1} - p_n - q_n.$$

Using the a_n and α_n, it easy to check that $u_0 = 0$ and also $u_1 = 0$. (To find u_1 you need r_4, and to compute this, it's best to use the formula (8.43) with $n = 1$; if you do so, you'll get $r_4 = 19$.) By Lemma 8.28 and Eq. (8.44), it follows that for $n \geq 2$, we have

$$u_n = 2(2n + 1)u_{n-1} + u_{n-2}.$$

Putting $n = 2$ and using that $u_0 = u_1 = 0$, we get $u_2 = 0$. Then putting $n = 3$, we get $u_3 = 0$. In fact, by induction, all the u_n are zero! This proves our result.   ∎

Finally, we can now prove the continued fraction expansion for e:

$$\langle 2; 1, 1, 4, 1, 1, \ldots \rangle = \lim \frac{r_n}{s_n} = \lim \frac{r_{3n+1}}{s_{3n+1}} = \lim \frac{p_n + q_n}{p_n - q_n}$$

$$= \lim \frac{p_n/q_n + 1}{p_n/q_n - 1} = \frac{\frac{e+1}{e-1} + 1}{\frac{e+1}{e-1} - 1} = \frac{\frac{e}{e-1}}{\frac{1}{e-1}} = e.$$

See [185] for another proof of this formula based on a proof by Charles Hermite (1822–1901). In the problems, we derive, along with other things, the following beautiful continued fraction for $\cot x$:

$$x \cot x = 1 + \cfrac{x^2}{3 - \cfrac{x^2}{5 - \cfrac{x^2}{7 - \cfrac{x^2}{9 - \cdots}}}}. \tag{8.45}$$

From this continued fraction, we can derive the beautiful companion result for $\tan x$:

$$\tan x = \cfrac{x}{1 - \cfrac{x^2}{3 - \cfrac{x^2}{5 - \cfrac{x^2}{7 - \cdots}}}}.$$

▶ **Exercises 8.7**

1. For all $n = 1, 2, \ldots$, let $a_n > 0$, $b_n \geq 0$, with $a_n \geq b_n + 1$. We shall prove that the following continued fraction converges:

$$\frac{b_1}{a_1+} \, \frac{-b_2}{a_2+} \, \frac{-b_3}{a_3+} \, \frac{-b_4}{a_4+} \, \ldots \tag{8.46}$$

Note that for the continued fraction we are studying, $a_0 = 0$. The Wallis–Euler recurrence relations (8.17) and (8.18) in this situation are (just replace b_n with $-b_n$)

$$p_n = a_n p_{n-1} - b_n p_{n-2} \, , \quad q_n = a_n q_{n-1} - b_n q_{n-2}, \quad n = 2, 3, 4, \ldots$$
$$p_0 = 0 \, , \quad p_1 = b_1 \, , \quad q_0 = 1 \, , \quad q_1 = a_1.$$

(i) Prove (via induction, for instance) that $q_n \geq q_{n-1}$ for all $n = 1, 2, \ldots$. In particular, since $q_0 = 1$, we have $q_n \geq 1$ for all n, so the convergents $c_n = p_n/q_n$ of (8.46) are defined.

(ii) Verify that $q_1 - p_1 \geq 1 = q_0 - p_0$. Now prove by induction that $q_n - p_n \geq q_{n-1} - p_{n-1}$ for all $n = 1, 2, \ldots$. In particular, since $q_0 - p_0 = 1$, we have $q_n - p_n \geq 1$ for all n. Dividing by q_n, we see that $0 \leq c_n \leq 1$ for all $n = 1, 2, \ldots$.

(iii) Using the fundamental recurrence relations for $c_n - c_{n-1}$, prove that $c_n - c_{n-1} \geq 0$ for all $n = 1, 2, \ldots$. Combining this with (ii) shows that $0 \leq c_1 \leq c_2 \leq c_3 \leq \cdots \leq 1$; that is, $\{c_n\}$ is a bounded monotone sequence and hence converges. Thus, the continued fraction (8.46) converges.

2. For all $n = 1, 2, \ldots$, let $a_n > 0$, $b_n \geq 0$, with $a_n \geq b_n + 1$. From the previous problem, it follows that given $a_0 \in \mathbb{R}$, the continued fraction $a_0 - \dfrac{b_1}{a_1+} \, \dfrac{-b_2}{a_2+} \, \dfrac{-b_3}{a_3}$ $+ \dfrac{-b_4}{a_4+} \, \ldots$ converges. Prove the following variant of the continued fraction convergence theorem (Theorem 8.14 on p. 628): Let $\xi_0, \xi_1, \xi_2, \ldots$ be a sequence of real numbers with $\xi_n > 0$ for $n \geq 1$ and suppose that these numbers are related by

$$\xi_n = a_n + \frac{-b_{n+1}}{\xi_{n+1}} \, , \quad n = 0, 1, 2, \ldots.$$

Then ξ_0 is equal to the continued fraction

$$\xi_0 = a_0 - \frac{b_1}{a_1+} \, \frac{-b_2}{a_2+} \, \frac{-b_3}{a_3+} \, \frac{-b_4}{a_4+} \, \frac{-b_5}{a_5+} \, \ldots.$$

Suggestion: Follow closely the proof of Theorem 8.14.

3. We are now ready to derive the beautiful cotangent continued fraction (8.45).

(i) Let $a > 0$. Then as we derived the identity (8.40) found in Theorem 8.25, prove that if we define

$$\eta_n = \frac{(a+n)F(a+n, -x)}{F(a+n+1, -x)} \, , \quad a_n = a + n, \, b_n = x, \quad n = 0, 1, 2, \ldots,$$

then

$$\eta_n = a_n + \frac{-b_{n+1}}{\eta_{n+1}}, \quad n = 0, 1, 2, 3, \ldots.$$

(ii) Using Problem 2, prove that for $x \geq 0$ sufficiently small, we have

$$\frac{a F(a, -x)}{F(a+1, -x)} = a - \frac{x}{a+1+} \frac{-x}{a+2+} \frac{-x}{a+3+} \frac{-x}{a+4+} \frac{-x}{a+5+} \cdots.$$
(8.47)

(iii) Prove that (cf. the proof of Proposition 8.22)

$$F\left(\frac{1}{2}, -\frac{x^2}{4}\right) = \cos x \quad , \quad x F\left(\frac{3}{2}, -\frac{x^2}{4}\right) = \sin x.$$

(iv) Now put $a = 1/2$ and replace x with $x^2/4$ in (8.47) to derive, for $x \neq 0$ sufficiently small, the beautiful cotangent expansion (8.45). Finally, relax and contemplate this fine formula!

4. **(Irrationality of** $\log r$**)** Using Theorem 8.26, prove that if $r > 0$ is rational with $r \neq 1$, then $\log r$ is irrational. In particular, one of our favorite constants, $\log 2$, is irrational.

8.8   Quadratic Irrationals and Periodic Continued Fractions

We already know (p. 226 in Section 3.8) that a real number has a periodic decimal expansion if and only if the number is rational. One can ask the same thing about continued fractions: What types of real numbers have periodic simple continued fractions? The answer, as you will see in this section, is the real numbers called quadratic irrationals.

8.8.1   Periodic Continued Fractions

The object of this section is to characterize continued fractions that "repeat."

Example 8.29 We have already encountered the beautiful continued fraction

$$\frac{1 + \sqrt{5}}{2} = \langle 1; 1, 1, 1, 1, 1, 1, 1, 1, \ldots \rangle.$$

We usually write the right-hand side as $\langle \overline{1} \rangle$ to emphasize that the 1 repeats.

Example 8.30 Another continued fraction that repeats is

$$\sqrt{8} = \langle 2; 1, 4, 1, 4, 1, 4, 1, 4, \ldots \rangle,$$

where we have an infinite repeating block of $1, 4$. We usually write the right-hand side as $\sqrt{8} = \langle 2; \overline{1, 4} \rangle$.

Example 8.31 Yet one more continued fraction that repeats is

$$\sqrt{19} = \langle 4; 2, 1, 3, 1, 2, 8, 2, 1, 3, 1, 2, 8, \ldots \rangle,$$

where we have an infinite repeating block of $2, 1, 3, 1, 2, 8$. We usually write the right-hand side as $\sqrt{19} = \langle 4; \overline{2, 1, 3, 1, 2, 8} \rangle$.

Notice that the above repeating continued fractions are continued fractions for expressions with square roots.

Example 8.32 Consider now the expression

$$\xi = \langle 3; 2, 1, 2, 1, 2, 1, 2, 1, \ldots \rangle = \langle 3; \overline{2, 1} \rangle.$$

Let $\eta = \langle 2; 1, 2, 1, 2, 1, 2, \ldots \rangle$. Then $\xi = 3 + \frac{1}{\eta}$, and

$$\eta = 2 + \cfrac{1}{1 + \cfrac{1}{2 + \cfrac{1}{1 + \cdots}}} \quad \Longrightarrow \quad \eta = 2 + \cfrac{1}{1 + \cfrac{1}{\eta}}.$$

Solving for η, we get a quadratic formula, and solving it, we find that $\eta = 1 + \sqrt{3}$. Hence,

$$\xi = 3 + \frac{1}{\eta} = 3 + \frac{1}{1 + \sqrt{3}} = 3 + \frac{\sqrt{3} - 1}{2} = \frac{5 + \sqrt{3}}{6},$$

yet another square root expression.

Consider the infinite repeating simple continued fraction

$$\xi = \langle a_0; a_1, \ldots, a_{\ell-1}, b_0, b_1, \ldots, b_{m-1}, b_0, b_1, \ldots, b_{m-1}, b_0, b_1, \ldots, b_{m-1}, \ldots \rangle \tag{8.48}$$

$$= \langle a_0; a_1, \ldots, a_{\ell-1}, \overline{b_0, b_1, \ldots, b_{m-1}} \rangle,$$

where the bar denotes that the block of numbers $b_0, b_1, \ldots, b_{m-1}$ repeats forever. Such a continued fraction is said to be **periodic**. When writing a continued fraction in this way, we assume that there is no shorter repeating block and that the repeating block cannot start at an earlier position. For example, we would *never* write

$$\langle 2; 1, 2, 4, 3, 4, 3, 4, 3, 4, \ldots \rangle \quad \text{as} \quad \langle 2; 1, 2, 4, \overline{3, 4, 3, 4} \rangle;$$

we simply write it as $\langle 2; 1, 2, \overline{4, 3} \rangle$. The integer m is called the **period** of the simple continued fraction. An equivalent way to define a periodic continued fraction is as an infinite simple continued fraction $\xi = \langle a_0; a_1, a_2, \ldots \rangle$ such that for some m and ℓ, we have

$$a_n = a_{m+n} \quad \text{for all } n = \ell, \ell + 1, \ell + 2, \ldots. \tag{8.49}$$

The examples above suggest that infinite periodic simple continued fractions are intimately related to expressions with square roots; in fact, these expressions are called quadratic irrationals, as we shall see in a moment.

8.8.2 Quadratic Irrationals

A **quadratic irrational** is, exactly as its name suggests, an irrational real number that is a solution of a quadratic equation with integer coefficients. Using the quadratic equation, we leave you to show that a quadratic irrational ξ can be written in the form

$$\xi = r + s\sqrt{b}, \tag{8.50}$$

where r, s are rational numbers and $b > 0$ is an integer that is not a perfect square (for if b were a perfect square, then $\sqrt{b}$ would be an integer, so the right-hand side of ξ would be rational, contradicting that ξ is irrational). Conversely, given *any* real number of the form (8.50), one can check that ξ is a root of the equation

$$x^2 - 2r x + (r^2 - s^2 b) = 0.$$

Multiplying both sides of this equation by a common denominator of the rational numbers $2r$ and $r^2 - s^2 b$, we can make the polynomial on the left have integer coefficients. Thus, a real number is a quadratic irrational if and only if it is of the form (8.50). As we shall see in Theorem 8.30 below, it would be helpful to write quadratic irrationals in a way we now explain. Let ξ take the form in (8.50) with $r = m/n$ and $s = p/q$, where $n, q > 0$. Assuming that $p \geq 0$, with the help of some mathematical gymnastics, we see that

$$\xi = \frac{m}{n} + \frac{p\sqrt{b}}{q} = \frac{mq + np\sqrt{b}}{nq} = \frac{mq + \sqrt{bn^2 p^2}}{nq} = \frac{mnq^2 + \sqrt{bn^4 p^2 q^2}}{n^2 q^2}.$$

Notice that if we set $\alpha = mnq^2$, $\beta = n^2 q^2$, and $d = bn^4 p^2 q^2$, then $d - \alpha^2 = bn^4 p^2 q^2 - m^2 n^2 q^4 = (bn^2 p^2 - m^2 q^2)(n^2 q^2)$ is divisible by $\beta = n^2 q^2$. Therefore, we can write every quadratic irrational in the form

$$\xi = \frac{\alpha + \sqrt{d}}{\beta}, \quad \alpha, \beta, d \in \mathbb{Z}, d > 0 \text{ is not a perfect square, and } \beta \mid (d - \alpha^2).$$

A similar argument shows that ξ has this same form in case $p < 0$. Using this expression as the starting point, we prove the following nice theorem, which gives formulas for the convergents of the continued fraction expansion of ξ.

Theorem 8.30 *Let $\xi = \frac{\alpha + \sqrt{d}}{\beta}$ be a quadratic irrational with complete quotients $\{\xi_n\}$ (with $\xi_0 = \xi$) and partial quotients $\{a_n\}$, where $a_n = \lfloor \xi_n \rfloor$. Then we can write*

$$\xi_n = \frac{\alpha_n + \sqrt{d}}{\beta_n},$$

where α_n and β_n are integers with $\beta_n \neq 0$, defined by the recursive sequences

$$\alpha_0 = \alpha, \quad \beta_0 = \beta, \quad \alpha_{n+1} = a_n \beta_n - \alpha_n, \quad \beta_{n+1} = \frac{d - \alpha_{n+1}^2}{\beta_n};$$

moreover, $\beta_n \mid (d - \alpha_n^2)$ for all n.

Proof We first show that all the α_n and β_n defined above are integers with β_n never zero and $\beta_n \mid (d - \alpha_n^2)$. This is automatic with $n = 0$. Assume that this is true for n. Then $\alpha_{n+1} = a_n \beta_n - \alpha_n$ is an integer. To see that β_{n+1} is also an integer, observe that

$$\beta_{n+1} = \frac{d - \alpha_{n+1}^2}{\beta_n} = \frac{d - (a_n \beta_n - \alpha_n)^2}{\beta_n} = \frac{d - a_n^2 \beta_n^2 + 2a_n \beta_n \alpha_n - \alpha_n^2}{\beta_n}$$

$$= \frac{d - \alpha_n^2}{\beta_n} + 2a_n \alpha_n - a_n^2 \beta_n.$$

By the induction hypothesis, $(d - \alpha_n^2)/\beta_n$ is an integer, and so is $2a_n \alpha_n - a_n^2 \beta_n$. Thus, β_{n+1} is an integer too. Moreover, $\beta_{n+1} \neq 0$, because if $\beta_{n+1} = 0$, then we must have $d - \alpha_{n+1}^2 = 0$, which shows that d is a perfect square, contrary to our condition on d. Finally,

$$\beta_{n+1} = \frac{d - \alpha_{n+1}^2}{\beta_n} \implies \beta_n = \frac{d - \alpha_{n+1}^2}{\beta_{n+1}} \implies \beta_{n+1} \mid (d - \alpha_{n+1}^2),$$

since β_n is an integer.

Lastly, it remains to prove that the $(\alpha_n + \sqrt{d})/\beta_n$ are the complete quotients of ξ. For each n let's put $\eta_n = (\alpha_n + \sqrt{d})/\beta_n$; we must show that $\eta_n = \xi_n$ for each n. Note that $\eta_0 = \xi = \xi_0$. For $n \geq 1$, observe that

$$\eta_n - a_n = \frac{\alpha_n + \sqrt{d}}{\beta_n} - \frac{\alpha_{n+1} + \alpha_n}{\beta_n} = \frac{\sqrt{d} - \alpha_{n+1}}{\beta_n},$$

where in the middle equality we solved $\alpha_{n+1} = a_n \beta_n - \alpha_n$ for a_n. Rationalizing and using the definition of β_{n+1} and ξ_{n+1}, we obtain

$$\eta_n - a_n = \frac{d - \alpha_{n+1}^2}{\beta_n(\sqrt{d} + \alpha_{n+1})} = \frac{\beta_{n+1}}{\sqrt{d} + \alpha_{n+1}} = \frac{1}{\eta_{n+1}} \implies \eta_n = a_n + \frac{1}{\eta_{n+1}}.$$

Using this formula plus induction on $n = 0, 1, 2, \ldots$ (recalling that $\eta_0 = \xi_0$) shows that $\eta_n = \xi_n$ for all n.   ∎

8.8.3   Quadratic Irrationals and Periodic Continued Fractions

After one preliminary result, we shall prove that an infinite simple continued fraction is a quadratic irrational if and only if it is periodic. Define

$$\mathbb{Z}[\sqrt{d}] = \{a + b\sqrt{d} \, ; \, a, b \in \mathbb{Z}\}$$

and

$$\mathbb{Q}[\sqrt{d}] = \{a + b\sqrt{d} \, ; \, a, b \in \mathbb{Q}\}.$$

Given $\xi = a + b\sqrt{d}$ in either $\mathbb{Z}[\sqrt{d}]$ or $\mathbb{Q}[\sqrt{d}]$, we define its **conjugate** by

$$\overline{\xi} = a - b\sqrt{d}.$$

> **Lemma 8.31** $\mathbb{Z}[\sqrt{d}]$ *is a commutative ring and* $\mathbb{Q}[\sqrt{d}]$ *is a field, and conjugation preserves the algebraic properties; for example, if* $\alpha, \beta \in \mathbb{Q}[\sqrt{d}]$, *then*
>
> $$\overline{\alpha \pm \beta} = \overline{\alpha} \pm \overline{\beta}, \quad \overline{\alpha \cdot \beta} = \overline{\alpha} \cdot \overline{\beta}, \quad \text{and} \quad \overline{\alpha/\beta} = \overline{\alpha}/\overline{\beta}.$$

Proof To prove that $\mathbb{Z}[\sqrt{d}]$ is a commutative ring we just need to prove that it has the same algebraic properties as the integers in that $\mathbb{Z}[\sqrt{d}]$ is closed under addition, subtraction, and multiplication; for more on this definition see our discussion on p. 48 in Section 2.3.1. For example, to see that $\mathbb{Z}[\sqrt{d}]$ is closed under multiplication, let $\alpha = a + b\sqrt{d}$ and $\beta = a' + b'\sqrt{d}$ be elements of $\mathbb{Z}[\sqrt{d}]$; then

$$\alpha\beta = (a + b\sqrt{d})(a' + b'\sqrt{d}) = aa' + bb'd + (ab' + a'b)\sqrt{d}, \tag{8.51}$$

which is also in $\mathbb{Z}[\sqrt{d}]$. Similarly, one can show that $\mathbb{Z}[\sqrt{d}]$ satisfies all the other properties of a commutative ring.

To prove that $\mathbb{Q}[\sqrt{d}]$ is a field, we need to prove that it has the same algebraic properties as the rational numbers in that $\mathbb{Q}[\sqrt{d}]$ is closed under addition, multiplication, subtraction, and division (by nonzero elements); for more on this definition, see our discussion on p. 75 in Section 2.6.1. For example, to see that $\mathbb{Q}[\sqrt{d}]$ is closed under taking reciprocals, observe that if $\alpha = a + b\sqrt{d} \in \mathbb{Q}[\sqrt{d}]$ is not zero, then

$$\frac{1}{\alpha} = \frac{1}{a + b\sqrt{d}} \cdot \frac{a - b\sqrt{d}}{a - b\sqrt{d}} = \frac{a - b\sqrt{d}}{a^2 - b^2 d} = \frac{a}{a^2 - b^2 d} - \frac{b}{a^2 - b^2 d}\sqrt{d}.$$

Note that $a^2 - b^2 d \neq 0$, since being zero would imply that $\sqrt{d} = a/b$, a rational number, which by assumption is false. Similarly, one can show that $\mathbb{Q}[\sqrt{d}]$ satisfies all the other properties of a field.

Finally, we need to prove that conjugation preserves the algebraic properties. For example, let's prove the equality $\overline{\alpha \cdot \beta} = \overline{\alpha} \cdot \overline{\beta}$, leaving the other properties to you. If $\alpha = a + b\sqrt{d}$ and $\beta = a' + b'\sqrt{d}$, then according to (8.51), we have

$$\overline{\alpha\beta} = aa' + bb'd - (ab' + a'b)\sqrt{d}.$$

On the other hand,

$$\overline{\alpha} \cdot \overline{\beta} = (a - b\sqrt{d})(a' - b'\sqrt{d}) = aa' + bb'd - (ab' + a'b)\sqrt{d},$$

which equals $\overline{\alpha\beta}$. ∎

The following theorem is named in honor or Joseph-Louis Lagrange (1736–1813).

Lagrange's theorem

Theorem 8.32 *An infinite simple continued fraction is a quadratic irrational if and only if it is periodic.*

Proof We first prove the "if" part, then the "only if" part.

Step 1: Let $\xi = \langle a_0; a_1, \ldots, a_{\ell-1}, \overline{b_0, \ldots, b_m} \rangle$ be periodic and let η be the repeating block in ξ:

$$\eta = \langle b_0; b_1, \ldots, b_m, b_0, b_1, \ldots, b_m, b_0, b_1, \ldots, b_m, \ldots \rangle = \langle b_0; b_1, \ldots, b_m, \eta \rangle.$$

Thus, $\xi = \langle a_0, a_1, \ldots, a_{\ell-1}, \eta \rangle$. The idea is to prove that η is a quadratic irrational, then deduce that ξ must be one as well. To prove that η is a quadratic irrational, we use Theorem 8.5 on p. 612 to write

$$\eta = \frac{\eta s_{m-1} + s_{m-2}}{\eta t_{m-1} + t_{m-2}},$$

where s_n/t_n are the convergents for η. Multiplying both sides by $\eta t_{m-1} + t_{m-2}$, we see that

$$\eta^2 t_{m-1} + \eta t_{m-2} = \eta s_{m-1} + s_{m-2} \implies a\,\eta^2 + b\,\eta + c = 0,$$

where $a = t_{m-1}, b = t_{m-2} - s_{m-1}$, and $c = -s_{m-2}$. Hence, η is a quadratic irrational. Now using that $\xi = \langle a_0, a_1, \ldots, a_{\ell-1}, \eta \rangle$ and again using Theorem 8.5 on p. 612, we obtain

$$\xi = \frac{\eta p_{m-1} + p_{m-2}}{\eta q_{m-1} + q_{m-2}},$$

where p_n/q_n are the convergents for ξ. Since η is a quadratic irrational, it follows that ξ is a quadratic irrational, since $\mathbb{Q}[\sqrt{d}]$ is a field from Theorem 8.31. Thus, we have proved that periodic simple continued fractions are quadratic irrationals.

Step 2: Now let $\xi = \langle a_0; a_1, a_2, \ldots \rangle$ be a quadratic irrational; we shall prove that its continued fraction expansion is periodic. The main idea is to show that the integers α_n and β_n of the complete quotients of ξ found in Theorem 8.30 are bounded, then invoke the pigeonhole principle. To this end, let ξ_n be the nth complete quotient of ξ. Then we can write $\xi = \langle a_0; a_1, a_2, \ldots, a_{n-1}, \xi_n \rangle$, so by Theorem 8.5 we have

$$\xi = \frac{\xi_n p_{n-1} + p_{n-2}}{\xi_n q_{n-1} + q_{n-2}}.$$

Solving for ξ_n, after a little algebra, we find that

$$\xi_n = -\frac{q_{n-2}}{q_{n-1}} \left(\frac{\xi - c_{n-2}}{\xi - c_{n-1}} \right).$$

By our lemma, conjugation preserves the algebraic operations, so

$$\bar{\xi}_n = -\frac{q_{n-2}}{q_{n-1}} \left(\frac{\bar{\xi} - c_{n-2}}{\bar{\xi} - c_{n-1}} \right). \tag{8.52}$$

If $\xi = (\alpha + \sqrt{d})/\beta$, then $\bar{\xi} - \xi = 2\sqrt{d}/\beta \neq 0$. Therefore, since $c_k \to \xi$ as $k \to \infty$, it follows that as $n \to \infty$,

$$\frac{\bar{\xi} - c_{n-2}}{\bar{\xi} - c_{n-1}} \to \frac{\bar{\xi} - \xi}{\bar{\xi} - \xi} = 1.$$

In particular, there is an $N \in \mathbb{N}$ such that for $n > N$, $(\bar{\xi} - c_{n-2})/(\bar{\xi} - c_{n-1}) > 0$. Thus, since $q_k > 0$ for $k \geq 0$, according to (8.52), for $n > N$ we have $\bar{\xi}_n < 0$. Writing $\xi_n = (\alpha_n + \sqrt{d})/\beta_n$ as shown in Theorem 8.30, we have

$$\xi_n - \bar{\xi}_n = 2\frac{\sqrt{d}}{\beta_n}.$$

Since $\xi_n > 0$ for $n \geq 1$ and since $-\overline{\xi}_n > 0$ for $n > N$, it follows that for $n > N$, we have $\beta_n > 0$. Now solving the identity $\beta_{n+1} = \frac{d - \alpha_{n+1}^2}{\beta_n}$ in Theorem 8.30 for d, we see that

$$\beta_n \beta_{n+1} + \alpha_{n+1}^2 = d.$$

For $n > N$, both β_n and β_{n+1} are positive; hence for such n we must have $0 < \beta_n \leq d$ and $0 \leq |\alpha_n| \leq d$. (For if either β_n or $|\alpha_n|$ were greater than d, then $\beta_n \beta_{n+1} + \alpha_{n+1}^2$ would be strictly larger than d, an impossibility, since the sum is supposed to equal d.) In particular, if A is the finite set

$$A = \{(\ell, m) \in \mathbb{Z} \times \mathbb{Z} \,;\, -d \leq \ell \leq d \,,\, 1 \leq m \leq d\},$$

then for the infinitely many $n > N$, the pair (α_n, β_n) is in the finite set A. By the pigeonhole principle, there must be distinct $j, k > N$ such that $(\alpha_j, \beta_j) = (\alpha_k, \beta_k)$. Assume that $j > k$ and let $m = j - k$. Then $j = m + k$, so

$$\alpha_k = \alpha_{m+k} \quad \text{and} \quad \beta_k = \beta_{k+m}.$$

Since $a_k = \lfloor \xi_k \rfloor$ and $a_{m+k} = \lfloor \xi_{m+k} \rfloor$, by Theorem 8.30 we have

$$\xi_k = \frac{\alpha_k + \sqrt{d}}{\beta_k} = \frac{\alpha_{m+k} + \sqrt{d}}{\beta_{m+k}} = \xi_{m+k} \quad \implies \quad a_k = \lfloor \xi_k \rfloor = \lfloor \xi_{m+k} \rfloor = a_{m+k}.$$

By our formulas for α_{k+1} and β_{k+1} from Theorem 8.30, we see that

$$\alpha_{k+1} = a_k \beta_k - \alpha_k = a_{m+k} \beta_{m+k} - \alpha_{m+k} = \alpha_{m+k+1},$$

and

$$\beta_{k+1} = \frac{d - \alpha_{k+1}^2}{\beta_k} = \frac{d - \alpha_{m+k+1}^2}{\beta_{m+k}} = \beta_{m+k+1}.$$

Thus,

$$\xi_{k+1} = \frac{\alpha_{k+1} + \sqrt{d}}{\beta_{k+1}} = \frac{\alpha_{m+k+1} + \sqrt{d}}{\beta_{m+k+1}} = \xi_{m+k+1}$$

$$\implies \quad a_{k+1} = \lfloor \xi_{k+1} \rfloor = \lfloor \xi_{m+k+1} \rfloor = a_{m+k+1}.$$

Continuing this process by induction shows that $a_n = a_{m+n}$ for all $n = k, k+1, k+2, k+3, \ldots$. Thus, by the definition of periodicity in (8.49), we see that ξ has a periodic simple continued fraction. ■

A periodic simple continued fraction is called **purely periodic** if it is of the form $\xi = \langle a_0; a_1, \ldots, a_{m-1} \rangle$.

Example 8.33 The simplest example of such a fraction is the golden ratio:

$$\Phi = \frac{1+\sqrt{5}}{2} = \langle \overline{1} \rangle = \langle 1; 1, 1, 1, 1, 1, \ldots \rangle.$$

Observe that Φ has the following properties:

$$\Phi > 1 \quad \text{and} \quad \overline{\Phi} = \frac{1-\sqrt{5}}{2} = -0.618\ldots \quad \Longrightarrow \quad \Phi > 1 \quad \text{and} \quad -1 < \overline{\Phi} < 0.$$

In the following theorem, Évariste Galois's.[8] (1811–1832) first publication (at the age of 17), we characterize purely periodic expansions as those quadratic irrationals having the same properties as Φ. (Don't believe everything you read about the legendary Galois; see [202]. See [235] for an introduction to Galois's famous theory.)

Galois' theorem

> **Theorem 8.33** *A quadratic irrational ξ is purely periodic if and only if*
>
> $$\xi > 1 \quad \text{and} \quad -1 < \overline{\xi} < 0.$$

Proof Assume that $\xi = \langle a_0; \ldots, a_{m-1}, a_0, a_1, \ldots, a_{m-1}, \ldots \rangle$ is purely periodic; we shall prove that $\xi > 1$ and $-1 < \overline{\xi} < 0$. Recall that in general, for a simple continued fraction $\langle b_0; b_1, b_2, \ldots \rangle$, all the b_n are positive after b_0. Thus, since a_0 appears again (and again, and again, $\ldots$) after the first a_0 in ξ, it follows that $a_0 \geq 1$. Hence, $\xi = a_0 + \frac{1}{\xi_1} > 1$. Now applying Theorem 8.5 to $\langle a_0; \ldots, a_{m-1}, \xi \rangle$, we get

$$\xi = \frac{\xi p_{m-1} + p_{m-2}}{\xi q_{m-1} + q_{m-2}},$$

where p_n/q_n are the convergents for ξ. Multiplying both sides by $\xi q_{m-1} + q_{m-2}$, we obtain

$$\xi^2 q_{m-1} + \xi q_{m-2} = \xi p_{m-1} + p_{m-2} \quad \Longrightarrow \quad f(\xi) = 0, \qquad (8.53)$$

where $f(x) = q_{m-1}x^2 + (q_{m-2} - p_{m-1})x - p_{m-2}$ is a quadratic polynomial. In particular, ξ is a root of f. Taking conjugates of the equation in (8.53), we see that $f(\overline{\xi}) = 0$ as well. Thus, ξ and $\overline{\xi}$ are the two roots of f. Now $\xi > 1$, so by the Wallis–Euler recurrence relations, $p_n > 0$, $p_n < p_{n+1}$, and $q_n < q_{n+1}$ for all n. Hence,

---

[8][From the preface to his final manuscript (Évariste died from a pistol duel at the age of 20)]. "Since the beginning of the century, computational procedures have become so complicated that any progress by those means has become impossible, without the elegance which modern mathematicians have brought to bear on their research, and by means of which the spirit comprehends quickly and in one step a great many computations. It is clear that elegance, so vaunted and so aptly named, can have no other purpose. ...Go to the roots of these calculations! Group the operations. Classify them according to their complexities rather than their appearances! This, I believe, is the mission of future mathematicians. This is the road on which I am embarking in this work." Évariste Galois (1811–1832).

$$f(-1) = (q_{m-1} - q_{m-2}) + (p_{m-1} - p_{m-2}) > 0 \quad \text{and} \quad f(0) = -p_{m-2} < 0.$$

By the intermediate value theorem, $f(x) = 0$ for some $-1 < x < 0$. Since $\overline{\xi}$ is the other root of f, we have $-1 < \overline{\xi} < 0$.

Assume now that ξ is a quadratic irrational with $\xi > 1$ and $-1 < \overline{\xi} < 0$; we shall prove that ξ is purely periodic. To do so, we first prove that if $\{\xi_n\}$ are the complete quotients of ξ, then $-1 < \overline{\xi}_n < 0$ for all n. Since $\xi_0 = \xi$, this is already true for $n = 0$. Assume that this holds for n; then

$$\xi_n = a_n + \frac{1}{\xi_{n+1}} \quad \Longrightarrow \quad \frac{1}{\overline{\xi}_{n+1}} = \overline{\xi}_n - a_n < -a_n \le -1 \quad \Longrightarrow \quad \frac{1}{\overline{\xi}_{n+1}} < -1.$$

The inequality $\frac{1}{\overline{\xi}_{n+1}} < -1$ shows that $-1 < \overline{\xi}_{n+1} < 0$ and completes the induction. Now we already know that ξ is periodic, so let us assume for the sake of contradiction that ξ is not purely periodic, that is, $\xi = \langle a_0; a_1, \ldots, a_{\ell-1}, \overline{a_\ell, \ldots, a_{\ell+m-1}} \rangle$, where $\ell \ge 1$. Then $a_{\ell-1} \ne a_{\ell+m-1}$, for otherwise, we could start the repeating block at $a_{\ell-1}$. Since

$$\xi_{\ell-1} = a_{\ell-1} + \langle \overline{a_\ell, \ldots, a_{\ell+m-1}} \rangle \quad \text{and} \quad \xi_{\ell+m-1} = a_{\ell+m-1} + \langle \overline{a_\ell, \ldots, a_{\ell+m-1}} \rangle, \tag{8.54}$$

it follows that

$$\xi_{\ell-1} \ne \xi_{\ell+m-1}. \tag{8.55}$$

From (8.54), it also follows that $\xi_{\ell-1} - \xi_{\ell+m-1}$ is an integer (which is equal to $a_{\ell-1} - a_{\ell+m-1}$, although this is not important). Since rational numbers are self-conjugate, we have

$$\overline{\xi}_{\ell-1} - \overline{\xi}_{\ell+m-1} = \xi_{\ell-1} - \xi_{\ell+m-1}.$$

Now we already proved that $-1 < \overline{\xi}_{\ell-1} < 0$ and $-1 < \overline{\xi}_{\ell+m-1} < 0$, from which we obtain

$$-1 < \overline{\xi}_{\ell-1} - \overline{\xi}_{\ell+m-1} < 1.$$

Thus, $-1 < \xi_{\ell-1} - \xi_{\ell+m-1} < 1$. However, $\xi_{\ell-1} - \xi_{\ell+m-1}$ is an integer, and since the only integer strictly between -1 and 1 is 0, it must be that $\xi_{\ell-1} = \xi_{\ell+m-1}$. However, this contradicts (8.55), and our proof is complete. ∎

8.8.4 Square Roots and Periodic Continued Fractions

Recall that

$$\sqrt{19} = \langle 4; \overline{2, 1, 3, 1, 2, 8} \rangle;$$

if you didn't notice the beautiful symmetry before, observe that we can write this as $\sqrt{19} = \langle a_0; \overline{a_1, a_2, a_3, a_2, a_1, 2a_0} \rangle$, where the repeating block has a symmetric part and an ending part twice a_0. It turns out that every square root has this nice symmetry property. To prove this fact, we first prove the following.

Lemma 8.34 *Let $\xi = \overline{\langle a_0; a_1, \ldots, a_{m-1} \rangle}$ be purely periodic.*

(1) $-1/\bar{\xi}$ is the reversal of ξ: $-1/\bar{\xi} = \overline{\langle a_{m-1}; a_{m-2}, \ldots, a_0 \rangle}$.
(2) If $\{\xi_n\}$ is the sequence of complete quotients, then $\xi_n = \xi_0$ if and only if n is a multiple of m.

Proof Writing out the complete quotients $\xi, \xi_1, \xi_2, \ldots, \xi_{m-1}$ of

$$\xi = \overline{\langle a_0; a_1, \ldots, a_{m-1} \rangle} = \langle a_0; a_1, \ldots, a_{m-1}, \xi \rangle,$$

we obtain

$$\xi = a_0 + \frac{1}{\xi_1} \; , \; \xi_1 = a_1 + \frac{1}{\xi_2} \; , \ldots, \; \xi_{m-2} = a_{m-2} + \frac{1}{\xi_{m-1}} \; , \; \xi_{m-1} = a_{m-1} + \frac{1}{\xi}.$$

Taking conjugates of all of these and listing them in reverse order, we find that

$$\frac{-1}{\bar{\xi}} = a_{m-1} - \bar{\xi}_{m-1} \; , \; \frac{-1}{\bar{\xi}_{m-1}} = a_{m-2} - \bar{\xi}_{m-2} \; , \ldots, \; \frac{-1}{\bar{\xi}_2} = a_1 - \bar{\xi}_1 \; , \; \frac{-1}{\bar{\xi}_1} = a_0 - \bar{\xi}.$$

Let us define $\eta_0 := -1/\bar{\xi}$, $\eta_1 = -1/\bar{\xi}_{m-1}$, $\eta_2 = -1/\bar{\xi}_{m-2}, \ldots, \eta_{m-1} = -1/\bar{\xi}_1$. Then we can write the previous displayed equalities as

$$\eta_0 = a_{m-1} + \frac{1}{\eta_1} \; , \; \eta_1 = a_{m-2} + \frac{1}{\eta_2} \; , \ldots, \eta_{m-2} = a_1 + \frac{1}{\eta_{m-1}} \; , \; \eta_{m-1} = a_0 + \frac{1}{\eta_0};$$

in other words, η_0 is just the continued fraction

$$\eta_0 = \langle a_{m-1}; a_{m-2}, \ldots, a_1, a_0, \eta_0 \rangle = \overline{\langle a_{m-1}; a_{m-2}, \ldots, a_1, a_0 \rangle}.$$

Since $\eta_0 = -1/\bar{\xi}$, our proof is complete. We leave the proof of *(2)* as an exercise. ∎

Let d be natural number, not a perfect square. Then by Theorem 8.30, the complete quotients ξ_n and the partial quotients a_n for $\sqrt{d}$ are determined by

$$\xi_n = \frac{\alpha_n + \sqrt{d}}{\beta_n} \; , \quad a_n = \lfloor \xi_n \rfloor,$$

where the α_n, β_n are integers satisfying the relations given in Theorem 8.30.

Theorem 8.35 *The simple continued fraction of $\sqrt{d}$ has the form*

$$\sqrt{d} = \langle a_0; \overline{a_1, a_2, a_3, \ldots, a_3, a_2, a_1, 2a_0} \rangle.$$

Moreover, $\beta_n \neq -1$ for all n, and $\beta_n = +1$ if and only if n is a multiple of the period of $\sqrt{d}$.

Proof Starting the continued fraction algorithm for $\sqrt{d}$, we obtain $\sqrt{d} = a_0 + \frac{1}{\xi_1}$, where $\xi_1 > 1$. Since $\frac{1}{\xi_1} = -a_0 + \sqrt{d}$, we have

$$-\frac{1}{\overline{\xi_1}} = -\left(-a_0 - \sqrt{d}\right) = a_0 + \sqrt{d} > 1, \qquad (8.56)$$

so we must have $-1 < \overline{\xi_1} < 0$. Since both $\xi_1 > 1$ and $-1 < \overline{\xi_1} < 0$, by Galois's theorem, Theorem 8.33, we know that ξ_1 is purely periodic: $\xi_1 = \langle \overline{a_1; a_2, \ldots, a_m} \rangle$. Thus,

$$\sqrt{d} = a_0 + \frac{1}{\xi_1} = \langle a_0; \xi_1 \rangle = \langle a_0; \overline{a_1, a_2, \ldots, a_m} \rangle.$$

On the other hand, from (8.56) and from Lemma 8.34, we see that

$$\langle \overline{2a_0; a_1, a_2, \ldots, a_m, a_1, a_2, \ldots, a_m, \ldots} \rangle = a_0 + \sqrt{d} = -\frac{1}{\overline{\xi_1}} = \langle \overline{a_m; \ldots, a_1} \rangle$$

$$= \langle \overline{a_m; a_{m-1}, a_{m-2}, \ldots, a_1, a_m, a_{m-1}, a_{m-2}, \ldots, a_1, \ldots} \rangle.$$

By the uniqueness of simple continued fraction expansions (Problem 8 on p. 636), we have $a_m = 2a_0$, $a_{m-1} = a_1$, $a_{m-2} = a_2$, $a_{m-3} = a_3$, and so forth. Therefore,

$$\sqrt{d} = \langle a_0; \overline{a_1, a_2, \ldots, a_m} \rangle = \langle a_0; \overline{a_1, a_2, a_3, \ldots, a_3, a_2, a_1, 2a_0} \rangle.$$

We now prove that β_n never equals -1, and $\beta_n = +1$ if and only if n is a multiple of the period m. Since $\beta_0 = +1$ by definition (see Theorem 8.30), we henceforth assume that $n > 0$. By the form of the continued fraction expansion of $\sqrt{d}$ we just derived, observe that for $n > 0$, the nth complete quotient ξ_n for $\sqrt{d}$ is purely periodic. In particular, by Galois's theorem, Theorem 8.33, we know that

$$\xi_n > 1 \quad \text{and} \quad -1 < \overline{\xi_n} < 0. \qquad (8.57)$$

Now for the sake of contradiction, assume that $\beta_n = -1$. Then the formula $\xi_n = (\alpha_n + \sqrt{d})/\beta_n$ with $\beta_n = -1$ and (8.57) imply that

$$1 < \xi_n = -\alpha_n - \sqrt{d} \implies \alpha_n < -1 - \sqrt{d} \implies \alpha_n < 0.$$

On the other hand, (8.57) also implies that

$$-1 < \overline{\xi}_n = -\alpha_n + \sqrt{d} < 0 \quad \Longrightarrow \quad \sqrt{d} < \alpha_n \quad \Longrightarrow \quad 0 < \alpha_n.$$

The contradictions $\alpha_n < 0$ and $\alpha_n > 0$ imply that $\beta_n = -1$ is impossible.

We now prove that $\beta_n = 1$ if and only if n is a multiple of the period m. Note that $\beta_n = 1$ if and only if $\xi_n = \alpha_n + \sqrt{d}$. Since $\sqrt{d} = a_0 + 1/\xi_1$, we see that $\xi_n = a_n + 1/\xi_1$, where $a_n = \alpha_n + a_0 \in \mathbb{Z}$. In particular, $\xi_{n+1} = \xi_1$. In conclusion, $\beta_n = 1$ if and only if $\xi_{n+1} = \xi_1$. Now if we put $\eta_k = \xi_{k+1}$ for all $k = 0, 1, 2, \ldots$, then observe that η_k is purely periodic and $\beta_n = 1$ if and only if $\eta_n = \eta_0$. By the previous lemma, $\eta_n = \eta_0$ if and only if n is a multiple of m. ∎

▶ **Exercises 8.8**

1. Find the canonical continued fraction expansions for

$$(a)\ \sqrt{29} \ , \quad (b)\ \frac{1 + \sqrt{13}}{2} \ , \quad (c)\ \frac{2 + \sqrt{5}}{3}.$$

2. Find the values of the following continued fractions:

$$(a)\ \langle 3; \overline{2, 6} \rangle \ , \quad (b)\ \langle \overline{1; 2, 3} \rangle \ , \quad (c)\ \langle 1; 2, \overline{3} \rangle \ , \quad (d)\ \langle 2; 5, \overline{1, 3, 5} \rangle.$$

3. Let $m, n \in \mathbb{N}$. Find the quadratic irrational numbers represented by

$$(a)\ \langle \overline{n} \rangle = \langle n; n, n, n, \ldots \rangle \ , \quad (b)\ \langle \overline{n; 1} \rangle \ , \quad (c)\ \langle \overline{n; n+1} \rangle \ , \quad (d)\ \langle m; \overline{n} \rangle.$$

4. Prove Part *(2)* of Lemma 8.34.

8.9 Archimedes's Crazy Cattle Conundrum and Diophantine Equations

Archimedes (287–212) was known to think in preposterous proportions. In *The Sand Reckoner* [171, p. 420], a fun story written by Archimedes, he concluded that if he could fill the universe with grains of sand, there would be approximately 8×10^{63} grains! According to Pappus of Alexandria (290–350), at one time Archimedes said (see [60, p. 15]), "Give me a place to stand, and I will move the Earth!" In the following, we shall look at a cattle problem proposed by Archimedes, whose solution involves approximately 8×10^{206544} cattle! If you feel moved to read more on Achimedes's cattle, see [18, 145, 167, 251, 266].

8.9.1 Archimedes's Crazy Cattle Conundrum

Here is a poem written by Archimedes to students at Alexandria in a letter to Eratosthenes of Cyrene (276–194 B.C.). (The following is adapted from [106], as written in [18].)

> Compute, O stranger! the number of cattle of Helios, which once grazed on the plains of Sicily, divided according to their color, to wit:
>
> (1) White bulls $= \frac{1}{2}$ black bulls $+ \frac{1}{3}$ black bulls $+$ yellow bulls
>
> (2) Black bulls $= \frac{1}{4}$ spotted bulls $+ \frac{1}{5}$ spotted bulls $+$ yellow bulls
>
> (3) Spotted bulls $= \frac{1}{6}$ white bulls $+ \frac{1}{7}$ white bulls $+$ yellow bulls
>
> (4) White cows $= \frac{1}{3}$ black herd $+ \frac{1}{4}$ black herd (here, "herd" $=$ bulls $+$ cows)
>
> (5) Black cows $= \frac{1}{4}$ spotted herd $+ \frac{1}{5}$ spotted herd
>
> (6) Dappled cows $= \frac{1}{5}$ yellow herd $+ \frac{1}{6}$ yellow herd
>
> (7) Yellow cows $= \frac{1}{6}$ white herd $+ \frac{1}{7}$ white herd
>
> He who can answer the above is no novice in numbers. Nevertheless, he is not yet skilled in wise calculations! But come consider also all the following numerical relations between the Oxen of the Sun:
>
> (8) If the white bulls were combined with the black bulls they would be in a figure equal in depth and breadth and the far stretching plains of Sicily would be covered by the square formed by them.
>
> (9) If the yellow and spotted bulls were collected in one place, they would stand, if they ranged themselves one after another, completing the form of an equilateral triangle.
>
> If thou discover the solution of this at the same time; if thou grasp it with thy brain; and give correctly all the numbers; O Stranger! go and exult as conqueror; be assured that thou art by all means proved to have abundant knowledge in this science.

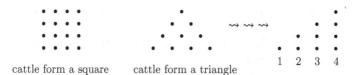

cattle form a square   cattle form a triangle

Fig. 8.7 With the dots as bulls, on the *left*, the number of bulls is a square number (4^2 in this case) and the number of bulls on the *right* is a triangular number ($1 + 2 + 3 + 4$ in this case)

To solve this puzzle, we need to turn it into mathematics! Let W, B, Y, S denote the numbers of white, black, yellow, and spotted bulls, respectively, and w, b, y, s for the number of white, black, yellow, and spotted cows, respectively.

The conditions (1)–(7) can be written as

$$(1) \ W = \left(\frac{1}{2} + \frac{1}{3}\right)B + Y \qquad (2) \ B = \left(\frac{1}{4} + \frac{1}{5}\right)S + Y,$$

$$(3)\ S = \left(\frac{1}{6} + \frac{1}{7}\right)W + Y \qquad (4)\ w = \left(\frac{1}{3} + \frac{1}{4}\right)(B + b),$$

$$(5)\ b = \left(\frac{1}{4} + \frac{1}{5}\right)(S + s) \qquad (6)\ s = \left(\frac{1}{5} + \frac{1}{6}\right)(Y + y),$$

$$(7)\ y = \left(\frac{1}{6} + \frac{1}{7}\right)(W + w).$$

Now how do we interpret (8) and (9)? We will interpret (8) as meaning that the number of white and black bulls should be a square number (a perfect square); see the left picture in Fig. 8.7. A **triangular number** is a number of the form

$$1 + 2 + 3 + 4 + \cdots + n = \frac{n(n+1)}{2},$$

for some n. Then we will interpret (9) as meaning that the number of yellow and spotted bulls should be a triangular number; see the right picture in Fig. 8.7. Thus, (8) and (9) become

(8) $W + B = $ a square number ,   (9) $Y + S = $ a triangular number.

In summary: We want to find *integers* W, B, Y, S, w, b, y, s (here we assume that there is no such thing as "fractional cattle") solving equations (1)–(9). Now to the solution of Archimedes's cattle problem. First of all, equations (1)–(7) are just linear equations, so these equations can be solved using simple linear algebra. Instead of solving these equations by hand, which will probably take a few hours, it would be best to use a computer. Doing so, you will find that in order for W, B, Y, S, w, b, y, s to solve (1)–(7), they must be of the form

$$W = 10366482\,k \ , \quad B = 7460514\,k \ , \quad Y = 4149387\,k \ , \quad S = 7358060\,k$$
$$w = 7206360\,k \ , \quad b = 4893246\,k \ , \quad y = 5439213\,k \ , \quad s = 3515820\,k,$$
$$(8.58)$$

where k can equal $1, 2, 3, \ldots$. Thus, in order for us to fulfill conditions (1)–(7), we would have at the very least, setting $k = 1$,

$$10366482 + 7460514 + 4149387 + 7358060 + 7206360 + 4893246$$
$$+\ 5439213 + 3515820 = 50389082 \approx 50 \text{ million cattle!}$$

Now we are "no novice in numbers!" Nevertheless, we are not yet skilled in wise calculations! To be skilled, we still have to satisfy conditions (8) and (9). For (8), this means

$$W + B = 10366482\,k + 7460514\,k = 17826996\,k = \text{a square number.}$$

Factoring $17826996 = 2^2 \cdot 3 \cdot 11 \cdot 29 \cdot 4657$ into its prime factors, we see that we must have

$$2^2 \cdot 3 \cdot 11 \cdot 29 \cdot 4657\, k = \text{a perfect square.}$$

Thus, we need $3 \cdot 11 \cdot 29 \cdot 4657\, k$ to be a square, which holds if and only if

$$k = 3 \cdot 11 \cdot 29 \cdot 4657\, m^2 = 4456749\, m^2$$

for some integer m. Plugging this value into (8.58), we get

$$
\begin{aligned}
W &= 46200808287018\, m^2\,, \quad B = 33249638308986\, m^2 \\
Y &= 18492776362863\, m^2\,, \quad S = 32793026546940\, m^2 \\
w &= 32116937723640\, m^2\,, \quad b = 21807969217254\, m^2 \\
y &= 24241207098537\, m^2\,, \quad s = 15669127269180\, m^2,
\end{aligned}
\tag{8.59}
$$

where m can equal $1, 2, 3, \ldots$. Thus, in order for us to fulfill conditions (1)–(8), we would have at the very least, setting $m = 1$,

$$
\begin{aligned}
& 46200808287018 + 33249638308986 + 18492776362863 + 32793026546940 \\
& + 32116937723640 + 21807969217254 + 24241207098537 \\
& + 15669127269180 = 2.2457\ldots \times 10^{14} \approx 2.2 \text{ trillion cattle!}
\end{aligned}
$$

It now remains to satisfy condition (9):

$$
\begin{aligned}
Y + S &= 18492776362863\, m^2 + 32793026546940\, m^2 \\
&= 51285802909803\, m^2 = \frac{\ell(\ell+1)}{2},
\end{aligned}
$$

for some integer ℓ. Multiplying both sides by 8 and adding 1, we obtain

$$8 \cdot 51285802909803\, m^2 + 1 = 4\ell^2 + 4\ell + 1 = (2\ell + 1)^2 = n^2,$$

where $n = 2\ell + 1$. Since $8 \cdot 51285802909803 = 410286423278424$, we finally conclude that conditions (1)–(9) are all fulfilled if we can find *integers* m, n satisfying the equation

$$n^2 - 410286423278424\, m^2 = 1. \tag{8.60}$$

This equation is commonly called a **Pell equation** and is an example of a Diophantine equation. As we'll see in the next subsection, we can solve this equation by simply (!) finding the simple continued fraction expansion of $\sqrt{410286423278424}$. The calculations involved are just sheer madness, but they can be done and have been done [18, 266]. In the end, we find that the smallest total number of cattle that satisfies (1)–(9) is a number with 206545 digits (!) and is equal to

> **Solution to Archimedes's crazy cattle conundrum:**
> $7760271406\ldots(206525$ other digits go here$)\ldots 9455081800 \approx 8 \times 10^{206544}$.

We are now skilled in wise calculations! A copy of this number is printed on 42 computer sheets, took 7 hours and 49 minutes computing time (in 1965, on massive IBM supercomputers), and was deposited in the unpublished mathematical tables of the journal *Mathematics of Computation*.

8.9.2 Pell's Equation

Generalizing the cattle Eq. (8.60), given a non-perfect-square natural number d, we call a Diophantine equation of the form

$$x^2 - d\,y^2 = 1 \tag{8.61}$$

a **Pell equation** and we seek *integer* solutions. The real number pairs (x, y) satisfying this equation define a hyperbola:

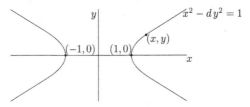

Thus, we are asking whether pairs of *integers* lie on this curve. Of course, $(x, y) = (\pm 1, 0)$ solve this equation. These solutions are called the **trivial solutions**; the other solutions are not so easily attained. We remark that Pell's equation was named by Euler after John Pell (1611–1685), although Brahmagupta (598–670) studied this equation a thousand years earlier [35, p. 221]. In any case, we shall see that the continued fraction expansion of $\sqrt{d}$ plays an important role in solving this equation. We note that if (x, y) solves (8.61), then trivially so do $(\pm x, \pm y)$, because of the squares in (8.61); thus, we restrict ourselves to the positive solutions.

Recall that the continued fraction expansion for $\sqrt{d}$ has the complete quotients ξ_n and partial quotients a_n determined by

$$\xi_n = \frac{\alpha_n + \sqrt{d}}{\beta_n} \quad , \quad a_n = \lfloor \xi_n \rfloor,$$

where α_n and β_n are integers defined in Theorem 8.30. The exact forms of these integers are not important; what is important is that β_n never equals -1 and $\beta_n = +1$ if and only if n is a multiple of the period of $\sqrt{d}$, as we saw in Theorem 8.35. The following lemma shows how the convergents of $\sqrt{d}$ enter Pell's equation.

Lemma 8.36 *If $d \in \mathbb{N}$ is not a perfect square and p_n/q_n denotes the nth convergent of $\sqrt{d}$, then for all $n = 0, 1, 2, \ldots$, we have*

$$p_n^2 - d\, q_n^2 = (-1)^{n+1} \beta_{n+1}.$$

Proof Since we can write $\sqrt{d} = \langle a_0; a_1, a_2, a_3, \ldots, a_n, \xi_{n+1} \rangle$ and $\xi_{n+1} = (\alpha_{n+1} + \sqrt{d})/\beta_{n+1}$, by (8.20) of Corollary 8.7, we have

$$\sqrt{d} = \frac{\xi_{n+1} p_n + p_{n-1}}{\xi_{n+1} q_n + q_{n-1}} = \frac{(\alpha_{n+1} + \sqrt{d})\, p_n + \beta_{n+1} p_{n-1}}{(\alpha_{n+1} + \sqrt{d})\, q_n + \beta_{n+1} q_{n-1}}.$$

Multiplying both sides by the denominator of the right-hand side, we get

$$\sqrt{d}(\alpha_{n+1} + \sqrt{d})\, q_n + \sqrt{d}\beta_{n+1} q_{n-1} = (\alpha_{n+1} + \sqrt{d})\, p_n + \beta_{n+1} p_{n-1}.$$

Equating the integers on each side and the coefficients of $\sqrt{d}$, we obtain the two equalities

$$dq_n = \alpha_{n+1} p_n + \beta_{n+1} p_{n-1} \quad \text{and} \quad \alpha_{n+1} q_n + \beta_{n+1} q_{n-1} = p_n.$$

Multiplying the first equation by q_n and the second equation by p_n and equating the $\alpha_{n+1} p_n q_n$ terms in each resulting equation, we obtain

$$dq_n^2 - \beta_{n+1} p_{n-1} q_n = p_n^2 - \beta_{n+1} p_n q_{n-1}$$
$$\implies \quad p_n^2 - d\, q_n^2 = (p_n q_{n-1} - p_{n-1} q_n) \cdot \beta_{n+1} = (-1)^{n+1} \cdot \beta_{n+1},$$

where we used that $p_n q_{n-1} - p_{n-1} q_n = (-1)^{n-1} = (-1)^{n+1}$ from Corollary 8.7. ∎

Next, we show that *all* solutions of Pell's equation can be found via the convergents of $\sqrt{d}$.

Theorem 8.37 *Let $d \in \mathbb{N}$ be not a perfect square, let p_n/q_n denote the nth convergent of $\sqrt{d}$, and let m be the period of $\sqrt{d}$. Then the positive integer solutions to*

$$x^2 - d\, y^2 = 1$$

are precisely the numerators and denominators of the odd convergents of $\sqrt{d}$ of the form $x = p_{nm-1}$ and $y = q_{nm-1}$, where $n > 0$ is any positive integer for m even and $n > 0$ is even for m odd.

Proof We prove our theorem in two steps.

Step 1: We first look for convergents (p_k, q_k) that make $p_k^2 - d\, q_k^2 = 1$. To this end, recall from Lemma 8.36 that

$$p_{k-1}^2 - d\, q_{k-1}^2 = (-1)^k \beta_k,$$

where β_k never equals -1 and $\beta_k = 1$ if and only if k is a multiple of m, the period of $\sqrt{d}$. Now $p_{k-1}^2 - d\, q_{k-1}^2 = 1$ if and only if $\beta_k = (-1)^k$. This holds if and only if $\beta_k = 1$ and k is even, which holds if and only if k is a multiple of m and k is even. Thus, $p_{k-1}^2 - d\, q_{k-1}^2 = 1$ if and only if $k = mn$, where $n > 0$ is any positive integer for m even and $n > 0$ is even for m odd.

Step 2: Our proof will be complete once we show that if $x^2 - d\, y^2 = 1$ with $y > 0$, then x/y is a convergent of $\sqrt{d}$. To see this, observe that since $1 = x^2 - d\, y^2 = (x - \sqrt{d}\, y)(x + \sqrt{d}\, y)$, we have $x - \sqrt{d}\, y = 1/(x + \sqrt{d}\, y)$, so

$$\left| \frac{x}{y} - \sqrt{d} \right| = \left| \frac{x - \sqrt{d}\, y}{y} \right| = \frac{1}{y\,|x + \sqrt{d}\, y|}.$$

Also, $x^2 = d\, y^2 + 1 > d\, y^2$ implies $x > \sqrt{d}\, y$, which implies

$$x + \sqrt{d}\, y > \sqrt{d}\, y + \sqrt{d}\, y > y + y = 2y.$$

Hence,

$$\left| \frac{x}{y} - \sqrt{d} \right| = \frac{1}{y\,|x + \sqrt{d}\, y|} < \frac{1}{y \cdot 2y} \implies \left| \frac{x}{y} - \sqrt{d} \right| < \frac{1}{2y^2}.$$

By Legendre's theorem 8.21, x/y must be a convergent of $\sqrt{d}$. ∎

The **fundamental solution** of Pell's equation is the "smallest" positive solution of Pell's equation; here, a solution (x, y) is **positive** means $x, y > 0$. Explicitly, the fundamental solution is (p_{m-1}, q_{m-1}) for an even period m of $\sqrt{d}$ or (p_{2m-1}, p_{2m-1}) for an odd period m.

Example 8.34 Consider the equation $x^2 - 3y^2 = 1$. Since $\sqrt{3} = \langle 1; \overline{1, 2} \rangle$ has period $m = 2$, our theorem says that the positive solutions of $x^2 - 3y^2 = 1$ are precisely $x = p_{2n-1}$ and $y = q_{2n-1}$ for all $n > 0$; that is, (p_1, q_1), (p_3, q_3), (p_5, q_5), ... Now the convergents of $\sqrt{3}$ are

n	0	1	2	3	4	5	6	7
p_n	1	2	5	7	19	26	71	97
q_n	1	1	3	4	11	15	41	56

In particular, the fundamental solution is $(2, 1)$ and the rest of the positive solutions are $(7, 4)$, $(26, 15)$, $(97, 56)$, ... Just to verify a couple of entries:

$$2^2 - 3 \cdot 1^2 = 4 - 3 = 1$$

and

$$7^2 - 3 \cdot 4^2 = 49 - 3 \cdot 16 = 49 - 48 = 1,$$

and one can continue verifying that the odd convergents give solutions.

Example 8.35 For another example, consider the equation $x^2 - 13\,y^2 = 1$. In this case, we find that $\sqrt{13} = \langle 3; \overline{1, 1, 1, 1, 6} \rangle$ has period $m = 5$. Thus, our theorem says that the positive solutions of $x^2 - 13y^2 = 1$ are precisely $x = p_{5n-1}$ and $y = q_{5n-1}$ for all $n > 0$ *even*; that is, (p_9, q_9), (p_{19}, q_{19}), (p_{29}, q_{29}), ... The convergents of $\sqrt{13}$ are

n	0	1	2	3	4	5	6	7	8	9
p_n	3	4	7	11	18	119	137	256	393	649
q_n	1	1	2	3	5	33	38	71	109	180

In particular, the fundamental solution is (649, 180).

8.9.3   Brahmagupta's Algorithm

Thus, to find solutions of Pell's equation we just have to find certain convergents of $\sqrt{d}$. Finding all convergents is quite a daunting task—try finding the solution (p_{19}, q_{19}) for $\sqrt{13}$—but it turns out that all the positive solutions can be found from the fundamental solution.

Example 8.36 We know that the fundamental solution of $x^2 - 3y^2 = 1$ is $(2, 1)$ and the rest of the positive solutions are $(7, 4)$, $(26, 15)$, $(97, 56)$, ... Observe that

$$(2 + 1 \cdot \sqrt{3})^2 = 4 + 4\sqrt{3} + 3 = 7 + 4\sqrt{3}.$$

Note that the second positive solution $(7, 4)$ to $x^2 - 3y^2 = 1$ appears on the right. Now observe that

$$(2 + 1 \cdot \sqrt{3})^3 = (2 + \sqrt{3})^2 \, (2 + \sqrt{3}) = (7 + 4\sqrt{3}) \, (2 + \sqrt{3}) = 26 + 15\sqrt{3}.$$

Note that the third positive solution $(26, 15)$ to $x^2 - 3y^2 = 1$ appears on the right. One may conjecture that the nth positive solution (x_n, y_n) to $x^2 - 3\,y^2 = 1$ is found by multiplying out

$$x_n + y_n \sqrt{d} = (2 + 1 \cdot \sqrt{3})^n.$$

This is in fact correct, as the following theorem shows.

Brahmagupta's algorithm

Theorem 8.38 *If (x_1, y_1) is the fundamental solution of Pell's equation*

$$x^2 - d\,y^2 = 1,$$

then all the other positive solutions (x_n, y_n) can be obtained from the equation

$$x_n + y_n\sqrt{d} = (x_1 + y_1\sqrt{d})^n \quad, \quad n = 0, 1, 2, 3, \ldots.$$

Proof To simplify this proof a little, we shall say that $\zeta = x + y\sqrt{d} \in \mathbb{Z}[\sqrt{d}]$ solves Pell's equation to mean that (x, y) solves Pell's equation; similarly, we say ζ is a positive solution to mean that $x, y > 0$. Throughout this proof we shall use the following fact:

$$\zeta \text{ solves Pell's equation} \quad\Longleftrightarrow\quad \zeta\overline{\zeta} = 1 \text{ (that is, } 1/\zeta = \overline{\zeta}). \qquad (8.62)$$

This is holds for the simple reason that

$$\zeta\overline{\zeta} = (x + y\sqrt{d})\,(x - y\sqrt{d}) = x^2 - d\,y^2.$$

In particular, if we set $\alpha := x_1 + y_1\sqrt{d}$, then $\alpha\overline{\alpha} = 1$, because (x_1, y_1) solves Pell's equation. We now prove our theorem. We can write $\alpha^n = x_n + y_n\sqrt{d}$ for some natural numbers x_n, y_n; then,

$$(x_n + y_n\sqrt{d})\,\overline{(x_n + y_n\sqrt{d})} = \alpha^n \cdot \overline{\alpha^n} = \alpha^n \cdot (\overline{\alpha})^n = (\alpha \cdot \overline{\alpha})^n = 1^n = 1,$$

from which, in view of (8.62), we conclude that (x_n, y_n) solves Pell's equation. Now suppose that $\xi \in \mathbb{Z}[\sqrt{d}]$ is a positive solution to Pell's equation; we must show that ξ is some power of α. To this end, note that $\alpha \leq \xi$, because $\alpha = x_1 + y_1\sqrt{d}$ and (x_1, y_1) is the smallest positive solution of Pell's equation. Since $1 < \alpha$, it follows that $\alpha^k \to \infty$ as $k \to \infty$, so we can choose $n \in \mathbb{N}$ to be the largest natural number such that $\alpha^n \leq \xi$. Define

$$\eta = \frac{\xi}{\alpha^n} = \xi \cdot (\overline{\alpha})^n,$$

where we used that $1/\alpha = \overline{\alpha}$ from (8.62). We shall prove that $\eta = 1$, which shows that $\xi = \alpha^n$ and completes our proof. To do so, we begin by observing that since $\mathbb{Z}[\sqrt{d}]$ is a ring (Lemma 8.31 on p. 672), we know that $\eta = \xi \cdot (\overline{\alpha})^n \in \mathbb{Z}[\sqrt{d}]$. Moreover, η solves Pell's equation, because

$$\eta\overline{\eta} = \xi \cdot (\overline{\alpha})^n \cdot \overline{\xi} \cdot \alpha^n = (\xi\overline{\xi}) \cdot (\overline{\alpha}\alpha)^n = 1 \cdot 1 = 1.$$

In particular, $\overline{\eta} = 1/\eta$. Note that $\alpha^n \leq \xi < \alpha^{n+1}$, so $1 \leq \eta < \alpha$. Now let $\eta = p + q\sqrt{d}$, where $p, q \in \mathbb{Z}$. Then since $\eta \geq 1$, we have $1/\eta > 0$, so

$$2p = (p + q\sqrt{d}) + (p - q\sqrt{d}) = \eta + \overline{\eta} = \eta + \frac{1}{\eta} > 0,$$

and

$$2q\sqrt{d} = (p + q\sqrt{d}) - (p - q\sqrt{d}) = \eta - \overline{\eta} = \eta - \frac{1}{\eta} = \frac{\eta^2 - 1}{\eta} \geq 0.$$

Thus, $p > 0$, $q \geq 0$, and $p^2 - dq^2 = 1$ (since η solves Pell's equation). Therefore, $(p, q) = (1, 0)$ or (p, q) is a positive (numerator, denominator) of a convergent of $\sqrt{d}$. However, we know that (x_1, y_1) is the smallest such positive (numerator, denominator), and that $p + q\sqrt{d} = \eta < \alpha = x_1 + y_1\sqrt{d}$. Therefore, we must have $(p, q) = (1, 0)$. This implies that $\eta = 1$ and hence $\xi = \alpha^n$. ∎

Example 8.37 Since $(649, 180)$ is the fundamental solution to $x^2 - 13y^2 = 1$, all the positive solutions are given by

$$x_n + y_n\sqrt{13} = (649 + 180\sqrt{13})^n.$$

For instance, for $n = 2$, we find that

$$(649 + 180\sqrt{13})^2 = 842401 + 233640\sqrt{13} \implies (x_2, y_2) = (842401, 233640),$$

which is much easier than finding (p_{19}, q_{19}).

There are many cool applications of Pell's equation explored in the exercises. Here's one of my favorites (see Problem 12): Every prime of the form $p = 4k + 1$ is a sum of two squares. This was conjectured by Pierre de Fermat[9] (1601–1665) in 1640 and proved by Euler in 1754. For example, 5, 13, 17 are such primes, and $5 = 1^2 + 2^2$, $13 = 2^2 + 3^2$, and $17 = 1^2 + 4^2$.

▶ **Exercises 8.9**

1. Brahmagupta (598–670) is reported to have said, "A person who can, within a year, solve $x^2 - 92y^2 = 1$ is a mathematician." Given that

$$\sqrt{92} = \langle 9; \overline{1, 1, 2, 4, 2, 1, 1, 18} \rangle,$$

find the fundamental solution to the equation $x^2 - 92y^2 = 1$. (Your calculation should end with the convergent $1151/120$.) You are now a mathematician!

---

[9] [In the margin of his copy of Diophantus's *Arithmetica*, Fermat wrote] "To divide a cube into two other cubes, a fourth power or in general any power whatever into two powers of the same denomination above the second is impossible, and I have assuredly found an admirable proof of this, but the margin is too narrow to contain it." Pierre de Fermat (1601–1665). Fermat's claim in this marginal note, later to be called "Fermat's last theorem," remained an unsolved problem in mathematics until 1995 when Andrew Wiles (b. 1953) finally proved it.

2. Find the fundamental solutions to the equations

$$(a)\ x^2 - 8y^2 = 1 \quad , \quad (b)\ x^2 - 5y^2 = 1 \quad , \quad (c)x^2 - 7y^2 = 1.$$

Using the fundamental solution, find the next two solutions.

3. **(Pythagorean triples)** (Cf. [186]) Here is a nice problem solvable using continued fractions. A **Pythagorean triple** consists of three natural numbers (x, y, z) such that $x^2 + y^2 = z^2$. For example, $(3, 4, 5)$, $(5, 12, 13)$, and $(8, 15, 17)$ are Pythagorean triples. (Can you find more?) The first example, $(3, 4, 5)$, has the property that the first two numbers are consecutive integers; here are some steps to find more Pythagorean triples of this sort.

   (i) Show that (x, y, z) is a Pythagorean triple with $y = x + 1$ if and only if

   $$(2x + 1)^2 - 2z^2 = 1.$$

   (ii) By solving the Pell equation $u^2 - 2v^2 = 1$, find the next three Pythagorean triples (x, y, z) (after $(3, 4, 5)$) where x and y are consecutive integers.

4. Are there infinitely many triples of *consecutive* natural numbers, each of which is a sum of two squares? Suggestion: Consider $n - 1, n, n + 1$ and try to write $n - 1 = k^2 + k^2$ and $n = \ell^2 + 0^2$.

5. **(Triangular numbers)** Here is another very nice problem that can be solved using continued fractions. Find all triangular numbers that are squares, where recall that a triangular number is of the form $1 + 2 + \cdots + n = n(n + 1)/2$. Here are some steps.

   (i) Show that $n(n + 1)/2 = m^2$ if and only if

   $$(2n + 1)^2 - 8m^2 = 1.$$

   (ii) By solving the Pell equation $x^2 - 8y^2 = 1$, find the first three triangular numbers that are squares.

6. In this problem we answer the following question: For which $n \in \mathbb{N}$ is the standard deviation of the $2n + 1$ numbers $0, \pm 1, \ldots, \pm n$ an integer? Here, the **standard deviation** of real numbers $x_1, \ldots, x_N$ is by definition the number $\sqrt{\frac{1}{N} \sum_{i=1}^{N} (x_i - \bar{x})^2}$, where $\bar{x}$ is the average of $x_1, \ldots, x_N$.

   (i) Show that the standard deviation of $0, \pm 1, \ldots, \pm n$ equals $\sqrt{\frac{1}{3}n(n + 1)}$.
   Suggestion: The formula $1^2 + 2^2 + \cdots + n^2 = \frac{n(n+1)(2n+1)}{6}$ from Problem 3b on p. 31 might be helpful.

   (ii) Therefore, we want $\frac{1}{3}n(n + 1) = y^2$, where $y \in \mathbb{N}$. If we put $x = 2n + 1$, prove that $\frac{1}{3}n(n + 1) = y^2$ if and only if $x^2 - 12y^2 = 1$, where $x = 2n + 1$.

   (iii) Now solve the equation $x^2 - 12y^2 = 1$ to answer our question.

7. The Diophantine equation $x^2 - d\,y^2 = -1$ (where $d > 0$ is not a perfect square) is also of interest. In this problem we determine when this equation has solutions. Following the proof of Theorem 8.37, prove the following statements.

   (i) If (x, y) solves $x^2 - d\,y^2 = -1$, $y > 0$, then x/y is a convergent of $\sqrt{d}$.
   (ii) The equation $x^2 - d\,y^2 = -1$ has a solution if and only if the period of $\sqrt{d}$ is odd, in which case the nonnegative solutions are exactly $x = p_{nm-1}$ and $y = q_{nm-1}$ for all $n > 0$ odd.

8. Which of the following equations have solutions? If an equation has solutions, find the fundamental solution.

   $$(a)\ x^2 - 2\,y^2 = -1 \quad , \quad (b)\ x^2 - 3\,y^2 = -1 \quad , \quad (c)\ x^2 - 17\,y^2 = -1.$$

9. Are there infinitely many pairs of *consecutive* natural numbers, the sum of whose squares is a perfect square? Suggestion: Find solutions to $n^2 + (n+1)^2 = k^2$.

10. (**A not-so-crazy cattle conundrum**) Compute, O stranger! the number of cattle of Helios, which once grazed on the plains of Sicily, divided into two groups, spotted and yellow. We have spotted bulls $= \frac{1}{2}$ yellow bulls, there are more than 200 and less than 7000 yellow bulls, and if the two groups separated and ranged themselves one after another, each group would form an equilateral triangle. If thou discover the solution of this at the same time; if thou grasp it with thy brain; and give correctly all the numbers; O Stranger! go and exult as conqueror; be assured that thou art by all means proved to have abundance of knowledge in this science. Suggestion: Put the number of spotted (respectively yellow) bulls equal to $m(m+1)/2$ (respectively $n(n+1)/2$).

11. In this problem we prove that the Diophantine equation $x^2 - p\,y^2 = -1$ always has a solution if p is a prime number of the form $p = 4k + 1$ for an integer k. For instance, since $13 = 4 \cdot 3 + 1$ and $17 = 4 \cdot 4 + 1$, it follows that $x^2 - 13y^2 = -1$ and $x^2 - 17y^2 = -1$ have solutions. Let $p = 4k + 1$ be prime.

    (i) Let (x_1, y_1) be the fundamental solution of $x^2 - p\,y^2 = 1$. Prove that x_1 and y_1 cannot both be even and cannot both be odd.
    (ii) Show that the case x_1 is even and y_1 is odd cannot happen. Suggestion: Write $x_1 = 2a$ and $y_1 = 2b + 1$ and plug this into $x_1^2 - p\,y_1^2 = 1$.
    (iii) Thus, we may write $x_1 = 2a + 1$ and $y_1 = 2b$. Show that $p\,b^2 = a\,(a+1)$. Conclude that p must divide a or $a + 1$.
    (iv) Suppose that p divides a; that is, $a = mp$ for an integer m. Show that $b^2 = m\,(mp + 1)$ and that m and $mp + 1$ are relatively prime. Using this equality, prove that $m = s^2$ and $mp + 1 = t^2$ for integers s, t. Conclude that $t^2 - p\,s^2 = 1$ and derive a contradiction.
    (v) Thus, it must be the case that p divides $a + 1$. Using this fact and an argument similar to the one in the previous step, find a solution to $x^2 - d\,y^2 = -1$.

12. (**Fermat's two squares theorem**) We shall prove that every prime of the form $p = 4k + 1$ can be expressed as the sum of two squares.

(i) Let $p = 4k + 1$ be prime. Using the previous problem and Problem 7, prove that the period of $\sqrt{p}$ is odd and deduce that $\sqrt{p}$ has an expansion of the form

$$\sqrt{p} = \langle a_0; \overline{a_1, a_2, \ldots, a_{\ell-1}, a_\ell, a_\ell, a_{\ell-1}, \ldots, a_1, 2a_0} \rangle.$$

(ii) Let η be the complete quotient $\xi_{\ell+1}$. Prove that $-1 = \eta \cdot \overline{\eta}$. Suggestion: Recall Lemma 8.34.

(iii) Finally, writing $\eta = (a + \sqrt{p})/b$ (why does η have this form?), show that $p = a^2 + b^2$.

8.10 Epilogue: Transcendental Numbers, π, e, and Where's Calculus?

It's time to get a box of tissues, because, unfortunately, our adventures through this book have come to an end. In this section we wrap up this book with a discussion on transcendental numbers and continued fractions.

8.10.1 Approximable Numbers

A real number ξ is said to be **approximable** (by rationals) to order $n \geq 1$ if there exist a constant C and infinitely many rational numbers p/q in lowest terms with $q > 0$ such that

$$\left| \xi - \frac{p}{q} \right| < \frac{C}{q^n}. \tag{8.63}$$

Observe that if ξ is approximable to order $n > 1$, then it is automatically approximable to $n - 1$; this is because

$$\left| \xi - \frac{p}{q} \right| < \frac{C}{q^n} \leq \frac{C}{q^{n-1}}.$$

Similarly, ξ approximable to every order k with $1 \leq k \leq n$. Intuitively, the approximability order n measures how closely we can surround ξ with "good" rational numbers, that is, rational numbers having small denominators. To see what this means, suppose that ξ is approximable only to order 1. Thus, there exist a constant C and infinitely many rational numbers p/q in lowest terms with $q > 0$ such that

$$\left| \xi - \frac{p}{q} \right| < \frac{C}{q}.$$

This inequality suggests that in order to find rational numbers very close to ξ, these rational numbers need to have large denominators to make C/q small. However, if ξ were approximable to order 1000, then there would exist a constant C and infinitely many rational numbers p/q in lowest terms with $q > 0$ such that

$$\left| \xi - \frac{p}{q} \right| < \frac{C}{q^{1000}}.$$

This inequality suggests that to find rational numbers very close to ξ, those rational numbers don't need to have large denominators, because even for small q, the large power of 1000 will make C/q^{1000} small. The following lemma shows that there is a limit to how close we can surround algebraic numbers by "good" rational numbers.

Lemma 8.39 *If ξ is a real algebraic number of degree $n \geq 1$ (so ξ is rational if $n = 1$), then there exists a constant $c > 0$ such that for all rational numbers $p/q \neq \xi$ with $q > 0$, we have*

$$\left| \xi - \frac{p}{q} \right| \geq \frac{c}{q^n}.$$

Proof Assume that $f(\xi) = 0$, where

$$f(x) = a_n x^n + a_{n-1} x^{n-1} + \cdots + a_1 x + a_0 = 0, \qquad a_k \in \mathbb{Z},$$

and that no such polynomial function of lower degree has this property. First, we claim that $f(r) \neq 0$ for every rational number $r \neq \xi$. Indeed, if $f(r) = 0$ for some rational number $r \neq \xi$, then we can write $f(x) = (x - r)g(x)$, where g is a polynomial of degree $n - 1$; we leave you to check that $g(x)$ has rational coefficients. Then $0 = f(\xi) = (\xi - r)g(\xi)$ implies, since $\xi \neq r$, that $g(\xi) = 0$. This implies that the degree of ξ is $n - 1$, contradicting the fact that the degree of ξ is n. Now for every rational $p/q \neq \xi$ with $q > 0$, we see that

$$0 \neq |f(p/q)| = \left| a_n \left(\frac{p}{q} \right)^n + a_{n-1} \left(\frac{p}{q} \right)^{n-1} + \cdots + a_1 \left(\frac{p}{q} \right) + a_0 \right|$$

$$= \frac{|a_n p^n + a_{n-1} p^{n-1} q + \cdots + a_1 p q^{n-1} + a_0 q^n|}{q^n}.$$

The numerator is a nonnegative integer, which cannot be zero, so the numerator must be ≥ 1. Therefore,

$$|f(p/q)| \geq 1/q^n \text{ for all rational numbers } p/q \neq \xi \text{ with } q > 0. \tag{8.64}$$

Second, we claim that there is an $M > 0$ such that

$$|x - \xi| \le 1 \quad \Longrightarrow \quad |f(x)| \le M|x - \xi|. \tag{8.65}$$

Indeed, note that since $f(\xi) = 0$, we have

$$f(x) = f(x) - f(\xi) \quad = a_n(x^n - \xi^n) + a_{n-1}(x^{n-1} - \xi^{n-1}) + \cdots + a_1(x - \xi).$$

Since

$$x^k - \xi^k = (x - \xi)\, q_k(x), \quad q_k(x) = x^{k-1} + x^{k-2}\xi + \cdots + x\,\xi^{k-2} + \xi^{k-1},$$

plugging each of these, for $k = 1, 2, 3, \ldots, n$, into the previous equation for $f(x)$, we see that $f(x) = (x - \xi)h(x)$, where h is a continuous function. In particular, since $[\xi - 1, \xi + 1]$ is a closed and bounded interval, there is an M such that $|h(x)| \le M$ for all $x \in [\xi - 1, \xi + 1]$. This proves our claim.

Finally, let $p/q \ne \xi$ be a rational number with $q > 0$. If $|\xi - p/q| > 1$, then

$$\left| \xi - \frac{p}{q} \right| > 1 \ge \frac{1}{q^n}.$$

If $|\xi - p/q| \le 1$, then by (8.64) and (8.65), we have

$$\left| \xi - \frac{p}{q} \right| \ge \frac{1}{M} |f(p/q)| \ge \frac{1}{M}\frac{1}{q^n}.$$

Hence, $|\xi - p/q| \ge c/q^n$ for all rational $p/q \ne \xi$ with $q > 0$, where c is the smaller of 1 and $1/M$. ∎

Let us form the contrapositive of the statement of this lemma: If $n \in \mathbb{N}$ and for all constants $c > 0$, there exists a rational number $p/q \ne \xi$ with $q > 0$ such that

$$\left| \xi - \frac{p}{q} \right| < \frac{c}{q^n}, \tag{8.66}$$

then ξ is not algebraic of degree n. Since a transcendental number is a number that is not algebraic of any degree n, we can think of a transcendental number as a number that can be surrounded arbitrarily closely by "good" rational numbers. This leads us to Liouville numbers, to be discussed shortly, but before talking about these special transcendental numbers, we use our lemma to prove the following important result.

Theorem 8.40 *A real algebraic number of degree n is not approximable to order n + 1 (and hence not to any higher order). Moreover, a rational number is approximable to order 1, and a real number is irrational if and only if it is approximable to order 2.*

Proof Let ξ be algebraic of degree $n \geq 1$ (so ξ is rational if $n = 1$). Then by Lemma 8.39, there exists a constant c such that for all rational numbers $p/q \neq \xi$ with $q > 0$, we have

$$\left| \xi - \frac{p}{q} \right| \geq \frac{c}{q^n}.$$

It follows that ξ is not approximable by rationals to order $n + 1$, because

$$\left| \xi - \frac{p}{q} \right| < \frac{C}{q^{n+1}} \implies \frac{c}{q^n} < \frac{C}{q^{n+1}} \implies q < C/c.$$

Since there are only finitely many integers q such that $q < C/c$; it follows that there are only finitely many fractions p/q such that $|\xi - p/q| < C/q^{n+1}$.

Let a/b be a rational number in lowest terms with $b \geq 1$; we shall prove that a/b is approximable to order 1. (Note that we already know from our first statement that a/b is not approximable to order 2.) From Theorem 8.9 on p. 617, we know that the equation $ax - by = 1$ has an infinite number of integer solutions (x, y). The solutions (x, y) are automatically relatively prime. Moreover, if (x_0, y_0) is any one integral solution, then all solutions are of the form

$$x = x_0 + bt \quad , \quad y = y_0 + at \quad , \quad t \in \mathbb{Z}.$$

Since $b \geq 1$, we can choose t large so as to get infinitely many solutions with $x > 0$. With $x > 0$, we see that

$$\left| \frac{a}{b} - \frac{y}{x} \right| = \left| \frac{ax - by}{bx} \right| = \frac{1}{bx} < \frac{2}{x},$$

which shows that a/b is approximable to order 1.

Finally, if a number is irrational, then it is approximable to order 2 from Legendre's approximation theorem 8.21 on p. 621; conversely, if a number is approximable to order 2, then it must be irrational, because rationals are approximable to order 1 and hence are not approximable to order 2 by the first statement of this theorem. ∎

Using this theorem, we can prove that certain numbers must be irrational. For instance, let $\{a_n\}$ be any sequence of 0's and 1's in which there are infinitely many 1's. Consider

$$\xi = \sum_{n=0}^{\infty} \frac{a_n}{2^{2^n}}.$$

We remark that ξ is the real number with binary expansion $a_0.0a_10a_20\ldots$, with a_n in the 2^nth decimal place and with zeros everywhere else. Now fix a natural number n with $a_n \neq 0$ and let $s_n = \sum_{k=0}^{n} \frac{a_k}{2^{2^k}}$, the nth partial sum of this series. Then we can write s_n as p/q, where $q = 2^{2^n}$. Observe that

$$\left|\xi - s_n\right| \le \frac{1}{2^{2^{n+1}}} + \frac{1}{2^{2^{n+2}}} + \frac{1}{2^{2^{n+3}}} + \frac{1}{2^{2^{n+4}}} + \cdots$$

$$< \frac{1}{2^{2^{n+1}}} + \frac{1}{2^{2^{n+1}+1}} + \frac{1}{2^{2^{n+1}+2}} + \frac{1}{2^{2^{n+1}+3}} + \cdots$$

$$= \frac{1}{2^{2^{n+1}}} \left(1 + \frac{1}{2^1} + \frac{1}{2^2} + \frac{1}{2^3} + \cdots\right) = \frac{2}{2^{2^{n+1}}} = \frac{2}{(2^{2^n})^2}.$$

In conclusion,

$$\left|\xi - s_n\right| < \frac{2}{(2^{2^n})^2} = \frac{C}{q^2},$$

where $C = 2$. Thus, ξ is approximable to order 2, and hence must be irrational.

8.10.2   Liouville Numbers

Numbers that satisfy (8.66) with $c = 1$ are special: A real number ξ is called a **Liouville number**, after Joseph Liouville (1809–1882), if for every natural number n there is a rational number $p/q \ne \xi$ with $q > 1$ such that

$$\left|\xi - \frac{p}{q}\right| < \frac{1}{q^n}.$$

In Problem 2, you'll prove that Liouville numbers are transcendental (see our discussion around (8.66)). Because this fact is so important, we state it as a theorem.

Liouville's theorem

> **Theorem 8.41**   *Every Liouville number is transcendental.*

Using Liouville's theorem, we can give many (in fact uncountably many; see Problem 4) examples of transcendental numbers. Let $\{a_n\}$ be a sequence of integers in $0, 1, \ldots, 9$ in which there are infinitely many nonzero integers. Let

$$\xi = \sum_{n=0}^{\infty} \frac{a_n}{10^{n!}}.$$

Note that ξ is the real number with decimal expansion

$$a_0.a_1 a_2 000 a_3 00000000000000000000 a_4 \ldots,$$

with a_n in the $n!$th decimal place and with zeros everywhere else. Using Liouville's theorem, we'll show that ξ is transcendental. Fix a natural number n with $a_n \neq 0$ and let s_n be the nth partial sum of this series. Then s_n can be written as p/q, where $q = 10^{n!} > 1$. Observe that

$$
\begin{aligned}
\left| \xi - s_n \right| &\leq \frac{9}{10^{(n+1)!}} + \frac{9}{10^{(n+2)!}} + \frac{9}{10^{(n+3)!}} + \frac{9}{10^{(n+4)!}} + \cdots \\
&< \frac{9}{10^{(n+1)!}} + \frac{9}{10^{(n+1)!+1}} + \frac{9}{10^{(n+1)!+2}} + \frac{9}{10^{(n+1)!+3}} + \cdots \\
&= \frac{9}{10^{(n+1)!}} \left(1 + \frac{1}{10^1} + \frac{1}{10^2} + \frac{1}{10^3} + \cdots \right) \\
&= \frac{10}{10^{(n+1)!}} = \frac{10}{10^{n \cdot n!} \cdot 10^{n!}} \leq \frac{1}{10^{n \cdot n!}}.
\end{aligned}
$$

In conclusion,

$$
\left| \xi - s_n \right| < \frac{1}{(10^{n!})^n} = \frac{1}{q^n},
$$

so ξ is a Liouville number and therefore is transcendental.

8.10.3 Continued Fractions and the "Most Extreme" Irrational of All

We now show how continued fractions can be used to *construct* transcendental numbers! This is achieved by the following simple observation. Let $\xi = \langle a_0; a_1, \ldots \rangle$ be an irrational real number written as a simple continued fraction and let $\{p_n/q_n\}$ be its convergents. Then by our fundamental approximation Theorem 8.18 on p. 640, we know that

$$
\left| \xi - \frac{p_n}{q_n} \right| < \frac{1}{q_n q_{n+1}}.
$$

Since

$$
q_n q_{n+1} = q_n (a_{n+1} q_n + q_{n-1}) \geq a_{n+1} q_n^2,
$$

we see that

$$
\left| \xi - \frac{p_n}{q_n} \right| < \frac{1}{a_{n+1} q_n^2}. \tag{8.67}
$$

Thus, the larger the partial quotient a_{n+1}, the more closely the rational number p_n/q_n approximates ξ. We use this observation in the following theorem.

Theorem 8.42 *For every function* $\varphi : \mathbb{N} \to (0, \infty)$, *there exist an irrational number* ξ *and infinitely many rational numbers* p/q *such that*

$$\left| \xi - \frac{p}{q} \right| < \frac{1}{\varphi(q)}.$$

Proof We define $\xi = \langle a_0; a_1, a_2, \ldots \rangle$ by choosing the a_n inductively as follows. Let $a_0 \in \mathbb{N}$ be arbitrary. Assume that $a_0, \ldots, a_n$ have been chosen. With q_n the denominator of $\langle a_0; a_1, \ldots, a_n \rangle$, choose (using the Archimedean ordering property of the natural numbers) $a_{n+1} \in \mathbb{N}$ such that

$$a_{n+1} q_n^2 > \varphi(q_n).$$

This defines $\{a_n\}$. Now defining $\xi := \langle a_0; a_1, a_2, \ldots \rangle$, by (8.67), for every natural number n we have

$$\left| \xi - \frac{p_n}{q_n} \right| < \frac{1}{a_{n+1} q_n^2} < \frac{1}{\varphi(q_n)}.$$

This completes our proof.   ∎

Using this theorem, we can easily find transcendental numbers. For example, with $\varphi(q) = e^q$, we can find an irrational ξ such that for infinitely many rational numbers p/q, we have

$$\left| \xi - \frac{p}{q} \right| < \frac{1}{e^q}.$$

Since for $n \in \mathbb{N}$, we have $e^q = \sum_{k=0}^{\infty} q^k / k! > q^n / n!$, it follows that for infinitely many rational numbers p/q, we have

$$\left| \xi - \frac{p}{q} \right| < \frac{\text{constant}}{q^n}.$$

In particular, ξ is transcendental.

To review two of our discussions in this section: (1) We can form transcendental numbers by choosing the partial quotients in an infinite simple continued fraction to be very large, and (2) transcendental numbers are irrational numbers that are "closest" to (good) rational numbers. With this in mind, we can think of infinite continued fractions with small partial quotients as being "far" from transcendental and hence are "far" from rational numbers. Since 1 is the smallest natural number, we can consider the golden ratio

$$\Phi = \frac{1 + \sqrt{5}}{2} = \langle 1; 1, 1, 1, 1, 1, 1, 1, \ldots \rangle$$

as being the "most extreme" or "most irrational" of all irrational numbers in the sense that it is the "farthest" irrational number from being transcendental or the "farthest" irrational number from rationals.

8.10.4 What About π and e and What About Calculus?

Above, we have already seen examples (in fact, uncountably many; see Problem 4) of transcendental numbers, and we even know how to construct them using continued fractions. However, those numbers seem in some sense to be "artificially" made. What about numbers that are more "natural," such as π and e? Are these numbers transcendental? In fact, these numbers do turn out to be transcendental, but the "easiest" proofs of these facts need the technology of calculus (derivatives) [174, 175]! This might give one reason (among many others) to take more courses in analysis in which the calculus is taught. **Advertisement** ☺: The book [146] is a sequel to the book you're holding, and in it is the next adventure through topology and calculus, and during our journey we'll prove that π and e are transcendental. (You might also want to look at the book [148] concerning Lebesgue's theory of integration.) However, if you choose to go on this adventure, we ask you to look back at all the amazing things that we've encountered during these past chapters, and all without using one single derivative or integral!

▶ **Exercises 8.10**

1. Given an integer $b \geq 2$, prove that $\xi = \sum_{n=0}^{\infty} b^{-2^n}$ is irrational.
2. Prove that Liouville numbers are transcendental.
3. Let $b \geq 2$ be an integer and let $\{a_n\}$ be a sequence of integers $0, 1, \ldots, b - 1$ in which there are infinitely many nonzero a_n. Prove that $\xi = \sum_{n=1}^{\infty} a_n b^{-n!}$ is transcendental.
4. Using a Cantor diagonal argument as in the proof of Theorem 3.35, prove that the set of all numbers of the form $\xi = \sum_{n=0}^{\infty} \frac{a_n}{10^{n!}}$, where $a_n \in \{0, 1, 2, \ldots, 9\}$ (and are not all zero), is uncountable. That is, assume that the set of all such numbers is countable and construct a number of the same sort not in the set. Since we already showed that all these numbers are Liouville numbers, they are transcendental, so this argument provides another proof that the set of all transcendental numbers is uncountable.
5. Going through the construction of Theorem 8.42, define a real number ξ such that if $\{p_n/q_n\}$ are the convergents of its canonical continued fraction expansion, then

$$\left| \xi - \frac{p_n}{q_n} \right| < \frac{1}{q_n^n}, \quad \text{for all } n.$$

Show that ξ is a Liouville number, and hence is transcendental.

Bibliography

1. S. Abbott, *Understanding Analysis*, Undergraduate Texts in Mathematics (Springer, New York, 2001)
2. A.D. Abrams, M.J. Paris, The probability that (a, b) = 1. College Math. J. **23**(1), 47 (1992)
3. E.S. Allen, The scientific work of Vito Volterra. Amer. Math. Mon. **48**, 516–519 (1941)
4. N. Altshiller, J.J. Ginsburg, Solution to problem 460. Am. Math. Mon. **24**(1), 32–33 (1917)
5. R.N. Andersen, J. Stumpf, J. Tiller, Let π be 3. Math. Mag. **76**(3), 225–231 (2003)
6. T. Apostol, Another elementary proof of Euler's formula for $\zeta(2k)$. Am. Math. Mon. **80**(4), 425–431 (1973)
7. T.M. Apostol, *Mathematical Analysis: A Modern Approach to Advanced Calculus* (Addison-Wesley Publishing Company, Inc., Reading, 1957)
8. R.C. Archibald, Mathematicians and music. Am. Math. Mon. **31**(1), 1–25 (1924)
9. J. Arndt, C. Haenel, *Pi—Unleashed*, 2nd edn. (Springer, Berlin, 2001). Translated from the 1998 German original by Catriona Lischka and David Lischka
10. R. Ayoub, Euler and the zeta function. Am. Math. Mon. **81**(10), 1067–1086 (1974)
11. B.S. Babcock, J.W. Dawson Jr., A neglected approach to the logarithm. Two Year College Math. J. **9**(3), 136–140 (1978)
12. D.H. Bailey, J.M. Borwein, P.B. Borwein, S. Plouffe, The quest for pi. Math. Intell. **19**(1), 50–57 (1997)
13. W.W. Ball, *Short Account of the History of Mathematics*, 4th edn. (Dover Publications Inc., New York, 1960)
14. J.M. Barbour, Music and ternary continued fractions. Am. Math. Mon. **55**(9), 545–555 (1948)
15. C.W. Barnes, Euler's constant and e. Am. Math. Mon. **91**(7), 428–430 (1984)
16. R.G. Bartle, D.R. Sherbert, *Introduction to Real Analysis*, 2nd edn. (Wiley, New York, 1992)
17. A.F. Beardon, Sums of powers of integers. Am. Math. Mon. **103**(3), 201–213 (1996)
18. A.H. Bell, The "cattle problem." By Archimedes 251 B.C. Am. Math. Mon. **2**(5), 140–141 (1885)
19. H.E. Bell, Proof of a fundamental theorem on sequences. Am. Math. Mon. **71**(6), 665–666 (1964)
20. J. Bell, On the sums of series of reciprocals. Originally published as De summis serierum reciprocarum, Commentarii academiae scientiarum Petropolitanae **7**, 123134 (1740) and reprinted in Leonhard Euler, Opera Omnia, Series 1: Opera mathematica, vol. 14, Birkhäuser, 1992. Bell's paper available at http://arxiv.org/abs/math/0506415 and Euler's text, numbered E41, is available at the Euler Archive, http://www.eulerarchive.org
21. W.W. Bell, *Special Functions for Scientists and Engineers* (Dover Publications Inc., Mineola, 2004). Reprint of the 1968 original
22. R. Bellman, A note on the divergence of a series. Am. Math. Mon. **50**(5), 318–319 (1943)

© Paul Loya 2017
P. Loya, *Amazing and Aesthetic Aspects of Analysis*,
https://doi.org/10.1007/978-1-4939-6795-7

23. P. Benacerraf, H. Putnam (eds.), *Philosophy of Mathematics: Selected Readings* (Cambridge University Press, Cambridge, 1964)

24. S.J. Benkoski, The probability that k positive integers are relatively r-prime. J. Number Theory **8**(2), 218–223 (1976)

25. L. Berggren, J. Borwein, P. Borwein, *Pi: A Source Book*, 3rd edn. (Springer, New York, 2004)

26. B.C. Berndt, Ramanujan's notebooks. Math. Mag. **51**(3), 147–164 (1978)

27. N.M. Beskin, *Fascinating Fractions* (Mir Publishers, Moscow, 1980). Translated by V.I Kisln, 1986

28. F. Beukers, A note on the irrationality of $\zeta(2)$ and $\zeta(3)$. Bull. Lond. Math. Soc. **11**(3), 268–272 (1979)

29. R.P. Boas, Tannery's theorem. Math. Mag. **38**(2), 64–66 (1965)

30. R.P. Boas Jr., *A Primer of Real Functions*, vol. 13, 4th edn., Carus Mathematical Monographs (Mathematical Association of America, Washington, DC, 1996)

31. J.M. Borwein, P.B. Borwein, Ramanujan, modular equations, and approximations to pi or how to compute one billion digits of pi. Am. Math. Mon. **96**(3), 201–219 (1989)

32. J.M. Borwein, *Pi and the AGM*, 4th edn., Canadian Mathematical Society Series of Monographs and Advanced Texts (Wiley, New York, 1998). Reprint of the 1987 original

33. R.H.M. Bosanquet, *An elementary treatise on musical intervals and temperament (London, 1876)* (Diapason press, Utrecht, 1987)

34. C.B. Boyer, Fermat's integration of X^n. National Math. Mag. **20**, 29–32 (1945)

35. C.B. Boyer, A. History, of Mathematics, 2nd edn. (Wiley, New York, *(With a foreword by Isaac Asimov* (Revised and with a preface by Uta C, Merzbach), 1991)

36. P. Bracken, B.S. Burdick, Euler's formula for zeta function convolutions: 10754. Am. Math. Mon. **108**(8), 771–773 (2001)

37. D. Bressoud, Was calculus invented in India? College Math. J. **33**(1), 2–13 (2002)

38. D. Brewster, *Letters of Euler to a German Princess on Different Subjects in Physics and Philosophy* (Harper and Brothers, New York, 1834). In two volumes

39. W.E. Briggs, N. Franceschine, Problem 1302. Math. Mag. **62**(4), 275–276 (1989)

40. T. J.I'A. Bromwich, An Introduction to the Theory of Infinite Series, 2nd edn. (Macmillan, London, 1926)

41. R.A. Brualdi, Mathematical notes. Am. Math. Mon. **84**(10), 803–807 (1977)

42. R. Bumcrot, Irrationality made easy. College Math. J. **17**(3), 243–244 (1986)

43. F. Burk, Euler's constant. College Math. J. **16**(4), 279 (1985)

44. F. Cajori, *A History of Mathematics*, 5th edn. (Chelsea Publishing Co., Bronx, 1991)

45. F. Cajori, *A History of Mathematical Notations* (Dover Publications Inc., New York, 1993). 2 Vol in 1 edition

46. B.C. Carlson, Algorithms involving arithmetic and geometric means. Am. Math. Mon. **78**, 496–505 (1971)

47. D. Castellanos, The ubiquitous π. Math. Mag. **61**(2), 67–98 (1988)

48. D. Castellanos, The ubiquitous π. Math. Mag. **61**(3), 148–163 (1988)

49. R. Chapman, Evaluating $\zeta(2)$ (1999)

50. R.R. Christian, Another completeness property. Am. Math. Mon. **71**(1), 78 (1964)

51. J.A. Clarkson, On the series of prime reciprocals. Proc. Am. Math. Soc. **17**(2), 541 (1966)

52. B. Cloitre, private communication

53. J.B. Conrey, The Riemann hypothesis. Notices Am. Math. Soc. **50**(3), 341–353 (2003)

54. F.L. Cook, A simple explicit formula for the Bernoulli numbers. Two Year College Math. J. **13**(4), 273–274 (1982)

55. J.L. Coolidge, The number e. Am. Math. Mon. **57**, 591–602 (1950)

56. Fr.G. Costa, Solution 277. College Math. J. **17**(1), 98–99 (1986)

57. R. Courant, H. Robbins, *What is Mathematics?* (Oxford University Press, New York, 1979). An elementary approach to ideas and methods

58. J.W. Dauben, *Georg Cantor* (Princeton University Press, Princeton, 1990)

59. R. Dedekind, *Essays on the Theory of Numbers, "Continuity and Irrational Numbers"* (Dover Publications Inc., New York, 1963)

60. E.J. Dijksterhuis, *Archimedes* (Princeton University Press, Princeton, 1987). Translated from the Dutch by C. Dikshoorn, Reprint of the 1956 edition
61. U. Dudley, *A Budget of Trisections* (Springer, New York, 1987)
62. W. Dunham, A historical gem from Vito Volterra. Math. Mag. **63**(4), 234–237 (1990)
63. W. Dunham, Euler and the fundamental theorem of algebra. College Math. J. **22**(4), 282–293 (1991)
64. E. Dunne, M. Mcconnell, Pianos and continued fractions. Math. Mag. **72**(2), 104–115 (1999)
65. E. Dux, Ein kurzer beweis der divergenz der unendlichen reihe $\sum_{r=1}^{\infty} 1/pr$. Elem. Math. **11**, 50–51 (1956)
66. P. Erdös, Uber die reihe $\sum p$. Mathematica Zutphen. B **7**, 1–2 (1938)
67. L. Euler, De progressionibus transcendentibus seu quarum termini generales algebraice dari nequeunt (on transcendental progressions that is, those whose general terms cannot be given algebraically), Commentarii academiae scientiarum Petropolitanae **5**, 36–57 (1738). Presented to the St. Petersburg Academy on November 28, 1729. Published in Opera Omnia: Series 1, vol. 14, pp. 1–24, Eneström Index is E19, and is available at EulerArchive.org
68. L. Euler, Variae observationes circa series infinitas (various observations about infinite series). Commentarii academiae scientiarum Petropolitanae **9**, 160–188 (1744). Published in Opera Omnia: Series 1, vol. 14, pp. 217–244, Eneström Index is E72, and is available at EulerArchive.org
69. L. Euler, Introductio in analysin infinitorum (Introduction to analysis of the infinite. Book I), vol. 1 (Springer, New York, 1988)
70. L. Euler, Introductio in analysin infinitorum (Introduction to analysis of the infinite. Book II), vol. 2 (Springer, New York, 1990)
71. H. Eves, Irrationality of $\sqrt{2}$. Math. Teacher **38**(7), 317–318 (1945)
72. H. Eves, *Mathematical Circles Squared* (Prindle Weber & Schmidt, Boston, 1972)
73. P. Eymard, J.-P. Lafon, *The number π* (American Mathematical Society, Providence, 2004). Translated from the 1999 French original by Stephen S. Wilson
74. C. Fefferman, An easy proof of the fundmental theorem of algebra. Am. Math. Mon. **74**(7), 854–855 (1967)
75. W. Feller, *An Introduction to Probability Theory and Its Applications*, vol. I, 3rd edn. (Wiley, New York, 1968)
76. W. Feller, *An Introduction to Probability Theory and Its Applications*, vol. II, 2nd edn. (Wiley, New York, 1971)
77. R. Fenn, The table theorem. Bull. Lond. Math. Soc. **2**, 73–76 (1970)
78. D. Ferguson, Evaluation of $π$. Are Shanks' figures correct? Mathematical Gazette **30**, 89–90 (1946)
79. W.L. Ferrar, *A Textbook of Convergence* (The Clarendon Press, Oxford University Press, New York, 1980)
80. J. Ferreirós, *Labyrinth of thought*, 2nd edn. (Birkhäuser Verlag, Basel, 2007)
81. S.R. Finch, *Mathematical Constants*, vol. 94 (Encyclopedia of Mathematics and its Applications (Cambridge University Press, Cambridge, 2003)
82. P. Flajolet, I. Vardi, Zeta function expansions of classical constants (1996). preprint
83. T. Fort, Application of the summation by parts formula to summability of series. Math. Mag. **26**(26), 199–204 (1953)
84. G. Fredricks, R.B. Nelsen, Summation by parts. College Math. J. **23**(1), 39–42 (1992)
85. R.J. Friedlander, Factoring factorials. Two Year College Math. J. **12**(1), 12–20 (1981)
86. J.A. Gallian, *Contemporary Abstract Algebra*, 6th edn. (Houghton Mifflin Company, Boston, 2005)
87. M. Gardner, *Mathematical Games (Scientific American* (Simon and Schuster, New York, 1958)
88. M. Gardner, *Second Scientific American Book of Mathematical Puzzles and Diversions* (University of Chicago press, Chicago, 1987). Reprint edition
89. J. Glaisher, History of Euler's constant. Messenger Math. **1**, 25–30 (1872)
90. E.J. Goodwin, Quadrature of the circle. Am. Math. Mon. **1**(1), 246–247 (1894)

91. R.A. Gordon, The use of tagged partitions in elementary real analysis. Am. Math. Mon. **105**(2), 107–117 (1998)

92. H.W. Gould, Explicit formulas for Bernoulli numbers. Am. Math. Mon. **79**(1), 44–51 (1972)

93. D.S. Greenstein, A property of the logarithm. Am. Math. Mon. **72**(7), 767 (1965)

94. R. Grey, Georg Cantor and transcendental numbers. Am. Math. Mon. **101**(9), 819–832 (1994)

95. L. Guilbeau, The history of the solution of the cubic equation. Math. News Lett. **5**(4), 8–12 (1930)

96. R.W. Hall, K. Josić, The mathematics of musical instruments. Am. Math. Mon. **108**(4), 347–357 (2001)

97. A.E. Hallerberg, Indiana's squared circle. Math. Mag. **50**(3), 136–140 (1977)

98. P.R. Halmos, *Naive Set Theory* (Springer, Heidelberg, 1974). Reprint of the 1960 edition. Undergraduate Texts in Mathematics

99. P.R. Halmos, *I Want to be a Mathematician* (Springer, Heidelberg, 1985). An automathography

100. G.D. Halsey, E. Hewitt, More on the superparticular ratios in music. Am. Math. Mon. **79**(10), 1096–1100 (1972)

101. G.H. Hardy, J.E. Littlewood, G. Pólya, *Inequalities*, Cambridge Mathematical Library (Cambridge University Press, Cambridge, 1988). Reprint of the 1952 edition

102. G.H. Hardy, E.M. Wright, *An Introduction to the Theory of Numbers*, 5th edn. (The Clarendon Press, New York, 1979)

103. V.C. Harris, On proofs of the irrationality of $\sqrt{2}$. Math. Teacher **64**(1), 19–21 (1971)

104. J. Havil, *Gamma (Princeton University Press, Princeton* (Exploring Euler's constant, With a foreward by Freeman Dyson, 2003)

105. K. Hayashi, Fibonacci numbers and the arctangent function. Math. Mag. **76**(3), 214–215 (2003)

106. T.L. Heath, *Diophantus of Alexandria: A Study in the History of Greek Algebra* (Cambridge University Press, England, 1889)

107. T.L. Heath, *The Works of Archimedes* (Cambridge University Press, England, 1897)

108. T.L. Heath, *A History of Greek Mathematics*, vol. I (Dover Publications Inc., New York, 1981). From Thales to Euclid, Corrected reprint of the 1921 original

109. A. Herschfeld, On infinite radicals. Am. Math. Mon. **42**(7), 419–429 (1935)

110. J. Hofbauer, A simple proof of $1 + 1 = 2^2 + 1 = 3^2 + \cdots = \pi^2/6$ and related identities. Am. Math. Mon. **109**(2), 196–200 (2002)

111. K. Hrbacek, T. Jech, *Introduction to Set Theory*, vol. 220, 3rd edn., Monographs and Textbooks in Pure and Applied Mathematics (Marcel Dekker Inc., New York, 1999)

112. P. Iain, *Science, Theology and Einstein* (Oxford University, Oxford, 1982)

113. F. Irwin, A curious convergent series. Am. Math. Mon. **23**(5), 149–152 (1916)

114. S.J.H. Jeans, *Science and Music* (Dover Publications Inc., New York, 1968). Reprint of the 1937 edition

115. D.J. Jones, Continued powers and a sufficient condition for their convergence. Math. Mag. **68**(5), 387–392 (1995)

116. G.A. Jones, $6 = /\pi^2$. Math. Mag. **66**(5), 290–298 (1993)

117. J.P. Jones, S. Toporowski, Irrational numbers. Am. Math. Mon. **80**(4), 423–424 (1973)

118. D. Kalman, Six ways to sum a series. College Math. J. **24**(5), 402–421 (1993)

119. E. Kasner, J. Newman, *Mathematics and the Imagination* (Dover Publications Inc., New York, 2001)

120. V.J. Katz, Ideas of calculus in Islam and India. Math. Mag. **68**(3), 163–174 (1995)

121. G.W. Kelly, *Short-cut Math* (Dover Publications Inc., New York, 1984)

122. A.J. Kempner, A curious convergent series. Am. Math. Mon. **21**(2), 48–50 (1914)

123. A.N. Khovanskii, *The application of continued fractions and their generalizations to problems in approximation theory, Translated by Peter Wynn* (P. Noordhoff N. V, Groningen, 1963)

124. S.J. Kifowit, T.A. Stamps, The harmonic series diverges again and again. AMATYC Rev. **27**(2), 31–43 (2006)

125. M.S. Klamkin, R. Steinberg, Problem 4431. Am. Math. Mon. **59**(7), 471–472 (1952)

126. M.S. Klamkin, J.V. Whittaker, Problem 4564. Am. Math. Mon. **62**(2), 129–130 (1955)

127. I. Kleiner, Evolution of the function concept: a brief survey. Two Year College Math. J. **20**(4), 282–300 (1989)
128. M. Kline, Euler and infinite series. Math. Mag. **56**(5), 307–314 (1983)
129. K. Knopp, *Infinite Sequences and Series* (Dover Publications Inc., New York, 1956). Translated by Frederick Bagemihl
130. R. Knott, Fibonacci numbers and the golden section. http://www.mcs.surrey.ac.uk/Personal/R.Knott/Fibonacci/
131. D.E. Knuth, *The Art of Computer Programming*, vol. 2 (Addison-Wesley Publishing Co., Reading, 1981). Seminumerical algorithms, Addison-Wesley Series in Computer Science and Information Processing
132. R.A. Kortram, Simple proofs for $\sum_{k=1}^{\infty} k^2 = \pi^2/6$ and $\sin x = x \prod_{k=1}^{\infty} (1 - x^2/k^2\pi^2)$. Math. Mag. **69**(2), 122–125 (1996)
133. M. Krom, On sums of powers of natural numbers. Two Year College Math. J. **14**(4), 349–351 (1983)
134. D.E. Kullman, What's harmonic about the harmonic series. College Math. J. **32**(3), 201–203 (2001)
135. R. Kumanduri, C. Romero, *Number Theory with Computer Applications* (Prentice-Hall, New Jersey, 1998)
136. M. Laczkovich, On Lambert's proof of the irrationality of π. Am. Math. Mon. **104**(5), 439–443 (1997)
137. E. Landau, *Foundations of Analysis* (Chelsea Publishing Company, New York, 1966). Translated by F. Steinhardt of the German-language book Grundlagen der analysis
138. S. Lang, *A First Course in Calculus*, 5th edn. (Addison-Wesley Publishing Co., Reading, 1964)
139. L.J. Lange, An elegant continued fraction for π. Am. Math. Mon. **106**(5), 456–458 (1999)
140. W.G. Leavitt, The sum of the reciprocals of the primes. Two Year College Math. J. **10**(3), 198–199 (1979)
141. D.H. Lehmer, Problem 3801. Am. Math. Mon. **43**(9), 580 (1936)
142. D.H. Lehmer, On arccotangent relations for π. Am. Math. Mon. **45**(10), 657–664 (1938)
143. D.H. Lehmer, M.A. Heaslet, Solution 3801. Am. Math. Mon. **45**(9), 636–637 (1938)
144. A.L. Leigh, Silver, Musimatics or the nun's fiddle. Am. Math. Mon **78**(4), 351–357 (1971)
145. H.W. Lenstra, Solving the Pell equation. Notices Am. Math. Soc. **49**(2), 182–192 (2002)
146. P. Loya, Amazing and aesthetic aspects of analysis: The celebrated calculus. In preparation
147. P. Loya, Fourier's art of analysis. In preparation
148. P. Loya, Lebesgue's remarkable theory of measure and integration with probability. In preparation
149. P. Loya, What is a number? A beginners guide to proofs and number systems. In preparation
150. N. Luzin, Function: part I. Am. Math. Mon. **105**(1), 59–67 (1998)
151. N. Luzin, Function: part II. Am. Math. Mon. **105**(3), 263–270 (1998)
152. R. Lyon, M. Ward, The limit for e. Am. Math. Mon. **59**(2), 102–103 (1952)
153. D. MacHales, *Comic Sections: The Book of Mathematical Jokes, Humour, Wit, and Wisdom* (Boole Press, Dublin, 1993)
154. A.L. Mackay, *Dictionary of Scientific Quotations* (Institute of Physics Publishing, Bristol, 1994)
155. E.A. Maier, On the irrationality of certain trigonometric numbers. Am. Math. Mon. **72**(9), 1012–1013 (1965)
156. E.A. Maier, I. Niven, A method of establishing certain irrationalities. Math. Mag. **37**(4), 208–210 (1964)
157. S.C. Malik, Introduction to Convergence (Halsted Press, a division of Wiley, New Delhi, 1984)
158. E. Maor, *E: The Story of a Number* (Princeton University Press, Princetown, 1994)
159. G. Markowsky, Misconceptions about the golden ratio. Two Year College Math. J. **23**(1), 2–19 (1992)
160. J. Mathews, Gear trains and continued fractions. Am. Math. Mon. **97**(6), 505–510 (1990)

161. M. Mazur, Irrationality of $\sqrt{2}$ (2004). private communication
162. J.H. McKay, The William Lowell Putnam mathematical competition. Am. Math. Mon. **74**(7), 771–777 (1967)
163. G. Miel, Of calculations past and present: the Archimedean algorithm. Am. Math. Mon. **90**(1), 17–35 (1983)
164. J.E. Morrill, Set theory and the indicator function. Am. Math. Mon. **89**(9), 694–695 (1982)
165. L. Moser, On the series, $\sum 1/p$. Am. Math. Mon. **65**, 104–105 (1958)
166. J.A. Nathan, The irrationality of e^x for nonzero rational x. Am. Math. Mon. **105**(8), 762–763 (1998)
167. H.L. Nelson, A solution to Archimedes' cattle problem. J. Recreat. Math. **13**, 162–176 (1980)
168. D.J. Newman, Solution to problem e924. Am. Math. Mon. **58**(3), 190–191 (1951)
169. D.J. Newman, Arithmetic, geometric inequality. Am. Math. Mon. **67**(9), 886 (1960)
170. D.J. Newman, T.D. Parsons, On monotone subsequences. Am. Math. Mon. **95**(1), 44–45 (1988)
171. J.R. Newman (ed.), *The World of Mathematics*, vol. 1 (Dover Publications Inc., Mineola, 2000). Reprint of the 1956 original
172. J.R. Newman (ed.), *The World of Mathematics*, vol. 3 (Dover Publications Inc., Mineola, 2000). Reprint of the 1956 original
173. J. Nickel, *Mathematics: Is God Silent?* (Ross House Books, Vallecito, 2001)
174. I. Niven, The transcendence of π. Am. Math. Mon. **46**(8), 469–471 (1939)
175. I. Niven, *Irrational Numbers, The Carus Mathematical Monographs, The Mathematical Association of America*, vol. 11 (Wiley, New York, 1956)
176. I. Niven, A proof of the divergence of $\sum 1/p$. Am. Math. Mon. **78**(3), 272–273 (1971)
177. I. Niven, H.S. Zuckerman, *An Introduction to the Theory of Numbers*, 3rd edn. (Wiley, New York, 1972)
178. J. Nunemacher, R.M. Young, On the sum of consecutive kth powers. Math. Mag. **60**(4), 237–238 (1987)
179. C.D. Olds, The simple continued fraction expansion of e. Am. Math. Mon. **77**(9), 968–974 (1970)
180. G.A. Osborne, A problem in number theory. Am. Math. Mon. **21**(5), 148–150 (1914)
181. T.J. Osler, The union of Vieta's and Wallis's products for pi. Am. Math. Mon. **106**(8), 774–776 (1999)
182. T.J. Osler, J. Smoak, A magic trick from Fibonacci. College Math. J. **34**, 58–60 (2003)
183. T.J. Osler, N. Stugard, A collection of numbers whose proof of irrationality is like that of the number e. Math. Comput. Edu. **40**, 103–107 (2006)
184. T.J. Osler, M. Wilhelm, Variations on Vieta's and Wallis's products for pi. Math. Comput. Edu. **35**, 225–232 (2001)
185. I. Papadimitriou, A simple proof of the formula $\sum_{k=1}^{\infty} k^{-2} = \pi^2/6$. Am. Math. Mon. **80**(4), 424–425 (1973)
186. L.L. Pennisi, Elementary proof that e is irrational. Am. Math. Mon. **60**, 474 (1953)
187. G.M. Phillips, Archimedes the numerical analyst. Am. Math. Mon. **88**(3), 165–169 (1981)
188. R.C. Pierce Jr., A brief history of logarithms. Two Year College Math. J. **8**(1), 22–26 (1977)
189. A.S. Posamentier, I. Lehmann, π: *A Biography of the World's Most Mysterious Number* (Prometheus Books, Amherst, 2004). With an afterword by Herbert A. Hauptman
190. R. Preston, The mountains of π, The New Yorker (1992). http://www.newyorker.com/magazine/1992/03/02/the-mountains-of-pi
191. G.B. Price, Telescoping sums and the summation of sequences. Two Year College Math. J. **4**(4), 16–29 (1973)
192. R. Redheffer, What! Another note just on the fundamental theorem of algebra. Am. Math. Mon. **71**(2), 180–185 (1964)
193. R. Remmert, Vom fundamentalsatz der algebra zum satz von Gelfand-Mazur. Mathematische Semesterberichte **40**(1), 63–71 (1993)
194. D. Rice, History of π (or pi). Math. News Lett. **2**, 6–8 (1928)
195. N. Rose, *Mathematical Maxims and Minims* (Rome Press Inc., Raleigh, 1988)

196. T. Rothman, Genius and biographers: the fictionalization of Evariste Galois. Am. Math. Mon. **89**(2), 84–106 (1982)
197. R. Roy, The discovery of the series formula for π by Leibniz. Gregory and Nilakantha. Math. Mag. **63**(5), 291–306 (1990)
198. W. Rudin, *Principles of Mathematical Analysis*, 3rd edn. (McGraw-Hill Book Co., New York, 1976). International Series in Pure and Applied Mathematics
199. W. Rudin, *Real and Complex Analysis*, 3rd edn. (McGraw-Hill Book Co., New York, 1987)
200. B. Russell, *The Philosophy of Logical Atomism* (Routledge, New York, 2010)
201. O. Sacks, *The Man Who Mistook His Wife for a Hat: And Other Clinical Tales* (Touchstone, New York, 1985)
202. H. Samelson, More on Kummer's test. Am. Math. Mon. **102**(9), 817–818 (1995)
203. E. Sandifer, How Euler did it. http://www.maa.org/news/howeulerdidit.html
204. N. Schaumberger, An instant proof of $e^{\pi} > \pi^{e}$. College Math. J. **16**(4), 280 (1985)
205. M. Schechter, Tempered scales and continued fractions. Am. Math. Mon. **87**(1), 40–42 (1980)
206. H.C. Schepler, A chronology of pi. Math. Mag. **23**(3), 165–170 (1950)
207. H.C. Schepler, A chronology of pi. Math. Mag. **23**(4), 216–228 (1950)
208. H.C. Schepler, A chronology of pi. Math. Mag. **23**(5), 279–283 (1950)
209. P.J. Schillo, On primitive Pythagorean triangles. Am. Math. Mon. **58**(1), 30–32 (1951)
210. F. Schuh, *The Master Book of Mathematical Recreations* (Dover Publications Inc., New York, 1968). Translated by F. Göbel
211. M.Ó. Searcóid, *Elements of Abstract Analysis, Springer Undergraduate Mathematics Series* (Springer, London, 2002)
212. P. Sebah, X. Gourdon, A collection of formulae for the Euler constant. http://numbers.computation.free.fr/Constants/Gamma/gammaFormulas.pdf
213. P. Sebah, X. Gourdon, A collection of series for π. http://numbers.computation.free.fr/Constants/Pi/piSeries.html
214. P. Sebah, X. Gourdon, The constant e and its computation. http://numbers.computation.free.fr/Constants/constants.html
215. P. Sebah, X. Gourdon, Introduction on Bernoulli's numbers. http://numbers.computation.free.fr/Constants/constants.html
216. P. Sebah, X. Gourdon, π and its computation through the ages. http://numbers.computation.free.fr/Constants/constants.html
217. L. Seidel, Ueber eine darstellung des kreisbogens, des logarithmus und des elliptischen integrals erster art durch unendliche producte. J. Reine Angew. Math. **73**, 273–291 (1871)
218. A.A. Shaw, Note on roman numerals. Nat. Math. Mag. **13**(3), 127–128 (1938)
219. G.E. Shilov, Elementary Real and Complex Analysis, English, Dover Publications Inc., Mineola (1996). Revised English edition translated from the Russian and edited by Richard A. Silverman
220. G.F. Simmons, *Calculus Gems* (Mcgraw Hill Inc., New York, 1992)
221. J.G. Simmons, A new look at an old function, $e^{i\theta}$. College Math. J. **26**(1), 6–10 (1995)
222. S. Singh, On dividing coconuts: a linear diophantine problem. College Math. J. **28**(3), 203–204 (1997)
223. D. Singmaster, The legal values of pi. Math. Intell. **7**(2), 69–72 (1985)
224. D. Singmaster, Coconuts: The history and solutions of a classic Diophantine problem. Ga?ita-Bhāratī **19**(1–4), 35–51 (1997)
225. W.S. Sizer, Continued roots. Math. Mag. **59**(1), 23–27 (1986)
226. D.E. Smith, *A Source Book in Mathematics*, vol. 1,2 (Dover Publications Inc., New York, 1959). Unabridged and unaltered republ. of the first ed. 1929
227. J. Sondow, Problem 88. Math Horizons **32**, 34 (1997)
228. H. Steinhaus, Mathematical Snapshots, English (Dover Publications Inc., Mineola, 1999). Translated from the Polish, With a preface by Morris Kline
229. I. Stewart, *Concepts of Modern Mathematics* (Dover Publications Inc., New York, 1995)
230. J. Stillwell, Galois theory for beginners. Am. Math. Mon. **101**(1), 22–27 (1994)

231. D.J. Struik (ed.), A Source Book in Mathematics, 1200-1800 (Princeton Paperbacks, Princeton University Press, Princeton, 1986). Reprint of the 1969 edition
232. M.V. Subbarao, A simple irrationality proof for quadratic surds. Am. Math. Mon. **75**(3), 772–773 (1968)
233. J.J. Sylvester, On certain inequalities relating to prime number. Nature **38**, 259–262 (1888)
234. F. Terkelsen, The fundamental theorem of algebra. Am. Math. Mon. **83**(8), 647 (1976)
235. C.J. Thomae, Einleitung in die theorie der bestimmten integrale, Verlag von Louis Nebert, Halle (1875). http://books.google.com/
236. H. Thurston, A simple proof that every sequence has a monotone subsequence. Math. Mag. **67**(5), 344 (1994)
237. C. Tøndering, Frequently asked questions about calendars (2003). http://www.tondering.dk/claus/
238. H. Turnbull, *The Great Mathematicians* (Barnes & Noble, New York, 1993)
239. H. Turnbull (ed.), The Correspondence of Isaac Newton, vol. II (Published for the Royal Society, Cambridge University Press, New York, 1960), pp. 1676–1687
240. University of St. Andrews, A chronology of pi. http://www-gap.dcs.st-and.ac.uk/~history/HistTopics/Pichronology.html
241. University of St. Andrews, Eudoxus of Cnidus. http://www-groups.dcs.st-and.ac.uk/~history/Biographies/Eudoxus.html
242. University of St. Andrews, A history of pi. http://www-gap.dcs.st-and.ac.uk/~history/HistTopics/Pithroughtheages.html
243. University of St. Andrews, Leonhard Euler. http://www-groups.dcs.st-and.ac.uk/history/Mathematicians/Euler.html
244. University of St. Andrews, Madhava of Sangamagramma. http://www-gap.dcs.st-and.ac.uk/history/Mathematicians/Madhava.html
245. D.J. Uherka, A.M. Sergott, On the continuous dependence of the roots of a polynomial on its coeficients. Am. Math. Mon. **84**(5), 368–370 (1977)
246. R.S. Underwood, R.E. Moritz, Solution to problem 3242. Am. Math. Mon. **35**(1), 47–48 (1928)
247. J.V. Uspensky, *Introduction to Mathematical Probability* (McGraw-Hill Book Co., New York, 1937)
248. A. van der Poorten, A proof that Euler missed...Apéry's proof of the irrationality of $\zeta(3)$. Math. Intell. **1**(4), 195–203 (1978/1979). An informal report
249. C. Vanden, Eynden, Proofs that $\sum 1/p$ diverges. Am. Math. Mon. **87**(5), 394–397 (1980)
250. I. Vardi, *Computational Recreations in Mathematica* (Addison-Wesley Publishing Company Advanced Book Program, Redwood City, 1991)
251. I. Vardi, Archimedes' cattle problem. Am. Math. Mon. **105**(4), 305–319 (1998)
252. P.G.J. Vredenduin, A paradox of set theory. Am. Math. Mon. **76**(1), 59–60 (1969)
253. A.D. Wadhwa, An interesting subseries of the harmonic series. Am. Math. Mon. **82**(9), 931–933 (1975)
254. M. Ward, A mnemonic for Euler's constant. Am. Math. Mon. **38**(9), 6 (1931)
255. A. Weil, *Number Theory* (Birkhäuser Boston Inc., Boston, 1984). An approach through history, From Hammurapi to Legendre
256. E. Weisstein, Dirichlet function. From MathWorld—A Wolfram web resource. http://mathworld.wolfram.com/DirichletFunction.html
257. E. Weisstein, Landau symbols. From MathWorld—A Wolfram web resource. http://mathworld.wolfram.com/LandauSymbols.html
258. E. Weisstein, Pi approximations. From MathWorld—A Wolfram web resource. http://mathworld.wolfram.com/PiApproximations.html
259. E. Weisstein, Pi formulas. From MathWorld—A Wolfram web resource. http://mathworld.wolfram.com/PiFormulas.html
260. B.R. Wenner, Continuous, exactly k-to-one functions on R. Math. Mag. **45**, 224–225 (1972)
261. J. Wiener, Bernoulli's inequality and the number e. College Math. J. **16**(5), 399–400 (1985)

262. E. Wigner, The unreasonable effectiveness of mathematics in the natural sciences. Commun. Pure Appl. Math. **13**, 1–14 (1960)
263. E. Wigner, *Symmetries and Reflections: Scientific Essays* (The MIT Press, Cambridge, 1970)
264. H.S. Wilf, *Generatingfunctionology*, 3rd edn. (A K Peters Ltd., Wellesley, 2006). http://www.cis.upenn.edu/wilf/
265. G.T. Williams, A new method of evaluating $\zeta(2n)$. Am. Math. Mon. **60**(1), 12–25 (1953)
266. H.C. Williams, R.A. German, C.R. Zarnke, Solution of the cattle problem of Archimedes. Math. Comp. **19**(92), 671–674 (1965)
267. A.M. Yaglom, I.M. Yaglom, *Challenging Mathematical Problems with Elementary Solutions*, vol. II (Dover Publications Inc., New York, 1987). Translated from the Russian by James McCawley, Jr., Reprint of the 1967 edition
268. H. Yang, Y. Heng, The arithmetic-geometric mean inequality and the constant e. Math. Mag. **74**(4), 321–323 (2001)
269. G.S. Young, The linear functional equation. Am. Math. Mon. **65**(1), 37–38 (1958)
270. R.M. Young, *Excursions in Calculus*, vol. 13 (The Dolciani Mathematical Expositions (Mathematical Association of America, Washington, DC, 1992)
271. D. Zagier, The first 50 million prime numbers. Math. Intell., 7–19 (1977/1978)
272. L. Zia, Using the finite difference calculus to sum powers of integers. College Math. J. **22**(4), 294–300 (1991)

Index

© Paul Loya 2017
P. Loya, *Amazing and Aesthetic Aspects of Analysis*,
https://doi.org/10.1007/978-1-4939-6795-7

Printed in the United States
By Bookmasters